AF334490

TRITICALE: TODAY AND TOMORROW

Developments in Plant Breeding

VOLUME 5

Triticale:
Today and Tomorrow

Edited by

HENRIQUE GUEDES-PINTO

*Departamento de Genética e Biotecnologia,
Universidade de Trás-os-Montes e Alto Douro,
Vila Real, Portugal*

NORMAN DARVEY

*Plant Breeding Institute,
University of Sydney,
Cobbity, Australia*

and

VALDEMAR P. CARNIDE

*Departamento de Genética e Biotecnologia,
Universidade de Trás-os-Montes e Alto Douro,
Vila Real, Portugal*

KLUWER ACADEMIC PUBLISHERS
DORDRECHT / BOSTON / LONDON

A C.I.P. Catalogue record for this book is available from the Library of Congress

ISBN 0-7923-4212-7

Published by Kluwer Academic Publishers,
P.O. Box 17, 3300 AA Dordrecht, The Netherlands.

Kluwer Academic Publishers incorporates
the publishing programmes of
D. Reidel, Martinus Nijhoff, Dr W. Junk and MTP Press.

Sold and distributed in the U.S.A. and Canada
by Kluwer Academic Publishers,
101 Philip Drive, Norwell, MA 02061, U.S.A.

In all other countries, sold and distributed
by Kluwer Academic Publishers Group,
P.O. Box 322, 3300 AH Dordrecht, The Netherlands.

Printed on acid-free paper

All Rights Reserved
© 1996 Kluwer Academic Publishers
No part of the material protected by this copyright notice may be reproduced or
utilized in any form or by any means, electronic or mechanical,
including photocopying, recording or by any information storage and
retrieval system, without written permission from the copyright owner.

Printed in the Netherlands

Contents

III - Germplasm

VIII - Regional Trials and Crop Management

LIST OF THE REFEREES

DON SALMON - Field Crop Development Centre, Alberta Agriculture, Food and Rural Development, Bag#47, Lacombe, Alberta, T0C 1S0, Canada.

EUGENE E. SAARI - CIMMYT, P. O. Box 5186, Kathmandu, Nepal.

GITTA OETTLER - University of Hohenheim, State Plant Breeding Institute, D-70593 Stuttgart, Germany.

HENRIQUE GUEDES-PINTO - Departamento de Genética e Biotecnologia, Universidade de Trás-os--Montes e Alto Douro, Apdo. 202, 5001 Vila Real Codex, Portugal.

J. C. MCLEOD - Semiarid Prairie Agricultural Research, P.O. Box 1030, Swift Current, SK, S9H 3X2, Canada.

KATHERINE COOPER - The University of Adelaide, Department of Plant Science, Waite Campus, Glen Osmond 5064, South Australia.

NORMAN DARVEY - University of Sydney, Plant Breeding Institute, Cobbity NSW 2570, Australia.

PERRY GUSTAFSON - United States Department of Agriculture, Agricultural Research Service, PGRU, University of Missouri, Columbia, MO 65211, USA.

P.J. KALTSIKES - Agricultural University of Athens, Department of Plant Breeding and Biometry, Athens, Greece, 118 55.

ROBIN JESSOP - University of New England, Department of Agronomy and Soil Science, Armidale, NSW 2351, Australia.

VALDEMAR CARNIDE - Departamento de Genética e Biotecnologia, Universidade de Trás-os-Montes e Alto Douro, Apdo. 202, 5001 Vila Real Codex, Portugal.

SPONSORS

Centro Internacional de Mejoramiento de Maiz y Trigo (CIMMYT)
EUCARPIA
Fundação Calouste Gulbenkian
International Triticale Association (ITA)
Junta Nacional de Investigação Científica e Tecnológica (JNICT)
SEMUNDO
Universidade de Trás-os-Montes e Alto Douro (UTAD)

INTRODUCTION

WELCOME AND NETWORK INFORMATION

On behalf of the International Triticale Association (ITA), I would like to welcome all of you to the joint EUCARPIA/ITA's Third International Triticale Symposium. It is wonderful to be in the beautiful city of Lisbon, in Portugal, and while I am sure we will all savour its atmosphere this evening and over the following days, I trust that you will go away from here feeling part of an Organisation that not just hosts International Conferences, as many other International Symposia do, but also is interested in co-ordinating Triticale Research and Communication at an International Level. I would like to remind you that the majority of us are members of the International Triticale Association and therefore have a voice in its future direction, so that when various proposals are brought forward for us to consider, we all share in the responsibility for directing our own paths.

One thing that most of you must be aware of, is the continuing worldwide decline in Agricultural funding, particularly at the CGIAR level, for International Research organisations such as CIMMYT or ICRISAT. This has in part coincided with a worldwide recession, and an increasing concern about the environment and sustainable agricultur. It would be rather ironic if cutbacks to plant breeding research and increased funding for "Save the Environment" type research, led to dramatic increases in death by starvation; no one would want this to happen, but I wish I had the confidence that those who make these decisions knew what they were doing, and be there to accept responsibility if they are proven wrong.

When it comes to the future of Triticale, the prognosis looks good if one looks at the increased area sown to triticale worldwide over the last 5-10 years. However, much of this recent growth has involved CIMMYT-derived germplasm, and hence the picture is not so rosy when one considers that CIMMYT funding has been significantly cut in the last year, that the triticale effort has been reduced and that triticale is looking rather vulnerable when it comes to discussing priority research areas. Hence, it may be somewhat fortuitous that one or more Co-operative International Research Networks will be established at this Symposium, so that materials, technology, information and research can be better co-ordinated, results disseminated and relevant germplasm be made more readily available. A model for the establishment of the ITA's first Research Co-operative on Germplasm Resources and other networks for "Communication", "Quality" and "Diseases" will be proposed. These networks are informal in nature, elect their own executives and co-ordinator(s) and may, from time to time, organise "Satellite Meetings" under the auspices of the ITA.

I conclude this brief address where I stated and acknowledge the significant input of the Local Organising Committee in making this Symposium possible, and I wish that all participants come away from here stimulated, challenged and with special memories and friendships that will linger well into the future.

Thank you.

Norman L. Darvey, Secretary, International Triticale Association

H. Guedes-Pinto et al. (eds.), Triticale: Today and Tomorrow, 3.
© *1996 Kluwer Academic Publishers. Printed in the Netherlands.*

TRITICALE: MILESTONES, MILLSTONES, AND WORLD FOOD

Calvin O. Qualset[1] and Henrique Guedes-Pinto[2]
[1]University of California, Davis, California, USA
[2]Universidade de Trás-os-Montes e Alto Douro, Vila Real, Portugal

Triticale is a synthesized crop, conceived with the idea of combining the best of both of its parents-wheat and rye. There was a long period of frustration and disappointment since the first sterile wheat-rye hybrids appeared in wheat fields until the time when the most appropriate type of hybrids were made. First octoploid triticale was produced from common hexaploid wheat x rye amphiploids. These did not fulfill the original dream for triticale. It was the idea that durum (tetraploid) wheat x rye hybrids may be more productive than octoploid types that renewed optimism that a truly new synthesized crop could be achieved. This third international symposium is a tribute to the validity of the that concept. The symposium upheld the excellence of the previous ones in Sydney, Australia in 1987 and Passo Fundo, Brazil in 1990. The global interest in triticale was obvious from the offering of 160 abstracts of original research and reviews to the symposium.

This is a unique point in the history of agriculture. Up to now we have inherited many crop plants whose evolutionary histories are only partially known. What we can deduce quite well are the biological factors that have been involved in the domestication of plants for human use. Much less is known about how early farmers shaped the phenotypes of crops to adapt them to meet their basic needs for food security. With triticale we can trace the human events that resulted in this new crop plant. That history was very nicely summarized by N. Jouve and we were so fortunate to have in our audience one of the conceptual founders of modern triticale. Professor Sanchez-Monge gave us the hope that the right way to utilize triticale would be trough hybridization of durum wheat and rye to make hexaploid triticale. History proved him correct. So, as triticale emerges as an important world crop we can know about the successes and failures in its culture and how it has come to be accepted as a food source for the rapidly expanding world population. This level of social, biological, and agricultural documentation about the evolution of a human food crop plant has not previously occurred. At least as far as we know from the archeological record, there were not countless local, regional, and international conferences about how to produce a new agricultural plant! This international symposium is a major landmark for triticale and we are fortunate that its broad scope has brought us up-to-date on new scientific achievements on the biology of triticale and on how the crop can be cultured and used.

5

H. Guedes-Pinto et al. (eds.), Triticale: Today and Tomorrow, 5–9.
© 1996 *Kluwer Academic Publishers. Printed in the Netherlands.*

The short history of triticale shows many important advances ("milestones") and still some deficiences ("millstones") which require attention from research, but also from marketing and social acceptance of the crop. "World food" is also included in the title because if we are to achieve anything of significance with triticale it should be for the betterment of the human condition in the face of global population expansion which will reach 10 billion people in the next century. To support this growth, Varughese reported that it has been estimated that the increase in consumption of cereals will be 3.5% per year. Recent productivity gains were estimated to be 2.4% per year, indicating that the rate of increase in grain yields per hectare must be dramatically accelerated. Triticale can contribute to narrowing this yield gap, because genetic improvements can still be made, perhaps at greater rates than in any other cereal crop. There are strong indications that triticale can be a component of the agricultural systems that will sustain this large population. Land area available for agriculture is not expanding, and in fact, should decrease in the near future, as the most fragile lands which are now used for crop production are returned to nonagricultural uses. Such uses will provide "ecosystem services", such as improved watersheds and reduced soil erosion, which in turn help increase crop productivity of the farmed lands. This suggests that agriculture will have to intensify production on the better lands and judiciously use those of lower quality. From the various papers presented during this symposium, it is evident that triticale has characteristics to fit into this sophisticated agricultural enterprise.

MILESTONES

- Triticale is advancing in area planted in some parts of the world. It was estimated that 1.2 million hectares were planted in 1990 and by 1994 the area had doubled to 2.4 million hectares. Schlegel reported that 120 varieties have been released in 35 countries.
- Recent data from CIMMYT with winter-planted, spring-type triticale in northern Mexico showed that the triticales outperformed common and durum wheats. This has been observed previously in other research. It was further shown that the rate of genetic gain in yield in the CIMMYT breeding program was 1.5% per year, exceeding the gains achieved in wheat breeding.
- Improved varieties are being released from a number of programs, thus plant breeding is progressing well, considering the limited resources devoted to this crop. Introduction of dominant gene expression for male sterility into triticale was reported by Venkatanagappa and Darvey which can aid recurrent selection in triticale and also provide a mechanism for hybrid triticale production. Although the permanent hybridity of wheat and triticale were viewed as sufficient to capture heterosis for these crops (Kaltsikes).
- Several presentations showed that triticale performs well in suboptimal environments, so it is a candidate as the crop of choice for such environments. Triticale varieties having strong tolerance to low soil pH and high concentrations of aluminum and to saline environments was reported. But as mentioned above, let us not fall into the trap of promoting the planting of triticale on lands where no crops should be grown. The range of adaptation of triticale in suboptimal environments is apparently greater than wheat and yields outpace rye.

- Grain quality has improved remarkably during the past decade for breeding. The kernel density and smoothness is approaches that of wheat, resulting in improved milling performance. Several workers reported that triticale has numerous human food uses and others pointed out the value of the grain and forage for animal diets.
- Triticale is a host for a number of diseases and other pests, but one author pointed out that, relative to wheat and rye, it is still a healthy crop.
- Triticale, in its early days of exploitation of the hexaploid type, had a narrow genetic background. This has been expanded in breeding programs through intercrossing new primary triticale with current varieties and continued intercrossing of secondary types. CIMMYT has established a world gene bank for triticale, reported at this symposium to have about 15.000 accessions. About 3.500 North American triticale genetic resources were assembled from several currently inactive breeding programs and evaluated for numerous traits. Similar collections should be assembled from the breeding programs of Europe.
- High fertility has been achieved in high-yielding triticales. The mating system is predominantly self-fertilization, but outcrossing can be rather high, requiring attention to isolation in seed production fields.
- Triticale has become established as having mixed genomes or as "pure" A, B, and R genomes. The substitution of D chromosomes for individual R chromosomes has resulted in improved performance, but, in the case of 6R replaced by 6D, Pfeiffer reported that while kernel density was increased there was a decrease in quality. Substitution of 1D into the A genome of triticale has given improved breadmaking quality of triticale (Kazman and Lelley). Triticale is proving to be a model for study of gene interactions among different genomic origins as shown for the *Nor* loci by Gustafson and Flavell.
- Rye has been proven to be an important gene resource for wheat as whole chromosome arms or small segments of rye chromosomes were recombined into wheat chromosomes. The spectacular success of the Veery family of wheat varieties produced by CIMMYT is an outstanding example of the contribution of rye genes for grain yield improvement. Triticale is a good bridging species for gene transfer to wheat.
- Substantial progress was reported for *in vitro* regeneration of plants from scutellum tissue. The first successful transformation by microprojectile bombardment was reported at this symposium by Zinmy and Lorz.
- RFLP genomic maps of hexaploid wheat and, to a lesser extent, for rye have been completed. More than, 1.000 markers are now mapped. Devos and Gale reported substantial conservation of linearity of the wheat and rye genomes. At the same time the widely known wheat rye chromosome translocations were confirmed, with 8 of 14R chromosome arms having wheat-rye translocations. This gives complications to the genome mappers, but provides important information for the plant breeders. Breeding progress can now be enhanced by using the molecular recombination map to tag genes for ease of selection for some traits.
- Plant height reduction has been a goal for breeders, but introduction of *Rht* genes has not shown great promise. Steady decrease in height has been achieved in the CIMMYT program. Walski has used dwarfing genes from rye and these have been shown to have recessive gene action.

8

- Triticale is exceptionally versatile in its end use possibilities as a food in bakery and extruded products and its value as a feed grain and forage has been widely accepted.
- The triticale model as a synthetic polyploid has given encouragement for the development of other new crops. Martin showed great progress with *Hordeum chilense* x wheat hybrids and others (not reported at this symposium) have found good potential for *Dasypyrum villosum* x wheat amphiploids.

MILLSTONES

In spite of all of the remarkable advances in triticale science and crop development there is still limited adoption of the crop. Many of the factors contributing to the greater success of triticale were discussed during the symposium and are briefly identified here.

- ***Markets.*** Traditional markets for wheat, barley, rye, and oat are well-established and reasonably reliable. Pricing is controlled in many countries. Will triticale remain a niche--market crop or will it enter the mainstream of world cereals markets?
- ***Marketing boards.*** Some countries have marketing boards that control the production and sale of commodities. Some local production can be enhanced for a crop outside of the marketing boards, but at the same time there is no support for developing a stable production base for an orphan crop.
- ***Official grade.*** Triticale is graded as livestock feed grain in some countries to its disadvantage as a food crop. This can be a disincentive for farmer adoption of the crop.
- ***Wheat.*** Triticale is most like wheat and can be used for the same products as wheat. Traditional uses and processes for milling adopted for wheat are not easily modified unless sufficient and consistent volume is available.
- ***Grain quality.*** Triticale is judged in comparison with wheat and acceptance is doubtful if it does not perform very similarly to wheat. Peña pointed out that at present it is most like a soft texture wheat and can be adapted for the same products as soft wheats. Quality is jeopardized by preharvest sprouting in some environments, thus reducing end-use potential. Weak gluten has been a definite deterrent to adoption of triticale; introduction of the high molecular weight alleles from the D genome promises to alter the end-use potential of triticale substantially. As a feed grain, some workers have noted a decline in protein quantity and quality (lowered lysine, for example) as the varieties were improved in agronomic characteristics.
- ***Genetic instability.*** Even though high fertility is observed in modern triticale varieties, the crop still is prone to reproductive stability as seen by the presence of aneuploids is seed production fields and as somaclonal variation in *in vitro* culture. Further genetic improvements in the stability of the reproductive sytem would be desirable.
- ***Diseases.*** Triticale's reputation as a healthy crop is likely to diminish as pathogens adapt to this crop. The rust epidemic in Australia is an example. More attention to disease biology and host resistance will be required in future research on triticale.
- ***Diminishing research efforts.*** Throughout the world there are very few research and crop development scientists devoted full-time to the advancement of triticale. Traditional crops generate R & D funds and these funds are appropriately used to

improve and protect those crops. Little allocation of public funds is dedicated to new crop development.

OUTLOOK

Triticale is well along the developmental pathway to becoming a much more significant world crop. Scientific advances are remarkably steady in spite of limited dedication of personnel and funds to the crop. Still a great limitation is consumer acceptance, which in turn is supported by lack of available product to bring triticale into the mainstream for food and feed uses.

Will the biological and agronomic superiority of triticale be recognized by widespread adoption? This is doubtful, given the current allocation of promotional resources to triticale. One can ask, **"What will be the fate of triticale when its enduse properties exactly mimic wheat and the field performance is clearly better than wheat?"** At that point triticale will replace wheat. That is a realistic outcome. The advances in triticale science reported at this symposium point to the potential for this outcome.

Definitely, triticale is no longer a mere scientific curiosity. A tremendous amount of work mainly achieved over the last 30 years on basic studies and breeding has lead to triticale reaching the dangerous age of the **end of youth** and the start of **maturity**: a fully-fledged and significant world cereal crop.

I - TRITICALE TODAY AND TOMORROW

TRITICALE: PRESENT STATUS AND CHALLENGES AHEAD

George Varughese
International Maize and Wheat Improvement Center (CIMMYT)
Apdo. Postal 6-641, 06600 Mexico, D.F. Mexico

Abstract

During the past three decades, the world has witnessed dramatic increases in the productivity of cereals, especially in developing countries. These increases, which were well above the growth of population and demand, were triggered by high-yielding varieties, higher inputs, appropriate policies, and expansion in area and irrigation. However, production growth now appears to be slowing down as several of these driving factors are approaching their limits. This reduced growth rate, combined with heightened concern for sustainability in agriculture and the environment, is making it increasingly difficult to keep up with the steady growth in demand for food (3% per annum for wheat in developing countries).

Current estimates indicate that triticales contribute more than 6 million metric tons per year to global cereal production. Thus, triticale production surpasses the production of crops like linseed, sesame, castor, and lentils and approaches that of chickpeas. During the 1986-92 period alone, the area under triticale more than doubled from 1 million to 2.4 million hectares.

Introduction

Despite small production on a global scale (1% of wheat), area expansion of triticale is taking place rapidly. This paper argues that the crop provides more opportunities that bread wheat in enhancing the productivity of cereals. The "D" genome is what made bread wheat to most important cereal in he world. The difference between triticale and bread wheat is that, in triticale, the "D" genome has been replaced by the "R" genome. When comparing the benefits and drawbacks of these two genomes, we find the "R" genome to be far superior to the "D" genome with respect to yield potential, disease resistance, tolerance to minor element deficiencies or toxicities, and phosphorus uptake efficiency. All these factors are extremely important in maintaining sustainable productivity gains in future. The only known defect to date that the "R" genome brings to this crop is inferior bread making quality. The scientific evidence we have today indicates that this negative trait can be easily modified by selectively incorporating the high molecular weight glutenin alleles into the "R" genome from the "D" genome. The paper concludes that riticale will be a vital crop for meeting the ever growing food demands of the world.

13

H. Guedes-Pinto et al. (eds.), Triticale: Today and Tomorrow, 13–20.
© 1996 *Kluwer Academic Publishers. Printed in the Netherlands.*

Global Cereal Situation and Emerging Trends

In 1992 world cereal production reached an all time record of 1.9 billion tons. Wheat, maize, and rice continue to be the three main cereal crops--combined accounting for 83% of global cereal production (Table 1).

Table 1. World Cereal Production in 1992 (1000 t).

Crop	Production	Percentage
Wheat	566,282	28.8
Maize	530,067	27.0
Rice paddy	526,360	26.8
Barley	165,037	8.4
Sorghum	69,245	3.5
Oats	33,884	1.7
Rye	29,254	1.5
Millet	28,908	1.5
Others	14,986	0.8
Total	1,964,023	100.0

Source: FAO Production Year Books.

Table 2. Growth rate of cereals in developing countries during the past three decades.

	1963-72			1973-82			1983-92		
	A	Y	P	A	Y	P	A	Y	P
Wheat	1.8	3.4	5.2	1.1	4.0	5.1	0.4	2.0	2.4
Maize	1.4	2.6	4.0	0.6	3.1	3.7	1.1	2.1	3.1
Rice	1.2	1.9	3.2	0.5	2.4	2.9	0.3	1.5	1.8
Other Cereals	-0.3	1.5	1.2	-1.5	2.2	0.7	0.0	-0.3	-0.3
All Cereals	0.9	2.4	3.3	0.1	3.0	3.1	0.4	1.6	2.0
All Cereals Global	0.3	3.0	3.4	0.3	2.2	2.5	-0.2	1.6	1.4

A = area; Y = yield; P = production.
Source: FAO Production Year Books.

During the past three decades, the world witnessed dramatic increases in the productivity of cereals and this phenomenon gained the popular term "Green Revolution". During the Green Revolution and the post-Green Revolution periods, productivity growth was well above population growth and also above demand growth. Developing countries, in general, faired better than the global average and wheat had the highest growth rate, followed by maize and then rice (Table 2).

Many factors contributed towards this dramatic change in productivity gains. The advent of high yielding, input-responsive, and input-efficient varieties was the primary trigger. As a consequence, utilization of inputs like fertilizer also registered rapid growth. Increases in area and irrigation also contributed substantially towards this expansion, more so during the 1960s and early 1970s [1].

Today, almost all the driving factors (i.e., land, irrigation, and fertilizer) responsible for the dramatic growth rates are reaching their limits and hence are showing either stagnation in productivity or in some instances a decline--the only exception being yield potential. In the 1960s, the global growth rate of production was 3.4% and in the 1980s it was only 1.4%. Almost all the high and medium potential areas are now saturated with modern high yielding varieties (Table 3). Thus, the primary contributing factor for the rapid advancement of productivity in the 1960s and 1970s, unfortunately, will not have the same level of impact in the future. Input use is reaching its optimum level and thus additional inputs are not going to be profitable. Expansion in area and irrigation will be minimal. So, the options for gains in the future must be different from those of the past.

Table 3. Percentage of wheat area sown by moisture regime in developing countries to semidwarf wheat varieties, 1990.

Wheat type	Percentage of area	
	Well watered	Dryland
Spring Bread Wheat	99	63
Spring Durum Wheat	82	33
Winter Bread Wheat	94	9
Winter Durum Wheat	100	0

Source: [2].

Evidence in the 1980s clearly indicates reductions in productivity growth (Table 2). This reduced growth rate--combined with heightened concerns for the environment and sustainability in agriculture--is making it increasingly difficult to keep up with the steady growth in the demand for food of around 3% per annum due to population and income increases. In order to continue along the path of supply matching demand, researchers have to provide a steady flow of new ideas, innovations, and inventions in creating more productive and at the same time sustainable and environmentally sound practices and cropping systems. Triticale definitely is one of these options.

Triticale--Its Current Status

Progress made in triticale improvement, to date, has been remarkable. In 1967 the late Prof. Arne Müntzing concluded his book on triticale with the following statement: "It can be expected that the new, manmade cereal, triticale, will definitely join the old cereals as food for the rapidly growing human populations and their domestic animals." The first commercial triticales were released two years after the publication of his book. Today, 25 years later, triticale is grown on more than 2.4 million hectares and contributes more than 6 million ton per year to global cereal production.

Triticale production surpasses the production of crops like linseed, sesame, castor, and lentils and approaches that of chickpeas. During the 1986-92 period alone, the area under triticale more than doubled from 1 to 2.4 million hectares (Table 4). Although production on a global scale is still small, expansion of area and productivity of triticale will continue because the crop provides better options than the cereals with which it competes.

Future of Triticale and its Role in Enhancing the Productivity of Cereals

During the 1961-90 period, consumption of cereals in developing countries grew at a rate of 3.5%. However, this trend is projected to slow down to 2.1% during the 1990-2005 period [3]. Current information indicates that the productivity gains for all the cereals during the past three decades in developing countries were only 2% per year. Even in the case of wheat (the crop with the highest growth rate), productivity gains of the past decade (2.4%) did not keep pace with the projected consumption demand (3%). Thus, it is imperative that, in order to satisfy the future demand, productivity must continue to grow while at the same time being sustainable.

The "D" genome made bread wheat the most important cereal in the world. The difference between triticale and bread wheat is that, in triticale, the "D" genome has been replaced by the "R" genome. *Triticum tauschii*, the ancestral parent of bread wheat that donated the "D" genome, is a grass while rye, the "R" genome parent, is a cultivated crop with many positive attributes that are not found in the "D" genome and also missing or deficient in the wheat crop. It is well documented that triticale has inherited most of these positive traits from its rye parent like better tolerance to drought and cold, superior resistance to many diseases, tolerance to minor element deficiencies or toxicities, and phosphorus uptake efficiencies. A recent study indicates that nitrogen accumulation at heading and physiologic maturity in triticale is higher than in bread wheat [4]. This difference in nitrogen accumulation is maximum at lower levels of N application and the advantage diminishes with higher doses. This would indicate that triticale may be a better crop for soils with low nitrogen fertility. All these positive traits will be needed by crops of the future to enhance productivity in a sustainable manner. Thus, triticale will be important for sustainability in agriculture and the environment and may even be a better alternative than wheat.

Increasing crop yield potential is one of the few means available to meet the food demands of the future. The short history of triticale indicates that it is definitely a better competitor than wheat in marginal environments [5]. Data accumulated for irrigated environments during the past few years show that triticale is catching up

Table 4. Estimation of triticale area in countries growing 1,000 hectares or more in 1986, 1991-92.

Country	Area in Hectares	
	1986	1991-1992
Algeria		10,000
Argentina	10,000	16,000
Australia	160,000	100,000
Austria	1,000	2,000
Belgium	5,000	10,000
Brazil	5,000	90,000
Bulgaria	10,000	100,000
Canada	6,500	2,000
Chile	5,000	10,000
China (Northeast/Heilongjiang)	25,000	1,500
Fed. Rep. Czechoslovakia		25,000
France	300,000	162,000
Germany	30,000	207,000
Hungary	5,000	5,000
India	500	
Italy	15,000	30,000
Kenya		8,000
Luxemburg	400	2,000
Mexico	8,000	3,000
Morocco		10,000
Netherlands	1,000	4,000
New Zealand		2,000
Poland	100,000	659,300
Portugal	7,000	90,000
Romania		20,000
South Africa	15,000	95,000
USSR	250,000	500,000
Spain	30,000	80,000
Sweden		1,000
Switzerland	5,000	11,000
Tanzania	400	
Tunisia	5,000	16,000
UK	16,000	16,000
USA	60,000	180,000
Total	1,075,800	2,467,800

Source: Dr. W.H. Pfeiffer, Wheat Program, CIMMYT.

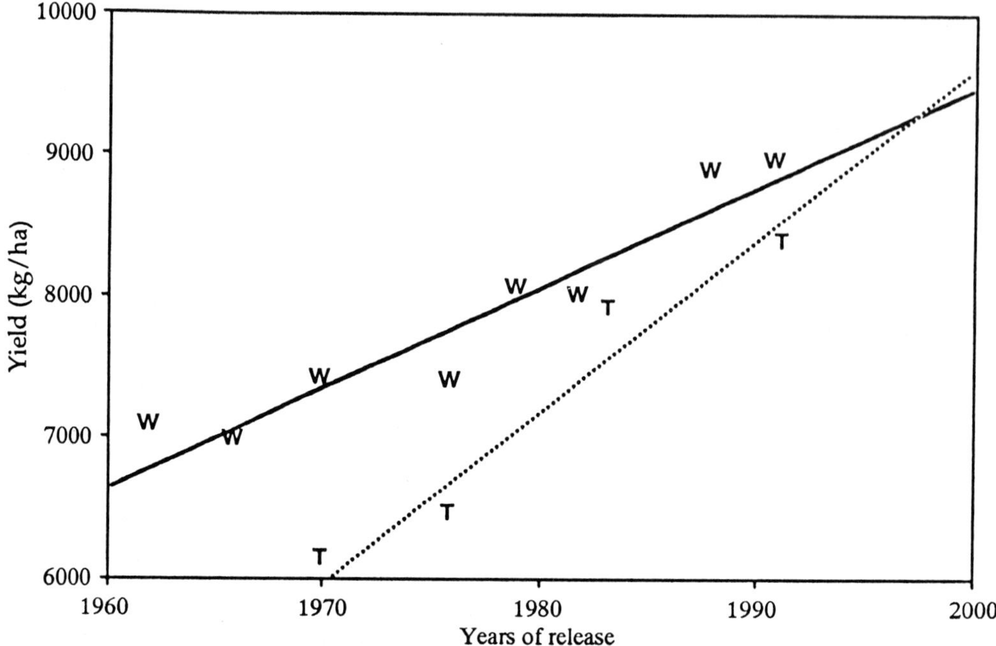

Figure 1. Historical yield gains of bread wheat (W) and triticale (T). Leading W and T varieties grown in optimally managed yield trials in northwestern Mexico; mean of years 1991-92 and 1992-93. Source: K.D. Sayre, pers. comm.

with wheat in yield potential and that it may eventually surpass wheat--even in optimal environments (Figure 1). These data are based on the average performance during 1991-92 and 1992-93 crop cycles of an historical set of varieties planted under optimal conditions at the CIANO experiment station in northwestern Mexico. Under optimal conditions, current yield gains through breeding are 71 kg/ha per year for bread wheat and 122 kg/ha per year for triticale. If these trends continue, triticale will have the same yield potential as that of bread wheat by 1997. Thus, from a yield potential point-of-view, triticale is definitely competitive with wheat.

To date, the only known defect that the "R" genome brings to triticale is inferior breadmaking quality. In several research institutions, attempts are underway to transfer the high molecular weight glutenin alleles from the "D" genome to the "R" genome. CIMMYT, with the help of Dr. A.J. Lukaszewski in California, looked at the contribution of all the D genome chromosomes through substitutions in the triticale background and found that 1D has the highest contribution towards breadmaking properties (unpublished). Based on this information, Dr. Lukaszewski induced 1D, 1R translocations in the triticale variety, Rhino. A few of these allele transfers (2+12 and 5+10) are at the disposal of CIMMYT but not yet available for general distribution because Dr. Lukaszewski has yet to publish the details. In the few confirmed transfers we have examined (Table 5), substantially improved breadmaking properties appear to be present. Although Rhino is an excellent genotype with respect to yield potential, test weight, disease resistance, and adaptation, it probably has the poorest breadmaking quality among all triticales. So, attempts are underway to combine these favorable alleles with triticales that already have better breadmaking properties. Once this is achieved, we are confident that the breadmaking properties of triticale will be comparable to those of bread wheat. This opens enormous opportunities for triticale to become a direct competitor with bread wheat in meeting the ever-growing food demands of the world.

Table 5. Milling and baking quality parameters of Rhino and Rhino with IRS IDL translocation. Average of 14 observations.

	TW	FY	FP	SDS	ALV
Rhino	76.8	72.7	8.50	5.60	102.5
Rhino IRS IDL (5+10)	75.8	70.7	8.94	7.98	143.6

TW = Test weight; FY = Flour yield; FP = Flour protein; SDS = Sedimentation; ALV = Alviogram volume.

Source: Unpublished data of A.J. Lukaszewski, W.H. Pfeiffer, and R.J. Peña.

In conclusion, evidence suggests that there are strains of triticale available today that are superior to wheat with respect to tolerance to abiotic stresses like drought and cold, phosphorus extraction ability, nitrogen uptake efficiency, and tolerance to minor element deficiencies and toxicities. By 1997, we will be able to add yield potential to this list. At the same time, evidence indicates that we can have triticale with breadmaking quality approaching acceptable levels to that of wheat. Thus, when the world is faced with the problem of slowing productivity growth for established crops like wheat, maize, and rice, this newcomer opens an important additional option for enhancing global cereal productivity.

References

1. Longmire JL. Longer term developments in the world cereals market: Looking towards 2000. Draft paper, CIMMYT Economics Program. CIMMYT. Mexico, D.F., 1987.

2. Byerlee D, Moya P. Impact of International Wheat Breeding Research in the Developing World, 1966-90. CIMMYT. Mexico, D.F., 1993.

3. CIMMYT. The World Wheat Situation: Current Developments and Emerging Trends. 1992/1993 CIMMYT World Wheat Facts and Trends, Part 2. CIMMYT. Mexico, D.F., 1993.

4. Ortiz-Monasterio JI, Sayre KD, Pfeiffer WH. Differences in nitrogen recovery among CIMMYT's bread wheats and complete and 2D(2R) substituted triticales. Triticale Topics 1993;11:6-9 (International Edition, New South Wales).

5. Varughese G. 1986. Triticale--a crop for marginal environments. In CIMMYT Research Highlights. 1985. CIMMYT. Mexico, D.F., 1986: 72-80.

TRITICALE - TODAY AND TOMORROW

Rolf Schlegel

Institute of Plant Genetics & Crop Plant Research

Corrensstr. 1, D-06466 Gatersleben, Germany

Abstract

Among three basic ploidy levels so far developed for triticale, the hexaploid type, *Triticale turgidocereale*, became the most important for breeding and agriculture. Although during the past three decades the acreage of triticale subsequently increased of about 2.4 Mio ha and more than 120 varieties are available in about 35 countries, the further progress requires some new strategic accomplishments. Additive as well as the combining effects of wheat and rye genomes were provable for phenotypic characters. They can be utilised in order to develop an improved crop plant compared to the established cereals either for low-input or marginal environment conditions. The rye genome bears the potential for better water use capacity, higher nutritional efficiency, disease resistance and adaptability, while the wheat component can contribute to essential yield and quality characters. Considering those prerequisites and depending on the specific national requirements of agriculture triticale can be bred for different end-use orientations (Pfeiffer 1993). A systematic broadening of genetic variability remains a main task. However, the recombination of wheat and rye genomes should be realised stepwise and avoiding heavy disturbances of the genetic and cytological balance of advanced secondary types.

Introduction

Notwithstanding the early enthusiastic and recently critical discussions on triticale, it remained a cereal plant with considerable future promise. While geneticists influenced the progress of research during the first 70 years, since the seventies the breeders have taken the initiative with all their accumulated knowledge, practical experience and determination. The examples of Russia, Poland and lately Germany demonstrate what a few breeders can perform over a comparatively short period. The specific advantages of triticale were utilised under different ecological environments in these countries.

Since the regular triticale symposia of the past three decades the regional cropping has been estimated. It corresponds with the time of systematic breeding activities. The Polish and the CIMMYT programmes might serve as a reflection of this development (Skovmand et al. 1984, Wolski 1987). The percentage of the world cereal production is still extremely small (< 1 %) but a subsequent increase of triticale acreage can be observed (Fig. 1). The introduction of new crops

H. Guedes-Pinto et al. (eds.), Triticale: Today and Tomorrow, 21–31.
© 1996 *Kluwer Academic Publishers. Printed in the Netherlands.*

in Europe such as potato, maize or sugar beets took much more time in order to reach the same acreage as presently accounts for triticale.

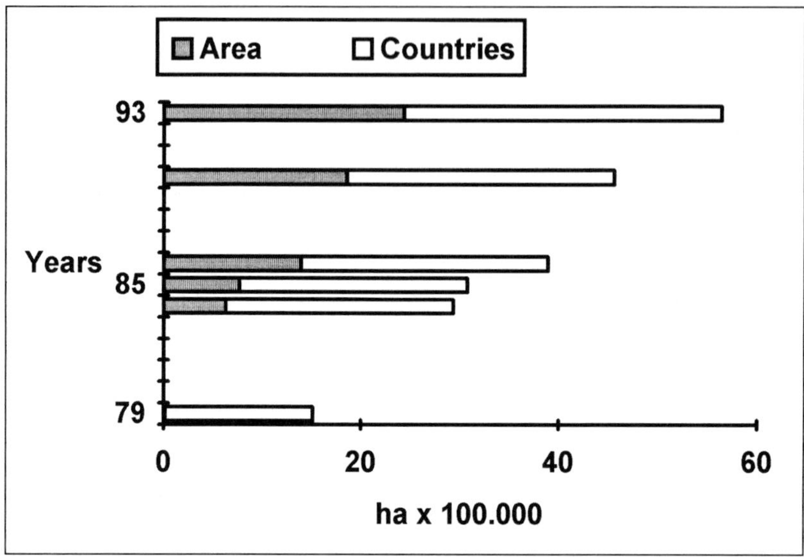

Fig. 1. World triticale area (ha x 100.000) and number of countries growing triticale (pooled of data from Anonymos 1980, 1984, 1985, 1988, 1990 and 1993).

The slow progress in triticale expansion is either a function of breeding input (capacities are small when they are compared with the capacities of established crops), the agronomic/economical interest, or a function of the potential of wheat-rye combination to give a promising new cereal. Depending on these three aspects one can assess advances in triticale either as very pessimistic or very optimistic. During this symposium let us talk about the realistic perspectives.
We know that we cannot expect a miracle plant that gives us all the advantages of two well-adapted species. This would contradict any biological logic and breeder's experience. However it seems possible to adjust a more productive cereal plant for less favourable conditions. Therefore we are at the parting of the ways - either to create a cereal competing with most advanced varieties of wheat, barley or maize under optimum growing conditions or to produce a plant for marginal sites with a reasonable yield, productivity and quality. By reason of the surplus production of developed countries and the ecological problems with high-input agriculture, the question seems to be decided in favour of the latter alternative, at least in the medium term.

What is triticale today?

The sophisticated techniques of inhibition of pro- and postgamous incompatibility, of embryo rescue, protoplast fusion, induced somaclonal variation and the recognition of wheat and rye genomes by means of differential staining and/or labelling procedures allowed the production of a wide range of hybrids which more or less could be cytologically or genetically stabilised if the selection pressure was strong enough. Besides the originally octoploid (AABBDDRR) and later hexaploid (AABBRR) triticales a few tetraploids (ABABRR) including several derivatives were created. After intensive discussion during the 2nd ITS in 1990 most participants agreed to taxonomically differentiate the three basic types of triticale with species-like names as follows (McKey 1990):

Genus:	*Triticum*
Section:	*Triticale*
Notospecies:	*Triticum krolowii (4x)*
	Triticum turgidocereale (6x)
	Triticum rimpaui (8x)

Meanwhile it is accepted that *Triticum turgidocereale* (6x) represents the most important genome combination, i. e. the addition of the complete RR genomes of rye to the complete AABB genomes of tetraploid wheat. Spontaneous or induced chromosome substitutions and translocations within one of the three basic types we should consider as special karyotypic modifications. Against former hopes they seem to remain the exceptions. Thus the attention of the further discussion should be focused on the complete hexaploid triticale.

Does the genome/gene composition of triticale offer new resources of recombination and selection?

Basically, recombination of genetic determinants (genes) which discriminate the parents (wheat and rye) can be expected in the hybrid. It requires the intimate association of the different genetic material that is achieved by nuclear fusion or meiosis. The recombination itself is realised either by chromosome assortment which leads to interchromosomal recombination of unlinked traits (the combination of whole genomes being just a special case of it), by intrachromosomal recombination of linked markers *via* crossing over or by gene conversion of one allele in the presence of the other allele resulting in nonreciprocal recombinants. The phenotypic mixture of traits in triticale confirms the above assumption.
Compared to established cereal plants, three possibilities of recombination are available in order to increase the variability of triticale:

(1) Recombination of complete genomes of wheat and rye
(2) Recombination of chromosomes within and between wheat and rye genomes
(3) Recombination of genes within and between wheat and rye genomes

The first approach is practised whenever a tetraploid wheat is combined with diploid rye. Each amphidiploid hybrid represents another recombinant when at least one parent is more or less heterozygous. This is given when allogamous rye is involved in the crosses. The utilisation of inbred lines or self-fertile rye resulted in the similar sort of hybrids, except with lower variation between the offsprings. Experiments demonstrated that they do not contribute to more stable meiosis (Müntzing 1956, Sanchez-Mong 1959, Lelley and Larter 1980), just as restoration of complete heterozygosity by producing F1 hybrids among them (Kaltsikes 1974). When the hybrid derives from a common cross followed by mitotic diploidisation, both the wheat and the rye complement become completely homozygous. Whether or not the rye genome is heterozygous its pairing reduction, univalency and meiotic instability are remarkably increased compared to common rye and to the wheat component (Schlegel et al. 1977, Schlegel and Schrader 1990, cf. Fig. 2). This also reduces the chance of allelic recombination within the rye genome. Aneuploids as a consequence of meiotic disturbances have substantial negative effects on important agronomic traits such as days to heading, tillering, plant height, spike length, number of spikelets per spike, floret fertility and thousand grain weight (Lelley 1992). In order to produce new variability only in this way, an enormous number of hybrids must be realised, without a guarantee (or with low probability) that one of them meets all breeder's objectives. The chance of high performing primary hybrids by preselection of parents, i.e. on the basis of good 'general genomic combining

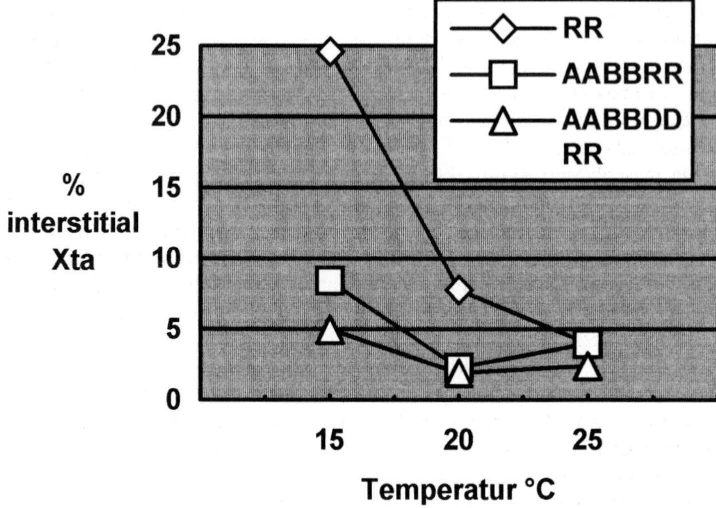

Fig. 2. The relationship between the percentage of interstitial chiasmata of the total number of chiasmata per rye genome and different temperatures during meiosis in diploid rye, hexaploid triticale and octoploid triticale

ability' (Lelley 1992), is still controversial (Oettler et al. 1991, Lelley 1992). The close correlation between the wheat component and the derived triticale point to strong additive effects of wheat on many morphological traits of triticale (cf. Table 1). Other qualitative characters including resistance to stem rust or ribosomal activity of rye are suppressed as well, depending on the strength of wheat genotype which is chosen (Kerber 1983, Gustafson et al. 1988, Kapila and Sethi 1994).

Notwithstanding, as found in numerous studies the fertility and seed quality of those hybrids is insufficient since they are still genetically unbalanced. However they can serve as important bridges between the gene pools of the donor species and secondary triticales or they can be themselves the starting-point of secondary plants.

Table 1. Phenotypic correlation between the wheat and rye parents and the primary triticale derived

Character	Wheat		Rye*
	T. turgidum*	T. durum**	S. cereale
Length of last internode	-	0.79	-
Awn length	-	0.25	-
Days to heading	0.90	0.76	0.67
Tillering	0.69	0.35	0.27
Height	0.99	0.82	0.06
Spike length	0.98	0.65	0.40
Spikelets/spike	0.97	0.67	0.48
Florets/spike	-	0.25	-
Floret fertility	0.49	0.10	0.19
TGW	0.63	-	0.10

* data modified after Lelley (1992); ** data modified after Wandelt (1988)

The second approach *per se* presupposes the presence of a primary or secondary triticale that allows either the intragenomic or intergenomic exchange of linkage groups by spontaneous or induced events. Whenever in a cross one of the A, B or R genomes become haploid or more than two non-homologous chromosomes are univalent, the chance of spontaneous chromosome substitutions arises. Depending on the selective importance and the genetic suitability they can be consciously or non-consciously selected. Both the 2D(2R) and the 6D(6A) substitutions found in advanced triticale arose in such a process. Of course, the same could also be experimentally performed by using defined aneuploids of wheat and rye. However, this would require a better understanding of the influence as well as substitution of individual chromosomes on the cytological and genetic balance of the hybrid product. At present, induced substitutions would be useful if defined wheat or rye chromosomes are available which determine monogenic or oligogenic characters with dominant expression. Since there is a simultaneous removal of a great number of mostly unknown genes located on the substituted chromosome, corresponding negative effects and interactions have to be considered likewise.

The third source of recombination, the recombination of alleles of different loci by crossing over, can primarily occur within the parental genomes realised by triticale x triticale, triticale x wheat and triticale x rye crosses. Homoeologous recombination between rye and wheat is extremely low and can usually happen when corresponding genomes are haploid throughout meiosis (Schlegel and Weryszko 1979). It is generated in (wheat x rye) x triticale crosses.

As studied by Lelley (1992) intragenomic recombination is provable in hexaploid triticale. However, the phenotypic effects are difficult to predict. It seems that morphological traits such as heading, tillering, height, spike length, spikelet number and TGW are predominantly

Table 2. Effects of 1RS.1BL chromosome translocation on yield components of the hexaploid wheats 'Borenos', 'Mikon', 'Bovictus' and/or corresponding hybrid populations (modified after Schlegel and Meinel 1994)

Character	Wheat varieties/populations						
	Bor 1BS	Mik 1BS	Bov 1RS	Bov/Bor 1BS	Bov/Bor **1RS**	Bov/Mik 1BS	Bov/Mik **1RS**
Spikes/plant	6.1	8.3	4.7	6.2	4.9*	5.3	5.4
Plant height (cm)	58.4	55.6	57.5	60.9	63.2	60.9	63.0
Spikelet number/spike	17.4	18.1	18.7	18.1	17.9	18.8	19.4
Seed number/spike	39.8	44.1	51.9	44.4	**49.6*****	46.6	**54.3*****
Seed number/spikelet	2.29	2.42	2.76	2.50	**2.76*****	2.51	**2.81****
Spike yield (g), calculated	1.69	1.71	2.32	1.85	2.22	1.98	2.38
TGW (g)	43.0	38.8	44.8	41.8	44.8**	42.6	43.8

* / **/*** significant at 5%, 1% and 0.1% level after *U*-test; only pairwise comparisons of Bov/Bor 1BS vs. Bov/Bor 1RS and of Bov/Mik 1BS vs. Bov/Mik 1RS) are indicated in this table;

determined by the wheat component, while, for example, the floret fertility is a matter of interaction of parental genomes, possibly triggered by genes of rye. The increased number of seeds per spikelet found in primary hexaploid and octoploid triticale (Wandelt 1988) and in 1RS.1BL wheat-rye translocations of hexaploid wheat (Schlegel and Meinel 1994; Table 2) might be caused by the same positive gene interaction. Another example shows that heat-shock proteins of wheat can be suppressed in triticale by genes of the rye genome resulting in a different pattern of stress reaction. Major repressive effects appear to be confined to chromosomes 1R and 3R (Somers and Filion 1990). In addition to the additive effects in triticale those interactions offer extra sources of selection which are not expressed in wheat or rye.

Variation coefficients for some yield characters of fully recombined genomes in comparison to recombination restricted to either wheat or rye genomes confirm the general expectation of greater recombination chances of three genomes of triticale instead of two in wheat or one in rye, although the mean performance of fully recombined triticale is significantly lower than that of partly recombined wheat genomes (Lelley 1992).

What are the advantages of triticale compared to competitive cereals?

Wheat characters dominate the plant growth habit of triticale (cf. Table 1). This might be one of the reasons why wheat breeders treated triticale similar as wheat, although there is quite a high degree of allogamy depending on the parents and weather conditions. Besides the vegetative vigour the spike of triticale became most interesting because of the increased number of florets per spike. If this heterotic character could be stabilised without grain shrivelling it could retain its value for breeding as it has in hexaploid wheat (cf. Table 2). Extra fertilisation of triticale does not compensate the partial failure of seed setting and grain filling. N fertilisation only tended to increase the number of spikes/m^2 and, to lesser extent, the number of seeds/spike (Kochhann et al. 1990).

What about the adaptability of triticale? Adaptation is defined as the ability of a genotype to produce consistently high yields of a range of environments. General and specific adaptation are just relative terms referring either to a broad or narrow response. Although this definition paraphrases very complex genotype x environmental interactions and precise measurements of adaptability are laborious, difficult and expensive, a first study of Pfeiffer and Fox (1991) and Fox et al. (1990) confirmed the common experience that (a) hexaploid triticale shows a slightly increased adaptability when it is compared with hexaploid wheat (see also Riede et al. 1991) and (b) a superior performance of complete-genomic hexaploid triticale when compared with 2D(2R) substituted types. Under most ecological conditions the completely genomic hexaploid triticales have adaptive advantages over the wide-spreaded 2D(2R) substituted types which one can ascribe to reduced genetic variability introduced by the rye genome as a balanced genetic system *per se*. As a consequence, the ratio of complete:substituted triticale strains of the CIMMYT breeding programme changed from 25:75 in 1984 to 90:10 in 1991 (Ortiz-Monasterio et al. 1993).

During the fifties it was intended in Canada to use the triticale as a bridge for transfer of resistance genes to leaf diseases of rye into durum wheat. It can still be an option but meanwhile it may serve as a bridge host for both wheat and rye diseases. Arseniuk and Czembor (1991) reported that triticale promotes the development of new pathogens which infect both wheat and rye. However, triticale is still relatively resistant to many foliar diseases affecting other cereal crops. In Central Europe most triticales are more resistant to rusts and powdery mildew than wheat and barley. Late infections are observed more frequently. Resistance to *Septoria* and *Fusarim* diseases remains an important breeding task.

The good nutritional quality remains one of the most important traits of triticale in feeding poultry and pigs (Belaid 1993, Leterme 1991). Generally, it exceeds the quality of rye. It is equal to wheat in poultry and better than barley in pig feeding (Wolski 1991). Among Polish triticales the crude protein content ranged from 9.4-15.3 % (at 11 % moisture), i.e. about 2 % higher than in rye and about equal to wheat. The mean lysine content amounts to 0.22 g/ 1g N (Rakowska et al. 1992). The digestibility measured in rats and pigs is within the same range of wheat and about 13 points higher than of rye. Compared with other cereals triticale has a better balance of essential amino acids, particularly lysine and methionine. Threonine is also found in relatively high levels compared to maize and soybean. It has an excellent nutritional profile, but it is still a by-product of breeding for human consumption.

Table 3. Relative starch amount in the dry matter of seeds of different wheats, ryes and two 1RS.1BL wheat-rye translocation lines (after Schlegel and Schnüber 1993)

Material		% starch	Total
Wheat:	Chinese Spring	55.3	
	Borenos	64.1	
	Fakon	64.3	
	Mikon	62.9	61.7
Rye:	Petkuser	52.4	
	King II	46.7	
	Inbred 361	42.4	47.1
Translocation:	Bovictus	64.6	
	Bov/Bor	63.8	
	Bov/Mik	65.0	64.5

Like each cereal, triticale shows a negative correlation between the fibre content and its energy value. The fibre as part of the seed structure cannot be totally removed but components such as lignins, which has the most detrimental influence on nutrient digestibility of pigs and poultry, can be reduced by selection.

Resorcinols as toxic alkaloids give triticale low palatability and affect animal feed intake. They are inherited from rye. Most of the new varieties contain smaller amounts of them. It ranged among 400 and 600 mg/g. The mean value is close to wheat while rye usually contains between 900 and 2000 mg/g (Rakowska et al. 1992). For initial triticale crosses rye genotypes with low resorcinols should be selected. The same is true for pentosans. The 5C-polysaccharides whose soluble fraction fills itself with water and becomes very sticky. Greater amounts in food decrease feed intake in chicken and piglets, hamper digestion and make droppings very sticky, resulting in inhibition of growth.

An investigation of the starch content in wheat and rye seeds showed that rye contains about 15 % less than hexaploid wheat. The mean value of 6.1 % of wheat was not provable modified when the chromosome arm 1RS was transferred to wheat (Table 3).

Antinutritional factors such as trypsin inhibitors seem to be less of a problem with new triticale cultivars. The protein-polysaccharide complexes did not limit growth of pigs, although the contents of trypsin, chymotrypsin inhibitors, water-soluble pentosans and sugars are higher in triticale than in wheat (Batterham et al. 1989). The concentration of trypsin inhibitor ranged among 0.6 and 1.6 that is close to wheat but much lower as in rye (Rakowska et al. 1992).

By improved test weights which reach already the level of rye and, in some strains, of wheat (72.3 kg/hl), the protein content was gradually reduced. Negative effects on baking quality must be expected due to the presence of secalins *Sec1, Sec2 and Sec3* located on chromosome arms 1RS, 2RS and 1RL, respectively (r=-0.42**), but also by the lack of glutenin (*Glu1*) and gliadin (*Gli1*) genes on wheat chromosome 1D, by 6D(6A) substitutions with the *Gli2* locus mapped on 6D, and by the rigorous selection for sprouting resistance, i.e. decreasing of amyolytic enzymes. However, re-introduction of alleles of the loci *Gli1-D1*, *Glu1-D1* and *Glu2-D1* into hexaploid triticale can strongly improve the baking quality (Kazman 1994, pers. comm.).

Comparisons between hexaploid triticale and hexaploid wheat show that triticale seems to have a better grain yield under soil residual nitrogen than wheat. The total nitrogen accumulated at heading and at physiological maturity is significantly higher in hexaploid triticale, particularly in complete types, than in wheat at lower levels of nitrogen dressing (0, 75 kg N/ha). At high N inputs (150, 300 kg N/ha) the differences between complete and substituted types are less distinctive.

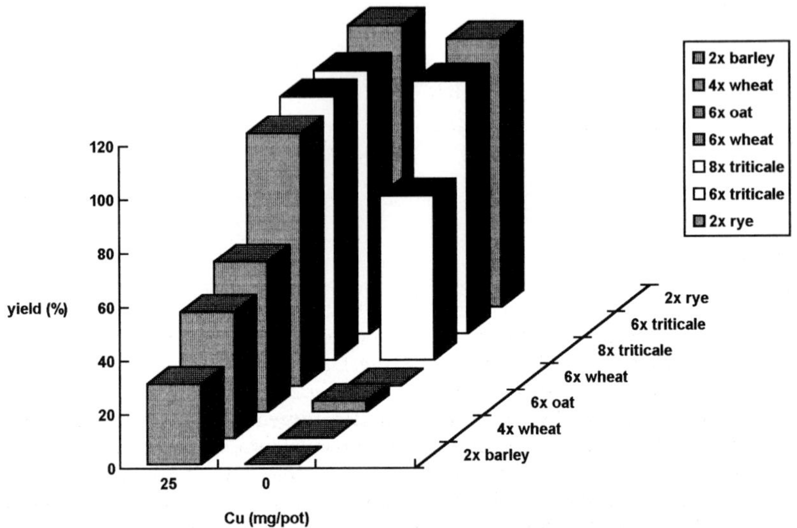

Fig. 3. Relative yields of diploid barley (Betzes), tetraploid wheat (Nodak), hexaploid oat (Sun II), hexaploid wheat (Holdfast), octoploid triticale (Holdfast-King II), hexaploid triticale (Bokolo), and diploid rye (King II) under copper deficient growth compared to sufficient supply of 50 mg/pot (= 100 %).

The data indicated that the larger amount of N accumulation in triticale is caused by a higher absorption capacity rather than by a longer period of growth (Ortiz-Monasterio et al. 1993).

Experiments of Camargo et al. (1990) demonstrated that all triticales tested showed also a high degree of tolerance to iron toxicity (10 mg/l Fe^{2+}) when they were compared with hexaploid wheats and barley. Additional to the comparably high aluminium tolerance (Aniol 1985, Carmargo et al. 1990) the triticale has great potential for drought and heat stress, with and without irrigation. Comparative studies showed that rye needs only 70-80 % of the water that wheat requires to produce the same amount of dry matter (Bushuk 1976). Under drought and cool growth conditions a competitive advantage over wheat was found as well.

Moreover, there are high broad sense heritability estimates (for example 0.89 for aluminium tolerance) which could facilitate directed intercrosses and selection for such sort of traits (Schlegel 1994). Triticale lines were already developed for Australian problem soils which grow more successful than wheat and barley on waterlogged acid soils (pH 5.4), on boron-toxic and manganese-deficient and zinc-deficient sites (Cooper 1991). The level of Zn and Mn efficiency corresponds to the level of diploid rye parent. Genes for Mn efficiency and boron tolerance were mapped on rye chromosome 2R (Graham 1987, Manyowa and Miller 1991). This indicates a dominant contribution by the rye genome. The same is true for copper efficiency (Fig. 3). A dominant gene or gene complex *(Ce)* was physically mapped on chromosome arm 5RL; linked to a dominant hairy neck character *(Ha1)*, to esterase loci *(Est6, Est7)* and probably to several RFLP markers (Schlegel et al. 1993, Kynast and Schlegel 1994, unpubl.). On the same chromosome arm there are genes which possibly control iron efficiency, at least the mobilisation of phytosiderophores.

The release of phytosiderophores is the main mechanism of cereal plants to acquire iron in the rhizosphere. In soils copper also gets easily mobilised by phytosiderophores and copper phytosiderophore chelates are somewhat preferentially taken up by the translocator for Fe

Table 4. Relative grain yields of 6x winter triticale on four classes of soil (modified after Wolski 1991)

| | Classes of soil | | | |
Material	very fertile	fertile	medium	light
Triticale	102	102	102	102
Wheat	104	102	101	90
Rye	89	96	95	101
Check (dt/ha)	61.2	58.4	48.8	44.8

phytosiderophores. Because of the similarities in the role of phytosiderophores in Fe and Cu acquisition and the fact that chromosome 5R was identified to carry the genes for Cu efficiency and for phytosiderophore synthesis - mugineic acid synthetase catalysing synthesis of MA from 2'-deoxymugineic acid and the gene for 3-hydroxymugineic acid synthetase catalysing synthesis of HMA from MA-, it has been concluded that iron and copper efficiency are controlled by a common mechanism. When different rye varieties are compared, those genotypes showing the highest copper and iron efficiency also represent the genotypes with the highest grain yield under Cu and Fe deficiency and with the highest HMA exsudation by the their roots (Schlegel 1992). Therefore, it is concluded that the rye genome contributes to better yield performance of triticale, particularly on sandy and acid soils with or without high level of free aluminium when it is tested with wheat. The data of Table 4 confirm this conclusion.

Final remarks

In 1988 we celebrated the centenary of triticale. It was in 1888 when a breeder, the German W. Rimpau, produced the fertile wheat-rye hybrid from 'Roter saechsischer Landweizen' x.'Schlanstedter Roggen' in order to combine the nondemanding features of diploid rye with the quality characters of hexaploid wheat. After several approaches during a long period breeders have returned to this target, namely to use the genetic potential of rye for increase the productivity of plants with wheat-related quality under marginal environmental conditions worldwide!

References

Aniol A 1985 Breeding triticale for aluminium tolerance Proc. 3rd EUCARPIA Meeting of Cereal Section on Triticale, Clermont-Ferrand (France), 573-582.
Anonymos 1980. Proc. 2nd EUCARPIA Meeting of Cereal Section on Triticale, Radzikow (Poland)
Anonymos 1984 Proc. 3rd EUCARPIA Meeting of Cereal Section on Triticale, Clermont-Ferrand (France)
Anonymos 1985 Proc. 1st Int. Triticale Symp., Sydney (Australia)
Anonymos 1988 4th EUCARPIA Meeting of Cereal Section on Triticale, Schwerin (Germany)
Anonymos 1990 Proc. 2nd Int. Triticale Symp., Passo Fundo (Brazil)
Anonymos 1993 Triticale Topics 11
Arseniuk E and Czembor H J 1991 Triticale diseases in central Poland in 1991. Triticale Topics 7, 2-4.

Batterham E S, Saini H S and Andersen L M 1989 The effect of mild heat on the nutritional value of triticale for growing pigs. Anim. Feed Sci. Techn. 26, 191-205

Belaid, A., 1993. Nutritive and economic value of triticale as a feed grain for poultry. Triticale Topics 11, 10-16

Bushuk W 1976 Rye: Production chemistry, and technology. St. Paul, Amer. Ass. of Cereal Chemists, Inc., 1-181.

Carmargo C E O, Felicio J C and Ferreira Filho A W P 1990 Abiotic stress, mineral toxicities, low nutrient availability, drought and heat. 2nd Int. Triticale Symp., Passo Fundo (Brazil), 164-170.

Cooper K V 1991 Breeding triticale for Australian problem soils. 2nd Int. Triticale Symp., Passo Fundo (Brazil), 188-195.

Fox P N, Skovmand B P, Thompson B K, Braun H J and Cormier R 1990 Yield and adaptation of hexaploid spring triticale. Euphytica 47, 57-64.

Graham R D 1987 Triticale: A cereal for micro nutrient deficient soils. Triticale Topics 1, 6-7.

Gustafson P J, Dera A R and Petrovic S 1988 Expression of modified rye ribosomal RNA genes in wheat. Proc. Nat. Acad. Sci., USA, 3943-3945.

Kaltsikes P J 1974 Univalency in triticale. Proc. Int. Symp., El Batan (Mexico),159-167.

Kapila R K and Sethi G S 1994 Expression of some rye (*Secale cereale*) traits in different triticale (x Triticosecale) x wheat (*Triticum aestivum* L.) hybrids. Cer. Res. Comm. 22, 27-32.

Kerber E R 1983 Suppression of rust resistance in amphidiploids of *Triticum*. Proc. 6th Int. Wheat Genet. Symp., Kyoto (Japan), 813-817.

Kochhann C H, Baier A C and Wiethölter S 1990 Harvest index, yield components and nitrogen content in triticale, wheat and rye. Proc. 2nd Int. Triticale Symp., Passo Fundo (Brazil), 71-73.

Lelley T 1992 Triticale, still a promise? Plant Breed. 109, 1-17.

Lelley T and Larter E N 1980 Meiotic regulation in triticale: interaction of the rye genotype and specific wheat chromosomes on meiotic pairing of the hybrids. Can. J. Genet. Cytol. 22, 1-6.

Leterme P 1991 Items on triticale from Belgium. Triticale Topics 7, 2

Manyowa N M and Miller T E 1991 The genetics of tolerance to high mineral concentrations in the tribe *Triticeae* - a review and update. Euphytica 57, 175-185.

McKey, J., 1990. Taxonomy of ryewheat. Proc. 2nd Int. Triticale Symp., Passo Fundo (Brazil), 36-40.

Müntzing A 1956 Studies on the properties and the ways of production of rye-wheat amphidiploids. Hereditas 25, 387-430.

Oettler G, Wehmann F and Utz H F 1991 Influence of wheat and rye parents on agronomic characters in primary hexaploid and octoploid triticale. Theor. Appl. Genet. 81, 401-405.

Ortiz-Monasterio R, Sayre K .D and Pfeiffer W H 1993 Differences in nitrogen recovery among CIMMYT's bread wheats and complete and 2D(2R) substituted triticales. Triticale Topics 11, 6-9

Pfeiffer W H 1993 Triticale improvement strategies at CIMMYT: Exploiting adaptive patterns and end-use orientation. Triticale Topics 11, 18-29.

Pfeiffer W H and Fox P N 1991 Adaptation of triticale. Proc. 2nd Int. Triticale Symp., Passo Fundo (Brazil), 54-63.

Rakowska M, Boros D and Gastrorowska M 1992 Quality traits of the grain of Polish varieties of triticale. Triticale Topics 8, 4-9.

Riede C E, Campos L A C and Fonseca Junior N S 1991 Proc. 2nd Int. Triticale Symp., Passo Fundo (Brazil), 79-85.

Sanchez-Mong E 1959 Hexaploid triticale. Proc. 1st Int. Wheat Genet. Symp., Winnipeg (Canada), 181-194.

Schlegel R 1992 Verbesserung der Mikronährstoff-Effizienz beim Weizen durch Introgression. Ber. 43. Arbeitstag. Saatzuchtleiter, Gumpenstein (Austria), 167-170.

Schlegel R. 1994 The genetic control of nutritional and stress characters in cereals. Plant and Soil, in press.

Schlegel R. and Weryszko E 1979 Intergeneric chromosome pairing in different wheat-rye hybrids revealed by the Giemsa banding technique and some implications on karyotype evolution in the genus *Secale* . Biol. Zbl. 98, 399-407.

Schlegel R and Schrader O 1990 Pairing restriction in homologous rye chromosomes of amphidiploid wheat-rye hybrids determined by genome dosage and temperature. Proc. 2nd Int. Triticale Symp., Passo Fundo (Brazil), 359-367.

Schlegel R. and Schnüber G 1993 No effect of chromosome arm 1RS on starch content of hexaploid wheat. Cer. Res. Comm. 21, 297-300.

Schlegel R. and Meinel A 1994 A quantitative trait locus (QTL) on chromosome arm 1RS of rye and its effect on yield performance of hexaploid wheat. Cer. Res. Comm. 22, 7-13.

Schlegel R., Bretschneider H and Viehweger S 1977 Über die Beziehungen zwischen Meioseverhalten und Aneuploidie bei hexaploidem Triticale. Arch. Züchtungsforsch. 7, 69-77

Schlegel R., Kynast R., Schwarzacher T, Römheld V. and Walter A 1993 Mapping of genes for copper efficiency in rye and the relationship between copper and iron efficiency. Plant and Soil 154, 61-65.

Skovmand B P, Fox P N and Villareal L R 1984 Triticale in commercial agriculture: progress and promise. Adv. Agron. 37, 1-45.

Somers D J and Filion W G 1990 Influence of the rye genome on heat-shock protein expression in triticale using rye addition lines. Proc. 2nd Int. Triticale Symp., Passo Fundo (Brazil), 332-337.

Wandelt W 1988 The influence of wheat and rye substitution effects on the performance of primary triticale. Tag.Ber. AdL (DDR) 266, 61-68.

Wolski T 1987 Twenty years of winter triticale breeding in the programme of Poznan plant breeders. Triticale Topics 2, 2-3

Wolski T 1991 Breeding triticale for production and stress resistance. Proc. EUCARPIA Cer. Sec. Meet., Schwerin (Germany), 189-193.

MARKETING - DILEMMA OR DREAM

Christopher Green
Semundo Limited, Cambridge, UK

Abstract

Agricultural policies over the last 40 years have had the primary purpose of increasing production. These policies have been supported by grower and market subsidies, which now increasingly bring a burden to the European budget. Todays economic environment is bringing a new focus to agricultural policies which are orientate more towards market needs than solely increasing production.
Triticale finds itself in an increasingly competitive cereals market. Will it remain a minor novelty cereal or does the species have the ability to become a main stream cereal. Why are we producing triticale and what to its future?
The problems of triticale may relate more to the marketplace than they do to the field or laboratory. This is the dilemma facing triticale.

Introduction

The question: "Why has triticale failed to live up to its early expectations?" has repeatedly been asked. This reflects the marketing dilemma of this crop. So what of the dream?

It was the civil rights campaigner, Martin Luther King who, in his clarion speech to the American people, made famous the words "I have a dream". On the 4th April 1968. I was at school, dreaming that I had passed my pending, final examinations, on that same day. Martin Luther King was assassinated. A hundred years before this, a young Scottish Botanist had the dream of bringing together the two species of wheat and rye. His dream was to unite the best characteristics of each of the parents into a new species. His scientific inquisitiveness offered a considerable challenge in those days. The realisation of this dream was the achievement and birth of triticale. His dream had therefore become a reality. With the scientific challenge achieved, a product was created; but what of the market?

H. Guedes-Pinto et al. (eds.), Triticale: Today and Tomorrow, 33–39.
© 1996 *Kluwer Academic Publishers. Printed in the Netherlands.*

To find some of the solutions to triticale marketing, we must understand a few marketing concepts.

The first finite concept to be appreciated is that **production is for consumption.** This is the fundamental principal of business. Products that are not consumed have no future. Whilst accepting that triticale is very much in its infancy as a world cereal, we should be clear at the onset; why it is being produced and for what market? For instance, take bread quality triticale; who is asking for it? Are the varieties being produced to satisfy a scientific or breeding goal, or to satisfy a new market?

The second concept, which I believe important, is the difference between selling and marketing. There are many wordy definitions for marketing, but coming from Ireland, I prefer my own simpler version, which is "Marketing is the creation of a preferential demand". The basic principal of marketing is that it requires a clear appreciation of consumers' 'wants'. It is necessary to understand the consumer markets, their structure, and the environment in which they operate and conduct their business. Factors which influence and affect this market, and the behaviour of the consumer must be understood.

Customers are the foundation to any business. The basic purpose of a business it to create and retain customers. Generally, customers only want to know what a product will do for them, therefore the development of any market must start with an understanding of the customers 'wants' and needs. In the case of the plant breeding industry, there is a chain of customers, from the seed merchant, to the adviser, the grower, the feed nutritionists, the feed producer and finally to the retail customer. For the chain to be complete, each must see the benefit of buying the product. Varieties which are produced without a clear and defined market are invariably sold or 'pushed' along this chain. On the other hand, products which are the result of consumer demand are invariably 'pulled' into the market by consumer demand. This is the subtle difference between selling and marketing. These two different approaches of bringing a product into the market are often referred to as the 'Push - Pull' strategies, where the selling approach is deemed to be the push strategy and a marketing approach the pull strategy.

PUSH STRATEGY

Plant Breeders → Seed Companies → Farmers → Consumers

In this strategy, a variety is introduced and sold to a seed company, who in turn will extol the varietal benefits to the farmer, who will produce the crop and then endeavour to find a buyer or consumer. This push strategy is aggressive by nature, in that each in the chain is pushing the other into action.

PULL STRATEGY

The pull strategy infers a marketing approach to commercialization. In this approach the consumers 'wants' have been identified and the product, or the variety, are developed to meet these needs.

From the diagrammatic representation of this strategy, it will be noted that this process responds to the demands or wants of the consumer. This strategy has created not only a product flow but an effective two-way communication. Unlike the Push strategy, which is a chain, this marketing approach is a continuous cycle which stimulates repeat business. It allows for the adoption of new ideas and facilitates better control of both planning and production. The essence of the pull strategy is that the consumer pulls the product into the market.

The market approach by Peter Kruse in Germany, provides a classic example of the success of adopting a pull strategy. Unfortunately all too often, triticale is being pushed into the market rather than being pulled.

Further differences between selling and marketing are shown in the following table:

SELLING	MARKETING
Emphasis is on the developed variety	Emphasis is on consumer 'wants'
Variety developed without any preconceived marketing plan	Breeding programme tailored to customer needs
Sales volume orientated	Profit orientated
Short term campaigns	Adaptable to meet changing 'wants'
The needs of the setter are more important than the needs of the buyer	The consumer is the final consumer, who ultimately directs the business

The third concept is that of a product life cycle. (See Figure 1) All products go through the same cycle of Introduction, Growth, Maturity and Decline. This principle can apply to a crop, as well as a variety. The time span in any one stage will vary considerably.

Figure 1 - TRADITIONAL LIFE CYCLE OF A PRODUCT - SALES AND PROFIT CURVES

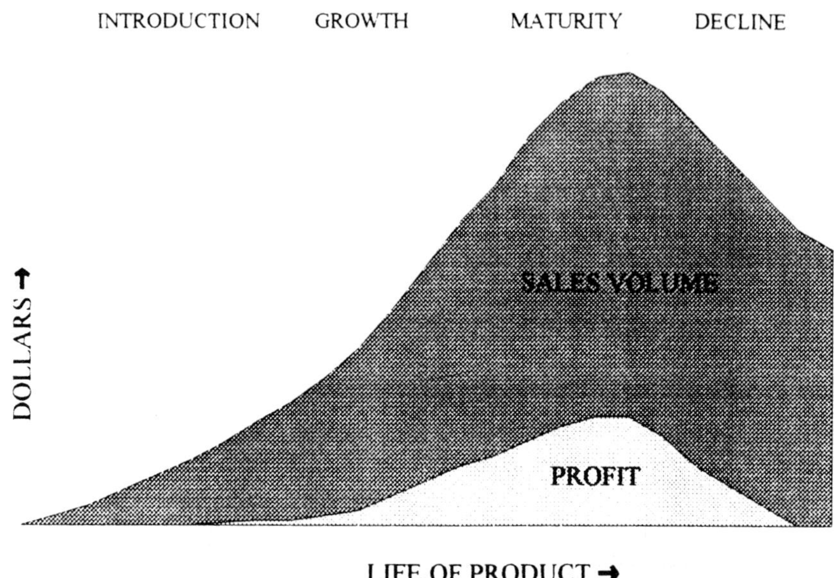

Whilst we accept that triticale, as a global cereal, is in its infancy and therefore in the introduction stage, nationally the picture may be different. In Germany and Poland, for instance, it is in its growth stage.

Bringing the last two concepts together, where a product is being introduced by consumer demand it will spend less time in the introductory phase. This may also mean lower support and marketing costs and a quicker pay-back time. The opposite applies for products being pushed into the market. If we adopt this concept to cereal breeding, then the need for continual introductions is clearly evident.

Turning now to the consumer, what makes them buy triticale? Rather than spend time dealing with the physical factors of price, quality, uniqueness, availability, competitiveness and useability, I want you to consider the sequence of events which lead to a buying decision being taken. This is reflected in the anacronim IADA, which stands for Attention, Interest, Desire and Action. (See Figure 2) Regardless of wether one is introducing triticale or a telecommunication system, the sequence of events is the same.

Figure 2 - THE AIDA PROCESS AND THE HIERARCHY OF EFFECTS MODEL

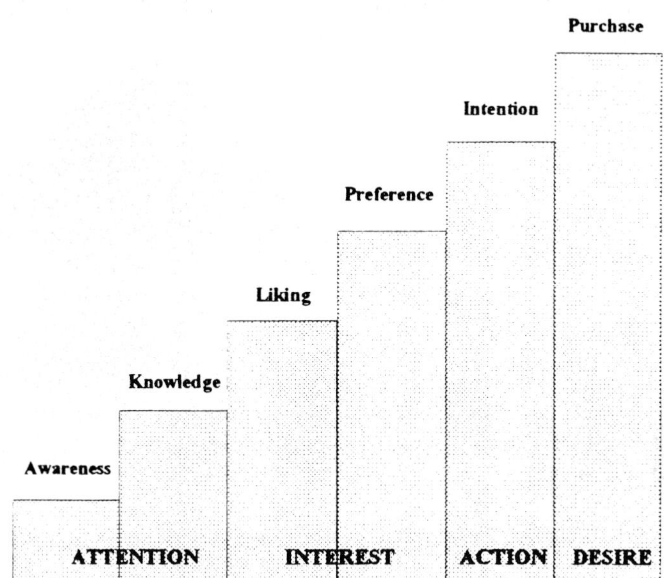

This can be developed further into the various steps which have to be pursued in order to ensure market effectiveness. These steps are awareness, knowledge, liking. preference, intentions and purchase.

APPLYING THESE CONCEPTS TO TRITICALE MARKETING

At the onset we must accept that the purpose of developing triticale and the breeding of new varieties is to create customers. This can be achieved through agronomic and quality improvements, which increase the competitiveness of triticale, but we should remember that the goal is to have our products consumed.

Triticale is at the beginning of its product life cycle. We may be expecting too much too soon. The prime movers for the crop are present at this conference and perhaps we are impatient for a breakthrough. From my perspective, I do not believe that there will be a dramatic breakthrough. I believe that political and economic factors will stimulate interest in triticale and in some countries growth will occur quite quickly. However, the evolutionary process of bringing this crop to the world market will be slow but steady. This will mean a prolonged introductory stage which may be better counted in decades rather than in years.

We must be clear of the market for triticale. The mass market is as a feed grain. Failure to achieve adequate penetration in this market sector will eventually result in failure in other secondary markets. The secondary or niche markets are best developed on the back of a sustained established market. Our focus must be to target triticale as a cost efficient quality feed grain.

We must better identify the customers for triticale and realise it is not only the person who makes the buying decision, but who is its customer. For instance, the agronomist and the animal nutritionist are important links in the chain, especially as they will act as information gatherers. Careful consideration must be given to our communication approach if we are to be successful in satisfying our customer needs.

Reflecting on the behaviour process of buyers, then the following can act as a guideline:

AWARENESS — The market must be made aware of triticale.

KNOWLEDGE — Simply being aware of triticale will not create a sale. The next step is to inform the selected target markets about the benefits and characteristics of triticale.

LIKING — A favourable attitude towards triticale must then be developed. Joint evaluation or testing can help at this stage.

PREFERENCE — Providing that triticale can do the job, is not sufficient in itself. The next stage is to engineer a preference for triticale. This requires extolling the benefits further and portraying the values to the consumer for the use of triticale.

INTENTION & PURCHASE — With the 'intend to buy' stage reached, the consumer then needs confidence that this decision will be a correct one. Only then are such factors like price, quality, availability and continuity of supply, important.

It is essential that there is a well-structured plan and that information and presentations are well targeted to those who can both influence and instigate buying decisions.

One problem with triticale is that there is too much attention being paid to technical and scientific issues, which insufficient effort on understanding the market and the consumer. This I believe is clearly reflected by the profile of delegates at this conference. Where are the consumers? The CABI database has revealed that there have been some 2,700 papers published on triticale since 1984 and out of the most recent 100, none have dealt with marketing issues.

There is the need to develop an educational campaign. There exists a great deal of scepticism about triticale. This is no doubt due to misinformation and perceived values and beliefs. The educational campaign should be designed not only to create a better awareness of the crop but to impart salient information across a wide spectrum of individuals and organisations. Only when interest has been stimulated will the crop move forward.

At this conference, I am preaching to the converted. For triticale to succeed, we must look outside our own sectorial interests and consider our consumers' needs both for product and information. There is a saying "If you want to feed the world then educate the people". Ladies and Gentlemen, let's stop dreaming and start marketing.

HOW GERMANY BECAME THE NUMBER 3 IN THE TRITICALE WORLD AND WHY I AM CONFIDENT ABOUT THE FUTURE DEVELOPMENT

Peter Kruse

KRUSE-SAATEN
Samengrosshandel
Schlosstr. 10-12, Spenge, Germany

Abstract

Growing Triticale has become more and more popular in Germany, and reached about 230.000 ha which is about 3,5% of the total cereal area. In some regions, for example in Westfalia, Triticale is the number 3 cereal crop with more than 8% of the area grown for cereal grain.

Several factors must be considered to be of decisive influence for this development.

LASKO as the first variety with a combination of good characters had the ability to attract growers' interest and surprise by excellent disease resistance and high yields.

An advantageous mix of sucessful growing and feeding of Triticale has been the basis of the sucess of Triticale in Germany.

THE CONVICTION

Twenty-five years ago, I started a cooperation with Prof. Wolski and his team. The beginning was limited to the rye varieties created by the leading rye breeding station in Poland. They proved to be well adapted to our climate and agriculture, and could be introduced in Germany with good success.

At this time, the LASKO variety of triticale was just ready for release in official trials which were started simultaneously in Poland and Germany in 1980. During the period of testing, there was a big discussion pro and contra triticale.

I tried to be objective, and very early observed and judged many trials in Germany and some neighbour countries. It was in trials under extreme conditions and on poor soils that triticale proved its superiority to other grain, especially to wheat. The scepticism with which I had approached triticale turned into conviction and enthusiasm. This proved to be necessary in the following years to surpass set-backs.

The conviction of triticale being a very special grain with a big future was the main fundamental for new and better ideas and creativity in marketing.

41

H. Guedes-Pinto et al. (eds.), Triticale: Today and Tomorrow, 41–43.
© 1996 *Kluwer Academic Publishers. Printed in the Netherlands.*

THE TECHNIQUE HOW TO GROW TRITICALE

Happily enough, we found out at the early stages that growing triticale needs a new technique in comparison with wheat and rye. Multiple trials were organized and observed. In addition, farmers' experiences were assembled. This information was the basis of very exact and detailed information published to the farmers and advisers, and the better technique was very rapidly accepted.

Because the new varieties following the first great success of LASKO, were highly different, a detailed growing technique was and continously has to be developed for every new variety released to the market.

Thus failures were avoided and farmers were given the chance to grow triticale successfully from the beginning.

HIGH QUALITY SEEDS

Introducing a totally new grain meant that highest care had to be given to seed production. To exclude admixtures in basic seed, fields have to be chosen with highest care, and by principal the first tankfilling from every field has to be dumped for feeding purposes, not for seed. In basic seed production a separate plant is being used for every single variety. The seeds are graded on bigger sieves than necessary, 2,25 to 2,5 mm as a standard, depending on the variety.

High attention is being given to a very thorough seperation on the cylinders, and the finish is being done on 2 different gravity seperators, one following the other. Highest quality basic seed has proven to serve as a solid basis to success.

INTELLIGENT USE OF A BETTER GRAIN

It has been a long way to develop a market for triticale grain. In the beginning, the farmers growing triticale had no other choice than to feed their animals, especially pigs. Fortunately enough, these farmers observed their pigs very thoroughly when feeding triticale, and found out that triticale feeds excellently. This encouraged officials to investigate the feeding values of triticale. They confirmed what farmers had observed, namely that triticale is a highly valuable grain for feeding animals, especially pigs, poultry and cattle. Eventually the feeding values were officially being described in the official lists issued by the German Agricultural Society. Further exact investigations followed, stating that triticale can be fed with a portion of 10 to 50% in feeding mixtures not only without disadvantages, but even with better successes than in compositions without triticale. High notes were given to triticale in the value replacing wheat, barley and even soymeal at a certain extent.

Today the feeding industry in Germany is aware that triticale is an excellent component for their mixtures, and they take advantage of the fact that triticale can be produced at lower costs than wheat. Thus triticale does not need any state intervention, but has been sold out throughout the years within a few months after the harvest.

NEW IDEAS FOR THE USE OF TRITICALE

Different approaches to introduce triticale grain for making bread have failed. The reason is that in Germany, the country with the greatest choice of different breads in the world, the present triticale varieties do not offer an advantage to the bread industry.

Amongst the many small farmers destilleries a couple of motivated specialists could be found who developed the technique to transform triticale into alcohol at an economical rate, and the authorities could be convinced to accept triticale as a raw material that may be used for the production of alcohol. This alcohol may be produced for technical purposes, and the first hectolitre of Triticale-Schnaps has carefully ripened in an oak barrel and is about to be released to the German public.

PROSPECTS FOR THE FUTURE DEVELOPMENT

With a surface of about 230.000 hectars, triticale takes about 3% of the total grain area in Germany. The low prices for grain to the farmers in Europe makes it necessary to save costs for seed, fertilisers and chemicals. Triticale can be sown at a rate as low as hybrid rye, but the seed is available at much lower prices. Triticale achieves higher yield with less fertiliser and much lower input of chemicals. Skilled farmers harvest high yields at low costs and make more money from triticale than with any other grain. The use of triticale for animal feeding has by far not reached the optimum, and new usages for triticale grain will be developed. There are areas in Germany, mainly in regions with a high portion of marginal soils and intensive meat production where triticale has reached a portion of 20% and more of the surface grown with grain. A 10% part of triticale of the total grain growing surface in Germany is a realistic aim that should be achieved until the end of this century.

Triticale is a young, healthy and strong plant with a rapid growth. With the special attention of a highly motivated group of experts, the future of triticale has just started to begin!

THE VERDICT ON TRITICALE - THE CASE FOR

Kath Cooper

University of Adelaide, Waite Campus, Glen Osmond, South Australia, 5064

Abstract.

Six points in favour of triticale are summarised. Thus triticale can: help feed the increasing world population; help facilitate the improvement of wheat; be used in farming practices to increase wheat yields; make a contribution to basic science and knowledge; help save our soil and help us be healthier and have more fun.

Presentation.

Many views have been expressed throughout the duration of this meeting : progress, problems, successes, disappointments, and I've been continually exposed to mood swings around me. My role is to present a brief summary of positive arguments for triticale; why we and the world need(s) and benefit(s) from triticale, and what avenues we should pursue to gain better recognition of and better use of this crop. The following exposition includes comments and information encountered during this symposium to which I've added experience from my own area of the world and my personal viewpoint.

What do I think of triticale? Triticale is an amazing success story. Its recent creation as a species aside, we must appreciate its very recent appearance as a crop plant and more pertinently when the first useful commercial varieties (not just unproven curiosities) appeared. In Australia the first adapted variety ("Satu", a selection from Armadillo) was released 15 years ago, in Germany it was ten, and in Sweden only four years ago that adapted varieties appeared. And in Sweden more triticale is currently grown than rye, which is one of their staple foods. Triticale has had to face tremendous political and cultural pressures, and compete in a world glutted with cereals. Despite these barriers and paucity of funding for breeding, research, development and promotion, triticale has succeeded in covering 2.5 million hectares of land across the world.

I'd like to present six major categories in which triticale provides positive benefits.

NUMBER ONE : TRITICALE CAN HELP FEED THE INCREASING WORLD POPULATION

George Varuguese reminded us at the start of the symposium, that with increasing world population the demand for cereals is increasing by 3% per annum. Most of the ways of meeting these needs have been exploited, notably by colonising new farming land, and by improving farming methods. In fact we're tending to move backwards in these respects. More and usually the best agricultural land is disappearing for housing development, and land degradation is increasing due to repeated cropping and lack of attendance to soil conservation methods. The threat of global warming makes the picture even grimmer. We

H. Guedes-Pinto et al. (eds.), Triticale: Today and Tomorrow, 45–47.
© 1996 *Kluwer Academic Publishers. Printed in the Netherlands.*

are going to be forced to be reliant on the poorer, marginal lands for agricultural production, and as has been repeatedly demonstrated in presentations at this meeting, triticale can be relied on to address this problem. I'd like particularly to note the oral presentation on salt tolerance by Koebner and the striking examples of tolerance to early drought on the poster by Mergoum of Morocco. Triticale grows superbly in a range of environments which are unsuitable for the traditional cereals, except for rye which triticale surpasses in nutritional and food processing quality attributes, and in yield in all but the worst environments. The mechanism of adaptation to these environments is complex and involves the interaction of many genes. An amphiploid is the quickest and simplest means of attending to this problem, and we should appreciate that we already have a well adapted amphiploid of a wheat x other species in triticale. It's widely adapted, gets fewer diseases, less often than wheat, and has adequate food processing qualities. Minor cultural changes may be necessary for widespread acceptance as a staple food, but this is not impossible as demonstrated in the past (e.g. acceptance of potatoes and rice in Europe). Triticale can be our insurance policy for raising cereal production in the future, as well as assisting marginal land dwellers now. Point number one alone should be sufficient to explain the need for triticale, but I' ll continue with two points relating to wheat, one of the most important cereal crops of the world today.

NUMBER TWO :   TRITICALE CAN FACILITATE THE IMPROVEMENT OF WHEAT

Triticale forms an extended gene pool for wheat improvement. Many individual genes for wheat improvement can be observed in triticale, and if the useful genes (usually from rye parentage) are expressed in triticale, there's a good chance they will be fully expressed in bread wheat. Attempts can then be made to transfer the genes. Triticale forms the ideal bridge between rye and wheat, and we should remember the improvements in wheat that came through crosses with triticale, such as the IB/IR and IA/IR translocations.

NUMBER THREE :   USE OF TRITICALE IN FARMING PRACTICES TO IMPROVE
                        WHEAT YIELDS

I'll give two examples from Southern Australia. Triticale is an excellent break crop for wheat, particularly where few alternative crop options exist. Triticale forms a good cleaning crop for root pathogens such as Cereal Cyst Nematode (*Heterodora avenae*)and Root Lesion Nematode (*Pratylenchus neglectus*) and is a useful crop following a build up of wheat rust inoculum, when a wheat rust epidemic is expected. In areas where drifting sand ridges are interspersed with adequate wheat growing "flats", growing triticale on the sandridges can halt the drift and conserve the better soils for wheat-growing.

NUMBER FOUR :   CONTRIBUTION TO BASIC SCIENCE AND KNOWLEDGE

Triticale provides a good system in which to study the action and interaction of genes and chromosomes. Triticales are relatively easy to make, and in this amphiploid it is possible to isolate the effects of chromosomes, parts of chromosomes or genes. Such studies would be much more difficult in natural polyploids such as wheat. Several presentations relating to this point were included in the symposium.

NUMBER FIVE :   TRITICALE CAN HELP SAVE OUR SOIL

Land degradation and environmental pollution are the biggest threats facing the world in the 21st century, and if we don't look after our soil and environment, the survival of our agriculture and ourselves will be in jeopardy. Triticale can be used to reduce land degradation in agriculture. Its early vigour, soil-binding and ground covering abilities enable it to reduce soil loss from wind erosion. A ground cover of triticale also reduces the effect of raindrops on the soil which otherwise cause compaction and leaching of nutrients, which as well as depleting the soil, pollutes the ground water. Triticale is already selected for these purposes in diverse locations around the world (e.g. the United States and Sweden).

We should be looking at farming methods which improve our soil and promote these together with triticale which can be recommended as a crop which helps increase the sustainability of agriculture. In South Australia, where we have soils which are shallow, low in organic matter, and which are easily erodible, it has been shown that incorporation of 3kg per hectare per year of straw into the soil can reverse the downward trend in fertility with continued cropping. Triticale is a good straw producer, and it can be well utilised for this purpose.

Indications are that triticale requires less applications of pesticides, and lower rates of herbicide applications for successful weed control, than wheat so triticale can be promoted as a more environmentally friendly cereal. If we can attend to the problem of requiring growth regulators to shorten the straw in more fertile environments, by using dwarfing genes or varieties with stiffer straw, then the situation looks even better.

NUMBER SIX : TRITICALE CAN HELP US BE HEALTHIER AND HAVE MORE FUN

Triticale is an ideal food for the more affluent or "western" societies, and for developing countries which are adopting, (usually the worst) aspects of the western diet. These societies tend to develop the so-called "diseases of affluence": cardiovascular disease, obesity, diabetes, bowl and breast cancers, constipation and diverticulitis, as they take up a diet low in fibre and high in fat. There is a growing body of knowledge demonstrating that high fibre wholegrain cereals are particularly useful in averting these diseases and indications that rye, and thus probably triticale is superior to wheat with relation to cancer reduction. Analysis of the fibre composition of triticale (Topping, personal communication) support the suggestion that triticale should be useful particularly in reducing obesity and diabetes.

People who are affluent tend to buy the foods they prefer, and which are easy to obtain and prepare, rather than particularly choosing foods which are necessarily the best for their health. Triticale is easy to use in the production of a wide range of home-baked products [1] and commercial production methods could be altered to suit triticale. Triticale is higher in fibre than wholegrain wheat and has a good mouth feel with respect to the amount of fibre it contains. Its sweet and nutty flavour means the salt and sugar contents of products made with triticale can be reduced whilst maintaining palatability, and this is a further point in triticale's favour.

At this point I'd like to demonstrate a new triticale product, a Finnish crispbread made by Vaasamills of Helsinki, Finland, which was shown to me last night. This product is indistinguishable from the rye crispbread but for its nuttier, less bitter flavour. Apparently the triticale crispbread is easier to make because the mixture requires less water, and the wheat equivalent is considerably more costly, requiring a specific quality of wheat to avoid a crispbread which is too hard, and also requiring the addition of fat and sugar to make an acceptable product.

Triticale can and should appear more widely in the food market. As well as increasing our physical health this move should increase the health of the growers' wallets. We can help move triticale along this path by researching and publicising how triticale influences disease, researching into the relevant quality properties of triticale. e.g. resistant starch, fibre, and by seeking to put triticale into relevant food products.

I trust you will forgive my omissions and simplifications in this short discussion. I hope you will add to it positive ideas of your own and will leave hopeful of continuing to work in triticale's favour.

Reference

1.	Cooper, K. The Australian Triticale Cookery Book. Savvas Publishing, 1985, 81 pp.

THE VERDICT OF TRITICALE - A CRITICAL VIEW

Tamas Lelley
Institute of Agronomy and Plant Breeding, Agricultural University, Vienna, Austria

Abstract

Despite intensive breeding efforts for the last 40 years the economic success of triticale falls much behind expectations. Two of the major reasons might be (1) lack of convincing economic advantage against endemic crops, and (2) breeding of triticale is proving to be more difficult than that of any other comparable selfpollinating species. It is postulated that the evolutionary divergence of wheat and rye might have already passed the threshold which would allow them to form a successful amphiploid combination. For future success of triticale breeding, a better understanding of the unique genetic structure of triticale is crucial.

The invitation to present my critical view on triticale was probably elicited by a review article I published in 1992 with the title *Triticale, still a Promise?* (Lelley 1992). In this article I reviewed my own 25 years of work with this fascinating plant which taught me so much about plant evolution, genetics and breeding. But the past 25 years of experiences led me too, to a somewhat ambiguous view about the future of this man-made plant. Today, I will go a little further in my critical assessment of triticale but with the main intention to provoke disagreement and discussion.

I'm aware and fully acknowledge the achievements of several outstanding triticale breeders in the development of new varieties. Indeed, many of the weaknesses of early triticale varieties were eliminated by breeding and today some two million hectares of triticale are being grown world wide. But the question is why only two million hectares?

H. Guedes-Pinto et al. (eds.), Triticale: Today and Tomorrow, 49–55.

© 1996 *Kluwer Academic Publishers. Printed in the Netherlands.*

Ten years ago a *realistic* and a *conservative* estimate was made about the future triticale-growing area by some highly respected triticale experts. The conservative estimate was set at an "absolute minimum" of about 8 million hectares in world agriculture (Skovmand et al. 1984). To me it appears that triticale growth is today still in a lag phase. But there are some other even more distressing questions.

Why have so many triticale breeding programs started with enthusiasm and then faded away? To me the most notable example is the triticale breeding program in Hungary led by Dr. Kiss. Why is the scientific research on triticale diminishing? Triticale research groups have dissolved, eminent scientists have pulled out of the field. Where is the promised breakthrough? The story of triticale is like a sinus curve with peaks of euphoria followed by disenchantment.

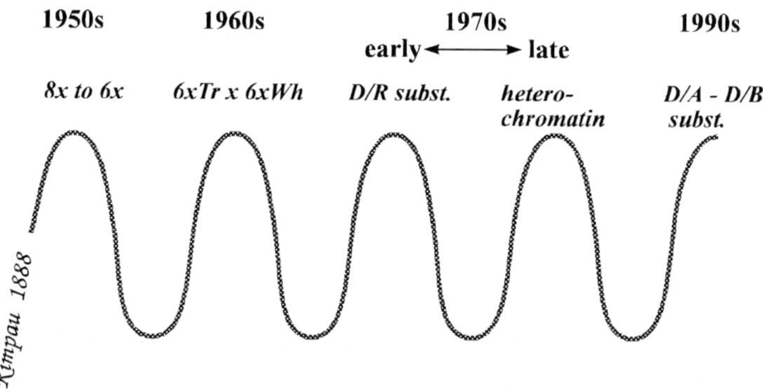

Fig. 1. Major events in recent history of triticale research giving new impetus to breeding efforts

Some experts are saying triticale´s major problem is lack of acceptance by farmers who are a most conservative type of people. Why did hybrid corn, triploid sugar beets, hybrid sunflowers, and hybrid rye, the seeds of which cost even more for the farmer than conventional seeding material, receive high acceptance? **May be, that triticale lacks the convincing economic advantage of those other new plants?**

We have to face the fact that the **real** breakthrough of triticale has not yet happened and probably it newer will. I venture to assert, that the combination of rye and wheat to a new amphiploid, which in nature never produced a competitive plant, **is an unfit one.**

As far as we are aware, natural amphiploids are composed of closely related species of the same genus. They possess some specific genetic mechanisms to control **diploid-like behavior** of the chromosomes and **disomic inheritance**. They are assumed to represent stable **intergenomic heterozygotes** expressing **permanent heterosis**. It is very likely that the evolutionary divergence of wheat and rye has already exceeded the threshold where their genomes still can form a competitive amphiploid. Chromosomes of wheat and rye pair only irregular, even in the absence of chromosome 5B. Structural differences are evident, considering the large difference in amount and distribution of heterochromatin in wheat and rye chromosomes. The rye genome contains 30% more DNA than the largest wheat genome (Bennett and Smith, 1976; wheat 1C 17.3, rye 8.3pg). The success of the spontaneous 1BL/1RS translocation in wheat breeding inversely indicates the lack of real compensation ability of the remaining six rye chromosomes or their individual arms.

We should recall that the idea of triticale was born in an era when knowledge of chromosomes or of the meaning of different breeding systems for the genetic constitution of a plant was far from established. Of course, the idea to combine the quality and yield of wheat with the durability and stress tolerance of rye was a very appealing one and some tantalizing plant phenotypes gave again and again a new impetus to continue breeding work.

Unfortunately, I never had the chance to breed triticale on a large scale. Thus, for a proper judgment I might lack some crucial field experiences. As a university worker in Germany, depending on successful project proposals for the next two or three years to survive, I tried to approach triticale from a scientific point of view. I acknowledged the fact that the combination of wheat and rye is the combination of two **contrasting breeding systems operating with different genetic mechanisms of adaptation.** The useful genetic variation in wheat is mainly additive, whereas in rye it appears mainly nonadditive. Since triticale is handled as an inbreeder, the power of the nonadditive genetic variance of the rye genome is lost. Rye can only contribute to the overall phenotype of the amphiploid either **through single genes or through nonadditive interactions with the wheat genome.**

To study the importance of interactions in triticale, I have produced primary triticales using pure lines of wheat and rye in an orthogonal way.

	inbred lines of rye		
	R₁	R₂	R₃ . . .
W₁	W₁+R₁	W₁+R₂	W₁+R₃
W₂	W₂+R₁	W₂+R₂	W₂+R₃
W₃	W₃+R₁	W₃+R₂	W₃+R₃

(4x wheat)

Fig. 2. Systematic production of primary 6x triticale lines using inbred lines of rye and pure lines of tetraploid wheat to test **genome combining ability.**

Assuming that the wheat and rye parents are homozygous the amphiploid should be a simple addition of the two.

This system allowed us to make comparisons of the wheat and rye parents and the amphiploids for many different characters, from cytological behavior to

52

agronomic traits like plant height, thousand grain weight, etc. (Jung et al. 1985, Jung and Lelley 1985b, Lelley and Gimbel 1989). To my regret, this approach was picked up by only a few triticale scientists, and is not being exploited, although it is probably the most suitable way to study systematically what is going on in triticale. Let me show you two examples of how this system works and what kind of results can be obtained by this approach.

Back in 1976, I produced several primary triticale genotypes using the tetraploid wheat variety *Langdon* and inbred lines of rye which I selected from the collection of the Plant Science Department of the University of Manitoba in Winnipeg. These two triticale genotypes on Fig. 3 (upper part) differ in their rye component, let them call *Langdon-A* and *Langdon-B*. The plants were sown at the same time. When *Langdon-A* with two months delay ultimately produced a few tillers, the number of set seeds was hardly enough to maintain the line. *Langdon-B* was not much more fertile as an ordinary primary triticale, but the plants looked very healthy and were vigorous. Somehow, we succeeded in producing a few F_1 hybrid seeds. The F_1 plants looked healthy and normal, like *Langdon-B,* and produced a few seeds. From these we obtained eight F_2 plants with 42 chromosomes which segregated exactly 2:6 (or 1:3) with respect to plant morphology (Fig. 3: lower part). It indicates that there is one **single gene** in the inbred rye line *A,* which doesn't have any visible deleterious effect on the inbred line itself, but which interacts with the genotype of *Langdon* and produces this strange phenotype of the primary triticale *Langdon-A* (Lelley unpublished). The second example of gene interaction between wheat and rye in triticale concerns hybrid necrosis. It is caused by the interaction of the well-known *Ne1* gene in tetraploid wheat and what we later called *Ner2* in the rye genome. *Ner2* is most probably a homoeoallele of *Ne2* in wheat (Jung and Lelley 1985a, Ren and Lelley, 1989a, 1989b).

The interaction between the wheat and rye genome in triticale ranges from complete lethality, i.e. incompatibility, indicated by certain wheat/rye combinations which we just could not obtain in our orthogonal schema even by repeated efforts, to genotypes like *Lasko, Presto* or *Beagle* being exceptionally successful combinations between two unknown wheat and rye genotypes in each. Unfortunately, we do not have a method to extract the rye and the wheat components of *Lasko* to look for the reason of success for this specific combination. But this **specific combining ability** of the wheat and rye genomes, decides the fate of a triticale genotype. As our data have shown, the **specific <u>genome</u> combining ability** is the major genetic source which determines the phenotype, especially the yield characteristics of a triticale plant (Jung et al. 1985, Jung and Lelley 1985b, Lelley and Gimbel 1989, Lelley 1992). From these results I would like to draw the following conclusions:

(1) Successful amphiploid combinations of wheat and rye are extremely rare specific genome combinations, probably more so than conventional diploid hybrids like in corn or in rye, with a further difference that in triticale **we do not have the means for testing the components for their general genome combining ability.** Actually, in natures vast nursery such a combination with the competitive ability of survival has not yet occurred, and it is probably only our ignorance (arrogance) which still makes us believe that we can do it better. And may be we can.

Fig. 3. Primary triticale genotypes obtained by combining the tetraploid wheat variety *Langdon* with two inbred lines of rye *A* and *B* (upper part). The primary triticale *Langdon-A* shows a weedy appearance and produces tillers with two months delay. Plants of *Langdon-B* look very healthy and are very vigorous. Crossing *Langdon-A* with *Langdon-B* yielded ultimately eight F_2 plants with 42 chromosomes. With respect to plant morphology they segregated exactly 2:6 (or 1:3, lower part).

54

(2) The delicate balance of a successful combination is extremely sensitive to disturbances caused by crossing it with another genotype. The more closely related the crossing partners are, the higher the chance of not disturbing too much the balance. We investigated the single seed descent progeny of several types of triticale crosses. We found that the distribution of F_2 genotypes in triticale is strongly skewed to the left in comparison with similar F_2 populations in wheat indicating the very narrow margin available for varietal improvement in a specific cross of triticale compared to wheat (Lelley 1990a, Lelley 1992).

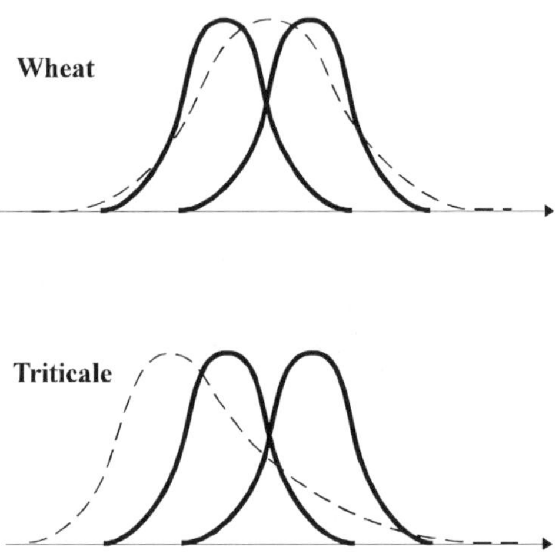

Fig. 5.   Distribution of F_2 genotypes (broken lines) in triticale is strongly skewed to the left in comparison with similar F_2 populations in wheat. Distribution of parents is shown by solid lines.

(3) It means of course on the other hand, that the more distant the crossing partners are, the less the chance for new improved types. The use of primary triticale for varietal improvement in triticale breeding is worse than using exotics in wheat breeding.

I believe that wheat/rye combinations with excellent agronomic performance are possible <u>but the effort</u> to find them is at present far greater than in a conventional wheat breeding program. The handling of genetic variation for varietal improvement in triticale is much more difficult than in wheat. I think triticale breeders should acknowledge the uniqueness of the genetic structure of the plant and should design specific breeding procedures for triticale (e.g. Lelley 1990b). Breeding should probably concentrate more on the slow improvement of the available best-adapted genotypes instead of creating new genetic variation. Hexaploid wheat started most likely by a very narrow genetic base and evolved initially by mutation. Triticale breeders should consider mutation breeding of the best available material. Through mutation we might be able to eliminate some still existing genes with deleterious intergenomic interactions. Crossing should be followed by several backcrossings to reestablish the original balance.

New chromosomal constructs such as D/A or D/B substitutions should be introduced in the best-adapted material by backcrossing programs. I acknowledge the fact that without believing in what we are doing it is hard to be successful. On the other hand, triticale has been around too long without the ever-expected breakthrough. This makes it increasingly difficult to attract further financial support for our research work. My personal interest in triticale is as alive as ever. I hope with my talk I have stirred enough emotions for a lively discussion on some basic issues concerning our common target of affection.

Literature

BENNETT, M.D., and J.B. SMITH, 1976. Nuclear DNA amounts in Angiosperms. Phil. Trans. roy. Soc. (Lond.) B **274**, 227-274.

JUNG, C., and T. LELLEY, 1985a. Hybrid necrosis in triticale caused by gene interaction in its wheat and rye components. Z. Pflanzenzüchtg. **94**, 344-347.

JUNG, C., T. LELLEY and G. RÖBBELEN, 1985. Genetic interactions between wheat and rye genomes in triticale. Part 1. Cytological results. Theor. Appl. Genet. **70**, 422-426.

JUNG, C., and T. LELLEY, 1985b. Genetic interactions between wheat and rye genomes in triticale. Part 2. Morphological and yield characters. Theor. Appl. Genet. **70**, 427-432.

LELLEY,T., and E.-M. GIMBEL, 1989. "Genome combining ability" of wheat and rye in triticale. Plant Breeding **102**, 273-280.

LELLEY, T., 1990a. The significance of "genome combining ability" of parental wheat and rye in triticale. Proc. 2nd Int. Triticale Symp. Passo Fundo (Brazil) 352-358.

LELLEY, T., 1990b. Breeding triticale: an alternative approach methods and first experiences. Proc. 2nd Int. Triticale Symp. Passo Fundo (Brazil) 282-285.

LELLEY, T., 1992. Triticale; still a promise? (A review). Plant Breeding **109**, 1-17.

REN, Z.-L., and T. LELLEY, 1989a. Hybrid necrosis in triticale and the expression of necrosis genes in allopolyploids. Theor. Appl. Genet. **77**, 742-748.

REN, Z.-L., and T. LELLEY, 1989b. Chromosomal localization of genes in the R genome causing hybrid necrosis in rye and triticale. Genome **33**, 40-43.

SKOVMAND, B., P.N. FOX, and R.L. VILLAREAL, 1984. Triticale in commercial agriculture: progress and promise. Advances in Agronomy **37**, 1-45.

TRITORDEUM: TRITICALE'S NEW BROTHER CEREAL

Antonio Martín, C. Martínez-Araque, D. Rubiales, and Juan Ballesteros

Instituto de Agricultura Sostenible (CSIC), Dpto. de Agronomía y Mejora Genética Vegetal, Apdo. 4084, 14080 Córdoba, Spain

Abstract

The amphiploids between *Hordeum chilense* and *Triticum* spp. (tritordeum) resemble their wheat parents agronomically. We have synthesized more than one hundred primary tritordeums for breeding a new cereal crop. From the results obtained on the first two cycles of selection, in which the yield level of tritordeum was raised to that of traditional wheat varieties, it is clear that tritordeum has the potential to become a crop when sufficient breeding effort is applied. Tritordeum is already been exploited as a bridge for introducing the genetic variability available in the section *Anisolepis* of the genus *Hordeum* into wheat.

Introduction

Interspecific and intergeneric hybridization is a common tool for widening the genetic basis of crops, and occasionally the result of such crosses are the starting point of new crops.

In the tribe Triticeae both outcomes have been obtained. Variation from wild species has been used successfully for breeding, mainly in wheat, but also a new crop, triticale, has been produced after crossing wheat and rye. Cultivated durum and bread wheats are natural amphiploids themselves.

Hordeum is another genus of agronomic interest of the tribe, and interspecific and intergeneric hybrids are abundant. Breeders have been interested in producing amphiploids between *Triticum* and cultivated barley, *H. vulgare*, since the beginning of the century and after the first success of Kruse (1973) many scientists obtained hybrids but have failed in producing fertile amphiploids. Some turned to wild *Hordeum* spp. and succeeded in producing fertile amphiploids (see Fedak, 1992 for a review). Among the amphiploids between *Triticum* and *Hordeum*, we think that *H. chilense* x *T. turgidum* (AABBH^{ch}H^{ch}) shows characteristics of a potential crop. We have called this amphiploid Tritordeum. The name X *Tritordeum* (Ascherson et Graebner) appeared for the first time as far as we know in a paper of 1902 referring to a cross between *Triticum repens* x *Hordeum secalinum*.

57

H. Guedes-Pinto et al. (eds.), Triticale: Today and Tomorrow, 57–72.
© 1996 *Kluwer Academic Publishers. Printed in the Netherlands.*

58

Tritordeum's mother, *Hordeum chilense*

Hordeum chilense Roem. et Schult is a native South American diploid wild barley included in the *Stenostachys* group (sect. *Anisolepis*). *H. chilense* occurs exclusively in Chile and Argentina and is extremely polymorphic both at the morphological and biochemical levels. Its distribution extends from 29° to 43° Latitude South and from sea level to 1.800 m of altitude. It is a weak perennial, although there are indications of genetic variation for this character. The species is found in wide range of edaphic conditions along lakeshores or wet areas. It is an element of natural pastures and is appreciated by cattle. The species is mainly autogamous but some accessions readily produce hybrids in nature and in experimental plots. The extreme case is represented by some accessions that are male sterile and for which outcrossing is obligatory.

After *H. vulgare* and *H. bulbosum*, we believe *H. chilense* is the species of the genus *Hordeum* with the highest potential for cereal breeding purposes, given its high crossability with other members of the Triticeae tribe and its agronomically interesting characteristics.

Crossability of Hordeum chilense

In Table 1 we present the species that have been crossed successfully with *H. chilense*. Most probably the whole tribe could be crossed with *H. chilense* and the variability of *H. chilense* could be exploited for breeding purposes if genetic recombination could be forced by whatever means.

The chromosome pairing in the meiosis in hybrids of *H. chilense* with other South American barleys is quite high but seed set is only obtained with *H. flexuosum* (Bothmer *et al.*, 1985), *H. patagonicum* ssp. *magellanicum* (Martín, unpublished). We can assume that after backcrossing to *H. chilense* the genetic variability of these two species would come available for tritordeum and for wheat breeding using tritordeum as a bridge.

The high compatibility of *H. chilense* with the genomes of *Triticum* species, giving rise to fertile and stable amphiploids, is surprising if we consider the chromosome instability observed in *H. vulgare* and *H. bulbosum* hybrids with *Triticum*. This compatibility seems to be especially high with *A. squarrosa* (*T. tauschii*). The amphiploid *H. chilense - A. squarrosa* is, until now, the only tetraploid amphiploid with *H. chilense* that is fertile (Table 2). The chromosomes of *H. chilense* are very similar in size to those of *Aegilops squarrosa*, although they differ in the nucleolar organizing regions; two pairs of satellited chromosomes in *H. chilense* and a small satellite in *A. squarrosa*. It seems that *A. squarrosa* and *H. chilense* are phylogenetically more closely related than expected or that postzygotic crossability barriers are absent in these two allopatric species. Nevertheless, there is no similarity at the chromosome pairing level. Crossing *H. chilense* with nuli 5B tetra 5D Chinese Spring we did not observe pairing between wheat and *H. chilense* chromosomes (Martín & Sánchez-Monge Laguna, 1980). The same result is observed in the hybrid *H. chilense* x *A. squarrosa* and X *Tritordeum* x *T. aestivum*.

Chromosome pairing is neither observed in *H. chilense* x *S. cereale* nor in *H. chilense* x *Dasypyrum villosum*. Nevertheless, in the hybrid *H. chilense - H. vulgare - A. squarrosa* ($H^{ch}H^{v}D$) some pairing is observed which could be attributed to *H. vulgare - H. chilense* chromosomes, if we assume there is no chromosome instability in this hybrid.

Table 1. Species which have been crossed successfully with *Hordeum chilense*

Species	Reference
Aegilops squarrosa	Martín, 1983
Dasypyrum villosum	Martín, unpublished
Hordeum bulbosum	Padilla & Martín, 1983b
H. brachyantherum	Bothmer & Jacobsen, 1986
H. californicum	Subrahmanyam, 1976
H. capense	Bothmer *et al.*, 1987
H. comosum	Bothmer & Jacobsen, 1986
H. cordobense	Bothmer *et al.*, 1985
H. depressum	Bothmer *et al.*, 1987
H. erectifolium	Bothmer & Jacobsen, 1986
H. euclaston	Bothmer *et al.*, 1985
H. flexuosum	Bothmer *et al.*, 1985
H. intercedens	Bothmer *et al.*, 1985
H. jubatum	Bothmer *et al.*, 1987
H. lechleri	Bothmer & Jacobsen, 1986
H. leporinum	Martín & Cubero, 1981
H. muticum	Bothmer *et al.*, 1985
H. patagonicum ssp. *magellanicum*	Martín, unpublished
H. pubiflorum	Bothmer & Jacobsen, 1986
H. pusillum	Bothmer *et al.*, 1985
H. roshevitzii	Bothmer & Jacobsen, 1986
H. secalinum	Bothmer *et al.*, 1987
H. stenostachys	Bothmer *et al.*, 1985
H. tetraploidicum	Bothmer *et al.*, 1987
H. vulgare	Sánchez-Monge Laguna & Martín, 1982
Secale cereale	Finch & Bennet, 1980
S. africanum	Finch & Bennet, 1980
S. montanum	Finch & Bennet, 1980
S. vavilovii	Finch & Bennet, 1980
X *Triticosecale* 6X	Schrader & Pohler, 1985
Triticum aestivum ssp. *vulgare*	Martín & Chapman, 1977
T. aestivum ssp. *sphaerococcum*	Martín *et al.*, 1987
T. monococcum	Martín, unpublished
T. turgidum conv. *durum*	Martín & Sánchez-Monge, 1980
T. turgidum ssp. *dicoccoides*	Padilla & Martín, 1983a
T. turgidum ssp. *georgicum*	Padilla & Martín, 1983a
T. turgidum conv. *polonicum*	Padilla & Martín, 1983a
T. turgidum ssp. *carthlicum*	Padilla & Martín, 1983a
T. turgidum conv. *turanicum*	Padilla & Martín, 1983a
T. timopheevi	Padilla & Martín, 1983a

Although it will be difficult to obtain introgression into wheat by recombination, we have observed high frequencies of translocation between the D and H^{ch} genomes from crosses between hexaploid tritordeum and bread wheat ($AABBDH^{ch}$) which facilitate the introgression of *H. chilense* into wheat and of *A. squarrosa* into tritordeum. Also, a high frequency of translocations has been observed among the progeny of tritordeum x triticale hybrids. These observations suggest that further efforts to incorporate this material into the tritordeum breeding programme would be worthwhile.

Octoploid, hexaploid and tetraploid tritordeums

Attempts to produce the amphiploid between *Triticum* spp. and *H. vulgare* have so far proved unsuccessful. However, fertile *Triticum - Hordeum* amphiploids (tritordeums) have been obtained with other wild barleys, as *H. bogdanii* (Kimber & Sallee, 1978), *H. californicum* (Fedak, 1992) and *H. bulbosum* (Rong & Wang, 1989).

In table 2 we present the amphiploids obtained to date using *H. chilense* as one of the parents. Among them, only the crosses *H. chilense* x *Triticum* spp. and *Aegilops squarrosa* yielded fertile plants.

The first fertile hybrid, after colchicine treatment, between *Hordeum chilense* and *Triticum aestivum* was obtained at the former Plant Breeding Institute in Cambridge (UK) by Martín and Chapman (1977). Miller and Chapman (1978) reported the amphiploid from the seed set obtained in that hybrid. The fertility of this amphiploid obtained from crossing *H. chilense* accession H1 and Chinese Spring was quite poor. However, the fertility was higher when other *H. chilense* accessions were used (Martín *et al.*, 1987; Martín, 1988).

Although some primary octoploid tritordeums have good fertility, they show poor initial growth and a high frequency of aneuploids that could limit their possible direct use as a crop. Nevertheless, octoploid tritordeums have become of great value for breeding hexaploid tritordeum.

Conversely, the primary hexaploid tritordeums show a low frequency of aneuploids, a wide variation in initial growth (from poor to good), and a good fertility, even when using H1 (Martín & Sánchez-Monge Laguna 1982). These and other favourable agronomic traits such as biomass, spikelets by spike, seed size, protein content, made us consider seriously the potential of this amphiploid as a possible new crop (Cubero *et al.*, 1986; Martín, 1988). There is prospect of using tritordeum in food industry (Alvarez *et al.*, 1992). Another interesting feature was the high osmotic adjustment of these tritordeum lines which was surprisingly much better than that of wheat. This trait could indicate high tolerance to drought (Gallardo & Fereres, 1989). Also, under NO_3^- nutrition, tritordeum generally exhibited higher levels of nitrate reductase activity than wheat and a quicker and more effective NO^{3-} extraction from the medium (Barro *et al.*, 1991). These traits suggest that it could be possible to obtain tritordeums with higher potential to extract N from soil than wheat.

Tetraploid tritordeum has been obtained, not from colchicine treated hybrids but directly by crossing tetraploid *H. chilense* and tetraploid *Aegilops squarrosa* (Cabrera & Martín, 1991). This amphiploid is quite fertile but of poor agronomic performance.

Tritordeum amphiploids including *H. chilense* and durum and bread wheat have been used to produce wheat - *H. chilense* chromosomes addition lines at both ploidy levels (Miller *et al.*, 1982; Fernández & Jouve, 1988). *H. chilense* addition lines on rye have also been obtained (Linde-Laursen *et al.*, 1993) by backcrossing the amphiploid *H. chilense* x rye with rye.

Table 2. Species crossed with *Hordeum chilense* giving rise to amphiploids.

Species	Fertility	Reference
Aegilops squarrosa	Self fertile	Cabrera & Martín, 1991
Agropyron cristatum	Sterile	Rubiales & Martín, unpublished
Agropyron desertorum	Sterile	Rubiales & Martín, unpublished
Hordeum bulbosum	Sterile	Cabrera & Martín, unpublished
H. vulgare	Sterile	Thomas & Pickering, 1985
Secale cereale	Sterile	Martín *et al.*, 1988
S. cereale-montanum	Sterile	Martín, unpublished
X *Triticosecale* 6X	Self fertile	Schrader & Pohler, 1985
Triticum aestivum ssp. *vulgare*	Self fertile	Martín & Chapman, 1977
T. aestivum ssp. *sphaerococcum*	Self fertile	Martín *et al.*, 1987
T. turgidum conv. *durum*	Self fertile	Martín & Sánchez-Monge Laguna, 1982
T. turgidum ssp. *carthlicum*	Female fertile	Ballesteros & Martín, unpublished
T. turgidum ssp. *dicoccoides*	Self fertile	Padilla & Martín, 1987
T. turgidum ssp. *georgicum*	Self fertile	Padilla & Martín, 1987
T. turgidum conv. *polonicum*	Self fertile	Padilla & Martín, 1987

Tritordeum breeding

The two first hexaploid amphiploids (HT22 and HT24) obtained were crossed with each other and selection applied following a pedigree method. The best six recombinant lines showed yield (800-1300 kg/ha) around 20% of that of commercial bread wheat, with a test weight (67-70 kg/Hl) similar to a good triticale and a high protein content (21-25%). The seeds were well-filled and plump.

One of these recombinant tritordeum lines, HT8 (6x), was crossed with a secondary tritordeum, HT9 (6x), obtained from a cross between hexaploid and octoploid tritordeums (HT22 (6x) x HT18 (8x)). A single-seed-descent scheme was followed with the progeny of this cross. The grain yield of the best six lines derived from this cross (2800-5600 kg/ha) was up to 80% of that of commercial wheat cultivars (average 7000 kg/ha), while total dry matter production was superior in some cases. The protein content (15-20%) although still higher than that of commercial wheats (13-15%) decreased when yield increased, as expected.

The response to selection indicates the high potential of this new species, especially when we consider the narrow genetic basis present at that moment; two *H. chilense* accessions, H1 and H7, and three wheat lines.

Among the very first decisions for breeding tritordeum, we decided to widen the genetic base of the amphiploid. We started to build a *H. chilense* collection using three approaches: i) requests to gene banks and individual scientists, ii) we paid for collections in Chile by Chilean colleagues and iii) we ourselves organized collection expeditions to Chile. The main sources of wheat germplam have been the crossing block of durum wheats of CIMMYT and ICARDA and traditional and modern Spanish cultivars of durum and bread wheats. Occasionaly non-cultivated diploids, tetraploids and hexaploid wheats have been used as parents. Hybrid plants have been obtained with 93 different *H. chilense* accessions. Fertile amphiploids have been obtained with 66 *H. chilense* accessions after colchicine treatment of the hybrids. Up to now 167 primary amphiploids have been obtained. One of them is a tetraploid, 115 are hexaploids and 51 are octoploids.

As we said previously, in the first cycle of breeding we followed a classical genealogical scheme. In the second cycle we used a single seed descent scheme to accelerate the programme. In a third cycle the primary tritordeums obtained each year, were crossed with the advanced lines and different selection schemes were applied: pedigree, single-seed descent and bulk. Among the priority traits to breed for were fertility, winter growth, free threshing, tough rachis, erect canopy, and plant stature. After these traits, resistance to biotic and abiotic stress factors, quality components, and yield are also being considered. We have obtained early flowering and near fully fertile lines, semidwarf and erect lines, free threshing lines, etc., but the obtention of a tritordeum line combining all the desired traits has proved more difficult than initially expected. Particularly difficult at present is the synthesis of a fully fertile semidwarf line. The same problem was found in semidwarf wheat and triticale breeding (Zillinsky, 1974). Triticale breeders solved the problem crossing hexaploid triticale with bread wheat and octoploid triticale with hexaploid triticale. We are following a similar strategy, crossing hexaploid tritordeum with bread wheat, with octoploid tritordeum and with hexaploid and octoploid triticale.

Actually backcrossing is the main tool in our programme. The best advanced tritordeum lines showing good fertility and early flowering are used as recurrent parents. These lines are crossed with primary tritordeums showing good agronomic traits and with semidwarf or free threshing advanced tritordeum lines. Most frequently the primary tritordeums are not used directly due to the lack of desired traits. Instead we are building up populations from multiple crosses among primaries, which are subsequently subjected to negative selection. The purpose is to maintain the genetic variation created and to increase the frequency of favourable alleles. The selected lines from this prebreeding scheme will be used in the backcrossing programme.

Comparison of tritordeum, wheat and triticale

From the beginning, primary hexaploid tritordeums showed unexpectedly good fertility, chromosome stability and a plant morphology similar to that of wheat (Martín & Cubero, 1981). In particular, the fact that the high protein content was not due to grain shrivelling as in primary triticales was interesting. These characters prompted us to study the agronomic potential of the new species.

Nevertheless, the evaluation of the potential of a plant as a future crop is difficult, given that the comparison currently cultivated amphiploids and primary amphiploids, that should be considered as raw material for breeding, is unfair. For this reason we designed a trial in which three primary hexaploid amphiploids (synthetic bread wheat (AABBDD), triticale (AABBRR) and tritordeum (AABBH^{ch}H^{ch})) were evaluated together with their durum wheat parents (AABB). This would allow us to assess whether we could expect that tritordeum would reach at least, the productivity of cultivated wheat and triticale, and might add any feature of interest for farmers, when sufficient breeding effort were applied.

Both the primary tritordeum and synthetic bread wheat were later, less fertile, had smaller kernels and lower biomass, yield and harvest indexes than their respective durum wheat parental line (Table 3). The protein contents of the synthetic bread wheat and that of the primary tritordeum were higher than those of their durum wheat parents. When harvest index was corrected, reducing mechanically the yield of the higher yielding lines by cutting part of the spike at heading, the protein content of synthetic bread wheat became lower than that of its wheat parent, whereas the protein content of the primary tritordeum was still higher than that of its durum wheat parent (17%). These data corroborated the potential of tritordeum to become a crop after sufficient breeding effort. In principle, primary tritordeum was at least as good as a primary bread wheat. The comparison with triticale was not completely fair because the triticale tested was not a primary in the same sense as the wheat and tritordeum tested were. The triticale segregated for endosperm proteins indicating that the primary triticale was a mixture of duplicated hybrids and could have, therefore, benefit from heterosis. In addition, triticale is the result of the fusion of two cultivated species while one of the parents of both tritordeum and wheat is a wild species.

Table 3 Characteristics of a primary tritordeum[1] (AABBH^{ch}H^{ch}), a synthetic bread wheat (AABBDD), a primary triticale (AABBRR), and their respective durum wheat (AABB) parental lines, under field conditions.

	Amphiploids			Durum Wheats		
	Tritordeum	Synthetic bread wheat	Triticale	parent ·of tritordeum	parent of bread wheat	parent of triticale
Days to flowering	160 a[2]	161 a	150 c	149 c	152 b	139 d
Biomass (kg/ha)	12500 c	11145 c	19618 a	16533 b	19130 a	14704 b
Plant height (cm)	89 c	100 b	189 a	106 b	197 a	100 b
Spikes/m^2	301 ab	349 a	273 bc	249 c	242 c	248 c
Yield (kg/ha)	1448 c	1438 c	4413 ab	3886 b	3834 b	4916 a
Harvest Index	11 c	13 c	22 b	23 b	20 b	33 a
1000 kernel weight (gr)	33 e	39 d	48 c	56 b	49 c	63 a
Protein content (%)	22 a	16 b	11 c	13 bc	12 c	12 c
Protein content after Harvest Index adjustment (%)	20 a	15 c	16 bc	17 b	17 b	16 b

[1] Trial with only one accession of each amphiploid.

[2] Letters in common per trait indicate that differences are not statistically significant (P<0.05, LSD).

Cytology and interspecific crosses

As we said previously, the pairing between *H. chilense* and wheat genomes is nill even in the absence of 5B chromosome. Thus, it is clear that anomalous meiosis or sterility caused by allosynthetic pairing will not be a problem in tritordeum. Nevertheless, the level of homologous chromosome pairing is lower in tritordeum than in the parental lines and this may cause aneuploidy especially in octoploids. Despite the breeding work of the last years a certain level of aneuploidy is still present in the advanced hexaploid lines.

In Table 4, some of the hybrids obtained from crosses involving tritordeum are shown. Tritordeums at octo- (AABBDDH^{ch}H^{ch}), hexa- (AABBH^{ch}H^{ch}) and tetraploid (DDH^{ch}H^{ch}) levels can be intercrossed readily. The hybrid hexa x octoploid (AABBH^{ch}H^{ch}D) and the reciprocal are fertile, while those with tetraploid (ABDH^{ch}H^{ch} and ABDDH^{ch}H^{ch}) are sterile. The best results in breeding tritordeum have been obtained by crossing octoploids and hexaploids. Among the possible crosses between tritordeum and wheat, the hybrid at the hexaploid level is the most valuable for breeding as a source of D/H^{ch} substitutions and translocations. Translocations D/H^{ch} have also been obtained from the hybrid AABBDRH^{ch}

H. vulgare (H^{v}H^{v}) crosses easily as female parent with tritordeum, although embryo rescue is necessary. Some plants have been obtained after crossing the hybrid *H. vulgare* X *Tritordeum* 6x (ABH^{ch}H^{v}) with durum wheat. Nevertheless, the chromosome number of these plants was lower than the expected 42 (AABBH^{ch}H^{v}), probably due to chromosome instability in the hybrid.

Table 4 Interspecific hybrids involving Tritordeum

Cross	Fertility	Reference
Aegilops squarrosa x X *Tritordeum* 4X	Sterile	Cabrera, unpublished
Hordeum vulgare x X *Tritordeum* 4X	Sterile	Martín *et at.*, 1995
Hordeum vulgare x X *Tritordeum* 6X	Female fertile	Martín *et al.*, 1995
Hordeum vulgare x X *Tritordeum* 8X	Sterile	Martín *et al.*, 1995
X *Triticosecale* 6X x X *Tritordeum* 6X	Female fertile	Fernández-Escobar & Martín, 1985
X *Triticosecale* 6X x X *Tritordeum* 8X	Sterile	Fernández-Escobar & Martín, 1988
X *Triticosecale* 8X x X *Tritordeum* 8X	Self fertile	Fernández-Escobar & Martín, 1989
Triticum aestivum x X *Tritordeum* 6X	Self fertile	Martín & Cubero, 1981
T. aestivum x X *Tritordeum* 8X	Self fertile	Miller *et al.*, 1982
X *Tritordeum* 4X x *T. monococcum*	Sterile	Rubiales & Martín, unpublished
T. turgidum x X *Tritordeum* 6x	Self fertile	Martín, unpublished
X *Tritordeum* 4X x *Secale cereale*	Sterile	Cabrera & Martín, 1992
X *Tritordeum* 6X x *Secale cereale*	Sterile	Martín, unpublished
X *Tritordeum* 8X x *Secale cereale*	Sterile	Martín, unpublished

The cytoplasm of Hordeum chilense

At present, the breeding programme of tritordeum is based on the introduction of variation, by synthesizing new amphiploids with a range of lines of *H. chilense* and wheat.

In all these crosses, *H. chilense* was the female parent, so primary hexaploid tritordeums should have *chilense* cytoplasm. To evaluate the effect of cytoplasm on agronomic traits, the genome of hexaploid tritordeum was introduced into the cytoplasm of *Triticum aestivum, T. turgidum* and *H. chilense* by repeated backcrossing to produce alloplasmic lines. This substitution did not greatly affect the characters studied, except yield per plant, which were lower with *T. turgidum* cytoplasm (Millán & Martín, 1992).

It is surprising that the lines with *H. chilense* cytoplasm were so similar to those with *T. aestivum* cytoplasm. Phylogenetically barley and wheat are more distant than wheat and rye (Vedel *et al.*, 1981), but triticale, also an amphiploid like tritordeum, has different characteristics with wheat and rye cytoplasm. In fact, hexaploid triticale in rye cytoplasm is no agricultural interest (Sánchez-Mongue & Soler, 1973). The results obtained with *H. chilense* cytoplasm are similar to those obtained with *H. vulgare*. Fedak (1978) did not find differences between morphologic traits in alloplasmic lines of durum wheat in *T. aestivum* and *H. vulgare* cytoplasms. We have obtained alloplasmic *T. turgidum* and *T. aestivum* in *H. chilense* cytoplasm and again we coud not observe any morphological difference.

The *H. chilense* cytoplasm tested (derived from only one line) seems to be appropriate for breeding tritordeum. Until now we have not considered cytoplasm as a factor in the breeding programme of tritordeum. However, no definitive or general conclusion can be made until more genotypes of tritordeum with *Triticum* and *H. chilense* cytoplasms have been compared.

Tissue culture

Tissue culture has been relatively successful in tritordeum. The regenerative capability of the inflorescence sheath-leaves of tritordeum is much higher than that of wheat (Barceló *et al.*, 1991). Cell suspensions and protoplast cultures established from immature embryos have given rise to green shoots and green plants (Barceló *et al.*, 1993). These achievements opened the possibility of transformation of tritordeum, which was obtained recently (Barceló *et al.*, 1994b) by particle bombardment of inflorescence cultures. Anther culture has also been established although its utilization in plant breeding is limited at present by the high frequency of albino plants produced (Barceló *et al.*, 1994a).

Resistance to diseases

Hordeum chilense is resistant to the barley and wheat brown rusts (*Puccinia hordei* and *P. recondita* f.sp. *tritici*, respectively). The reaction of tritordeum to these pathogens is very similar to that of the wheat parent, which is resistant to the barley brown rust and susceptible to the wheat brown rust (Rubiales *et al.*, 1991).

There are several mechanisms of resistance to brown rusts operative in *H. chilense*. Urediospore germlings of most rust fungi must seek and recognize leaf stomata, where appressorium formation should be triggered. Mechanisms that hamper the ability to find and to penetrate stomata had not previously been documented in close relatives of wheat. We detected however a very low appressorium formation by brown rust fungi on some *H. chilense* lines (Rubiales & Niks, 1992a). The poor recognition of the signal triggering the appressorium differentiation over the stomata seems to be due to a prominent wax covering on the stomata opening. This low appressorium formation is frequent in *H. chilense* and has also been detected in other *Hordeum* accessions (*H. brachyantherum*, *H. parodii*, and *H. secalinum*). All the accessions of the other wild Triticeae gave rather high appressorium induction rates (Rubiales *et al.*, 1995a).

Unfortunately this avoidance mechanism is not significantly expressed in tritordeum and transfer to wheat may not be feasible. *H. chilense* may be useful in the study of mechanisms that play a role in directional germ tube growth and appressorium differentiation. Studies are being performed to determine the inheritance of this avoidance mechanism. Trisomics of *H. chilense* accession H17, bearing this mechanism, are being developed which would allow the identification of the chromosome/s responsible/s for such avoidance. Crosses between *H. chilense*, *H. brachyantherum*, *H. parodii* and *H. secalinum* accessions, bearing this mechanism, with *H. vulgare* are being attempted in order to study the feasibility of exploitation of this avoidance in barley breeding.

H. chilense lines may be susceptible to *P. recondita* f.sp. *agropyrina*, to wheat and barley yellow rusts (*P. striiformis* ff.spp. *tritici* and *hordei*, respectively) (Rubiales & Niks, 1992b) and to the wheat stem rust (*P. graminis*). Tritordeum behaves in all instances as the wheat parent, being susceptible to the wheat yellow and stem rust, and resistant to the barley yellow rust and to the agropyron brown rust (although one tetraploid tritordeum, was susceptible as was its *Triticum tauschii* parent) (Rubiales *et al.*, 1991, 1993b).

Thus, with regard to rust diseases tritordeum should be considered to behave as a wheat, with the problems and the breeding strategy being those of a conventional wheat breeding programme. Emphasis should be given to the resistance already present in the wheat. Therefore, the selection of the wheat lines is crucial when planning the synthesis of new tritordeums. The effect of the resistance to rust fungi present in *H. chilense* on the resistance of tritordeum is minimal. However, when screening a large collection in the field over years, disease severities tended to be lower in tritordeum than in wheat for brown rust, but slighly higher for yellow rust. There were indications that the tritordeums from particular *H. chilense* (H7 and H17) lines tended to show lower yellow rust severities than the tritordeums from other *H. chilense* lines (H1 and H12). So, although the *H. chilense* resistance is generally suppressed by the wheat genome, there might still be a small, but useful, role of the *H. chilense* background in the response of tritordeum.

The resistance of *H. chilense* to rust fungi is suppressed by the wheat genome, but not by the rye genome (Rubiales *et al*, 1993c). Suppression of rust resistance due to intergenomic interactions has previously been reported in cereals. Suppressors may possibly be eliminated through induction of mutations as has been reported for the 7DL suppressor locus of wheat 'Canthatch' (Kerber, 1991) that inhibited resistance to stem rust.

H. chilense is resistant to the wheat powdery mildew (*Erysiphe graminis tritici*). There is a quantitative contribution to the resistance of tritordeum (Rubiales *et al.*, 1993a). Field trials showed that disease severity was lower in tritordeum than in its wheat parent. Detached leaf inoculations with several isolates showed that there was a considerable reduction in the mildew colonies in tritordeum although the infection type was that of the wheat parental line. The tetraploid tritordeum had 20x reduction of infection frequency in comparison with its wheat parent; hexaploid tritordeums had a 7-14x reduction; and octoploid tritordeums a 1.5-2.5x reduction. This quantitative nonhypersensitive resistance in tritordeum contributed by *H. chilense* fits the definition of Partial Resistance given by Parlevliet (1978). However, we did not study its genetic basis, and its durability can only be known after widespread use in agriculture (Jonhson, 1984).

H. chilense and tritordeum were resistant (infection type 0) to the barley and rye powdery mildew (*E. graminis* ff.ssp. *hordei* and *secalis*, respectively) isolates studied.

Both *H. chilense* and tritordeum are resistant to *Septoria tritici*. Resistance to *S. tritici* in wheat has been associated with late flowering and tall plant stature. Inoculations both in seedlings under control conditions and in mature plants in the field indicated that all hexaploid tritordeums were inmune to this septoria whatever their stature and lateness. The seedling test included both durum and bread wheat isolates. Octoploid tritordeums displayed some septoria development but considerably less than their respective bread wheat parental line. In octoploid tritordeum the resistance to septoria was associated with lateness but not with tall plant stature (Rubiales *et al.*, 1992).

The main interest in tritordeum breeding is in the hexaploid lines. All hexaploid lines tested, representing a wide genetic background, have proven to be resistant to *S. tritici* in the field. Also seedlings tested with isolates from very different origins were completely resistant. Therefore, we do not expect that septoria leaf blotch will be a problem in tritordeum.

H. chilense and tritordeum may be susceptible to *S. nodorum*. Nevertheless, seedlings and field experiments showed that some tritordeums were more resistant than their wheat parental lines, although others were not. In general tritordeum as a crop is more resistant than wheat (Rubiales *et al.*, 1995b). There was a slight dilution of the resistance at higher ploidy level, but not as clear as for *S. tritici*. As for *S. tritici*, tritordeum resistance to *S. nodorum* is associated with lateness but not with tall plant stature. It should therefore be possible to breed short tritordeum lines with resistance to both septoria diseases.

Tritordeum as a crop can be regarded as being susceptible to *Fusarium culmorum*, but some tritordeums were more resistant than their wheat parent. The contents of ergosterol (estimates the fungal biomass) and the phytotoxic mycotoxin deoxinivalenol (DON) showed that the level of resistance to colonization by *Fusarium* is on average higher in tritordeum than in wheat. Some *H. chilense* genotypes (H7, H17, H56, H61) enhanced the wheat resistance to *F. culmorum* in its tritordeum offspring, others (H1, H11, H12, H13) did not (Rubiales *et al.*, 1995b). As there does not seem to be specialization of resistance for *F. culmorum*, *F. graminearum* or *F. nivale*, we expect a similar response of tritordeum to these other pathogenic fusariums. This is supported by preliminary data on the reaction of a few tritordeum lines to *F. graminearum* (Fedak, pers. comm.).

H. chilense and tritordeum are resistant to common bunt (*Tilletia* spp.). No infected spikes of *H. chilense* or hexaploid tritordeum were found in bunt field tests at Córdoba. Octoploid tritordeum lines did, however, display some bunted spikes but much less than their respective bread wheat parental line (average of 3x reduction in % of infected spikes) (Rubiales, submitted).

H. chilense is also known to possess resistance to the aphids *Diuraphis noxia* (Clement & Lester, 1990) and *Schizaphis graminum* (Castro et al., 1994), to the nematode *Meloidogyne naasi* (Person-Dedryver *et al.*, 1990), to the smuts *Ustilago nuda* and *U. tritici* (Nielsen, 1987), to *Drechslera teres*, to *Rhynchosporium secalis* (Rubiales, unpublished) and tolerance to salt (Foster, *et al.*, 1990). Susceptibility in *H. chilense* has been reported to *Typhula ishikariensis* (Bothmer & Hagberg, 1983), and to *Pseudocercosporella herpotrichoides* (Rubiales, unpublished).

Acknowledgements

We are gratefully indebted to Prof. J.I. Cubero who encouraged the development of this programme. In addition to the authors of this review Dr. E. Sánchez-Monge Laguna, J. Riado, M.J. Jiménez-Alvear and Dr. J.A. Padilla have contributed significantly to the development of tritordeum. We are gratefully indebted to the ETSIAM and the CIDA of Córdoba, for allowing the use of their facilities; to the CICYT (projects AGR89-0552, AGF92-0184 and AGF95-0964-C02-01) for the financial support; and to Drs. R.E. Niks and P.A. Lazzeri for the critical reading of the manuscript.

References

Alvarez, J.B., J. Ballesteros, J.A. Sillero & L.M. Martín, 1992. Tritordeum: a new crop of potential importance in the food industry. Hereditas 116: 193-197.

Barceló, P., A. Cabrera, C. Hagel & H. Lörz, 1994a. Production of doubled-haploids plants from tritordeum anther culture. Theor. Appl. Genet. 87: 741-745.

Barceló, P., C. Hagel, D. Becker, A. Martín & H. Lörz, 1994b. Transgenic cereal (*Tritordeum*) plants obtained at high efficiency by microprojectile bombardment of inflorescence tissue. The Plant Journal 5(4): 583-592.

Barceló, P., P.A. Lazzeri, A. Martín & H. Lörz, 1991. Competence of cereal leaf cells. I. Patterns of proliferation and regeneration capability in vitro of inflorescence sheath leaves of barley, wheat and tritordeum. Plant Sci. 77:243-251.

Barceló, P., P.A. Lazzeri, P. Hernández, A. Martín & H. Lörz, 1993. Morphogenic cell and protoplast cultures of tritordeum. Plant Sci. 88: 209-218.

Barro, F., A.G. Fontes & J.M. Maldonado, 1991. Organic nitrogen content and nitrite reductase activities in tritordeum and wheat grown under nitrate or ammonium. Plant and Soil 135: 251-256.

Bothmer, R. von, J. Flink & T. Landström, 1987. Meiosis in *Hordeum* interspecific hybrids. II. Triploid hybrids. Evolutionary Trends in Plants, Vol. 1(1): 41-49.

Bothmer, R. von & A. Hagberg, 1983. Pre-breeding and wide hybridization in barley. In: *Pre-breeding in relation to genebanks*. EUCARPIA workshop, Beograd-Zemun, Yugoslavia, 1983. pp. 41-53.

Bothmer, R. von & N. Jacobsen, 1986. Interspecific crosses in *Hordeum*. Pl. Syst. Evol. 153: 49-64.

Bothmer, R. von, M. Kotimäki & Z. Persson, 1985. Genome relationships between eight diploid *Hordeum* species. Hereditas 103: 1-16.

Cabrera, A. & A. Martín, 1991. Cytology and morphology of the amphiploid *Hordeum chilense* (4x) x *Aegilops squarrosa* (4x). Theor. Appl. Genet. 81: 758-760.

Cabrera, A. & A. Martín, 1992. A trigeneric hybrid between *Hordeum*, *Aegilops* and *Secale*. Genome 35: 647-649.

Castro, A.M., L.M. Martín, A. Martín, H.O. Arriaga, N. Tobes & L.B. Almaraz, 1994. Screening for Greenburg Resistance in *Hordeum chilense* Roem et Schult. Plant Breeding, in press

Clement, S.L. & D.G. Lester, 1990. Screening wild *Hordeum* species for resistance to russian wheat aphid. Cereal Res. Comm. 18: 173-177.

Cubero, J.I., A. Martín, T. Millán, A. Gómez-Cabrera & A. de Haro, 1986. Tritordeum: a new alloploid of potential importance as a protein source crop. Crop Sci. 26: 1186-1190.

Fedak, G., 1978. Progeny of barley-wheat intercrosses. Barley Genet. Newsl. 8: 34-35.

Fedak, G., 1992. Intergeneric hybrids with *Hordeum*. In: P.R. Shewry (Ed.), Barley: Genetics, Biochemistry, Molecular Biology and Biotechnology, CAB International, UK, pp. 45-70.

Fernández, J.A. & N. Jouve, 1988. The addition of *Hordeum chilense* chromosomes to *Triticum turgidum* conv. *durum*. Biochemical, karyological and morphological characterisation. Euphytica 37: 247-259.

Fernández-Escobar, J. & A. Martín, 1985. A trigeneric hybrid from Triticale x Tritordeum. Z. Pflanzenzüchtg 95: 311-318.

Fernández-Escobar, J. & A. Martín, 1988. A hybrid between hexaploid triticale and octoploid tritordeum. Cereal Res. Comm. 16: 45-51.

Fernández-Escobar, J. & A. Martín, 1989. A self-fertile trigeneric hybrid in the Triticeae involving *Triticum*, *Hordeum* and *Secale*. Euphytica 42: 291-296.

Finch, R.A. & M.D. Bennet, 1980. Mitotic and meiotic chromosome behaviour in new hybrids of *Hordeum* with *Triticum* and *Secale*. Heredity 44: 201-209.

Foster, B.P., M.S. Philips, T.E. Miller, E. Baird & W. Powell, 1990. Chromosome location of genes controlling tolerance to salt (NaCl) and vigour in *Hordeum vulgare* and *H. chilense*. Heredity 65: 99-107.

Gallardo, M. & E. Fereres, 1989. Resistencia a la sequía del Tritordeo (*Hordeum chilense* x *Triticum turgidum*) en relación a la del trigo, cebada y triticale. Invest. Agr.: Prod. Veg. Vol. 4: 361-375.

Johnson, R., 1984. A critical analysis of durable resistance. Ann. Rev. Phytopathol. 22: 309-330.

Kerber, E.R., 1991. Stem rust resistance in 'Canthatch' hexaploid wheat induced by a nonsuppressor mutation on chromosome 7DL. Genome 34: 935-939.

Kimber, G. & P.J. Sallee, 1978. An amphiploid of *Triticum timopheevi* x *Hordeum bogdanii*. In: *Cytogenetics and Crop Improvement Symposium*, Varanasi, India, pp. 98-136.

Kruse, A., 1973. *Hordeum* x *Triticum* hybrids. Hereditas 73: 157-161.

Linde-Laursen, I.B., Schrader, O. & Zerneke F., 1993. Chromosomal constitution of rye (*Secale cereale*) *Hordeum chilense* addition lines. Hereditas 119: 21-29.

Martín, A., 1983. The cytology and morphology of the hybrid *Hordeum chilense* x *Aegilops squarrosa*. The Journal of Heredity 74: 487.

Martín, A., 1988. Tritordeum: the first ten years. Rachis 7: 12-15.

Martín, A. & V. Chapman, 1977. A hybrid between *Hordeum chilense* and *Triticum aestivum*. Cereal Res. Comm. 5: 365-368.

Martín, A. & J.I. Cubero, 1981. The use of *Hordeum chilense* in cereal breeding. Cereal Res. Comm. 9: 317-323.

Martín, A., T. Millán & J. Fernández-Escobar, 1988. Morfología y citología del híbrido y anfiploide *Hordeum chilense x secale cereale*. Ann. Aula Dei 19: 135-142.

Martín, A., J.A. Padilla & J. Fernández-Escobar, 1987. The amphiploid *Hordeum chilense* x *Triticum aestivum* ssp. *sphaerococcum*. Variability in octoploid tritordeum. Plant Breeding 99: 336-339.

Martín, A, D. Rubiales, J.M. Rubio & A. Cabrera, 1995. Hybrids of *Hordeum vulgare* and tetra-, hexa-, and octoploid tritordeum (amphiploid *H. chilense x Triticum* spp). Hereditas 123: 175-182.

Martín, A. & E. Sánchez-Monge, 1980. A hybrid between *Hordeum chilense* and *Triticum turgidum*. Cereal Res. Comm. 8: 349-353.

Martín, A. & E. Sánchez-Monge Laguna, 1980. Effects of the 5B system on control of pairing in *Hordeum chilense* x *Triticum aestivum* hybrids. Z. Pflanzenzüchtg 85: 122-127.

Martín, A. & E. Sánchez-Monge Laguna, 1982. Cytology and morphology of the amphiploid *Hordeum chilense* x *Triticum turgidum* conv. *durum*. Euphytica 31: 262-267.

Millán, T. & A. Martín, 1992. Effects of *Hordeum chilense* and *Triticum* cytoplasms on agronomical traits in hexaploid tritordeum. Plant Breeding 108: 328-331.

Miller, T.E. & V. Chapman, 1978. The amphiploid of *Hordeum chilense* x *Triticum aestivum*. Cereal Res. Comm. 6: 351-352.

Miller, T.E., S.M. Reader & V. Chapman, 1982. The addition of *Hordeum chilense* chromosomes to wheat. In: C. Broertjes (Ed.), *Induced Variability in Plant Breeding*, Pudoc, Wageningen, pp. 79-81.

Nielsen, J., 1987. Reaction of *Hordeum* species to the smut fungi *Ustilago nuda* and *U. tritici*. Can. J. Bot. 65: 2024- 2027.

Padilla, J.A. & A. Martín, 1983. New hybrids between *Hordeum chilense* and tetraploid wheats. Cereal Res. Comm. 11: 5-7.

Padilla, J.A. & A. Martín, 1983. Morphology and cytology of *Hordeum chilense* x *Hordeum bulbosum* hybrids. Theor. Appl. Genet. 65: 353-355.

Padilla, J.A. & A. Martín, 1987. Cytology, fertility and morphology of amphiploids *Hordeum chilense* x tetraploid wheats (Tritordeum). Plant Breeding 99: 295-302.

Parlevliet, J.E., 1978. Race-specific aspects of polygenic resistance of barley to leaf rust *Puccinia hordei*. Neth. J. Plant Pathol. 84: 121-126.

Person-Dedryver, F., J. Jahier, & T.E. Miller, 1990. Assessing the resistance to cereal root-knot nematode, *Meloidogyne naasi* in a wheat line with the added chromosome arm 1HchS of *Hordeum chilense*. J. Genet. & Breed. 44: 291-296.

Rong, J & L. Wang, 1989. Selection of wheat barley octoploid. Hereditas (Beijing) 11: 1-4.

Rubiales, D. & R.E. Niks, 1992a. Low appressorium formation by rust fungi on *Hordeum chilense* lines. Phytopathol. 82: 1007-1012.

Rubiales, D. & R.E. Niks, 1992b. Histological responses in *Hordeum chilense* to brown and yellow rust fungi. Plant Pathol. 41: 611-617.

Rubiales, D., J. Ballesteros & A. Martín, 1991. The reaction of X *Tritordeum* and its *Triticum* spp. and *Hordeum chilense* parents to rust diseases. Euphytica 54: 75-81.

Rubiales, D., J. Ballesteros & A. Martín, 1992. Resistance to *Septori tritici* in *Hordeum chilense* x *Triticum* spp. amphiploids. Plant Breeding 108: 281-286.

Rubiales, D., J.K.M. Brown & A. Martín, 1993a. *Hordeum chilense* resistance to powdery mildew and its potential use in cereal breeding. Euphytica 67: 215-220.

Rubiales, D., R.E. Niks, R.G. Dekens & A. Martín, 1993b. Histology of the infection of tritordeum and its parents by cereal brown rust. Plant Pathology 42: 93-99.

Rubiales, D., R.E. Niks & A. Martín, 1993c. Genomic interactions in the resistance to mildew and rust fungi in hybrids and amphiploids involving the genera *Triticum*, *Hordeum* and *Secale*. Cereal Res. Comm. 21: 187-194.

Rubiales, D., M.C. Ramírez & R.E. Niks, 1995a. Avoidance to leaf rust fungi in wild relatives of cultivated cereals. Euphytica, in press.

Rubiales, D., P. Nicholson, C.H.A. Snijders & A. Martín, 1995b. Reaction of tritordeum to *Septoria nodorum* and *Fusarium culmorum*. Euphytica, in press.

Sánchez-Monge, E. & C. Soler, 1973. Wheat and Triticale in rye cytoplasm. Proc. IV Int. Wheat Genet. Symp. 387-390.

Sánchez-Monge Laguna, E. & A. Martín, 1982. *Hordeum chilense* x *Hordeum vulgare* hybrids. Z. Pflanzenzüchtg 89: 115-120.

Schrader, O. & W. Pohler, 1985. Seed set from a colchicine-treated trigeneric hybrid of the cross *Hordeum chilense* (2x) x Triticale (6x). Cereal Res. Comm. 13: 63-69.

Subrahmanyam, N.C. 1976. Interspecific hybridization, chromosome elimination and haploidy in *Hordeum*. B.G.N. 6: 69-70.

Thomas, H.M. & R.A. Pickering, 1985. Comparisons of the hybrid *Hordeum chilense* x *H. vulgare*, *H. chilense* x *H. bulbosum*, *H. chilense* x *Secale cereale* and the amphiploid of *H. chilense* x *H. vulgare*. Theor. Appl. Genet. 69: 519-522.

Vedel, F., F. Quertier, Y. Cauderon, F. Dosba & G. Doussiault, 1981. Studies on maternal inheritance in polyploid wheats with cytoplasmic DNAs as a genetic marker. Theor. Appl. Genet. 59: 239-245.

Zillinsky, F.J., 1974. The development of triticale. Advances in Agronomy 315-349.

A RETROSPECTION ON TRITICALE

Enrique Sánchez-Monge
Polytechnical University, Madrid, Spain

Abstract

This lecture presents a personal retrospection into the work done on triticale since 1947 in Spain. The review reports in a condensed and autobiographic way the sucessive steps developed by the author. They started at the Aula Dei Experimental Station in Zaragoza and continued uninterruptedly until now at the Polytechnical University of Madrid. At the beginning, it was suggested that triticale could be an interesting cereal for the Spanish drylands. Early, was shown that hexaploid is the optimal ploidy level for vigour and productivity. Different projects were designed to improve ear size, fertility, endosperm quality and seed smoothness. Numerous line of hexaploid triticale were obtained varying the plasmagenic environment by means of irradiation of ears during flowering. They were designed labourious experiments to obtain alloplasmic forms by backcross substitution. Occasionally, changes of cytoplasm induced significant modifications in chromosome stability, morphological characters, yield, endosperm proteins, male-sterility, etc.. Different male sterile lines were essayed for the production of triticale hybrid seed.

THE EARLY STEPS

The history of triticale has already been narrated by several authors [1,2,3,4,5,6] and, as most of you are familiar with these publications, what I am going to offer you in this lecture is a personal retrospection, a look back, into the Spanish work on triticale since 1947, and with reference, only, to the hexaploid triticale.

My interest in triticale started during my stay in 1946 in the laboratory of Dr. Levan at the Sveriges Utsädesförenig of Svalöv, Sweden, and specially through ocassional contacts with Professor Müntzing of the Institute of Genetics of the University of Lund. Professor Müntzing was working on octoploid triticale at that time.

After my return to Spain in 1947 I joined the staff off a new Centre of Agricultural Research, the Aula Dei Experimental Station of the Superior Council of Scientific Research. This Centre was founded and directed by a well-known agronomist, Ramón Esteruelas, my mentor, to whom I devote a warm memory on this ocassion.

At the Aula Dei Station I took charge of the task of breeding cereals: wheat, barley, oats and rye. Very soon Dr. Joe Hin Tjio, a very skilful cytogeneticist, also joined our staff: a researcher who had been my fellow at the Levan laboratory.

H. Guedes-Pinto et al. (eds.), Triticale: Today and Tomorrow, 73–81.
© 1996 *Kluwer Academic Publishers. Printed in the Netherlands.*

A very big collection of winter cereals was formed at Aula Dei. The tetraploid and hexaploid wheats of the Iberian Peninsula were the best represented in the collection by a number of entries. Improved varieties of wheat from different countries were also included in the collection, and also several genotypes of Müntzing octoploid triticales from Sweden.

At that moment we were playing with the idea that triticale could be an interesting cereal for our drylands in Spain, by combining into a single species the baking quality of wheat and the rusticity of rye.

Both, the octoploid triticales and the hexaploid wheats coming from Northern Europe, proved to be of too late maturity for our environmental conditions and were very sensitive to black rust.

After the failures in the trials with octoploid triticales, a programme for obtaining new ones, more suitable for Spain, was prepared, using Iberian material of wheat and rye as parents. But during the discussion of this programme the visit to Aula Dei of Professor C.A. Jorgensen, from the University of Copenhagen took place. After observing the great variability of genotypes of tetraploid wheat from Spain and Portugal, Professor Jorgensen suggested to us the idea of using this tetraploid wheat material as female parents for obtaining hexaploid triticale. At that moment we agreed with the ideas of Prof. Jorgensen and Dr. Levan about the possible existence of an optimal ploidy level for vigour and productivity within each taxonomic group of plants. This optimal could be hexaploidy for wheat and for triticale also.

THE USE OF RYE INBRED LINES

Another idea taken into account for the hexaploid triticale programme was that, if the new cereal obtained was a self-pollinator, as was its female parent, the rye component, coming from an allogamous species, could induce a lack of vigour and flower fertility, as a consequence of homozygosis.

In another programme at the Aula Dei Station we were working to obtain inbred lines of rye with high self-fertility [7,8]. After selfing for several years rye plants from different Spanish populations, we obtained several lines which were relatively self-fertile, and it was decided to use these lines as male parents in the production of new triticales.

The experiments were conducted between 1950 and 1955 [9]. Crosses in the open field were made between 64 genotypes of tetraploid wheat as female parents and 82 rye lines as male parents. The number of cross combinations was 305 and the total number of flowers pollinated was 26,126.

From these crosses only 54 gave rise to at least one plant with 21 chromosomes.

After colchicine treatments, by means of capping cut tillers of potted 21 chromosome seedlings, 9 new types of 42-chromosome triticales were obtained.

Dr. Tjio and myself presented a short paper about hexaploid triticale at the IX International Congress of Genetics that took place in Lago Como, Italy in 1953, but our paper did not raise too much interest.

The interest of the world-wide cereal agricultural research on hexaploid triticale started probably during the First International Wheat Genetics Symposium that took place in Winnipeg, Canada, in 1958.

At this Symposium I was invited to present a lecture about "Hexaploid triticale" [10]. After the lecture and during the discussion I was asked about my opinion about the future of this cereal as a new agricultural crop, and my answer was an emphatic statement of my belief in a promising future.

Moreover, during the Symposium, the participants had the opportunity to visit a "living herbarium" with more than 800 wheat varieties from 40 countries, sown in small plots at the campus of the University of Manitoba. Among this material were three of our triticales. The vigour, health and ear size of these triticales was the cause of great general interest.

After the Winnipeg Symposium I was transferred to take charge of the direction of the Maize Breeding Centre of the National Institute of Agricultural Research (INIA) in Madrid, where it was not possible, at least officially, to devote time, money or experimental fields to triticale. But a grant from a private foundation, the "Fundación Juan March", allowed us to work on triticale at a small scale. Our main purpose at that time was to improve the smoothness of the triticale kernels.

The researchers devoted to obtaining and breeding triticale had already received at this time a lesson on humility. At the begining we had the intention to "create" a cereal as good a bread-maker as wheat and as rustic as rye, but we had really obtained a cereal which was more suitable for the manufacturing of animal feeds and, for human consumption, more in the form of chapatis than in the form of bread.

Another experiment made in Spain was about the use of triticale as a prop for vetches. The association of triticale and common vetch proved to be very satisfactory for hay making.

TRITICALE SEED SHRIVELLING

The problem of triticale seed shrivelling was treated in another different way, by starting a process of plasmagenic mutagenesis that perhaps it is interesting to describe.

The initial idea was that, as the hexaploid triticales were always obtained with wheat as the female parent, the cytoplasm of the amphiploid would carry wheat plasmagenes only, whereas its nucleus was composed of 2/3 of wheat chromosomes and 1/3 of rye chromosomes. The latter could perhaps malfunction, especially during mitosis of endosperm formation, due to finding themselves in a different plasmagenic environment.

For the experiment we used triticales with good floral fertility and ear size derived from Spanish wheats and self-fertile inbred lines of rye. In potted plants, 3-5 ears were emasculated and irradiated immediately with a dose of 1500-3000 r in a gamma field. After irradiation the emasculated ears were fertilized using pollen of unirradiated sister plants of the same triticale line.

From a total 77 progenies obtained it was possible to select 5 with slightly smoother seeds.

Any difference between the original lines and the selections must therefore have been due to dominant nuclear mutations or to plasmagenic mutations.

I order to ascertain if the improved endosperm quality was due to plasmagenic influence, reciprocal crosses were made between the 5 selections and their original lines.

The endosperm quality was estimated in the parents and crosses, and the results indicated a slight plasmagenic influence on the character. We concluded that the irradiation and posterior restoration of the nucleus could be one of the ways of improving seed smoothness by inducing favourable plasmagenic changes [11].

The material obtained in this last experiment was the origin of the variety "Cachirulo", the first triticale variety registered in Spain [12].

At that time I was promoted to the post of director of the Cereal Breeding Center of the INIA and so my work on triticale could then come out into the open.

With the variety Cachirulo, an experiment in gradual milling was conducted by Prof. García Olmedo and coworkers [13] at the Cereal Breeding Centre. By means of the gradual milling, fractions were obtained with very high protein content that could be used as additives in the manufacture of animal feeds.

CYTOPLASM EFFECT AND HYBRID TRITICALE

Another line of research was obtaining triticales in rye cytoplasm in order to compare isogenic lines with wheat and with rye cytoplasm [14].

As the donor of cytoplasm we used a tetraploid rye variety, "Gigantón", obtained in 1952 [15] and still in cultivation in Spain. The hexaploid triticales used as male parents were our own varieties (JM and T varieties) and some lines selected from segregating material sent to us by Prof. B.C. Jenkins from the University of Manitoba (JK lines).

Obtaining triticale in rye cytoplasm was made by backcross substitution [16]:

$$Secale\ cereale\ 4n \times Triticale^b$$

where b is the number of backcrosses (b=8 was reached, Table 1).

TABLE 1- TYPES OF SUBSTITUTION BACK-CROSSES

Secale cereale 4n x Triticale^b (b = 8)	(b = 8)
(*Aegilops ovata* x *Triticum aestivum*^{b}) x Triticale$^{b'}$	(b = 11, b' = 7)
[(*Ae. ovata* x *T. dicoccum*^{b}) x *T. turgidum*$^{b'}$)] x Triticale$^{b''}$	(b = 8, b' = 4, b'' = 2)
(*Ae. caudata* x *T. aestivum*^{b}) x Triticale$^{b'}$	(b = 12, b' = 3)
[(*Ae. caudata* x *T. aestivum*^{b}) x *T. turgidum*$^{b'}$] x Triticale$^{b''}$	(b = 7, b' = 4, b'' = 2)
(*T. timopheevi* x *T. aestivum*^{b}) x Triticale$^{b'}$	(b = 16, b' = 5)
(*T. timopheevi* x *T. turgidum*^{b}) x Triticale$^{b'}$	(b = 12, b' = 4)

Chromosome substitution in these backcrosses was relatively quick, and we obtained progenies with 42 somatic chromosomes and 21 bivalentes at meiosis as early as after the fourth backcross. Practically all the plants from the fifth backcross on showed cytological characteristics similar to triticale in wheat cytoplasm, i.e. 42 somatic chromosomes and 21 bivalents at meiosis.

The hexaploid triticale lines in rye cytoplasm showed different degrees of flower fertility. Full fertility was much more frequent than male sterility, which was found only in two lines (Table 2).

The high level of fertility shown by two of the lines obtained, JM-130 and JM-135 led us to make a comparative study between these two lines and their counterparts with wheat cytoplasm. The results of the comparison are shown in Tables 3 and 4.

TABLE 2 - FLOWER FERTILITY IN HEXAPLOID TRITICALE LINES
WITH RYE CYTOPLASM

Male-steriles	*S. cereale* $4n=$ x JM - 78^4) x T-236^2 (*S. cereale* $4n=$ x JM - 130^5) x JK-138
5 to 30% fertility	(*S. cereale* $4n=$ x JM - 130^5) x JK-138^2 (*S. cereale* $4n=$ x JM - 130^5) x JK-124 (*S. cereale* $4n=$ x JM - 74^6)
30 to 60% fertility	(*S. cereale* $4n=$ x JM-130^5) x JK-124^2 (*S. cereale* $4n=$ x JM - 78^6)
60 to 100% fertility	(*S. cereale* $4n=$ x JM - 130^5) x JM- 87 (*S. cereale* $4n=$ x JM- 130^5) x JM-139
Fertility 100%	(*S. cereale* $4n=$ x JM - 130^7) (*S. cereale* $4n=$ x JM - 135^7) (*S. cereale* $4n=$ x JM-130^5) x JK-147^2 (*S. cereale* $4n=$ x JM-130^5) x JK-5/1^2 (*S. cereale* $4n=$ x JM-130^5) x JK-5/6^2 (*S. cereale* $4n=$ x JM-130^5) x JK-122^2 (*S. cereale* $4n=$ x JM-130^5) x T-202

The only significant difference among the cytological characters was for the number of micronuclei per tetrad, which was higher with wheat cytoplasm (Table 3).

Two of the components of yield, "ears per plant" and "spikelets per ear" were consistently higher with wheat cytoplasm (Table 4). Flower fertility was higher in one line and lower in the other with the cytoplasm change, but the differences were not significant. Thousand kernel weight was always significantly higher for rye cytoplasm. Grain yield and protein per hectare were superior for wheat cytoplasm.

TABLE 3 - COMPARISON OF THE MEIOTIC STABILITY OF TRITICALE LINES WITH WHEAT AND RYE CYTOPLASM

LINE	JM-130		JM-135	
Cytoplasm	Wheat	Rye	Wheat	Rye
Univalents per PMC	1.46	1.46	1.08	1.20
Bivalents per PMC	20.27	20.27	20.46	20.40
Micronuclei per tetrad	0.97	0.67	0.85	0.52

TABLE 4 - COMPARISON FOR MORPHOLOGICAL CHARACTERS AND YIELD BETWEEN TRITICALE LINES WITH WHEAT AND RYE CYTOPLASM

CHARACTER	WHEAT VS. RYE CYTOPLASM	
	JM-135	JM-135
Plant height	5% higher*	4% higher*
Tillering	32% higher*	21% higher*
Maturity date	identical	identical
Flag leaf lenght	12% lower*	5% lower
Flag leaf breadth	4.6% higher	1.1% higher
Flag leaf weight per cm2	12% higher*	identical
Ears per plant	56% higher**	39% higher**
Spikelets per ear	5.5% higher*	0.9% higher
Glume length	6.4% lower**	identical
Glume breadth	identical	1.8% lower
Rachis length	2.7% higher	3.1% lower
Rachis internode length	0.9% higher	4.2% lower
Kernel length	5.9% lower**	0.8% higher
Chlorophyll *a* content	15% higher*	0.7% higher
Chlorophyll *b* content	13 % higher*	identical
Flower fertility	2.9% lower	5.8% higher
1000 kernels weight	10.7% lower**	3.2% lower*
Kernel protein content	8.4% higher*	6.3% lower
Yield	42.8% higher*	43.6% higher*
Protein per Ha	54.8% higher	34.6% higher

* Significant at the 5% level; ** Significant at the 1% level
The values for rye cytoplasm were taken as 100 for comparison.

The results seemed to indicate a lack of practical value of hexaploid triticale with rye cytoplasm, but no definitive conclusions could be made until more isogenic triticale lines with both cytoplasms had been compared.

Another line of experiments was initiated in order to obtain hybrid seed of triticale by means of cytoplasmic male-sterility and restoration. The big size of the anthers of hexaploid triticale and its good pollen production could make triticale more apt than wheat for the production of hybrid seed.

Male sterile lines were obtained by means of substitution backcrosses, using as female parents alloplasmic wheats with *Aegilops ovata, Ae. caudata* and *Triticum timopheevi* cytoplasms. The male parents were our JM, JK, T and Cachirulo derived lines and also CIMMYT lines. The types of backcrosses made are recorded in Table 1.

With the *Aegilops ovata* cytoplasms only the first of the two types of substitution backcrosses (Table 1) could be extended beyond the second backcross. The fourth backcross was obtained with the line JM-74 and the cross combination:

$$(\textit{Ae. ovata} \text{ x } \textit{Triticum aestivum}^6) \text{ x Triticale}^4$$

From this material the chromosomes of other triticale lines have been easily transfered to *ovata* cytoplasm, and we reached the seventh substitution backcross. All the pollinations made with JM, JK and Mexican lines gave only male-sterile progenies. The effects of *ovata* cytoplasm on morphology seem to be less drastic than in wheat and there was only a slight reduction in height and a small difference in earliness.

Two types of substitution backcrosses were also used for the *Ae. caudata* cytoplasm and only the first one gave viable seeds after three backcrosses. This was with the line JM-130 in the combination:

$$(\textit{Ae. caudata} \text{ x } \textit{T. aestivum}^{12}) \text{ x Triticale}^3$$

TABLE 5- TRITICALE LINES USED IN SUBSTITUTION BACK-CROSSES WITH
T. TIMOPHEEVI CYTOPLASM WITH TWO CYTOPLASM DONORS

CYTOPLASM DONOR	*T. timopheevi* x x *T. aestivum*[b]	*T. timopheevi* x *T. turgidum*[b]
Triticale lines (all giving male sterile progenies)	Armadillo 130 Armadillo 135 Armadillo 1524 Cachirulo 1642 Cachirulo 1643 Cachirulo 1647 Cachirulo 1810 JK-5/10 JM-78 JM-87 JM-130 JM-135 T-236	Armadillo 130 Armadillo 135 Armadillo 1524 Bronco 90 Cachirulo 1463 Cachirulo 1464 Cachirulo 1495 Rossner

With other triticale lines (JM, JK and Cachirulo) no seed was obtained.

Both types of backcrosses used for the *timopheevi* cytoplasm gave viable seeds and the fifth backcrosses was reached. Values of the b exponent from 6 to 13 (Table 1) gave good results for the backcrosses:

$$(\textit{T. timopheevi} \text{ x } \textit{T. aestivum}^b) \text{ x Triticale}^5$$

and values from 6 to 12 for the backcrosses:

$$(T.\ timopheevi \ \ \text{x} \ \ T.\ aestivum^b\ )\ \text{x Triticale}^4$$

All the progenies obtained (Table 5) were completely male sterile.

Thus, obtaining male sterile lines was not a problem, but no fertile alloplasmic lines were found that could act as restorer parents for obtaining the hybrid seed.

This is, more or less, a short history of our work till my retirement.

TABLE 6 - AREA DEVOTED TO TRITICALE CULTIVATION

País	*Ciclo*	*Superficie* $(10^3 ha)$
Germany	Winter	62
Algeria	Spring	7
Argentine	Spring	16
Australia	Spring	100
Austria	Winter	2
Belgium	Winter	8
Brazil	Spring	35
Bulgaria	Winter	10
Burundi	Spring	2
Canada	Winter + Spring	6
Chile	Spring	7
China	Spring + Winter	2
Spain	Spring	80
France	Spring + Winter	300
Hetherlands	Winter	4
Hungary	Winter	1
Italy	Spring	30
Kenya	Spring	5
Luxemburg	Winter	2
Moroccos	Spring	1
Mexico	Spring	4
New Zealand	Spring + Winter	2
Poland	Winter	750
Portugal	Spring	70
United Kingdom	Winter	16
Romania	Winter	13
South Africa	Spring + Winter	85
Sweden	Winter	1
Switerzand	Winter	10
Tunisia	Spring	50
ex-URSS	Winter	250
USA	Winter + Spring	200

To finish my lecture I should like to show you the last figures that I was able to obtain about the extension of triticale cultivation in the world (Table 6).

Looking at these figures we are tempted to say that something has been achieved due to the efforts of triticale researchers all over the world and that a little bit of this is owing to the efforts of the Spanish team.

References

1 - Larter, E.N., 1973. A review on the historical development of triticale. Symp. Bioch. Nutr. Triticale.

2 - Müntzing, A., 1973. Historical review of the development of triticale. Proc. Int. Triticale Symp.: 13-30.

3 - Müntzing, A., 1979. Triticale. Results and problems. Suppl. 10 J. Plant Breed.: 1-103.

4 - Sánchez-Monge, E., 1973. Development of triticales in Western Europe. Proc. Int. Triticale Symp.: 31-39.

5 - Cauderon, Y., 1981. Origine et évolution des triticales. Ind. Cereales (10): 3-9.

6 - Royo, C., 1992. El triticale.

7 - Ramirez, D. 1956. La selección de la fertilidad en el centeno. An. Aula Dei 4: 208-211.

8 - Lacadena, J.R., E. Sánchez-Monge and L.M. Villena, 1969. Selection for self--fertility in rye inbred lines. An. Aula Dei 10: 846-855.

9 - Sánchez-Monge, E., 1956. Studies on 42-chromosome triticale. The production of amphiploids. An. Aula Dei 4: 191-207.

10 - Sánchez-Monge, E., 1958. Hexaploid triticale. Proc. First Int. Wheat Genet. Symp.: 181-194.

11 - Sánchez-Monge, E., 1968. Improvement of endosperm quality in triticale. Proc. Third Int. Wheat Genet. Symp.: 371-372.

12 - Sánchez-Monge, E., 1969. La saga del "Cachirulo". An. Aula Dei 10: 795-799.

13 - Garcia Olmedo, F., J.M. Vallejo and S. de la Plaza, 1970. Millering and utilization of hexaploid triticale. Proc. V World Cong. Cereals Bread 3: 183-187.

14 - Sanchez-Monge, C. Soler, 1973. Wheat and triticale with rye cytoplasm. Proc. Fourth Int. Wheat Genet. Symp.: 387-390.

15 - Tjio, J.H., E. Sánchez-Monge and M. Alvarez Peña, 1973. Centenos tetraploides españoles. Agricultura 22: 138-140.

16 - Kihara, H., 1964. Restoration and substitution of nucleus. Proc. IX Int. Cong. Genetics: 900-902.

17. J.H. Tjio, 1974. Note on 42-chromosome triticale. Proc. IX Int. Cong. Genetics: 718.

LESS KNOWN STUDIES ON TRITICALE IN CENTRAL AND EASTERN EUROPE

Anton Tajnsek and Ivan Kreft
Biotechnical Faculty, Agronomy Dept., Ljubljana, Slovenia

Abstract

Many scientists in Central and Eastern Europe have done work on developing wheat-rye hybrids and studying triticale. Their results, however, have not been published in English and therefore they are less well known outside their countries. At the beginning of this century, shortly after the role of chromosomes in heredity was discovered, important research on sterile and fertile wheat-rye hybrids by E. von Tschermark and F. Jesenko was done in Vienna and Ljubljana. In 1920, after G. D. Karpechenko showed the possibility of restoring the fertility in distant hybrids by T. K. Meister worked on wheat-rye hybrids, while N. V. Tsytsyn studied wheat-Agropyron hybrids and wheat-Elymus crosses. Further distant hybridisation was encouraged by the works of N. I. Vavilov. A. R. Zhebrak, P. M. Zhukovsky and several other Russian, Belarussian and Ukrainian scientists studying different wheat alloploids. Development and research on triticale carried out by V. E. Pisarev was of great importance. Later works of other Central and East European authors are also reviewed and some of the breeding results in production of triticale cultivars presented.

Introduction

In the literature it is well known, that Wilson described in 1875 an almost sterile wheat-rye hybrid, and that Rimpau in 1891 obtained a fertile wheat-rye hybrid.

Less known are the achievements of other authors from Central and Eastern Europe from the beginning of this century, as they published mostly in Russian, German, French and some other languages. These languages are less well understood by most modern scientists.

Since they contributed their share to the current level of knowledge about triticale it is worth making a survey of their contribution available in the English language.

Early achievements on wheat rye crosses and early research and breeding of triticale, including some of the results from Central and Eastern Europe, were reviewed among others by Love and Craig (1), Bleier (2), Meister (3),Isenbeck and Rosenstiel (4), Müntzing (5), Kreft (6), Müntzing (7), Sulima (8) Rigin and Orlova (9), Fedorova (10). The review of Bleier (2) in German is very extensive and detailed for that time.

H. Guedes-Pinto et al. (eds.), Triticale: Today and Tomorrow, 83–87.
© 1996 *Kluwer Academic Publishers. Printed in the Netherlands.*

Early wheat-rye hybrids of Fran Jesenko in Vienna and in Ljubljana

First interspecific crosses in Vienna were reported and discussed by Tschermak (11, 12). At the beginning of the 20th century the Slovenian scientist Fran Jesenko started his work on wheat-rye hybrids in the research group of Erich von Tschermak in Vienna.
Jesenko was born in 1875 in Škofja Loka near Ljubljana (Slovenia). In 1900 Jesenko obtained his doctor's degree at Vienna University (Austria). He spent long periods in Egypt, France (Paris) and Sweden (Uppsala and Stockholm) as a botanist. From 1909 F. Jesenko was a research assistant at the experimental station of von Tschermak (13, 14). In 1913 he acquired the title of university lecturer for plant production and horticulture. During the 1914-1918 world war his scientific work was interrupted and plant material lost. After the war Jesenko worked as a lecturer at Zagreb University (Croatia) and in 1921 became professor of botany at Ljubljana University. In Ljubljana F. Jesenko started his work on wheat-rye hybrids again, as well as on *Triticum-Aegilops* hybrids. Some field trials were performed at Beltinci experimental station with F. Mikuž, who was later appointed as first professor of genetics and plant breeding at Ljubljana University. Prof. Fran Jesenko died in 1932 in a high-mountain accident, while on a botanical excursion.
Besides several papers concerning plant physiology, F. Jesenko published two papers on his wheat-rye crosses (15, 16). Jesenko successfully crossed different varieties of bread wheat with rye varieties, *Triticum dicoccoides* with rye and with *Secale montanum*, and *Triticum dicoccoides* with rye and with the hybrid of *Secale montanum* and *Secale cereale*. By crossing Mold Squarehead and Petkus Rye he obtained a perennial wheat-rye hybrid. This hybrid was backcrossed with wheat and a fertile perennial plant was obtained. F. Jesenko studied inheritance of some characters, cytology and morphology of the hybrids. He suggested that the reduced fertility in wheat-rye hybrids is not only due to irregular chromosome distribution, but also possibly connected with deviating flower morphology. He studied in detail morphology of the spikelets and the cytology of the tetrads and pollen grains.
Research of F. Jesenko on wheat-rye hybrids was of some importance for further triticale studies and breeding. His results were cited by later reports of many authors (1, 2, 3, 4, 7, 8, 12, 17, 18, 19, 21, 22, 23, 24, 25, 26, 28).

Triticale breeding in Russia and Ukraine

In 1920, after G. D. Karpechenko showed the possibility of restoring the fertility in distant hybrids by doubling the chromosome number, the development of wheat-rye and other amphiploids was promoted.
Further distant hybridisation was encouraged by the work of N. I. Vavilov. A. R. Zhebrak, P. M. Zhukovski and several other Russian, Belarussian and Ukrainian scientists studying different wheat alloploids.
In Saratov (Russia) octoploid triticale breeding started in 1917 (8). Results were reported by Meister (19), Meister and Meister (27) , Meister and Tyumyakoff (28), Levitsky and Benetskaya (21), and Avdulov (29). Later breeding of octoploid triticale started as well in Ukraine at Belocerkovsk experimental station (30, 31). In the meantime, G.D. Karpechenko studied his alloploid *Raphanobrassica*, obtained from spontaneous chromosome doubling of hybrids.
The first hexaploid triticale was obtained in the USSR in 1932-1933 by A.I. Derzhavin (8, 32) by crossing *Triticum durum* with *Secale montanum*.
After the discovery of colchicine treatment, the first octoploid triticale was obtained in 1940 by

Navalihina. In 1941 V.E. Pisarev obtained the first spring lines of octoploid triticale, and in 1945 first winter forms (8, 33).

It seems that research of triticale in Russia was strongly repressed in the USSR in the period from about 1936 until about 1955. Georgy Karlovich Meister tragically lost his life in prison in 1937, and Georgy Dmitriyevich Karpechenko shared a similar fate in 1942, when he was only 43 years old (the second author obtained this information in the Museum of General Genetics in Moscow in July 1989). Early Russian triticale have been used in triticale breeding in Sweden (34, 35) and probably in some other western countries as well.

Studies on triticale fertility

Nakao (36) was the first to analyse cytologically wheat-rye hybrids: He found abnormalities, including the fact that sometimes 5 or 6 cells resulted from one pollen mother cell. Similar abnormalities and pollen grains of different size were found two years later by Jesenko (16). Meiotic behaviour was studied in more detail by Kihara (37), Meister (38) and other scientists.

Quality studies of wheat interspecific crosses

Fedorova (39) studied aneuploidy and fertility in octoploid triticale and found chromosome numbers from 49 to 55. Some fertile aneuploids had higher yields than euploids..

In 1985 Fedorova (10) published a review of the achievements and prospects of triticale breeding in Russia and in other countries. As a result of the achievements of Ukrainian Scientific Institute of Crop Science, triticale 'AD 206', outyielding wheat cultivar 'Mironovskaya 808' , was registered in 1977.

Brandenstein et al. (40) studied dependence of primary triticale genotypes on rye and wheat nuclear genomes and the wheat cytoplasm. They found that *per se* performance of the wheat and rye parents is of little use for the triticale performance, because parental genomes and their interactions significantly influenced the expression of most traits.

Triticale starchy endosperm and aleurone ultrastructures were studied by scanning electron microscopy by Kreft (6) and starch grain and protein bodies similar to those in wheat were found.

Breeding of commercial triticale cultivars

Study of triticale started in Bulgaria in 1964. The largest number of genotypes used in the breeding programme originated from Russia and Mexico. In 1984 the first triticale cultivar was selected and recognized ('Vihren'), and another some years later ('Peshenk').

In former Yugoslavia triticale breeding work started in 1970. The first cultivar was registered in 1970 ('Kg-20'). Since that time have been a further seven cultivars of triticale inscribed in the official list of varieties. All of them have been created by Macedonian and Serbian breeders.

In Italy a group of breeders at CNEN began an intensive triticale breeding programme in 1970 (41). This work resulted in the first Italian triticale cultivar 'Mizar' in 1980.

In Hungary home-breed cultivars with a yield of more than 7 t/ha and a protein content of more than 13% were achieved alongside great success in breeding of wheat at the end of the 1980s.

Slovenia does not have its own program of breeding commercial triticale cultivars. There are 9 cultivars listed on the list of varieties inherited from the former Yugoslavia but only 'clercal' is produced for animal feed.

Literature

(1) Love HH, Craig,WT. Fertile wheat-rye hybrids. J. Heredity 1919;10(5):195-207.

(2) Bleier H. Genetik und Cytologie teilweise und ganz steriler Getreidebastarde. Bibliographia Genetica 1928;4:321-400.

(3) Meister NG. Formoobrazovatel'nyj process rzhano-pshenichniykh gibridov shenichnogo tipa.
In: Meister GK, red. Rzhano-pshenichnie gibridi u processe ih izuchenija i spolzovanija dlja selekcii. Seljhozgiz, Moskva, 1936: 15-141.

(4) Isenbeck K, Rosenstiel K. Die Züchtung des Weizens. P. Parey Berlin und Hamburg, 1950:529.

(5) Müntzing, A. Experiences from work with octoploid and hexaploid ryewheat (Triticale). Biol. Zbl. 1972;91:69-80.

(6) Kreft I. Genetika novega žita tritikale - od raziskav do uporabe v praksi. In: Spominski zbornik BF UL . Ljubljana, 1975:33-39

(7) Müntzing, A. Triticale; Results and Problems. V. P. Parey. Berlin und Hamburg,, 1979: 103.

(8) Sulima Ju.G.Tritikale, dostishenya, problemy, perspektivy. Stiinca, Kishiniev. Triticale, Studies and Breeding; 1976 Jul 3-7; Leningrad: Eucarpia Cereal Section, Intern. Symp.: 250

(9) Rigin BV, Orlova IN. Pshenichno-rzhanye amfidiploidy. Leningrad, Kolos, 1977: 300.

(10) Fedorova TN. Novoy zernovoy kul'ture - tritikale - 20 let. Genetika t. 1985;21:151-190.

(11) Tschermak E. Die Kreuzungzüchtung des Getreides und die Frage nach den Ursachen der Mutation. Monatshefte für Landwirtschaft 1908;1:4-12.

(12) Tschermak E. Über seltene Getreidebastarde. Beitrege zur Pflanzenzüchtung, 1913;III:49-61.

(13) Tschermak - Seysenegg, Erich von. Leben und Wirken. Verlag Paul Parey. Berlin und Hamburg, 1958: 196

(14) Kreft I. Fran Jesenko (1875-1932) and studies on his partly fertile wheat-rye hybrids Interspecific Hybridization in Plant Breeding. Proc. 8th Congr. EUCARPIA; 1977; Madrid: 41-42.

(15) Jesenko F. Sur un hybride fertile entre *Triticum sativum* (blé Mold-Squarehead) et *Secale cereale* (seigle de Petkus); 1911; Paris. IV Conf. Int. Génét.: 301-311.

(16) Jesenko F. Über Getreide-Speziesbastarde (Weizen-Roggen). Z. für induktive Abstammungs- und Vererbungslehre 1913; 10(4):311-326.

(17) Jesenko F.Wheat x Rye Hybrids. J. Heredity 1915;6:47.

(18) Leighty CE. Natural wheat-rye hybrids of 1918. J. Heredity 1920;11(3):129-137.

(19) Meister GK. Natural hybridization of wheat and rye in Russia. J. Heredity 1921;12(10):467-471

(20) Meister N, Tyumyakoff NA. Rye-wheat hybrids from reciprocal crosses. J. of Genetics 1928;20(2):233-247.

(21) Levytsky GA, Benetskaya GK. Citologiya pshenichno-rzhanykh amfidiploidov. Trudy vsesoyuznoyo s'ezda po genetike, selektsii, semenovodstvu i plemenomu zhivotnovodstvu, tom 2. Genetika; 1930; Leningrad.

(22) Florell VH. A genetic study of wheat x rye hybrids and back crosses. J. Agr. Res. 1931;42:315-339.

(23) Schiemann E. Entstehung der Kulturpflanzen. Berlin, 1932: 377.

(24) Meister NG. Rzhano-pshenichnye gibridy. In: Meister GK, red. Sbornik statey po selektsii i semenovodstvu. Saratov: Saratovskoe oblastnoe izdatel'stvo, 1937: 34-43.

(25) Forlani R. Il frumento, aspetti genetici del miglioramento della cultura granaria.

Monografie di Genetica Agraria. Pavia, 1954: 315.

(26) Korić S, Korić M. Kako nastaju nove sorte poljoprivrednog bilja. Zagreb,1970:326

(27) Meister GK, Meister NG. Rzhano-pshenichnye gibridy. Moskva, 1923. (Cited from Sulima, 1976).

(28) Meister NG, Tyumyakoff NA. Zhurnal opitnoy agronomii Yugo-Vostoka. Saratov, T. 4. vip. I. 1927:87-97. (Cited from Sulima, 1976).

(29) Avdulov NP. Povedeniye rzhanopshenichnykh amfidiploidov v skreshchi-vanyakh. Citologicheskiye issledovanija F1 *T. secalotricum saratoviense* x *S. ceereal.* In: Raboty po citologii kulturnykh rasteniy. Moskva-Leningrad, 1937:127-135. (Cited from Rigin and Orlova, 1977).

(30) Lebedev VN. Estestvenny rzhano-pshenichny gibridy na Belotserkovskoy selektsionnoy stantsii. Byul. Belotserk. selekts. stantsii, No I, s. ll4-120. (Cited from Sulima, 1975; and Rigin and Orlova, 1977).

(31) Lebedev VN. Novye yavleniya v pshenichno-rzhanykh gibridah. Tr. Ukr. NII sah. prom., Kiev, 1932.

(32) Derzhavin AI. Rezul'taty rabot po vyvedeniyu mnogoletnikh sortov pshenitsy i rzhi. (Raising perennial wheats and rye).- Izd-vo AN SSSR, 1938; No 3:663-665. (Cited from Rigin and Orlova, 1977).

(33) Pisarev VE. (1947): Izmenchivost potomstva amfidiploidov "yarovaya pshenitsa x yarovaya rozh". Doklady VASHNIL, vip. 1947; 12:40-47.

(34) Müntzing, A. Some recent results from breeding work with ryewheat. In: Recent Plant Breeding Research Svalöf 1946 - 1961. Almqvist & Wiksell. Stockholm, 1963: 167-178.

(35) Merker A. Triticale - a new cereal crop. In: Svalöf 1886 - 1986, Research and Results in Plant Breeding. Ed. G. Olsson, Lts förlag, Stockholm, 1986: 132-139.

(36) Nakao M. Cytological studies on the nuclear division of the pollenmothercells of some cereals and their hybrids. Journ. Coll. Agricult. Tohoku-Imp. Univ. Sapporo 4 1911; 173-190 (Cit. from Bleier, 1928).

(37) Kihara H. Cytologische und genetische Studien bei Wichtigen Getreidearten mit besonderer Rücksichtsnahme auf die Verhältnisse der Chromosomen und die Sterilität in den Bastarden. Memoirs Ser. B.1 1924; Vol. I (Cit. from Bleier, 1928).

(38) Meister GK, Meister NG. Wheat rye hybrids. Contr. from the Saratow Agric. Exp. Station. 1924; (Cit. from Bleier, 1928).

(39) Fedorova TN. Aneuploidija i fertil'nost' u Triticale. Genetika 1984; 20:274-283.

(40) Brandenstein C, Melchinger AE, Oettler G. Einfluß des Weizen- und Roggen-genoms sowie des Weizencytoplasmas auf agronomische Merkmale bei primären Triticale. Bericht über die Arbeitstagung 1992. Arbeitsgemeinschaft der Saatzuchtleiter. Vereinigung österreichischer Pflanzenzüchter; 1992 Nov 24-26; Gumpenstein.

(41) Mosconi C, Rossi L. Caratteristiche agronomiche del triticale Mizar. Agrario 1981: 16205-16207

II- CYTOGENETICS

TRITICALE GENOMIC AND CHROMOSOMES' HISTORY

Nicolas Jouve[1] & Consuelo Soler[2]
[1]Department of Cell Biology and Genetics, University of Alcalá de
.Henares, Campus, 28807-Alcalá de Henares (Madrid) Spain
[2]Department of Plant Breeding, C.I.T., I.N.I.A., La Canaleja, Alcalá de
Henares (Madrid) Spain

Abstract

The present report reviews the classical subjects in the history of triticale, that have been condensed and centered from the cytogeneticist's view point. It was early demonstrated that in respect to practical breeding, chromosome stability and fertility, hexaploids represent the optimal ploidic level. From the beginning, the cytogeneticist paid attention to the question of chromosome instability and meiotic irregularities in triticale. Several theories attributed the occurrence of univalents to different factors. Those that have deserved the most attention among researchers and breeders are reviewed: genetic wheat-rye interactions, time of duration of meiosis, and presence of telomeric heterochromatin in the rye chromosomes. The main cytogenetic strategies to broaden the germplasm are reviewed: obtention of triticales with different cytoplasms, secondary triticales following hybridization programs, tetraploid forms, wide crossing; new forms with variability produced by somaclonal or androgenic variation through *in vitro* culture.

Introduction

The possibility that a new species can arise by spontaneous hybridization between two preexisting forms of plants was pointed out by the father of biological nomenclature, Linnaeus, in 1760. This eminent naturalist described a number of new species, which he declared had been produced by hybrid generation, even though he did not accept an evolutionary concept of species. Common wheat, *Triticum aestivum* L. was born in this manner in the Mediterranean region and in southwestern Asia through two sets of crossings. According to Cauderon polyploid wheats clearly have

H. Guedes-Pinto et al. (eds.), Triticale: Today and Tomorrow, 91–118.
© 1996 *Kluwer Academic Publishers. Printed in the Netherlands.*

more possibilities than their diploid relatives [1]. This situation explains plant breeders' major interest in polyploidy as an important factor of evolution.

Interspecific and intergeneric hybridization has become a conventional method for the systhesis of new species in applied cytogenetics. The first artificial amphiploid, *Nicotiana digluta* (2n=72), was obtained in 1925, by crossing *N. glutinosa* (2n=24) and *N. tabacum* (2n=48) [2]. Two years later, was synthesized the amphiploid *Raphanobrassica* after chromosome doubling of the hybrid between *Raphanus sativus* and *Brassica oleracea* [3]. The plant breeders received an inestimable aid to promote artificial amphiploidy with the discovery of the chromosome-doubling properties of colchicine [4,5]. Since then, interspecific and intergeneric hybridizations have been undertaken for the artificial synthesis of new species of crops.

Triticale is the term coined by Lindschau and Oehler [6] for amphidiploid (2n=8x=56) *Triticum x Secale* hybrids, created to add the winter hardiness of rye to wheat. Müntzing [7] applied this term as the "generic name" that has been used since 1936 by cytogeneticists and breeders. Although a large number of amphiploids have been synthesized during the last fifty years, triticale is often considered to be the only example of a successful synthetic crop developed by this method [8].

The first attempts to produce an artificial hybrid between wheat *(Triticum aestivum* L.) and rye (*Secale cereale* L.) were reported more than a century ago [9]. The F_1 hybrids produced vigorous but sterile offspring. The first notice of the spontaneous chromosome doubling of a wheat-rye F_1 hybrid resulting in a partially fertile amphiploid was reported fifteen years later by Rimpau [10].

From that time, the short history of triticale underwent two main surges. The first happened at the Agricultural Experimental Station of Saratov in Russia, in the early 1920's. There, wheat breeders used rows of rye to separate from each other the experimental plots of different winter wheat varieties to avoid intervarietal hybridization. They were surprised by the presence of thousands hybrid plants among the progenies of certain wheat varieties [11]. The spontaneous wheat-rye hybrids were sterile but could give descendants by backcrossing to either wheat or rye parentals. The second surge came with the use of colchicine as an efficient method to double the number of chromosomes.

Towards the establishment of the optimal ploidic-level

A critical question that occupied some time in cytogenetic and breeding work with triticale was the comparative results of two basic types: octoploids and hexaploids. The **octoploids** are 2n=8x=56 and have the genome constitution AABBDDRR. They are produced by chromosomal doubling in hybrids of hexaploid wheat (2n=6x=42; AABBDD) and rye (2n=14; RR). The **hexaploids** are 2n=42 and have the genome constitution AABBRR. They are made by crossing *durum* wheat (2n=4x=28; AABB) and rye (2n=14; RR).

Müntzing directed his efforts towards obtaining octoploid triticales at the Plant Breeding Station in Svalof, Sweden [7,12-18]. He synthesized and improved a considerable collection of octoploid triticale lines. He gave his own opinion on this material at the Eucarpia Meeting on triticale held in St Petersbourg in 1973: *"(octoploids)…are still far from perfect. Fertility is reduced and the degree of straw-stiffness is not sufficient. For that reason our octoploid triticale-types tend to lodge, if they receive high amounts of nitrogen. In yield trials with relatively low amounts of nitrogen, these triticale-strains may give a higher yield than wheat. In other trials, however, where the amounts of nitrogen and soil conditions are optimal for wheat, the best triticale-types only reach 70 to 80 per cent of the yield of bread wheat"*. Opposite to these unfavourable characteristics, many octoploid forms obtained by Müntzing in Sweden, and others by Pissarev in Russia [19], have been very appreciated. They were as frost-resistant as rye, indifferent to soil fertilizers, presented a high protein content in grain and were resistant to different diseases. However, they were never extensively used in agriculture because of their reduced fertility, extreme meiotic instability and high aneuploid frequency. Moreover, octoploid triticale lines produce a certain frequency of aneuploids in bulk populations as well as in the progeny of euploid plants [20-26]. Another undesirable defect was the different degree of grain shrivelling, the cause of which is not yet totally known.

Most scientists thought that these defects could be overcome at the hexaploid level. O'Mara crossed a *durum* wheat with rye to obtain hexaploid triticales at the University of Missouri [27-28]. Sánchez-Monge, at the Agricultural Station of Aula Dei in Zaragoza (Spain), was the first to start a breeding program with hexaploid triticale. The objective was very clear: to improve the problems existing at the octoploid level [29-37]. At the First International Wheat Genetic Symposium held at Manitoba in Canada, he affirmed: *"The underlying idea is to obtain a new cereal with the milling and baking qualities of wheat combined with the drought resistance and the ability of rye to grow on poor soils… As a working hypothesis we assumed that the number of 56 chromosomes in the triticale obtained from the cross between common wheat and rye would be too high and therefore the attainment of 42-chromosome triticale from crosses between tetraploid wheat and rye should give a polyploid level that could be optimum for this cereal"*. In 1969, he released a variety for production named 'Cachirulo' that was the first hexaploid triticale to be marketed in the world. This cultivar has been considered a prototype of synthetic crop triticale.

Almost simultaneously, Kiss began a program on hexaploid triticale at Martonvasar in Hungary. He developed lines that in 1969 reached a growing surface of 40,000 Ha [38-39]. In Japan, Nakajima also synthesized many hexaploid triticales [40]. A great impulse towards the current interest in triticale was given at the University of Manitoba, Winnipeg, in Canada [41-45]. The history of the work on hexaploid triticale at this University was carefully detailed by Shebeski at the Meeting on Breeding Triticale held in 1979 in Radzikow (Poland) [46].

During the 1950s and 1960s, research work on octoploid and hexaploid triticales was extended. The results of many authors suggested that between octoploids and hexaploids and in respect to practical breeding, chromosome stability and fertility, hexaploids represent the optimal ploidic level [20,21,29,32,33,47-49,50]. The clear advantages of hexaploid relative to octoploid triticales were summarized by Shebeski who pointed out that "...*the synthesized hexaploid triticales were far more vigorous than either of the component parent species,....the synthesized bread wheats, although more vigorous than the* Aegilops squarrosa *parent, were far less vigorous than their tetraploid* Triticum *parents. This led to the natural conclusion the D genome was a poor combiner with the tetraploid wheats for productive capacity, whereas the R genome was an excellent combiner*". Hexaploids are superior to octoploids in their field performance and in their general potential as new crop species [51].

When hexaploid triticales proved to be superior, interest in the commercial use of octoploid forms quickly declined. However, the octoploid triticales have sustained an indirect interest as source material for improving hexaploid triticales. Guedes-Pinto, in Portugal, has been intensely activite during last years in obtaining new primary octoploid triticales. These are useful as raw material in hybridization to incorporate new wheat and rye germplasm in triticale [52,53]. The yield, quality, plant height, leaf crude protein content and other parameters have also been evaluated for breeding 8x-triticale as a forage crop in northern Portugal [54]. At present, octoploids are mainly cultivated by farmers in the People's Republic of China [55].

Other ploidic levels that have been studied are decaploid and tetraploid. Both ploidy levels have been questioned. The **decaploid** triticales ($2n=10x=70$; genome formula AABBDDRRRR) emerged in the early 1950s by crossing octoploid triticale with diploid rye, followed by chromosome doubling of the resulting pentaploid hybrid [56]. These forms were of no value because they had poor vigor and fertility, irregular meiosis and usually reverted to lower chromosome numbers. The **tetraploid** triticales have received progressively more attention ever since their appearance in the early 1970s. These triticales will be treated later in the section related with new triticale forms obtained by crossing.

The question of chromosome instability and meiotic irregularities

The literature on triticale cytology has been profuse. The first review entitled "*The Cytogenetics of Triticale*" was published by O'Mara in 1953 [28]. From then, other reviews have been published that cover practically all the applied and basic cytogenetic progress in triticale [8,42,57-64].

From the beginning, the cytogeneticists and breeders paid attention to the question of chromosome instability and meiotic irregularities in triticale. The possible environmental effects on the degree of meiotic instability in triticale has been considered [65,66]. Müntzing [56] pointed out that the higher the ploidic level in triticale, more frequent the meiotic irregularities. He considered that the ratio between numbers of different genomes influenced the course of meiosis. Shkutina and Khvostova, who carried out a cytological investigation in a series of lines with different ploidic levels, observed more irregularities in the octoploids than in the hexaploids [67]. They suggested an incompatibility between wheat and rye genomes and the inactivation of single loci on rye chromosomes as the main causes of the meiotic instability.

An important objective of much cytogenetic work in triticale concerns the origin of the univalents and open bivalents (Fig. 1). It was shown early that, because of the considerable failure to pair in homologous chromosomes, univalents were common in newly established triticale lines [14,28,33,68]. Lelley analyzed three hexaploid triticale lines and their hybrids in all possible combinations [69]. He observed a complete pairing in the F_1 hybrids and in the parents up to late diakinesis, and discussed the role of desynapsis in pairing failure at metaphase I.

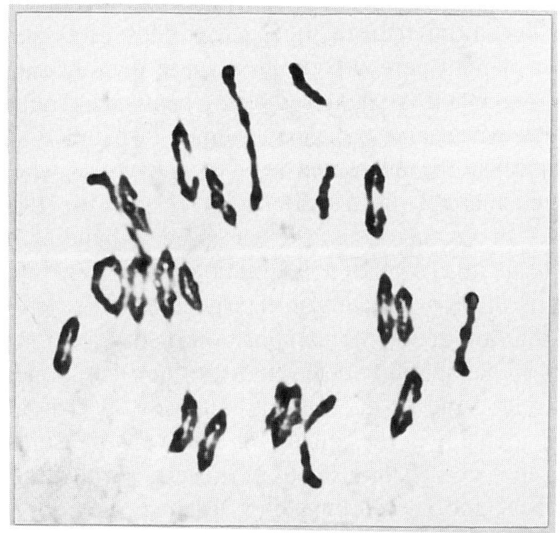

Fig. 1.- A metaphase I of the 6x-triticale 'Cachirulo' showing 16 ring bivalents + 4 open bivalents + 2 univalents produced by desynapsis.

Other authors also observed that univalents and open bivalents resulted from desynapsis or early terminalization of chiasmata after diakinesis [70,71]. Desynapsis is a universal characteristic of triticale and shows a remarkable stability over a wide range of temperatures. It is considered that univalency is a stable expression of some basic incompatibility within the hybrid nucleus. Several theories have attributed the occurrence of univalents at first metaphase on meiosis in triticale to different factors. We will consider the factors that have deserved the most attention among researchers and breeders: genotype and genetic wheat-rye interactions, time of duration of meiosis, and presence of telomeric heterochromatin in the rye chromosomes.

Genetic interactions between wheat and rye genome

The importance of the genotype for a satisfactory control of meiotic pairing in triticale was discussed by Riley and Miller [68]. Observation of differences in the number of univalent chromosomes in different triticale lines cultivated under the same environmental conditions suggested the importance of the genotype. On the other hand, the level of chromosome instability is quite heritable in triticale. Accordingly, the same triticale line with either a high or a low degree of univalency has shown similar results when studied by different authors under different environmental conditions. For instance, different authors separately found many univalents (5.75/CMP and 6.1/CMP, respectively) in the rather instable 8x-triticale 'Rimpau' [12,72]. It has been suggested that the reduction of pairing intensity, and consequently the failure of its maintenance in triticale is under polygenic control. Therefore the interaction between the wheat and rye genotypic components is conclusive [26,69,70,74-80].

It has been demonstrated that in hybrids having different doses of wheat and rye chromosomes, the wheat and rye chromosomes influence chromosomal pairing [81-83]. Lelley [79,84,85] studied the cytological and morphological expression of interactions between wheat and rye genomes in hexaploid triticale lines. The lines used were developed using genetically pure lines of both parental species. It was demonstrated that the polygenic balance can break down when rye is placed on a wheat background. A phenomenon of interference between wheat and rye chromosomes in triticale has been also discussed [86]. This would cause the decreased rye-rye and wheat-wheat homologous pairing observed at metaphase I in triticale with respect to their corresponding rye and wheat parents. In another study, it was used C-banding in meiosis as an approach to analyzing wheat and rye genome interactions in triticale [87]. It was observed that pairing intensity is more dependent on intergenomic interaction between the rye and wheat genomes than on genotypic composition of the genomes. Meiosis was investigated in four primary hexaploid triticale lines, in their component two tetraploid wheat and two rye parents, and in the hybrids obtained by crossing within each ploidic level. In each new triticale line, the pairing for wheat chromosomes was moderately reduced and pairing for rye chromosomes was very significantly reduced, when compared with the wheat and rye parents. For instance, the *durum* wheat 'Edmore', the rye 'Snoopy' and the 6x-triticale 'Edsnoo' derived from them, presented the following mean number of chiasmata per pollen mother cell: 'Edmore' 15.4 and 14.5 for the A- and B-genome chromosomes, respectively; 'Snoopy', 14.0 for the R-genome chromosomes; and 'Edsnoo' 14.1, 13.9 and 11.7, for the A-, B- and R-genomes, respectively. On the other hand, the hybridization between hexaploid triticales that differed either in the rye or wheat components, revealed that homozygosity or heterozygosity in the wheat complement did not affect the behavior of the rye chromosomes. These always exhibited a pairing reduction in triticale. All these data suggest a strong negative intergenomic interaction between the rye genome and the wheat genome, which mainly influences the meiotic pairing of the rye chromosomes.

Is there an inter-genomic disharmony in time of duration of meiosis?

The allocycly, or genomic disharmony of wheat and rye genomes, has been noticed as another cause of meiotic instabilility in triticale [67]. It was observed genomic disharmony with respect to the nucleolar organization during meiotic prophase, and independent behavior of the rye genome in triticale. This genome would form a separate nucleolus at prophase, organize its own spindle at metaphase I and separate independently at anaphase I. The explanation for this genomic disharmony has been given in terms of the different requirements of the time available to rye and wheat genomes for pairing. It has been suggested that the higher meiotic stability in hexaploid triticale can be explained by the longer duration of its meiosis (20 hours in octoploids and 37 hours in hexaploid triticales), which is more compatible with that of rye (51 hours) [88,89]. However, Roupakias and Kaltsikes [90] suggested that within 6x-triticale itself there is no correlation between the time taken to complete meiosis and the degree of univalency.

Telomeric heterochromatin of the rye chromosomes

Kernel shrivelling and univalency have been considered the most important problems of synthetic amphiploid triticales. Both effects have been correlated in part with the presence of the large terminal heterochromatin blocks on the rye chromosomes [91-95]. C-banding has been the most useful method for identifying chromosomes in rye, wheat and triticale during the last years (Fig. 2a and b). The blocks of constitutive heterochromatin are mainly located at the telomeres of the rye chromosomes, and at intercalary positions on the wheat chromosomes. Accordingly, a series of studies have analyzed chromosome constitution and the amount of telomeric heterochromatin in the rye chromosomes of secondary triticale lines [96-101]. The heterochromatin distribution pattern is rather constant. However, the rye chromosomes display intra- and inter-varietal variability in their open-pollinated rye parents [102]. Because of this, the usefulness of the C-banding technique to identify rye chromosomes was questioned [103]. Besides C-banding, several other techniques are being used, such as fluorescent banding [104], improved C-banding techniques [64] and fluorescent *in situ* hybridization (FISH) [105-107] (Fig 3). Also, isozymes can be used as markers to analyze the presence of certain chromosomes in segregant triticale plants [108-109]. The telomeric heterochromatin in rye chromosomes has been correlated with certain morphologically and cytologically undesirable characters. Kaltsikes [61] noticed that the presence of rye heterochromatin in triticale causes aberrant nuclei in a coenocytic endosperm. The production of aberrant nuclei leads to the formation of cavities in the endosperm and grain shrivilling at maturity. Varguese and Lelley studied the Giemsa karyotype of the rye chromosomes in four triticale lines with different degrees of grain shrivilling [99]. They observed no significant differences in the C-banding pattern between lines with plump and well-filled seeds and lines with shrivelled grains.

The capacity of C-heterochromatin to influence univalency and aneuploidy in triticale has been widely discussed [110-117]. The existence of effects of both arm length and amount and distribution of heterochromatin of each chromosome pair in triticale was also studied in 6x-triticale x *durum* wheat progeny [118,119].

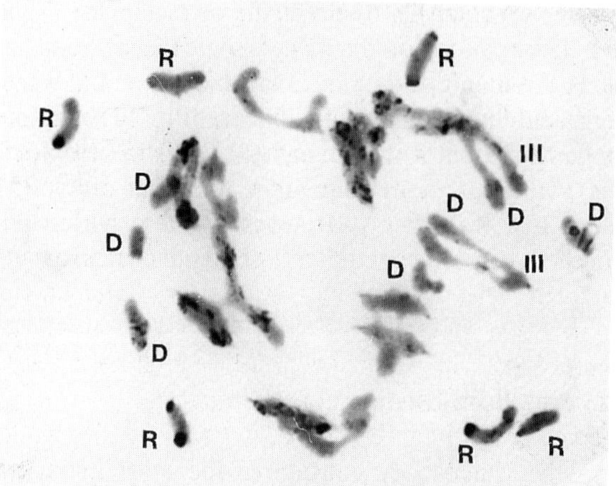

Fig 2.- a) A somatic mataphase stained by Giemsa C-banding. b) Metaphase I of F_1 hybrid 6x-triticale cv. 'Cappelli' *ph1c/ph1c* x *T. aestivum* cv. 'Chinese Spring' *ph1a/ph1a* using C-banding to distinguish A, B, R and D-genome chromosomes. It shows 12 A+B genome bivalents, 7 R genome univalents, 5 D genome univalents, and 2 A- or B+D genome trivalents.

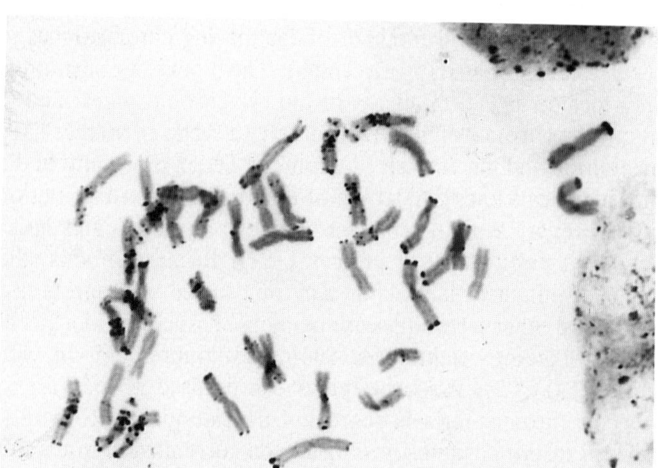

The absence of the telomeric heterochromatin in rye chromosomes is associated with an improvement in their meiotic stability. Thus, Kaltsikes and Gustafson on a triticale background, combined chromosomes 4R and 6R with and without heterochromatin in the telomeres [120]. They found that the absence of heterochromatin, improved meiotic stability. Soler et al. studied chromosome constitution and meiotic behavior of a progeny segregating for missing heterochromatin in the telomeres of the 3RS arm [121]. Chromosome pair 3R showed significantly more chiasmata in plants in which the heterochromatin was either deleted (- -) or present

(+ +) in both telomeres than in the heterozygous (-+) ones. As a conclusion, the absence of telomeric heterochromatin could have a beneficial effect on meiotic chromosome pairing in triticale. Thus, rye stocks lacking heterochromain in the telomeres should be selected for use in the production of new triticale lines.

The hypothesis that rye chromosomes are responsible for the disturbances seen at meiosis has driven speculation on the co-adaptation of wheat and rye genomes in triticale. Badaeva et al. found that many variations occurred in heterochromatin when chromosomes passed from parentals to the amphiploid [122]. They suggested that the heterochromatin distribution in rye would have a significant role in deciding the viability of new triticale lines. In another paper, it was investigated the distribution and characterization of heterochromatin in rye chromosomes in a series of cultivars, primary triticales, and new lines derived by crossing between common wheat and triticale [123]. The crossings contained materials that had changes from the parentals to the amphiploid and over generation time. The changes consisted in two tendencies: one to lose intercalary heterochromatin and the other to modify the amount of telomeric heterochromatin. Other results suggest that the same cytotypes are preferentially selected for rye chromosomes in triticales from rye parents. A high level of intervarietal polymorphism was detected in the banding pattern. A total of 126 bands were characterized and 39 belonged to the R genome. The ratio of polymorphic bands per genome was 74% (30/39) for the rye genome. The modifications can involve translocations, deletions or other structural aberrations [96,100,124], and are transmitted via male and female gametophytes [96,97]. As we will mention later, similar chromosome aberrations are produced in tissue culture [125-129].

The possibility that changes in the rye heterochromatin content in triticale may be induced by either the environment or selective pressure has been discussed by Gustafson et al. [130]. The incorporation of heterochromatin modifications into the germ-line cells and the systematic tendency to their loss or increase over generation time, seems to suggest some role for heterochromatin in selecting a more balanced nucleotype in triticale. According to Appels et al [131], the variations in heterochromatin may have mechanical or genetic consequences, especially in modulating gene activity. If so, the changes in heterochromatin could be the chromosomal manifestation of the most cryptic changes at molecular level.

The nature of the changes that occur in the heterochromatin pattern is an open question. Some explanations have been suggested: underreplication of DNA, molecular drive, allosyndetic end-by-end pairing during meiosis and mobile elements. Progress in understanding the particular role(s) of heterochromatin in triticale and rye depends on further cytogenetic, genetic and molecular analyses. The use of *in situ* hybridization could, perhaps, contribute new information on the molecular nature of the rye heterochromatin [132].

Broadening the genetic basis of triticale

At the end of fifties it seemed clear that hexaploid triticales offered better possibilities than octoploids, and a very intense activity was initiated to broaden the germplasm available in breeding programs. In triticale there was less research done on the use of physical and chemical mutagenesis to induce the desirable variability [133,134] than on cytogenetic procedures. The main cytogenetic strategies followed to increase available variability in triticale breeding programs are: obtention of triticales with different cytoplasms; obtention of secondary triticales following hybridization programs; obtention of new triticale forms by wide crossing; obtention of new forms by *in vitro* culture. We will pay attention to all these tendencies.

Triticale with different cytoplasms

Synthetic amphiploid triticale is ideally obtained with the cytoplasm of either of its parental species. Since crossing wheat with rye is easier when wheat is used as the female parent, triticale usually carries cytoplasmic information from the wheat parent.

The harmonic coexistence of the nucleus and cytoplasm is an important factor that was early considered in the production of new forms of triticale [19,34,36,39,47,135]. The major interest in this field was developed during the sixties. It was showed that triticales with hexaploid wheat cytoplasm have more favorable agronomic traits than isogenic ones with tetraploid wheat cytoplasm [136-139]. Sánchez-Monge at the INIA (Spain), was the first to start a breeding program of production of alloplasmic triticales using the nuclear restoration and substitution method developed by Kihara [140]. He produced triticales with different cytoplasms: *Secale cereale* L., *Aegilops ovata* L., *Ae. caudata* L., and *Triticum timopheevi* Zhuk [36,141,142]. Production of these alloplasmic forms was considered of interest for two reasons. First, a triticale with a different cytoplasm may be a better agricultural crop than the same one produced with wheat cytoplasm. Secondly, a change of cytoplasm may induce male-sterility, and if male-sterile lines of triticale could be obtained, they can be used for the production of triticale hybrid seed. The cytoplasm of *Ae. ovata*, *Ae. caudata* and *T. timopheevi* was strongly male-sterilizing for hexaploid triticale and could be used for hybrid seed production [141].

Cytoplasmic-nuclear interaction as a factor that might be responsible for univalent formation was first suggested by Sisodia and McGinnis [135]. Other authors observed that triticales having the extracted AABB component of hexaploid wheat presented improved meiotic stability. Moreover, several lines of 6x-triticale, with hexaploid and tetraploid wheat cytoplasms presented significant differences in the number of univalents [66,136]. Studies using Isogenic forms of hexaploid triticale with rye and wheat cytoplasms, and F_1 interlinear hybrids with different cytoplasms, seem to support the general assumption of lack of effects on the level of chromosome pairing at first metaphase [70,143,144]. However, significant differences are found in the behavior at later stages of meiosis between materials that have an identical nuclear genotype constitution but different cytoplasms.

The influence of *Ae ovata, T. timopheevi* and *S. cereale* cytoplasm in the meiotic behavior of wheat and rye chromosomes, using conventional chromosome staining [145] and a C-banding method [146], was also investigated. The results suggested a differential effect of the source of cytoplasm on the particular stability of the chromosomes of parental species. It has also been suggested that alien cytoplasms could be of interest in programs to improve hexaploid triticale. However, a comparative morphological study of some lines of triticale with the wheat and rye cytoplasms, showed that the rye cytoplasm lacked of practical value for the production of male-sterility in triticale. As a consequence, the use of this cytoplasm was not considered advisable for improving the agronomic performance of 6x-triticale [142]. The cytological study showed a similar mean number of univalents per PMC in both forms (1.1-1.46 in the form with 4x wheat cytoplasm and 1.2-1.46 in the form possessing rye cytoplasm), but the number of micronuclei per tetrad was lower in the triticale with rye cytoplasm (0.56-0.67) than in the form possessing 4x wheat cytoplasm (0.85-0.97) [142,147].

Secondary triticales

The term "secondary" was applied by Kiss [48] to describe stable hexaploid derivatives obtained by intercrossing the octoploid and hexaploid triticales. Jenkins reported the selection of improved hexaploid triticales from crosses between hexaploid triticale and bread wheat [44]. Many secondary triticales were very promising and showed high yields, good nutritional quality, and other agronomically valuable properties [19,47-59,51,148]. Kiss and Videki crossed hexaploid triticale with a series of different rye varieties and observed more bivalents in the hybrids than in hybrids of hexaploid wheat with rye [49]. The F_1 hybrids were backcrossed twice with rye and three groups of plants were obtained: i) triticale-like individuals with moderate fertility; ii) rye-like plants with poor fertility, and iii) F_1-like, very sterile plants with an intermediate type.

It could be affirmed that the arrival of the secondary triticales marks the beginning of a new period in the genomic and chromosome triticale story. According to Merker, the superiority of secondary hexaploids could be explained along three different lines: genic, cytoplasmic and chromosomal [149]. The genic explanation is based on the assumption that the A and B genomes of hexaploid and tetraploid wheats, joined in the F_1 hybrids to produce secondary triticales, maintain tangible genetic differences that are very profitable in breeding programs. The differences are due to the long period of divergent evolution behind both species [17,66]. The effects of cytoplasmic changes have been considered in the preceding section in this review.

To consider the factors of the superiority of secondary triticales at the chromosomal level, requires examining the substitution of rye chromosomes by D genome chromosomes. Zillinsky and Borlaug, based on plant morphology and physiology, presumed that the 'Armadillo' triticales, obtained at CIMMYT in México, were the result of a spontaneous outcross between triticale and bread wheat [51]. Gustafson and Zillinsky proved the presence of chromosome 2D in the substitution of 2R in an 'Armadillo' line [150]. The discovery of this chromosome substitution was an important event in the history of triticale. Since then on a new distinction among triticales became necessary: the one between "complete" and "substitute" lines.

Complete lines retain all the chromosomes of the rye parent unchanged. They tend to be more productive under stressful conditions and are considered the triticales of choice for marginal environments, because of their robustness and resistance to difficult environmental conditions. A marked progress with complete triticales has been carried out in Poland by Wolski and his colleagues during the last years. They produced very good varieties for sandy and acidic-sols, that are aluminium-tolerant, and good in the difficult climatic conditions of Poland and other central-European regions [151-154].

On the other hand, substituted triticales present one or more rye chromosomes replaced by homoeologous of the D genome. With exceptions, substituted triticales perform better than complete ones under nonstressed conditions. The C-banding staining technique for chromosomal heterochromatin has proved to be very valuable in triticale, since it allows identification of the seven pairs of rye chromosomes [155] (Fig 1b). The chromosome composition of 50 lines of hexaploid triticale and triticale wheat intercross lines from the CIMMYT breeding program were investigated by Merker using the C-banding technique. He found that the third genome can present a wide range of combinations of rye- and D-genome chromosomes [92]. Most of the advanced secondary lines with good agronomic characteristics have a mixed chromosome composition. This work was extended to many other triticale lines by other authors [156-161], who observed that most CIMMYT lines contained D-R substitutions. Some chromosome constitutions examined earlier in the seventies were later re-examined with improved C-banding techniques [130]. These studies confirmed the high frequency of the 2D(2R) substitution, but demonstrated that other supposed substitutions were actually modified versions of rye chromosomes. Other authors used the analysis of pairing configurations in triticale x wheat hybrids to detect chromosome substitutions. The use of ditelocentric or wheat-rye addition lines is very useful in identifying substituted chromosomes [162].

New triticale forms obtained by crossing

There are several available crosses that increase triticale germplasm or obtain forms with new chromosomal constitutions, modifications or recombinations. They include the following hybridization methods and results:
-8x-triticale x 8x-triticale: Recombined 8x-triticales.
-6x-triticale x 6x-triticale: Recombined 6x-triticales.
-8x-triticale x 6x-triticale, followed by backcrosses to 6x-triticales: Recombined secondary 6x-triticales.

-6x-triticale x *T. aestivum*, followed by selfing: Secondary 6x-triticales with possible
 D(R) chromosome substitutions; 4x-wheats on *T. aestivum* cytoplasm; 6x-wheats
 with possible R(D) chromosome substitutions.
-6x-triticale x *T. aestivum*, followed by backcrossing with 6x-triticale: Secondary 6x-
 triticales with possible chromosome D(R) substitutions.
-6x-triticale x *T. turgidum*, followed by selfing or backcrossing with 6x-triticale:
 Recombined 6x-triticales.
-6x-triticale x 2x-*Secale cereale*, followed by selfing: 4x-triticales.
-8x-triticale x 4x-triticale: 6x-triticales with possible A(D), B(D) substitutions.
-6x-triticale x 4x-triticale: Secondary 4x-triticales.
-6x or 8x-triticale x Related species, followed by backcrossing with triticale: triticales
 with introgression from alien species.

Crossing triticale with either of its parental species, or with other plant material
is the simplest method for transferring additional genetic information from wheat or rye
to triticale. The genetic variation of 6x-triticale has been increased in recent years by
genetic recombination in both wheat and rye components, and by D(R) chromosome
substitutions and translocations. All these phenomena often occur during meiosis in
hybrids involving primary or secondary 6x- and 8x-triticales, or crossed by common
wheat [63]. This kind of crossing is also useful for introducing rye chromosome
substitutions and/or segments into wheat. Ren et al karyotyped more that a thousand
plants of the BC_1, F_2 and F_3 populations of three octoploid triticale x common wheat
hybrids using Giemsa C-banding [163]. The 14 chromosome arms of rye and the A-,
B- and D-genome chromosomes of wheat were involved in the wheat/rye
translocations, 48.57% of the identified wheat/rye translocations occurred
between homoeologous and 51.43% between non-homoeologous chromosomes. The
study suggests that sufficient numbers of wheat/rye translocations can be produced
for wheat breeding by crossing octoploid triticale and common wheat. Crosses between
6x-triticale x bread wheat have been also utilized to obtain adaptative advantages by
rye/wheat chromosome substitutions in the genomic background of bread wheat [164].

The germplasm variation in triticale is still very limited due to the extensive use
of 'Armadillo' and another good varieties in breeding programs [165]. The basis of
triticale breeding is genetic variability, and this depends on recombination and genetic
assortment. The utilization of crosses involving triticale and wheat is very helpful in
achieving these objectives. The work of several authors who have demonstrated
differences in the recombination frequency in the same chromosome regions between
triticale and its parental species is quite interesting. Very recently, has been studied the
frequency and distribution of recombination along the chromosome in a diploid rye and
in a 6x-triticale derived from it [166]. In rye, the total recombination frequency in five
segments of chromosome 1R was 93.7%. Recombination was concentrated in the distal
regions of both chromosome arms and was infrequent in the proximal regions. In 6x-
triticale the total recombination frequency in the same chromosome was reduced to
51.7%. Similar results were found in chromosomes 1R and 2R [167]. As a
conclusion, it seems clear that genetic background influences not only the total amount
of recombination, but also its distribution along the length of the chromosome.

The introduction of alien genetic variation by homoeologous pairing, chromosome translocation or substitutions, has been essayed through different crosses. Table 1 shows the results of a series of investigations on meiotic pairing in different hybrids involving triticales with different ploidic levels and other forms. Evidence for homoeologous pairing between wheat and rye (A-R, B-R, or D-R) and univalent distribution in hybrids has previously been reported [111,113,168-172].

Table 1.- *Meiotic behaviour at first metaphase in hybrids of Triticale and other Triticeae*

Hybrid	2n=	Chromosome/genome constitution		Chromosome associations				Reference
				I	II	III /	IV-V	Xata
6x-Triticale * *Triticum aestivum*	42	AABBDR	14.6	13.2	0.31			[170]
	42	AABBDR	12.2	14.9	0.01		29.6	[111]
6x-Triticale * *T. aestivum Ph1/ph1*	42	AABBDR	23.7	11.90	1.50		24.00	[172]
6x-Triticale * *T. aestivum ph1/ph1*	42	AABBDR	13	10.65	2.03 /.38		24.31	[172]
6x-Triticale * *T. turgidum*	35	AABBR	7.9	13.5	0.01			[215]
6x-Triticale * *Agropyron intermedium*	42	ABRE	27.7	6.9	0.17		7.6	[216]
8x-Triticale * *Secale cereale*	35	ABDRR	22.6	5.2	0.02			[217]
	32	ABDRR	22.6	4.6	0.02		5.4	[218]
8x-Triticale x 4x-*S. cereale*	42	ABDRRR	14.8	6.9	4.34		22.1	[219]
8x-Triticale x *S. montanum*	35	ABDRRm	15.6	8.1	1.00		10.9**	[219]
6x-Triticale * *S. cereale*	28	ABRR	12.2	7.8	0.02		12.3	[220]
	28	ABRR	9.3	9.3			15.2	[221]
	28	ABRR	14.1	6.6	0.19			[113]
	28	ABRR	10.1	8.7	0.1			[222]
	28	ABRR	13.6	7.1	0.04		10.5	[218]
6x-Triticale x *S. montanum*	28	ABRRm	11.5	6.8	0.79		10.0**	[219]
6x-Triticale (*dicoccoides*) * *S. cereale*	28	ABRR	19.6	4.1	0.05		9.1	[223]
6x-Triticale * 4x-*S. cereale*	35	ABRRR	9.8	6.9	4.34		17.5**	[219]
	35	ABRRR	6.1	9.2	3.3			[222]
6x-Triticale (*montanum*) * *S. cereale*	35	ABRRRm	11.6	7.2	3.07		17.2**	[219]
4x-Triticale * *S. cereale*	21	(AxBy)RR	7.2	6.8	0.02		12.4	[218]
8x-Triticale * 8x-*Tritordeum*	56	AABBDDRHch	17.6*	18.8	0.22		33.24**	[224]
6x-Triticale * 6x-*Tritordeum*	42	AABBRHch	14.4	13.8			24.4	[225]
anfipl.(*T.aes-Agrop*) * *S. cereale*	35	ABDER	22.9	5.7	0.26			[226]
* 4x-*S. cereale*	42	ABDERR	19.0	10.9	0.37			[226]
* 6x-Triticale	49	AABBDER	22.2	13.1	0.14			[226]
* 4x-Triticale	49	ABDE(AB)RR	11.9	13.4	0.74			[226]
H. parodii * Triticale-6x	28	HABR	25.9	6.8	1.10		13.1	[227]
H. jubatum * 6x-Triticale	35	H1H2ABR	34.7*	0.11			0.11	[228]
S. cereale * 6x-Triticale	28	ABRR	11.9	8.0	0.02			[216]

* = most favorable pairing amongst hybrids differing in the genotype of parentals.
**= data indicated this way were not reported in the original paper cited but calculated according to the mean number of the different meiotic configurations given by the authors.

As early as 1957, Riley and Chapman observed that rye chromosomes either do not pair with wheat chromosomes, or do so only with extremely low frequency [173]. The diploid-like behavior at meiosis of amphiploids, such as hexaploid and tetraploid wheats and triticale, is the result of balanced interaction between several genes that influence chromosome pairing. The major effect is due to the gene *Ph1* that is located on chromosome arm 5BL of wheat [174]. Different mutations, *ph1b* and *ph1c* have respectively been obtained in common wheat [175] and *T. turgidum* [176]. These mutations can be used to promote homoeologous pairing within wheat genomes and between these and the genomes of related species. A primary hexaploid triticale carrying the *ph1c* mutation of *T. turgidum* cv. 'Cappelli' was produced in Italy [177]. The 6x-triticale *ph1c* was crossed with common wheat cv 'Chinese Spring' containing the mutant allele *ph1b* [178]. The meiotic behavior of hybrids having different doses of the mutant alleles was investigated using the Feulgen staining and C-banding methods [178,179]. D-genome chromosomes, which were clearly distinguished by their small size and unbanded response to Giemsa staining, were increasingly more likely to pair with the homoeologous chromosomes of the A and B genomes in the absence of normal *Ph1* gene. However, the wheat-rye associations were not enhanced when one or two mutant alleles were present (Fig 1c). During many years the major cytogenetic work in triticale was directed towards the development of primary and secondary hexaploid triticales. Interest in obtaining new chromosome combinations began in the early seventies. Chaudry in Canada, attempted to produce a new ploidic possibility: the "tetraploid" triticales [179]. However, he failed to obtain this material using direct diploid wheat x rye crossing. Krolow used indirect methods and obtained the first tetraploid triticales (2n=4x=28; (A-B)(A-B)RR, BBRR, AARR...). These were initially produced through the cross 6x-triticale x 2x-rye [180-182]. The interest in these forms is based on their higher proportion of rye genetic material which could favour stronger expression of rye adaptative characteristics, like winter-hardiness and acid, sandy soil tolerance [183,184]. The obtention of all the theoretically possible combinations of the A- and B-genome chromosomes has been discussed. The use of C-banding has considerably facititated the identification of chromosome constitutions. For instance, Bernard et al obtained a tetraploid triticale with the chromosome constitution 1B, 2A, 3B, 4A, 5B, 6B, 7A and all seven rye chromosome pairs [185].

Distant hybridization and autoallohexaploid triticales have been successfully used to introduce genetic material (D-genome chromosomes) from *T. aestivum* into tetraploid triticale as demonstrated using C-banding. Thus, Lehmann et al. tested a sample of 63 karyotyped plants from five lines, and identified six plants with complete D-genome chromosomes in homoeologous groups 4, 6 and 7. In addition, two plants had a monosomic 7DS/7AL translocation [186]. The total frequency of plants with D-genome chromosomes in the sample was 12.7%. Badaev et al. analysed 13 forms of tetraploid triticale using a C-banding technique [187]. All forms contained 14 R-genome chromosomes, and 69 combinations of A-, B- and D-genome chromosomes from wheat. The following general features were found for the formation of the synthetic karyotype of 4x-triticale: (i) the homoeologous groups were stabilized at different rates; (ii) "heterozygotes" (AB) were selected against; and (iii) chromosomal constitution of some homoeologous groups exerted an effect on the rate and direction of stabilization in other groups. According to Badaev et al the most probable variants of the wheat genome in 4x triticale are 1B, 2B, 3A, 4A, 5A, 6A, 7B and the reverse. Selection occurred at the level of homoeologues under the effect of environmental factors [187].

More recently, Hohmann obtained F_1 hybrids with the genome constitution ABDERR (2n=6x=42) or ABDE(AB)RR (2n=7x=49) [188]. These forms are derived from crosses between either an octoploid *Triticum aestivum/Thinopyrum elongatum [Elymus elongatus]* amphiploid and tetraploid rye (AABBDDEE X RRRR) or an autoallohexaploid triticale (AABBDDEE X (AB)(AB)RRRR). Backcrossing the F_1 to tetraploid triticale (AB)(AB)RR and selfing for several generations, Hohmann obtained euploid plants with 28 chromosomes. Among them were different tetraploid karyotypes with different chromosome constitutions: (i) without any detectable D-genome chromosome from *T. aestivum* or any E-genome chromosome from *T. elongatum;* (ii) lines (ABD)(ABD)RR tetraploids with one-to-three disomic substitutions of D-genome chromosomes for A- or B-genome chromosomes without any disomic substitution of E-genome chromosomes what so ever; and (iii) plants with minor differences in chromosome C-banding in homoeologous groups 1, 5, and 6 suggest the possibility of translocations between the A-, B-, D-, and E-genome chromosomes.

Today, the use of tetraploid triticale is mainly seen as a tool for improving hexaploid and octoploid triticale [62,180]. According to Lukaszewski and Gustafson tetraploid triticales are extremely interesting from both the practical and the theoretical points of view [64]. However, the genetic basis of the available material is too narrow to estimate its real agronomic potential. Comparing performances of materials with different genetic origins would help give the discussion on the possible practical value of tetraploid triticale a better basis [189,190].

Hexaploid triticale has also been crossed with some species from different genera including *Aegilops, Agropyron,* and artificial amphiploids (*T. aestivum-Agropyron*; *T.aestivum-H. chilense*; *T.turgidum-H. chilense*...). According to the origin and intergenomic homoeology, the related species used for triticale breeding could be used for alien genetic transfer by means of recombination or translocation. The simplest method to achieve interspecific recombination is to exploit the allosyndetic pairing that can be shared between homoeologous chromosomes in interspecific hybrids [191]. The results on the possibilities of this kind of transfer are shown in Table 1. Bernard and Gay [192] obtained the amphiploid from *Ae. ventricosa* and rye (2n=42, and genome formula DD M^{v}M^v RR). This amphiploid was crossed as male and female with hexaploid and octoploid triticales and rye. The new materials joined the genomes from different origins in a haploid or a diploid condition (A B D M^v RR. A B M^v DD RR and D M^v RR). All these materials showed a high asyndesis rate, even with the R genome chromosomes. Some hybrids presented a fairly regular meiosis and even produced some seeds. From these many plants were produced by embryo culture and they were fertile. This is a very promising way to obtain new triticales that incorporate genes for disease resistance from *Ae. ventricosa.* The introgression into tetraploid triticale of chromosomes from *Agropyron elongatum* (2n=14; genomes EE), and the wheat D-genome has been also attempted [183]. Krolow crossed an amphiploid *T. aestivum x A. elongatum* (2n=8x=56; AABBDDEE) with tetraploid rye (2n=4x=28; RRRR). The F_1 hybrid (ABDERR) was backcrossed twice to tetraploid triticale. A series of lines carrying different D-genome chromosomes but lacking E-genome chromosome were recovered from the segregant progeny. This result has been interpreted as a negative compensation of *A. elongatum* chromosomes in competition with wheat chromosomes [193]. The distant hybridization between triticale and related species perhaps offer us the best possibilities for developing new materials with very promising potential for triticale breeding. The potential utility of "allo-autopolyploid" triticales (2n=56; AABBRRRR) has also been demonstrated. They can be produced by one of two steps: obtention of hexaploid x 4x-rye F_1 hybrid (6x-triticale x 4x-rye; 2n=28; ABRR), or colchicine duplication of the hybrids [194].. Some allo-autoploids presented a pronounced level of diploidization. Triticale-like forms with new chromosome constitutions or mixogenomes can be expected to be produced by the use of these materials and they will probably be of interest for breeding purposes.

Obtention of new forms by *in vitro* culture

Another effective method to increase genetic variation in triticale is *in vitro* culture. The first plant embryos obtained from somatic tissues were achieved in carrot (*Daucus carota*) in 1958 [195,196]. *In vitro* culture is an all-embracing term that describes the growth and manipulation of plant explants, single cells (protoplasts) or complete organs, under more-or-less defined nutrient conditions. This technique offers different ways to obtain genetic variation in a very short time. Variability in tissue culture has been described at all levels of the process from *callus* formation to plant regeneration. Somaclonal variation refers to the variation arising in material that was cultured *in vitro* and underwent a *callus* phase [197]. Considerable attention has been devoted to somaclonal variation because of the genetic consequences of variability and its possible sexual propagation in regenerated plants. On the other hand, *in vitro* culture of haploid gametophytes (pollen through culture) offers the possibility of obtaining haploid regenerants that are useful for rapid production of homozygous lines by artificial or spontaneous chromosome doubling. Androgenesis by anther culture of F_1 hybrids is the technique of choice to avoid the time-consuming conventional methods that require several selfing generations after the hybridization in order to obtain homozygous lines. The variations can affect the number and/or structure of chromosomes in plants regenerated by *in vitro* culture of either somatic tissues or reproductive cells.

Nakamura and Keller using inflorescence-derived *callus* cultures on different media, examined the culture conditions to optimize triticale plant regeneration [198]. The *in vitro* embryo rescue response of hybrid embryos of different ages and sizes of bread wheat x 6x-triticale, crosses was recently studied [199]. It was found that response is very dependent on the parental wheat genotype. The association of genes and culture conditions with plant regeneration in somatic embryogenesis of 6x-triticale and in embryo rescue for primary triticale production is being investigated [200,201]. Many studies have been designed to provide information on culture systems, and increase tissue efficiency for regeneration and genetic variability by somaclonal variation in triticale [53,127,190,202-206]. Armstrong et al. observed variations, mainly involving R-genome chromosomes, in plants regenerated from immature triticale embryo culture [207]. The frequency of these variations increased with the duration of *in vitro* culture. All these works served to demonstrate that somaclonal variation can be induced and so be used as a tool for producing genetic variation in triticale. Plants are often produced that are outside the range of the control population, and some of them are better than the best control plants [127,129,208,210].

Bebeli and Kaltsikes demonstrated that telomeric heterochromatin influences the amount and kind of variation found in regenerated families of hexaploid triticale [208]. They used two sets of nearly isogenic lines arising from the cultivars 'Drira' and 'Rosner', that differed in the presence or absence of certain telomeric heterochromatin blocks. They also concluded that the influence of heterochromatin on the production of somaclonal variation is modified by both the chromosome on which it is borne and by the residual genotype. It was also demonstrated that somaclonal variation, which was studied for two generations under field conditions, was transmitted via the sexual cycle [210]. The work of Kaltsikes and the group at the Agricultural Faculty of Athens during last years adds a new role to C-heterochromatin. The multiple negative effects of heterochromatin in triticale have extensively been studied by Prof Kaltsikes [90,91,110,114,120]

Another important application of *in vitro* culture techniques in triticale breeding programs concerns anther culturing to produce androgenic material. Bernard developed an efficient method to obtain many haploid progeny from which homozygous lines can be derived using colchicine treatment or spontaneous duplication [211,212]. The genetic basis of androgenesis in triticale was discussed by Charmet and Bernard [213] who assumed that this phenomenon is under control of nuclear genes with additive, non-additive and cytoplasmic effects. It was observed that many plants, which had been regenerated by anther culture from previously obtained triticale F_1 hybrids do not show the euploid number of chromosomes. Most of the chromosome modifications observed in androgenic plants were losses or gains of whole chromosomes or telosomes. However, aneuploidy and chromosome rearrangements do not involve the *in vitro* technique itself but rather the choice of the material in triticale. Anther culture could be a valuable method to create homozygous lines from F_1 hybrids [80]. The best chance to reveal the potential ability of androgenesis corresponds to hybrids between parentals with the largest genetic and even chromosome constitution differences (for instance, triticale x wheat hybrids). Genetical analysis of microspore-derived triticale plants also revealed changes in biochemical markers [214]. All these results offer a world of possibilities to obtain new variability and new selected lines using dihaploid plants that carry new gene combinations.

Acknowledgements. The authors ackowledge the financial assistance received from Instituto Nacional de Investigaciones Agrarias (INIA), Comision Asesora de Investigación Científica y Técnica (CAICYT) and Comisión de Investigación Científica y Técnica (CICYT) of Spain, in the form of successive research grants (N° PB82/1558 from CAICYT, 1984-87; N° 5764 from INIA 1984-87; N° 7689 from INIA 1987-91; N° PB86-054 from CICYT 1988-90; N° PB89/0209 from CICYT 1991-93; and N° AGR91-0191 from CICYT 1992-1994).

References

1. Cauderon Y. Alloploidy. In *Interspecific Hybridization in Plant Breeding* Sánchez-Monge, E, and García-Olmedo, F, eds..Proc of the 8th Congress of Eucarpia, Madrid, Spain 1978.

2. Clausen RE, Goodspeed TH. Interspecific hybridization in *Nicotiana*. II. A tetraploid *glutinosa-tabacum* hybrid. Genetics 1925;10:278-84.

3. Karpechenko GD. Polyploid hybrid of *Raphanus sativus* L. and *Brassica oleracea* L. Bull Appl Bot Genet Pl Breed 1927;17:398-410.

4. Blakeslee A. De'doublement du nombre de chromosomes chez les plantes par traitement chimique. CR Acad Sci Paris 205 1937:476-79.

5. Eigsti O. A cytogenetical study of colchicine effects in the induction of polyploidy in plants. Prc Nat Acad Sci 1938;24:56-63.

6. Lindschau M, Oehler E. Untersuchungen am konstant intermediaren additiven Rimpauschen Weizen-Roggen-Bastarden. Der Zuchter 1935;7:228-33.

7. Müntzing A. Über die Entstehungsweise 56-chromosomiger Weizen-Roggen Bastarde. Züchter 1936;8:188-91

8. Gupta PK, Reddy VBK. Cytogenetics of Triticale -A man made cereal. In "Chromosome Engineering in Plants: genetics, breeding, evolution" PK Gupta and T Tsuchiya, eds. Elsevier, Part A 1991:335-59.

9. Wilson AS. Wheat and rye hybrids. Edinburgh Bat Sac Trans 1876;12:286-88.

10. Rimpau W. Keuzungsprodukte landwirthschaftlicher Kulturpflanzen. Lanw Jbüch 1891;20:335-71.

11. Meister GK. Natural hybridization of wheat and rye in Russia. J Hered 1921;12:467-70.

12. Müntzing A. Studies on the properties and the ways of production of rye-wheat amphidiploids. Hereditas 1939;25:387-430.

13. Müntzing A. Experiences from work with induced polyploidy in cereals. Svalöf 1886-1946. In "History and Present Problems" Akerman et al, eds. Lund 1948:324-37.

14. Müntzing A. Cytogenetic studies in rye-wheat Triticale. Proc 1st Int Wheat Genet Symp Tokyo. Cytologia supp 1957:51-56.

15. Müntzing A. Some recent results from breeding work with rye-wheat. In "Recent Plant Breeding Research" Akerberg A, and Hayberg A, eds. Svalof 1946-1961, 1963.

16. Müntzing A. Cytogenetic and breeding studies in Triticale. Proc 2nd Int Wheat Genet Symp Lund. Hereditas supp 1966;2:291-300.

17. Müntzing A. Experiences from work with octoploid and hexaploid rye-wheat Triticale. Biol Zentralbl 1972;91:69-80.

18. Müntzing A. Some results from cytogenetic studies and breeding work in triticale. Proc on the Eucarpia Section Cereals Symp on Triticale Leningrad 1975:70-73.

19. Pissarev V. Diferent approaches in Triticale breeding. Proc 2nd Int Wheat Genet Symp Lund. Hereditas supp 1966;2:279-90.

20. Krolow KD. Aneuploidie und Fertilität bei amphidiploiden Weizen-Roggen- Bastarden Triticale. I. Aneuploidie und Selektion auf Fertilität bei oktoploiden Triticale-Formen. Z Pflanzenzüchtg 1962;48:177-96.

21. Krolow KD. Aneuploidie und Fertilität bei amphidiploiden Weizen-Roggen- Bastarden Triticale. II .Aneuploidie und Fertilitätsuntersuchungen an einer oktoploiden Triticale-Form mit starker Abregulierungstendenz. Z Pflanzenzüchtg 1963;49:240-42

22. Krolow KD. Cytologische untersuchungen an kreuzungen zwischen 8x und 6x Triticale. I.

Untersuchungen an den Eltern, an der F_1 und der F_2. Z Pflanzenzüchtg 1969;62:241-71.

23. Pieritz WJ. Untersuchungen über de ursachen der Abeuploidie bei amphidiploiden weizen- roggen- bastarden und über die funktionsfähigkeit ihrer männlichen und weiblichen gameten. Z Pflanzenzüchtg 1966;56:27-69.

24. Weimarck A. Cytogenetic behaviour in octoploid Triticale. I. Meiosis, aneuploiDy and fertility. Hereditas 1973;74:103-18.

25. Weimarck A. Kernel size and frequency of euploids in octoploid Triticale. Hereditas 1975;80:69-72.

26. Weimarck A. Cytogenetic behaviour in octoploid Triticale. II. Meiosis with special reference to chiasma frequency and fertility in F_1 and parents. Hereditas 1975;80:121-30.

27. O'Mara JG. Fertility in allopolyploid. Rec Genet Sci Am 1948;17:52.

28. O'Mara JG. The cytogenetics of Triticale. Bot Rev 1953;19:587-605.

29. Sánchez-Monge E. Studies on 42-chromosome Triticale. I. The production of the amphiploids. Ann Est Exp Aula Dei 1956;4:191-207.

30. Sánchez-Monge E. Crossability of tetraploid wheat species with cultivated rye. Wheat Inf S e r v 1956;3:30.

31. Sánchez-Monge E. Fertility in triticale. Wheat Inf Serv 1956;3:29.

32. Sánchez-Monge E. Hexaploid triticale. Proc 1st Int Wheat Genet Symp Manitoba 1958:181- 94

33. Sánchez-Monge E. Improvement of endosperm quality in Triticale. Proc 3rd Int Wheat Genet $\mathbb{S}$ 1968:371-72.

34. Sánchez-Monge E. La saga del Cachirulo. Ann Est Exp Aula Dei 1969;10:795- 99.

35. Sánchez-Monge E. Hexaploid triticale with different cytoplasms. Ann INIA. Ser Prod Veg 1973;3:37-43.

36. Sánchez-Monge E. Development of triticales in western Europe. Proc Symp Triticale El Batán, México 1974:31-39.

37. Sánchez-Monge E, Tjio JH. Note on 42 chromosome Triticale. Caryologia supp. 1954;2:748.11.Meister GK. Natural hybridization of wheat and rye in Russia.J Hered 1921;12:467- 70.

38. Kiss A, Redei G. Kisèrletek buza-rozs hibridek triticale eloallitasara. Növénytermeles 1952;1.

39. Kiss A. A hexaploid triticale nemesitési problémai. MTA Agrartudományi Közlemények 1971;30:187-96.

40. Nakajima G. Genetical and cytological studies in breeding of amphiploid types between *Triticum* and *Secale*. I, The external characters and chromosomes of the fertile F_1 *T. turgidum* n=14 x *S. cereale* n=7 and its F_2 progenies. Jap J Genet 1958;25:139-48.

41. Larter EN. Triticale. Agric Inst Can Rev 1968;23:12-15.

42. Larter EN, Tsuchiya T, Evans LE. Breeding and cytology of Triticale. Proc 3rd Int Wheat Genet Symp Canberra 1968;213-21.

43. Larter EN, Shedeski LH, McGuinnis RC, Evans LE, Kaltsikes PJ. Rosner. A hexaploid triticale cultivar. Can J Plant Sci 1970;50:122-24.

44. Jenkins BC. History of the development of some presently promising hexaploid triticales. Wheat Inf Serv 1969;28:18-20

45. Jenkins BC. Hexaploid triticale: Past, present and future. Proc Int Symp on Triticale Leningrado, 1975:26-29.

46. Shebeski LH. Triticale in retrospect and prospect. Hodowla Rosl Aklim Nasienn 1980;24:279-85.

47. Kiss A. Neue richtung in der Triticale-züchtung. Z Pflanzenzüchtg 1966;55:309- 29.

48. Kiss A. Kreuzungsversuche mit Triticale. Der Zuchter 1966;36:249-55.

49. Kiss A, Videki L. Development of secondary hexaploid triticales by crossing Triticale with rye. Wheat Inf Ser 1971;32:17-20.

50. Pissarev V, Zhilkina MD. Triticale x 2n=42. Genetika 1967;4:3-12.

51. Zillinsky FJ, Borlaug NE. Progress in developing Triticale as an economic crop. CIMMYT Res Bull 1971;17:27.52. Guedes-Pinto H, Carnide O, Carnide VP. New primary 8x-Triticales for Portugal. Broteria Genet 1984;V LXXX:136-46.

52. Guedes-Pinto H, Carnide O, Carnide VP. New primary 8x-Triticales for Portugal. Broteria Gen 1984;V LXXX:136-46.

53. Guedes-Pinto H, Carnide O, Leal F. Segmentation effect on immature spike on triticale *calli*. In "Plant Aging: Basic and Applied Approaches" R. Rodriguez et al, eds Plenum Press, New York 1990: 361-65.

54. Carnide O, Guedes-Pinto H. Forage apitude of primary 8x-Triticale compared with rye and wheat progenitors. Porc 2nd Int Triticale Symp Passo Fundo, Brazil 1990:536-41.

55. Bao WK . Evaluation of primary strains in breeding work of octoploid triticale. Eucarpia meeting on Triticale Clermont Ferrand, France:1984.

56. Múntzing A. Mode of production and properties of a Triticale-Strain with 70 chromosomes. Wst Inf Serv 1955;2:1-12.

57. Tsuchiya T. Cytogenetics in hexaploid Triticale. Wheat Newsletter 1969;15:10- 17.

58. Briggle LW. Triticale -a review. Crop Sci 1969;9:197-02.

59. Scoles GJ, Kaltsikes PJ. The cytology and cytogenetics of Triticale. Z. Pflanzenzüchtg 1974;73:13- 43.

60. Müntzing A. Triticale, results and problems. Verlag Paul Parey. Berlin 1979.

61. Kaltsikes PJ. Univalency in Triticale. In "Triticale". Proc of an International Symposium El Ban, Mexico 1974:59-167

62. Gupta PK, Priyadarsham PM. Triticale: present status and future prospects. Adv Genet 1982;21:255- 45.

63. Gustafson JP. Cytogenetics of Triticale. In "Cytogenetics of crop plants" MS Swaminathan, P K Gupta, U Sinha, eds.1982:228-50. Macmillan India Ltd.

64. Lukaszewski AJ, Gustafson JP. Cytogenetics of Triticale. Plant Breeding Reviews 1987;3:41-94.

65. Boyd WJR, Sisodia NS, Larter EN. A comparative study of the cytological and reproductive behaviour of wheat and triticale subjected to two temperature regimes. Euphytica 1970;19:470-497.

66. Thomas JB, Kaltsikes PJ. Genotypic and cytological influences on the meiosis of hexaploid Triticale. Can J Genet Cytol 1972;4:889-98.

67. Shkutina FM, Khvostova VV. Cytological investigation of the 42-chromosome wheat-rye amphidiploids. Theor Appl Genet 1971;41:109-19.

68. Riley R, Miller TE. Meiotic chromosome pairing in Triticale. Nature 1970;227:82-83.

69. Lelley T. Desynapsis as a possible source of univalents in Metaphase I of Triticale. Z Pflanzenzüchtg 73:249-58.

70. Jouve N, Soler C, Saiz G. Cytoplasmic influence on the meiosis of 6x-Triticale. Z Pflanzenzüchtg 1977;78:124-34.

71. Chen C, Qualset, CO, Stanford EH. Meiotic studies of secondary 42- chromosome triticales. Bot Bull Academia Sinica 1977;18:89-99.

72. Berg KH, Oehler E. Untersuchungen über die cytogenetik amphidiploider weizen- roggen- bastarde. Züchter 1938;10: 226-38.

73. Lelley T. Triticale breeding. A new approach. In "Genetics and Breeding of Triticale" Proc Eucarpia meeting Clermont Ferrand, France 1985:135-144.

74. Rajora A, Sareen PK, Chowdhury JK. Cytological studies in Triticale hexaploide Lart., *Triticum durum* L., and *Secale cereale* L. Z Pflanzenzüchtg 1979;83:127- 132.

75. Pohler W, Kistner G, Kison HU, Szigat G. Meioseuntersuchungen an Triticale. V. Meioseverhalten, pollenvitalität und fertilität von triticale-F^1-bastarden und deren eltern. Biol Zb 1978;97:453-70.

76. Lelley T, Larter EN. Meiotic regulation in triticale. Interaction of the rye genotype and specific wheat chromosomes on meiotic pairing in the hybrids. Can J Genet Cytol 1980;22:1-6.

77. Oettler G. The influence of the wheat and rye genome on the performance of primary triticale. In "Genetics and Breeding of Triticale". Proc Eucarpia meeting Clermont Ferrand, France 1985:125-34.

78. Guedes-Pinto H, Rangel-Figueriedo T, Carnide O. Aneuploidy in high yielding 6x triticale. Cerl Res Comm 1984;12:229-35.

112

79. Jung C, Lelley T, Robbelen G. Genetic interactions between wheat and rye genome in triticale. 1. Cytological results. Theor Appl Genet 1985;70:422-26.
80. Charmet G, Bernard S, Bernard M. Origin of anuploid plants obtained by anther culture in triticale. Can J Genet Cytol 1986;28:444-52.
81. Lacadena JR. Introduction of alien variation into wheat by gene recombination. I. Crosses between mono-V 5B *Triticum aestivum* L. and *Secale cereale* L and *Aegilops columnaris* Zhuk. Euphytica 1967;16:221-30.
82. Naranjo T. Análisis del comportamiento meiótico en diversas combinaciones trigo-centeno. Ph D Thesis Univ Complutense of Madrid. Spain 1978.
83. Naranjo T, Palla O. Genetic control of meiotic pairing in rye *Secale-cereale*. Heredity 1982;48:57-62. 84. Jung C, Lelley. Cytological and morphological expression of interactions between wheat and rye genomes in triticale. Eucarpia meeting on Triticale Clermont Ferrand, France 1985:145-52.
85. Jung C, Lelley T. Genetic interactions between wheat and rye genomes in triticale. 2. Morphological,and yield characters. Theor Appl Genet 1985;70:427- 32.
86. Fominaya A, Orellana J. Does differential C-heterochromatin content affect chromosome pairing in octoploid triticale? Heredity 1988;61:167-73.
87. Galindo C, Jouve N. C-banding in meiosis. An approach to the study of genome interactions in Triticale. Genome 1989;32:1074-78.
88. Bennett MD, Chapman V. Riley R. The duration of meiosis in pollen mother cells of wheat, rye and triticale. Proc Roy Soc London B 1971;178: 259-75.
89. Bennett MD, Kaltsikes PJ. The duration of meiosis in a diploid rye, a tetraploid wheat and the hexaploid Triticale derived from them. Can J Genet Cytol 1973;15:671-79.
90. Roupakias DG, Kaltsikes PJ. Independence of duration of meiosis and chromosome pairing in hexaploid triticale. Can J Genet Cytol 1977;19:345-54.
91. Thomas JB, Kaltsikes PJ. The genomic origin of the unpaired chromosomes in triticale. Can J Genet Cytol 1976;18:687-700.
92. Merker A. Chromosome composition of hexaploid triticale. Hereditas 1975;80:41- 52.
93. Merker A. The cytogenetic effect of heterochromatin in hexaploid triticale. Hereditas 1976;83:215-22.
94. Bennett MD. Heterochromatin, endosperm nuclei and grain shrivelling in wheat- rye genotypes. Heredity 1977;39: 411-419.
95. Gustafson JP, Bennett MD. The effect of telomeric heterochromatin from *Secale cereale* L on triticale x*Triticosecale* Wittmack I. The influence of several blocks of telomeric heterochromatin on early endosperm development and kernel characteristics at maturity. Can J Genet Cytol 1982;24:83-92.
96. Lukaszewski AJ, Apolinarska B, Gustafson JP, Krolow KD, Chromosome pairing and aneuploidy in tetraploid triticale. I. Stabilized karyotypes. Genome 1987;29:554-61.
97. Lukaszewski AJ, Apolinarska B, Gustafson JP, Krolow KD. Chromosome pairing and aneuploidy in tetraploid triticale. II. Unstabilized karyotypes. Genome 1987;29:562-69.
98. Ziauddin A, Kasha AJ. Giemsa C-band identification of rye chromosomes in some advanced lines of winter triticale. Can J Genet Cytol 1982;24:721-27.
99. Varghese JP, Lelley T, Origin of nuclear aberrations and seed shrivelling in triticale: a re-evaluation of the role of C-heterochromatin. Theor Appl Genet 1983;66:159-67.
100. Gustafson JP, Lukaszewski AJ, Bennett MD. Somatic deletion and redistribution of telomeric heterochromatin in the genus *Secale* and in Triticale. Chromosoma 1983;88:293-98.
101. Lukaszewski AJ. Mapping the D-genome from bread wheat for hexaploid triticale breeding. Proc Int Triticale Symp Sydney, Australia 1986:53-62.
102. Giraldez R, Cermeño MC, Orellana J. Comparison of C-banding pattern in the chromosomes of inbred lines and open pollinated varieties of rye. Z. Pflanzenzüchtg 1979;83:40-48.
103. May CE, Appels R. Rye chromosome translocation in hexaploid wheat: a re- evaluation of loss of heterochromatin from rye chromosomes. Theor Appl Genet 1980;56:15-23.
104. Schlegel R, Metz G, Mettin D. Rye cytology, cytogenetics and genetics -current status. Theor Appl

Genet 1986;72:721-34.

105. Schwarzacher T, Leitch AR, Bennett MD, Heslop-Harrison JS. *In situ* localization of parental genomes in a wide hybrid. Ann Bot 1989;64:315-24.

106. Heslop-Harrison JS, Leitch, AR, Schwarzacher T. The physical organization of interphase nuclei. In "The chromosome" Heslop-Harrison JS, Flavell RB, eds. Oxford BIOS 1993: 178-82 and 221-32.

107. Cuadrado A, Jouve N. Mapping and organization of highly-repeated DNA sequences by means of simultaneous and sequential FISH and C-banding in *xTriticosecale*. Chromosome Res 1994;2:331-38.

108. Jouve N, Bernardo A, García M, García P, Soler C. C-banding and isozyme markers to analyze the segregation of rye chromosomes in the progenies of triticale x wheat hybrids. In "Genetic Manipulation in Plant Breeding" Horn et al., eds. Walter and Gruyter, Co. New York 1986:163-65.

109. Bernardo A, Luengo P, Jouve N. Chromosome constitution in G_2 and G_3 progenies of 6x-triticale x *T. turgidum* L hybrids. Euphytica 37 1988:157-66.

110. Roupakias DG, Kaltsikes PJ. The effect of telomeric heterochromatin on chromosome pairing of hexaploid triticale. Can J Genet Cytol 1977;19:543-48.

111. Schlegel R, Zaripoba Z, Shchapova AI. Further evidence on wheat-rye chromosome pairing in F_1 triticale x wheat hybrids. Biol Zentralbl 1980;99:585- 90.

112. Naranjo T, Lacadena JR. C-banding pattern and meiotic pairing in five rye chromosomes of hexaploid triticale. Theor Appl Genet 1982;61:233-37.

113. Soler C, Montalvo D, Jouve N. Secondary association and univalent chromosomes in hybrids of hexaploid triticale and rye and wheat. J Hered 1980;71:408-10.

114. Kaltsikes PJ, Lukaszewski AJ, Gustafson JP. The effect of telomeric heterochromatin on chromosome pairing in several wheat-*Secale* hybrids. Proc 6th Int Wheat Genet Symp Kyoto Japan 1983:885-88.

115. Miazga D, Chrzastekk M. The identification of rye univalents by means of Giemsa technique. Cer Res Comm 1984;12:107-109.

116. Schlegel R, Huelgenhof E. Heterochromatin alterations in chromosomes of hexaploid triticale and their effects on meiotic pairing behaviour. Proc Eucarpia meeting Clermont Ferrand, France 1985:35-47..

117. García P, Soler C, Jouve N. New germplasm for triticale breeding: cytogenetic studies in segregant progenies of the cross 6x-triticale x *Triticum aestivum* L. An Aula Dei 1988;19:169-78.

118. Bernardo A, Díaz F, Jouve N. Chromosome factors affecting pairing in progenies of 6x-triticale x *Triticum turgidum* L ssp *turgidum* conv. *durum* Desf.. Heredity 1988;60:455-61.

119. Bernardo A, García M, Jouve N. The effect of *Secale cereale* L. heterochromatin on wheat chromosome pairing. Genetica 77 1988:89-95.

120. Kaltsikes PJ, Gustafson JP. Factors affecting chromosome pairing in Triticale. Proc Int Symp Genetic Approaches to Crop Impreovement, Karachi, Pkistan 1982:234-35.

121. Soler C, García P, Jouve N. Meiotic expression of modified chromosome constitution and structure in x*Triticosecale* Wittmack. Heredity 1990;65:21-28.

122. Badaeva ED, Badaev NS, Bolsheva, NL, Zelenin, AV. Chromosome alterations in the karyotype of triticale in comparison with the parental forms. i. Heterochromatin regions of R genome chromosomes. Theor Appl Genet 1986.;72:518-23.

123. Jouve N, Galindo C, Mesta M, Diaz F, Albella B, García P, Soler C. Changes in triticale heterochromatin visualized by C-banding. Genome 1989;32:735-42.

124. Lukaszewski AJ, Gustafson JP. Translocaion and modification of chromosomes in triticale x wheat hybrids. Theor Appl Genet 1983;64:239-48.

125. Lapitan NL, Sears RS, Gill BS. Translocations and other karyotypic structural changes in wheat x rye hybrids regenerated from tissue culture. Theor Appl Genet 1984;68:547-54.

126. Jordan MC, Larter EN. Somaclonal variation in triticale x*Triticosecale* Wittmack cv, Carman. Can J Genet Cytol 1985;27:151-57.

127. Bebeli PJ, Kaltsikes PJ. Somaclonal variation in agronomic traits of isogenic lines of Triticale. In

CIMMYT Proc 2nd Int Triticale Symp México, CIMMYT: 1991.

128. Kaltsikes PJ, Bebeli PJ. The effect of rye telomeric heterochromatin on the nature and size of variance in regenerates families of hexaploid triticale. J Genet Breed 1992;46:359-62.

129. Kaltsikes PJ, Bebeli PJ. Somaclonal variation causes changes in the interrelationships between traits in hexaploid triticale. Japan J Breed 1993;43:45- 51.

130. Gustafson JP, Lukaszewski AJ, Robertson K. Chromosome substitution and modifications in hexaploid triticale: a reevaluation. Eucarpia meeting on Triticale Clermont Ferrand, France 1985:15- 27.

131. Appels R., Gustafson JP, May CE, Structural variation in the heterochromatin of rye chromosomes in triticale. Theor Appl Genet 1982;63: 235-44.

132. Heslop-Harrison JS. The molecular cytogenetics of plants. Journal of Cell Sci 1991;100:15-21.

133. Giorgi B. A male-sterile mutant of durum wheat. II. Induction of haploid plants in wheat and triticale. preliminary results. Cer Res Comm 1991;19:267-68.

134. Wolski T, Pojmaj MS, Sawicka EJ. Evaluation of short triticale mutants for hybrid breeding. Cer Res Comm 1991;19:261-66.

135. Sisodia NS, McGinnis RC. Importance of hexaploid wheat germplasm in hexaploid triticale breeding. Crop Sci 1970;10:161-62.

136. Larter EN, Hsam SLK. Performance of hexaploid triticale as influenced by source of cytoplasm. Proc 4th Int Wheat Genet Symp Columbia, MO, USA 1973:245-51.

137. Hsam SLK, Larter EN. Influence of source of wheat cytoplasm on the synthesis and plant characteristics of hexaploid triticale. Can J Gen Cytol 1974;16:333-40.

138. Hsam SLK, Larter EN. Influence of source of wheat cytoplasm on the nature of proteins in hexaploid triticale. Can J Genet Cytol 1974;16:529-37.

139. Hsam SLK, Larter EN. Quantitative relationships of cellular-protein, RNA, and nuclear-histone in hexaploid triticale as influenced by source of wheat cytoplasm. Can J Genet Cytol 1974;16:619-25.

140. Kihara H. Substitution of nucleus and its effects on genome manifestation. Cytologia 1951;16:177-93.

141. Sánchez-Monge E. Hexaploid triticale with different cytoplasms. Proc Eucarpia Triticale Symposium Leningrado, USSR 1975:175-180.

142. Sánchez-Monge E, Soler C. Wheat and Triticale with rye cytoplasm. Proc 4th Int Wheat Genet Symp Columbia, Missouri 1973:387-90.

143. Kiss A, Trefas GS. The effect of cytoplasms in triticale breeding. Proc 4th Int Wheat Genet Symp Columbia, Missouri 1973:233-36.

144. Lelley T. Genetic control of pairing of rye chromosomes in triticale. Z Pflanzenzüchtg 1975;75:24-29.

145. Jouve N, Soler C. Influence of the cytoplasms of *Triticum timopheevi* Zhuk and *Aegilops ovata* L. in the meiosis of hexaploid triticale. Cer Res Comm 1978;6:235-240.

146. Jouve N, Montalvo D, Soler C. Cytogenetic analysis of 6x-triticale with different cytoplasms. Hodowla Rosl Aklim Nasienn 1980;24:323-26.

147. Soler C. Estudio comparativo de un alohexaploide artificial, *Triticum x Secale* Triticale sobre los citoplasmas de las especies parentales. An INIA Ser Prod Veg 1975;5:9-82.

148. Nakajima G, Zennyozi A. Cytogenetics of wheat and rye hybrids. Seiken Ziho 1966;18:39-48.

149. Merker A. Chromosome substitutions, genetic recombination and the breeding of hexaploid triticale. Wheat Inf Serv 1976;41-42:44-48.

150. Gustafson JP, Zillinsky FJ. Identification of D-genome chromosomes from hexaploid wheat in a 42 chromosome triticale. 4th Int Wheat Genet Symp Columbia, MO, USA 1973:225-31.

151. Wolski T, Tymienicka E. The present state and main problems in winter triticale breeding in Laski and Choryn Exp. Stations. Hod Rosl Aklim i Nas 1980;24:475- 86.

152. Wolski T, Tymienicka E. Stan obecny i perspektwy ulepszenia ozimego Triticale Stacjach Poznanskiej Hodowli Roslin. Post Nauk Roln 1982;5:3-26.

153. Wolski T, Maczinska L, Tymienicka E. Winter triticale varieties from the Choryn and Laski Experimental Stations. Eucarpia meeting on Triticale Clermont Ferrand, France 1985:487-96.

154. Tymieniecka E, Wolski T, Madra M. Breeding of winter triticale for improvement of agronomic value. Eucarpia meeting on Triticale Clermont Ferrand, France 1985:445-54.

155. Merker A. A Giemsa technique for rapid identification of chromosomes in Triticale. Hereditas 1973;75:280-82.

156. Rogalska S. Chromosome constitution of plants of selected lines of secondary hexaploid triticale. Hodowla Rosl Aklim Nasienn 1978;24:357-64.

157. Pilch J. Rye chromosome constitution and the amount of telomeric heterochromatin of the widely and narrowly adapted CIMMYT hexaploid triticales. Z. Pflanzenzüchtg 1981;87:56-68.

158. Pilch J. Analysis of the rye chromosome constitution and the amount of telomeric heterochromatin in the widely and narrowly adapted hexaploid triticales. Theor Appl Genet 1981;60:145-49.

159. Lukaszewski AJ, Apolinarska B. The chromosome constitution of hexaploid winter triticale. Can J Genet Cytol 1981;23:281-85.

160. Seal A. C-banded chromosomes in wheat and triticale. Theor Appl Genet 1982;63:39-47.

161. Sandha GS, Grewal KD, Satija CK. Study of R-D chromosome substitutions and their effect in triticale. Crop Improv 1984;11:119-22.

162. Gupta PK, Balyan HS, Fedak G. A study of D/R substitutions in some spring triticales using wheat ditelocentrics. Proc 7th Int Wheat Genet Symp Cambridge, UK 1988:297-301.

163. Ren Z, Lelley T, Robbelen G. Translocations of chromosomes in octoploid triticale x common wheat hybrids Acta Genet Sin 1991;18: 228-34.

164. Plaha P, Sethi GS. Sdaptative advantage to 6R chromosome of rye in the genomic background of bread wheat. Cer Res Comm 1993;21:2-3.

165. Skovmand B, Fox PN, Villarreal PJ. Triticale in commercial agriculture: progress and promise. Adv Agric 1984;37:1-45.

166. Lukaszewski AJ. A comparison of physical distribution of recombination in chromosome 1R in diploid rye and in hexaploid triticale. Theor Appl Genet 1992;83:1048-53.

167. Kaltsikes PJ, Lukaszewski AJ, Gustafson JP. Cross-over frequencies in chromosomes 1R and 2R of rye *Secale cereale* L.. Proc Int Triticale Symp. The Australian Institute of Agricultural Sciences, Sydney 1991:321-26.

168. Sánchez-Monge E, Sánchez-Monge E jr. Meiotic pairing in wheat-triticale hybrids. Z. Pflanzenzüchrg 1977;79:122-33.

169. Soler C, Montalvo D, Jouve N. Introducción de variación genética en trigo y triticale mediante hibridación de triticale con *Triticum aestivum* L. Anales INIA, Ser Agr 1982;21:95-108.

170. Jouve N, Montalvo D, Soler C. C-banding in cytogenetics of 6x-triticale x *Triticum aestivum* 1 hybrids. Z Pflanzenzüchtg 1982;88:311-21.

171. Jouve N, Montalvo D, Soler C. Distribution of univalents in the meiosis and chromosomal analysis of the progeny of 6x triticale x common wheat hybrids. Eucarpia meeting on Triticale Clermont Ferrand, France 1985:227-37.

172. Jouve N, Giorgi B. Analysis of induced homoeologous pairing in hybrids between triticale *ph1* mutant and *Triticum aestivum* L. Can J Genet Cytol 1986;28:696- 700.

173. Riley R, Chapman V. Haploids and polyhaploids in *Aegilops* and *Triticum*. Heredity 1957;11:195-207.

174. Riley R, Chapman V. Genetic control of the cytologically diploid behaviour of hexaploid wheat. Nature 1958;182:713-15.

175. Sears ER. An induced utant with homoeologous pairing in common wheat. Can J Genet Cytol 1977;19:585-93.

176. Giorgi B. A homoeologous pairing mutant isolated in *Triticum durum* cv Cappelli. Mutat Breed News 1978;11:4-5.

177. Giorgi B. Origin, behaviour and utilization of a *Ph1* mutant of durum wheat, *Triticum turgidum* L. var *durum*. Proc 6th Int Wheat Genet Symp Kyoto, Japan 1983:1033-40.

178. Giorgi B, Ceoloni C. A *ph1* hexaploid triticale: production, cytogenetic behaviour and use for intergeneric gene transfer. Eucarpia meeting on Triticale Clermont Ferrand, France 1985;105-117.

179. Chaudry MN. Synthesis of tetraploid triticale. Ph D Thesis Univ Complutense of Manitoba, Winnipeg, Canada 1968.

180. Krolow KD. 4x triticale production and use in triticale breeding. Proc 4th Int Wheat Genet Symp Columbia, MO, USA 1973:691-96.

181. Krolow KD. Research work with 4x-triticale in Germany Berlin. Proc Int Triticale Symp El Batán, Mexico 1974:51-60.

182. Krolow KD. Selection of 4x triticale from the cross 6x-triticale x 2x rye. Proc Eucarpia Triticale Symposium Leningrado, USSR 1975:114-122.

183. Krolow KD. New aspects for the use of 4x-triticale 2n=28 in triticale development. Proc 6th Int Wheat Genet Symp Kyoto, Japan 1983:903-08.

184. Lapinski B, Apolinarska B. Polish work on 4x-triticale. Eucarpia meeting on Triticale Clermont Ferrand, France 1985:261-66.

185. Bernard M, Bernard S, Saigne B, Tetraploid triticales: investigations on their genome and chromosome constitution. Eucarpia meeting on Triticale Clermont Ferrand, France 1985:277-88.

186. Lehmann C, Hohmann U, Krolow KD. Tetraploid triticale with D-genome chromosomes from *Triticum aestivum* produced with autoallohexaploid triticale. Cereal Res Comm 1991;19:469-76.

187 .Badaev NS, Badaeva ED, Dubovets NI, Bolsheva, NL Bormotov VE,. Zelenin AV. Formation of a synthetic karyotype of tetraploid triticale. Genome 1992;35:311-17.

188. Hohmann U. Stabilization of tetraploid triticale with chromosomes from *Triticum aestivum* ABDABDRR 2n = 28. Theor App Genet 1993;86:356-64

189. Baum M, Lelley T. A new method to produce 4x triticales and their application in studying the development of a new polyploid plant. Plant Breed 1988;100: 260- 67.

190. Lehmann C, Krolow K. Variability of morphological traits, fertility and yield- related characters of tetraploid triticale. Cer Res Comm 1993;21:75-81.

191. Martin A, Jouve N. Cytogenetics of F_1 and their progenies. In "Distant Hybridization of Crop Plants". Kalloo G, Chowdhury, JB, eds. Monogr Theor Appl Genet 16. Springer Verlag.Berlin.1992:82-105.

192. Bernard S, Gay G. Introduction of *Aegilops ventricosa* germplasm into hexaploid triticale. Eucarpia meeting on Triticale Clermont Rerrand, France 1985:221-24.

193. Krolow KD, Lukaszewski AJ, Gustafson PJ. Preliminary results on the incorporation of D- and E-genome chromosomes into 4x-triticale. Eucarpia meeting on Triticale Clermont Ferrand, France 1985:289-95.

194. Guedes-Pinto H, Mello-Sampayo T. Allo-autopolyploid triticales AABBRRRR: I. Origin, behaviour and propects. Eucarpia meeting on Triticale Clermont Ferrand, France 1985:205-13.

195. Steward FC. Growth and development of cultivated cells. III. Interpretations of the growth from free cell to carrott plant. Amer J Bot 1958;45:709-13.

196. Reinert J. Morphogenese und ihre kontrolle an gewebekulturen aus carotten. Naturwissenshaten 1958;45:344-45.

197. Larkin PJ, Scowcroft WR. Somaclonal variation. A novel source of variability from cell cultures for plant improvement. Theor Appl Genet 1981;60:197-214.

198. Nakamura C, Keller WA. Plant tegeneration from inflorescence cultures of hexaploid triticale. Plant Sci Let 1982;24:275-80.

199. Kapila RK, Sethi GS. Genotype and age effect on *in vitro* embryo rescue of bread wheat x hexaploid triticale hybrids. Plant Cell Tiss Org Culture 1993;35:287- 91.

200. Reddy VD, Reddy GM. Genetic basis of plant regeneration in hexaploid triticale. Euphytica 1993;70:17-19.

201. Sirkka A, Immonen T. Comparison of callus culture with embryo culture at different times of embryo rescue for primary triticale production. Euphytica 1993;70:185-90.

202. Stolarz A, Lörz H. Somatic embryogenesis *in vitro* manipulation and plant regeneration from immature embryos of hexaploid triticale x *Triticosecale* Wittmack. Z Pflanzenzüchtg 1986;96:353-62.

203. Guedes-Pinto H, Carnide O. Plantlets regeneration from *in vitro* culture of 6x- triticale immature spikes. Ciencia Biologica 1987;supl 12 5A:208.

204. Guedes-Pinto H, Carnide O, Alpoim F. *Calli* induction from triticale immature spikes. Vortr Pflanzenzüchtg heft 1989;15-1:7-14.

205. Bebeli PJ, Karp A, Kaltsikes PJ. Plant regeneration and somaclonal variation from cultured immature embryos of sister lines of rye and triticale differeing in their content of heterochromatin. 1. Morphogenetic response. Theor Appl Genet   1988;75: 929-936.

206. Felföldi K, Purnhause L. Induction of regeneration callus cultures from immature embryos of 44 wheat and 3 triticale cultivars. Cer Res Comm 1992;20:273-77.

207. Armstrong KC, Nakamura C, Keller WA. Karyotype instability in tissue culture regenerants of triticale *xtriticosecale* Wittmack cv. Welsh from 6-month-old *callus* cultures. Z. Pflanzenzüchtg 1983;91:233-45.

208. Bebeli PJ, Kaltsikes PJ. The effect of rye telomeric heterochromatin on the nature and size of variance in regenerated families of hexaploid triticale. J Genet Breed 1992;46: 359-62.

209. Bebeli PJ, Kaltsikes PJ. Somaclonal variation causes changes in the inter- relationships between traits in hexaploid triticale. Japan J Breed 1992;43:45-51.

210. Bebeli PJ, Kaltsikes PJ, Karp A. Field evaluation of somaclonal variation in rye lines differing in telomeric heterochromatin. J Genet Breed 1993;47: 15-22.

211. Bernard S. *In vitro* androgenesis in hexaploid triticale: determination of physical conditions increasing embryod and green plant production. Z Plflanzenzüchtg 1988;85:308-321.

212. Bernard S. Direct embryogenesis and plant production through *in vitro* androgenesis in triticale and wheat. Vortr Pflanzenzüchtg 1989;15:7-15.

213. Charmet G, Bernard S. Diallel analysis of androgenic plant production in hexaploid triticale x*Triticosecale* Wittmack. Theor Appl Genet 1984;69:55-61.

214. González JM, López LA, Bernard S, Jouve N. Prolamin analysis of progenies from androgenetic plants of triticale. Plant Breed 1993;111:42-48.

215. Jouve N, Bernardo A, Soler C. Hybrids 6x-triticale x *Triticum turgidum* L and the obtention of its F_2 and BC_1 progenies. Cer Res Comm 1984;12:223-28.

216. Gupta PK, Fedak G. Variation in induction of homoeologous pairing among chromosomes of 6x *Hordeum parodii* as a result of three triticale x*Triticosecale* Wittmack cultivars. Can J Genet Cytol 1986;28:420-25.

217. Gupta PK, Priyadarshan PM, Misra AK. Cytogenetic studies in triticales: I. Cytology of F_1 and F_2 hybrids involving rye. Proc 6th Int Wheat Genet Symp Kyoto Japan 1983: 909-13

218. Gupta PK, Priyadarsham PM. Analysis of meiosis in Triticale (x*Triticosecale* Wittmack) x rye (*Secale cereale* L.) F_1 hybrids at three ploidy levels. Theor Appl Genet 1987;73:893-98.

219. Miller TE, Riley R. Meiotic chromosome pairing in wheat-rye combinations. Genet Iber 1972;24:241-50.

220. Bernard S, B. Saigne B. Étude cytogénétique de F_1 et F_2 (Triticale hexaploide xSeigle): Mise en évidence d'une inhibition du systeme diploidisant du Blé tendre. Ann Amelior Plantes 1977;27:737-48.

221. Jouve N, Montalvo D. Meiotic behaviour in hybrids beteen 6x-triticale and *Secale cereale* L. Proc 8th Congr Eucarpia Madrid, Spain 1977:191-97.

222. Naranjo T, Lacadena JR, Giraldez R. Interaction between wheat and rye genomes on homologous and homoeologous pairing. Z Pflanzenzüchtg 1979;82:289-305.

223. Jouve N, Diez N, Rodriguez M. C-banding in 6x-triticale x *Secale cereale* hybrid cytogenetics. Theor Appl Genet 1988;57:75-79.

224. Fernández-Escobar J, Martin A. Morphology, cytology and fertility of a trigeneric hybrid from triticale x tritordeum. Euphytica 1989;42:291-96

225. Fernández JA, Jouve N. Meiotic pairing in hybrids of 6x-triticale and the amphiploid *Hordeum chilense* x *Triticum turgidum* conv *durum*. J Hered 1985;76:63-64.

226. Vos DJ. Introgresion of material from *Agropyron elongatum* (2n=14) into triticale. In "Sakamoto S (ed.). Proc 6th Int Wheat Geneti Symp, Jyoto, Japan 1983:897-902.

227. Gupta PK, Fedak G. Intergeneric hybrids between x*Triticosecale* cv Welsh n=42, and three genotypes of *Agropyron intermedium* 2n=42. Can J Genet Cytol 1986;28:176-79.

228. Gupta PK, Fedak G. Intergeneric hybrids between *Hordeum jubatum* and triticale (x*Triticosecale* Wittmack). Genome 1987;29:671-73.

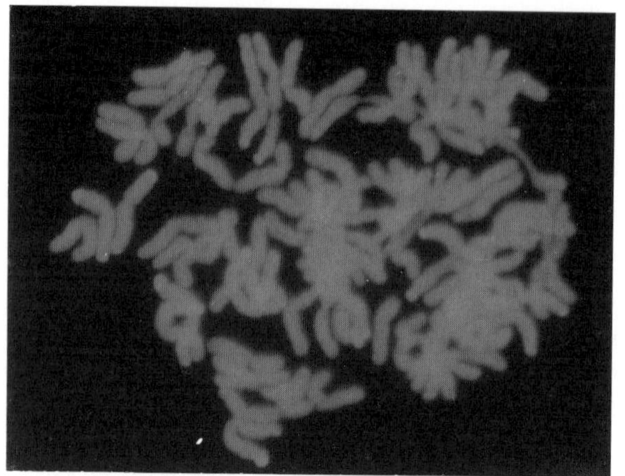

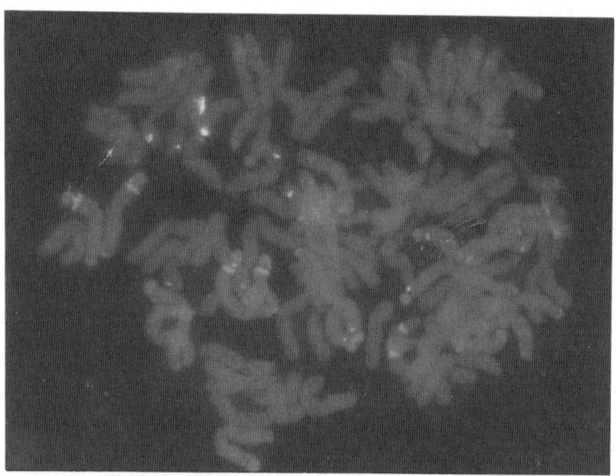

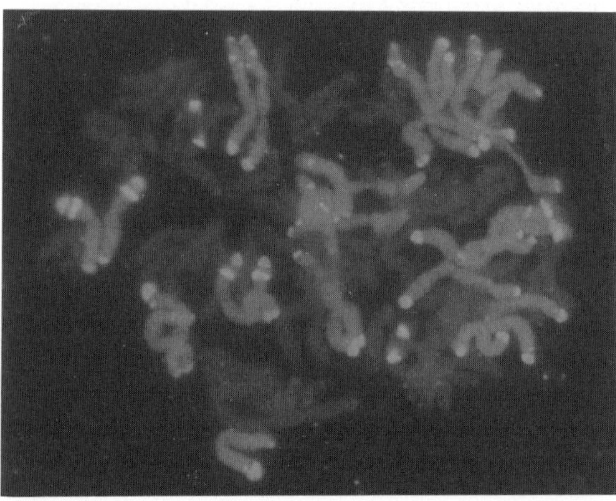

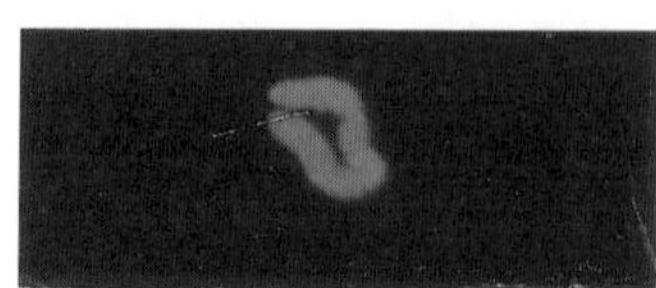

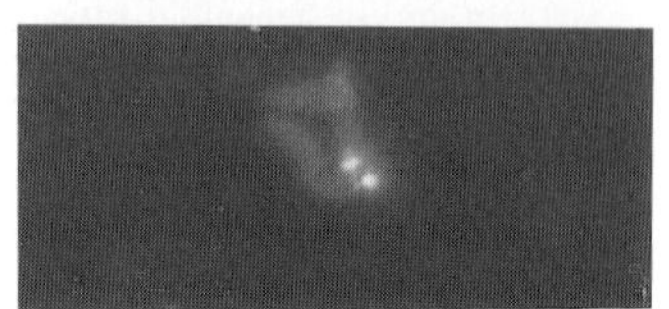

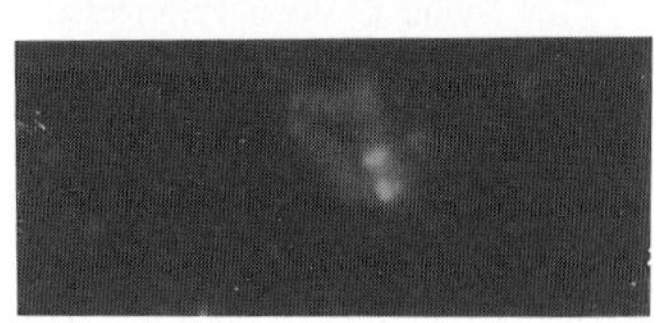

Fig 3.- A late metaphase plate of the Spanish triticale line 'Tajuña' having the 2D/2R chromosome substitution after mixed fluorescence *in situ* hybridization. **From head to foot at left**: DAPI counterstain. Digoxigenin labelled pTa794 probe (5S rDNA) detected by fluorescein conjugate (green). Genomic *in situ* using total DNA from rye and the hybridization sites to the rhodamine labelled pTa71 probe (25S-5.8S-18S rDNA of NOR, isolated from wheat) (red), where rye chromosomes are shown as reddish enhanced fluorescence. **Above right**: a detail of chromosome 1B (photographs given by our colleague Angeles Cuadrado).

CONTROL OF NUCLEOLAR EXPRESSION IN TRITICALE*

John Perry Gustafson[1], and Robert B. Flavell[2] 1. U.S.
Department of Agriculture-ARS, Plant Genetics Resources Unit,
and Plant Science Unit, University of Missouri, Columbia, USA
and 2. John Innes Centre, Norwich, United Kingdom

Abstract

Ribosomal rRNA genes present in the nucleolar organizing region
(NOR) in hexaploid wheat (*Triticum aestivum* L. em Thell.) are known to
be expressed at different levels and some loci exhibit full or partial
dominance over others. Nucleolar dominance is best seen in wide-hybrids
like hexaploid triticale (X *Triticosecale* Wittmack) where the wheat loci
dominate over the rye (*Secale cereale* L.) NOR locus. A correlation
appears to exist between the methylation of cytosine residues and the
expression of a rRNA locus suggesting that a dominant locus would have
a much reduced methylation pattern over a recessive locus The effect of
rye NOR expression in tetraploid triticales containing various wheat and
rye NOR loci was studied. The results showed that when both wheat
chromosomes 1B and 6B were present, the rye NOR locus was methylated
and suppressed. When either 1B or 6B were present, some minor rye
activity was seen, and that a slightly higher degree of activity was
observed in the absence of 1B versus 6B. When both 1B and 6B were
absent, the rye locus was expressed as in rye. The data indicates that
either one of the wheat rRNA loci will suppress rye rRNA loci.

Introduction

When a rye genome (*Secale cereale* L.), gene, or gene complex has
been placed into a wheat (*Triticum* ssp.) background, agronomically
useful expression of rye characters is not always realized and the
expression level appears generally skewed, being lower than that of the
mid-parent. It would be helpful to triticale (X *Triticosecale* Wittmack)
researchers to have an understanding of the problems associated with

---

Names are necessary to report factually on available data; however, the USDA neither
guarantees nor warrants the standard of the product, and use of the name by USDA
implies no approval of the product to the exclusion of others that may also be suitable.

H. Guedes-Pinto et al. (eds.), Triticale: Today and Tomorrow, 119–125.
© 1996 *Kluwer Academic Publishers. Printed in the Netherlands.*

genome interactions.

The most important source of alien introgression into present-day triticale breeding programs involves combining of new sources of rye chromosomes with wheat (*T. durum* L. and *T. aestivum* L. em Thell.) genomes. Variability in rye gene expression, when present in a wheat background has been well documented and has been shown to be totally unpredictable (1, 2). Some rye disease resistance genes present in a triticale are expressed at the same level as in the rye parent (3). Other rye genes are expressed at an intermediate level of that in either the rye or the wheat parents. Finally, there are rye disease resistance genes that have no detectable level of expression when in a triticale (4). However, a recent study showed repressed expression of specific wheat heat-shock proteins in triticale, where the triticale exhibited a heat-shock protein pattern similar to rye (5).

A well studied example of gene suppression in wide hybrids is in the expression of the nucleolar organizing region (NOR) ribosomal RNA (rRNA) genes (6). Studies of triticale and wheat-rye addition lines using various techniques (i.e. counting nucleoli, silver staining, and *in situ* hybridization) have detected little or no rye NOR locus activity (7-10).

DNaseI sensitivity studies (11) indicated that the quantity and specific sequence of an NOR locus plays a role in preferential rRNA gene transcription. Gene number, promoter binding site number and the kinetics between the site and various transcription factors appear to work in determining transcription. Suppression of rye rRNA expression in a wheat background has been attributed to the larger and more numerous ribosomal DNA spacer regions (containing transcription start sites) present in wheat. It was proposed that through basic competition, the region with the more repeats, more promoter sites, acquires transcription factors readily and is preferentially transcribed (12).

Genes of cereal NORs are known to show abundant methylation at the CpG sites in CCGG regions as assayed by *HpaII* restriction digests (13), but occasionally a site is unmethylated. In a small proportion of wheat rRNA genes, a CCGG site is unmethylated in the promoter region, and these unmethylated sites appear more often in loci displaying a greater transcriptional activity. Therefore, these sites are thought to be components of the active fraction of the wheat NOR producing rRNA (6).

Martini and Flavell (11) showed that chromosome 1B of wheat contributes twice the amount of rDNA chromatin into the nucleolus as does chromosome 6B, even though 6B has almost three times the number of rRNA genes. This indicates that 1B is semidominant over 6B, despite the smaller number of rRNA genes of the former. However, levels of expression can and do change from variety to variety depending on the number and specific sequence of spacers and genes present on each chromosome. Through unequal crossing-over and gene conversion, extensive varietal polymorphism has been observed in the spacer region of *T. aestivum* (14), which affects the expression level of the NOR. Using tetraploid triticales, *in situ* hybridization detected dispersion of the rye

NOR and hence NOR activity; an approximate 20% decrease of activity
was noted when chromosome 1B was present. A 60% decrease in rye NOR
activity was seen in the presence of only 6B and a 60% decrease of rye
activity was seen when both 1B and 6B were present (Gustafson, unpubl.
data). These results further indicate that whether or not 1B is
semidominant over 6B depends very much on the quantity and specific
sequence of rDNA spacers and genes present in the wheat being tested.

A study was conducted utilizing 5-azacytidine treated root tips in
order to establish whether or not the rye NOR was methylated when in a
wheat background (15). Since 5-azacytidine inhibits DNA methylation,
any rye NOR activity in triticale would indicate that it had been previously
suppressed by being methylated. This analysis showed that there were
more countable nucleoli present in the triticales that had been treated
with 5-azacytidine. It was suggested that these new nucleoli were derived
from reactivating the rye NOR locus that had been previously methylated.

The purpose of the present study was to examine the methylation
pattern the rye NOR locus of several tetraploid triticales containing
various complements of the wheat NOR chromosomes 1B and 6B, by using
methylation sensitive restriction enzymes. This would determine the
effect of the presence or absence of wheat NORs on the methylation pattern
of the rye NOR and relate these methylation patterns to expression of the
rye NOR.

Materials

The tetraploid triticale stocks used were created in the 1970's by Drs. K.D.
Krolow, Freie Universitat Berlin, Germany (4x14 #113), and A.J.
Lukaszewski, B. Lapinski, and B. Apolinarska, Poznan, Poland (P1X #5,
P1C6 #24 and P1U #59)(Table 1).

Table 1.   Chromosome composition of the tetraploid triticales, and the
Chinese Spring (CS) wheat and D. zlote rye controls.

name	rye complement	wheat complement	NOR location		
D. zlote	7"		1R		
P1X#5	7"	1A,2A,3A,4B,5A,6A,7A	1R		(1A)
P1C6#24	7"	1A,2A,3A,4A,5B,**6B**,7A	1R	6B	(1A)
P1U#59	7"	1B,2A,3A,4B,5B,6A,7B	1R	1B	
4x14#113	7"	1B,2A,3B,4A,5B,**6B**,7B	1R	1B	6B
CS	-	1B,2B,3B,4B,5B,**6B**,7B	1B		6B

parentheses indicate possible minor NOR regions

The chromosome constitutions were identified using the standard C-band
analysis. The pedigree of 4x14 #113 is unknown, but the three triticales
from Poland (P1X#5, P1C#24, P1U#59) are somewhat inter-related (16).
The tetraploid triticales were obtained from backcrossing a hexaploid
triticale or an ABR haploid to diploid rye. The rye D. zlote and the wheat
Chinese Spring were present as representative controls.

122

Methods

Plants were grown in a greenhouse under approximately 16 hr of light , during the summer, at the John Innes Centre, England. Young tissue (4-6 leaf stage) was homogenized by grinding in liquid nitrogen with acid washed sea sand. The ground tissue was immediately transferred to a tube containing the following extraction buffer: 100 mM Tris-HCl pH 8, 100mM NaCl, 50 mM EDTA pH 8, 2% SDS and 0.1mg/ml proteinase K was added immediately before use. After mixing, the tissue and buffer mixture was placed at 37°C for 1 hour, with additional mixing every 20 minutes. The DNA was extracted following standard procedures. The sample was RNAse A treated, re-extracted and re-precipitated. Two micrograms of each DNA sample were digested, according to the manufacturer's instructions, with *Taq*I or*Taq*I plus *Hpa*II. *Hpa*II digests were carried out first followed by the *Taq*I digests because the latter requires a higher incubation temperature. The DNA digests were electrophoresed overnight at 50V in 1% agarose gels, or until the loading dye migrated 2/3 the length of the gel and the DNA was transferred to nitrocellulose. Cloned DNA from the NOR spacer region of rye or wheat (pScR4-T1 or pTa71, respectively) was radioactively labeled using the random-primed method. Hybridizations were carried out overnight at 65°C in a roller bottle oven. Blots were washed with a final stringency of 0.2X SSC, 0.1% SDS, 65°C.

Results and Discussion

The observed decrease in band intensity from the non-*Hpa*II-digested lane to the *Hpa*II digested lane would be an indicator of sequences being unmethylated and transcribed. This decrease in band intensity would be accompanied by an increase of intensity or the appearance of a new band migrating at a lower molecular weight. In contrast, methylated sequences would show little or no differences in banding patterns between the *Hpa*II- and non-*Hpa*II-digested lanes (Fig. 1).

The wheat and rye spacer region DNA probes used showed a high degree of homology to their respective genomic DNA restriction fragments, especially the major rye band present in the rye and all triticale *Taq*I digested DNA blots. Since the triticales did not contain the same wheat composition, differences in their banding patterns were observed when probed with the wheat spacer region probe pTa71. The banding patterns showed that the triticale containing wheat chromosomes 1B, 6A and the triticale with 1A, 6B came from different wheat backgrounds.

Figure 1. A Schematic diagram showing the characteristic banding patterns on a Southern blot of both methylated and unmethylated rye NORs when hybridized with the rye specific NOR spacer region (pScR4-T1).

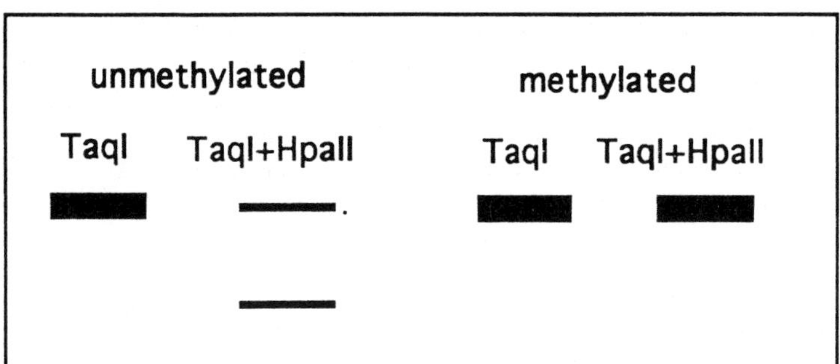

When probed with the rye spacer region probe (pScR4T1), the triticale containing no major wheat NOR loci, had similar band intensities as the rye control for both the *Hpa*II digested and non-*Hpa*II-digested lanes. This indicated that the rye NOR locus in the triticale with no major wheat NOR loci was active in a manner similar to that observed in the rye control. The triticale containing only the wheat 6B NOR locus, the chromosome with a greater number of rRNA genes, showed a faint banding differentiation in the +*Hpa*II digest. While in the +*Hpa*II digest of the triticale containing only the wheat 1B locus the band differences in the higher molecular weight area were faint and a lower molecular weight band appeared. The wheat 1B locus contains a longer intergenic spacer than 6B and this is considered to be the deciding factor in preferential transcription for this region as compared to the 6B locus. The triticale containing both 1B and 6B showed almost no digestion with *Hpa*II. The similarity of the banding patterns indicated that the rye control D. zlote was probably one of the parents used in creating the three tetraploid triticales from Poland.

Flavell (17) described a competition model between NORs for a protein that can bind to intergenic regulatory parts of rRNA genes described. Along with NOR loci competition, methylation must also be considered a transcriptional regulator. It is likely that any rye NORs containing unmethylated CCGG sites, contribute to the active rRNA fraction. Using 5-azacytidine, Vieira et al. (15) showed that wheat nucleolar dominance in wheat-rye hybrids was probably due to methylation of rRNA genes or their regulators.

In the present experiment, employing triticales containing various wheat and rye NORs, the proportion of rye rRNA spacers that can be cleaved by *Hpa*II is extremely small in the presence of wheat NORs. When both wheat NORs were absent, more rye rDNA spacers were cleaved by *Hpa*II than when either the 1B or 6B loci were present. This agrees with reduced methylation being correlated with significant expression of the rye NOR locus. The data clearly indicate either one or both of the wheat NOR loci can suppress expression of the rye NOR locus through methylation. Though varietal differences will definitely show variable NOR expression levels, there will be a marked and predictable loss of rye NOR activity as seen by decreased sensitivity to DNAse I treatment and increased methylation at these sites. Without studying the effects of several different wheats showing differences in their NOR loci, no conclusions can be made about wheat NOR loci at 1B showing dominance over the NOR located on chromosome 6B. However, it was apparent that the presence of both major wheat NORs was coupled with methylation and suppression of rye NOR activity when in a wheat background.

Acknowledgments

We would like to thank K. Houchins and M. O'Dell for their technical assistance. JPG would like to thank the Underwood Foundation for financial support.

References

1. Gustafson, J.P. Cytogenetics of triticale. In Swaminathan, Gupta and Sinha, eds., Cytogenetics of Crop Plants, Macmillan India Ltd. pp. 225-250. 1982.

2. Lukaszewski, A.J., and Gustafson, J.P. Cytogenetics of triticale. In Janik, J. (ed.) Plant Breeding Reviews 5:41-94. 1986.

3. McIntosh, R. A., and Singh, S. J., Rusts - real and potential problems for triticale, *in: "Proc. Intl. Triticale Symposium,"* N. L. Darvey, ed., Australian Institute of Agricultural Science, pp. 199. 1987.

4. Morrison, R. J., Larter, E. N., and Green, G. J., The genetics of resistance to Puccinia graminis tritici in hexaploid triticale, *Can. J. Genet. Cytol.*, 19, 683- ? . 1977.

5. Somers, D.J., Gustafson, J.P., and Filion, W.G. The influence of the rye genome on expression of heat-shock proteins in triticales. Theor. Appl. Genet. 34:845-848. 1992.

6. Flavell, R.B., M. O'Dell, P. Sharp, E. Nevo and A. Beiles. Variation in the intergenic spacer of ribosomal DNA of wild wheat, Triticum dicoccoides, in Israel. Mol. Biol. Evol. 3(6):547-558. 1986.

7. Darvey, N.L., and Driscoll, C.J. Nucleolar behaviour in Triticum. Chromosoma 36:131-139. 1972.

8. Appels, R., Moran, L.B., and Gustafson, J.P. The structure of DNA from the rye (Secale cereale) Nor-R1 locus and its behaviour in wheat backgrounds. Can. J. Genet. Cytol. 28:673-685. 1986.

9. Gustafson, J.P., Dera, A.R., and Petrovic, S. Expression of modified rye ribosomal RNA genes in wheat. Proc. Natl. Acad. Sci., USA. 85:3943-3945. 1988.

10. Martini, G., O'Dell, M., and Flavell, R.B. Partial inactivation of wheat nucleolus organizers by nucleolus organizers from Aegilops umbellulata. Chromosoma 84:687-700. 1982.

11. Flavell, R.B., O'Dell, M., and Thompson, W.F. Regulation of cytosine methylation in ribosomal DNA and nucleolus organizer expression in wheat. J. Mol. Biol. 204:523-534. 1988.

12. Flavell, R., J. Jones, D. Lonsdale and M. O'Dell. Higher plant genome structure and the dynamics of genome evolution. Advances in Gene Technology: Molecular Genetics of Plants and Animals. 1983.

13. Appels, R. and J. Dvorak. Relative rates of divergence of spacer   and gene sequences within rDNA region of species in the Triticeae: Implications for the maintenance of homogeneity of a repeated gene family. Theor Appl Genet 63:361-365. 1982.

14. Vieira, R., Queiroz, A., Morais, L., Barao, A., Mello-Sampayo, T., and Viegas, W. 1R chromosome nucleolus organizer activation by 5-azacytidine in wheat x rye hybrids. Genome 33:702-712. 1990

15. Lukaszewski, A.J., Apolinarska, B., Gustafson, J.P., and Krolow, K.D. Chromosome constitution of tetraploid triticale. Z. Pflanzenzuchtg. 93:222-236. 1984.

16. Flavell, R.B., Sardana, R., Jackson, S., and O'Dell, M. The molecular basis of variation affecting gene expression: evidence from studies on the ribosomal RNA gene loci of wheat. In Stadler Genetics Symposium. Ed. J.P. Gustafson. Plenum Press, New York. pp. 419-430. 1990.

17. Vieira, R., T. Mello-Sampayo and W. Viegas. Genetic control of   1R nucleolus organizer region expression in the presence of wheat genomes. Genome 33:713-718, 1990.

EXPRESSION OF 1R rDNA *loci* IN TRITICALE: GENETIC AND DEVELOPMENTAL CONTROLS

Wanda Viegas[1], Manuela Silva[1], Nuno Neves[1], Alexandra Castilho[1], Augusta Barão[1], Álvaro Queiroz[1], J.S. (Pat) Heslop-Harrison[2], Leonor Morais-Cecílio[1], Luis Amado[1], Margarida Delgado[1], Vitor Carvalho[1]
[1]Departamento de Botânica e Engenharia Biológica, Instituto Superior de Agronomia, Lisboa, Portugal
[2]Karyobiology Group, John Innes Centre, Colney, Norwich, United Kingdom

Abstract

Using the silver staining technique, which allows the identification in metaphase cells of rDNA *loci* active in the previous interphase, it was observed the consistent expression of 1R NOR in hexaploid triticale cultivars with a 2R/2D chromosome substitution, confirmed through genomic *in situ* hybridization using rye total genomic DNA and different cloned probes of repetitive sequences. A control mediated by non-nucleolar chromosomes on the suppression of rye NOR in triticale was therefore detected since it was possible to observe the expression of those *loci* in the presence of all wheat major NORs.

It was also established that the 1R NOR expression pattern in complete hexaploid triticale cultivars is a developmental-dependent process, since: (i) the epigenetic pattern of 1R NOR inactivation in the presence of wheat genomes is syncronously established in embryos and endosperms between the 4th and the 5th day after fertilization; (ii) the influence of induced DNA demethylation on that pattern, through the incorporation of 5-azacytidine, is also dependent on the seed developmental stage; and (iii) the rye NOR suppression pattern is maintained until meiosis being erased during this process.

Introduction

In wheat (*Triticum aestivum* Thell) x rye (*Secale cereale* L.) hybrids and in the artificial amphiploid triticale (X *Triticosecale* Wittmack) a nucleolar dominance of wheat is observed, resulting in an almost total inactivation of rDNA from rye origin [1-4]. It has moreover been extensively documented that the suppression of rye NOR in distinct wheat backgrounds is mediated by the presence of any of the wheat major nucleolar organizing regions [5-7].

H. Guedes-Pinto et al. (eds.), Triticale: Today and Tomorrow, 127–134.

© 1996 *Kluwer Academic Publishers. Printed in the Netherlands.*

The process of wheat nucleolar dominance has also been considered as a phenomenon characteristic of all plant developmental stages and consequent of a competition for limiting factors essential for rDNA transcription [8]. Although the importance of the structure of rDNA spacers for that competitive process, as became clear from the analysis of *X. laevis* rDNA [9, 10], there was not, until now, any indication about the mapping of potential genes for any of those limiting elements or other factors influencing those interactions.

Through the evaluation of nucleolar activity in hexaploid triticale cultivars with the 2R/2D chromosome substitution it was possible to observe for the first time the expression of 1R rDNA *loci* in the presence of all wheat major NORs. The further analysis of the expression patterns of those *loci* in distinct cell types of complete hexaploid triticale cultivars, associated with the evaluation of modifications induced by DNA hypomethylation in different stages of seed maturation, disclosed important aspects of nucleolar dominance observed in this species.

It was therefore unravelled that the key features in determining the epigenetic pattern of rye NOR are mediated by non-nucleolar chromosomes and restricted to specific developmental stages.

Material and Methods

The following plants were used: (i) hexaploid triticale, X *Triticosecale* Wittmack cvs. Juanilho, Drira, Bacum and Rosner kindly supplied by the ENMPE, Portugal. Triticale cultivars Drira and Juanilho are euploids (2n=2x=42, with full chromosome sets from genomes A, B and R); cultivars Bacum and Rosner have a chromosome substitution (2n=6x=42, AABBRR(2R)2D, i.e. the pair of chromosomes 2 from rye substituted by the pair of homoeologous chromosomes 2D of hexaploid wheat.; (ii) bread wheat *Triticum aestivum* (L.) em. Thell cv. Chinese Spring (2n=6x=42, AABBDD); and (iv) rye *Secale cereale* L. cv. Centeio do Alto (2n=2x=14, RR) from the North of Portugal.

Cytological evaluation of nucleolar activity. For the evaluation of nucleolar activity somatic cells, pollen mother cells and microspore cells were selected and further treated for silver staining impregnation accordingly with the methods already described [11]. The number of silver positive satelite NORs (Ag-SAT-NORs) per metaphase cell was scored.

Analysis of chromosome constitution. Non-radioactive *in situ* hybridization using (i) rye genomic probe to distinguish the parental origin of triticale chromosomes, (ii) pTa71 to allocate the rRNA genes, and (iii) the repetitive sequence pSc119.2 for individual rye chromosome identification was used as previously described [12, 13].

Embryos and endosperms in distinct developmental stages. Wheat plants were transferred to environment-controlled cabinets before being emasculated and were further pollinated with rye pollen. From each spike 3 to 10 day post-fertilized ovaries were excised, treated for silver staining and dissected to pulled apart young embryos and endosperms [14].

5-Azacytidine treatments. (I) - During seed germination: seeds of several triticale cultivars were germinated in the presence of a 10 µg/ml 5-AC solution, following the standard procedures of nucleolar activity evaluation by silver staining and methylation pattern analysis by Southern hybridization technique. (II) - During embryogenesis: after controlled pollination a 10µl/ml 5-azacytidine (5-AC) solution was injected in each maturing seed for three consecutive days at three distinct stages: (i) - injections at the 2nd, 3rd and 4th days after fertilization; (ii) - injections at the 5th, 6th and 7th days; (iii) - injections at the 8th, 9th and 10th days. In the resulting seeds the nucleolar activity was evaluated in root tip cells by silver staining [15].

Analysis of methylation patterns of intergenic sequences from rDNA loci. The methylation pattern of intergenic sequences from rDNA *loci* was evaluated by Southern transfer and hybridization using the ECL system (Amersham). DNAs from control and 5-AC treated triticale plants [15] were digested to completion with Apa I (methylation sensitive) and Dra I. The probes used were pTa71 (wheat rDNA unit, [16]) and pScR4-T1 (rye rDNA intergenic sequence, [6]).

Results and Discussion

Nucleolar activity on complete and 2R/2D substitution triticale cultivars. Through the analysis of the nucleolar activity on somatic cells of four distinct cultivars of hexaploid triticale a marked difference was observed between complete triticale cultivars and those carrying the chromosome substitution 2R/2D (Table 1). In fact considering the frequencies of metaphase cells from root tips with different number of silver positive subterminal NORs (Ag-SAT-NORs) it can be observed that cvs. Juanilho and Drira show 86% of metaphase cells with 4 major satellite NORs belonging to chromosomes 1B and 6B as identified by silver staining and *in situ* hybridization, which was expected as 1R NOR is usually suppressed in wheat backgrounds [1-4].

In triticale cultivars with the chromosome substitution 2R/2D however, one extra pair of Ag-SAT-NORs was observed in most metaphase cells. Through *in situ* hybridization with rye genomic DNA, pTa71 and pSc119.2 probes, the extra pair of major NORs was allocated to chromosomes 1R and it also confirmed the chromosome substitution 2R/2D present in these cultivars.

Table 1 - Frequency of root-tip metaphase cells with different numbers of Ag-SAT-NORs in complete and substituted (2R/2D) triticale cultivars

X *Triticosecale*	Genome constitution	% of metaphase cells with different numbers Ag-NORs on SAT-chromosomes*	
		2-4	5-6
cv. Drira	AABBRR	94	6
cv. Juanilho	6x=2n=42	96	4
cv. Bacum	AABBRR(2R)2D	4	96
cv. Rosner	6x=2n=42	12	88

* SAT-chromosomes: 1B, 6B and 1R

An identical pattern of 1R rDNA expression in both substitution triticales is a strong indication that the expression of ribosomal genes from rye origin, usually silent in complete triticales, is mainly mediated by the absence of chromosome 2R. A clear influence of the non-nucleolar rye chromosome 2 on the process responsible for the establishment of the epigenetic pattern of 1R NOR inactivation in wheat backgrounds is therefore demonstrated. Any potential effect of the presence of 2D can not be considered, as in octoploid triticales the common inactivation of 1R rDNA *loci* is observed [5].

When is established the 1R NOR inactivation pattern? The identification of the developmental stage during which occurs the establishment of the 1R NOR inactivation pattern in wheat-rye F1 hybrids occurs, was evaluated through the analysis of embryos and endosperms from the 3rd to the 7th day after fertilization (Table 2, [14]). At the 3rd and 4th days after fertilization all metaphase cells from embryos had three distinct subterminal Ag-NORs, allocated on 1B, 6B and 1R chromosomes. From the 5th to the 7th days after fertilization only two Ag-NORs could however be observed on metaphase cells of all embryos analysed. An analogous reduction was observed on metaphase cells from endosperms since until the 4th after fertilization 95% of the cells presented five Ag-NORs, contrasting with the four Ag-NORs observed on most of the cells (96%) after that stage.

The frequencies of metaphase cells with different numbers of silver staining SAT-NORs scored in both structures demonstrated that the 1R NOR inactivation pattern is simultaneously established in those structures, between the 4th and the 5th days after fertilization when 30-45 cells were present in embryos and 1000-3000 cells in the endosperms.

Table 2: 1R rDNA *loci* expression patterns in developing seeds of wheat-rye F$_1$ hybrids.

Seed structures	n° of days after fertilization				
	3rd	4th	5th	6th	7th
Embryo (ABDR)	ACTIVE **3** Ag-SAT-NORs		INACTIVE **2** Ag-SAT-NORs		
Endosperm (AABBDDR)	ACTIVE **5** Ag-SAT-NORs		INACTIVE **4** Ag-SAT-NORs		

The commitment of de novo methylases to specific developmental stages. The comparative analysis of rye rDNA patterns of cytosine methylation between control and 5-azacytidine (5AC) germinated triticale plants revealed an additional ApaI/DraI fragment from rye origin only on the DNA from the 5AC treated plants [15]. That molecular analysis confirmed the correlation previously suggested [4, 8] between methylation patterns of ribosomal intergenic sequences and the rye NOR expression patterns observed by silver staining.

The developmental dependence of wheat nucleolar dominance was further confirmed through the analysis of nucleolar activity on cells of root tips from triticale seeds injected with 5AC in three distinct stages (Table 3): on cells from root tips of seeds injected from the 2nd-4th days or 5-7th days after fertilization the usual pattern of 1R NOR inactivation was observed since 96% of cells presented four Ag-SAT-NORs; this contrasts with the analysis of root tip cells from seeds injected from the 8-10th days after fertilization, where 68% of them had more than four Ag-SAT-NORs.

Table 3 - Frequency of metaphase cells with different numbers of Ag-SAT-NORs in root tips of triticale cv. Juanilho from seeds untreated or injected with 5AC during distinct periods after fertilization.

Injection of 5-AC after fertilization	% of metaphase cells with different numbers Ag-SAT-NORs*	
	2-4	5-6
untreated	96	4
2, 3 and 4th days	96	4
5, 6 and 7th days	96	4
8, 9 and 10th days	32	68

* SAT-chromosomes 1B, 6B and 1R

Modifications of DNA methylation patterns by the incorporation of 5AC should then be induced after the 7th day of fertilization in order to be maintained until later developmental stages. The *de novo* methylases responsible for the establishment of the epigenetic 1R rDNA inactivation pattern should therefore be present or active only until that stage.

Meiotic reprogramming of 1R rDNA loci.   The comparative study of nucleolar activity between somatic root tip cells and pollen mother cells of complete triticale cultivars confirmed that the inactivation pattern of 1R rDNA *loci* established until the 7th day after fertilization is further inherited since in both types of cells only wheat NORs are usualy active.

The analysis of post-meiotic cells demonstrated however that wheat nucleolar dominance is erased during meiosis (Table 4): in the first post-meiotic mitosis three SAT-NORs were clearly observed on at least 96% of the metaphase cells analysed demonstrating the consistent expression of 1B, 6B and 1R rDNA *loci*.

Table 4 - Frequencies of pollen mother cells and microspores of complete triticale cultivars with 1R NOR active.

Types of cells	Number of Ag-SAT-NORs*	Cultivars	
		Drira	Juanilho
Pollen mother cells (AABBRR)	5-6	5%	7%
Microspores (ABR)	3	100%	96%

* SAT-chromosomes: 1B, 6B and 1R

Conclusions

The importance of rye chromosome 2 on the establishment of the process of wheat nucleolar dominance in triticale can be interpreted as the mapping of gene(s) of any of the suggested limiting factors essential for rDNA transcription [17] or of other also unknown controls than the ones underlying the competition phenomenon.

Our results indeed seem to postulate new controls, hierachically higher than the competition of rDNA *loci* for limiting factors essential for transcription, strongly suggested by the following aspects: first, the induced erasing of the 1R NOR suppression pattern, through the incorporation of 5-AC after the 7th day of fertilization and its maintenance in further stages of plant development, clearly demonstrates that any competition phenomenon underlying that process is restricted to that specific developmental stage. On the other side, the normal reversion during meiosis of the epigenetic 1R NOR inactivation pattern, maintained until the pre-meiotic interphase, should be associated to a programmed demethylation of the spacer sequences previously responsible for the preferential inactivation of those *loci* in somatic tissues.

The above mentionated controls of wheat nucleolar dominance in triticale, discloses, thus far, that whatever the molecular basis for the supression of 1R rDNA *loci* should be, it should also include controls mediated by non-nucleolar chromosomes and by developmental dependent processes, which can probably be influenced by changes in chromatin structure and/or modification of cromatin spatial domains. This closer knowledge on new controls of wheat nucleolar dominance in triticale will certainly contribute for the development of better strategies for genetic or epigenetic manipulations in other analogous systems.

Acknowledgements: Financial support was provided by Junta Nacional de Investigação Científica e Tecnológica (JNICT), Portugal.

References

1. Orellana J, Santos JL, Lacadena JR, Cermeño, MC. Nucleolar competition analysis in *Aegilops ventrucosa* and its amphiploids with tetraploid wheats and diploi rye by the silver-staining procedure. Can J Genet Cytol 1983, 26, 34-9.

2. Lacadena JR, Cermeño MC, Orellana J, Santos JL. Evidence for wheat-rye nucleolar competition (amphiplasty) in triticale by silver staining procedure. Theor Appl Genet 1984a; 67:207-13.

3. Lacadena JR, Cermeño MC, Orellana J, Santos JL. Analysis of nucleolar activity in *Agropyron elongatum*, its amphiploid with *Triticum aestivum* and the chromosome addition lines. Theor Appl Genet 1984b;68:75-80.

4. Vieira R, Queiroz A, Morais L, Barão A, Mello-Sampayo T, Viegas W. 1R chromosome nucleolus organizer region activation by 5-azacytidine in wheat x rye hybrids. Genome 1990;33:707-12.

5. Cermeño MC, Orelana J, Santos JL, Lacadena JR. Nucleolar organizer activity in wheat, rye and derivatives analysed by a silver staining procedure. Chromosome 1984; 89:370-6.

6. Appels R, Moran LB, Gustafson JP. The structure of DNA from the rye (*Secale cereale*) NOR R1 locus and its behaviour in wheat backgrounds. Can J Genet Cytol 1986; 28:673-85.

7. Gustafson JP, Dera AR, Petrovic S. Expression of modified rye ribosomal genes in wheat. Proc Natl Acad Sci USA 1988; 85:3943-5.

8. Flavell RB, O'Dell MP, Vicente M, Sardana R., Barker RF. Phil Trans R Soc London 1986, ser B, 314,385-97.

9. Moss T. A transcriptional function for the repetitive ribosomal species in *Xenopus laevis*. Nature 1983, 302, 223-8.

134

10. Reeder RH, Roah JG, Dunaway M. Spacer regulation of *Xenopus* ribosomal gene transcription: competition in oocytes. Cell 1983, 35, 449-56.
11. Silva M, Queiroz A, Neves N, Barão A, Castilho A, Morais-Cecílio L., Viegas W. Reprogramming of rye rDNA in triticale during microsporogenesis. Chrom. Res. 1995; 3 (in press).
12. Leitch AR, Scharzacher T, Jackson D, Leitch IJ. A pratical guide to *in situ* hybridization. 1993. in press.
13. Schwarzacher T, Abanthawat.Jonsson K, Harrison GE, Islam AKMR, Jia JZ, King IP, Leitch AR, Miller TE, Reader SM, Rogers WJ, Shi M, Heslop-Harrison JS. Genomic *in situ* hybridization to identify alien chromosomes and chromosome segments in wheat. Theor Appl Genet 1992a; 84:778-86.
14. Castilho A., Queiroz A., Neves N., Barão A., Silva M., Viegas W. The developmental stage of inactivation of rye origin rRNA genes in the embryo and endosperm of wheat x rye F1 hybrids. Chrom. Res. 1995; 3:169-174.
15. Neves N., Heslop-Harrison J.S., Viegas W. rRNA gene activity and control of expression mediated by methylation and imprinting during embryo development in wheat x rye hybrids. Theor. Appl. Genet. 1995; 91: 529-533.
16. Gerlach WL, Bedbrook JR. Cloning and characterization of ribosomal RNA genes from wheat and barley. Nucl Acid Res 1979; 7:1869-85.
17. Reeder RH. Mechanisms of nucleolar dominance in animal and in plants. J Cell Sci 1985, 101, 201-6 326.

THE EXPRESSION OF THE NUCLEOLUS ORGANIZER REGIONS (NORs) IN BREAD WHEAT X RYE HYBRID LINES CARRYING A 1RS/1BL CHROMOSOME TRANSLOCATION

Silva, Joaquim P.[1]; Vieira, Rita[2]; Jouve, Nicolas[3] and Mello-Sampayo, Tristão[4]

[1]University of Beira Interior, Dep. of Chemistry, Covilhã, Portugal; [2]University of Coimbra, Inst. of Botany, Coimbra, Portugal; [3]University of Alcalá de Henares, Dep. of Cell Biology and Genetics, Alcalá de Henares (Madrid), Spain; [4]Estação Nacional de Melhoramento de Plantas, Elvas, Portugal

ABSTRACT

In wheat-rye hybrids NOR activity is regulated by genes carried in both genomes which interact. The amphiplasty is a well known example of this phenomenon which causes inactivation of 1R NORs in triticale and in 1R(1B) rye-wheat substitution lines [3].

It was found that the inactivity of 1RS can be avoided in cases in which 1RL is missing such as in 1RS wheat addition lines and 1BL/1RS translocation lines or under the effect of 5-azacytidine, a demethylating agent that was employed in certain cases on root tip cells [8]. The absence of 1BS and the interaction of genes located between the 1BL and 1RS arms are also of pertinence. However, the 1RS NOR activity in wheat containing the 1BL/1RS translocation and also carrying the 1R(1D) substitutions poses several problems which will be discussed. In a 2n=42 wheat plant which was also a 1BL/1RS (1B) and 1R(1D) double disomic substitution line, NORs of 6B, 1BL/1RS and 1R were simultaneously active.

INTRODUCTION

Plants carry large numbers of ribosomal genes (rDNA) that are repeated from one or few hundreds to several thousands at the nucleolar organizing regions (NORs). Chromosomes 1B and 6B are responsible for the production of the major nucleoli

H. Guedes-Pinto et al. (eds.), Triticale: Today and Tomorrow, 135–139.
© 1996 *Kluwer Academic Publishers. Printed in the Netherlands.*

while chromosomes 1A and 5D show only a minor nucleolar activity. In triticale, rye NOR activity has been reported to be visually undetectable [1]. This was interpreted as nucleolar competition (amphiplasty) leading to the preferential inactivation of rDNA genes in one of the parental genomes, which was induced by the presence of genomes of different species[2]. This amphiplasty effect seems to result from the interaction between wheat and rye genes involving the long and short arms of rye and wheat chromosomes of the homeologous group 1. As a matter of fact, there is no amphiplasty in the absence of the long arm of the 1R chromosome, as described for some 1AL/1RS, 1BL/1RS and 1DL/1RS translocation lines and in hexaploid monotelosomic 1RS addition line (2n=43) [3]. On the contrary consistent amphiplasty exists in 1R(1B) disomic substitution lines such as hexaploid "Zorba" line [3].

In this work the interaction between wheat and rye genomes in three lines of wheat-rye hybrids was studied. The results have confirmed so far the existence of a specific amphiplasty elimination in 1BL/1RS translocated chromosomes.

Further results disclose the evidence of an extended influence of the 1BL/1RS translocated chromosomes. It was our purpose to reveal these effects for the first time and in more detail in the following description.

MATERIALS AND METHODS

In this work we used the following wheat-rye hybrid lines: 1) Anza/S149, a 1R(1D) substitution line [4] derived from a cross involving the hexaploid wheat "Anza" and the triticale 6A250; 2) Glennson 81, a hexaploid wheat from CIMMYT (México) containing a 1BL/1RS translocation and 3) 7841-34-4, a F_2 hybrid line between Glennson 81 and Anza/S149, containing the 1BL/1RS translocation and the 1R(1D) substitution from their parents [5,6].

Seeds were germinated in the dark on filter paper soaked either with the tap water or with a 0.1 µg/ml solution of 5-Azacytidine in Petri dishes at 20 ± 1 ^{o} C for 2-3 days. The filter paper solutions were replaced daily. Root tips (1 cm long) were detached and immersed for 4 hours in a saturated solution of 1-bromonaphtalene at 20 ± 1 ^{o}C, to maximize the mitotic index and they were fixed in FAA (a solution of 37% formaldehyde-glacial acetic acid - 50% ethanol, 1:1:18) for 2-3 days at 4 ^{o}C . 5-Azacytidine treatment as referred by [7], were performed in order to confirm the hypomethylation effect in 1RS NOR eventual inactivation in 1R(1D) chromosome substitutions.

Silver nitrate impregnation technique was used to selectively stain NORs in metaphase. Fixed root tips were immersed in a 2% aqueous solution of $AgNO_3$ at 60 ^{o}C, overnight. They were developed in a solution of 10% formaldehyde: 1% hydroquinone (1:1) for 3-5 minutes and then squashed in 45% acetic acid.

The major NORs we refer to are located at the short arms of chromosomes 6B, 1B and 1R.

RESULTS AND DISCUSSION

The number of Ag-NORs observed per cell in water was 4 Ag-NORs for Anza/S149 (Table I), as expected since in the presence of chromosomes 1B and 6B, the NOR of 1RS in the substitutive 1R(1D) chromosomes was inactive [3]. With 5-azacytidine the 1RS NOR become active because of the hypomethylation effect on the amphiplasty [8]. In its turn 6 NOR were expressed in 7841-34-4 (Table I and Figure 1) suggesting an inhibition of amphiplasty in 1R. As a matter of fact the expression of 1RS NOR was not suppressed either in the 1BL/1RS translocation or in the alien 1R complete chromosomes as described by [9].This suggests that the NOR recovery effect in 1BL/1RS chromosomes when 1B was absent became extensive to the 1RS NOR in the complete 1R substitutive chromosomes. That is why 7841-34-4 shows normally 6 major NORs at metaphase when 1B chromosomes were absent. In conclusion the three major Ag-NORs that were seen as expressing their activity at the mitotic metaphase of the 7841-34-4 plant must be those of chromosomes 6B, translocated 1BL/1RS and 1R, the latter two of which were the replacements for the chromosomes 1B and 1D, which were absent in the plant.

It was therefore concluded that the NOR carried by the complete chromosome 1R became activated due to the presence of the translocated chromosome 1BL/1RS, which has itself an active NOR in 1RS.

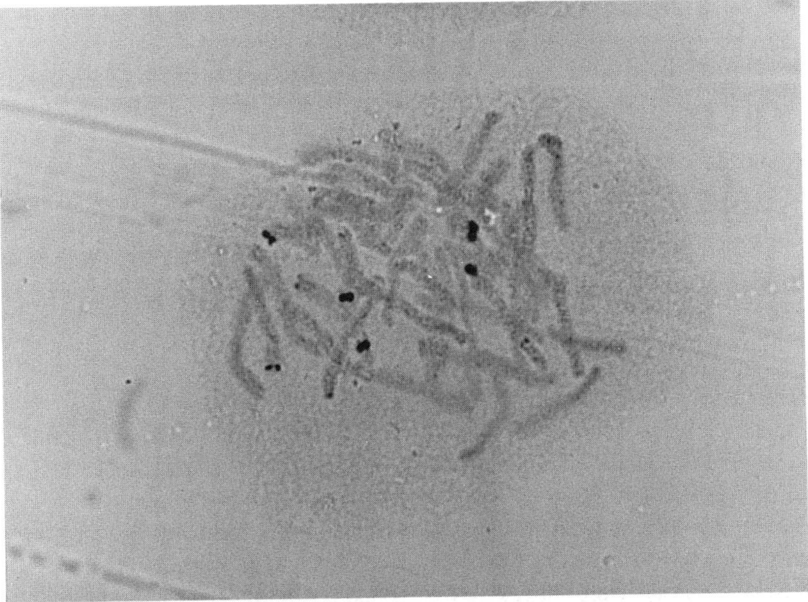

Figure 1 - Visualization of Ag-NORs in C-metaphase by silver nitrate impregnation in a 7841-34-4 (19"+(1RS/1BL)"+1R"-1B"-1D") plant with no 5 - Azacytidine treatment (6 major active NORs).

WITHOUT 5 - AZACYTIDINE

Line	Genotype (2n=42=21")	N. of NORs expected per cell	N. of Ag-NORs observed per cell	N. of metaphase cells	N. of interphase cells with 1-7 nucleolus (%) *							N. of interphase cells
					1	2	3	4	5	6	7	
Glennson 81	20"+(1RS/1BL)"-1B"	4 (6B, 1RS)	4	648	30.8	22.2	20.9	11.8	7.1	6.3	0.9	2593
Anza/S 149	20"+1R"-1D"	6 (1B, 6B,1R)	4	640	29.2	22.5	22.1	12.6	7.3	5.6	0.7	2573
7841-34-4	19"+(1RS/1BL)"+1R"-1B"-1D"	6 (6B, 1RS,1R)	6	730	24.9	18.6	18.4	16.7	13.4	6.9	0.3	2927

WITH 5 - AZACYTIDINE

Line	Genotype (2n=42=21")	N. of NORs expected per cell	N. of Ag-NORs observed per cell	N. of metaphase cells	N. of interphase cells with 1-8 nucleolus (%) *								N. of interphase cells
					1	2	3	4	5	6	7	8	
Glennson 81	20"+(1RS/1BL)"-1B"	4 (6B, 1RS)	4	532	36.6	23.5	22.5	14.3	2.0	0.6	0.5	-	2128
Anza/S 149	20"+1R"-1D"	6 (1B, 6B, 1R)	6	702	28.6	18.4	17.8	14.5	11.1	7.2	1.5	0.9	2809
7841-34-4	19"+(1RS/1BL)"+1R"-1B"-1D"	6 (6B, 1RS, 1R)	6	680	28.8	18.1	18.2	14.5	11.1	7.1	1.4	0.8	2756

* The number of nucleoli in interphase cell could not correspond exactly to the number of visibly active NORs as seen in metaphase because several nucleoli could be included in a single NOR.

Table I - Number of Ag-NORs expected and observed, and number of nucleoli observed in metaphase and interphase cells of root tips, respectively, for seeds germinated in either water or 5-azacytidine solution.

Acknowledgements. The authors are greatly indebted to Dr. Oscar A. Sequeira for his frendly help in making possible the functioning of our lab into the installations of the Department of Phytopathology of the Natl. Agricultural Station and to Fernando Cardoso from INETI for his colaboration on the typing of the manuscript.

REFERENCES

1. Lacadena J R, Cermeño M C, Orellana J, Santos J L. Evidence for wheat rye nucleolar competition (amphiplasty) in triticale by silver-staining procedure. Theor Appl Genet 1984; 67:207-213.

2. Rieger R, Nicoloff H, Anastossova-Kristeva M. Nucleolar dominance in interspecific hybrids and translocation lines - a review. Biol Zentralbl 1979; 48:385-398.

3. Vieira R, Queiroz A, Morais L "et al". Genetic control of 1R nucleolus organizer region expression in the presence of wheat genomes. Genome 1990b; 33:713-718.

4. Gustafson J P. Evaluation of a 1R(1D) substitution in wheat. Proc 7th Int Wheat Genet Symp, Institute of Plant Science Research. Cambridge. U K 1988: 193-196.

5. Gustafson J P, Bittel D C. An effective triticale manipulation system Problems and uses. Proc 2nd Int Triticale Symp Passo Fundo, Brasil 1990.

6. Bittel D C, Fominaya A, Jouve N, Gustafson J P. Dosage response of rye genes in a wheat background I. Selection of lines with multiple copies of 1RS. Plant Breeding 1992; 108:283-289.

7. Flavel R B, O'dell M, Thompson N F. Cytosine methylation of ribosomal RNA genes and nucleolus organizer activity in wheat. Kew Chromosome Conference II George Allen and Unwin, 1983: 11-17.

8. Vieira R, Queiroz A, Morais L "et al". 1R chromosome nucleolus organizer region activation by 5-Azacytidine in wheat x rye hybrids. Genome 1990a; 33: 707-712.

9. Gustafson, J.P. Flavell, R.B. Control of nucleolar expression in triticale.Abst. III International Triticale Symposium (Eucarpia- International Triticale Association), Lisbon, Portugal, 1994.

CAN BREAD-MAKING QUALITY BE INTRODUCED INTO HEXAPLOID TRITICALE BY WHOLE-CHROMOSOME MANIPULATION?

Ebrahim M. Kazman[1] and Tamàs Lelley[2]

[1]Institute of Agronomy and Plant Breeding, Georg-August-University, D-37075 Göttingen, Germany,

[2]Institute of Agronomy and Plant Breeding, University of Agriculture, A-1180 Vienna, Austria

Abstract

Synthetic 6x triticale lines with various D-genome chromosomes, derived from octoploid x tetraploid triticale crosses were grown in the field and analyzed for characters related to bread-making quality and yield. Considerable variability was found for each character. Grain protein content of the majority of lines was markedly higher than that of the wheat and triticale cultivars included in the experiments as controls. Grain yield of some of the lines was as high as that of the controls. Sedimentation volumes ranged from 33 to 103% of the control wheat cultivar exhibiting medium bread-making quality. No correlation was found between sedimentation volume and protein content, and between sedimentation volume and grain yield. The presence of chromosome 1D, in combination with 1B, invariably increased the sedimentation volume. Ranking for positive effect of homoeologous group 1 chromosomes on sedimentation was 1D>1B>1R>1A. These preliminary data suggest that through the introduction of chromosome 1D, especially as a 1D(1A) substitution, the bread-making quality of hexaploid triticale can be improved considerably, approaching the level of wheat. Effects of different alleles at *Glu-D1* locus are discussed.

Introduction

Despite of the relatively good agronomic performance of hexaploid triticale, its production area is only very slowly expanding. One of the main reasons for this slow increase is its limited use for human consumption, which can be improved if bread-making quality could be introduced into this cereal.

In wheat, bread-making quality to a great extent is determined by the amount and quality of its gluten. Several genes are involved in this complex trait. The allelic variation at *Glu-1* locus, which is widely studied, is often reported to be associated with bread-making quality. Most important was the

H. Guedes-Pinto et al. (eds.), Triticale: Today and Tomorrow, 141–148.
© 1996 *Kluwer Academic Publishers. Printed in the Netherlands.*

allelic variation at the *Glu-D1* locus on chromosome 1D [1,2,3,4]. Rogers et al. [5] showed that an increase in dosage of 1D drastically improved the bread-making quality in 'Chinese Spring'.

To utilize the D-genome gene-pool for the improvement of 6x triticale, different methods have been described. Introducing D-genome chromosomes into 6x triticale while keeping its rye genome intact, is probably the best way to combine the bread-making quality with the hardiness of rye. This can be achieved by substitution of the D-genome chromosomes for their homoeologues of the A- and/or B-genomes [6,7]. Octoploid x tetraploid triticale crosses, first formulated by Krolow [8], appear to be the simplest way of creating such substitutions [9,10]. One of its most likely limitations, i.e. the requirement for several selfing generations to achieve karyotype stabilization, can be overcome by backcrossing the F_1 hybrid to the 8x triticale parent followed by selfing. By using this crossing scheme, we produced a large number of stabilized 6x triticale lines containing a complete rye genome and various D-chromosomes [11]. The objective of the present study was to investigate the potential of such hexaploid forms for the improvement of bread-making quality in triticale.

Materials and Methods

Hexaploid triticale lines containing various D-genome chromosome substitutions for their A-and/or B-genome homoeologues (synthetic 6x triticale) were developed from octoploid x tetraploid triticale crosses. Their production and karyotype analyses have been described earlier [11]. In 1992/1993, seed of 280 lines in BC_1F_5 and F_6 generations, were directly sown in the field in Göttingen, Germany. Each line was represented by a double-row with two replications. In each row, 20 seeds were sown. Two 6x triticale cultivars and three wheat cultivars were also included in the trial as controls. The triticale cultivars consisted of the French cultivar, 'Clercal', which has been the parent of the 4x triticale used in the production of the synthetic 6x triticales, and the polish cultivar, 'Lasko'. The following wheat cultivars were used:
a) 'Kolibri'- spring wheat with medium to high bread-making quality (A6).
   This cultivar was originally used for the production of one of the 8x
   triticales, the other parent of most of the present synthetic 6x lines.
b) 'Greif' - winter wheat with medium bread-making quality (B4)
c) 'Bussard' - winter wheat with the highest bread-making quality (A9)

Of the synthetic triticale lines investigated, 61 were known for their karyotype and 89 were known for their allelic composition encoding HMW glutenin subunits. Due to the small amount of seed harvested, the following analyses were carried out for the quality assessment: A random of 132 lines were analyzed for their gluten quality using Zeleny-sedimentation test. Grain protein content was measured using Near Infrared Reflectance Spectroscopy (NIRS). In addition, the protein content of 85 lines was determined as N x 5.7, using Dumas digestion. Since a high correlation was found between both methods (r = 0.94), only the data of NIRS are presented. Duplicate determinations were made for each flour sample and analysis.

Results and Discussion

The frequency distribution of lines with different protein contents and sedimentation volumes is shown in Fig. 1a,b. Grain protein ranged from 9.6 to 17.5%. The majority of the synthetic triticale lines showed a markedly higher protein content than the wheat and triticale cultivars included in the experiments as controls, even in those cases where their grain yield surpassed those of the wheat cultivars (data not shown). The relative grain yield ranged from 0.9 to 186.9% when compared to that of the triticale cultivar 'Clercal' (Fig. 2a).

While the two triticale cultivars showed very low sedimentation volumes, several synthetic triticale lines reached a value similar to two of the wheat cultivars (Fig. 1a and Table 3). The sedimentation volumes ranged from 15 to 47ml, i. e. 32.9 to 103.3% of the volume of the medium-quality wheat 'Kolibri'.

No correlation did exist between sedimentation volume and protein content, and between the sedimentation volume and grain yield (Fig. 2a,b). The results suggest that the large sedimentation volume in some of the synthetic 6x triticale lines resulted from the qualitative changes in the protein composition, caused by specific chromosome/allele. Similar results have been reported earlier by Kolster et al. [4], Rogers et al. [5] and Carillo et al. [12].

Regardless of the rest of the karyotype, chromosome 1D significantly increased the sedimentation volume (Table 1). However, this positive effect of chromosome 1D on sedimentation was obvious only if chromosome 1B was present. This was found by analyzing a complete set of lines in which D-chromosome substitutions and additions were present in the homoeologous group 1 (Table 2). The loss of chromosome 1B in a 1D(1B) substitution had an adverse effect on sedimentation volumes, resulting in very low values. Our results agree with those reported by Rogers et al. [5]. They found that an increase in dosage of chromosome 1D positively changed the bread-making quality of 'Chinese Spring', whereas the loss of chromosome 1B had a negative effect. Such interaction between *Glu-B1* on chromosome 1B and *Glu-D1* on chromosome 1D is also reported by Odenbach and Mahgoub [2].

Table 1 Mean effect of chromosome 1D on sedimentation volume (ST) and grain protein content (%) of 74 synthetic 6x triticale lines, known for the presence or absence of chromosome 1D

Karyotype	No. of lines	ST (ml)	Protein content (%)
with disomic 1D	38	38.7[a]	14.2
with monosomic 1D	10	35.0[a]	14.0
without 1D	26	29.3[b]	14.4

Values with different letters are significant at p = 0.05

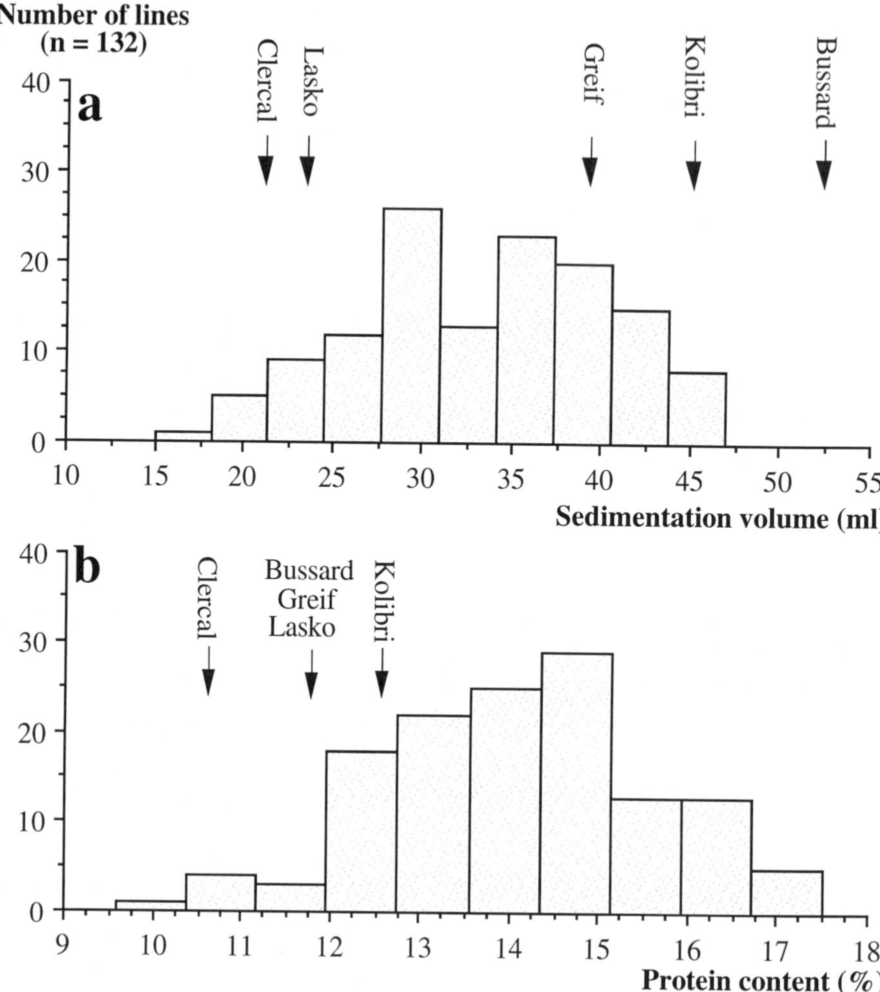

Figure 1 Frequencies of synthetic 6x triticale lines for (a) sedimentation volume and (b) protein content. Arrowheads point to the values measured for the wheat cultivars 'Greif', 'Kolibri' and 'Bussard' and for the triticale cultivars 'Clercal' and 'Lasko'

The 6x-synthetic triticale lines with the largest sedimentation volumes are listed in Table 3. Most of them carry a 1D(1A) substitution. For four lines the karyotype is not known and their SDS-PAGE, so far, did not reveal the subunits encoded by *Glu-D1*. These lines are being further investigated for a possible translocation or variation in gliadins and/or low-molecular weight glutenins. The disomic 1D addition lines, which ranged among the most fertile and highest yielding ones (data not shown), also showed a high sedimentation volume comparable to the wheat cultivar 'Greif'. Surprisingly, lines with 1D(1R) substitution realized relatively low sedimentation volumes, lower than that of the 1D(1A) substitution lines (Table 2). Accordingly, in the present study the ranking of the homoeologous group 1 chromosomes for positive effect on sedimentation volume in Triticale was 1D>1B>1R>1A.

Table 2 Sedimentation volume (ST) of a set of synthetic 6x triticale lines with single 1D(1A/1B/1R) disomic substitutions, disomic 1D addition (+1D") and lines with no D chromosomes (Null-D)

Karyotype	No. of lines	ST (ml) Mean	ST (ml) Min. - Max.
1D(1A)	4	41.5	37.5 - 43.5
1D(1B)	2	28.8	27.0 - 30.5
1D(1R)	2	34.5	34.0 - 35.0
+1D"	4	37.4	36.0 - 38.5
Null-D	7	29.1	26.0 - 34.5

Table 3 Chromosome constitution and HMW glutenin alleles at *Glu-B1* and *Glu-D1* of some synthetic 6x triticale lines with a sedimentation volume (ST) similar to that of the two control wheat cultivars, 'Greif' and 'Kolibri'

Karyotype	No. of lines	HMW-Glutenin allele *Glu-B1*	HMW-Glutenin allele *Glu-D1*	ST (ml) Min. - Max.
1D(1A)	3	7+8	2+12	42.0 - 43.5
1D(1A),2D(2B),3D(3A)	2	7+8	2+12	40.5 - 43.0
1D(1A),2D(2B),4D(4A),5D(5A)	2	7+8	2+12	43.0 - 44.5
1D(1A),2D(2B),6D6A	1	7+8	2+12	46.5
1D(1A),3D(3A)	2	7+8	2+12	41.5 - 41.5
1D(1A),3D(3A),4D(4B),7D(7B)	1	7+8	2+12	44.0
1D(1A),5D(5A)	2	7+9	5+10	42.0 - 46.0
Disomic 1D addition	1	7+8	2+12	38.5
not known	5	7+8	2+12	40.0 - 43.5
not known	2	7+9	5+10	44.0 - 46.5
not known	4	7+8	x	38.5 - 47.0
Controls				
'Clercal'		7+8	x	21.0
'Greif' (B4)		7	2+12	39.0
'Kolibri' (A6)		7+9	5+10	45.5
'Bussard' (A9)		7+9	5+10	52.5

x = no subunits present

146

Payne et al. [1], Odenbach and Mahgoub [2] Moonen et al. [13], and Rogers et al. [5], found more positive association of bread-making quality with the $Glu\text{-}D1_{5+10}$ allele on chromosome 1D than with its counterpart, $Glu\text{-}D1_{2+12}$. Data on the sedimentation volumes of 10 of the synthetic triticale lines carrying $Glu\text{-}D1_{5+10}$ - four of them are presented in Table 3 - showed no significant increase in the sedimentation volumes over the lines carrying $Glu\text{-}D1_{2+12}$ allele.

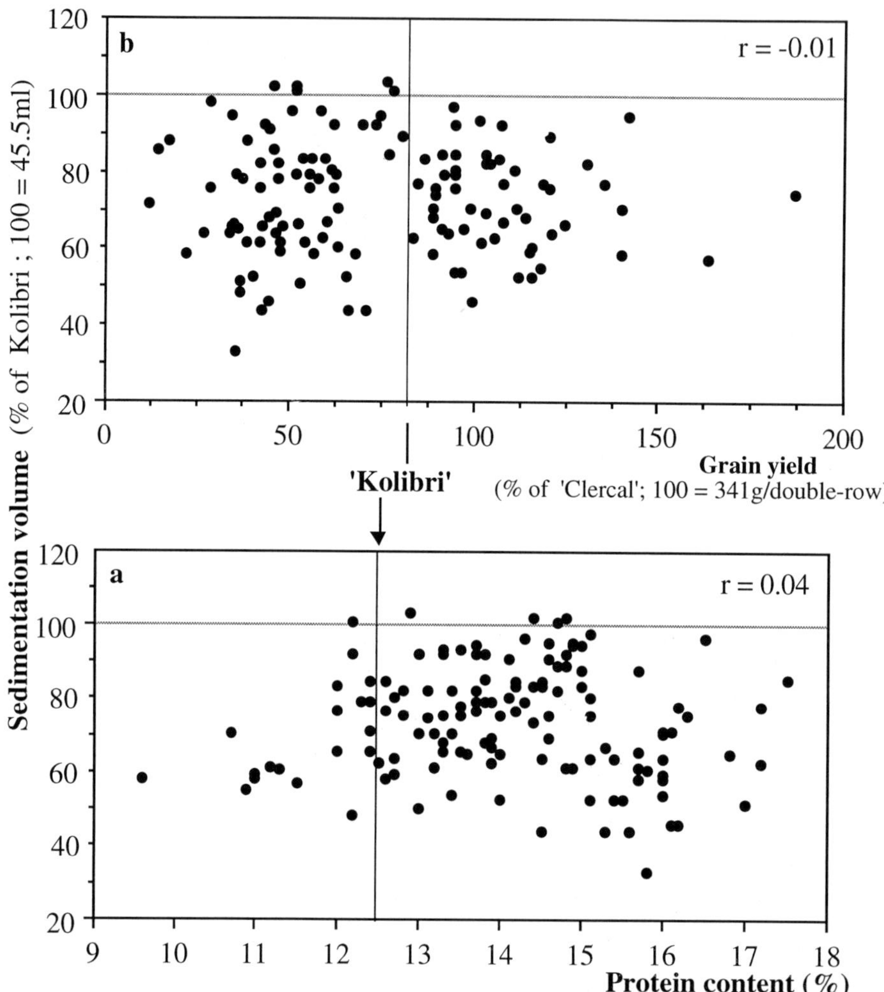

Figure 2 Relationship between (a) sedimentation volume (expressed as a percentage of the volume of the wheat cultivar Kolibri) and grain yield (expressed as a percentage of the triticale cultivar Clercal); (b) sedimentation volume and protein content (%). A sample of 132 synthetic 6x triticale lines was considered.

Conclusion

The results of this study have revealed a significant association of chromosome 1D, especially as a 1D(1A) substitution, to the sedimentation volume in 6x triticale. No differences in sedimentation volume were found between triticale lines carrying the alleles $Glu\text{-}D1_{5+10}$ or $Glu\text{-}D1_{2+12}$. Lines with a high sedimentation volume were shown to be high yielding and high in protein content, as well. Thus the inclusion of the whole chromosome 1D in breeding programs should have a positive impact on triticale. By crossing 1D(1A) substitution lines to an existing 6x triticale cultivar and using SDS-PAGE for screening, chromosome 1D can easily be transferred and incorporated in further, probably better backgrounds. On the other hand such lines can be crossed to octoploid triticales - carrying a desirable wheat component - to create allelic variation, especially at the $Glu\text{-}D1$ locus. Their original karyotype can be reconstructed by proper screening.

Acknowledgements

Special appreciation is extended to Dr. K. Brunckhorst of the F. von Lochow-Petkus GmbH for conducting the quality analyses. We also thank Prof. Dr. G. Röbbelen for his support and Dr. D. Stelling for his critical comments on the manuscript. This research was financially supported by the Deutsche Forschungsgemeinschaft.

References

1. Payne PI, Nightingale MA, Krattiger AF, Holt LM. The relationship between HMW glutenin subunit composition and the bread making quality of British-grown wheat varieties. J Sci Food Agric 1987;40:51-65.

2   Odenbach W, Mahgoub ELS. Relationship between HMW glutenin subunit composition and the sedimentation value in reciprocal sets of inbred backcross lines derived from two winter wheat crosses. Miller TE, Koebner RMD, editors. Proc. 7th Int. Wheat Genetics Symposium; Cambridge, England 1988; 987-991.

3   Rogers WJ, Payne PI, Harinder K. The HMW glutenin subunit and gliadin compositions of German-grown wheat varieties and their relationship with bread-making quality. Plant Breeding 1989;103:89-100.

4   Kolster P, van Eeuwijk FA , van Gelder WMJ. Additive and epistatic effects of allelic variation at the high molecular weight glutenin subunit loci in determining the bread-making quality of breeding lines of wheat. Euphytica 1991;55:277-285.

5   Rogers WJ, Law CN, Sayers EJ. Dosage effects of homoeologous group 1 chromosomes upon the bread-making quality of hexaploid wheat. Miller TE, Koebner RMD, editors. Proc. 7th Int. Wheat Genetics Symposium; Cambridge, England 1988;1003-1008.

6.   Larter EN, Noda K. Some characteristics of hexaploid triticale substitution lines involving the A-, B-, and D-genome chromosomes of wheat. Can J Genet Cytol 1981;23:679-689.

7.   Hohmann U. Direct use of hexaploid wheat in the production of recombined hexaploid triticale. Miller TE, Koebner RMD, editors. Proc. 7th Int. Wheat Genetics Symposium; Cambridge, England, 1988;303-308.

8.   Krolow K-D. 4x triticale production and use in triticale breeding. Proc. 4th Int. Wheat Genetics Symposium, Missouri Agricultural Experiment Station, Columbia, Missouri 1973;691-696.

9.   Bernard M, Gay G, Saigne B. Study of the fertility and chromosome behaviour of 3 successive generations obtained following crosses between octoploid and tetraploid triticale. In: Bernard M, Bernard S, editors. Genetic and breeding of triticale. Proc 3rd Eucarpia Meeting on Triticale, Clermont-Ferrand, France 1985;245-257.

10. Lukaszewski AJ, Apolinarska B, Gustafson JP. Introduction of the D-genome chromosomes from a bread wheat into hexaploid triticale with complete rye genome. Genome 1987;29:425-430.

11. Kazman E, Lelley T. Rapid incorporation of D genome chromosomes into A- and/or B genomes of hexaploid triticale. Plant Breeding 1994 (in press).

12. Carrillo JM, Rousset M, Qualset CO, Kasarda DD. Use of recombinant inbred lines of wheat for study of associations of high-molecular-weight glutenin subunit alleles to quantitative traits. I. Grain yield and quality prediction tests. Theor Appl Genet 1990;79:321-330.

13. Moonen JH, Kescheepstra AU, Graveland A. Use of the SDS-sedimentation test and SDS-polyacrylamide electrophoresis for screening breeder's samples of wheat for bread-making quality. Euphytica 1982;31:667-690.

INTERGENOMIC INTERACTION IN WHEAT X RYE HYBRIDS NON-ADDITIVE EXPRESSION OF GLIADINS AND ISOZYMES

Vitor Carvalho and Wanda Viegas
Secção de Genética e Melhoramento Vegetal; Instituto Superioı
de Agronomia, Lisboa, Portugal

ABSTRACT

To successfully introduce new genes in well established cultivars it is important to understand how to overcome genomic interactions that could modify the expected genetic expression in the new background.

When wheat and rye genomes are brought together several authors have pointed out that the suppression of different kind of genes (wheat leaf rust genetic resistance (Quinones *et al.*, 1972) or rye NOR suppression (Lacadena *et al.*, 1984; Viegas *et al.*, 1994)) can vary.

In this study, the non-additive expression of biochemical markers in wheat x rye F_1 hybrids, namely of phosphatase and glucose phosphate isomerase isozymes and of gliadins, using progenitors that do not show polymorphism on the studied traits, is demonstrated
Through isoelectric focusing it was possible to observe that some bands present in one of the progenitors were absent in the hybrid zimograms of alkaline phosphatase as well of glucose phosphate isomerase. The missing bands were, however, present in either the wheat or in the rye zimograms.

Through acidic PAGE of gliadins it was observed also that some bands present in the progenitors were absent in the hybrid zimogram.

INTRODUCTION

Isozyme and endosperm protein genes have been considered as useful genetic markers and they have been utilized for gene pool evaluations, cultivar identification, determination of the presence of alien chromatin, and also as chromosome markers for directed genetic manipulation.

Although they are widely utilized, their usefulness as genetic markers is conditioned by their expected codominant expression. Actually, several authors have pointed out, that the suppression of the expression of different kinds of genes could happen when different genomes are brought together, as happens with wheat leaf rust genetic resistance suppression by the rye genome in a triticale background (Quinones *et al.*, 1972) or with rye NOR

H. Guedes-Pinto et al. (eds.), Triticale: Today and Tomorrow, 149–154.
© 1996 *Kluwer Academic Publishers. Printed in the Netherlands.*

suppression (Darvey and Driscoll, 1972; Orellana *et al.*, 1983; Lacadena *et al.*, 1984; Vieira *et al.*, 1990; Viegas *et al.*, 1994) in wheat x rye F_1 hybrids.

Galili and Feldman (1984) also suggested the existence of controlling genes on the D genome are responsible for suppression of expression of some gliadin genes as tetraploid wheats (genomes AABB) extracted from hexaploid wheat lines (genomes AABBDD) exhibited electrophoretic patterns not present in the hexaploid "progenitor". This phenomenon therefore suggested that intergenomic gene interactions could be involved in the control of the expression of those endosperm proteins also.

In this work it is intended to show the occurrence of non-additive expression of several biochemical genetic markers in wheat x rye F_1 hybrids, namely for alkaline phosphatase and glucose phosphate isomerase isozymes and for gliadins patterns, using progenitors that do not show polymorphism for the studied traits.

MATERIALS AND METHODS

PLANT MATERIAL

Triticum aestivum **L. em Thell. cv. Chinese Spring (CS):** hexaploid wheat cultivar (AABBDD genome).

Secale cereale **L. cv. Centeio do Alto (CA):** rye cultivar (RR genome) with genes for gliadin like proteins located in the $1R^S$ chromosome arm.

CS x CA: F1 hybrid between the previous cultivars.

GLIADIN (GLI) DETECTION (adapted from Bushuk and Zillman, 1978)

Gli proteins were extracted with 70% ethanol from single kernels, and stained with 0,06% Coomassie Brilliant Blue R-250 in 12,5% trichloroacetic acid after electrophoresis carried out in a Phast System apparatus (Pharmacia ™) during 200 Vh, using Phast Gel Gradient 8-25 with discontinuous sodium lactate - glycine buffer system (gel buffer: sodium lactate 10Mm, glicine 10Mm, pH 3.08; electrode buffer: sodium lactate 10mM, glicine 50mM, pH 2.8)

GLUCOSE 6-PHOSPHATE ISOMERASE (GPI) DETECTION

GPI isozymes were extracted from single kernels with 0,1M Tris-HCl buffer pH 7.5, the macerate being centrifuged at 8000 g, 10 min. and the supernatant directly used for electrophoresis. GPI activity was detected following the tetrazolium method adapted from Pasteur *et al.* (1988), after electrophoresis carried out in a Phast System apparatus (Pharmacia ™), using Phast Gel IEF 5-8 and following the respective separation program (PhastFile100).

ALKALINE PHOSPHATASE (AkPh)

AkPh isozymes were extracted from single kernels with 0,1M NaCl pH 5.5, the macerate being centrifuged at 8000 g, 10 min. and the supernatant directly used for electrophoresis. AkPh activity was detected following the diazonium method adapted from Pasteur *et al.* (1988), after electrophoresis carried out in a Phast System apparatus (Pharmacia ™), using Phast Gel IEF 4-6.5 and following the respective separation program (PhastFile100).

RESULTS AND DISCUSSION

The electrophoretic patterns for glucose phosphate isomerase isozymes (Fig. 1a), gliadins (Fig. 1b), and alkaline phosphatase isozymes (Fig. 1c) for *Triticum aestivum* cv. Chinese Spring, *Secale cereale* cv. Centeio do Alto and their F1 hybrid are presented. The patterns obtained were characteristic of each genotype and no variability between samples was detected, under the conditions utilised.

Through isoelectric focusing and acidic PAGE it was possible to observe that some of the bands present in at least one of the progenitors were absent in the hybrid pattern for alkaline phosphatase and glucose phosphate isomerase as well as for gliadins (Fig. 1a, c and b arrowheads). The missing bands were, however, present either in the wheat or in the rye zimograms. For GPI and Gli the bands not present in the hybrid pattern were exclusively from rye origin. For AkPh the missing bands were both of wheat and rye origin. The patterns of genetic markers studied in the F1 wheat x rye hybrid were not therefore additive, some bands present in at least one of the progenitors were absent in the hybrid.

An analogous result has already been described for minor gliadin and glutenin subunits (Bietz *et al.*, 1975) and for acid phosphatase (Graça *et al.*, 1991) where it was found that 10 rye bands and 2 wheat bands where missing in the wheat x rye F1 hybrid pattern and that new monomers associations could not account for the disappearance.

The existence of a gene suppression mechanism underlaying the allopolyploid wheat formation due to the interactions between the different genomes and consequent diploidization of genes coding for endosperm storage proteins was already proposed, following the observations by Galili and Feldman (1983).

The occurrence of gene suppression seems to be the mechanism mediating the disappearance of specific bands from each progenitor in the hybrid as no simultaneous detection of new bands was observed which could indicate that the undetected ones were involved in any potential new association.

CONCLUSIONS

From the presented results it must be pointed out:

The existence of intergenomic suppression on marker genes usually considered as presenting codominant expression (isozymes and storage proteins), in wheat x rye F1 hybrids;

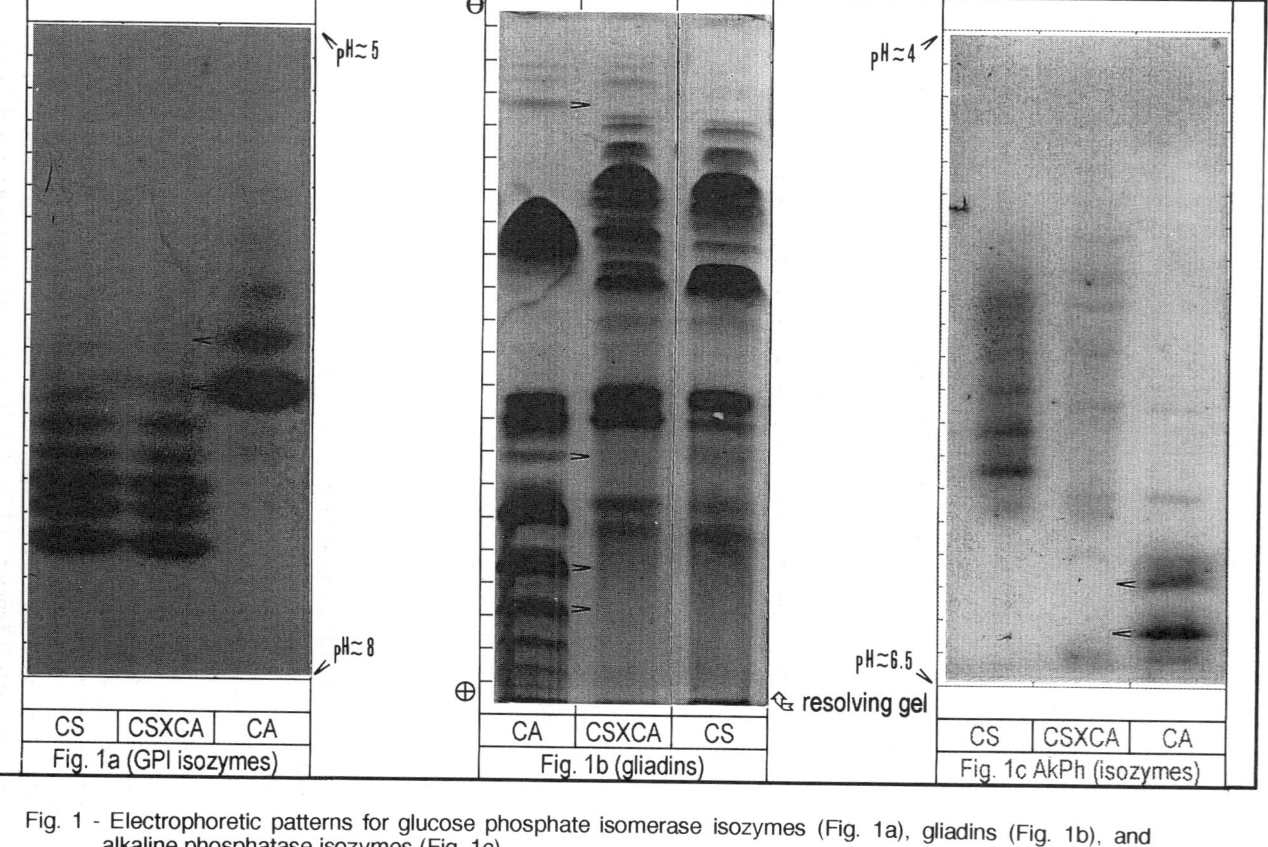

Fig. 1 - Electrophoretic patterns for glucose phosphate isomerase isozymes (Fig. 1a), gliadins (Fig. 1b), and alkaline phosphatase isozymes (Fig. 1c). **CS**: *Triticum aestivum* cv. Chinese Spring. **CA**: *Secale cereale* cv. Centeio do Alto. **CS x CA**: F$_1$ hybrid. The > indicate missing bands in the hybrid but present in at least in one of the progenitors

The suppression of gene expression could happen in genes from both parents (Fig. 1c) or be directed towards one of the parents (Fig. 1a, 1b - preferential suppression of the expression of genes from rye).

For this later case, it could be suggested that an entire chromosome region on 1R it is silenced following genome interactions in F1 wheat x rye hybrids, at least between *Nor-R1* (Lacadena *et al.*, 1984) and *Sec-1* passing through *GPI-R1,* for a distance of about 40 arbitrary units of genetic distance, as referred by Baum and Appels (1991). In this case gene silencing could be due to a topologic effect, resulting from genome interaction.

The mechanisms underlying this suppression of gene expression could be related to the difficulties in expressing transferred genes. Consequently, to successfully introduce new genes in well established cultivars it is needed to understand how overcome this genomic interaction that could modify the expected genetic expression in the new background.

The utilisation of marker genes that could be subjected to gene suppression will interfere with the gene screening on gene transfer and with cultivar certification but, on the other hand, these marker genes could be easily used as "flags" for gene suppression activity over the newly transferred DNA.

ACKNOWLEDGEMENTS

The first author was financied by a Junta Nacional de Investigação Científica grant

REFERENCES

Baum, M., and R.Appels; 1991. *Chromosoma* **101**:1-10

Bietz, J.A., K.W.Shepherd, and J.S.Wall; 1975. *Ceral Chem.* **52**:513-32

Bushuk, W. and Zillman, R.R.; 1978. *Can. J. Plant Sci.,* **58**: 505-515.

Darvey, N.L. and Driscoll, C.J.; 1972. Chromosoma **36**:131-9.

Graça, I., V.Carvalho, W.S.Viegas; 1991. XXVI Jorn.Luso Espanholas Genét. p. 208.

Galili, G. and Feldman, M.;1983. *Theor.Applied.Genet.* **64**:97-101

Galili, G. and Feldman, M.; 1984. *Can J. Genet. Cytol.,* **26** : 651-656.

Lacadena, J.R.; M.C.Cermeño; J.Orellana and S.L.Santos; 1984. *Theor.Appl.Genet..* **67**:207

Pasteur, N., G.Pasteur, F.Bonhomme, J.Catalan, and J.Britton-Davidian; 1988. Practical Isozyme Genetics. Ellis Horwood, Chichester.

Phast System Separation Technique File nº 100, Pharmacia LKB Biotechnology.

Quinones, M.A.; E.N.Larter and P.J.Samborski; 1972. *Can.J.Genet.Cytol.* **14**:495

Vallejos, C.E.; 1983. *in* S.D.Tanksley e T.J.Orton; Isozymes in Plant Genetics and Breeding, Part A. p. 469. Elsevier Science Publishers.

Viegas, W., M. Silva, N. Neves, A. Castilho, A. Barão, A. Queiroz, J.S. Heslop-Harrison, L. Morais, L. Amado, M. Delgado, V. Carvalho; 1994. Expression of 1R rDNA *loci* in triticale: genetic and developmental controls. (Kluwer Academic Publishers B.V.) (in press).
Orellana, J., J.L.Santos, J.R.Lacadena, M.C.Cermeño; 1983. *Can.J.Genet.Cytol.* **26**:34-9
Vieira, R., A.Queiroz, L.Morais, A.Barão, T.Mello-Sampayo, W.Viegas; 1990. *Genome* **33**:707-12

COMPARATIVE ANALYSIS OF TELOMERIC HETEROCHROMATIN OF RYE CHROMOSOMES IN RYE AND TRITICALE BY FISH

Angeles Cuadrado & Nicolás Jouve
Dept. Cell Biology and Genetics, University of Alcalá de Henares 28871-
Alcalá de Henares, Spain.

Abstract

The repetitive DNA composition in the telomeres of rye chromosomes in two rye lines and three triticale varieties is carried out using a sensitive sequential and successive fluorescence *in situ* hybridization (FISH). The rye lines: 'Petkus', and 'Imperial', and the 6x-triticale lines: 'Badiel', Tajuña' and '0-22-80', were comparatively investigated. Repetitive DNA sequences of 120 bp, 480 bp and 610bp from rye were studied. The 5S rDNA sequence from wheat was used to identify the metacentric chromosomes. Interstitial hybridization sites detected correspond to only one sequence 120bp or 480bp. Concerning the telomeric repetitive DNA sequences our results demonstrate: i) the high polymorphism that exits for the composition and arrangement in the telomeres of open-pollination rye lines; and, ii) the complex organization in the telomeres of rye chromosomes in triticale. These consist of arrays of multiple family sequences. The number and order of repeated sequences varied between and within varieties.

Introduction

The presence of repeated DNA in rye heterochromatin was first reported by Appels et al (1978). Jones and Flavell (1982a) using radioactive *in situ* hybridization confirmed that heterochromatin consists of repetitive DNA sequences, and that these are arranged in tandem in chromosomes of rye. Bedbrook et al. (1980) described four different repeated DNA sequences in *Secale*. Jones and Flavell (1982b) also studied the structure, amount and chromosomal localization of these sequences in the species of the genus *Secale*. Two

155

H. Guedes-Pinto et al. (eds.), Triticale: Today and Tomorrow, 155–163.

© 1996 *Kluwer Academic Publishers. Printed in the Netherlands.*

repeated sequence families of 120 bp and 480bp occupy predominantly the telomeres, and some major interstitial sites in rye chromosomes. The 120bp repeat unit is very conserved in the *Triticeae* and shows cross-hybridization to chromosomes of wheat (Rayburn and Gill, 1985). Jones and Flavell (1982b) also suggested that although the amounts of repeated sequences at each site can vary within and between rye varieties, the sites are relatively invariant.

C-banding has been the most useful method of chromosome identification in cereals. Giemsa or Leishman staining techniques give slightly different patterns for heterochromatin distribution in the chromosomes of these species. Although C-banding staining techniques prove that heterochromatin distribution is rather constant in rye chromosomes, Giraldez et al. (1979) reported intravarietal and intervarietal variation in open-pollinated cultivars. Moreover, the main heterochromatic blocks show clear modifications when wheat and rye chromosomes are joined in an amphiploid (Darvey and Gustafson, 1975; Badaeva et al., 1986; Jouve et al., 1989). All these works reported the quantitative modification of C-bands in the telomeres and interstitial sites, when passing from diploid rye to the hexaploid Triticale. These changes in heterochromatin could be the chromosomal manifestation of the most cryptic changes at molecular level.

It has been suggested that the heterochromatin distribution in rye must play a significant role in determining the viability of new triticale lines. Accordingly, the absence of the telomeric heterochromatin in rye chromosomes has been associated with an improvement in their meiotic stability (Kaltsikes et al., 1983; Soler et al 1990). Thus, rye stocks lacking heterochromatin in the telomeres should be very useful for producing new triticale lines. The aim of this work is the analysis of the repetitive DNA composition in the telomeres of rye chromosomes as an approach to understand their molecular variation in the synthetic amphiploid. The analysis is carried out in two rye lines and three triticale varieties using a sensitive sequential and successive fluorescence *in situ* hybridization (FISH).

Materials and methods

Seeds of two different rye lines: 'Petkus', and 'Imperial', and three hexaploid triticale lines: 'Badiel', Tajuña' and '0-22-80', formed the plant material used in this work. The rye lines have a well-known karyotype and piercing blocks of constitutive heterochromatin (Driscoll and Sears, 1971; Darvey and Gustafson, 1975; Sybenga, 1983; Mukai et al., 1992). Samples of the triticale lines were kindly given by Dr Soler.

The following probes containing different repetitive DNA family sequences from *Secale cereale* L. were used for *in situ* hybridization: pSc119.2, a sub-clone of the 120 bp family isolated by McIntyre et al. (1990); pSc74, that contains the 480 bp family isolated by Bedbrook et al. (1980); pSc34 that contains the 610 bp family obtained by Bedbrook et al. (1980),; and pTa794, which comprise a complete 410bp 5S rDNA gene unit from wheat (Gerlach and Dyer, 1980). These DNA probes were kindly supplied by Dr J.P. Gustafson (Univ Missouri, Columbia, MO, USA), and Drs J.S. Heslop-Harrison and R Koebner, from the John Innes Institute (Norwich, UK). The total genomic DNA from *Secale cereale* cv. 'Petkus' was used as a probe to distinguish rye and wheat chromosomes in triticale. The probes pSc119.2, pSc34, and pSc74 were labelled with Rhodamine-4-dUTP or Digoxigenin-11-dUTP, using the polymerase chain reaction (PCR) to spawn massively the labelled sequence. The total genomic DNA from rye 'Petkus' was labelled with Rhodamine-4-dUTP. Chromosome and probe denaturation, and the *in situ* hybridization followed to the method of Heslop-Harrison et al. (1991). A programmable thermal controller (PT-100, M.J. Research Inc.) was used for denaturation of spread chromosome preparations on slides prior to *in situ* hybridization.

Results and discussion

The arrangement of repetitive sequences in the telomeres of rye and tritcale lines are registered in Table 1. We will describe separately the most relevant results.

THE REPETITIVE DNA IN RYE LINES

All rye chromosomes showed strong telomeric hybridization sites to the 120bp repetitive sequence (Fig 1 F to I). Moreover, the probe pSc119.2 marked interstitial sites distributed on the 1R, 4R, 5R, 6R, and 7R chromosomes. The 480 bp sequence hybridized to telomeric regions of all chromosomes. Moreover, this sequence hybridized to interstitial sites on the long arm of the submetacentric chromosomes 5R and 6R. The probe pSc34 (610 bp repeat unit) hybridized exclusively to telomeric sites, and marked predominantly all arms, with the exceptions of 1RL and 6RL in 'Imperial', 7RL in 'Petkus' and 3RL in both rye lines.

The quantitative distribution of 120bp, as judged by differences in intensity and distribution of the hybridization sites, showed polymorphism within and between both rye lines. However, the comparisons of the intervarietal hybridization patterns reveal that the position of this family is relatively invariant. Quantitative and qualitative variation for the distribution of the 480bp repeated sequences was detected between and within homologous

Table 1.- The arrangements of the repetitive sequences in the telomeres.

Number of successive telomeric repetitive sequences	DNA sequences and order (from proximal to distal)	*Secale cereale* i = 'Imperial' p = 'Petkus'		6x-Triticale b = 'Badiel' o = '0-22-80' t = 'Tajuña'	
One	-120		$5RL^{ip}$	$4RS^{bot}$ $6RS^{b}$	$5RL^{bot}$ $6RL^{t}$
	-480		$4RL^{p}$	$6RL^{o}$	
	-610		$6RL^{i}$	$6RL^{b}$	
Two	-120-480		$2RL^{ip}$	$1RL^{o}$ $3RL^{bot}$ $4RL^{ot}$	
	-480-120		$6RL^{p}$	$4RL^{b}$	
Three	-120-480-610	$1RS^{i}$ $2RS^{i}$ $3RS^{i}$ $4RS^{ip}$ $6RS^{ip}$ $7RS^{p}$	$1RL^{i}$ $3RL^{ip}$ $7RL^{p}$	$2RS^{o}$ $5RS^{t}$	
	-120-610-480	$3RS^{i}$ $7RS^{i}$	$7RL^{p}$	$1RS^{o}$ $2RS^{b}$ $3RS^{bot}$ $6RS^{t}$ $7RS^{bot}$	
	-480-120-610	$1RS^{p}$			$6RL^{b}$
	-480-610-120			$1RL^{o}$ $4RL^{ot}$ $7RL^{bo}$	
	-480-120-480			$2RL^{b}$ $7RL^{bo}$	
Four	-480-120-480-610	$2RS^{b}$			
	-120-480-610-120			$5RS^{o}$	
	-120-480-120-480				$2RL^{o}$
	-120-610-480-610			$1RS^{b}$ $6RS^{o}$	
Five	-120-480-610-120-480	$5RS^{p}$			
	-120-480-120-480-610	$5RS^{i}$			
	-120-480-610-480-120			$5RS^{b}$	
	-480-610-120-480-610				$1RL^{b}$
Six	-480-610-120-480-610-120				$1RL^{t}$

chromosomes in both rye lines.

THE REPETITIVE DNA IN TRITICALE LINES

The repetitive DNA sequences (120 bp and 480 bp) hybridized to the telomeres and interstitial regions in all chromosomes in the three triticale lines analyzed (Fig.1 A to E) The seven rye chromosomes showed a clear hybridization signal with the 120 bp repeat unit. In addition, interstitial hybridization was visible in chromosome arms 1RS, 1RL, 4RL, 5RL, 6RS, 6RL, 7RS and 7RL. Furthermore, the 480bp sequence appeared at interstitial sites in the long arm of the most heterobrachial chromosomes (5R and 6R). Chromosomes 1R and 7R in 'Tajuña', 1R, 2R and 7R in 'Badiel', and 7R in '02280' showed interstitial hybridization to pSc74 near the telomere of the long arm. All of these interstitial bands are the most effective features for chromosome identification in the present material. The simultaneous hybridization with the pSc119.2 and pTa794 probes effective resulted in distinguishing the rye metacentrics 2R and 3R. These chromosomes did not show interstitial sites with probe pSc119.2. A clear signal on 3RS of the 5S rRNA gene locus is useful as an identifying feature. The third metacentric chromosome of rye (7R) was easily distinguished from 2R and 3R by using their interstitial hybridization to the pSc119.2.

Our observed results confirmed the hybridization of the 120 bp sequence on 4A, 5A and the seven B-genome chromosomes, as previously reported by Mukai et al. (1993). Besides this, we observed a new hybridization site on the telomere of chromosome arm 7BL. The B-genome in the three triticale lines showed multiple pSc119.2 hybridization sites. These include the positions previously described for the pSc119.2 in the chromosomes of genome B in 'Chinese Spring' (Rayburn and Gill, 1985; Mukai et al., 1993).

FEATURES OF REPETITIVE DNA IN RYE AND TRITICALE

Regarding the arrangement of the repetitive sequences in the interstitial regions, all hybridization sites detected corresponded to only one sequence 120bp or 480bp. This result was common in all rye and triticale lines investigated.

The analysis of telomeric repetitive DNA sequences in rye and triticale seemed to demonstrate two specific properties: i) the high polymorphism that exits for the composition and arrangement in the telomeres of open-pollination rye lines; and, ii) the complex organization that is usually observed in the telomeres of rye chromosomes in triticale. These consist of arrays of multiple family sequences.

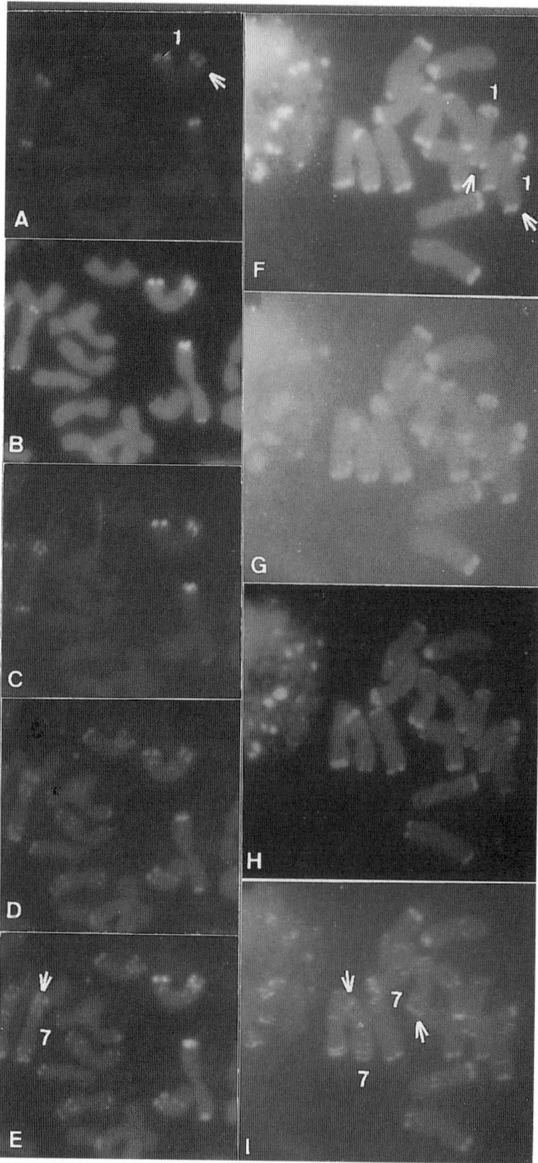

Fig. 1. Multicolor result of *in situ* hybridization in a partial cell of the triticale line 'Tajuña (A to E), and a metaphase plate of rye 'Imperial' (F to I). In A, E, F and I have been marked the telomeres 1RL and 7RL to illustrate the arrangement of the three repeat sequences in both photographic sequences: A) Hybridization using digoxigenin labelled pSc34 DNA probe (610 bp) detected with fluorescein conjugate (green). B) Hybridization using the rhodamine labelled pSc74 (480 bp) visualized as pink on bluish chromosomes by double-exposure with two different filters, for DAPI and red. C) Simultaneous visualization of hybridization chromosome sites to probes pSc34 (red), and pSc74 (yellow-green); D) hybridization sites to the digoxigenin labelled pSc119.2 (green). E) Simultaneous visualization of hybridization chromosome sites to probes pSc34, pSc74 and pSc119.2. F) Hybridization using the rhodamine labelled pSc34 visualized as pink on bluish chromosomes by double-exposure with two different filters, for DAPI and red. G) Simultaneous hybridization to the digoxigenin labelled pSc34 (red) and pTa794 (5S rDNA). H) Hybridization using the rhodamine labelled pSc74 (pink) on bluish chromosomes by double-exposure with two different filters, for DAPI and red; I) Hybridization using digoxigenin labelled pSc119.2 detected with fluorescein conjugate (green).

The polymorphism in the telomeres of rye chromosomes in 'Petkus' and 'Imperial', could be explained as being due to the phenomena of deletion and/or amplification of the repetitive sequences. When two or more different blocks of repetitive DNA were found in the telomeres, the 120 bp and 480 bp sequences were usually present in proximal and distal sites, respectively for most the telomeres (2RS, 3RS, 3RL, 6RS, and 7RS). The repetitive sequences were arranged in tandem in the telomeres. The tandem formed by the repetitive sequences of 120-480-610bp, from proximal to distal position, was the most frequently observed (1RS, 1RL, 2RS, 3RS, 3RL, 4RS, 6RS, 7RS and 7RL). This organization confirms the position of blocks suggested by Jones and Flavell (1982b) for the telomeres of rye chromosomes. However, the same repeat sequences were found in different orders in other telomeres or as polymorphisms for the same telomere. The number and order of repeated sequences varied between and within varieties and were more complex in triticale lines than in the rye cultivars (Fig 1). In sum, it is not convenient to generalize on the arrangement of the adjacent blocks of the repetitive families within telomeric heterochromatin in rye chromosomes.

The most complex arrangements consits of arrays of five or six repetitive families organized in tandem. They were found in the telomeres of chromosome arms 5RS ('Imperial', 'Petkus' and 'Badiel') and 1RL ('Badiel' and 'Tajuña'). Some of these organizations imply the repetition of arrays of two or three repeat sequences in the same order. This is the case of the organization 480-610-120, that was single in the telomeres of 1RL in triticale '0-22-80' and duplicated in 1RL in triticale 'Tajuña'.

The outlined results demonstrate that simultaneous and successive FISH, using multiple DNA probes on a single metaphase, is a powerful technique for molecular mapping repetitive sequences in cereals. Its application in triitcale seems to explain that modificaions frequently observed in the telomeric heterochromatin, should be due to phenomena of amplification or deletion of blocks or group of blocks of the repetitive sequences.

Acknowledgements. The authors thank CICYT of Spain for the award of a grant (N^c AGR91-0191), and Dr J.S. Heslop-Harrison for expert assistance.

References

Appels R, Driscoll C, Peacock WJ. Heterochromatin and highly repeated DNA sequences in Rye (*Secale cereale* L.). Chromosoma 1978;70 67-89.

162

Badaeva ED, Badaev NS, Bolsheva NL, Zelenin AV Chromosome alterations in the karyotype of triticale in comparison with the parental forms. i. Heterochromatin regions of R genome chromosomes. Theor Appl Genet 1986;72:518-23.

Bedbrook JR, Jones J, O'Neil, M, Thompson RD, Flavell RB. Molecular characterization of telomeric heterochromatin in *Secale* species. Cell 1980;19:545-60.

Darvey NL, Gustafson JP. Identification of rye chromosomes in wheat-rye addition lines and triticales by heterochromatin bands. Crop Sci 1975;15:239-43.

Driscoll CJ, Sears ER. Individual addition of the chromosomes of 'Imperial' rye to wheat. Agron Abstr 1971;.6.

Gerlach WL, Dyer TA. Sequence organization of the repeating units in the nucleus of wheat that contain 5S rRNA genes. Nucleic Acid Res 1980;8:4851-65.

Giraldez R, Cermeño MC, Orellana J. Comparison of C-banding pattern in the chromosomes of inbred lines and open pollinated varieties of rye. Z. Pflanzenzüchtg 1979; 83:40-48.

Heslop-Harrison JS, Schwarzacher T, Anamthawat-Jónsson K, Leitch AR, Shi M, Leitch IJ. *In situ* hybridization with automated chromosome denaturation. Technique-A Journal of Methods in Cell and Molecular Bioloby 1991; 3:109-16.

Jones JDG, Flavell RB, The mapping of highly-repeated DNA families and their relationships to C-bands in chromosomes of *Secale cereale*. Chromosoma 1982a;86: 595-612.

Jones JDG, Flavell RB. The structure, amount and chromosomal localization of defined repeated DNA sequences in species of the genus *Secale*. Chromosoma 1982b;86: 613-41.

Jouve N, Galindo C, Mesta M, Díaz F, Albella B, García P, Soler C. Changes in triticale heterochromatin visualized by C-banding. Genome 1989; 32:735-42.

Kaltsikes PJ, Lukaszewski AJ, Gustafson JP. The effect of telomeric heterochromatin on chromosome pairing in several wheat-*Secale* hybrids. Proc 6th Int Wheat Genet Symp Kyoto Japan 1983:885-88.

McIntyre CL, Pereira S, Moran LB, Appels R. New *Secale cereale* (rye) DNA derivatives for the detection of rye chromosome segments in wheat. Genome 1990; 33: 635-40.

Mukai Y, Friebe B, Gill BS. Comparison of C-banding patterns and in situ hybridization sites using highly repetitive and total genomic rye DNA probes of ,Imperial, rye chromosomes added to ,Chinese Spring, wheat. Jap J Genet 1992;67:71-83.

Mukai Y, Nakahara, Y, Yamamoto M. Simultaneous discrimination of the three genomes in hexaploid wheat by multicolor fluorescence *in situ* hybridization using total genomic and highly repeated DNA probes. Genome 1993;36:489-94.

Rayburn AL, Gill BS. Use of biotin-labelled probes to map specific DNA sequences on wheat chromosomes. J Heredity 1985;76:78-81.

Rayburn AL, Gill BS. Isolation of a D-genome specific repeated DNA sequence from Aegilops squarrosa. Plant Mol Biol Rep 1987; 4:102-09.

Soler C., García P,Jouve N. Meiotic expression of modified chromosome constitution and structure in xTriticosecale Wittmack. Heredity 1990;65:21-28.

Sybenga J Rye chromosome nomenclature and homoeology relationships. Z Pflanzenzüchtg 1983; 90:97-304.

PRODUCTION AND FERTILITY OF HEXAPLOID PRIMARY TRITICALES

Gert Kleijer and Aldo Fossati

Station Fédérale de Recherches Agronomiques de Changins (RAC), Rte de Duillier, case postale 254, CH-1260 Nyon, Switzerland

Abstract

Broadening the genetic base of a triticale breeding programme is a necessity to ensure its sustainability. One way is the production of primary triticales. We tried to obtain short, fertile hexaploid triticales with a satisfactory grain formation. For this purpose, short durum wheats were crossed with short rye lines as well as with some long local Swiss rye varieties. Average results of a five year crossing programme show on 16026 emasculated flowers a seed set of 26%. 76,5% of these seeds had an embryo from which 22,5% germinated. Chromosome doubling with colchicine succeeded for 57% of the obtained plants. Plant height of the primary triticales varied from 65 cm to 175 cm. The lines were split into two groups, one with the interesting short triticales (spt) with plant height ranging from 65 cm to 115 cm, and the other with the less interesting long triticales (lpt) ranging from 125 cm to 175 cm. The average fertility of the spt and lpt was 1.69 and 1.83 grains per spikelet respectively. Within the spt the whole range of fertility, from nearly sterile with 0.45 grains per spikelet to fully fertile with 2.68 grains per spikelet, was observed. The average note for grain formation was 6.4 for the spt and 5.4 for the lpt. A comparaison between durum wheat, rye and their doubled hybrids showed that of the observed characters (plant height, thousand kernel weight, fertility and grain formation) grain formation of the hybrids was very poor and the difference between the note of the hybrids (5.9) and the parents (3.4) was highly significant. As grain formation is a problem in a triticale breeding programme only primary triticales with a fairly good grain formation should be taken as parents. A number of the spt produced, combined good fertility and a rather good grain formation and are thus very interesting to our breeding programme.

Introduction

Broadening the genetic base of a triticale breeding programme is an absolute condition of its maintenance and success. Triticale being a synthetic crop is not as diverse as naturally evolved crops. The production of primary triticales by crossing wheat and rye is one of the various approaches of broadening the genetic base of triticale. The cross between *durum* wheat and rye has been carried out several times and the success depends on the genotypes used [1-5]. Due to a lack of differentiation during early embryo and endosperm development, embryo rescue has

H. Guedes-Pinto et al. (eds.), Triticale: Today and Tomorrow, 165–171.
© 1996 *Kluwer Academic Publishers. Printed in the Netherlands.*

to be carried out [6]. Seed set is not the only important factor in the production of wheat/rye amphidiploids. The survival of embryos when rescued *in vitro* and the success of colchicine treatment for doubling chromosome number are also relevant [4,7]. The evaluation of agronomic characteristics has not been carried out very often and information is scarce [8,9].

The aim of our breeding programme is the creation of short, high yielding hexaploid triticale with good grain formation [10]. Selection of primary triticales for grain formation, fertility and plant height is useful in order to avoid introducing too much negative characters in the breeding programme. In this paper, the production of primary hexaploid triticales as well as some of their agronomic characteristics will be presented.

Material and methods

During five years (1988-1992) of primary triticale production, 37 different *durum* wheats have been used. Most of them (22) originated from F_2 lines received from CIMMYT/Mexico and selected at RAC during four to five years for plant height and winter hardiness, five lines originated from the U.S.A., four varieties from Italy, and three from Hungary. Except for three lines obtained from Gatersleben (Germany), all *durum* lines were semi-dwarf. The rye parents originate from Switzerland (9), Italy (4), Russia (1), and U.S.A. (1) and were used for their winter hardiness. All these lines were tall. Short lines obtained from the Botanical Garden in Warsaw, Poland (11), short inbred lines from the Agricultural University of Wageningen, The Netherlands (6), and some German varieties (4) with intermediate plant height were also used. The sending of all these lines is gratefully acknowledged. The crosses, embryo rescue and colchicine treatment were carried out as described earlier [11].

Primary triticales were grown in the field during 1992 and 1993 on 1.5 m² plots, one replication, under normal tillage practices. The occurring outcrosses were removed. Fertility was measured by counting the number of grains per spikelet for five ears of each plot. Grain shrivelling was estimated by a score from 1 to 9, where 1 represents good grain formation and 9 completely shrivelled. The comparison between the wheat and rye parents with their hybrids (in total 42) was carried out on the field in 1993 using small plots with three randomized replications. For some observations correlation analyses have been carried out as well as Student's t-test.

Results and discussion

The average results of the crosses over the 5 year period are presented in table 1.

Average seed set varied from 10 to 39.6% over the years and from 0 (one parent) to 59.4% for the *durum* wheat parents. Only five *durum* wheat parents showed a seed set lower than 10%. The seed set, related to the rye parents, varied from 2% to 58%. Crossability depends thus on the genotype of *durum* wheat and rye which confirms the investigations carried out earlier [2-5,7,11]. The percentage of grains with an embryo varied from 36.9% to 84.8% over the years and 11.9 to 94.3 for the different *durum* wheat parents. Crossability and embryo development in the grain seemed to be influenced by environment [5]. This was confirmed by our study which showed important differences between years, but we used different genotypes for the crosses each year, so part of the variation may have been genotypic. The correlation coefficient between the percent seed set and embryo formation was low, r = 0.35, but significant at P = 0.05. Earlier studies showed that seed set and embryo development were independent of each other [4]. Germination of the embryos on an artificial medium was rather low, especially in 1989 (11.7%) with a maximum of 38.5% in 1992. The colchicine treatment succeeded for 57% of obtained plants.

Table 1. Average crossability of *durum* wheats with ryes and response to colchicine treatment

Year	Number *durum* wheat parents	Number rye parents	Number flowers pollinated	% Seed set	% Grains with embryo	Plantlets recovered as % of embryos cultured	Doubled plants as % of total plants
1988	6	8	2628	10.0	84.8	27.3	42.6
1989	6	9	3539	39.6	80.2	11.7	51.5
1990	10	8	3428	32.0	79.4	24.8	44.9
1991	9	8	3210	31.9	81.3	30.8	76.2
1992	7	10	3257	11.9	36.9	38.5	46.5

One hundred and sixty five primary triticales (Tcp) were analyzed for some agronomic characters in 1993 and some of them also in 1992. Tcp were sown for the second to fourth time depending on the year of production. Some outcrosses occurred so plots were not always homogenous. Measurements were carried out on the most frequent phenotypes in a plot. Plant height varied from 65 cm to 175 cm. In order to study the most interesting lines more in detail, the Tcp were split into two groups : the short < 120 cm and the tall > 120 cm. The scores for grain formation of these two types of lines are presented in figure 1. The average score for the tall triticales was 5.4 and for the short ones 6.4. The difference was highly significant (t-test). The number of grains per spikelet of these primary triticale lines are presented in figure 2. The average fertility for the short lines was 1.69 and for the tall 1.83 grains per spikelet. This difference was significant at the 5% level. Within the short primary triticales, the whole range of fertility, from nearly sterile (0.45 grains per spikelet) to fully fertile (2.68 grains per spikelet) was observed. Grain formation was noted for 125 Tcp in 1992 (average 5.1) and 1993 (average 6.0). The correlation between years was rather low but significant (r = 0.45, P < 0.01) The differences in grain formation and fertility between 1992 and 1993 were highly significant indicating an environmental effect.
A special trial was carried out to compare the parents with their hybrids. Average results for grain score, plant height, fertility and thousand kernel weight are presented in table 2.
Thousand kernel weight of the primary triticales was at the same level as for the secondary triticales. Fertility was intermediate between *durum* wheat and rye parents. Plant height showed an important variation. The average was lower than that of the standard secondary triticales. The dwarfing genes of the *durum* wheat parents and the rye parents expressed themselves in the primary triticales. The average grain formation was bad and was the only character in which the primary triticales were worse than the parents. Grain formation is thus the most striking problem in primary hexaploid triticale. A more important grain shrivelling in hexaploid than in octoploid primary triticales was observed [8].

168

Fig.1   Distribution of score for grain formation of the short (average 6.4) and tall (average 5.4) primary triticales.

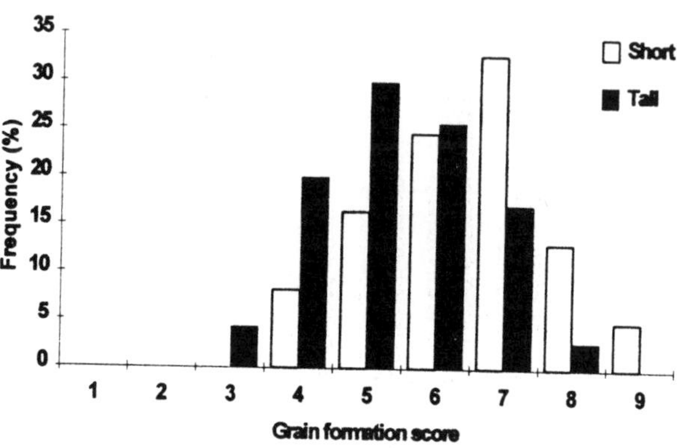

Fig.2   Distribution of fertility, measured as number of grains per spikelet of the short (average 1.69) and tall (average 1.83) primary triticales.

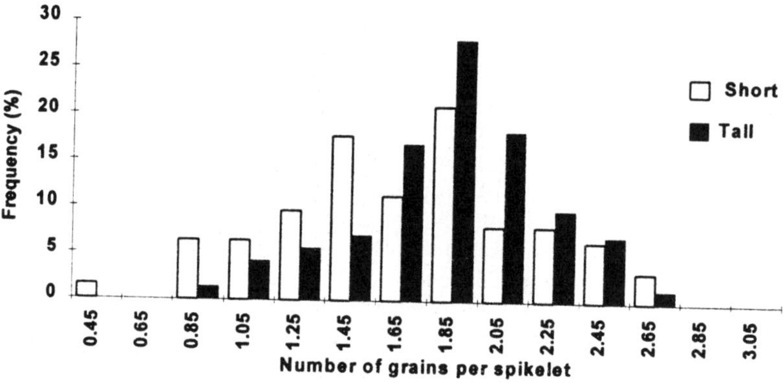

Fig.1   Distribution of score for grain formation of the short (average 6.4) and tall (average 5.4) primary triticales.

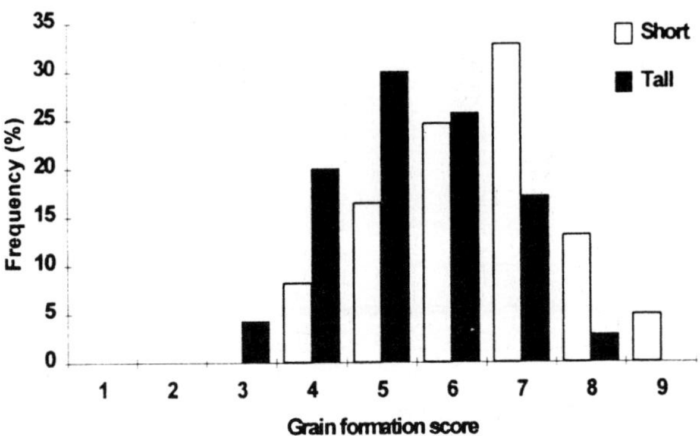

Fig.2   Distribution of fertility, measured as number of grains per spikelet of the short (average 1.69) and tall (average 1.83) primary triticales.

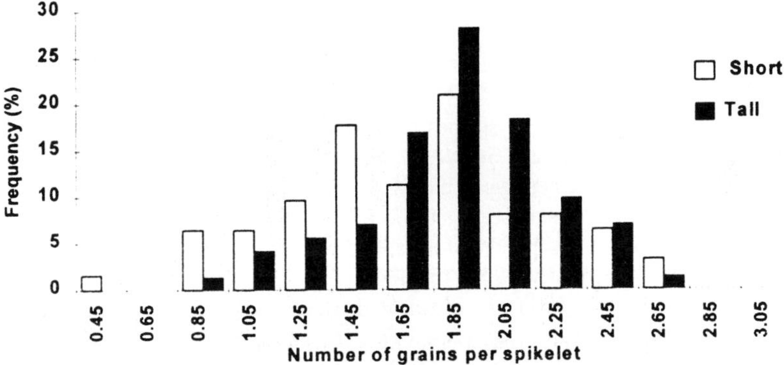

Table 2. Averages of some agronomic characteristics of primary triticales, their parents and a secondary triticale standard

	Grain score	Plant height (cm)	Fertility (grains per spikelet)	Thousand kernel weight (g)
durum wheat	3.29	84.4	2.56	49.4
Primary triticale	5.88	115.0	1.95	55.7
Rye	3.57	116.0	1.60	38.4
Secondary triticale	3.58	123.0	2.42	55.8

Conclusions

Although important genotypic differences in seed set occurs, the production of primary triticales is possible with most of the *durum* wheat parents.

The success of the production of wheat/rye hybrids depends not only on seed set but also on embryo formation, survival of embryos when rescued and chromosome doubling. One of the most important problems of primary triticales is grain formation. Our results show that selection of primary triticales for short plant length, high fertility and acceptable grain formation is possible.

References

1. Jalani BS, Moss PJ. The effects of species, polyploidy and embryo transplantation on the crossability between *Triticum* and *Secale*. Z Pflanzenzüchtg 1981;86: 286-297.
2. Oettler G. Effect of parental genotype on crossability and response to colchicine treatment in wheat-rye hybrids. Z Pflanzenzüchtg 1982; 88: 322-330.
3. Verzea M. Crossability of tetraploid and hexaploid wheats with diploid rye. Proc 2nd Int. Triticale Symp Cimmyt Mexico 1991: 157-160.
4. Balatero CH, Darvey NL. Influence of selected wheat and rye genotypes on the direct synthesis of hexaploid triticale. Euphytica 1993; 66: 179-185.
5. Sirkka A, Immonen T, Varughese G, Pfeiffer WH, Mujeeb-Kazi A. Crossability of tetraploid and hexaploid wheats with ryes for primary triticale production. Euphytica 1993; 65: 203-210.
6. Keyworth S, Larter EN. Embryo and endosperm development in wheat-rye hybrids. Cereal Res Comm 1979; 7: 135-144.
7. Oettler G. Parental effects on crossability, embryo differentiation and plantlet recovery in wheat X rye hybrids. Euphytica 1984; 33: 233-239.
8. Oettler G. The influence of the wheat and rye genome on the performance of primary triticale. In Genetics and Breeding of Triticale Eucarpia meeting Clermont-Ferrand, France 1989, INRA, Paris 1985: 125-134.
9. Oettler G, Wehmann F, Utz HF. Influence of wheat and rye parents on agronomic characters in primary hexaploid and octoploid triticale. Theor Appl Genet 1991; 81: 401-

405.
10. Fossati A, Fossati D, Kleijer G. Triticale breeding in Switzerland. In Proc III International Triticale Symposium 13-17 June 1994, Lisbon, Portugal (this volume)
11. Kleijer G. Les triticales primaires. Rev Suisse Agric 1992; 24: 38.
12. Taira T, Larter EN. Factors influencing development of wheat/rye hybrid embryos *in vitro*. Crop Sci 1978; 18: 348-350.

STUDY OF HYBRIDS BETWEEN 2n=6X=42 and 2n=6X=42 TRITICALE FORMS. I. INVESTIGATIONS IN F$_2$

Stoyan M. Tsvetkov
Institute of Wheat and Sunflower "Dobrudja"
General Toshevo, Bulgaria

Abstract

In triticale crosses of triticale of the same ploidy level (2n=6X=42 x 2n=6x=42) a large diversity of most different hybrid derivates is developed as a result of the character pre--combinations. It is claimed that the morphogenetic process is very restricted in some combinations.

Our investigations shows that the great diversity of hybrid forms in phenotypic groups from the segregation with a very low fertility (1-12 kernels in a spike) and such ones with a very fertility (over 61 kernels in a spike) is an index showing that the selection directed to a high productivity is efficient and necessary even in F$_2$.

Introduction

In the same ploid crosses of triticale (2n=6x=42 x 2n=6x=42) a large diversity of most different hybrid derivatives is developed as a result of the character pre-combination between the parent forms and following generations. According to Shulindin (3) the morphogenetic process is very restricted in some combinations, while in some it is running rapidly and practically approaches to the interspecific crossings.

In the foreign and national special literature the studies in triticale hybridization 2n=42 x 2n=42 are very scanty regarding the systematization of the product from segregation and their fertility in F$_2$. The lack of such scientific information is still considered as a serious obstacle for improving of the selection efficiency in the development of promising forms with a production importance.

The present study has the aim to make certain more definite conclusions for the character of hybrids segregation and fertility of the same ploid level crossings in F$_2$.

Material and Methods

The investigations on the segregation character in F$_2$ were done in two basic groups at the single combinations. Hybrid combinations were studied on a reciprocal basis with the participation of the short-stemmed triticale T-AD-17-B-with a resistant to lodging and very long spikes (4) in the first group. In the second the hybrid combinations are also on a reciprocal basis, but with the assistance of Mexican high fertile triticales (5).

The hybrid derivatives from the segregation in F$_2$ were distributed into 7 basic phenotypic groups based on a series of morphologic characteres of the spike (with characters of ♀, with characters of ♂, intermediate type, with new characters, durum type,

H. Guedes-Pinto et al. (eds.), Triticale: Today and Tomorrow, 173–178.
© 1996 *Kluwer Academic Publishers. Printed in the Netherlands.*

spelta type, compactum type). The fertility was determined on the average of one spike and floret depending on the systematic belonging of the groups produced from the segregation. The cytological analyses of the meiosis in PMC were conducted according to the established techniques (2).

Results

SEGREGATION OF THE HYBRID DERIVATIVES IN SYSTEMATIC CHARACTERS OF THE SPIKE

The experimental data shows that the number of the hybrid derivatives from the phenotypic group with new characters is most significant in the segregation of the same ploid level in F_2. The percentage of the phenotypic groups taking part with $\female$ and $\male$ like morphology of the spike is considerable less. The same are close to the parent forms according to morphological characters of the spike and the stem, but do not coincide with them. The number of the hybrid plants from the groups with intermediate and durum like morphology of the spike and durum type is insignificant. The percentage of the hybrid derivatives from the phenotypic groups of spelta and compactum like morphology of the spike is also very slight.

FERTILITY

In contrast to the investigations from abroad, our work studied the fertility of the hybrids from the same ploid level in phenotypic groups from the segregation in hybrid combinations in F_2.

From the data in Table 1 it is seen that both in hybrids involving the short--stemmed triticale T-AD-17-B, and in others with the participation of Mexican triticales the highest average fertility of a spikelet in the spike is found in the phenotypic groups with $\female$ and $\male$ like morphology of the spike. These two groups in average fertility of the spike are presented with a high diversity, and these are over 60 grains in one spike.

The hybrid plants from the group with new characters of a average fertility per spikelet in the ear are ranked intermediate to that one with $\female$ and $\male$ like morphology of the spike. The group here is also presented by a high hybrid diversity including both plants with unsatisfactory (1-12) kernels and such ones with a very good (over 61) kernels.

Of all hybrid plants studied in the F_2 segregation with lowest average fertility per spikelet in an ear, these of the phenotypic groups of spelta, compactum and to a certain degree of durum-like morphology of the spike are distinguished (Table 1). According to Muntzing et al (1), the appearance of hybrid plants with a very low grain formation is a result of significant disturbances of the meiosis development in PMC.

POTENTIAL OPPORTUNITIES FOR SELECTION OF PRODUCTIVITY

In the hybrid combinations including the short-stemmed triticale T-AD-17-B (2n=42) and these of the Mexican triticales (2n=42) the common relationship was determined by reducing the plant height in the hybrids with 51-90 grains showed a spike decrease, but

Table 1 - Average fertility within groups from the segregation in F_2 in the same ploid crossing of triticale (2n=42 x 2n=42)

Cross	Plants Studied Nº	Degree of fertility per spike								grain weight per spike, g
		percent						average		
		1-12	13-24	25-36	37-48	49-60	over 61	1 spike	1 spikelet	
With participation of short-stemmed triticale T-AD-17-B (2n=42)										
With characteres of ♀	119	0.84	0.84	3.36	10.08	19.33	65.54	60.7	2.0	3.1
With characteres of ♂	97	-	-	4.12	8.25	14.43	73.20	75.5	2.1	3.8
Intermediate type	52	5.77	9.62	11.54	9.62	11.54	51.93	49.7	1.8	2.5
New characteres	957	1.57	2.19	5.43	10.45	19.02	61.23	65.5	2.1	3.2
Durum type	82	2.44	9.76	18.28	14.63	28.05	26.83	45.7	1.5	2.2
Spelta type	19	26.32	21.5	5.26	15.79	21.05	10.53	29.6	1.1	1.3
Compactum type	6	-	-	33.33	16.67	16.67	33.33	31.7	1.1	1.6
With participation of mexican triticales (2n=42)										
With characteres of ♀	431	1.16	4.64	4.64	16.47	25.99	45.94	65.8	2.3	3.1
With characteres of ♂	330	1.21	3.03	7.58	13.03	29.70	45.15	55.9	2.2	2.8
Intermediate type	171	2.91	7.02	9.36	12.87	22.81	45.03	53.7	1.7	2.5
New characteres	748	4.01	8.02	10.03	14.17	22.99	40.51	60.8	1.9	3.4
Durum type	50	10.00	14.0	8.00	16.00	22.00	28.00	45.5	1.5	1.7
Spelta type	18	16.67	27.8	11.11	22.22	5.56	5.56	38.1	1.1	1.1
Compactum type	13	-	7.69	15.38	7.69	23.08	46.15	69.5	1.6	3.6

with the special feature, that a possibility exists even in a low degree for a plant selection with a height from 51 to 70 cm in combination with a high productivity. The highest number of hybrids with a spike productivity of 51 to 80 grains is produced also in the two hybrid combinations studied with a plant height in the ranges from 91 to 120 cm (Table 2).

MEIOSIS CHARACTERISTICS IN PMC

The disturbances in the same ploid level hybrids of triticale in F_2 are referred to cells with univalents in M_I and M_{II}, cells with lagging chromosomes and bridges in A_I and A_{II}, tetrads with micronuclei and poliads.

The date in Table 3 indicate that the average percent of the disturbances of the hybrid combinations studied in F_2 varies from 75,23 to 80,98% compared to 90,18 and 92,57% for the some combinations in F_1 and 33,42 and 52,99% for the initial parent forms. This data shows that a stability process occurs in F_2, although the rate is more slowly, expressing itself in a gradual elimination of a part of the univalents.

Conclusion

The great diversity of hybrid forms in phenotypic groups from the segregation with a very low fertility (1-12 kernels in a spike) and such ones with a very high fertility (over 61 kernels in a spike) is an index showing that the selection directed to a high productivity is efficient and necessary even in F_2.

The percent of meiosis disturbances in PMC of the hybrid combinations studied in F_2 varies from 75,23 to 80,98% compared to 90,18 and 92, 57% for the same in F_2, and for the parent initial forms it is 33,42 and 52,99%. These data indicate that a process of stability occurs in F_2, though at a more slow rate, which is expressed in a gradual elimination of a part of the univalents.

References

1. Muntzing, A., N. J. Hrishi & C. Tarcowski. Reversion to haploid in strains of hexaploid and octaploid triticale. Hereditas 1963;49: 78-90.
2. Nikolov, Chr. & St. Daskalov. Cytological technique. ASN, Sofia: Academy Press, 1966: 1-177.
3. Shulindin, A.. Interspecific hybridization of triticale. Plant Breeding and Seed Production 1977; 2: 14-15.
4. Tsvetkov, St.. Breeding of winter (2n=6x=42) triticales in bulgaria, dissertation, Institute for Wheat and Sunflower, 1982, 1-602.
5. Zillinsky, F & N. E. Borlaug. Triticale research in Mexico. Agr. Sci. Review, 1971; 9, 4: 28-35.

Table 2 - Potentialities for a selection on productivity in F_2 depending on the kernel number in a spike and the plant height in the same ploid crossing of triticale (2n=42 x 2n=42)

Number of grains per spike	Plants studied Nº	Plants with 51 to 110 grains per spike	Plant height (cm)							Over 120
			51-60	61-70	71-80	81-90	91-100	101-110	111-120	
			percent (%)							
With participation of short-stemmed triticale T-AD-17-B (2n=42)										
Below 50	1332	984	0.23	1.35	2.18	3.90	3.90	2.18	3.00	7.96
51-60		(73.8%)	0.08	0.38	1.05	2.63	2.78	2.48	1.95	5.11
61-70			0.08	0.23	1.05	2.10	1.95	2.55	2.93	7.88
71-80			-	0.15	0.90	1.95	1.95	2.33	3.45	8.48
81-90			-	-	0.38	0.83	1.43	1.65	2.63	4.65
91-100			-	-	0.15	0.30	0.83	0.53	1.20	2.93
101-110			-	-	-	0.30	0.15	0.15	0.15	1.20
111-120			-	-	-	0.15	0.15	0.15	-	0.68
Over 120			-	-	-	0.15	-	-	-	0.15
With participation of mexican triticales (2n=42)										
Below 50	1985	1553	-	0.36	1.13	3.17	6.80	7.98	7.83	8.08
51-60		(79.4%)	0.05	0.15	0.66	1.18	3.48	3.79	4.25	5.73
61-70			-	0.05	0.41	1.43	2.61	3.84	4.30	5.73
71-80			-	0.20	0.26	0.92	1.43	2.46	3.53	4.45
81-90			0.05	-	0.15	0.46	0.97	1.33	1.53	2.46
91-100			-	0.05	0.10	0.31	0.62	1.02	1.07	1.59
101-110			-	-	-	-	-	0.51	0.51	2.46
111-120			-	-	-	-	-	0.10	0.05	0.26
Over 120			-	-	0.05	-	-	0.05	-	0.05

Table 3 - Degree and characteristics of the meiosis disturbance of PMC in the hybridization of the same ploid crossings of triticale (2n=42 x 2n=42)

Cross	Phases	Cells Studied Nº	Cells with disturbance						
			With uni valents	With laggin chrom.	With bridges	With micro nuclei	With polyads	Total number of cells	% distur-bance
AD Nº6 x	M_I	548	421	-	-	-	-	412	75.18
AD-Winter	A_I	777	-	599	16	-	-	615	79.15
	M_{II}	433	312	-	-	-	-	312	70.43
	A_{II}	133	-	115	6	-	-	121	90.98
	T	420	-	-	-	333	-	333	79.29
AD-Winter	M_I	410	335	-	-	-	-	335	87.71
x AD Nº6	A_I	465	-	333	5	-	-	338	72.69
	M_{II}	263	187	-	-	-	-	187	71.10
	A_{II}	148	-	128	1	-	-	129	87.16
	T	414	-	-	-	376	-	376	90.82
AD Nº8 x	M_I	408	297	-	-	-	-	297	72.79
AD-Winter	A_I	326	-	247	2	-	-	249	76.38
	M_{II}	200	145	-	-	-	-	145	72.50
	A_{II}	260	-	202	3	-	-	205	78.85
	T	316	-	-	-	240	-	240	75.95
AD-Winter	M_I	279	222	-	-	-	-	222	79.57
x AD Nº8	A_I	328	-	259	-	-	-	259	78.96
	M_{II}	244	193	-	-	-	-	193	79.10
	A_{II}	216	-	186	2	-	-	188	87.04
	T	642	-	-	-	521	1	522	81.31

TETRAPLOID TRITICALE FROM *AEGILOPS SQUARROSA* L. X *SECALE* L. SPP

Adoración Cabrera[1], Isabel Domínguez[1], Diego Rubiales[2], Juan Ballesteros[2] and António Martín[2]

[1]Dept. de Genética, ETSIAM, Córdoba, Spain

[2]Instituto de Agricultura Sostenible, Córdoba, Spain

Abstract

Tetraploid triticales have been obtained by direct crossing of *Aegilops squarrosa* (4x) with *Secale cereale* (4x) and *S. cereale-montanum* (4x), and by chromosome doubling of the hybrid *A. squarrosa* (2x) x *S. cereale* (2x). Vigourous spring growth is the most interesting agronomic trait of this amphiploid in field tests. Hybrids between octoploid and tetraploid triticales have also been obtained in order to increase the genetic variability of the later. This hybrid (DDRRAB) is self sterile but grain were obtained after backcrossing with tetraploid triticale.

Introduction

During the last decades major effort in triticale breeding has been directed toward the development of hexaploid and octoploid triticales. At the tetraploid level, triticales have been also obtained, opening new possibilities for the improvement of 6x and 8x triticales. The obtention of tetraploid triticales was achieved by Krolow [1] crossing hexaploid triticale with diploid rye followed by selfing the F1 hybrid. These tetraploid triticales were found to have a complete set of rye chromosomes and a variable number of chromosomes from the A and B genomes [2]. Tetraploid triticales originated from crosses between diploid progenitors of cultivated wheat and rye have been also produced. A sterile amphiploid was obtained by colchicine treatment of the hybrid between *Triticum monococcum* and *Secale cereale* [3]. Amphiploids between *A. squarrosa* and *S. cereale* have been obtained using different strategies. Tetraploid regenerants were obtained from callus induced on immature inflorescences of the hybrid between *T. tauschii* and *S. cereale* [4], by colchicine treatment of the hybrid [5] and crossing the tetraploid plants for each parent [6]. In the present paper we present preliminary results about the morphological characteristics, meiotic behaviour, fertility and agronomical performance of the amphiploid between *A. squarrosa* and *S. cereale* and the effort on widening the genetic basis of tetraploid triticale.

179

H. Guedes-Pinto et al. (eds.), Triticale: Today and Tomorrow, 179–182.
© 1996 *Kluwer Academic Publishers. Printed in the Netherlands.*

180

Material and Methods

Tetraploid *A. squarrosa* (syn. *T. tauschii*) (DDDD, 2n=4x=28) was used as female. The male parentals were tetraploid *S. cereale* ($R^c R^c R^c R^c$, 2n=4x=28) and tetraploid "perennial rye" [7], *Secale cereale* X *Secale montanum* ($R^c R^c R^m R^m$, 2n=4x=28). At the diploid level, *A. squarrosa* and *S. cereale* cv. Petkus were also crossed. One or two days after emasculation, spikes of *A. squarrosa* were pollinated with *S. cereale*. Gibberelic acid, at a concentration of 75 ppm, was applied to pollinated florets 24 h after pollination by means of a syringe. Three weeks later, the embryos were excised under sterile aseptic conditions and cultured in orchid agar. The embryos were kept in the dark at 10° C until germination and then grown under continuous light. When the cultured embryos reached the three-leaf stage, they were transplanted into pots. Young plants in tillering stage were treated with colchicine at 0.3 % using the capping technique. For somatic chromosome counts, root tips of adult plants were fixed in alcohol:acetic acid (3:1) and stained by the Feulgen procedure. For meiotic studies, immature inflorescences with pollen mother cells (PMC) at metaphase I stage were fixed and stained by the same procedure.

Results

DIRECT SYNTHESIS OF TETRAPLOID TRITICALES

Embryos were obtained by pollinating *A. squarrosa* (4x) with tetraploid rye and adult plants were stablished. Somatic chromosome counts revealed that the plants had the expected chromosome number of 2n=4x=28. The difference in chromosome size between the two parental species clearly proved that the plants were true amphiploids, with 14 *Aegilops* and 14 of the larger *Secale* chromosomes. The amphiploids with $R^c R^c R^c R^c$ were fertile and vigorous. A mean of five seeds per spike were obtained from the synthesized $DDR^c R^c$ amphiploid plants. A high frequency (more than 90%) of the plants in the progenies of the amphiploids were euploid. Cytological analysis of PMCs showed 60% of the cells with 14 bivalents. No cells with less than twelve bivalents were observed.

The results obtained by the meiotic analysis of the $DDR^c R^c$ amphiploids showed that no homoeology existed between the parental genomes. No trivalents or quadrivalents were observed in PMCs. These results are in agreement with the chromosome pairing observed in the *A. squarrosa* X *S. cereale* hybrid [4] and in the $H^{ch}DR$ hybrid [8].

A high frequency of bivalents formation has been found in PMCs observed. A mean of 13.45 II + 1.1 I per cell was observed after three cycles of recombination, indicating a high meiotic regularity in the amphiploid. Data given by other authors for chromosome pairing at meiosis in the amphiploid were 11.7 II + 4.1 I [5] and 10.05 II + 7.9 I [6].

These differences in frequencies of bivalent formation may be caused by differences on the specific genotypes employed to produce the amphiploids. Rye is know to have a polygenic system controlling the extent of pairing between homoeologous chromosomes of wheat [9]. The highest frequency of bivalents formation in the $DDR^c R^c$ amphiploid observed in the present report may be also the result of selection for fertility, if it is associated with meiotic regularity.

The morphology of the plants is intermediate between both parents. Spikes, although similar to rye in general aspect, present wheat-like spikelets with tough rachis. Some characteristics of *Secale*, such as hairy neck, allogamous reproductive behaviour and seed shape were also present in the amphiploids. Tetraploid triticale in field trials, at maturity, resembles a cultivated cereal, with upright growth habit, more similar to rye than to *A. squarrosa*. Although showing some susceptibility, the DDR^{c}R^c amphiploid is more resistant than its female parent to the wheat brown and yellow rusts, and than the rye parent to the rye brown rust. This triticale is however very resistant to both the wheat and rye powdery mildews [10].

This tetraploid triticale has also been subjected to tissue culture showing an exceptionally high regeneration capacity [11].

The amphiploid was selfed for three generations resulting in increased fertility. There were differences in the fertility of the plants after the first generation of selfing. Some of the plants were completely sterile. However, most of the plants were fertile and some of them produced more than two hundred seeds per plant. One reason for the lack of fertility of some of these DDR^{c}R^c plants may have been manifestation of the self-incompatibility mechanism of the rye parent, as indeed was found in the DDR^{c}R^c regenerants obtained by tissue culture of the DR^c plants [4].

In field conditions, plants showed prostrated growth habit during winter, dark colour and poor early vigour, typical of cereals with moderate vernalization requirements. This results in good ground cover followed with high vigour in spring, quite normal behaviour of cereals adapted to Mediterranean conditions. If the seed set could be increased by selection for improved fertility, this triticale might be of interest as a combination of forage green manure usage and grain harvest would be feasible. If a plant breeding programme is to be carried out we need to wide the genetic basis. With this objective we initially explore four approaches:

SYNTHESIS OF NEW TETRAPLOID TRITICALES CROSSING TETRAPLOID PARENTS

As we said previously tetraploid triticale could be used as a forage crop. With this idea in mind it is attractive to incorporate in the breeding programme the character of perennial growth habit, and perennial rye [7] is the source available. The amphiploid DDR^{c}R^m was obtained last year pollinating tetraploid *A. squarrosa* with the tetraploid *S. cereale x S. montanum*. It seems to have vernalization requirements. The anthers are poorly developed and in this first year is not setting seeds. It can be however vegetatively multiplied and will be vernalized for the next season. We are also pollinating it with DDR^{c}R^c but until now unsuccessfully.

SYNTHESIS OF TETRAPLOID TRITICALES CROSSING DIPLOID PARENTS AND DOUBLING CHROMOSOME NUMBER

The hybrid *A. squarrosa* x *S. cereale* treated with colchicine gives some seeds. We have not counted chromosome numbers yet. Hopefully these seeds shall be amphiploids. Backcrossing the treated DR^c hybrid with tetraploid DDR^{c}R^c did not produced any seed until now.

CROSSING TETRAPLOID TRITICALE AND BREAD WHEAT AND BACKCROSSING WITH TETRAPLOID TRITICALE

Tetraploid triticale crosses readily with *Triticum aestivum*. The hybrid plant grows vigorously but is completely sterile even backcrossed with tetraploid triticale. More effort should be devoted to this approach, since the result obtained are based only in crosses between the first DDR^{c}R^c obtained and Chinese Spring.

CROSSING TETRAPLOID AND OCTOPLOID TRITICALES AND BACKCROSSING WITH TETRAPLOID TRITICALE

Hybrid seeds from two different octoploids triticales (8x) pollinated with tetraploid triticale (4x) were obtained easily. Embryo rescue was not necessary. The hybrid plants grew vigorously and resembled phenotypically the male parental. The meiotic pairing was much lower than expected. Never 14 bivalent from the DDR^cR^c genomes were observed. This factor could be the origin of the low fertility. The hybrid is self sterile and only two grains have been obtained until now after hundreds of backcrosses with pollen from the tetraploid triticale parental. Nevertheless, this approach should be explored further.

Acknowledgments

We are greatly indebted to the Spanish C.I.C.Y.T. (Projects AGF92-0999 and AGF92-0184) for the financial support, and to the C.I.D.A. of Córdoba for allowing the use of its facilities.

References

1.   Krolow KD. 4x Triticale, production and use in Triticale breeding. Proc. Int. Wheat Genet. Symp., 4th, 1973, Columbia, Missouri, pp. 237-243.
2.   Gustafson JP & Krolow KD. A tentative identification of chromosomes present in the tetraploid triticale based on heterochromatin banding patters. Can. J. Cytol. 1978;20:199-204.
3.   Sodkiewicz W. Amphiploid *Triticum monococcum* L. x *Secale cereale* L. (AARR)- a new form of tetraploid triticale. Cereal Res. Commun. 1984;12:35-40.
4.   Fedak G. Cytogenetics of tissue culture regenerated hybrids of *Triticum tauschii* x *Secale cereale*. Can. J. Genet. Cytol. 1984;26:382-386.
5.   Bernard S & Bernard M. Creating new forms of 4x, 6x and 8x primary triticale associating both complete R and D genomes. Theor. Appl. Genet. 1987;74:55-59.
6.   Kawakubo J. & Taira T. Intergeneric hybrids between *Aegilops squarrosa* and *Secale cereale* and their meiotic chromosome behaviour. Plant Breed. 1992;109: 108-115.
7.   Reimann-Philipp R. Perennial spring rye as a crop alternative. J Agronomy & Crop Science 1986;157:281-285.
8.   Cabrera A & Martín A. A trigeneric hybrid between *Hordeum*, *Aegilops* and *Secale*. Genome 1992;35:647-649.
9.   Dvorák J. Effect of rye on homoeologous chromosome paring in wheat X rye hybrids. Can. J. Genet. Cytol. 1977;19:549-566.
10.   Rubiales D., Niks RE, & Martín A. Genomic interactions in the resistance to mildew and rust fungi in hybrids and amphiploids involving the genera *Triticum*, *Hordeum* and *Secale*. Cereal Res. Commun. 1993;21:187-194.
11.   Canalejo AL. Regeneración *in vitro* y transformación genética en anfiploides de triticíneas [dissertation]. Córdoba (España): Universidad de Córdoba, 1994.

MOLECULAR CYTOGENETICS ANALYSIS OF TRITICALE X TRITORDEUM F₁ HYBRIDS

José Lima-Brito[1], Henrique Guedes-Pinto[1] and J.S.Heslop-Harrison[2]

[1]Department of Genetics and Biotechnology, University of Trás-os-Montes e Alto Douro, 5000 Vila Real, Portugal
[2]Karyobiology Group, John Innes Centre, Colney Lane, NR4 7UH Norwich, England

Abstract

The parental origin of all chromosomes present in Triticale X Tritordeum (AABBRH^{ch}, 2n=6X=42) F₁ hybrids was established in metaphases using fluorescence *in situ* hybridization technique with cloned, repetitive DNA probes and total genomic DNA. Total genomic rye DNA can be used as a probe to label the seven rye-origin chromosomes strongly and uniformly along their lengths and telomeric regions. The seven *Hordeum chilense*-origin chromosomes and the 28 wheat-origin chromosomes were labelled weakly and almost undetectably, respectively. In contrast, when total genomic *Hordeum chilense* DNA was used as a probe, under the same conditions, the seven *H. chilense* and rye-origin chromosomes were strongly labelled and the wheat-origin chromosomes showed a weak *in situ* hybridization signal. Cloned and repetitive DNA probes of rye (pSc200) and *H. chilense* (pHcKB6) enabled all their own-species chromosomes to be identified in the F₁ hybrids. The *in situ* hybridization patterns analysis of several probes used on prophase and interphase nuclei showed that the different parental origin chromosomes did not lie randomly, but were located in one to three different domains. The fluorescence *in situ* hybridization technique is a useful molecular cytogenetic method on the identification of the different parental origin chromosomes present in these hybrids not only in metaphase but also at other stages of cell cycle.

Introduction

Different trigeneric (*Triticum* , *Hordeum* and *Secale*) hybrids have been produced in *Triticeae* from crosses between triticale (X *Triticosecale* Wittmack) and tritordeum (X *Tritordeum* Ascherson et Garbner) [1,2,3]. These will be used to make substitution lines or recombinant lines with improved agronomic performance.

In the present study we identify the parental origin of all chromosomes of a triticale x tritordeum F₁ hybrid either in metaphases or in prophases and interphases, using *in situ* DNA:DNA hybridization with cloned, repetitive DNA probes and total genomic DNA.

183

H. Guedes-Pinto et al. (eds.), Triticale: Today and Tomorrow, 183–188.
© 1996 *Kluwer Academic Publishers. Printed in the Netherlands.*

184

Material and methods

We used F₁ hybrids AABBRH^{ch} (2n=42) between 6x triticale advanced line UTAD17/85 (with the genome constitution AABBRR) and 6x tritordeum advanced line HT67 (AABBH^{ch}H^{ch}), following [4] for root-tip spreads preparation and [5] for *in situ* hybridization and probe detection.

The probes used in hybridization were:

Total genomic DNA from *Hordeum chilense* and *Secale cereale* cv. "Petkus" labelled with digoxigenin and pHcKB6 [6] and pSc200 [7] labelled with biotin. *Triticum aestivum* cv. "Chinese Spring" DNA was used as blocking DNA to reduce cross hybridization.

The slides were counterstained with DAPI (4'-6' diamidino-2-phenylindole).

Results

CHROMOSOME IDENTIFICATION

Figures 1 and 3 show metaphase chromosomes of the F₁ hybrid (AABBRH^{ch}, 2n=42) after staining and *in situ* hybridization.

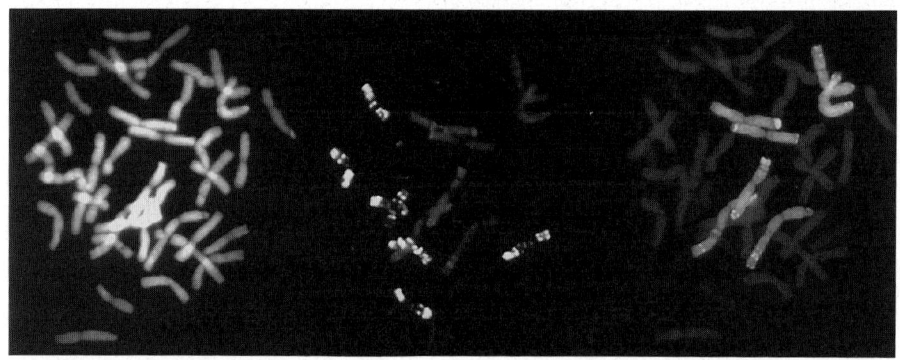

FIGURE 1 - Root tip metaphase of the F₁ hybrid triticale x tritordeum counterstained with DAPI (blue; left), and probed with pHcKB6 (red; centre) and total genomic rye DNA (green; right).

Triticale x tritordeum nuclei contain 42 chromosomes, seven of *H.chilense* -origin, seven of rye-origin and 28 of wheat-origin. The genomic origin of H^{ch} and R chromosomes could be distinguished. Total genomic rye DNA strongly and uniformly labels chromosomes with stronger regions coincident with terminal heterochromatin. The wheat and *H. chilense* -origin chromosomes were weakly stained.

The repetitive probe pHcKB6 enables the identification of *H. chilense* chromosomes with a characteristic pattern of hybridization (Figure 2). Some minor hybridization signal is present on wheat-origin chromosomes.

Each *H. chilense* chromosome arm has a characteristic *in situ* banding pattern which enables its identification and may be useful to find breakpoints in translocations or identify chromosomes in substitution or addition lines.

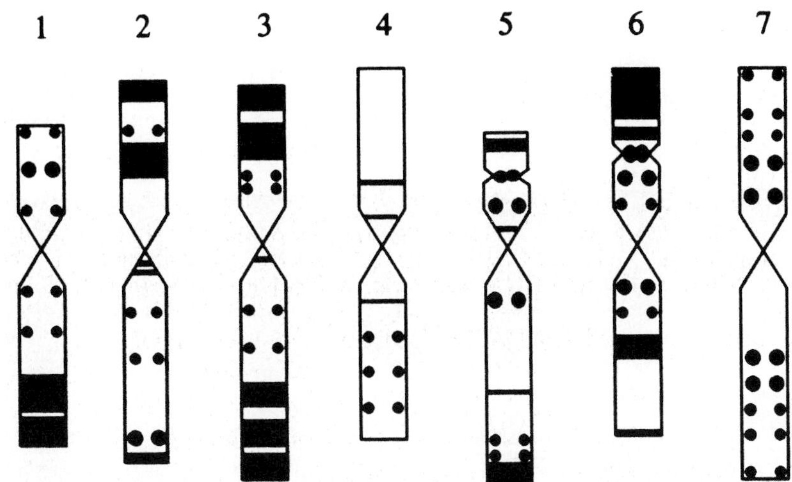

FIGURE 2 - Karyotype of *H. chilense* probed with pHcKB6.

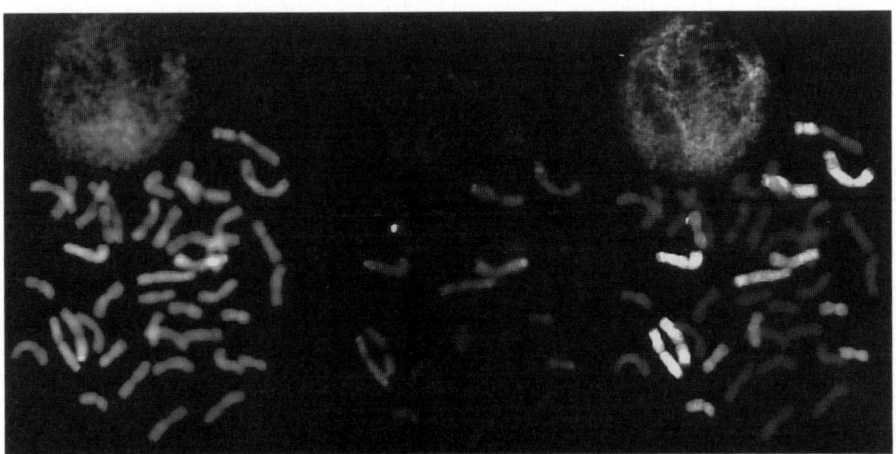

FIGURE 3 - Root tip metaphase of the F₁ hybrid triticale x tritordeum after staining with DAPI (blue; left) and *in situ* hybridization with pSc200 (red; centre) and total genomic *H. chilense* DNA (green; right).

The 42 chromosomes show uniform DAPI staining with some brighter fluorescence at heterochromatin bands. Total genomic *H. chilense* DNA labelled brightly the small *H. chilense*- origin chromosomes and the rye-origin chromosomes as well, excluding the subtelomeric heterochromatin. The wheat-origin chromosomes are weakly labelled but in some experiments characteristic bands were seen. The repetitive probe pSc200 gave a characteristic hybridization pattern in the telomeric and subtelomeric regions of rye-origin chromosomes and the *H. chilense* and wheat-origin chromosomes were not labelled.

NUCLEAR ARCHITECTURE

Figures 4 and 5 show prophases and interphase nuclei of the F1 hybrid triticale x tritordeum after staining and *in situ* hybridization.

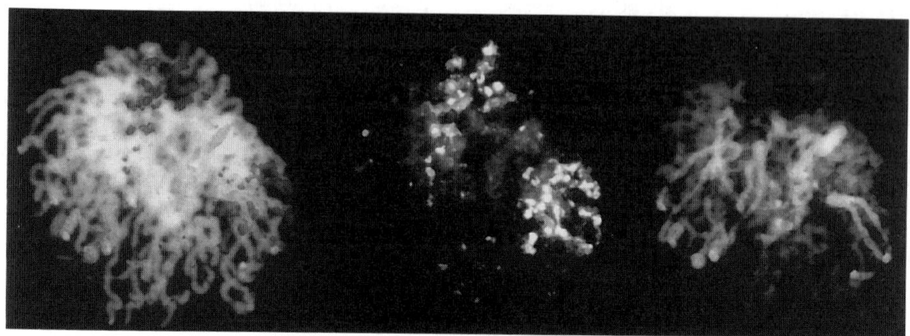

FIGURE 4 - Root tip prophase of the F1 hybrid triticale x tritordeum after staining with DAPI (blue; left) and simultaneous *in situ* hybridization with pHcKB6 (red;centre) and total genomic rye DNA (green; right).

Analysis of hybridization patterns of the probes to prophase nuclei revealed that the chromosomes from the three genera were lying in differents domains in most cells although some *H. chilense* -rye genome intermixture occurs. The *H. chilense* and rye-origin chromosomes were clustered in one or three domains, particularly in the earliest prophases (Fig.4).

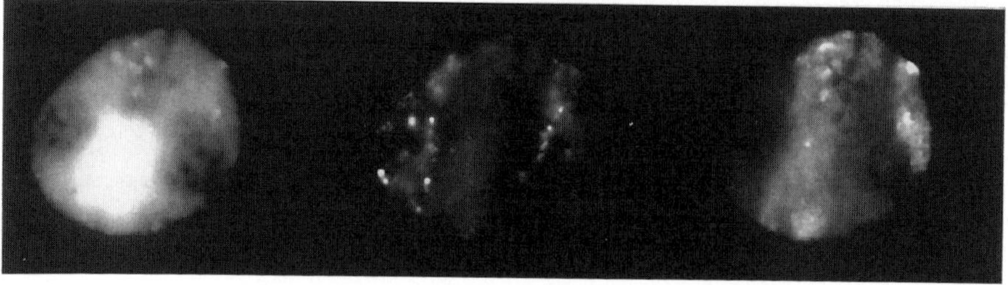

FIGURE 5 - Root tip interphase of the F1 hybrid triticale x tritordeum stained with DAPI (blue; left) and probed with total genomic rye DNA (green; centre) and pHcKB6 (red, right).

At interphase, different domains were also visible although the axes of the chromosomes could not be distinguished (Fig.5).

Discussion

Total genomic DNA *in situ* hybridization is a valuable method to identify chromosomes originating from different genomes in hybrids (see [8]).

No translocations were detected in the lines (Figures 1 and 2), although even small interchanges between rye, *H. chilense* and wheat-origin chromosomes would have been identified with a combination of rye and *H. chilense* genomic DNA probe and pHcKB6.

The high level of cross hybridization detected between *H. chilense* and rye chromosomes when we used *H. chilense* as a probe was surprising especially since there was negligible rye hybridization to *H. chilense*-origin chromosomes (Figure 1).

As genomic *H. chilense* hybridizes weakly to the wheat-origin chromosomes a repetitive sequence from *H. chilense* may be useful for the identification of wheat chromosomes and it would be worthwhile to isolate and clone the sequence involved.

Both pHcKB6 (Figure 1) and pSc200 (Figure2) revealed their specificity to the *H. chilense* and rye-origin chromosomes giving different but characteristic *in situ* hybridization patterns.

Conclusion

In situ hybridization with both cloned, repetitive DNA probes and total genomic DNA probes enables the identification of most chromosomes in metaphase, prophase and interphase nuclei of Triticale X Tritordeum F_1 hybrids. It will be useful for both the characterization of substitution or recombinant lines produced in such crosses and for the selection of desirable chromosome combinations in the lines.

Acknowledgements

This work was supported by Junta Nacional de Investigação Científica e Tecnológica (JNICT) grant BD 2661/93-IE.

REFERENCES

1. Fernández-Escobar J, Martin A. A hybrid between hexaploid Triticale and octoploid Tritordeum. Cer. Res. Comm. 1988; 16(1-2): 45-51.

2. Fernández-Escobar J, Martin A. A self-fertile trigeneric hybrid in the *Triticeae* involving *Triticum*, *Hordeum* and *Secale*. Euphytica 1989; 42: 291-296.

3. Lima-Brito J, Guedes-Pinto H. Tritordeum x Triticale crossability studies in dialelic crosses. Book of Abstracts of XXVIII Jornadas Luso-Espanholas de Genética, Faro, Portugal 1993, p.125. (*in press*).

4. Schwarzacher T, Leitch AR, Bennet MD, Heslop-Harrison JS. *In situ* localization of parental genomes in a wild hybrid". Ann. Bot. 1989; 64: 315-324.

5. Schwarzacher T, Leitch AR, Heslop-Harrison JS. DNA-DNA *in situ* hybridization-methods for light microscopy. In: Harris N, Oparka KJ, eds. Plant Cell Biology: a Pratical Approach. Oxford: 1RL/Oxford University Press, 1994: 127-155.

6. Anamthawat-Jónsson K, Heslop-Harrison JS. Isolation and characterization of genome-specific sequences in *Triticeae* species. Mol. Gen. Genet 1993; 240: 151-158.

7. Vershinin A, Schwarzacher T, Heslop-Harrison JS. The large scale genomic organization of repetitive DNA families at the telomeres of rye chromosomes. 1996 (Submitted).

8. Schwarzacher T *et al*. Genomic in situ hybridization to identify alien chromosomes and chromosome segments in wheat". Theor. Appl. Genet. 1992; 84: 778-786.

DIFFERENT CHROMOSOME COMBINATIONS ON TETRA- AND HEXAPLOID LEVEL FROM HYBRIDS OF TETRAPLOID RYE × TETRAPLOID TRITICALE

Barbara Apolinarska
Institute of Plant Genetics, Polish Academy of Sciences
ul. Strzeszyńska 34, 60-479 Poznań

Abstract

Different chromosome combinations at the tetra- and hexaploid level from the backcross hybrids 4x rye × 4x triticale with 4x triticale and 4x rye in the F_2 generation were obtained. (1) Rye-cytoplasmic tetraploid triticale with 8 different chromosome constitutions in wheat genome. In the wheat genome, all chromosomes from A and B genomes, except the chromosome 3A, were represented. The ratio of wheat chromosomes in genomes A to B was from 1:6 (A_1B_6) to 4:3 (A_4B_3). (2) Lines of tetraploid rye with additional wheat chromosomes. Addition chromosomes in the lines of tetraploid rye were all chromosomes from the A genome, except chromosome 3A, and from B genome only the chromosomes 3B and 7B. The chromosome 3A occurred in a translocation form 3AS/5AL. (3) Lines of tetraploid rye with A/R and B/R substitutions. In the substitution lines occurred monosomics of chromosomes from A genome – 1A, 4A, 5A and 7A, and from B genome – 2B, 3B and 7B. Only two chromosomes from genome A – 1A and 5A occurred in the disomic form. (4) Substitution-additive lines of tetraploid rye with wheat chromosomes. (5) Allohexaploids with ABRRRR genome constitution. A similar karyotype (one chromosome combination) – 1B, 2A, 3B, 4B, 5B, 6A and 7B – occurred in the wheat genome of all plants. (6) Tetraploid rye.

Introduction

Tetraploid triticale with various chromosome combinations in wheat genome is an ideal form to extend genetic variation in triticale and rye. A suitable manipulation of crossing between tetraploid and octoploid triticale, between tetraploid triticale and di- and tetraploid rye leads to the obtaining of numerous substitution lines of 6x triticale [1, 2], addition lines of diploid rye [3], addition, substitution, addition-substitution and translocation lines of tetraploid rye.

189

H. Guedes-Pinto et al. (eds.), Triticale: Today and Tomorrow, 189–194.
© 1996 *Kluwer Academic Publishers. Printed in the Netherlands.*

Material and Methods

A line of tetraploid rye derived from twin embryos [Sulinowski, unpublished] was crossed with different lines of tetraploid triticale. Plants of F_1 generation were backcrossed to both 4x triticale and 4x rye. The progeny of backcross hybrids was exposed to open pollination. In the BC_1-F_2 progeny 119 plants were analysed by the c-banding method according to the technique of Lukaszewski and Gustafson [4].

Results and Discussion

Different lines of tetraploid triticale derived from simple and complex crosses of hexaploid triticale with rye [5-13] or with tetraploid wheat [13] having a mixed wheat genome (AB) or a pure genome B [10] are not so important in breeding. The best forms obtain about 60% of hexaploid triticale yield [8] or hardly exceed 50% yield of the standard cultivar Malno [12], whereas tetraploid triticale can be successfully used in chromosome engineering for manipulation of chromosomes on different ploidy levels. Thus, it has been used for D genome introduction into hexaploid triticale instead of some chromosomes of the A and/or B genomes. As a result of crossing between octoploid and tetraploid triticale followed by plant selection at the hexaploid level in the F_5 and F_6 generations, Lukaszewski et al. [1] and Apolinarska [2] obtained 140 lines of hexaploid triticale with 70 different D/A and D/B substitutions, while simultaneously preserving a complete rye genome.

Table 1. Stabilized and unstabilized karyotypes of rye-cytoplasmic tetraploid triticale

Homeologous Group							Lines
1	2	3	4	5	6	7	
AA	AA	BB	AA	AA	BB	BB	1
AB	AA	BB	AA	AA	BB	BB	1
AA	AA	BB	AA	BB	AA	BB	13
AA	AA	BB	AA	BB	BB	BB	8
AA	AA	BB	BB	AA	BB	AA	1
AA	AA	BB	BB	BB	BB	BB	1
AA	BB	BB	BB	AA	AA	AA	1
BB	AA	BB	AA	BB	BB	BB	1
BB	AA	BB	A	AB	BB	BB	1
BB	BB	BB	AA	BB	BB	BB	1

Using backcross hybrids and an inbred rye line, Thiele et al. [3] derived addition lines of diploid rye with rye cytoplasm and additional wheat chromosomes.
Backcross hybrids of tetraploid rye with tetraploid triticale were a source of different chromosome combinations at the tetra- and hexaploid level.

Twenty nine plants of secondary tetraploid triticale containing rye cytoplasm with eight different chromosome constitutions in wheat genome were obtained from backcross hybrids of tetraploid rye × tetraploid triticale with 4x triticale (Table 1). Two plants – one in the first and another in the fifth homeologous group – had a nonstabilized chromosome constitution. Besides that, two plants were aneuploid, i.e. having 27 chromosomes, with 13 rye chromosomes and the missing chromosome 4R. One of the plants had 15 rye and 13 wheat chromosomes, the chromosome 4A having a monosomic form. The wheat genome of tetraploid triticale contained all chromosomes of the A and B genomes except the chromosome 3A, and the chromosome ratio between A and B genomes ranged from 1:6 (A_1: B_6) to 4:3 (A_4: B_3). No complete A and B genomes were found.

Hexaploid (AABBRR) and octoploid (AABBDDRR) triticale have a stable genome constitution, except for a few substitution and translocation lines. In the case of tetraploid triticale derived from hybrids of hexaploid triticale or from tetraploid wheat and rye, different chromosomes from the A and B genomes may occur in wheat genome. In 105 analysed lines of wheat-cytoplasmic tetraploid triticale, 48 different karyotypes were obtained from 128 theoretically possible combinations of chromosomes constitutions of wheat genome on a tetraploid level [14, 6, 7, 8, 10, Łapiński, unpublished]. In spite of such diversity, only three karyotypes out of eight chromosome constitutions of secondary rye-cytoplasmic tetraploid triticale were in common with wheat-cytoplasmic 4x triticale lines [15]. Karyotypes of the discussed lines were also different from the karyotype of rye-cytoplasmic 4x triticale obtained from backcross hybrids to 6x triticale and 2x rye [16].

Despite many efforts, the lines of wheat-cytoplasmic tetraploid triticale did not satisfy breeders' requirements. It happenes so probably because of many conditions. One of them is the specificity of wheat genome consisting of chromosomes from different genomes. On the other hand, however, the equilibrium of rye and wheat genomes, which occurs in 4x triticale, and thereby a larger expression of rye genomes as compared to hexaploid or octoploid triticale, probably require a replacement of wheat cytoplasm with rye cytoplasm. But the first rye-cytoplasmic tetraploid triticale obtained by Melz and Schlegel [16] was not characterized by more favourable traits in comparison to 4x triticale with wheat cytoplasm, and even in the case of the chromosome conjugation analysis it showed a considerably larger number of univalents than tetraploid triticale with wheat cytoplasm [17].

The lines of tetraploid triticale obtained from backcross hybrids of tetraploid rye with tetraploid triticale were characterized by an unsatisfactory fertility, which was also varying and ranged from 0 to 31 grains per spike. The seeds were rye-like and quite well filled. This trait distinguished the discussed triticale from that obtained by Melz and Schlegel [16].

Tetraploid triticale may play a particularly significant role in obtaining additive, substitution and translocation lines of diploid and tetraploid rye.

Thiele et al. [3] as a result of a double backcross of rye-cytoplasmic tetraploid triticale with an inbred line of diploid rye derived the first additive line of diploid rye with rye cytoplasm and additional chromosome 6B. The plants showed almost normal development with a small reduction of tillering and a decreased plant height.

The first additive lines of diploid rye with six different chromosomes, but with wheat cytoplasm were obtained by Schlegel [18] and Schlegel et al. [19]. These lines were derived from backcross hybrids of wheat × rye with rye, and probably wheat cytoplasm caused a considerably weaker development of plants.

Tetraploid triticale played a very significant role in hybrids with tetraploid rye, since it was used for the obtaining of tetraploid rye lines with additional wheat chromosomes.

In the F$_2$ generation of the backcross hybrids tetraploid rye × tetraploid triticale with rye, nine additive lines of tetraploid rye were obtained. Their chromosome number ranged from 29 to 31. Additive were all chromosomes from A genome, except 3A, and only 3B and 7B from B genome. The chromosome 3A occurred in a translocation form 3AS/5Al (Table 2). In all additive plants the chromosomes were monosomics, except two plants. In one of them there occurred a pair of additive chromosomes 5A, and in another – a pair of translocation chromosomes 3AS/5Al.

From the same combination of backcross hybrids of tetraploid rye × tetraploid triticale with rye, the substitution lines of tetraploid rye were obtained. In 30 substitution lines of 4x rye the following wheat chromosomes occurred: 1A, 4A, 5A, 6A, 7A – from A genome and 2B, 3B, 7B – from B genome. Only two chromosomes – 1A and 5A were disomics, the chromosome 1A occurring in one plant and 5A – in two plants. In 28-chromosome plants of tetraploid rye the number of substitutions ranged from 1 to 3. In four substitution lines there occurred monosomics of the translocated chromosomes 3AS/5AL, 3AS/5BL, 7AS/5Al (Table 2). Plants with A/R and B/R substitutions significantly deviated from 4x rye – a maternal form and additive lines in the rate of development. In the initial period of development they strongly tillered, but eared about a month later than 4x rye.

Table 2. Frequency of A and B genome chromosomes the lines of tetraploid rye

Genom	No. of chromosomes in		
	Addition lines	Substitution lines	Substitution-addition lines
1A	5	6	4
2A	1		
3A			
4A	3	4	1
5A	5	8	3
6A	1	2	1
7A	3	7	5
Total A	18	27	14
1B			
2B		1	
3B	1	2	
4B			
5B			
6B			
7B	2	2	1
Total B	3	5	1
3AS/5AL	3	2	1
7AS/7RL		1	
3AS/5BL		1	
Total trans-locations	3	4	1

Out of 29-30-chromosome backcross hybrids of tetraploid rye × tetraploid triticale with rye a group of substitution-additive plants was separated in F_2. Four 29-chromosome plants were obtained. Three of them had two monosomic wheat chromosomes in their karyotype – 1A and 7A, and one plant had three wheat chromosomes – two normal, 1A and 7A, and one plant had three wheat chromosomes – two normal, 1A and 7A, and one translocation – 3AS/5AL. Two plants were 30-chromosome: in one of them there occurred the combination 2×5A, 6A, 7B, whereas in another – 4A, 5A and 7A. Substitution-additive plants resembled 4x rye in their development and morphology.

Completely unexpected in the backcross hybrids of tetraploid rye × tetraploid triticale with rye were 42-chromosome plants – allohexaploids with the genome constitution ABRRRR. A similar karyotype (one chromosome combination) – 1B, 2A, 3B, 4B, 5B, 6A and 7B – occurred in wheat genome of all plants.

Allohexaploids probably arose from unreduced gametes as a result of irregularities in meiotic divisions. It is conceivable that the maternal form – 4x rye, which was derived from twin embryos of diploid rye, had an influence on the formation of allohexaploids. That line had probably genetic tendencies to produce forms with different ploidy. Twin embryos capable to produce forms with different ploidy, were also used in grass polyploidization [20].

Allohexaploids have the genome structure identical to that of autoallohexaploids and can be used according to the suggestion of Krolow and Lukaszewski [15], Lehmann and Hohmann [21] – for introduction of D genome into hexaploid triticale in the form of D/A and D/B substitutions. Quite a large group (15 plants) in the backcross hybrids of tetraploid rye × tetraploid triticale with rye were constituted by tetraploid rye.

Addition, substitution and translocation lines are widely used in genetic and breeding studies. These lines serve to transmit genetically alien material from related species and thus play an important role in the extension of genetic variation.

Genetic and breeding works focused on the introduction of desirable genes of rye into wheat. An example can be the translocation 1BL/1RS which widely occur in many wheat varieties. Few additive lines of diploid rye [18, 19, 3] as well as additive, substitution and translocation lines of tetraploid rye may initiate studies on the introduction of desirable wheat genes into rye [22].

References

1. Lukaszewski A.J., Apolinarska B. Introduction of the D-genome chromosomes from bread wheat into hexaploid triticale. Genome 1987; 29: 425-430.
2. Apolinarska B. Substytucje D/A i D/B w pszenżycie heksaploidalnym (D/A and D/B substitutions in hexaploid triticale). Zeszyty Probl. Inst. Hod. Akl. Rośl. Radzików 1990; 57-64.
3. Thiele V., Buschbeck R., Seidel A., Melz. G. Identification of the first rye cytoplasmic rye - wheat - addition using L A P - isozymes. Plant Breed. 1988; 101: 250-252.
4. Lukaszewski A.J., Gustafson J.P. Translocations and modifications of chromosomes in triticale × wheat hybrids. Theor. Appl. Genet. 1983; 64: 239-248.
5. Krolow K.-D. Research work with 4x-Triticale in Germany (Berlin). Proc. Int. Symp. El Batun, Mexicco, 1973: 51-60.

6. Lukaszewski A.J., Apolinarska B., Gustafson J.P., Krolow K.-D. Chromosome constitution of tetraploid Triticale. Z. Pflanz. 1984; 93: 222-236.

7. Bernard M., Bernard S., Saigne B. Tetraploid triticales; investigations on their genome and chromosome constitution. Proc. on Eucarpia. Meet. on Triticale 1984, July 2-3; Clermont-Ferrand France. 1985; 277-288.

8. Hohmann U. Cytology and fertility of primary and secondary tetraploid triticale and advanced populations. Proc. Eucarpia Meet. on Triticale 1984, July 2-3. Clermont-Ferrand, France. 1985; 267-276.

9. Sabeva Z. Obtaining and investigation of tetraploid Triticale forms. Cer. Res. Com. 1985; 13: 71-76.

10. Baum M., Lelley T. A new method to produce 4x Triticales and their application in studying the development of a new polyploid plant. Plant Breed. 1988; 100: 260-267.

11. Dubovets N.I., Badaev N.S., Bolsheva N.L., Badaeva E.D., Shcherbakova A.M., Bormotov V.E., Zelenin A.V.. Regularities of karyotype formation in tetraploid Triticale. Cer. Res. Com. 1989; 17: 253-257.

12. Łapiński B., Budzianowski G. Tetraploid Triticale breeding in the P B A E S Malyszyn. Biul. Inst. Hod. Akl. Rośl. 1992; 183: 119-124.

13. Apolinarska B. Stabilization of ploidy and fertility level of tetraploid triticale obtained from four cross combinations. Genetica Polonica 1993; 34: 121-131.

14. Gustafson J.P., Krolow K.-D. A tentative identifications of chromosomes present in tetraploid triticale based on heterochromatic banding patterns. Can. J. Genet. Cyt. 1978; 20: 199-204.

15. Krolow. K.-D., Lukaszewski A.J. Tetraploid Triticale – a tool in hexaploid Triticale breeding. Genet. Manip. in plant Breed. 1986: 105-118.

16. Melz G., Schlegel R. Production and cytogenetical analysis of a rye-cytoplasmic tetraploid Triticale. Plant. Breed. 1987; 98: 200-204.

17. Lukaszewski A.J., Apolinarska B., Gustafson J.P., Krolow K.-D. Chromosome pairing and aneuploidy in tetraploid triticale. I. Stabilized karyotypes. Genome 1987; 29: 554-561.

18. Schlegel R., First evidence for rye – wheat additions. Biol. Zbl. 1982; 101: 641-646.

19. Schlegel R., Kynast R., Schmidt J.-C. Alien chromosome transfer from wheat into rye. Genet. Manip. in Plant Breed. 1986: 129-136.

20. Sulinowski S., Wiśniewska H., Sękowska K. . Frequency of spontaneous polyploids in Lolium perenne and Festuca pratensis. Eucarpia Fodder Crop Section. "The Utilization of Genetic Resources in Fodder Crop Breeding". 1982: 55-59.

21. Lehmann C., Hohmann U., Krolow K.-D. Tetraploid triticale with D-genome chromosomes from Triticum aestivum produced with autoallohexaploid triticale. Cer. Res. Com. 1991; 19: 469-476.

22. Schlegel R., Kynast R. Wheat chromosome 6B compensates genetical information of diploid rye, Secale cereale L. Seventh Int. Wheat Genetics Symp. Cambrigde. 1988: 421-426

CYTOLOGICAL ANALYSIS OF F₁ AEGILOPS HYBRIDS WITH TRITICOSECALE

Daniela Gruszecka, Czeslaw Tarkowski, Grazyna Stefanowska,
Krystyna Marciniak
Institute of Genetics and Plant Breeding, Agricultural University
Lublin, Poland

Abstract

The aim of the work was to analyse the degree of chromosome conjugation and the vitality of pollen of hybrids F_1 originating from the hybridization of Triticum, Secale and Aegilops.

In MI hybrid plants of a secondary hexaploid triticale with Ae. squarrosa 2x, single bivalents were found. On average, 0.74 chiasmata fell to one cell. A relatively low degree of conjugation occurred in hybrids of triticale with Ae. crassa 4x (on average 2.4-4.2 chiasmata/cell) and in hybrids of triticale with Ae. juvenalis 6x and Ae. triaristata 6x (4.3-5.8 chiasmata/cell). In combinations with Ae. crassa 4x, Ae. juvenalis 6x, and Ae. triaristata 6x only single multivalents were found.

The vitality of pollen was low. It ranged from 16.3% (triticale with Ae. crassa 4x) to 42.5% (triticale with Ae. juvenalis 6x). Hybrids with Ae. juvenalis 6x showed a considerably higher pollen vitality than hybrids with Ae. triaristata 6x.

Cytogenetic studies have shown that there is possibility of introgression of genes from Aegilops to triticale.

Introduction

The purpose of the crossings between various species and genera is to create new plant genotypes, which could be practically used in a wider range than the basic forms, because of the proper conjunction of the features. The Aegilops species do not take much place in the breeding programme, but still they constitute a highly valuable source of resistance genes against pests and diseases concerning mainly the cultivated cereal [1- 4].

H. Guedes-Pinto et al. (eds.), Triticale: Today and Tomorrow, 195–201.

© 1996 Kluwer Academic Publishers. Printed in the Netherlands.

The aim of the study was to analyse the degree of chromosome conjugation and the vitality and size of pollen grains of F_1 hybrids originating from the hybridization of Triticum, Secale and Aegilops species.

Materials and Methods

The F_1 plants of eight combinations of Triticosecale with Aegilops were investigated (from 1 to 3 plants per one combination, total number of plants - 19). The hexaploid triticale components (n=3x=21=ABR) were received at the Institute of Genetics and Plant Breeding, Agricultural University in Lublin: (Lanca x L506/79) x CZR 142/79, (Jana x Tempo) x Jana, Panda x Dańkowskie Złote. Goat grasses: Ae. squarrosa (n=x=D), Ae. crassa (n=2x=DM^{cr}), Ae. juvenalis n=3x=DC^{U}M^J) and Ae. triaristata (n=3x=C^{u}M^{t}M^{t2}) were received from the collection. The 14-day old F_1 germs were grown in vitro on the Murashige and Skoog's nutrient.

In the pollen mother cells (PMC) stained by the orcein method, the stage of the metaphase I and tetrads (50 cells per plant) was analysed. The following were determined: the frequency of univalents, the number and kind of bivalents and multivalents, the number of chiasmata per cell and the number of micronuclei per tetrad.

Applying the staining by the methylene blue solution, the vitality of pollen was estimated (over 500 grains from each plant), classifying it into one of the following three groups: pollen totally (75-100% vol.) and partly (25-75% vol.) filled with cytoplasm was regarded as a vital one, and an empty pollen (0-25% vol.) as an unfertile pollen. From among total 60 measurements of pollen grains (length and width), each analysed group mentioned above comprised twenty measurements.

Results

F_1 crossing the goat-grasses with secondary hexaploid triticale the hybrid plants were obtained with the frequency of 3.7% per the pollinated flower (3186 flowers had been pollinated and 118 embryos were received), of the chromosome number of 2n=28 (with Ae. squarrosa), 2n=35 (with Ae. crassa) and 2n=42 (with Ae. juvenalis and Ae. triaristata).

The meiotic analyses have shown very low chromosome pairing rates (Table 1). The bivalents in the metaphase I were most often seen as rod, rarely as rings. Single multivalents (tri-and quadrivalents) were found in the combinations of triticale with Ae. crassa, Ae. juvenalis or Ae. triaristata. There may possibly exist a homologous, as well as homoeologous pairing between the triticale and goat-grass chromosomes (Photos.1-3). F_1 the hybrids of triticale with Ae. squarrosa at MI on average, 26.9 univalents and single bivalents; 0.74 chiasmata per cell were observed (Table 1). In the hybrids of triticale with Ae.crassa there were, on average, 28.3-31.2 univalents and 2.4-4.2 chiasmata per cell. Higher number of univalents and lower number of chiasmata were observed in the combinations of Ae. crassa x (Panda x Dańkowskie Złote). Relatively low degrees of pairing also appeared in the hybrids of triticale and hexaploid forms of goat-grasses. As the result of back-crossing it was found that the hybrid with the maternal Ae. triaristata form had a slightly higher degree of pairing, compared to the combination, in which this hybrid was the pollinator.

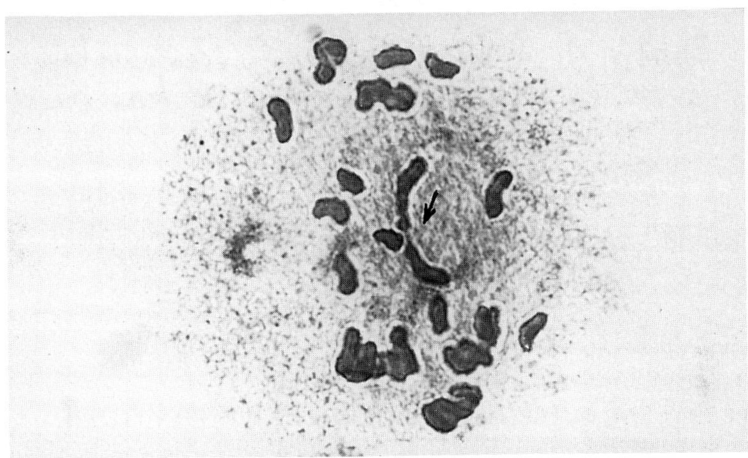

Photo.1. M I-26$_I$ +1$_{II}$, F$_1$(Ae. squarrosa 2x x [(Lanca x L 506/79) x CZR 42/79])

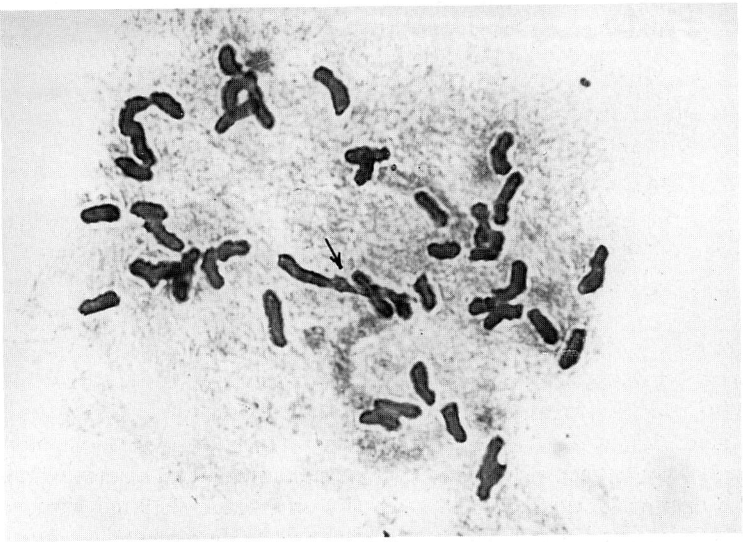

Photo.2   M I - 40$_I$ +1$_{II}$, F$_1$[(Panda x Dańkowskie Złote.) x Ae. triaristata 6x]

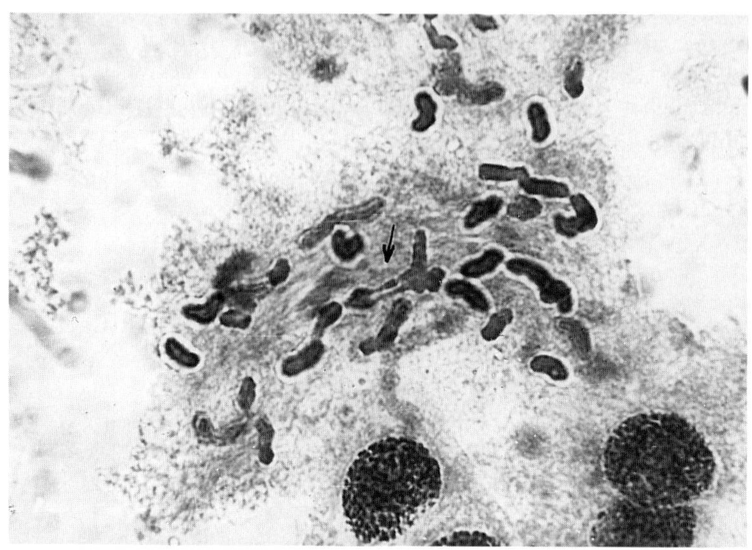

Photo.3 M $_I$ - 32$_I$ + 1$_{III}$, F$_1$ {Ae. crassa 4x x [Lanca x L 506/79) x 142/79]}

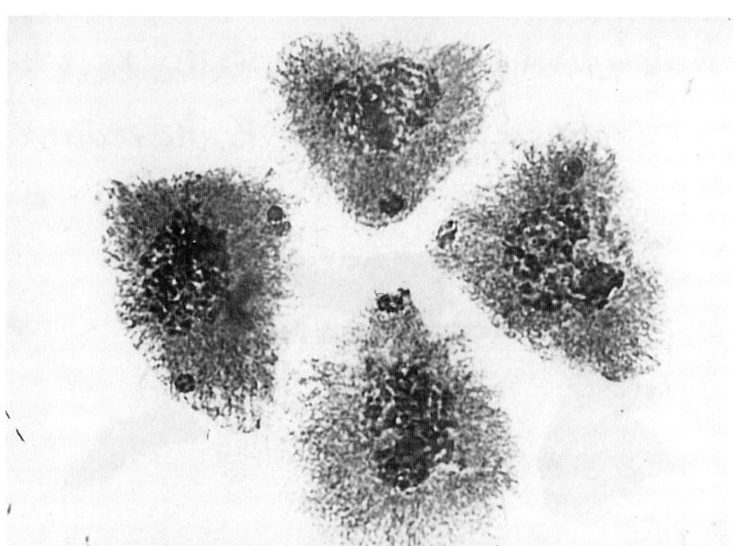

Photo.4 Tetrad - 10 micronuclei, F$_1$ (Ae. juvenalis 6x x [(Jana x Tempo) x Jana])

In the hybrid plants, the smallest number of micronuclei in the tetrads occurred in the combination with Ae. squarrosa (average value - 3.9), while the highest number with Ae. triaristata (average number 10.1) (Table 1, Photo 4/. No relationship between the number of univalents and the number of micronuclei was found.

Table 1. Analysis of meiotic division of F₁ hybrids (Aegilops x Triticosecale) (mean values).

No.	Crossing combination	F$_1$ chromosome number	Univalents/ PMC*	Bivalents/PMC		Multivalents/PMC		Chiasmata/ PMC	Micronuclei/ tetrad
				rods	rings	trivalents	quadrivalents		
1.	Ae.squarrosa 2x x [{Lanca x L506/79} x CZR 142/79]	28	26.9	0.16	0.23	-	-	0.74	3.9
2.	Ae. crassa 4x x [{Lanca x L506/79} x CZR 142/79]	35	28.3	2.50	0.70	0.04	0.02	4.20	8.2
3.	Ae. crassa 4x x {Panda x Dańk. Złote.}	35	31.2	1.80	0.10	-	-	2.40	6.4
4.	Ae. juvenalis 6x x [{Lanca x L506/79} x CZR 142/79}	42	33.0	3.40	0.80	0.20	0.02	5.30	7.1
5.	Ae. juvenalis 6x x [{Jana x Tempo} x Jana]	42	34.8	3.30	0.40	-	-	4.70	7.1
6.	Ae. triaristata 6x x {Panda x Dańk. Złote}	42	33.1	3.60	0.60	0.14	0.02	5.20	10.1
7.	[{Jana x Tempo} x Jana x Ae. juvenalis 6x	42	34.2	3.30	0.50	0.04	-	5.80	8.2
8.	{Panda x Dańk. Złote} x Ae. triaristata 6x	42	35.7	3.50	0.60	-	-	4.30	7.3

*PMC - pollen mother cell.

Table 2. Analysis of vitality (%) and size of pollem grains (μ) of F1 hybrids (Aegilops x X Triticosecale) (mean values).

| No. | Crossing combination | Pollen grains (%) | | | Pollen grain size (μ) | | | | |
| | | Fully filled | Partly filled | Empty | Vitality (ful-ly+part-ly filled) | Vital pollen (fully+partly filled) | | Average size | |
						length	width	length	width
1.	Ae.squarrosa 2x x [{Lanca x L506/79} x CZR 142/79]	6.8	10.1	83.1	16.9	57.2	52.6	55.9	50.1
2.	Ae. crassa 4x x [{Lanca x L506/79} x CZR 142/79]	5.9	10.3	83.7	16.3	57.1	53.1	54.5	50.2
3.	Ae. crassa 4x x {Panda x Dańk. Złote}	31.3	8.7	60.0	40.0	66.3	61.1	62.5	57.7
4.	Ae. juvenalis 6x x [{Lanca x L506/79} x CZR 142/79}	13.6	16.4	70.0	30.0	59.3	54.3	57.5	53.1
5.	Ae. juvenalis 6x x [{Jana x Tempo} x Jana]	24.4	17.8	57.8	42.2	58.3	54.7	57.4	53.4
6.	Ae. triaristata 6x x {Panda x Dańk. Złote}	7.5	12.2	80.4	19.6	58.4	53.6	57.0	52.5
7.	[{Jana x Tempo} x Jana x Ae. juvenalis 6x	22.7	19.8	57.5	42.5	60.1	55.8	58.2	56.9
8.	{Panda x Dańk. Złote} x Ae. triaristata 6x	8.6	10.2	81.2	18.8	55.5	50.6	53.4	48.7

The values of pollen vitality analysis ranged from 16.3% or the combination of Ae. crassa x [(Lanca x L 506/79) x CZR 142/79] to 42.5% for the combination of [(Jana x Tempo) x Jana] x Ae. juvenalis (Table 2). Among the hybrids of triticale with hexaploid goat-grasses, the forms with Ae. juvenalis had considerably greater vitality (30.0-42.5% on average) than those with Ae. triaristata (18.8-19.6%). The hybrid combinations of the highest pollen fertility also had the greatest dimensions: e.g., 60.1/55.8 u for the combination [(Jana x Tempo) x Jana] x Ae. juvenalis and 66.3/61.1 u for the combination Ae. crassa x (Panda x Dańkowskie Złote).

Conclusion

The occurrence of bivalents and multivalents in the metaphase I of the F hybrid meiosis points to homologous and homoeologous pairing between the chromosomes of triticale and Aegilops. It also points to a possible introgression of genes from Ae. crassa 4x, Ae. juvenalis 6x and Ae. triaristata 6x into triticale 6x.

References

1. Asay K.H.: Breeding Potentials in Perennial Triticeae Grasses. International Symposium (Sweden); 1991.
2. Bernard S., Gay G.: Introduction of Aegilops ventricosa germplasm into hexaploid triticale. Genetics and breeding of Triticale, EUCARPIA Meeting, Clermont-Ferrand (France); 1984 July 2-5; INRA. Paris: 1985: 221-225.
3. Gupta P.K., Fedak G.: Wide crosses for possible improvement of 6x Triticale (Triticosecale Wittmack). Proceedings of International Triticale Congress, Sydney: 1986: 464-468.
4. Stefanowska G., Tarkowski C., Gruszecka D., Prażak R.: Preliminary results on crossability between Triticum, Triticosecale and Aegilops species. XIII EUCARPIA Congress, Auger (France); 1992 July 6-11; Book of Poster Abstracts: 271-272.

C-BANDING CHARACTERISTICS OF RYE CHROMOSOMES IN WINTER HEXAPLOID TRITICALE CROSSED WITH MAIZE *[ZEA MAYS]*

Rogalska Stanislawa Maria
Department of Plant Genetics and Breeding
Agricultural University Poznan, Wojska Polskiego, Poznan, Poland

Abstract

In order to stablished C-banding the patterns rye chromosomes of eight winter 6x-triticale were studied. Those 6x triticale will be used for the obtaintion of double haploids. Low variability was observed for heterochromatin bands on rye chromosomes, that differed mainly on size and prominence of telomeric bands.

Introduction

The C-banding patterns of the rye chromosomes in eight winter, hexaploid triticales were observed. The aim of the experiment was to establish C-banding karyotypes for the triticale forms crossed with maize to facilitate further investigations of the chromosomes in derived doubled haploid plants. The triticale forms analyzed were Lasko, Presto, Prego, Vero, SZD1740, SZD1834, SZD3801 and MAH1590. Chromosomes were stained with Giemsa following the procedure of Darvey and Gustafson [1] with some modifications. These were: pretreatment in ice water overnight; slides were immersed into absolute ethanol for one hour only. C-banding of Imperial rye chromosomes were used as the standard [2]. The triticale forms used in this study had 14 rye and 28 wheat chromosomes.

Variability in heterochromatin banding was found among the triticales and lines studied. In general, Lasko, Presto, Prego and SZD1740 appeared to have less prominent telomeric heterochromatin bands compared with the rest of studied triticale forms [Fig. 1]. Cultivar Vero exhibited modification of the 1R chromosomes pair. There were reduced telomeric bands on the satelite and a big, prominent band in the NOR [Nucleolar Organising Region]. All rye chromosomes in this cultivar have big, prominent telomeric bands. In the rest of analyzed forms, the 1R chromosomes have similar banding patterns to each other and to the pattern in rye as well. The 2R chromosome pairs did not appear to differ widely among the studied forms. The same situation applied to chromosome pairs 3R, 4R and 5R. The chromosome 6R pair appeared to have some variability in number of

H. Guedes-Pinto et al. (eds.), Triticale: Today and Tomorrow, 203–205.

© 1996 *Kluwer Academic Publishers. Printed in the Netherlands.*

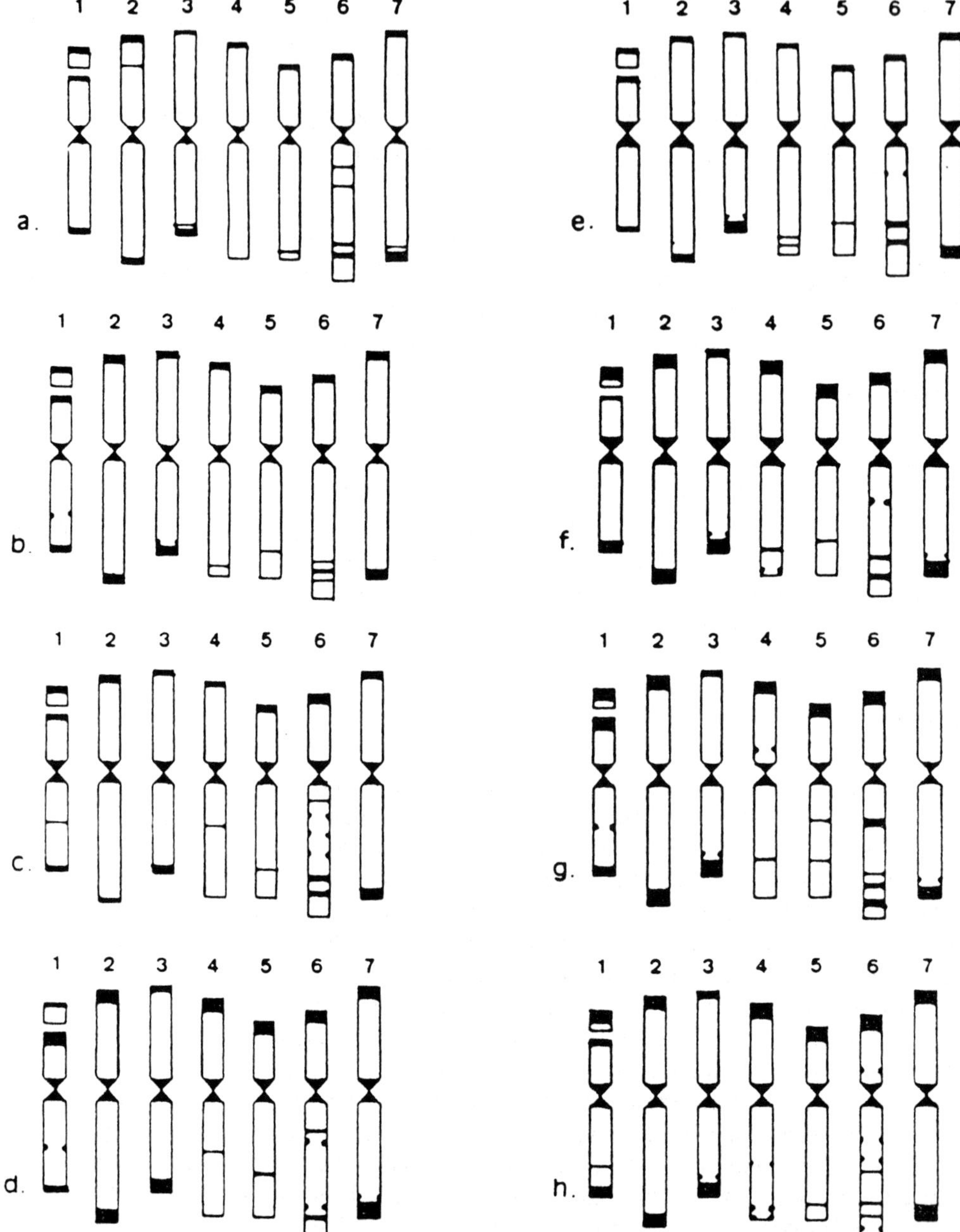

Fig. 1. Karyotypes of C-banding patterns of chromosomes of rye of investigated triticale forms: a-Lasko; b-Presto; c-Prego; d-Vero; e-SZD1740; f-SZD1834; g-SZD3801; h-MAH1590.

interstitial bands on longer arm. In all triticales chromosome 7R appeared to have both telomeric bands with a less prominent band on short arms. In summary, all rye chromosomes of triticales were very similar to those of Imperial rye.

Conclusion

From this study we concluded that the triticale forms do not show much variability in the presence and number of heterochromatin bands on rye chromosomes. They differed mainly with respect the size and prominence of telomeric bands.

References

1. Darvey, N.L., Gustafson, J.P., Identification of rye chromosomes in wheat-rye addition lines and triticale by heterochromatin bands. Crop Sci. 1975, 15: 239-245.
2. Sybenga, J., Rye chromosome nomeclature and homoeology relationships. Workshop report. Zeitschrift. f. Pflanzchtg. 1983, 90: 297-304.

CYTOLOGICAL DISORDERS IN HEXAPLOID TRITICALE (*TRITICOSECALE* WITTMACK). ASSOCIATION AND STABILITY BETWEEN TWO INDEXES.

Adriana Juárez ; Adriana Ordóñez ; Ricardo Maich ; Diana Manero de Zumelzú.
Facultad de Ciencias Agropecuarias.Universidad Nacional de Córdoba.Argentina.

Abstract

Cytologic heterogeneity of triticale (*Triticosecale* Wittmack) has been well studied at both varietal and intravarietal levels.The aim of this work was to determine the degree of association between meiotic index (MI) or percentage of normal tetrades and post-tetrades index (PTI),namely percentage of abnormal microspores.Eight hexaploid triticale sublines fron two different varieties,Don Frank (DF) and Jenkins TA 203 (J) were evaluated during two years.Statistical analysis showed a dissociation between indexes in J variety.PTI index would be most suitable for intravarietal selection based on cytological aspects.

Introduction

Hexaploid triticale (*Triticosecale* Wittmack) is a man developed alopoliploid,and like most intergeneric poliploids it exhibits irregular meiosis.This crop has been cytologically well studied considering meiotic irregularities as a whole [1 , 4]; especially for the important role they may play on fertility [5]; as well as its cytological stability at both varietal and intravarietal levels [2].
Among the parameters used to cytologically identify triticales are meiotic index (MI) [5] and rate of healthy microspores (without micronuclei) [2],or post-tetrades index (PTI) as it will be known in the present study.
According to bibliography both would be equally suitable to cytological characterization of varieties or experimental triticale strains.
The purpose to this study was to define the degree of association between MI and PTI stability.

Material and Methods

Two hexaploid triticale varieties, Don Frank (DF) and Jenkins TA 203 (J) , were intravarietaly investigated as both appeared cytologically stable through the years.DF displays constant high values of cytological health whereas J shows low values [2 , 3].
Studies were carried out on eight sublines,four for each of the two varieties studied.The materials were cultivated during the 1989-1990 period in the Experimental Farm of the Facultad de Ciencias Agropecuarias - Córdoba - Argentina,according to a completely randomized block design with two repetitions.Two row plots of 1.30 m long,0.20 m apart and using seeding rate of 100 seeds m2 were

H. Guedes-Pinto et al. (eds.), Triticale: Today and Tomorrow, 207–209.
© 1996 *Kluwer Academic Publishers. Printed in the Netherlands.*

used.Four of five plants per plot were analized removing their principal spikes to estimate MI and PTI indexes,recording 50 and 200 tetrades and microspores respectively.Abnormal tetrades are those in which,at least,one cell contain micronuclei.The inmature spikes were collected and treated according to the technique of Ochoa de Suárez *et al* (1987) [4].Correlation coefficients between sublines within varieties for the two years were estimated.

Results and discussion

Significant positive correlations (p≤ 0.01) between MI and PTI indexes during the two years for DF variety (1989 ,r = 0.99;1990 r = 0.96) were observed; however,not for Jenkins TA 203 variety.When each one was cytologically characterized through the corresponding sublines,(Figs. 1,2), a higher range of variation for MI when compared with PTI was observed.

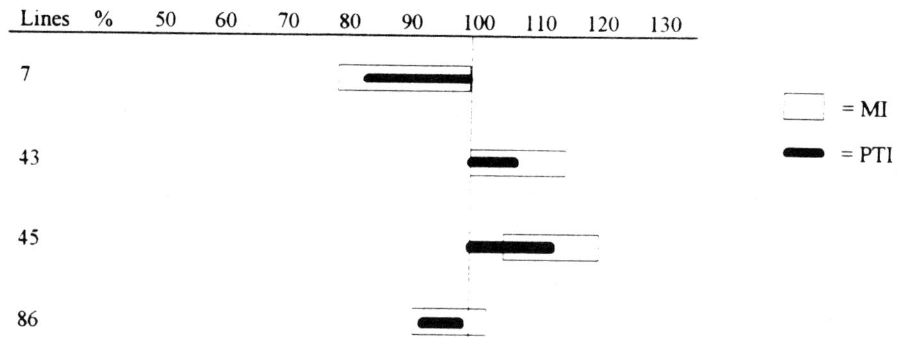

Figure 1: Range of variation of the Meiotic Index (MI) and
Post-Tetrades Index (PTI) as media percentage (= 100)
for Don Frank´s (DF) sublines evaluated
during 1989/90.

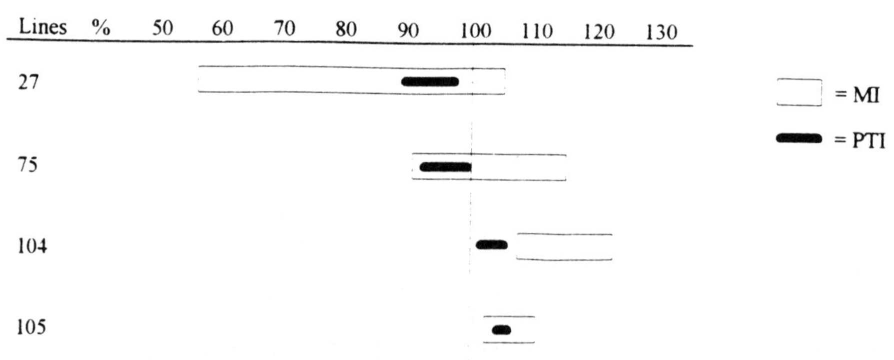

Figure 1: Range of variation of the Meiotic Index (MI) and
Post-Tetrades Index (PTI) as media percentage (= 100)
for Jenkin´s TA 203 (J) sublines evaluated
during 1989/90.

Based on correlation coefficients between MI and PTI it is not possible to confirm that both are indistincts to identify cytologically hexaploid triticale varieties,since a dissociation between them was observed within the most irregular variety (J).
Intravarietal characterization (as regards varietal media) in J variety through the use PTI indicates that this index would be more efficient that MI to carry out intravarietal selection based on cytological characters talking account that sensitivity to the environment was higher for MI when compared with PTI.
PTI is an index that should be taken into account not only for triticale varieties (of similar patterns of behaviour to J) but also with those intergeneric hybrids product of artificial crossings,a process by which two or more different genomes are incorporated to the same organism.

References

1. Falċao,TMMA , B. de Moraes Fernandes, M , Zanettini, M H B . Genotypic and Environmental Effect on Meiotic Behaviour and the Influence of Chromosomal Abnormalities of Hexaploid Triticale *(Triticosecale* Wittmack). Rev Brasil Genet 1981 ; IV , 4 : 611 - 624.

2. Manero de Zumelzú,D , Ordóñez,A , Maich,R , Torres,L .Meiotic Irregularities in Hexaploid triti cale *(Triticosecale* Wittmack). Variability Analysis. Cereal Res Comm 1992 ; 20 (1-2) : 125-130.

3. Manero de Zumelzú,D , Maich,R , Juárez,A .Cytological Disorders in Hexaploid Triticale (*Tritico secale* Wittmack).Association and stability between two indexes.Cytologia 1995 ; 60 : 303-305.

4. Ochoa de Suárez,B , Manero de Zumelzú,D , Macchiavelli,R .Aberraciones meióticas en triticales hexaploides. Rev Cs Agrop 1987 ; V : 135-144.

5. Szpiniak de Ferreira,B . Relación entre fertilidad e índice meiótico en 8 cultivares de triticale (*Triti cosecale* Wittmack) . Mendeliana 1983 ; VI , 1 : 43-54.

GELELECTROPHORETIC GLIADIN PATTERNS OF EUPLASMIC AND ALLOPLASMIC PRIMARY TRITICALE AND THE CORRESPONDING WHEAT PARENTS

Thomas Günther [1], Claus-Ulrich Hesemann [2] and Gitta Oettler [3]

[1] Research Centre Biotechnology and Plant Breeding,
[2] Institute of Genetics,
[3] State Plant Breeding Institute, University of Hohenheim, Germany

Summary

This study was conducted to determine the influence of wheat cytoplasms on storage protein patterns in wheat and triticale. Euplasmic and alloplasmic triticale had been produced by crossing *T. durum* and *T. aestivum*, having nine different cytoplasms, with four rye inbred lines. The gliadin patterns of these euplasmic and alloplasmic primary hexa- and octoploid triticale and the corresponding wheat parents were biochemically analysed, using the acid PAGE as electrophoretic technique. Comparing the gliadin patterns of the euplasmic and alloplasmic wheat parents it was found that the variation of the quantitative expression in the gliadin patterns was amplified in the background of alien wheat cytoplasms. This variation was reduced in alloplasmic triticale, especially in hexaploids. The quantitative expression of rye secalin bands in triticale was influenced by the wheat genome. From this extensive study, it can be concluded that alien wheat cytoplasms produce quantitative and qualitative variations in the expression of gliadin patterns.

Introduction

The performance of a triticale genotype (x *Triticosecale* Wittmack) is determined by the harmonic cooperation of the nuclear genes of the wheat and rye parents with the genomes of the organelles in the wheat cytoplasm. Up to now, the effects of the nuclear genomes were mainly investigated in connexion with a improvement of the performance. It is known that the plasmon influences the expression of many characters, extends the genetic variability and contributes to a greater adaptability [1, 2]. But the number of publications about plasmon effects is very limited.

The analyses of this project were conducted to get information about the influences of the genome and plasmon on the expression of electrophoretic prolamin patterns in primary triticale and the corresponding parents. Prolamins together with glutenins are important components for baking and cooking properties in cereals. Besides, the electrophoretic separation of the prolamines has proved to be a useful technique for characterization of genotypes of different cereal species and also for genetical studies [3-6]. As method of separation the acid polyacrylamid gelelectrophoresis (PAGE) is especially suitable [7]. We developed a high resolving, horizontal acid PAGE with a pore gradient and a simple buffer system based on available methods [8, 9]. This type of PAGE was used for simultaneous and

H. Guedes-Pinto et al. (eds.), Triticale: Today and Tomorrow, 211–216.
© 1996 *Kluwer Academic Publishers. Printed in the Netherlands.*

clear separation between prolamins of rye, wheat, and triticale in the same gel. The separation was carried out using flour extracts and for fine analysis also investigations of single grains.

Materials

For the production of the investigated primary euplasmic and alloplasmic triticales, the wheat parents *T. aestivum* ('Jubilar'), *T. durum* lines (D8, D12) and as *Secale cereale* parents the inbred lines L283, L290, L301, L304 were taken as genome donors. The alien cytoplasms derived from *T. aestivum, T. spelta, T. macha, T. compactum, T. sphaerococcum, T. durum, T. turgidum, T. dicoccum* und *T. dicoccoides*. The plant material was cultivated in 1990 (wheats), 1991 (hexaploid triticales) and in 1992 (octoploid triticales) at Hohenheim. Of each primary triticale seeds of three colchicine-treated homozygous plants were analysed.

Methods

PREPARATION OF THE PROBES

Gliadin extraction from meal: A sample of 20 kernels was ground by hand using a mortar and pestle. 25 mg of this meal was incubated overnight with 150 µl of 25% 2-chloroethanol. After centrifugation, the clear supernatant was taken for electrophoretic separation.

Gliadin extraction from single kernel: From a sample of 20 kernels, each kernel was weighted and then incubated with six times its weight of 25% 2-chloroethanol overnight. The kernels were then crushed in the Eppendorf tube and again incubated overnight. After centrifugation the clear supernatant was taken for electrophoretic separation.

PREPARATION OF THE GRADIENT GEL

A 0.5 mm thick polyacrylamid gradient gel was polymerized on a supporting film using a 12.5 x 26 cm glass plate with slot formers and a gradient maker purchased from Pharmacia Biotech (Freiburg). The gel included a stacking gel with T: 5%, C: 4% and a separation gel with T: 12-20%, C: 3%. After polymerization the gel was washed in dest. water for one hour to remove the ammonium persulfate and the gel buffer and then incubated in 10% glycerol solution for 30 min. The gel could be stored for several days in a refrigerator.

REHYDRATION BUFFER

Before electrophoresis the gel was rehydrated for one hour in a buffer containing 0.1% glycin and 3% urea adjusted to pH 2.7 with glacial acetic acid.

ELECTROPHORESIS BUFFER

The electrophoresis buffer was a sodium lactate-lactic buffer. In 1800 ml dest. water 0.9 ml of 50% lactate solution was diluted and adjusted to pH 3.1 with lactic acid and then filled up to 2l volume with dest. water.

ELECTROPHORESIS

The electrophoresis system was purchased from Pharmacia Biotech (Freiburg) consisting of a horizontal Multiphor II electrophoresis unit, MultiDrive XL power supply, and Multitemp II kryostat. Synthetic cleaning cloths were used as electrode wicks, which were necessary for the contact between electrophoresis buffer and gel. Of each probe 4 µl were deposited with a pipette into the slots. The power supply was programmed with the following maximal values: 1st phase (entry): 500 V, 15 mA, 30 W, 15 min; 2nd phase (separation): 900 V, 30 mA, 30 W, 185 mAh. Under this condition the duration of electrophoresis was 5-6 h.

FIXATION AND STAINING

After electrophoresis the gel was incubated in a 10% acetic acid containing 0.04% Coomassie blue R-250 for 50 min at 50°C. The gel was destained in 10% acetic acid for 2 h to remove the background. Thereafter the gel was prepared for drying in 10% glycerol solution overnight.

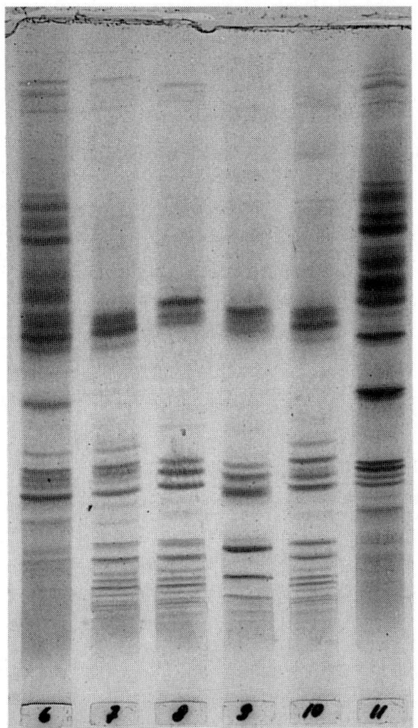

Figure 1: Electrophoregram of the primary triticale D12 x L283 (6), rye inbred lines L283, L290, L301, L304 (7-10) and the *durum* line D12 (11).

Results and discussion

With the electrophoresis method developed for this study, it was possible to separate in a high resolution in the same gel the prolamins of triticale and the corresponding parents in (Fig.1). Therefore, we could compare directly their protein patterns. Due to the high resolution, we were able to detect also very small variations in gene expression. This was important for the analysis of cytoplasmic effects on the prolamin pattern.

WHEAT PARENTS

In the first step, the extracts from the meal were compared. The analysis of the electrophoregrams showed little differences between some alloplasmic wheats and their corresponding euplasmic lines. Nevertheless, all alloplasmic triticales demonstrated the typical gliadin pattern of their genome donors. The mentioned variations in protein patterns of the hexaploid wheats consisted quantitative changes of a few bands only, whereas in the tetraploids quantitative and also qualitative differences appeared. Besides the different expression of the same protein band, we defined also the appearance and disappearance of a single band as quantitative variations. Changing mobilities of individual bands were interpreted as qualitative variations. Individual kernel analyses were conducted to get more information about these variations. The gliadin patterns of some alloplasmic varieties showed an extensive polymorphism. No or only very small differences were observed between the electrophoregrams from single kernels of the euplasmic wheats. The kernels of the wheat line D8 and cultivar Jubilar had only one protein pattern, whereas those of line D12 showed two patterns. The variations involved the γ-gliadins and the ω-gliadins. Comparing the alloplasmic lines, we found the greatest polymorphism in the electrophoregrams of genotypes with the genome from *durum* line D12 and the smallest with the genome from the hexaploid wheat cultivar Jubilar. Within all alloplasmic wheats the gliadin patterns of the lines containing cytoplasm of *T. turgidum* showed an exceptional high polymorphism. In the electrophoregrams of (*T. turgidum*) 'Jubilar' and the line (*T. turgidum*) D8 we found three and in those of the line (*T. turgidum*) D12 ten different gliadin patterns. In contrast, all investigated wheats with the cytoplasm derived from *T. compactum* had the same protein patterns as the euplasmic wheats. Investigating also the electrophoregrams of single kernels from the same spike, we found different patterns depending on genome and cytoplasm. The kernels of some spikes showed almost all the different gliadin patterns which were found in a sample of several plants.

Interpreting these results one might think of the phenomena of outcrossing if a trait shows polymorphism. This explanation is not very probable because wheat is a selfpollinator. Furthermore, the assumption of cross-pollination cannot explain the remarkable differences between alloplasmic and euplasmic wheats. Therefore, we assume that alien cytoplasms influence the gliadin patterns. We propose that components in the cytoplasm can regulatively switch on gliadin alleles in the genome, which were normally switched off in the euplasmic wheat. The existence of multiple alleles from gliadin coding genes was frequently reported. Metakovsky presented a catalogue of such gliadin alleles in 350 common wheat cultivars [10].

PRIMARY TRITICALE

Euplasmic triticale: The protein patterns of the investigated primary triticale of both ploidy levels contained the prolamin bands of both parents, gliadins of the wheat and secalins of the rye. But individual areas of bands showed variation of the quantitative expression compared with the parents. On the one side, some of the ω-gliadins were clearly weakened, on the other side, the γ-gliadins were expressed unchanged or only slightly weaker. The extent of the weakening in single bands was different and dependent on the wheat parent. The ω-bands of the wheat parent D8 were very faint in comparison to D12 and 'Jubilar' in the corresponding triticale. As a rule the α- and β-gliadin patterns showed no changes.

The quantitative expression of the secalin patterns in octoploid triticales was in contrast to the hexaploids. In the octoploid triticales, the γ-secalins were present only as weakened bands. However, in the hexaploids these bands were clearly expressed in similar quantitative proportions to the other secalins compared with the corresponding rye inbred lines. All investigated triticale contained only a trace of the HMW secalins. Apart from the measurable variation of the protein content in the progeny of a single cross, the different expression of the ω-secalins in relation to gliadin bands was interesting and conspicuous. This effect was found in octoploid triticale and very frequently in cross combinations of the rye line L290. This result demonstrates the influence of the wheat genome on the expression of distinct rye proteins.

Alloplasmic triticale: In all alloplasmic triticale we found the same phenomena of different expression of gliadin bands as in the euplasmic triticales. But surprisingly we could not prove the same great polymorphism of the gliadin bands as in the alloplasmic wheat parents. The analyses of single kernels showed widely homogeneous prolamin patterns. In all triticales, the gliadin patterns predominantly of the euplasmic wheats were manifested. In some progenies of hexa- and octoploid triticale we have additionally established variations of single gliadin bands at the quantitative and qualitative level, but in these cases we could not find a relation to the gliadin pattern of the wheat parents.

We are of the opinion that the observed 'quietening' of the gliadin patterns could only in part be explained by the homozygosity of the produced primary triticale. Interactions between the plasmon and the rye genome could have suppressed the regulative influence of the plasmon. We hope that we can provide further and more detailed data of this phenomena when the extensive investigations are finished using also densitometrical analyses of the prolamin bands. But a complete solution of the effective mechanisms could only be achieved, if further detailed experimental investigations are carried out. Our experience shows that investigations of the prolamin patterns represent a very suitable system, in general, for analyses of interactions between wheat plasmon and wheat and rye genome within a triticale. We are convinced that more information about the relation between both genomes and the interactions with the plasmon provides valuable scientific findings in connexion with future triticale breeding programs.

Acknowledgements

The project was supported by the Vater und Sohn Eiselen - Stiftung, Ulm.

References

1.   Sage GCM. Nucleo-cytoplasmic relationships in wheat. Adv Agron 1976;28:267-300.

2.   Mathias RJ, Fukui K, Law CN. Cytoplasmic effects on the tissue culture response of wheat (Triticum aestivum) callus. Theor Appl Genet 1986;72:70-75.

3.   Bushuk W, Zillman RR. Wheat cultivar identification by gliadin electrophoregrams. I. Apparatus, method and nomenclature. Can J Plant Sci 1978;58:505-515.

4.   Du Cros DL Wrigley CW. Improved electrophoretic methods for identifying cereal varieties. J Sci Food Agric 1979;30:785-794.

5.   Dal Belin Peruffo A, Pallavicini C, Varanini Z, Pogna, NE. Analysis of wheat varieties by gliadin electrophoregrams I.Catalogue of electrophoregram formulas of 29 common wheat cultivars grown in Italy. Genet Agr 1981;35:195-208.

6.   Metakovsky EV, Kudryavtsev AM, Iakobashvili ZA, Novoselskaya AYu. Analysis of phylogenetic relations of durum, carthlicum and common wheats by means of alleles of gliadin coding loci. Theor Appl Genet 1989;77:881-887.

7.   Van de Weghe L. Comparative study of electrophoretic methods for cultivar identification of wheat and triticale. Seed Sci & Technol 1991;19:41-50.

8.   Skeritt JH, Martinuzzi O, Metakovsky EV. Chromosomal control of wheat gliadin protein epitopes: analysis with specific monoclonal antibodies. Theor Appl Genet 1991;82:44-53.

9.   Federmann G, Goecke EU, Steiner AM. Der elektrophoretische Nachweis von Weichweizen (Triticum aestivum L.) in Dinkelmehlen (Triticum spelta L.). Getreide Mehl und Brot 1992;46(10):309-312.

10. Metakovsky EV. Gliadin allele identification in common wheat II. Catalogue of gliadin alleles in common wheat. J Genet & Breed 1991;45:325-344.

DEVELOPMENT OF A DOMINANT MALE-STERILE TRITICALE (*MS3*) FOR RECURRENT SELECTION AND HYBRID SEED PRODUCTION

Shoba Venkatanagappa & Norman L. Darvey
University of Sydney, Plant Breeding Institute, Cobbitty, NSW, Australia.

Abstract

With the reports of dominant male sterility (DMS) in wheats such as cv. Taigu (*Ms2*) and cv. Selkirk (*Ms3*) interest has developed in the transfer of the male sterility genes into triticales for utilization in breeding programs. This paper presents the steps involved in the manipulation of chromosome 5A (*Ms3*) of cv. Selkirk to incorporate relevant marker genes into it via recombination so as to enable the identification of the seeds prior to sowing or plants prior to anthesis. Furthermore, the synthesis of dominant male-sterile triticale and its application and relevance to triticale breeding are described and discussed.

Introduction

Breeding efforts over the past decades have enabled the successful extraction and isolation of triticales with several valuable characteristics such as disease resistance, high lysine, stress tolerance and increase in yields [1]. However, undesirable characteristics, such as seed shrivelling, plant height (tallness), late maturity, low test weight and low flour yield (as compared to wheat) have limited their widespread use as a commercial grain crop [2,3]. These characteristics may be controlled either by qualitative or quantitative genes which may be linked. Therefore, the improvement of triticales may not necessarily be achieved via conventional breeding techniques alone [3]. Hence, this necessitates the supplementation of traditional breeding methodologies with alternative breeding strategies which not only eliminate the cost of labor associated with the production of hybrids through manual emasculation but also provide a feasible means to obtain several crosses from each generation thus permitting novel recombinations and breakage of adverse linkages.

Following the reports of the dominant male sterility (DMS) genes in wheats, *Ms2* in cv. Taigu [4], and *Ms3* in cv. Selkirk [5], has developed in the transfer of the male sterility genes into triticales, for utilization in breeding programs. Ji and Deng [6,7] reported the production of octoploid triticales using the Taigu wheat and four rye lines namely, Jingzhou, Snoopy 2, Snoopy 3 and Lanzhou. Production of triticales, with increased seed plumpness via recurrent selection using dominant male-sterile triticales was reported [3]. However, the DMS allele was not found to be linked to any phenotypically identifiable

217

H. Guedes-Pinto et al. (eds.), Triticale: Today and Tomorrow, 217–224.
© 1996 *Kluwer Academic Publishers. Printed in the Netherlands.*

marker gene and hence the progeny obtained could not be identified prior to anthesis in the field [8]. Therefore, manipulation of the chromosome carrying the DMS allele is needed so as to establish a permanent linkage between the male sterility allele and an appropriate marker that can be used to identify either the male-sterile seeds prior to sowing or male--sterile plants prior to anthesis.

Liu and Deng [10] proposed that a linkage between the *Ms2* allele in Taigu DMS wheat and the blue aleurone color gene on chromosome 4E of *Thinopyrum* (*Agropyron*) species be established to enable the identification of sterile plants. Darvey [11] proposed a 2-step method by which a translocation chromosome 5AL.5AS-5RS carrying *Ms3* on chromosome segment 5AS could be obtained in the first step. This could be further modified via centric fusion with chromosome 4E from a *Thinopyrum* species carrying the blue aleurone gene *Ba*. The resulting chromosome would be 4EL.5AS-5RS with linkage between *Ms3* and *Ba*. It was anticipated that the translocated arm would not pair with a complete 5AS arm.

This paper describes the following:

1. Steps required to manipulate chromosome 5A carrying the DMS allele so as to obtain a translocation chromosome carrying the relevant marker genes.

2. Synthesis of dominant male-sterile triticales and their relevance and application for triticale breeding.

Materials and Methods

EXPERIMENT 1

This study was undertaken with the aim of achieving the recombination between chromosomes 5A (*Ms3*) and 5R (Step 1). Four breeding lines namely *Triticum aestivum* cv. Selkirk, *ph1b* (5B) mutant in cv. Chinese Spring, a line monosomic for chromosome 5B in cv. Chinese Spring and a 5R (5A) disomic substitution in cv. Chinese Spring were used in the crossing procedure. During the crossing procedure, chromosome 5A from cv. Selkirk and chromosome 5R were maintained as monosomics and were expected to pair and recombine in the presence of the homozygous *ph1b* mutant. Several types of male--sterile gametic types were expected from the female parent. The following were considered to be of relevance to this crossing procedure.

Type a) Those without the 5RL chromosome arm; these either carried a non-recombinant chromosome 5A or a recombinant 5AL-5RS translocation chromosome.

Type b) Those with the 5RL chromosome arm; these either carried non-recombinant chromosome 5R or a recombinant 5RL-5AS translocation chromosome.

A schematic diagram of steps involved in the manipulation of chromosome 5A of cv. Selkirk has been presented in Figure 1.

Figure 1. Proposed steps for the manipulation of chromosome 5A of cv. Selkirk carrying the *Ms3* allele to enable the incorporation of marker gene *Ba* (blue aleurone) and segments of chromosome arms 5RS and 4EL.

Step 1.

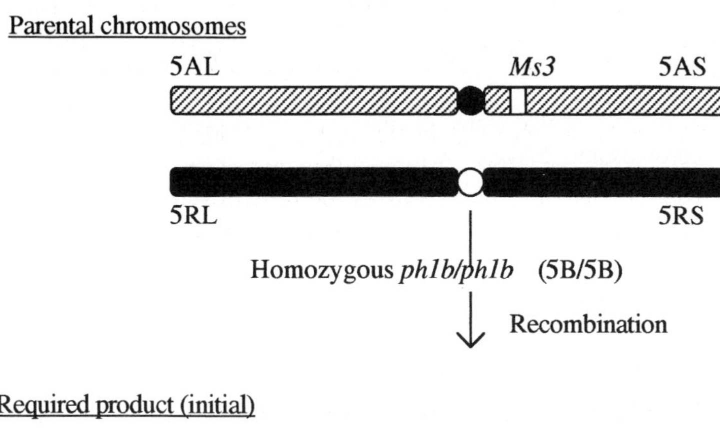

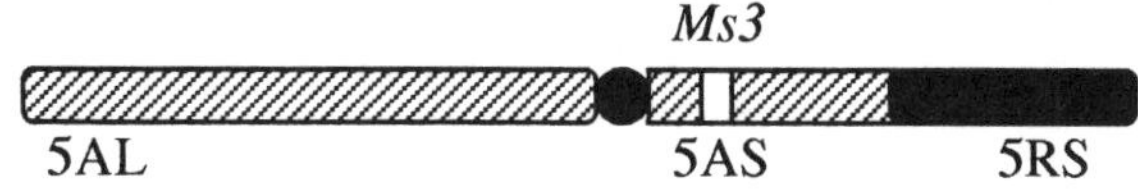

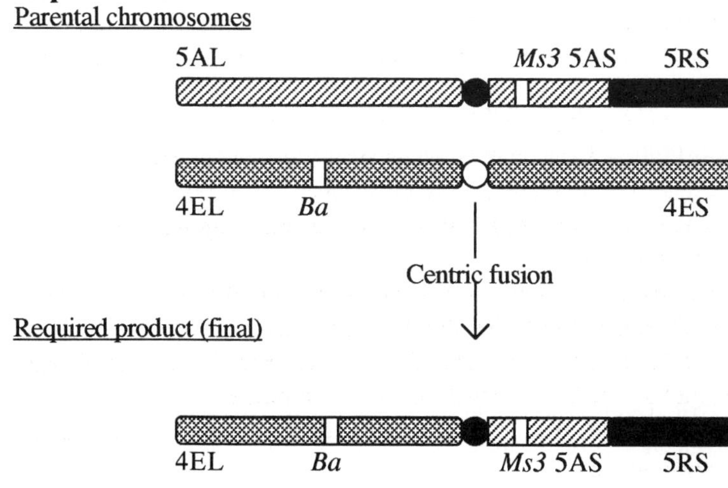

The DMS -Selkirk plants segregated into male-sterile and male-fertile plants. Standard crossing techniques were used for hybridization. The presence of indehiscent sterile anthers [whitish to pale yellow in color, 5] and presence of a 'hairy neck' due the hairy peduncle gene *Hp* on chromosome 5R [12,13] were used as morphological markers. Cytological analyses were conducted to check the chromosome numbers. Iso-electric focusing was conducted to detect the presence of genes β - *Amy-A2* on chromosome arm 5AL [14,15] and β - *Amy-R2* on chromosome 5R [16] coding for β - amylase, and the genes *Ibf-A1* and *Ibf-R1* coding for Iodine-binding factor [17]. A rye specific probe [pAW173, 18] was used to detect rye DNA sequences in the DNA of the leaves of plants with probable translocations between chromosomes 5A and 5R.

EXPERIMENT 2

The crossing procedure for the synthesis of dominant male-sterile triticales is presented in Figure 2.

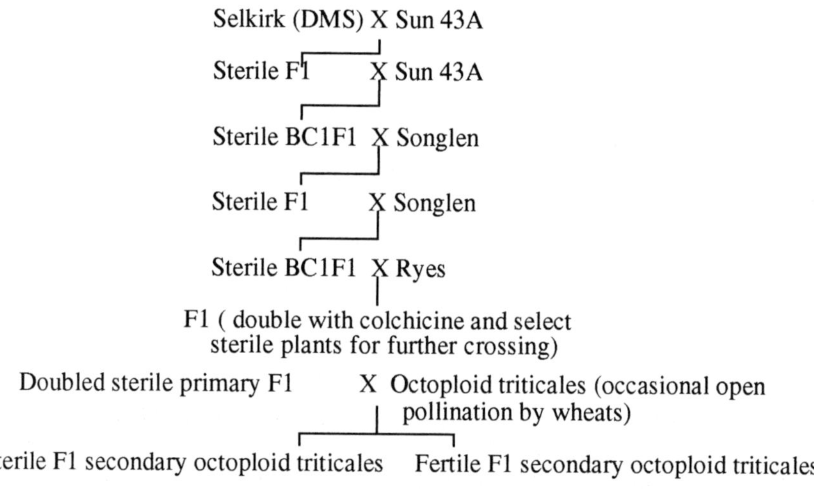

Figure 2. Schematic representation of crossing procedure for the synthesis of dominant male-sterile triticales

Cultivar Selkirk was chosen as the DMS source and the four selections from *Secale cereale* rye and one selection each from *S. montanum, S. vavillovii and S. silvestris* were selected as rye sources. The embryos of the wheat-rye amphihaploids were harvested 15-20 days after pollination and aseptically cultured on modified Norstog medium [19]. The plants were treated with colchicine [20]. The florets of sterile doubled plants were pollinated with the primary octoploid triticales UC115 (C. Qualset, University of California, Davis) or Black Pucara x Extra Early (EE) rye (R. Metzger, The University of Oregon, Corvallis) or allowed to cross-pollinate with the fertile sibs of DMS wheats. The chromosome numbers of these seeds were either 56 (due to pollination with octoploid triticale) or 49 (due to pollination with wheat). The plants obtained from these seeds were further crossed with two octoploid selections Black Pucara x EE rye and 8A 90 (University of Manitoba collection).

Results

EXPERIMENT 1-STEP 1

Two out of 501 seeds carried morphological and genetic markers were consistent with the presence of a translocated chromosome containing chromosome arm 5AL and a part of chromosome arm 5RS. However only one was sterile indicating *Ms3*. This apparent translocation will need to be further confirmed by crossing with euploid wheat and the resulting progeny will segregate into fertiles and steriles. The steriles should have the translocated chromosome. Analyses by C-banding and *in-situ* hybridization of mitotic chromosomes would reveal the presence of the translocated chromosome. The presence of the terminal C-band on the chromosome would confirm the presence of a terminal segment of the chromosome arm 5RS and the absence of the interstitial band located close to the end of the long arm of chromosome 5R would reveal its absence[21,22]. The length of the rye segment would be assessed by *in-situ* hybridization. One further plant had 2n=41+t', hairy neck, and bands for *β-Amy-R1* but not for *Ibf-R1*. This collection of markers is thus most consistent with a recombination event having taken place on chromosome arm 5RL.

EXPERIMENT 2

Several wheat-rye hybrids (a total of 128) were treated with colchicine and several of these upon recovery, segregated into fertiles (47) and genetically male-steriles (33) as indicated by their anther morphology. Sterile plants, homozygous for the DMS allele *Ms3,* were either crossed with octoploid triticales or open pollinated by wheat (Figure 2). The resulting male-sterile plants were crossed with octoploid triticales.

Discussion

Once the centric-fusion chromosome 4EL-5AS-5RS carrying the *Ms3* and *Ba* alleles is produced, a breeder will be able to utilize this chromosome as an alien addition in hybrid breeding programs. The transfer of this chromosome to triticales will enable its direct use in hybrid triticale breeding, since the short arm of the centric-fusion chromosome would pair with chromosome arm 5RS in triticale. This is prefarable to its use as a monosomic alien addition in wheat, which results in meiotic instability, as the alien chromosome segment would not pair with any homoeologous wheat segment. Such an instability would limit the utilization of the DMS systems along similar lines to Driscoll's XYZ system using the recessive male sterile gene *ms1c* [23]. Therefore the ideal translocation chromosome for hybrid would comprise the relevant chromosome arms/segments bearing the alleles *Ms3, Ba* and/or *Rht 12* [allele conferring dwarfness located on chromosome 5AL, 24] and be stably transmitted during meiosis. Development of a translocation chromosome with segments of four different chromosomes as drawn below would be worthy of consideration.

Such a chromosome with the required linkage relationship and terminal homology to chromosome 5A on both arms could be more stable than the monosomic alien addition chromosome and form 21 bivalents at meiosis when crossed with other wheat lines in the hybrid breeding system.

The stable expression of male sterility in the presence of the rye genome (as shown in Experiment 2) demonstrates that chromosome carrying the male sterility allele *Ms3* could be transferred to triticales. However, in order to improve the stability during meiosis in triticales, further modification of the centric-fusion chromosome as diagrammed below should be considered.

Triticales carrying a relevant centic-fusion chromosome they can be used in breeding schemes as follows

1. Germplasm development using hexaploid wheats (with crossability genes) and cereal rye.
2. To assist with crossing among triticales

    Ms3/ms3 triticales x *ms3/ms3* triticales
               ↓
    50% fertiles   :   50% steriles

3. For recurrent selection procedures [4].

The benefits of dominant male-sterile system for triticale breeding would be as follows:

1. Labor costs
2. Apply cross-breeding strategies as well as traditional self-pollinating methods to triticale breeding. Since two out of three genomes ie., B & R are derived from cross-bred species, the relevance of applying cross-breeding methodologies to triticale is evident.
3. Breakage of established linkage groups and promote increased recombinations

4. Enhanced germplasm base- DMS wheats could be used for rapid production of new wheat-rye amphihaploids (4x) leading to hexaploid triticale production.

5. Population improvement via enforced cross-pollination involving selection of fertile and sterile plants in each generation and allowing intercrossing among the best of these.
- a form of recurrent selection with the added benefits of applying different selection pressures (eg, rust, seed quality and acid soil tolerance etc.) in each generation.

References

1.   Gustafson, J.P. and D.C. Bittel. (1990). An effective triticale manipulation system: Problems and uses. In: Proc. 2nd Int. Triticale Symp. Passo Fundo, Brazil.-pp. 278-281.

2.   Skovmand, B., P.N. Fox and R.L. Villareal. (1984). Triticale in commercial agriculture: progress and promise. Advances in Agronomy.37:1-45.

3.   Sun,Y. and C. Wang.(1990) Study on utilization of triticale male sterility in breeding. In: Proc. of the Second Int. Triticale Symp. Brazil, 1990. pp. 131-135.

4.   Deng, J.Y. and Z. Gao. (1987). Determination of a dominant male sterile gene and its prospect in practice. In: Taigu genic male-sterile wheat. J.Y. Deng. (ed.). Elsevier. Amsterdam, Oxford, New York, Tokyo. pp. 7-19.

5.   Franckowiak, S.S. Mann, and N.D.Willams. (1976). A proposal for hybrid wheat utilizing *Aegilops squarrosa* L. cytoplasm. Crop Sci. 16:725-728.

6.   Ji. F., and J.Y. Deng. (1985). Further study on the inheritance of genic male-sterile wheat of Taigu and the production of the dominant male-sterile octoploid triticale. Sci.Sinica XXVIII (6): 608-617.

7.   Ji. F., and J.Y. Deng. (1987). Utilization of the dominant male sterile Ta 1 gene in triticale breeding program. In: Int.Triticale Symp. Sydney, 1986. (Supp). Darvey, N.L (ed). 31: The Australian Institute of Agricultural Science, Australia. pp. 77-81.

8.   Sorrells, M.E. and S.E. Fritz. (1982). Application of a dominant male-sterile allele to the improvement of self-pollinated crops. Crop Sci. 22:1033-1035.

9.   Sorrells, M.E. and S.E. Fritz. (1982). Application of a dominant male-sterile allele to the improvement of self-pollinated crops. Crop Sci. 22:1033-1035.

10.   Liu, B. and J.Y. Deng. (1986). A dominant gene for male sterility in wheat. Plant Breed. 97:204-209.

11.   Darvey, N.L. (1987). Dominant male sterility in hybrid wheat and triticale production. In: Int. Triticale Symp. (Supp.). N.L. Darvey (ed.) 31: Australian Institute of Agricultural Science, Sydney, Australia. pp. 82-85.

12.   Chang, T.D., G. Kimber and E.R. Sears. (1973). Genetic analysis of rye chromosomes added to wheat. In: Proc. 4th Int. Wheat Genet. Symp. Mo. USA.1973. E.R. Sears and L.M.S. Sears. (eds.). Agric. Expt. Stn. University of Missouri. pp. 151-153.

13.   Chang, T.D. (1975). Mapping of the gene for hairy peduncle (Hp) on rye chromosome 5R. Can. J. Genet. Cytol. 17:127-128.

14.   Ainsworth, C.C., M.D. Gale and S. Baird. (1983). The genetics of β-amylase isozymes in wheat. I. Allelic variation among hexaploid varieties and intra-chromosomal gene locations. Theor. Appl. Genet. 66: 39-49.

15.   Sharp, P.J., S. Desai, M.D. Gale. (1988). Isozyme variation and RFLPs at the β-amylase loci in wheat. Theor. Appl. Genet.76: 691-699.

16.   Ainsworth, C.C., T.E. Miller and M.D. Gale. (1987). α -amylase and β-amylase homoeoloci in species related to wheat. Genet. Res. Camb. 49: 93-103.

17.   Liu, C.J. and M.D.Gale. (1989). Ibf-1(Iodine binding factor), a highly variable marker system in the Triticeae. Theor. Appl.Genet. 77:233-240.

18.   Rogowsky, P.M., S. Manning, J-Y. Liu, P. Langridge (1991). The R173 family of rye-specific repetitive DNA sequences: a structural analysis. Genome 34: 88-95.

19.   Taira, T. and E.N. Larter. (1978). Factors influencing development of wheat-rye hybrid embryos in vitro. Crop Sci. 18: 348-350.

20.   Taira, T., Z.Z. Shao, H. Hamawaki and E.N. Larter. (1991). The effect of colchicine as a chromosome doubling agent for wheat-rye hybrids as influenced by pH, method of application, and post-treatment environment. Plant Breeding. 106:329-333.

21.   Gill, B.S. and G. Kimber. (1974). The giemsa C-banding karyotype of rye. Proc. Nat. Acad. Sci. 71(4):1247-1249.

22.   Darvey, N.L. and J.P. Gustafson. (1975). Identification of rye chromosomes in wheat-rye addition lines and triticale by heterochromatin bands. Crop Sci. 15:239-243.

23.   Rao, M.K., K. Uma Devi and A. Arundhati. (1990). Applications of genetic male sterility in plant breeding. Plant Breed.105:1-25.

24.   Sutka, J. and G. Kovacs. (1987). Chromosomal location of dwarfing gene *Rht12* in wheat. Euphytica: 521-523.

CMS SYSTEM IN HEXAPLOID TRITICALE

Roman Warzecha, Krystyna Salak-Warzecha,
Zygmunt Staszewski

Plant Breeding and Acclimatization Institute, Radzików, Poland

Abstract

The male sterility system developed in the present study is a result of interaction of a nuclear gene (s) of hexaploid triticale and *Triticum timopheevi* cytoplasm. It was found that the majority of tested triticale accessions were restorers within this system. The winter Polish cultivar MALNO and the spring CIMMYT cultivar ZEBRA 31 acted as nonrestores and were used to develop male sterile lines of triticale. Also winter cultivars PRESTO and PURDY, as well as six advanced spring triticale lines, which showed non-restoring character, are used to breed their male sterile analogues.

Introduction

The study initiated in Poland three decades ago has resulted in the introduction of the new cereal crop, hexaploid triticale, into our agriculture. All winter and spring triticale cultivars released till now were bred by the use of conventional breeding methods. The development of a new breeding method seems to be an important factor for further improvement of triticale. Breeding of hybrid triticale has an increasing importance.

The possibility of commercial production of hybrid triticale is determined at least by two essential factors: sufficient heterosis in grain yield and a system facilitating crossing of parental components.

High heterosis in grain yield and other important agronomic traits in triticale (performance of hybrids vs. parents or hybrids vs. check cultivar) were reported by several authors [1,4,7,8,10]. Hybrids outyielding the check cultivar up to 30% were obtained in our own study (Warzecha et al., unpublished).

Development of a system useful for hybrid seed production is associated with

225

H. Guedes-Pinto et al. (eds.), Triticale: Today and Tomorrow, 225–232.
© 1996 *Kluwer Academic Publishers. Printed in the Netherlands.*

research on alloplasmic triticale. The replacement of triticale's own (*Triticum turgidum* or *Triticum aestivum*) cytoplasm by "alien" cytoplasm from related species (wheats or Aegilops) may induce its pollen sterility [2,4,6,9].

The aim of the present study was to develop the gene-cytoplasmic (CMS) system of male sterility in hexaploid triticale: male sterile, nonrestorer and restorer lines.

Materials and Methods

The initial maternal material - donor of *T. timopheevi* cytoplasm - consisted of two triticale lines CMH75.993   and CMH80.1150 obtained from CIMMYT Special Germplasm Development Program.   Hexaploid triticale cultivars and strains of Polish and foreign origin were used as pollen parents of hybrids. Polish accessions were obtained from Plant Breeding "DANKO" and IHAR Breeding Station at Małyszyn. The foreign ones were obtained from CIMMYT, Mexico and from Triticale Gene Bank of Agricultural University in Lublin, Poland. The names of pollen parents used in the study are given in pedigree of hybrids in table 1 and 2.

In 1986 materials were planted in greenhouse. The maternal lines were emasculated and pollinated with pollen of different hexaploid triticales. The obtained F_1 progenies were grown and evaluated for their pollen fertility. Several of male sterile segregants were selected from these progenies. These plants were again crossed to the pollen parents. At the beginning of the study only two triticales - spring ZEBRA 31 and winter MALNO -crossed to male sterile plants produced hybrids, which showed high degree of male sterility. However no one of these hybrid was completely male sterile. The selected hybrids were subsequently backcrossed to their pollen parents ZEBRA 31 and MALNO to produced male sterile analogues. They served as the testers of restoring ability of hexaploid spring and male sterile winter triticale accessions in our study.

The following research procedure was applied to evaluate hybrids. F_1 hybrids were planted in greenhouse. Seedlings of winter progenies were vernalized before planting in cold room under artificial light at 4° C through 42 days. The fertility of hybrids was evaluated visually at anthesis. The plants were grouped in three classes according to their fertility: MF - male fertile, PMF - partially male fertile and MS - male sterile. MF plants produced normal size, dehiscend anthers which shed pollen and set seeds in bagged spikes like the male parents of hybrids. The fertility of PMF plants was reduced and they set much less seeds in bagged spikes than MF plants (1-10 seeds per one spike). MS plants formed reduced size, undehiscend anthers, which did not shed pollen and their bagged spikes did not set any kernels. The anthers of MS plants contained pollen grain usually inviable in 90-100%, empty, unstained in acetocarmine. Only a few viable pollen grains could be observed occasionally by microscope in squashed anthers of MS plants.

Results

The results concerning fertility of the hybrids are collectively shown in Table 1. These hybrids having *T. timopheevi* cytoplasm consisted mainly of MF plants and in part some PMF plants. The results indicate that restoring genotypes were much more frequent than non-restoring ones among tested triticale accessions.

Among winter triticale accessions carrying restorer genes for *T. timopheevi* cytoplasm are cultivars released in Poland: LASKO, BOLERO, ALMO, UGO, VERO, NEMO, MONIKO, MORENO, TEWO. Also many Polish winter triticale strains acted as restores: MAGO, ALGO, LAD 285, CHD 1190, MAH 1490, MAH 1389.

All Polish spring triticale accessions which have been tested till now were restorers for MS plants with *T.timopheevi* cytoplasm. These were cultivars JAGO, MAJA and advanced lines from Malyszyn Breeding Station MAH 7695-9/1, MAH 14158-1, MAH 15551-1, MAH 13354-3, MAH 14272-1, MAH 9429-4, MAH 13523-1, MAH 13297-10, MAH 14285-1. Restoring ability was also found in advanced spring lines no. 25, 103 and 127 from CIMMYT, Canadian cultivar ROSNER and U.S. strains FLORIDA 201, ARK 2301 and A 876.

Since restoring genotypes are very common among Polish and foreign accessions of hexaploid triticale, we have focused our research on development of nonrestorer genotypes and their male sterile analogues. In this research nonrestorer genotypes were backcrossed to MS plants as recurrent pollen parents. The results are shown in Table 2.

In BC_5 male sterile analogues of the winter cultivar MALNO, numbered 118, 125 and 126, only MS plants were observed in 1993. These strains tested in BC_6 were also completely male sterile in 1994.

BC_5 male sterile analogues of the spring cultivar ZEBRA 31 produced still some MF and PMF plants (9.0% - 27.0%), but in 1994 BC_6 analogues of ZEBRA 31 showed increase of male sterility.

Polish winter cultivar PRESTO acted as a nonrestorer genotype: its BC_3 analogue was completely male sterile in 1993. In 1994 we observed some fertile plants in BC4 analogue of this cultivar. Winter cultivar Purdy, which is Polish-Dutch bred, showed non-restoring ability and was used to derive its male sterile analogue.

It was found that CIMMYT spring triticale germplasm can serve as the source of nonrestorer genes for CMS-timopheevi system in triticale. Advanced lines no. 50, 97, 15, 39, 41, 59 used as the pollinators for MS plants produced completely or partially male sterile F_1 and BC_2 progenies (Table 2). Nonrestorer character of these lines was confirmed in 1994 by testing their BC_1 and BC_3 analogues.

In 1994 the subsequent backcrosses were made to produce male sterile analogues of all selected nonrestorer genotypes. Also the first experimental spring and winter triticale hybrids developed on male sterile lines of cv. MALNO and ZEBRA 31 as well as on male fertile lines of the same cultivars (produced by hand emasculation of the spikes) are being tested.

Table 1

Fertility of F_1 progenies resulting from crossing of MS-*timopheevi* triticale and restoring genotypes

Hybrids pedigree	Number of plants		
	Total	MF + PMF	MS
<u>winter type</u>			
MS Malno x LAD 285	10	10	0
MS Malno x CHD 1190	15	15	0
MS Malno x MAH 1490	19	19	0
MS Malno x MAH 1389	14	14	0
MS Malno x Lasko	20	20	0
MS Malno x Bolero	20	20	0
MS Malno x Almo	20	20	0
MS Malno X Ugo	20	20	0
MS Malno x Nemo	20	20	0
MS Malno x Moniko	20	20	0
MS Malno x Moreno	18	17	1
MS Malno x Tewo	19	18	1
MS Malno x Mago	20	20	0
MS Malno x Algo	20	20	0
<u>spring type</u>			
MS Zebra 31 x Jago	20	20	0
MS Zebra 31 x Maja	20	20	0
MS Zebra 31 x MAH 7695-9/1	18	18	0
MS Zebra 31 x MAH 14158-1	13	12	1
MS Zebra 31 x MAH 15551-1	18	18	0
MS Zebra 31 x MAH 13354-3	11	11	0
MS Zebra 31 x MAH 14272-1	10	10	0

Table 1 cont.

Hybrids pedigree	Number of plants		
	Total	MF + PMF	MS
MS Zebra 31 x MAH 9429-4	18	18	0
MS Zebra 31 x MAH 13523-1	11	11	0
MS Zebra 31 x MAH 13297-10	20	20	0
MS Zebra 31 x MAH 14285-1	14	14	0
MS Zebra 31 x Rosner	20	18	2
MS Zebra 31 x Florida 201	20	20	0
MS Zebra 31 x ARK 2301	20	19	1
MS Zebra 31 x A 876	20	20	0
MS Zebra 31 x line no. 25 [1]	14	14	0
MS Zebra 31 x line no. 103 [2]	17	17	0
MS Zebra 31 x line no. 127 [3]	15	15	0

Pedigrees: 1 - DF 99//Cin "S"/Bgl "S"/3/Hare 268;
2 - Dgo "S"/Tesmo "S"//Pika "S";
3 - Df "S" (Pft 80137)/Zebra 32//H 507. 71A/2 * Bgl "S".

Table 2

Fertility of the progenies resulting from crossing of MS-*timopheevi* triticale and non-restoring genotypes.

Hybrids pedigree	Generation	Number of plants		
		Total	PMF + MF	MS
118 MS x Malno-5-1	BC_4	49	0	49
121 MS x Malno-9-1	BC_5	61	11	50
124 MS x Malno-9-1	BC_5	23	1	22
125 MS x Malno-9-1	BC_5	76	0	76
126 MS x Malno-9-1	BC_5	40	0	40
150 MS x Zebra 31	BC_5	158	31	127
152 MS x Zebra 31	BC_5	53	11	42
153 MS x Zebra 31	BC_5	13	2	11
154 MS x Zebra 31	BC_5	33	3	30
156 MS x Zebra 31	BC_5	17	3	14
157 MS x Zebra 31	BC_5	37	10	27
130 MS x Presto	BC_3	20	0	20
123 MS x Purdy	F_1	18	0	18
138 MS x line no. 50[1]	BC_2	18	0	18
142 MS x line no. 97[2]	BC_2	12	0	12
150 MS x line no. 59[3]	F_1	12	1	11
153 MS x line no. 41[4]	F_1	18	5	13
154 MS x line no. 15[5]	F_1	14	9	5
154 MS x line no. 39[6]	F_1	19	1	18

Pedigrees: 1 - Bat "S"/Whale "S"; 2 - Pft 80413/Df "S"//Yogur "S"/3/ Lynx "S";
3 - Mst "S"/Caborca 79; 4 - Crg//Au/Dove "S"/3/Pft 8138;
5 - Civet "S"/Dakold 97; 6 - Zebra 31//H 507.71A/2*Bgl "S".

Discussion

The male sterility system developed in the present study is a result of interaction of nuclear gene (s) of hexaploid triticale and *T. timopheevi* cytoplasm. This system, according to Kaul [3], should be classified as gene-cytoplasmic type of male sterility.

It is widely known that *T. timopheevi* cytoplasm was used worldwide for induction of male sterility in bread wheat (*T. aestivum*). The programs of breeding wheat hybrids were unsuccessful mainly due to insufficient availability of the restorer gene sources. In hexaploid triticale the situation is reverse: the majority of accessions are restorers for MS-*timopheevi* system. This was proved in the present study. Also the other authors [2,4] found that transfer of *T. timopheevi* cytoplasm to hexaploid triticale resulted much more frequently in full male fertility than in male sterility. In the study of Cauderon et al. only one cultivar out of ten, namely ROSNER, induced male sterility in *T.timopheevi* cytoplasm. In our study we were unsuccessful to breed male sterile analogue of cv. ROSNER, because this cultivar caused restoring of male fertility. It was presumably due to differences in sources of *T. timopheevi* cytoplasm.

Since restorers for MS-*timopheevi* system are very rare in wheat accessions, we are of the opinion that in hexaploid triticale the rye genome is responsible for fertility restoration in male sterile triticale with *T. timopheevi* cytoplasm.

Also an important question is whether "alien" *T. timopheevi* cytoplasm does cause any negative effects on agronomic characters of hexaploid triticale. The authors mentioned above [2,4] found that this cytoplasm was one of the most prospective for triticale breeding. The potential of *T. timopheevi* cytoplasm for triticale improvement was also well documented and discussed in the paper of Nalepa and Jenkins [5]. In our study, which of course has to be continued, we developed the genetic system, which makes the improvement of many important traits more realistic by breeding of triticale hybrids.

Acknowledgments

This research has been financed since 1993 by the Committee of Scientific Research within the grant KBN no. 5 S301 049 04.

Abbreviations: CMS = gene-cytoplasmic male sterility; **MS** = male sterile;
MF = male fertile; **PMF** = partially male fertile.

References

1. Barker TC, Varughese G. Combinig ability and heterosis among eight complete spring hexaploid triticale lines. Crop Sci. 1992; 32: 340-344.

2. Cauderon Y, Cauderon A, Gay G, Roussel J. Alloplasmic lines and nucleo-cytoplasmic interactions in triticale. Genetics and breeding of triticale, Eucarpia meeting. Clermont-Ferrand (France), 2-5 July 1984. INRA Paris, 1985: 177-192.

3. Kaul MLH. Male Sterility in Higher Plants. Monographs on Theor. Appl. Genet.10. Springer-Verlag Berlin Heidelberg, 1988.

4. Nalepa S. Hybrid Triticale: Present and future. Proceedings of the Second International Triticale Symposium. Passo Fundo, Rio Grande do Sul, Brasil, 1990: 402-407.

5. Nalepa S, Jenkins BC. The influence of T. timophevi cytoplasm on endosperm development in triticale. Genetics and breeding of Triticale, Eucarpia meeting, Clermont-Ferrand (France), 2-5 July 1984. INRA Paris, 1985: 193-201.

6. Sanchez-Monge E. Hexaploid triticale with different cytoplasm. Proc. Int. Sym. Leningrad, 1975: 175-180.

7. Spiss L. Perspektywy wyhodowania heterozyjnych mieszańców pszenżyta. (Perspectives of developing heterotic hybrids of triticale) Hodowla Roślin i Nasiennictwo. Biul. Branż. 1990; 2-3: 1-3.

8. Spiss L., Góral H. Efekt heterozji w mieszańcach z karłowymi mutantami pszenżyta. (Heterotic effect in hybrids of dwarf mutants of triticale). Hod. Rośl. Aklim. 1990; 3-4: 33-37.

9. Tsunewaki K, Iwanaga M, Maekawa M, Tsuji S. Production and characterization of alloplasmic lines of a triticale 'Rosner'. Theor. Appl. Genet. 1984; 68: 169-177.

10. Wolski T. Wstępne badania efektów heterozji pszenżyta. (A preliminary study of heterotic effect in triticale). Conference on triticale breeding, Poznań 1988 (Manuscript).

GENETIC CONTROL OF NON-STARCH POLYSACCHARIDES IN WHEAT RYE ADDITION LINES

Malgorzata Cyran[1], Maria Rakowska[1] and Danuta Miazga[2]

[1] Institute of Plant Breeding and Acclimatization, Radzikow, Blonie, Poland

[2] Institute of Genetics and Plant Breeding, Agricultural University, Akademicka 15, Lublin, Poland

Abstract

Content of dietary fiber (DF) as well as content and composition of non-starch polysaccharides (NSP) were evaluated in the disomic, wheat rye addition lines and octoploid triticale compared to their parental species - wheat (Grana) and rye (Dankowskie Zlote). A large variation in the content of NSP and DF fractions was observed in wheat rye addition lines, especially in the soluble fractions. The level of soluble NSP found in 3RS (0.61% of grain) was significantly lower ($P<0.05$), than in wheat (0.91%) whereas in 3R and 1R (0.88%) it was similar to that of wheat. The highest level of soluble NSP was found in 2R (2.13%) and 5R (1.65%), although 5R had an intermediate level of total NSP (8.57%), while in 2R (9.34%) the highest level of NSP was observed in both NSP fractions. A major differences in sugar composition of soluble NSP have been found in 2R and 5R, where the glucose polymers were the main constituent (57.3% and 46.7%, respectively) in contrast to the rest of wheat rye addition lines, octoploid triticale, wheat and rye, where arabinoxylans (AX) were found to be prevailing among NSP constituents (on average 58.0%, 65.2%, 57.1% and 65.0%, respectively). In the addition lines 6R, 6RL, 4R and octoploid triticale the highest content of soluble AX has been found (0.87%, 0.78%, 0.78% and 0.82% of the grain), although it was twofold lower, than the respective value for rye (1.91%). On the other hand these samples exhibited an intermediate level of soluble NSP. The octoploid triticale demonstrated the highest content of total AX (7.69%), exceeding that of rye (7.02%) as well as in addition lines 4R, 6R and 6RL the high, approximate that of rye content of total AX has been observed (7.45%, 7.28% and 6.84%). It seems that soluble dietary fiber (SDF) analysis might be used as preliminary screening test of cell wall components in wheat rye addition lines, although to detailed study more precisely analysis of NSP by gas chromatography is recommended.

Abbreviations

AX - arabinoxylan; CV - coefficient of variation; DF - dietary fiber; IDF - insoluble dietary fiber; NSP - non starch polysaccharides; SDF - soluble dietary fiber; TDF - total dietary fiber.

H. Guedes-Pinto et al. (eds.), Triticale: Today and Tomorrow, 233–239.

© 1996 *Kluwer Academic Publishers. Printed in the Netherlands.*

Introduction

Dietary fiber (DF) is derived primarily from plant cell walls, that are resistant to hydrolysis by the digestive enzymes of monogastrics (Trowell, 1974). The main components of DF are non-starch polysaccharides (NSP), which form a mixture of polymers associated with proteins, polyphenolics and lignin. The NSP may be divided into soluble and insoluble fractions, but in fact they act as a complex. Their beneficial role in the intestinal regulation and metabolism in man is stressed (Topping, 1991). On the other hand in animal feeding, especially soluble-NSP have been considered as the main antinutritive factor in rye, diminishing the efficiency of feed utilisation (Ward & Marquardt, 1987). Both NSP fractions otherwise influence baking quality of rye (Kühn & Grosch, 1989) and wheat (Shogren *et al.*, 1987). The NSP of each plant material has its own, unique spectrum of sugars. Detailed analysis of these fractions allows more accurate predictions of the possible physiological, technological and nutritional role of the NSP from various plants, which might be of interest to the rye and triticale breeders.

In connection to that, the content and composition of NSP in wheat rye addition lines with 1R to 6R chromosomes were studied.

Materials and methods

The material studied consisted of: wheat-cv. Grana, rye-cv. Dankowskie Zlote, octoploid triticale (Grana x Dankowskie Zlote, 2n=8x=56) and disomic, wheat-rye addition lines -1R", 2R", 3R", 3RS", 4R", 5R", 6R" and 6RL". All cereals were grown under the same soil and climatic conditions.

Soluble dietary fiber (SDF) and insoluble dietary fiber (IDF) components were isolated by gravimetric method based on enzymic digestion as described by Asp *et al.* (1983). According to this procedure total dietary fiber (TDF) was calculated from the sum of SDF and IDF. Content of the DF fractions was assayed without correction for ash and protein.

DF fractions isolated underwent acid hydrolysis in 1N TFA at 125°C for 1h. Aldononitrile acetate derivatives of neutral sugars were obtained according to McGinnis (1982). The samples were analyzed using Varian 3700 gas chromatograph, equipped with hydrogen flame ionization detector on a DB-WAX wide bore capillary column. The correction factors used for the calculation of individual monosaccharides in samples account for losses during hydrolysis and derivatization and for response factors on the gas chromatographo. Allose was used as an internal standard. The values obtained were expressed as polysaccharides by multiplying pentoses by 0.88 and hexoses by 0.90.

Analysis of variance was performed according to MSTAT (microcomputer statistical program for data analysis : Michigan State University).

Results and discussion

CONTENT OF DIETARY FIBRE (DF)

A high coefficient of variation (CV=33.4%) for the SDF analysis of wheat rye addition lines was observed with the range from 1.32% to 11.71% of the grain (Table 1).

Table 1. Contents of soluble (SDF) and total (TDF) dietary fiber in wheat-Grana, rye-
-Dankowskie Zlote, octoploid triticale (Grana x Dankowskie Zlote) and eight
wheat-rye addition lines, in % of whole grain

Genotype[*]	SDF	TDF
	%	
Grana-wheat	2.22 a[***]	12.1 a
3RS[**]	1.32 b	12.7 a
3R	2.35 a	13.1 a
6R	2.82 ac	17.9 b
1R	3.59 cd	17.3 b
4R	4.02 de	19.5 bc
octoploid triticale	4.76 ef	20.6 c
6RL	5.04 f	19.5 bc
5R	9.40 g	23.7 d
2R	11.71 h	25.5 d
Dankowskie Zlote-rye	7.98 i	20.6 c
CV%	33.4	10.9

* Genotypes of wheat rye addition lines and octoploid triticale are ranked on the basis of their SDF
contents.
** Disomic addition of short arm of Dankowskie Zlote rye chromosome 3 to Grana wheat.
*** Values followed by the same letter are not significantly different by multiple range test
(P<0.05).

The level of SDF found in 5R and 2R (9.40% and 11.70%, respectively) was
significantly higher (P<0.05), than that of rye (7.98%). The lowest level of SDF was
found in 3RS (1.32%), while in 3R (2.35%) it was similar to wheat (2.22%). In the
rest of samples (6R, 1R, 4R, 6RL and octoploid triticale) an intermediate SDF content
was found. It was significantly lower (P<0.05), than those of rye, 5R and 2R. No
significant differences (P<0.05) were found in TDF content between 3RS, 3R and
wheat (12.7%, 13.1% and 12.1%, Table 1), while in 6RL, 4R and octoploid triticale
(19.5%, 19.5% and 20.6%, respectively) it was similar to that of rye (20.6%). The
highest level of TDF was observed in 5R and 2R (23.7% and 25.5%), it was
significantly higher (P<0.05), than the respective value for rye. In 1R and 6R an
intermediate level of TDF was found, significantly different (P<0.05), than those of
rye and wheat. The TDF contents obtained in our experiment for wheat and rye include

236

protein bound to DF and ash, therefore are slightly higher, than these reported by Pettersson & Åman (1987), 10.4% and 16.5% for wheat and rye respectively. The variation (CV=10.9%) in tested material, in terms of TDF content was evidently lower compared to that for SDF content.

CONTENT OF NSP

NSP may be divided into cellulose and non-cellulosic polysaccharides. Data presented in this paper concern the non-cellulosic NSP obtained by hydrolysis with 1N TFA. Therefore differences between DF and NSP values are mainly connected with contents of cellulose and lignin, also important constituents of DF isolated by enzymic--gravimetric procedure.

The highest CV (58.1%) was obtained for the content of soluble NSP. In 3R, 1R and wheat (0.89%, 0.88% and 0.91% of grain) the same content of soluble NSP has been found, while it was significantly lower (P<0.05) in 3RS (0.61%) likewise in the case of SDF analysis (Figure 1). Although soluble NSP content found in 5R and 2R (1.65% and 2.13%) was evidently lower, than that of rye (2.94%) in contrast to SDF and TDF values observed in both samples. These results indicate, that 5R and 2R are richer in cellulose and lignin than in non-cellulosic NSP. Soluble NSP values of rye grain found in our experiment were consistent with the results reported by Bengtsson (1992). The total NSP comprised 17.2% and 28.4% of soluble NSP in wheat and rye, respectively, while in addition lines they comprised from 9.0% in 3RS to 14.1% in octoploid triticale to 22.8% in 2R. The content of total NSP (Figure 2) followed the same pattern of change, but less variation (CV=13.6%) was observed.

COMPOSITION OF NSP

Arabinoxylans (sum of arabinose and xylose) and glucose polymers were the main NSP constituents. The major differences in composition of soluble NSP was found in 2R and 5R (Figure 1), where arabinoxylans constituted only 30.0% and 37.6% of NSP in contrast to the rest of the wheat-rye addition lines, octoploid triticale, rye and wheat, where they constituted 58.0%, 65.2%, 65.0% and 57.1%, respectively. Soluble glucose polymers were the main constituents of NSP in 2R and 5R instead of arabinoxylans (57.3% and 46.7%), whereas in the other samples they constituted about 25%. These NSP glucose polymers may correspond to ß-glucans and resistant starch. In the addition lines 6R, 6RL, 4R and octoploid triticale the highest content of soluble AX was observed (0.87%, 0.78%, 0.78% and 0.82% of the grain), although it was twofold lower, than the respective value for rye (1.91%). This observation agreed with the results of Thiele *et al.* (1989), obtained in an experiment on sparrows. In the soluble NSP fraction, galactose constituted on average 14.4% in wheat-rye addition lines and octoploid triticale, 15.4% in wheat, while only 4.8% in rye. Soluble mannose constituted 4.4% in rye and about 2.1% in the remaining samples. No major differences have been observed in total NSP composition of wheat-rye addition lines and octoploid triticale, which was an intermediate between their parental species (Figure 2), although the octoploid triticale demonstrated the highest content of total AX (7.69% of the grain), exceeding that of rye (7.02%) as well as in the addition lines 4R, 6R and 6RL. The high approximate that of rye content of total AX has been found (7.45%, 7.28% and 6.84%).

soluble NSP (% of whole grain)

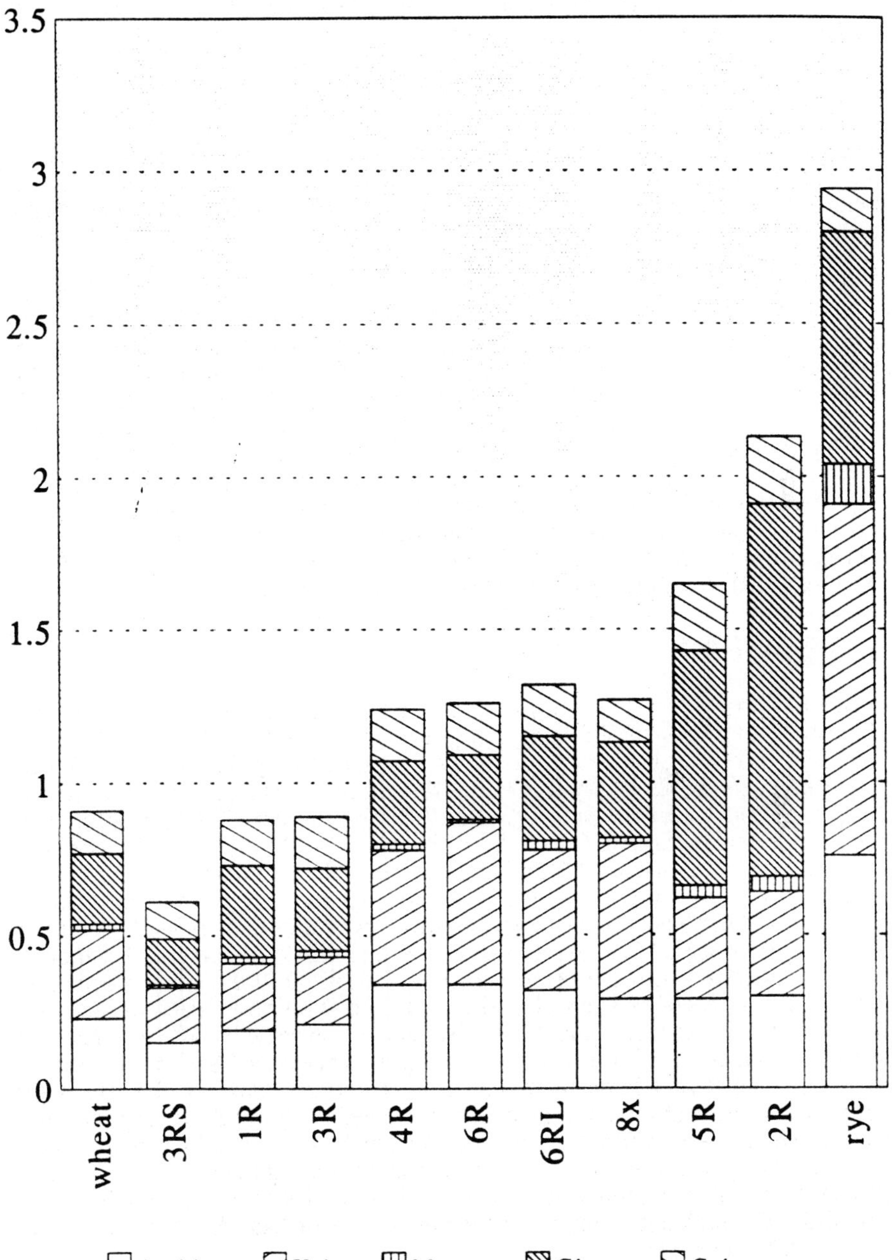

Figure 1. Content and composition of soluble NSP (% of whole grain) in wheat rye addition lines and octoploid triticale (8x) compared to their parental species.

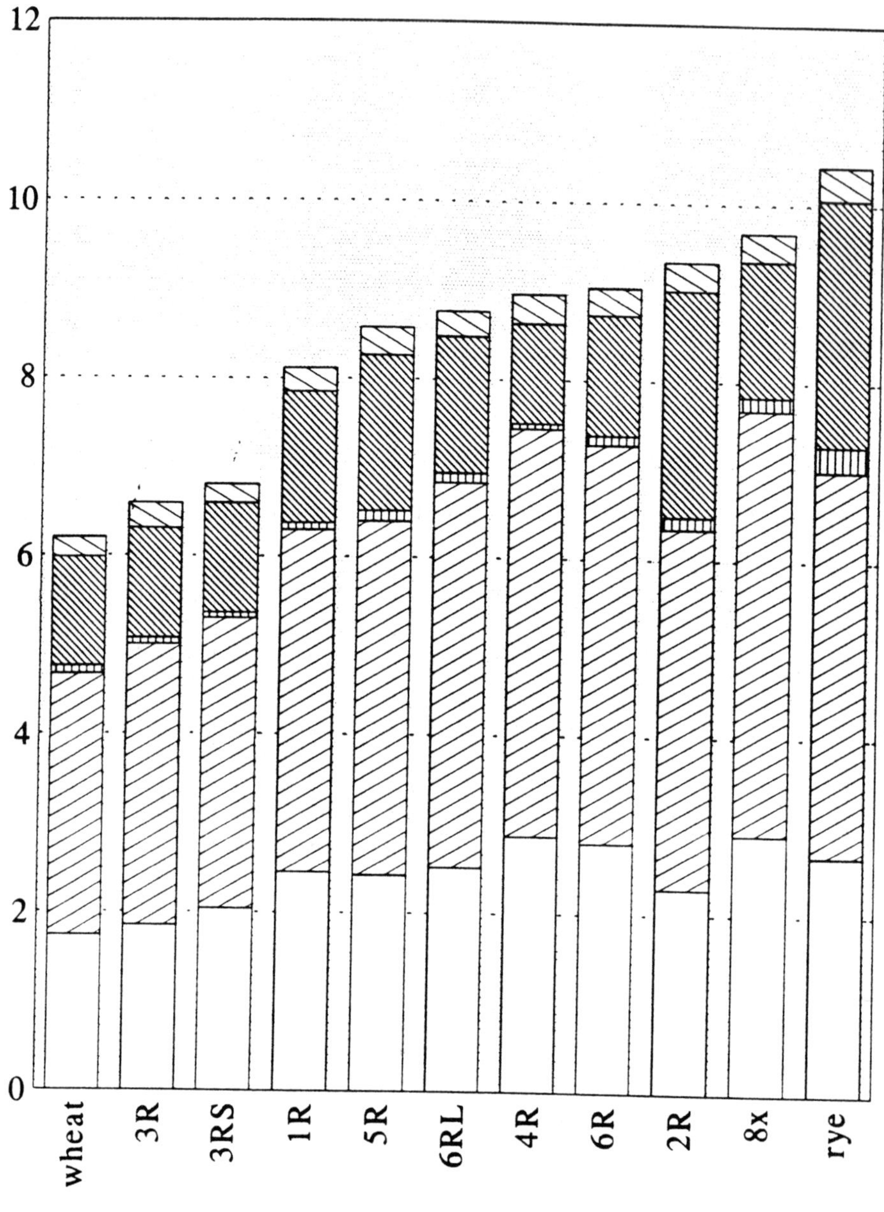

Figure 2. Content and composition of soluble NSP (% of whole grain) in wheat rye addition lines and octoploid triticale (8x) compared to their parental species.

Conclusion

The detailed analysis of DF fractions and NSP composition of wheat-rye addition lines allowed the recognition of chromosomes controling synthesis of the rye fibre components. The obtained data might be of the interest in breeding rye or triticale with lower viscous arabinoxylan contents.

References

Asp, N.G., C.G. Johansson, H. Hallmer and M. Siljeström, 1983. Rapid enzymatic assay of insoluble and soluble dietary fiber. J. Agric. Food Chem. 31: 476-482.

Bengtsson, S., R. Andersson, E. Westerlund and P. Åman, 1992. Content, structure and viscosity of soluble arabinoxylans in rye grain from several countries. J. Sci. Food Agric. 58: 331-337.

Kühn, M.C. and W. Grosch, 1989. Baking functionality of reconstituted rye flours having different nonstarchy polysaccharide and starch contents. Cereal Chem. 66: 149--154.

McGinnis, G.D., 1982. Preparation of aldononitrile acetates using N-methylimidazole as catalyst and solvent. Carbohydr. Res. 108: 284-292.

Pettersson, D. and P. Åman, 1987. The variation in chemical composition of triticales grown in Sweden. Acta Agric. Scand. 37: 20-26.

Shogren, M.D., S. Hashimoto and Y. Pomeranz, 1987. Cereal pentosans: Their estimation and significance. II. Pentosans and breadmaking characteristics of hard red winter wheat flours. Cereal Chem. 64: 35-38.

Thiele, V., G. Melz and W. Flamme, 1989. Genetic analysis of rye (*Secale cereale* L.). Location of gene Anu controlling an anti-nutritive factor. Arch. Züchtungsforsch. 19; 239-241.

Topping, D.L., 1991. Soluble fiber polysaccharides: Effects on plasma cholesterol and colonic fermentation. Nutr. Rev. 49: 195-203.

Trowell, H., 1974. Definitions of fibre. Lancet 1: 503.

Ward, A.T. and R.R. Marquardt, 1987. Antinutritional activity of a water-soluble pentosan rich fraction from rye grain. Poultry Sci. 66: 1665-1674.

III - GERMPLASM

TRITICALE GERMPLASM: A BASIC TOOL FOR PLANT BREEDING

Bent Skovmand, Paul N. Fox, Wolfang H. Pfeiffer, and George Varughese
International Maize and Wheat Improvement Center (CIMMYT)
Apdo Postal 6-641, 06600 Mexico, D.F. Mexico

Abstract

In 1990, the Wheat Genetic Resource Section of the CIMMYT Wheat Program[1] established a world collection of triticale in response to a resolution of the 2nd International Triticale Symposium held in Brazil in 1990. The CIMMYT triticale collection now has about 15,000 accessions and consists of primary and secondary triticales of spring, winter and facultative types. Most triticale lines, that have entered CIMMYT's International Nursery Program and germplasm, which has been utilized in CIMMYT crosses, are maintained in the collection and we are attempting to collect all released cultivars. Passport and characterization information is being loaded into CIMMYT's International Wheat Information System (IWIS), which stores information on bread wheat, durum wheat, triticale, and rye and manages pedigrees, selection histories, and will store performance data and genetic information. This system will analyze genetic and cytoplasmic diversity with performance data to define sub sets for better exploitation by breeders. CIMMYT collections are held in trust.

Introduction

Understanding the sources of genetic diversity and inter-relations between cultivated crops and their wild relatives has rapidly expanded in the latter half of the 20th century following the pioneering work of N.I. Vavilov. This enhanced awareness has produced the recognition that genetic resources are our heritage to be preserved, protected, and made available without restrictions.

Manifestations of the importance of conservation of genetic resources include:
  · The establishment of the FAO Commission on Plant Genetic Resources.
  · Ongoing conservation efforts of international agricultural research centers (IARCs) such as IBPGR and crop-oriented centers such as CIMMYT.
  · The Convention on Biological Diversity.

---

[1] The mandated crops of the CIMMYT Wheat Program are bread wheat, durum wheat and triticale.

243

H. Guedes-Pinto et al. (eds.), Triticale: Today and Tomorrow, 243–252.
© 1996 *Kluwer Academic Publishers. Printed in the Netherlands.*

The establishment of a world collection of triticale in 1990 by the Wheat Genetic Resources Section of CIMMYT, in response to a resolution of the 2nd International Triticale symposium in Brazil [1], demonstrates our conviction that genetic diversity is fundamental to sustaining triticale production. Further, the development of the International Wheat Information System (IWIS) indicates CIMMYT's commitment to the interchange of information about germplasm on an international scale.

Germplasm exchange catalyzed a revolution in plant breeding. A similar revolutionary effect can be expected from the free exchange of information related to germplasm, adding value to germplasm and strengthening bonds between institutions and between scientists. Positive dynamic feedback between genetics, conventional and molecular, and environmental information will provide new insights into crop adaptation.

The Triticale World Collection

The Wheat Genetic Resources Section supports triticale improvement by:

- Acquiring essential germplasm,
- Maintaining collections of selected germplasm representative of all significant germplasm pools,
- Identifying and documenting useful genetic variation,
- Transferring variation into useful genotypes through pre-breeding,
- Distributing accessions and information freely throughout the world.

Presently, the collection contains about 15,000 triticale accessions. The triticale collection consists of primary triticales, produced by crossing wheat and rye, and secondary triticales, derived from intercrosses between triticales, and includes spring, winter, and facultative triticales. Most triticale lines that have entered CIMMYT's International Nursery Program and germplasm, which has been utilized in the CIMMYT crossing program, are maintained in the collection. Recently, an assembly of North American triticales was added to the collection [2]. We are attempting to collect all released triticale cultivars and we invite national programs to submit such material for inclusion in the World Collection.

The Wheat Germplasm Bank interacts with national agricultural research systems (NARSs) by:

- Acting as a backup for national collections,
- Satisfying any request for germplasm and by requesting specific germplasm from a national program to be included in the Bank,
- Asking for assistance in verifying the identity of accessions,
- Requesting evaluations that cannot be done in Mexico.

By serving as a backup for national program collections, including working collections, the Bank provides insurance against loss of germplasm. Collections can be stored, either as part of the CIMMYT collection or as special holdings that will not be available for distribution. We anticipate that NARSs will become increasingly active partners in the evaluation of wheat genetic resources.

The Triticale World Collection is maintained in The Wheat Germplasm Bank, which became operational in 1981 [3]. The Bank also contains over 80,000 accessions of diploid, tetraploid, and hexaploid wheats and their wild relatives, spanning more than 50 years of breeding and collection activities. A small rye collection is maintained to support triticale development. It is an active collection--stored at -2°C to maintain seed for 40 to 50 years. We are planning to add a long-term storage facility, which will maintain a temperature of -18°C to maintain viability for up to 100 years.

THE LEGAL STATUS OF THE WHEAT COLLECTION

The CGIAR policy on genetic resources has been formulated to ensure unrestricted availability of germplasm to *bona fide* users and states that the IARC collections are held in trust. The legal concept of trusteeship indicates that the germplasm collections are maintained and managed by CIMMYT on behalf of beneficiaries. Many of the accessions have been donated by individual countries or institutions or collected in agreement with countries concerned on the general understanding that these accessions would be freely available and that the germplasm would be conserved and used for research. CIMMYT is responsible for maintaining the germplasm and for ensuring the safety of the genetic resources and prevent their destruction and misappropriation.

DOCUMENTATION USING THE WHEAT GERMPLASM BANK SYSTEM (WGBS)

The Wheat Germplasm Bank System (WGBS) is a database system that manages passport, characterization, evaluation, and logistical data for the wheat and triticale collection.

WGBS is an integral part of IWIS and uses the pedigree identifiers of the Wheat Pedigree Management System (WPMS). This allows data interchange among the breeders, the international nursery section, and the collection. When germplasm is submitted to the Bank, existing evaluation data will immediately be available for bank use.

The specific functions of the WGBS include:

- Entering and updating passport data of accessions;
- Producing field books for introductions, regenerations, and characterizations;
- Producing lists for seed shipment;
- Assigning storage locations in the active and base collections;
- Monitoring storage dates, seed viability, and seed stocks and recognize materials that require regeneration;
- Filing data obtained from introduction and regeneration nurseries;
- Facilitating selection of accessions based on passport data and/or specific traits.

When IWIS is completed, WGBS will function as a seamless component.

The International Wheat Information System

While costs of generating field and laboratory data are increasing, costs of data storage, management, and analysis are decreasing rapidly. Conservative estimates indicate that, throughout the world, national agricultural research systems collectively invest more than US$1 million per year in field plot management of the germplasm received from the CIMMYT Wheat Program alone. However, information technology is required to put the data generated by such major investments to work in crop improvement. One of many advantages of data integration will be ease of use of data from gene banks and laboratories in the planning of crosses in breeding.

Integrating the CIMMYT system for unique identification of germplasm with work of the USDA Plant Genome Research Program and other groups will make it possible for an International Wheat Information System to store, query, and disseminate data on wheat germplasm held by many countries.

THE CIMMYT MODEL FOR SELF-POLLINATED CROPS

Until the CIMMYT Wheat Program developed a strategy based on unique identification of germplasm, information generated by different sources, e.g., national trials, international trials, laboratories, and germplasm banks, could not be integrated around the germplasm to which it pertained. Data "islands" were formed.

The International Wheat Information System (IWIS) will enable data integration and be based on three major components:

> • The Wheat Pedigree Management System (WPMS) has been completed and uniquely identifies germplasm by cross identification (CID) and selection identification (SID) numbers. Coefficients of parentage are generated from pedigrees, based on a computer program of T.S. Cox.

> • The completed Wheat Germplasm Bank System (WGBS).

> • The Wheat Data Management System (WDMS), which will interface with WPMS and WGBS and consider three types of data categories--Genetic (G), Genotype x Environment--including disease surveys--(GE, or phenotypic) and Environmental (E)--and the interrelations between these categories.

IWIS will link to Geographic Information Systems (GIS) through E data and to genetic mapping initiatives through G data.

REMOVING BARRIERS TO ASSOCIATION

In the bulletins of the International Spring Wheat Yield Nurseries (ISWYN) the wheat 'Siete Cerros' was represented 13 different ways and one had to know a lot about wheat (and a little Spanish) to realize that these 13 designations, with the associated data, pertain to one wheat line. Such multiple synonyms and also the use of the same name for different cultivars create barriers to association. WPMS overcame these barriers, by uniquely identifying wheat germplasm, and is the core of the International Wheat Information System. Costly and unnecessary repetition of evaluations, such as industrial quality tests, is eliminated by unequivocal identification. WPMS is a database and repository of information on genealogies and selection histories, using the Purdy/USDA system for cross notation.

Breeders commonly retain some genealogical data in their heads, typically back to the grandparents of a variety. WPMS can provide such information at any specified level, for example grandparents, great grandparents or as far back as data exist. For example, "Copi", a recently developed triticale, contains breeding contributions from at least 25

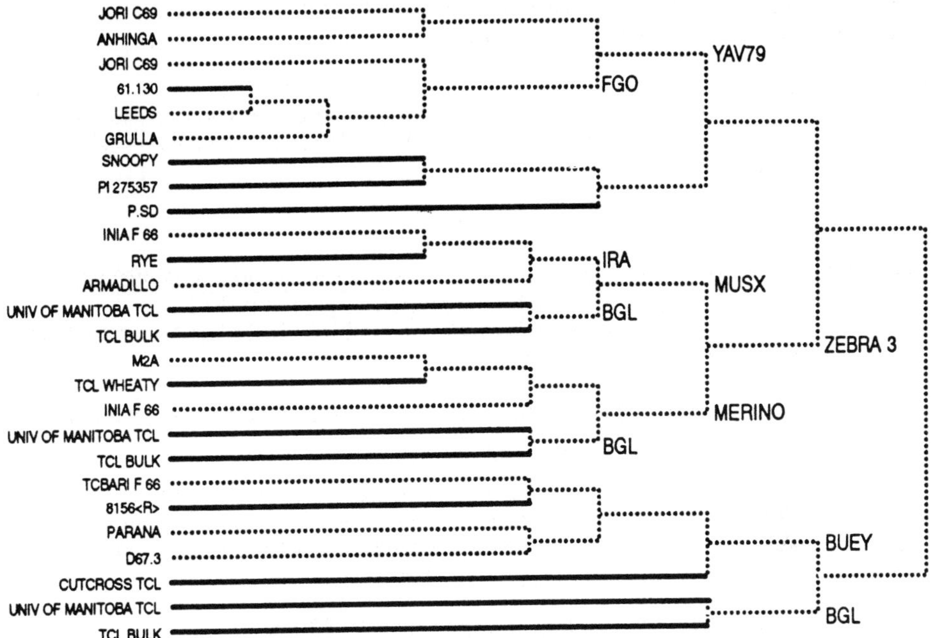

Figure 1. Partial genealogy for triticale line COP1 going back 5 levels.

countries (Table 1) and is traceable to 76 landraces, 9 triticale parents, 5 rye parents, 15 durum, and 47 bread wheat parents. There are no landrace cultivars in triticale as it is a manmade crop, however, these nine parents have unknown pedigrees and are treated as landraces in the genealogies. Figure 1 is a genealogical dendrogram for Cop1. Pedigrees may be extended with the discovery of additional information on ancestors.

CROSSING DATA FRONTIERS

IWIS will integrate information from different sources around the germplasm to which it pertains, providing a secure, flexible system for data storage and facilitating powerful associations between genetic information and performance data. This interface was seldom crossed because of problems in association of data from different sources. Database technology makes cross referencing this information feasible and efficient.

The scope of IWIS embraces data from international trials, national trials, WGBS, industrial quality and pathology laboratories, and research in molecular biology. All genes in the "Catalogue of Gene Symbols for Wheat" will be included, along with RFLPs and other molecular markers. IWIS will be specifically designed to interface with databases such as GrainGenes which stores genetic maps.

DATA MANAGEMENT AND EXPLORATION

IWIS could list instances where a variety, to be released by a national program, occurs in the database and flag all susceptible disease reactions reported internationally for the

Table 1. Landrace cultivars of bread wheat, durum wheat, and rye and triticales of unknown ancestry in the triticale line 'Copi', their origin, and theoretical genetic contribution.

NAME	CROP	ORIGIN	THEORETICAL GENETIC CONTRIBUTION
TCL BULK	TRITICALE	MEXICO	18.75
UNIV OF MANITOBA TCL	TRITICALE	CANADA	18.75
OUTCROSS TCL	TRITICALE	MEXICO	12.50
P. SD	RYE		6.25
CAPPELLI	DURUM	ITA	4.10
MINDUM	DURUM	USA	3.95
PI275357	DURUM		3.13
SNOOPY	RYE	MEX	3.13
CARAVACA 1	DURUM	SPAIN	2.05
TCL WHEATY	TRITICALE	MEXICO	1.56
RYE	RYE		1.56
TCL PERS	TRITICALE	MEXICO	1.46
HEITI	BREAD	PAKISTAN	1.44
LANGDON 379	DURUM	USA	1.36
KENYA 324	BREAD	KENYA	1.27
MARROQUI 588	BREAD	MOROCCO	.98
TCL DIC	TRITICALE	MEXICO	.97
TAGANROG	RYE	RUSSIA	.78
CANDIAL	DURUM	ARGENTINA	.78
61.130	DURUM	USA	.78
AKAGOMUGHI	BREAD	JAPAN	.72
RED FIFE	BREAD	CANADA	.70
STEINWEDEL	BREAD	AUSTRALIA	.70
BOUTEILLE	DURUM	ALGERIA	.49
TCL DUR	TRITICALE	MEXICO	.49
TCL DUR CRLT	TRITICALE	MEXICO	.49
TCL DUR GIZA	TRITICALE	MEXICO	.49
TREMEZ MOLLE	DURUM	PORTUGAL	.49
HARD RED CALCUTTA	BREAD	INDIA	.48
ALFREDO CHAVEZ 6.21	BREAD	BRAZIL	.43
POLYSSU	BREAD	BRAZIL	.43
KANRED	BREAD	USA	.41
IUMILLO	DURUM	SPAIN	.41
DURUM NO2E	DURUM	ARGENTINA	.39

Table 1. Continued.

RYE UNK	RYE		.39
ZENATI BOUTEILLE	DURUM	ALGERIA	.39
ZENATI	DURUM	ALGERIA	.39
GAZA	BREAD	EGYPT	.38
TURKEY RED	BREAD	USA	.37
VERNAL EMMER	DURUM	USA	.36
REITI	BREAD	ITALY	.36
KENYA BF4-3B.10.V.1	BREAD	KENYA	.29
RED EGYPTIAN	BREAD	S. AFRICA	.29
ORO	BREAD	USA	.28
SQUAREHEAD	BREAD	USA	.28
MARIA ESCOBAR	BREAD	ARGENTINA	.23
ETOILE DE CHOISY	BREAD	FRANCE	.20
KENYA 338	BREAD	KENYA	.20
MAHMOUDI 981	BREAD	TUNESIA	.20
KENYA C9906	BREAD	KENYA	.20
FULTZ	BREAD	USA	.18
DARUMA	BREAD	JAPAN	.18
YAROSLAV	DURUM	USA	.18
CHINO	BREAD	ARGENTINA	.14
BARLETA	BREAD	ARGENTINA	.14
BARRIGON YAQUI	BREAD	MEXICO	.14
BLOUNT"S LAMBRIGG	BREAD	USA	.14
TURKEY	BREAD	USA	.11
FORTYFOLD	BREAD	USA	.11
MEDITERRANEAN	BREAD	USA	.10
IMPROVED FIFE	BREAD	USA	.10
BISKRI	BREAD	TUNESIA	.09
AMERICANO 44D	BREAD	URUGUY	.09
BARLETA 7D	BREAD	ARGENTINA	.09
ZEEUWSE WITTE	BREAD	NETHERLAND	.08
INDIAN G	BREAD	INDIA	.07
SASKATCHEWAN FIFE	BREAD	CANADA	.05
PURPLE STRAW	BREAD	AUSTRALIA	.05
PELON 33C	BREAD	URUGUY	.04
AMERICANO 25E	BREAD	URUGUY	.04
TACUR TIPO-125	BREAD	MEXICO	.04
FIFE	BREAD	POLAND	.04
WHITE NAPLES	BREAD	AUSTRALIA	.04
SICILIAN SQUAREHEAD	BREAD	FRANCE	.03
ETAWAH	BREAD	INDIA	.02
ST 464	BREAD	ETHIOPIA	.01

variety. Alternatively, it could identify sites where bread wheat, durum wheat, and triticale trials have been grown in the same season with less than 350 mm of available water.

Breeders lament the fact that genetic databases are not at their fingertips to plan crosses. In general, wheat scientists require better access to raw data, results of analyses, and interpretative summaries as well as identification of novel or under-utilized germplasm. IWIS is designed to provide this information by linking initiatives in many countries through CIMMYT's WPMS.

Long-term trends in genetic diversity were compared using coefficients of parentage for wheat in the Pakistani Punjab and in the Yaqui Valley of Mexico. Trends in adoption by farmers were more important in determining diversity within these agricultural areas than the number of varieties released. An analysis of coefficients of parentage of a single ISWYN indicated extensive pedigree variation among entries, many of which are significantly divergent from US germplasm. Of particular interest, the CIMMYT wheats 'Bagula' and 'Milan' demonstrated adaptation to the Pacific Northwest and have pedigrees divergent from US cultivars. These wheats meet quality parameters of the Pacific Northwest and add to genetic diversity. IWIS will allow international researchers to make similar determinations about diversity in their triticale germplasm.

The power of WPMS has been seen in unexpected ways. For example, investigations suggest that cytoplasmic diversity in CIMMYT bread wheats is restricted and that one of the dominant cytoplasms is tetraploid. Cytoplasmic variation of triticale will be investigated in the near future.

CIMMYT will increasingly distribute data with nurseries to improve the selection efficiency of clients. For example, there is much widely usable information on industrial quality generated before germplasm is distributed in an international trial. In the long-term, extensive genetic profiles will be distributed with seed.

IWIS will enable better use of race-specific rust reactions that are routinely generated in USA, Canada, South Africa, Australia, and other countries. In parallel, CIMMYT is embarking upon a more directed phase of data collection, especially from developed countries where many important genetic data of international importance are generated but not disseminated. IWIS aims to be a clearinghouse to make data work better for all wheat and triticale researchers.

GENES AND THE ENVIRONMENT DYNAMIC FEEDBACK

Here is an example of how data integration might function. If we knew the genetics of boron tolerance and could select contrasting entries, tolerant and intolerant, for this trait in our extensive performance databases (e.g., the 30 years' data accumulated by the ISWYNs) and could then extract international performance data on them, preliminary inferences about the distribution of boron related problems could be made. Similarly, we can examine the performance of known nematode resistant lines relative to susceptible ones across the world and achieve a rapid bio-assay for the pathogen.

Conversely, environmental knowledge facilitates characterization of germplasm. Mapping the occurrence of problems will aid selecting tolerant germplasm, using a database, by identifying top performers across environments subject to a specific stress. Soil mapping will indicate the most probable areas of the world for locating genetic tolerances and resistances to nutrient stresses. GIS may be used to identify disease "hot-spots". The intensity and frequency of specific diseases, or insect vectors, can be predicted from long-term meteorological records. For example, temperature, relative humidity, dew point and dew duration, and wind direction and velocity all have important and direct influences on wheat rust epidemiology. So-called neutral marker genes may prove to have adaptive significance if their influence can be gauged from extensive performance data.

PUTTING IT ALL TOGETHER AND TOUCHING BASES

The IWIS project aims to transfer functions to platform independence and public domain software for universal access, recognizing that PC systems are essential for widespread use by the NARSs. Alternatives are being actively investigated with the Integrated Information Management Laboratory at Texas A & M University. With a view to minimizing duplication, CIMMYT is in contact with other groups developing or planning crop databases.

Most users will require the specialties of the house, those more innovative functions based on unique germplasm identification. These are tools for data input, verification and querying using the following:

- Generation and maintenance of CIDs and SIDs for germplasm,
- Pedigree expansion at any desired level for which information exists,
- Origins of cytoplasms,
- Synonym and nomenclature service,
- Coefficients of parentage,
- Genetic profiles of germplasm,
- International performance profiles of germplasm,
- GIS.

Limited options for experimental design and breeding stationery will be available for users requiring a total system. However, IWIS will not concentrate on areas where excellent software is available (e.g., generation of stationery, experimental design and statistical analysis) and aims to export data and link with available commercial software wherever possible.

Conclusion

Genetic diversity is fundamental to sustaining triticale production in the future. Collection and maintenance of diversity are our starting point. From there we are applying advances in information technology to maximize efficient use of genetic diversity in breeding. The IWIS software being developed at CIMMYT is directly applicable to self-pollinated crops (or any species with a biparental mating system) and will be transferred to a PC environment for scientists around the world.

References

1. Darvey NL. Germplasm creation centers and banks for spring, winter and facultative triticale. Proceedings, 2nd International Triticale Symposium. Passo Fundo, Rio Grande do Sul, 1-5 Oct. 1990. 1991: 14-19.

2. Qualset CO, Furman BJ, Skovmand B, Wesenberg DM. Triticale variation. Agronomy Abstracts 1990;191.

3. Skovmand B, Varughese G, Hettel GP. Wheat Genetic Resources at CIMMYT: Their Preservation, Enrichment, and Distribution. CIMMYT, Mexico, D.F., 1992.

THE CONTRIBUTION OF RYE GERMPLASM TOWARDS CEREAL IMPROVEMENT

A. Mujeeb-Kazi, MDHM William, RL. Villareal, RJ. Peña, S. Rajaram, MN. Islam-Faridi, L. Gilchrist and G. Varughese
International Maize and Wheat Improvement Center (CIMMYT), México, D.F., MEXICO

ABSTRACT

Alien introgression methodologies have elucidated the usage of distantly or closely related species through intergeneric and interspecific hybridization strategies. Of the various annual and perennial Triticeae species, *Secale cereale* (2n=2x=14, RR) is one that expresses a well demonstrated potential for cereal improvement with over five million hectares being cultivated to the rye translocation wheats possessing the 1AL/1RS or 1BL/1RS chromosome translocations. We now attribute a 4.3 percent yield advantage to the 1BL/1RS translocation as observed for present analyses of random F_2-derived F_6 lines emerging from the *Triticum aestivum* cvs. Nacozari/Seri 82 combination. Stringent evaluation of the contribution of 1BL/1RS translocation is anticipated from yield analyses of bread wheat (Seri 82, 1BL/1RS) and durum wheat (Altar 84, 1B) germplasm in which chromosomes 1B or 1BL/1RS respectively have been substituted by eight backcrosses. We have initially inferred that the adverse bread-making quality is not an exclusive function of the 1BL/1RS translocation. The evaluation of the critical genetic substitution stocks produced by backcrossing to Seri 82 will shed further light upon the contribution of 1BL/1RS translocation towards bread making. Additional utilization of rye comes from germplasm exhibiting improved copper efficiency (5AS/5RL or 4BL/5RL), cereal cyst nematode (6BS/6RL) and Karnal bunt resistance (disomic additions 4R and 6R). Transfers and introgressions from these rye sources are being made into some CIMMYT spring wheats. Advanced derivatives from triticale x *T. aestivum*; with a translocation homozygote; are a source of *Septoria tritici* resistance. Detection of rye presence has utilized fluorescent *in situ* hybridization (FISH) as a diagnostic.

INTRODUCTION

Exploiting alien genetic variability within the Triticeae relatives for wheat improvement is associated with genetic introgressions through intergeneric and interspecific hybridization methodologies. The interspecific approach has a demonstrated advantage because of the genetic proximity of the various species with the genomes of wheat that enable rapid recombinational exchanges. The alien variation of the more distant relatives, however, is complex to incorporate into wheat and requires an array of cytogenetic based inputs. To a considerable extent the choice of the alien species becomes a factor in achieving success for intergeneric areas and usually alien diploids receive priority, of which *Secale cereale* is a potential choice. A utilization form of *S.*

H. Guedes-Pinto et al. (eds.), Triticale: Today and Tomorrow, 253–260.
© 1996 *Kluwer Academic Publishers. Printed in the Netherlands.*

cereale in agriculture is so vividly apparent through X *Triticosecale* where both its hexaploid and octoploid forms have rendered significant impact. This status provides logic that rye may offer a more specific means for improving wheat preferably through cryptic DNA introgressions. The concept is encouraged by the tremendous impact that the 1RS chromosome arm has made globally through release and cultivation of bread wheats that possess the 1BL/1RS or 1AL/1RS translocations. Impact of the 1BL/1RS translocation appears to be greater, and consequently has warranted a more stringent assessment linked with yield, genetic and quality evaluations. Over 5 million hectares grown to such wheats. The 1AL/1RS wheats however, have yet to reach major land occupancy. Other rye chromosomes with the potential of significant addressing agricultural constraints are 4R, 5R and 6R that have shown promise in initial evaluations. Large scale exploitation of exchanges between these rye chromosomes and those of wheat is forthcoming. We describe in this presentation our wheat germplasm that has significantly benefitted from rye introgressions. This germplasm as complemented by supportive methodology and diagnostics is further elucidated as it impinges upon our futuristic research endeavors.

Translocations involving the 1RS chromosome arm
1. *1BL/1RS translocated wheats.* Approximately fifty percent of the advanced breeding lines of CIMMYT's bread wheat breeding program possess the 1BL/1RS translocation. The genes *Lr26, Sr31, Yr9* and *Pm8* are mapped on the 1RS arm (McIntosh 1983). Though several countries possess virulences for these biotic stress genes, cultivars with the translocation continue to be significant contributors for wheat production. A survey in Great Britain indicated that among the 84 wheat cultivars grown in 1980 to 1985, 10.7% had the 1BL/1RS translocation (Payne et al. 1987). Similarly, 22.8% of the 57 cultivars recommended for cultivation in West Germany in 1983 to 1984 possessed the 1BL/1RS translocation (Rogers et al. 1989). In 1989 USA wheat nurseries the 1BL/1RS contribution was 7.1% (Lukaszewski 1990). Pakistani wheats show a heavy dependence of the 1RS translocation with approximately 75% of the varietal releases since 1980 possessing the 1B/1R translocation. In their national yield trials with three categories; normal duration, short duration and rainfed; 1B/1R wheat entries occupied percentages of 40.0, 61.5 and 84.6 respectively (Ter-Kuile et al. 1991). A CIMMYT report indicates that 41 cultivars with the 1BL/1RS translocation were released in 22 countries (Villareal and Rajaram 1988). This presence of the 1BL/1RS translocation is quite high. The status remains theoretically prone to the dogma of genetic vulnerability as a consequence of the narrow genetic base contributed by the 1RS chromosome arm from Petkus rye. Though several countries already possess virulences for the rust and mildew genes located on 1RS, breeding programs still continue to exploit such translocated wheats contending that the 1BL/1RS exchange embodies yield advantages, stability and wide adaptability (Rajaram et al. 1983). The ensuing adverse quality has been debated by Peña et al. (1990). To further the findings of Rajaram et al. (1983) 1B and 1BL/1RS cultivars of diverse origins were yield tested (Villareal et al. 1994). Genotypes with the 1BL/1RS translocation had 2.2% higher above-ground biomass yield, 2.1% more spikes/m^2, 1.12g higher 1000-grain weight and 0.8 kg/hl higher test weight. The non 1BL/1RS genotypes had 0.8% higher harvest index and 0.6 cm longer spikes. The 1BL/1RS group headed 2.5 days later than the 1B group. A similar test of yield and agronomic characteristics was conducted on 28 individual random F_2-derived F_6 lines, 14 with and 14 without the 1BL/1RS translocations from the Nacozari (1B)/ Seri 82 (1BL/1RS) cross under optimum and reduced irrigation conditions. Each

experimental plot consisted of 8 rows, 5 m long and 20 cm apart grown in randomized complete blocks of three replications. Plant height, days to flowering and physiological maturity, grain yield, above-ground biomass at maturity, harvest index, yield components and test weights were determined. Yield advantages of 4.3 (Table 1) and 4.2% have been statistically measured on the 1BL/1RS genotypes under optimum and reduced irrigation conditions respectively. Moreover, under both conditions, genotypes with the 1BL/1RS chromosome translocation had higher above-ground biomass yield, more grains/spike, higher 1000-grain weight, were shorter and possessed longer flowering dates. The superiority of the translocation group on grains/m^2 was only expressed under optimum environment. On the other hand, higher harvest index, longer spike length, grainfilling period and days to physiological maturity of the 1BL/1RS genotypes were only detected under limited irrigation trial. Germplasm now developed for cultivar Seri 82 where the chromosome 1B substitution for 1BL/1RS shall elucidate the 1RS contribution effects more stringently. Following eight backcrosses of the 1B, 1BL/1RS heterozygote with Seri 82 (1BL/1RS, 1BL/1RS) with a selfing, led to identifying Seri 82 (1B substituted) and Seri 82 (1BL/1RS extracted) lines. The latter will ellucidate the contribution to yield and yield components of the 1BL arm as a consequence of recombination. The original cv. Seri 82 will be the third test group in yield and quality evaluations that are currently underway.

Table 1. Optimum irrigation trial elucidating details of yield and its components in F_2 -derived F_6 lines of the *T. aestivum* from Nacozari/Seri 82 cross.

	1BL/1RS	1B	Adv. (%)
Grain yield	6266	6006	4.3 *
Biomass at maturity	15.2	14.7	3.4 *
Grains/m^2	15906	15634	1.7 *
Grains/spike	44.3	42.6	4.0 **
1000-grain weight	40.19	39.87	0.8 *
Test weight	79.8	78.8	1.3 ***
Plant height	91.6	94.1	3.2 ***
Spike length	9.9	10.1	3.0 *
Days to flowering	80.3	79.4	0.9 **

* , ** and *** indicates significance at 0.05, 0.01 and 0.001 levels of probability respectively.

2. *1BL/1RS Contributions to Quality.* The presence of the 1BL/1RS translocation in bread wheat has been attributed with reduced dough mixing tolerance and dough stickiness. This influence, however, has not been unequivocally demonstrated (Peña et al. 1990). Amaya et al. (1991) in a subsequent study found that out of 18 1BL/1RS translocation wheats three did not show stickiness. Dough stickiness, at low- and high dough mixing speed, has further also been observed even in normal wheats (Amaya et al. 1991). Non-sticky wheats, regardless of their 1BL/1RS status, possessed stronger gluten with a longer mixing time than wheats with some degree of stickiness

(Table 2). Recent studies involving random F_2 derived F_6 lines from the Nacozari (1B, 1B)/ Seri 82 (1BL/1RS, 1BL/1RS) cross substantiate our contention that dough stickiness not be exclusively associated with 1BL/1RS translocated wheats. It appears this may be a function of the genetic background interlinked with the growing environment.

Table 2. Mean values for quality characteristics of wheats, irrespective of their 1B and 1BL/1RS status, grouped according to their stickiness characteristic under high speed (185 rpm) mixing [a]

| | | | | Alveograph | | Bread making | |
| | | | | | | | |
Class [b]	1B	1BL/1RS	FP [c,d] %	Wx10^{-4}J	P/G	Mix (min)	LV (ml)
NS	7	1	11.0[a]	337[a]	5.4[a]	2.8[a]	702[a]
SS	12	11	10.4[a]	263[b]	6.5[a]	1.9[b]	688[a]
S	2	6	10.5[a]	224[b]	6.5[a]	1.7[b]	675[a]

a: adopted from Amaya et al. (1991)
b: NS = nonsticky; SS = slightly sticky; S = sticky.
c: FP = flour protein; W = gluten strength; P/G = tenacity/extensibility ratio; Mix = mixing time; LV = loaf volume.
d: mean values followed by the same letter are not significantly different (alpha = 0.05).

Since the 1BS chromosome arm is replaced by 1RS in the 1BL/1RS translocated wheats, gluten proteins controlled at *Gli-B1* (omega- and gamma-gliadins) and *Glu-B3* (low M_r glutenins) loci are lost because of their location on the 1BS arm. The 1RS chromosome arm leads to the introduction of rye proteins controlled at the *Sec-1a* (40-K gamma-secalins), *Sec-1b* (omega-secalins) and *Pr-3* (55-K secalin) loci. How these protein composition changes effect gluten functionality in the translocated wheats is as yet not well understood; but Dhaliwal and McRitchie (1990) have suggested a shift in the proportions of polymeric and monomeric proteins that may influence the adverse quality of the translocation wheat.
Our recent completion of substituting chromosome 1B for 1BL/1RS in Seri 82 through eight backcrosses is anticipated to shed more insight into the influence on baking quality of the 1RS chromosome segment.

3. *1AL/1RS translocated wheats*. The 1AL/1RS translocation originating for green-bug resistance (Sebesta and Wood 1978) in *T. aestivum* cv. Amigo is another translocation being exploited for bread wheat improvement. The 1RS arm (contributed

by Insave F.A. rye) provides resistance to powdery mildew, leaf rust, stem rust, green bug and wheat curl mite. 1AL/1RS wheats generally, exhibit superior performance under dryland conditions. The overall frequency of 1AL/1RS entries in the 1989 major USA wheat nurseries was 4.3% (Lukaszewsky 1990). Replicated yield trials conducted at CIMMYT, Mexico during the 1992-1993 wheat cycles involving 85 random F_2-derived F_6 lines from three 1AL/1RS x 1A bread wheat crosses under optimum and reduced irrigation conditions found that the 1AL/1RS translocation increased grain yield, above-ground biomass, spikes/m^2 and test weight. In addition, 1AL/1RS cultivars showed higher 1000-grain weight under optimum irrigation condition (Table 3), and required longer time to mature physiologically and for grainfilling under one irrigation treatment as compared to the 1A genotypes. A recent study comparing quality characters of 1BL/1RS and 1AL/1RS wheat rye-translocation lines showed that SDS sedimentation volumes and mixograph tolerance scores were higher among 1AL/1RS lines than among the 1BL/1RS sister lines.

Table 3. 1AL/1RS optimum irrigation trial showing yield and yield component details involving 85 random F_2-derived F_6 lines from three 1AL/1RS x 1A bread wheat crosses.

	1AL/1RS	1A	Adv. (%)
Grain yield	5585	5348	4.4 ***
Biomass at maturity	14.8	14.3	3.5 **
Spikes /m^2	474	441	7.5 ***
Grains/spike	31.4	33.6	7.0 ***
1000-grain weight	40.2	38.5	4.4 ***
Test weight	80.8	79.5	1.6 ***
Spike length	9.2	9.5	3.3 ***

** and *** indicate significance at 0.01 and 0.001 levels of probability respectively.

Additional rye chromosome contributions and utilization.
a) *The 5AS/5RL translocation.* In wheat cultivation, lack of copper leads to plant growth that expresses various degrees of sterility. Pot culture data (Graham 1975, 1978) attributed the fertility trait restoration to a single dominant gene on chromosome 5R long arm linked with the *Hp* gene on the same arm. The 5AS/5RL stock was obtained from T.E. Miller of IPSR, Norwich, UK, and is being transferred through backcrossing into 10 CIMMYT spring wheats, a few of which also posssess the 1BL/1RS translocation. Field testing sites for this germplasm after another 4 backcrosses shall be in Kenya and Australia. Another source demonstrating high copper efficiency is the 4BL/5RL translocation (Schlegel et al. 1991)

b) *The 6BS/6RL translocation.* Cereal cyst nematode (CCN) resistant germplasm needs have surfaced in Australia, India and Turkey. The necessity may also emerge in other areas. Until now, resistant sources have been identified in *Aegilops squarrosa*, *Ae. variabilis* and *S. cereale* of which the latter is pertinent to this report. Chromosome 6R of rye has been associated with CCN resistance (Asiedu et al. 1990) and a translocation stock (6BS/6RL; from L.A. Rayburn, Univ. of Illinois, USA) is a potential tester germplasm stock for eventual transfer to wheat cultivars.

c) *Chromosome 4R and 6R of rye.* Disomic addition lines 4R and 6R of Imperial rye in *T. aestivum* cv. Chinese Spring (source: Late E.R. Sears) were identified as resistant sources to Karnal bunt (KB); *Tilletia indica* (K.S. Gill pers. comm.). The additions have since been transferred to an agronomically superior but KB susceptible wheat WL711 by H.S. Dhaliwal in Ludhiana, India. Upon receiving this germplasm and in our attempts to increase it for our tests we have recovered di- and mono-telocentric plants for both 4R and 6R additions that may add towards additional gene location data associated with KB resistance.

d) *Septoria tritici resistance.* Use of triticale germplasm as a bridge for effecting rye transfers is a potent means to improve wheats and warrants additional exploitation. Through limited germplasm testing in Toluca, Mexico that comprises of inoculation of tillers with a mixture of five isolates, once weekly over a three week interval, resistant wheat germplasm has been identified that includes rye DNA presence. The advanced lines originated from hexaploid triticale x bread wheat crosses. The hexaploid triticale source (M$_2$A/CML) possessed 42 chromosomes; 28 AABB + 12RR+ one pair involving a D and R genome chromosome exchange. The rye segment is introgressed terminally on the long arm of a wheat D genome chromosome. The fluorescent *in situ* hybridization diagnostics used comprised of durum AABB DNA for blocking with biotin labelled rye (RR) and digoxigenin labelled *Ae. squarrosa* (DD) DNA as probes. There was no rye presence observed in the wheat cultivars Nyu Bay and Maringa (MRNG). Both advanced lines derived from M$_2$A/CML and wheat cultivar crosses exhibited good resistance expression (Table 4). In the M$_2$A/CML//2* Nyu Bay a homozygous 1BL/1RS translocation was present; origin of which can be positively speculated as a novel source. The 1BL/1RS translocation was also present in the M$_2$A/CML// Nyu Bay/3/CMH72A.576/MRNG line that in addition possessed the D and R genome chromosomal exchange exhibited by the M$_2$A/CML triticale. The contribution to *S. tritici* resistance of the rye exchanges is under further study.

Table 4. *Septoria tritici* evaluation of some Triticale and Wheat germplasm.

Line and pedigree	Septoria evaluation [a]
Maringa (MRNG)	3.2
Nyu Bay	6.4
M$_2$A/CML	5.3
CEP/76314-04	
M$_2$A/CML//2*Nyu Bay	5.2
CMH80A.1257-1B-2Y-2B-1Y-2B-0Y	
M$_2$A/CML//Nyu Bay//3/CMH72A.576/MRNG	3.2
CMH82.961-4Y-1B-3Y-5B-0Y	
Kauz (Susceptible bread wheat)	9.9

a Double digit scale at milk stage of the grain. First digit indicates height of infection on a 1 to 9 scale and second digit the severity. Score of 9 relates to top of plant and extreme susceptibility (see Kauz).

CONCLUSION

We demonstrate in this presentation the potential of *S. cereale* chromosomal sources in wheat improvement. Most of the exchanges are Robertsonian (centric break-fusion) types and their diagnostics is readily accomplished by C-banding, isozyme, as well as *in situ* hybridization methodologies. It is when the exchanges are of the R and D nature as present for the *S. tritici* material, or when cryptic rye DNA introgressions occur, that molecular cytogenetic analysis shall add its meaningful strength. We also feel that in the future, use of cryptic alien rye introgressions will become more enhanced through the use of the *ph* locus (Asiedu et al. 1989) and FISH diagnostics.

References
1. McIntosh RW, 1983. A catalogue of gene symbols. In: Proc. 6th IWGS, Kyoto, Japan, pp.1197-1255.
2. Payne PI, Nightingale MA, Krattiger AF, Holt LM. The relationship between HMW glutenin subunit composition and the bread-making quality of British-grown wheat varieties. J.Sci. Food. Agric. 1987;40:51-65.
3. Rogers WJ, Payne PI, Harinder K. The HMW glutenin subunit and gliadin compositions of German-grown wheat varieties and their relationship with bread-making quality. Plant Breed. 1989;103:89-100.
4. Lukaszewski AJ. Frequency of 1RS.1AL and 1RS.1BL translocations in United States wheats. Crop Sci. 1990; 30:1151-1153.
5. Ter-Kuile N, Jahan Q, Hashmi N, Aslam M, Vahidy AA, Mujeeb-Kazi A. 1B/1R translocation wheat cultivars detected by A-Page electrophoresis and C-banding in the 1990 National Uniform Wheat Yield Trial in Pakistan. Pak. J. Bot. 1991; 23:203-212.
6. Villareal RL, Rajaram S. Semidwarf bread wheats; Names, parentages, pedigrees, and origins. CIMMYT, Mexico, D.F., Mexico. 1988.
7. Rajaram S, Mann CE, Ortiz-Ferrara G, Mujeeb-Kazi A. Adaptation, stability and high yield potential of certain 1B/1R CIMMYT wheats. Proc. 6th IWGS, Kyoto, Japan, 1983:613-621.
8. Peña RJ, Amaya A, Rajaram S, Mujeeb-Kazi A. Variation in quality characteristics associated with some spring 1B/1R translocation wheats. J. Cereal Sci. 1990; 12:105-112.
9. Villareal RL, Mujeeb-Kazi A, Rajaram S, Del Toro, E. Associated effects of chromosome 1B/1R translocation on agronomic traits in hexaploid wheat. Breeding Science 1994;44:7-11.
10. Amaya A, Peña RJ, Zarco-Hernandez J, Rajaram S, Mujeeb-Kazi A . Quality (bread making) characteristics of normal (1B/1B) and translocation (1B/1R) wheats varying in dough stickiness character at two mixing speeds. Cereal Foods World 1991; 36:701.
11. Dhaliwal AS, MacRitchie F. Contributions of protein fractions to dough handling properties of wheat-rye translocation cultivars. J. Cereal Sci. 1990;12:145-149.
12. Sebesta EE, Wood EA, Jr. Transfer of greenbug resistance from rye to wheat with X-rays. In: 70th Annu Meet Am Soc Agron, Chicago, IL, 1978; Abstr, 61-62.

13. Graham RD. Male sterility in wheat plants deficient in copper. Nature 1975; 254:514-515.
14. Graham RD. Tolerance of triticale, wheat and rye to copper deficiency. Nature 1978; 271:542-543.
15. Asiedu R, Mujeeb-Kazi A, Ter-Kuile N. Marker assisted introgression of alien chromatin into wheat. In: Mujeeb-Kazi A,Sitch LA, (ed) Review of Advances in Plant Biotechnology, 1985-1989. 2nd Int Symp on Genet Manipulation in Crops. México DF, México and Manila, Philippines, CIMMYT and IRRI, 1989;133-144.
16. Schlegel R, Werner T, Hülgenhof E. Confirmation of a 4BL/5RL wheat-rye chromosome translocation line in the wheat cultivar 'Viking' showing high copper efficiency. Plant Breed. 1991;107:226-234.
17. Asiedu R, Fisher JM, Driscoll CJ. Resistance to *Heterodera avenae* in the rye genome of triticale. Theor. Appl. Genet. 1990;79:331-336

ASSEMBLY AND ANALYSIS OF A NORTH AMERICAN TRITICALE GENETIC RESOURCE COLLECTION

Calvin O. Qualset[1], Bonnie J. Furman[1], John H. Heaton[1], Bent Skovmand[2] and Darrell M. Wesenberg[3]

[1]University of California, Davis, California, USA
[2]CIMMYT, Int., Mexico City, Mexico
[3]USDA, National Small Grains Collection, Aberdeen, Idaho, USA

Abstract

More than 3000 triticale accessions were received as 12 collections from Canada, Mexico, and USA and grown in the field at Davis, CA in 1991 and 1992. Several of the collections were in danger of being lost due to retirement of the originating breeders, but a very high proportion of the accessions were rescued and seed provided for the USDA and CIMMYT gene banks. The collection contains mostly hexaploid secondary types and predominantly spring growth habit. The entire collection was characterized for several spike and growth characters and a subsample of 838 accessions was analysed by canonical analysis including both qualitative and quantitative characters. Diversity was examined using the Shannon-Weaver index and individual and composite coefficients of variation. Subcollections from Manitoba, Missouri, and California (Jenkins) had high diversity, while those from Oregon and Mexico showed the least diversity. Clustering analysis showed that the CIMMYT and California (Jenkins) collections were distinct and the Univ. of Calif. collections were intermediate as expected since those two gene pools had been hybridized in a breeding program. Cluster analysis showed distinct groupings of the complete and substitutional hexaploid types. Further analysis is planned so that a core subsample may be designated for use in initial screening of the collection for desired traits.

Introduction

Triticale is a product of modern plant breeding, and as such does not have a wealth of landraces from which to draw upon for use in plant breeding. Genetic resources of triticale are therefore triticale itself and its progenitor wheat and rye relatives. So in this sense triticale has a vast genepool from which genes may be extracted. Figure 1 shows the interrelationships of wild and cultivated species that can be exploited through conventional hybridization to produce triticale or to introduce genes into present-day triticale varieties. The history of modern triticale shows that noncultivated tetraploid wheats were key to the discovery of agronomically exciting new forms that resulted in the development of modern triticale. The gains made in transforming triticale from a raw amphiploid to an agronomically acceptable crop plant were hard won. Thousands of progenies from numerous primary triticales hybridized with bread wheat and other primary triticale were grown and selected in breeding programs.

H. Guedes-Pinto et al. (eds.), Triticale: Today and Tomorrow, 261–267.

© 1996 *Kluwer Academic Publishers. Printed in the Netherlands.*

Through several cycles of such hybridization and selection slow, steady gains were made in kernel quality, grain yield, and reduced plant height. On the other hand, rapid gains were made in reproductive capacity resulting in high floret fertility and reduced cytological irregularities.

Relatively few breeding programs (compared to wheat or barley) took up triticale breeding as a primary activity. In North America several of the programs have reduced their activity or have been eliminated. It became apparent much of the genetic resource base for triticale was at risk. We believed it was important for the future development of triticale that the gains of previous breeders be captured and retained in perpetuity for future use. Therefore, we initiated a program to assemble the North American triticale genetic resources from as many sources as possible and regenerate them in a common environment. The resulting North American Triticale Genetic Resources Collection (NATGRC) was transferred to the U.S. National Small Grains collection and the Wheat Gene Bank, CIMMYT, Mexico. The work was done at Davis, California where the mild winter of the Mediterranean-type climate was suitable for growing both winter and spring types.

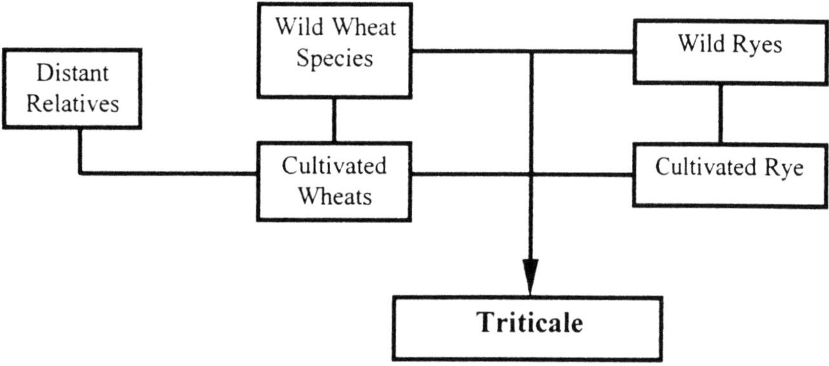

Figure 1. Triticale genetic resources showing lines of hybridization for producing triticale or introgressing genes from progenitor species.

The North American Triticale Genetic Resources Collection

Collections were obtained from 13 cooperators and all but one of them had viable seed and were successfully regenerated. After two seasons, 3,084 accessions had been regenerated. The contributions from Canada included 398 from University of Manitoba (E.N. Larter), 162 from the Plant Gene Resources Center (B. Fraleigh), and 16 from the breeding programs in Saskatchewan (G. McLeod) and Alberta (D. Salmon). The U.S. collections included 508 from the National Small Grains Collection at Aberdeen, Idaho (D.W. Wesenberg); 447 from the private collection of B.C. Jenkins, Salinas, California ; 301 from the USDA, Columbia, Missouri (J.P. Gustafson), 61 from the USDA, Corvallis, Oregon (R.J. Metzger); 250

from the University of California, Davis (C.O. Qualset); and 68 from the University of Florida (R.D. Barnett). From the CIMMYT program in Mexico (B. Skovmand) about 2,000 accessions were received and this was reduced to 873 after retaining representatives from large groups of closely related lines.

Methods

Plantings were made in November 1989 and 1990 using a single replicate of single-row plots in the first year and 2-row plots in the second year. The plots were 2.5 m in length with 0.6 m between rows. Twenty-five seeds per row were planted. The accessions were grouped according to their origins. Four reference (check) varieties were systematically included at 40-plot intervals. The check varieties included Anza common wheat; 6TA 204 and Juan, representing complete-type triticale; and Tetal 'S,' representing the substituted-type triticale. Standard fertilization and irrigation practices for triticale production were used.

Two spikes per accession were bagged preanthesis to ensure self-pollination. Data were collected in the field as follows: heading date, plant height at maturity, BYD visual symptom score (1989) (1 none, 9 severe), awn length (0, 1, 3, 6, and 9 cm), ploidy (6x or 8x based on general morphology), spike type (octoploid; 'Armadillo,' representing substituted triticale; 'Beagle,' representing complete triticale; compact, clavate or dense rachis types, and 'other' for types not fitting the previous categories), hairy neck (pubescence on the peduncle below the spike, 1 none, 5 heavy) glume pubescence (0 absent, 3 heavy), brittle and nonbrittle rachis, glume and awn color (black, brown, white). Flag leaf blade length and width were measured on a subsample of every fifth accession. At maturity the bagged spikes were collected as a future seed source and 5 unbagged spikes were collected from each plot for determination of kernels/spike, spike length, spikelet number/spike, kernels/spikelet, rachis internode length (spike density). Analysis of spike characters was completed on a systematic subsample of 838 accessions and the check varieties.

The quantitative data were analyzed first as replicates for the check varieties to detect field gradients. Minor gradients were found and the data for the accessions were adjusted according to the performance of the check varieties. This adjustment was successful in reducing variation due to the field environment and the accessions were analyzed on the basis of adjusted values. Frequency distributions, means, variances, and coefficients of variation (CV) were computed for each group of accessions and for the whole population. Principal component analysis was used for detecting patterns of variation among the groups of accessions for quantitative traits. The Shannon-Weaver diversity statistic (H') was computed for accession groups and for the whole collection.

Qualitative Traits

Detailed results for each of the subcollections will not be given here, but the overall frequencies are briefly summarized to give a general view of the extent of variation found in the NATGRC. Most of the materials were hexaploid (91%) and secondary types (98%) having spring growth habit (94%).

The spike types showed 30% of the Armadillo or substituted type, and 38% of the Beagle or complete type. The remaining included 13% of the club, dense or clavate types and 10%

miscellaneous types that were intermediate or very distinct from the above types. The hairy neck trait ranged widely, with the following frequencies of the five classes: 20, 28, 18, 12, and 22% for the classes 1 through 5. Glume pubescence showed a high frequency of low or nearly glabrous types (66%), 23% intermediate, and 11% with very heavy pubescence. Awn length was a rather variable character for scoring. The length was measured on spikelets from the center of the spike. Very few, as expected, short-awned or awnless types were found (<1% awnless and 8% short-awned (1 to 3 cm)). There were two distinct groups of the long-awned types with 31% about 6 cm and the predominant group (60%) having awns 9 cm or longer. Glume and awn color could be scored separately because some accessions had dark-colored awns and light-colored lemmas or glumes. The frequencies of classes for glume and awn color, respectively, were 84 and 76% white, 16 and 14% brown, and 0.2 and 10% black. For arid production areas the brittle rachis phenotype is extremely important. Most of the brittle rachis types have been eliminated in current breeding lines, but a few were found in this collection (6%) and the remaining were nonbrittle, but there was variation among the brittle types that was not scored.

The only disease reaction data obtained was a score for barley yellow dwarf in the first year. Triticale has been observed to be a source of BYD resistance previously, so there is interest in locating sources of resistance that could be used in triticale breeding programs. In this collection only 3% scored 1 or 2 (highly tolerant) and 40% showed tolerance (score 3 or 4). On the susceptibility side 47% scored 5 or 6 and only 10% were highly susceptible scoring 7, 8, or 9.

Shannon-Weaver diversity indices were computed for each trait and over 10 traits for each of the subcollection groups. The resulting H' values showed interesting differences among the groups. Four of the collections could be classified as rather uniform and four were distinctly more variable. The low-variability group included USDA/Corvallis (0.96), CIMMYT (1.05), Canada/PGRC (1.15), and California/UCD (1.16). The USDA/Corvallis group represented breeding lines selected for similar type and the low variability was expected. The CIMMYT group had a preponderance of two types, Armadillo and Beagle, thus contributing to more uniformity in that group. The Canada/PGRC group was surprisingly low in diversity, probably because the collection included advanced breeding lines and varieties from the Canadian programs. The UCD group included primarily progenies selected from CIMMYT x Jenkins materials and therefore was more uniform. As for the most diverse types, the Manitoba collection was the highest with H' = 1.56. The next most diverse group was from the USDA/Missouri collection (1.35) provided by Gustafson. That collection included substantial materials from Manitoba and CIMMYT. The USDA collection had H' = 1.32, and it included a wide range of old and new triticale types. The Jenkins group (1.22) had a wider range of spike types than the other collections and was the primary source of awnless and short-awned types.

Quantitative Traits

The collection showed a wide range of variation for heading time among the spring-type accessions, reflecting photoperiod sensitivity and insensitivity. For the whole collection, including the winter types, the range in heading time was 56 days; for the spring types the range was about one-half of that amount. The CIMMYT group was clearly the earliest (15 days past 31 March), being almost entirely photoperiod insensitive types and the USDA/Corvallis group was the latest (41 days past 31 March), being winter type accessions. Plant height varied from 48 to 220 cm with a mean of 125 cm, with the USDA/Corvallis group being the shortest at 93 cm and the CIMMYT group also short at 102 cm. The USDA/NSGC group was the tallest at 158 cm. Flag leaf blade area was greatest for the CIMMYT accessions (mean, 54 cm^2) with all other groups having flag leaf areas of 43 cm^2 or less.

Spike traits were also highly variable as expected. The minimum, maximum, and mean values for the whole collection were: number of spikelets/spike, 18, 50, and 30; spike length, 67, 205, and 43 mm; number of kernels/spike, 2, 163, and 79; number of kernels/spikelet, 0.1, 4.9, and 2.6; and spike density, 2.4, 9.3, and 4.8 mm/rachis internode. The USDA/Corvallis group had long spikes, with the highest number of kernels/spike (122), while the CIMMYT group had the highest fertility with 3.4 kenels/spikelet.

Mean coefficients of variation over all traits were used as an index of diversity for the whole collection and the subgroups. For the traits examined, the mean CVs were 46% for heading date, 22% for flag leaf blade area, 18% for plant height, 15% for number of spikelets/spike, 14% for spike length, 34% for number of kernels/spike, 34% for number of kernels/spikelet, and 15% for spike density. The large CV for heading date was a result of having both winter and spring types included and the large CVs for kernel number was influenced by a large range in reproductive fertility in this collection. For overall diversity, the results were practically the same as observed for the qualitative traits with the Shannon-Weaver index. In the low diversity group for H', the mean CVs were 15, 16, 17, and 20% for the CIMMYT, USDA/Corvallis, UCD, and Canada/PGRC, respectively. CVs in the group with high H' were 19, 22, 25, and 28% for California/Jenkins, USDA/Missouri, Canada/Manitoba, and USDA/NSGC, respectively.

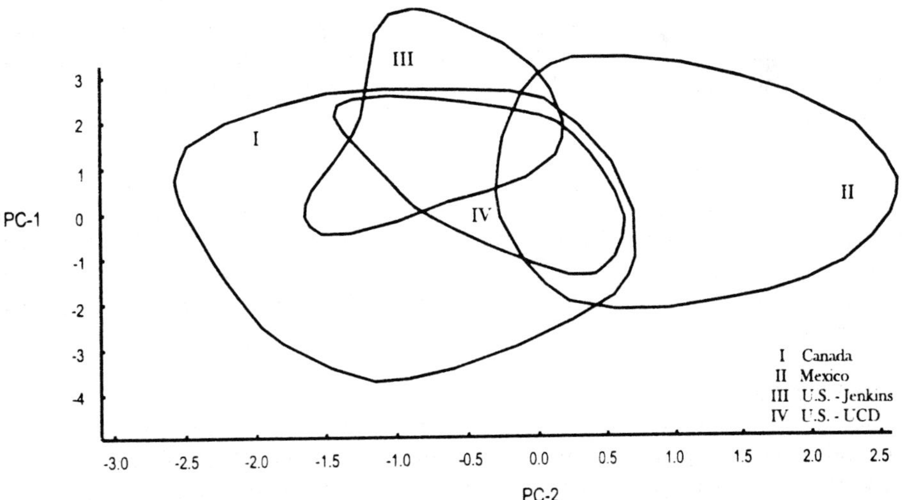

Figure 2. Principal component representation of the range of accessions from each of four triticale accession groups, based on quantitative characters identified in the text.

Principal components analysis for the quantitative traits illustrates some of the relationships among the various groups of accessions (Figure 2). In the figure, four groups of accessions are shown. The Canada group includes accessions from the three sources. This most closely represents the basic source population for North American triticale because the highly fertile types were discovered in the Manitoba program in the late 1950s and early 1960s. The CIMMYT group was developed from the Manitoba germplasm. The Jenkins materials represent materials that were developed in the Manitoba program when B.C. Jenkins was the plant breeder there and it represents his independent breeding efforts from the mid 1960s until his retirement in the mid 1980s. The California (UCD) materials were largely an outcome of intercrossing the Jenkins and the CIMMYT germplasm. Figure 2 shows the substantial divergence of the CIMMYT accessions from the Canadian ones, a lesser divergence of the Jenkins materials from the Canadian group, and intermediacy of the UCD materials with the Jenkins and CIMMYT groups. Thus, the distributions revealed by the clusters obtained by principal component analysis have meaning based on plant breeding history.

Discussion

The North American Triticale Genetic Resources Collection has been assembled and distributed to the U.S. National Small Grains Collection and to the CIMMYT Wheat Gene Bank. The data collected represent a partial characterization of the collection. Still, the analysis of the data revealed interesting relationships among the various sources of materials and provides the basis for breeders to select parental lines. The full database and more complete analysis of the data are still under development. The collection includes some redundancy of genotypes. These can be identified with careful data analysis and grow-out of putative duplicates. This is a considerable effort and if the redundancy is low it is perhaps better to leave the redundant accessions in the collection. Besides being a costly exercise, there may be minor, but important differences among the closely related accessions. These lines also provide useful hidden replication within a collection so that when evaluation and regeneration plantings are made comparable data may be expected from the redundant lines. In addition, the lines occurring more than once in a collection are most likely those which have been found to have useful traits in different programs. Those are the ones that an evaluator of a collection would like to receive. They might not appear in a subsample of a collection, as if a core collection were to be designated from the NATGRC.

It is hoped that the NATGRC contains most of the valuable germplasm of triticale from North America, but surely some important breeding lines and primary materials have not been included. It is urged that developers of such materials send them to a gene bank. Likewise, we urge that triticale workers in Europe and elsewhere take care to incorporate their materials into a secure gene bank. Triticale was not considered for conservation in the early days of the IBPGR cum IPGRI because there were no landraces. Now we can say that there are modern landraces and these should be conserved for triticale breeding and research—and for wheat breeding as well.

Acknowledgments

This collection could not have been assembled without the generosity and cooperation of the original developers, special thanks to Ed Larter, Bob Metzger, and Charles Jenkins who have taken formal career retirements, but remain active triticale enthusiasts. Frank Zillinsky, also retired, left a legacy at CIMMYT which was largely captured in this effort. Financial support was provided by Pioneer Hi-Bred International, CIMMYT, the USDA National Plant Germplasm System, National Small Grains Collection at Aberdeen, ID, the California Crop Improvement Association, and the California Genetic Resources Conservation Program.

ESTABLISHMENT OF A NETWORK FOR GENETIC RESOURCES

Norman L. Darvey[1], Gitta Oettler [2] and Wolfgang H. Pfeiffer[3]
University of Sydney[1]Universität of Hohenheim[2],CIMMYT[3]

Abstract

Systematic germplasm creation and deployment is the basis for the establishment of the ITA's first 'Co-operative Network'. Membership of the network will normally require an active contribution by an individual or organisation in the development of germplasm resources. Membership will also be afforded to organisations with a mandate for the development of regionally adapted germplasm. The germplasm centre at CIMMYT is willing to take a leadership role in the documentation, maintenance and distribution of newly developed germplasm. Germplasm will be freely interchanged among members of the co-operative, and all materials held in the CIMMYT collection will be available on request. The management committee will be subject to election by participating members of the co-operative during a meeting to be held at each International Triticale Symposium. The management committee is responsible for approving and amending membership lists and for co-ordinating various aspects of germplasm research. The genetic resources will initially include primary triticales, selected progenies from primary (or amphihaploid) times primary or secondary triticale crosses, standard wheats and ryes for triticale production or standard triticales with special attributes (eg. good GCA, disease resistance, spontaneous doubling genes for haploid producing systems, good green/albino ratio for anther culture derived regenerants, male sterility for germplasm deployment and recurrent selection, etc.). The network objectives, methodologies (strategies for germplasm production and evaluation/deployment), data and communication bases and contributions of the authors are presented along with membership and nomination lists for completion by intending participants.

Introduction

The breeding and development of all crops depends on many factors, the most important of which are the germplasm base, the genetic distance between raw or primary germplasm and cultivated germplasm, the international exchange of materials, the number of breeding programs, the skills of the individual breeder and the cross linkages to other disciplines, including agronomic, disease and quality testing.

As a relatively new crop, triticale has become adapted to a number of environmental niches; however its further expansion and wider scale acceptance depends, to a large extent, on its

H. Guedes-Pinto et al. (eds.), Triticale: Today and Tomorrow, 269–274.
© 1996 *Kluwer Academic Publishers. Printed in the Netherlands.*

increasing grain and forage production and the utilisation and competitiveness of the grain in traditional markets that have already become established for other grains with respect to the human food chain and stockfeed utilisation.

Considering the enormous genetic diversity among *Triticum* and *Secale* species, the possibility for genetic improvement with the aid of cytogenetics and biotechnology seems unlimited; however, triticale is in a catch-22 situation, for if all the financial resources are directed at plant breeding, the yield of triticale will gradually move ahead of its main competitor, namely bread wheat - but this will not improve its utilisation - if, on the other hand, the main financial resources are directed to end-product utilisation, there is the risk that yields will decline relative to the bread wheats, and hence triticale will remain non-competitive with the bread wheats, except in their environmental niches.

Hence triticale will struggle to expand on the international scene unless it either finds a key place in sustainable agriculture, its financial research inputs are increased or international researchers pool their resources, in order to reduce duplication and thus increase the emphasis on key research areas. The establishment of Collaborative International Research Networks is seen as a means of achieving the latter objective, and this should be effective provided that the objectives are clearly defined, there is open interchange of materials within the co-operatives and all parties benefit from the achievement of these objectives.

While quality improvement of the grain for end product utilisation is almost certainly the most important short to medium-term research goal, it is somewhat inevitable that the conservative nature of the grains processing industry in developed countries will not adopt triticale of equal product consistency as other grains in the short-term, unless it matches the industrial specifications of existing grains. This is because there is little flexibility in mechanised systems to shift production specifications for different cereal grains or grain mixtures. However, the benefits to developing countries or countries with more flexible processing systems, would be significant.

It is the view of the authors that the longer-term future of triticale depends to a large extent on the expansion and wise utilisation of the germplasm base. The starting point, therefore, for the establishment of a "Genetic Resources Network" is to develop sensible strategies for the creation, assessment and distribution of new germplasm.

Networking methodology

The issue of combining ability between wheat and rye, and between primary and secondary triticales, is at the core of future germplasm development. This is not a static process since many centres in the world are involved in the breeding of wheats (4x and 6x), ryes and triticales. Hence the CIN for Germplasm Resources has been initially established at three centres for winter rye, spring rye and wheat/triticale breeding (University of Hohenheim, University of Sydney and CIMMYT, respectively). Hence the traditional breeding of the rye and wheat components of triticale and special pre-breeding of these parents for triticale production can be effectively carried out. The latter includes the assessment of combining abilities and the development of parent stocks with special attributes (eg. crossability, acid soil tolerance, disease resistance, etc.).

The evaluation of newly produced germplasm *per se* and its combining ability with other triticales likewise needs to be assessed. Basic measurements of biomass production, fertility and seed quantity and quality need to be recorded as well as performance of F_1s (heterotic potential), doubled-haploids, or backcross F_1 progenies via the use of standard triticale tester stocks. This information will provide a genetic profile for each newly produced triticale. This might also be followed by biochemical and molecular analyses to provide information on quality parameters as well as genetic divergence. The total information which also includes pedigrees and details of disease and insect tolerance or resistance would then be stored in a database and nurseries prepared each year for general or specific attributes.

The storage and cataloguing of all new germplasm combinations and the distribution of nurseries would also be essential components of the Genetic Resources Network.

In addition to newly developed germplasm of various ploidy levels or selected progenies from "primary x secondary" crosses, the germplasm resources network will also include triticale germplasm showing special attributes such as high GCA, SCA, protein or amino acid components, gluten properties, tolerance to diseases, insects or mineral toxicities etc. Although a cytogenetics co-operative may be instigated at a later date, a special network of cytogenetic and transformation stocks would also be established within the germplasm resources network. The cytogenetic stocks would include chromosome substitution or translocation lines as well as aneuploid series, alien addition lines and useful genetic stocks for chromosome pairing manipulations.

A policy on the genetic ownership of stocks will need to be addressed.

The germplasm centre at CIMMYT has indicated a willingness to take a leadership role in the documentation, maintenance and open distribution of genetic resources. While the majority of germplasm would be openly available, some members may wish to limit the exchange of specific items of germplasm to "Research Members" of the Network for a period of say three to five years. In this case, the germplasm would be stored at a different location and distributed according to the wishes of the owners until such time as the ownership period for the material has elapsed. While the open distribution of the majority of germplasm would be the normal requirement of all research members of the co-operative, it is recognised that the cost of germplasm production is expensive and may be funded by private interests which demand ownership or priority use or plant variety rights in exchange for germplasm release. Open exchange within the Research Network therefore expands the utilisation of restricted germplasm while ensuring that all germplasm becomes openly available through the germplasm centre at CIMMYT once the ownership period has elapsed. Further debate and resolution of this issue would be appropriate during the first meeting of Network members during the Symposium.

The following section outlines the contributions to be made by the authors and invites all interested parties wishing to make a contribution to the Network to participate in the Network as Research or Associate Members. The Network will be responsible for electing its Executive, comprising one or two Co-ordinators and a maximum of five Executive Researchers. Membership forms and Nomination forms are attached.

1. UNIVERSITY OF SYDNEY GERMPLASM CENTRE

The primary emphasis at the Plant Breeding Institute, Cobbitty, is the development of genetic stocks for efficient germplasm production. These include standard <u>tetraploid wheats</u> of acceptable agronomic type with good CA with rye, good crossability with rye, high survival of wheat-rye embryos on tissue culture media and lines with spontaneous doubling or self-fertility genes for some of the newly-produced wheat-rye combinations.

<u>Standard tester triticales</u> derived by self-fertility of haploids via meiotic restitution, and carrying genes for anther culture compatibility have also been produced [1]. A triticale with the above characters plus high green:albino ratio among anther culture regenerants has recently been isolated (Tabatabaie and Darvey, unpublished). A dominant male-sterile triticale for ease of crossing, backcrossing and recurrent selection procedures has also been produced[2]

The tetraploid wheat base includes both Australian and CIMMYT durum wheats; the rye base for wider scale testing against the standard durums includes advanced spring lines from the University breeding program and spring derivatives from crosses between Australian germplasm and European hybrid rye cultivars. The hexaploid triticale testers are CIMMYT-type cultivars (eg. backcrosses to Juanillo and Rhino).

A range of tetraploid wheat germplasm can also be tested by producing F_1s between these and the standard durum tester lines. These F_1s have to date shown many of the attributes of the standard durum parents, including good crossability and embryo culturability in crosses with rye.

2. UNIVERSITY OF HOHENHEIM GERMPLASM CENTRE

At the State Plant Breeding Institute (Landessaatzuchtanstalt), Hohenheim, the emphasis has been on the production of 6x and 8x primary triticales. Exclusively winter rye inbred lines or their hybrids were used as male parents. The wheat base includes German winter wheat cultivars. and durum wheats of diverse origin. Some of these triticales have been produced from parents having special attributes such as dwarfing genes or Aluminium tolerance. Orthogonal sets of 6x and 8x alloplasmic triticales with various wheat cytoplasms have also been synthesized.

All primary triticales are homozygous lines and seed is produced in isolation plots or by bagging. Agronomic characters were assessed for some triticales, and triticale suitability of some rye lines is known [3,4]. A number of 8x triticales will be tested for their GCA and SCA with secondary triticales.

The secondary winter triticale lines are derivatives of our own primary triticales, Polish and EUCARPIA germplasm and to a lesser extent CIMMYT materials.

3. CIMMYT GERMPLASM CENTRE

The primary objective of the CIMMYT germplasm centre has been the development of a sensible balance between germplasm diversity and genetic progress for value-added traits. Targeted crosses among adapted elite progenitors in tactical breeding activities ensure short term progress; but this needs to be counterbalanced against the threat of genetic vulnerability, particularly in a man-made crop. Hence, a strategic crop enhancement program has to emphasise the generation and maintenance of genetic diversity, but also carefully balance diversity objectives required to ensure long term progress against the relatively narrow genetic variability necessary to achieve short term breeding goals. Trait oriented expansion of the genetic base in triticale (Tcl) and parent building are separated from tactical breeding activities and addressed via the development of special trait populations (STPs).

S.T.P. Development Facilitates:

(1) the utilisation of diversity by combining adapted sources with a range of unadapted sources
(2) the use of different breeding methodologies
(3) the evaluation and quantification of alternate gene sources
(4) assessment of progress
and
(5) a close focus on research objectives

In triticale, the lack of evolutionary diversity may be overcompensated by a spectrum of possibilities to introduce variability.

Spring and winter wheat and rye gene pools are accessed through direct interspecific (bread wheat x Tcl) and intraspecific (winter Tcl x spring Tcl) crosses, and the production of octoploid (8X) and hexaploid (6X) primary Tcls followed by primary x secondary Tcl crosses. 8X x 6X crosses guarantee an influx of cytoplasmic variability. Genetic systems from alien species eg., *Triticum tauschii* are transferred into Tcl via bread wheats carrying alien introgressions, while complete x 2D(2R) Tcl crosses continue to be an important medium to introgress genetic variability - the two gene-pools are distinct.

Advances in embryo rescue techniques, culture media and particularly the use of tissue culture and cloning with multiple plant regeneration has drastically increased the success rates of primary triticale production. Further, "breeder-friendly" primary triticales with better meiotic stability have evolved from crosses with modern wheat and rye germplasm [5,6,7,8,9].

References

1. Balatero, CH. Studies on the synthesis and androgenesis of hexaploid triticale. MSc.thesis. 1992.

2. Venkatanagappa S, NL Darvey. Development of a dominant male-sterile triticale (*Ms 3*) for recurrent selection and hybrid seed production. Proceedings of the Third International Triticale Symposium; 1994; Lisbon, Portugal (This proceedings).

3. Oettler, G, F Wehmann, HF Utz. Influence of wheat and rye parents on agronomic characters in primary hexaploid and octoploid triticale. Theor Appl Genet 1991; 81:401-405.

4. Geiger, HH, G Oettler, R Marker, HF Utz, F Wehmann. Evaluating rye inbred lines *per se* and in euplasmic and alloplasmic crosses for triticale suitability. Crop Sci 1993; 33:725-729.

5. Immonen, AST. Amino acid medium for somatic embryogenesis from immature triticale (*X Triticosecale* Wittmack) embryos. Cer Res Comm 1993; 21(1):51-55.

6.   Immonen, AST. Comparison of callus culture with embryo culture at different times of embryo rescue for primary triticale production. Euphytica 1993; 70:185-190.

7.   Immonen, AST, G Varughese, W.H Pfeiffer, A Mujeeb-K. Crossability of tetraploid and hexaploid wheats with ryes in triticale primary production. Euphytica 1993; 65:203-210.

8.   Immonen, AST. Effect of karyotype on somatic embryogenesis from immature riticale (*X Triticosecale* Wittmack) embryos. Plant Breeding 1992; 109:116-122.

9.   Immonen, AST, WH Pfeiffer. 1991. Callus culture in production of primary triticales: comparison of media and hormones. p. 196. In Agronomy Abstracts. ASA, Madison, WI USA.

DWARFING GENES OF WHEAT AND RYE AND ITS EXPRESSION IN TRITICALE

Andreas Börner and Jens Plaschke
Institut für Pflanzengenetik und Kulturpflanzenforschung, Gatersleben, Germany

Abstract

Primary octoploid triticale have been produced by crossing GA_3 sensitive tall wheats with a GA_3 insensitive dwarf (*ct2*) rye as well as a GA_3 insensitive dwarf (*Rht3*) wheat with a GA_3 sensitive tall rye. After colchicine treatment and vegetative multiplication of the hybrids, grains were harvested and cytologically checked. Then a GA_3 seedling test was conducted. Whereas the triticales carrying the dominant *Rht3* dwarfing gene of wheat showed the expected GA_3 insensitivity, the triticales with the recessive *ct2* dwarfing gene were GA_3 sensitive. Final plant heights were as expected from the results of the GA_3 test. It was suggested that the expression of *ct2* is affected by genomic interactions between wheat and rye.

Introduction

The exploitation of triticales in high yielding environments became possible only with a widespread adoption of new semi-dwarf varieties. Major genes for reduced height (*Rht* genes) from wheat have been employed in many triticale breeding programmes. The utilisation of these semi-dwarfing genes was advanced by the fact, that the dwarf phenotype was associated with insensitivity to exogenously applied gibberellic acid and, therefore, easy to select even at the seedling stage [1]. By using the GA_3 insensitivity for genetical studies two loci were identified on chromosomes 4B and 4D, respectively. The 4B locus is shared by *Rht1*, *Rht3* [2] and two alleles, identified in the varieties 'Saitama 27' and 'Besostaya mutant' [3]. For the locus on chromosome 4D, which can be used in hexaploid triticale only by substitution or translocation to another chromosome in the A, B or R genome, three alleles were described. These are *Rht2* [2], *Rht10* [4] and one allele identified in 'Ai-bian 1a', a subline of the *Rht10* carrier 'Ai-bian 1' [5]. In addition to the wheat genepool, promising dominant semi-dwarfing genes of rye are being used as parents in triticale breeding programmes [6].

Studying the GA_3 response of rye differing in plant height [7,8,9] four semi-dwarfing mutants ('Moskowskij Karlik', 'Gülzow kurz', 'R18', 'Dwarf C') were found, which showed a reduced sensitivity to GA_3. For 'Moskowskij Karlik' and 'Gülzow kurz' it was already known that their semi-dwarf character was determined by the recessive genes *ct2*, located on chromosome 5R and *ct1*, located on 7R, respectively [10,11]. Furthermore it was found that there is an additional allele on both the 5R locus ('Moskowskij Karlik' and 'R18') and the 7R locus ('Gülzow kurz' and 'Dwarf C'), respectively [9]. Both loci have been RFLP mapped [12,13].

275

H. Guedes-Pinto et al. (eds.), Triticale: Today and Tomorrow, 275–280.

© 1996 *Kluwer Academic Publishers. Printed in the Netherlands.*

Knowing the importance of GA_3 insensitive *Rht* genes from wheat, which were also shown to have a major effect on plant height in triticale [14,15], the aim of the present paper was to study the expression of the GA_3 insensitive dwarfing gene *ct2* in octoploid wheat-rye hybrids.

Materials and Methods

EXPERIMENT 1

Octoploid triticale lines were produced by using the hexaploid wheat variety 'Chinese Spring' ('CS'), known to be GA_3 sensitive [16] as the female parent and the GA_3 insensitive rye semi-dwarf 'Moskowskij Karlik' ('MK', *ct2*) as the male parent. For chromosome doubling haploid F_1 plants were immersed in 0.25% colchicine for 12 hours at room temperature when they had reached the EC 24 stage. After chromosome counting (root-tip squashes) and multiplication 10 triticale lines (T1 ... T10) were studied together with the two parental lines.

The GA_3 test was applied as described by [8]. Seedlings were treated with GA_3 and grown at 20 $^{\circ}$C until the three leaf stage, when the distance between the stem base and the end of the second leaf sheath was measured. Usually 15 seedlings per genotype were scored along with 10 control plants, that were grown under the same conditions but omitting GA_3. In addition 10 plants per genotype were grown in the greenhouse for measuring the final plant height.

EXPERIMENT 2

Four octoploid triticales (T11, T12, T13, T14) were obtained by crossing the GA_3 insensitive wheat line 'Bersee iso Rht3' ('B3'), a near isogenic line of the wheat variety 'Bersee' carrying *Rht3*, with the GA_3 sensitive rye strain 117R (*Ct2*) and 'MK' (GA_3 insensitive, *ct2*), respectively (T12, T13). Furthermore two crosses were made between 'Bersee' ('B', GA_3 sensitive, *rht*) or 'Bersee iso Rht2' ('B2', GA_3 insensitive, *Rht2*), respectively, and 'MK' (T10, T14). The procedures for the colchicine treatment, GA_3 test and plant height measurements were the same as in experiment one. For the statistical analysis the *U*-test was applied in both experiments.

Results

EXPERIMENT 1

The results of experiment 1 are summarised in figure 1. After GA_3 application both parents could be clearly distinguished. Whilst the absolute seedlings length of 'MK' (*ct2*) was about 50 mm (120% compared to the untreated control), 'CS' reached a length of 115 mm (203% response). The patterns of the 10 triticale lines were very similar to that of the GA_3 sensitive wheat parent. The means of the absolute seedlings lengths after GA_3 treatment and the percentage of response were about 120 mm and 190%, respectively.

The results obtained for final plant height are in accordance with the data of the GA_3 seedlings test. Again both parents could be clearly discriminated and all of the triticale lines outgrew 'CS' by on average 20 cm.

EXPERIMENT 2

In experiment 2 different combinations of dwarfing genes of wheat and rye were analysed (Figure 2). The results obtained for the triticale line T11, carrying the *ct2* gene of rye only, are in accordance with those described in experiment 1. The GA_3 response of this line was similar to that of the GA_3 sensitive wheat parent and the final plant height again was even higher. The line T12, having the wheat *Rht3* gene, showed the expected reduction in GA$_3$ response, which was, however, still significant in comparison to the untreated control, and also a reduced final plant height.

Of special interest are the lines T13 and T14, carrying the rye *ct2* gene in combination with the semi-dwarfing wheat genes *Rht3* or *Rht2*, respectively. Both lines showed a complete insensitivity to exogenously applied GA$_3$ as well as the greatest reduction in plant height. For the latter trait the difference to T12, having an *Rht* gene but not *ct2*, was significant.

Experiment 1

GA Response

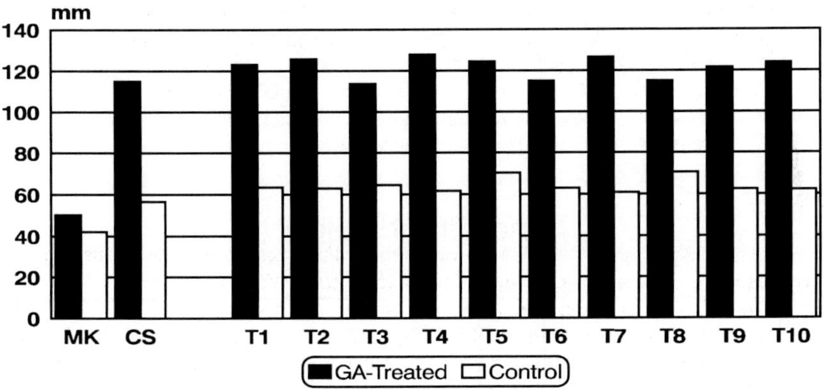

Final Plant Height

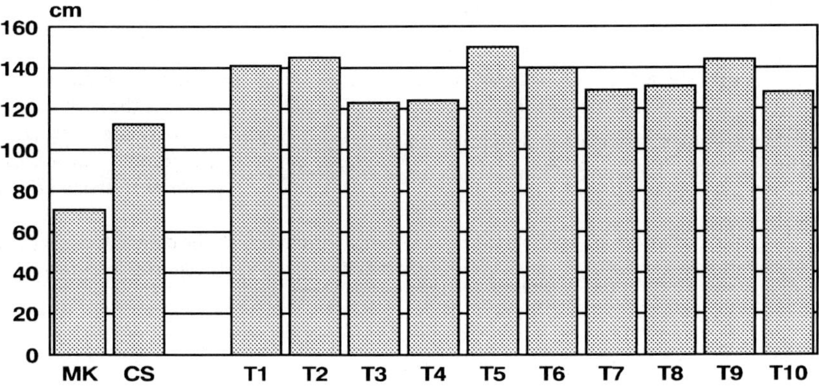

Figure 1: GA$_3$ seedlings response and final plant height of 'Chinese Spring' ('CS'), 'Moskowskij Karlik' ('MK') and ten derived octoploid triticales (T1 ... T10)

Experiment 2

GA Response

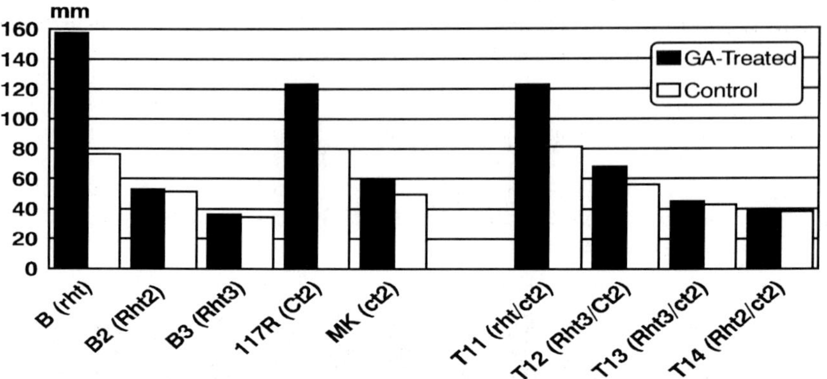

Final Plant Height

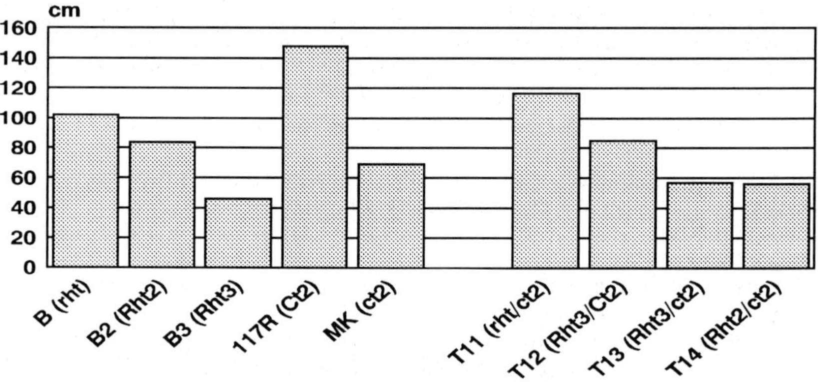

Figure 2: GA_3 seedlings response and final plant height of 'Bersee' ('B'), 'Bersee iso Rht2' ('B2'), 'Bersee iso Rht3' ('B3'), '117R', 'Moskowskij Karlik' ('MK') and four derived octoploid triticales (T11 ... T14)

Discussion

Triticale, the hybrid between wheat and rye provides a good opportunity to study intergeneric genomic interaction between both species at the level of gene expression. For the GA_3 insensitive semi-dwarfing genes of wheat *Rht1* and *Rht3* it was known that they are expressed in hexaploid triticale and therefore could be used successful in triticale breeding [14,15]. This was confirmed for octoploid triticale by the present results with the dominant wheat gene *Rht3* showing activity in line T12.

However, for the recessive gene *ct2*, which determines GA_3 insensitivity and reduced height in diploid rye no expression was found in triticale. Altogether 11 octoploid triticales showed similar results, being GA_3 sensitive and tall.

The results support the conclusion, that in terms of phenotypic markers such as disease resistance or morphological characters normally only dominant rye genes are expressed in wheat-rye amphiploids [17]. A successful exploitation of recessive rye semi-dwarfing genes, including *ct2* (or *ct1*) in triticale seems to be impossible. The utilisation of rye as a genepool for short straw will be limited to dominant semi-dwarfing genes. This may cause problems as only two out of ten described semi-dwarfing genes of rye show a dominant inheritance [18]. Neither of these genes are linked with GA_3 insensitivity [7]. Conversely, a dominant dwarfing gene which was GA_3 insensitive and expressed in triticale was described [19].

Interestingly, in combination with *Rht3* or *Rht2*, the latter is less potent in height reduction, the *ct2* gene has an effect on both GA_3 insensitivity and final plant height in triticale and therefore seems to be expressed to some extent.

References

1. Gale MD, Gregory RS. A rapid method for early generation selection of dwarf genotypes in wheat. Euphytica 1977;26:733-9.
2. Gale, MD, Youssefian S. Dwarfing genes in wheat 1. In: GE Russell editor. Progress in Plant Breeding. London: Butterworths and Co., 1985: 1-35.
3. Worland AJ, Petrovic S. The gibberellic acid insensitive dwarfing gene from the wheat variety Saitama 27. Euphytica 1988;38:55-63.
4. Börner A, Mettin D. The genetic control of gibberellic acid insensitivity of the wheat variety Ai-Bian 1. Proc 7th Int. Wheat Genet. Symp. IPSR, Cambridge, 1988: 489-92.
5. Börner A, Lehmann CO, Mettin D et al. GA-insensitivity of 'Ai-bian 1a'. Ann. Wheat Newsletter 1991;36:59-60.
6. Wolski T. Breeding of triticale for production and stress resistance. Vortr. Pflanzenzüchtg. 1991;20:189-93.
7. Börner A, Melz G. Response of rye genotypes differing in plant height to exogenous gibberellic acid application. Arch. Züchtungsforsch. 1988;18:79-82.
8. Börner A. Genetical studies of gibberellic acid insensitivity in rye (*Secale cereale* L.). Plant Breeding 1991;106:53-7.
9. Börner A, Melz G, Lenton JR. Genetical and physiological studies of gibberellic acid insensitivity in semidwarf rye. Hereditas 1992;116:199-201.
10. Sturm W, Müller HW, 1982. Localisation of the recessive gene of the Moscow dwarf mutant short straw character in *Secale cereale* L. Cytol. and Genet. 1982;16:13-7 (russ.).
11. De Vries JN, Sybenga J. Chromosomal location of 17 monogenically inherited morphological markers in rye (*Secale cereale* L.) using the translocation tester set. Z. Planzenzüchtg 1984;92:117-39.
12. Plaschke J, Börner A, Xie DX, Koebner RMD, Schlegel R, Gale MD. RFLP mapping of genes affecting plant height and growth habit in rye. Theor. Appl. Genet. 1993;85:1049-54.
13. Plaschke J, Korzun V, Koebner RMD, Börner A. Mapping of the GA_3-insensitive dwarfing gene *ct1* on chromosome 7 in rye. Plant Breeding. 1995;114:113-6.
14. Gregory RS. A technique for identifying major dwarfing genes and its application in a triticale breeding programme. Hodowla Roslin 1980;24:407-18.
15. Gregory RS. A technique for identifying major dwarfing genes in triticale. Z. Planzenzüchtg. 1984;92:177-84.
16. Gale MD, Marshall GA. Insensitivity to gibberellin in dwarf wheats. Ann. Bot. 1973;37:729-35.

17. Zeller FJ, Hsam SLK. Broadening the genetic variability of cultivated wheat by utilizing rye chromatin. Proc. 6th Int. Wheat Genet. Symp., Kyoto, Japan, 1983: 161-73.
18. Melz G, Schlegel R, Thiele V. Genetic linkage map of rye (*Secale cereale* L.). Theor. Appl. Genet. 1992;85:33-45.
19. Jlibene M, Gustafson JP. The identification of a gibberellic-acid-insensitive gene in *Secale cereale*. Plant Breeding 1992;108:229-33.

TRANSFER OF GENES Rht 1, Rht 2, AND Rht 3 FROM WHEAT TO TRITICALE

Czesław Tarkowski, Daniela Gruszecka,
Jadwiga Bichta,Krzysztof Kowalczyk
Institute of Genetics and Plant Breeding,
Agricultural University
Akademicka 15, 20-934 Lublin, Poland

Abstract

In order to introduce Rht genes from wheat to triticale a series of intercrossings of the variety Maris Widgeon with triticale variety Presto has been carried out. Seeds of F_1 plants as well as seeds of further generations were sown, using a single-seed sowing method, in spacing 20 x 10 cm and the height of plants was measured.

The F_1 was sterile to a high degree. Only single grains were set in the ear. In the F_2 a segregation of plant height has occured. Mean height of the variety Presto was 90.8 cm. The combination with Rht1 was slightly lower. The selection of the lowest plants (75.0 cm) has been performed. The plants with Rht2 gene were more differentiated than plants in former series. The lowest plants have about 60 cm. In combination with Rht3 the plant height was about 45 cm. Similarly the segregation occurred in combination with genes Rht2+3. Plant fertility was also low in F_2. The lowes fertility was observed in combination with Rht2, because together with the transfer of gene the substitution of chromosome 4D has also occurred. Similarly the sterility was very high in combination with Rht2+3.

Introduction

The aim of the study was to transfer the genes of dwarfness from wheat to triticale. Hitherto the dwarfness genes have been transferred from the dwarf rye L 506 to triticale [1]. Basing on these materials, two new semidwarf varieties, Debo and Dalo, have been breeded by the DANKO Co., Poland [2].

H. Guedes-Pinto et al. (eds.), Triticale: Today and Tomorrow, 281–284.
© 1996 *Kluwer Academic Publishers. Printed in the Netherlands.*

Materials and Methods

Wheat variety Maris Widgeon, which contains dwarfness genes, has been selected for the hybridization. The dwarfness genes were introduced to triticale variety Presto. This variety is considerably high and on good soils it undergoes lodging at higher rates of nitrogen fertilization. Since few years the Presto variety is being grown both in Poland and abroad. In the years 1990-1993 a series of crossings of wheat having Rht1, Rht2, and Rht3 genes with triticale plants from Presto variety were performed. Also the first backcross has been carried out. The seeds were sown according to a single-seed sowing method, in spacing 20 x 10 cm. The measurements were taken after the harvest.

Results

The F_1 hybrids appeared to be uniform as far as morphology is concerned. The spikes were long and, to a high degree, sterile. Only single grains were found in spikelets. In the F_2 segregation has occurred, both in respect to the height of plants and to other features. The spike fertility was differentiated depending on the crossing combination. The best seed appearance occurred in the combination Presto x Maris Widgeon Rht1, while the weakest was in case of Presto with Rht2. The hybrid plants were luxurious and produced many unfertile stems. The highest number of productive tillers was in the first backcross (Table 1).

Table 1. Plant height and number of tillers in F_2

Crossing combination	Sample size	Plant height /cm/	Number of prod. tillers	Number of unfert.stems
F_2 Presto x Maris Widgeon Rht1	28	89.4 (109-75)	6.32 (14-2)	1.46 (9-0)
F_2 Presto x Maris Widgeon Rht2	35	76.6 (115-60)	5.03 (11-2)	2.46 (7-0)
F_2 Presto x Maris Widgeon Rht3	24	71.4 (97-45)	4.67 (11-1)	2.42 (6-0)
F_2 Presto x Maris Widgeon Rht2+3	36	74.6 (100-48)	4.42 (12-1)	2.56 (10-0)
Presto	20	90.8 (100-83)	3.30 (6-1)	0.15 (1-0)
LSD at P=0,05		9.1	1.76	1.77
F_2 Presto x Maris Widgeon Rht1+2	3	64.7 (80-57)	5.00 (8-2)	1.00 (1-1)
(Presto x M.W. Rht1) x Presto	3	62.7 (71-52)	3.67 (6-1)	1.00 (2-0)
(Presto x M.W.Rht2) x Presto	1	89.0	9.00	0.00

The height of plants was also differentiated depending on the hybrid combination. The height of Presto variety plants reached about 90.8 cm, but in a very dense plant stand they reached 120 cm. The hybrids of F_2 generation of the Presto x Maris Widgeon Rht1 were slightly shorter than the Presto variety plants. The height of plants of this combination varied considerably, ranging from 75 to 109 cm (i.e. 89.4 cm average). Some of them were much taller than Presto plants.

The combination of Presto with wheat having the Rht2 genes significantly varied from the combination with Rht1 genes. It was caused by the fact that the Rht2 gene occured in the 4D chromosome. Therefore, together with this gene the substitution of 4D chromosome has occurred. The plants from this combination appeared to be shorter than the Presto variety plants. The average height reached 76.6 cm, and the variation was wide, ranging from 60 to 115 cm. The lowest and the most fertile plants were selected. The disturbances of meiosis in this combination were high, while the fertility was low. It can be assumed that it was caused by the substitution of the 4D chromosome instead of chromosome 4A
or 4B.

In the Presto x Maris Widgeon Rht3 combination the plants were low. The average plant height was 71.4 cm, and the variation ranged from 45 to 97 cm. In the fourth combination Presto has been crossed with wheat having Rht2+3 genes. The plant height was slightly higher than in the former combination, reaching 74.6 cm on average. Within this group, as in the former ones, the segregation in respect to plant height has been recorded. The fluctuations ranged from 48 to 100 cm, depending on the plant.

In the F_2 the backcrossing has been carried out. The hybrids were the maternal plants, while Presto plants were paternal forms. The plants obtained in the crossing combination (Presto x Maris Widgeon Rht1) x Presto appeared to be of the triticale type of 62.7 cm high.

The backcrossing and the selection of low plants is being continued. The plants after the first backcrossing, as well as those in F_2 are very similar to the Presto variety. However, it seems that total restitution of the Presto variety genotype is not possible. Beside the Rht genes from wheat, also other genes will be transferred, as a result of homoeologous conjugation of chromosomes and exchange of genes between genomes AB of Maris Widgeon wheat and AB of Presto triticale variety.

Conclusions

The studies carried out allow us to draw the following conclusions.
1. A high segregation has been found in respect to plant height and great variability in F_2 generation.
2. The shortest plants were observed in the combination of Presto with Maris Widgeon Rht3.

References

1. Tarkowski C, Masłowski J, Subda H. The influence of chromosomes in triticale on wheat properties (in Polish). Biul. IHAR 1993; 187: 45-46.
2. Wolski T, Czerwińska E, Ceglińska A. Breeding of triticale in Danko Company. Scientific Workshop on the Agrotechnics and Consumption of Triticale; 1994 March 2-3; Puławy: Institute of Soil Science and Plant Cultivation, IUNG, 1994: 19-32.

BEHAVIOUR OF TRITICALE GERMPLASM SELECTED AT ENMP BY INTROGRESSION OF WINTER GENES ONTO SPRING GENOTYPES

Benvindo Maçãs; José Coutinho; Francisco Bagulho;
Conceição Gomes and Nuno Pinheiro
Estação Nacional de Melhoramento de Plantas, Elvas, Portugal

Abstract

In Portugal, almost all cultivated triticales are spring types based on CIMMYT germplasm selections. In spite of the decrease of cultivated area, due to CAP policies, farmers still grow triticales in their lands . The traditional use of this species was grain for feed. Now, new purposes can be considered, as grazing and silage. So,changes on some morphophysiological characters in the spring triticales currently used, are required . The introduction of day-length insensitive genes in all spring CIMMYT germplasm has led to the development of continuous growth cycle genotypes. This material, when planted early in the conditions of South Portugal, reaches the reproductive phase earlier in Winter when late frosts can occur. The novel breeding approach, in Estação Nacional de Melhoramento de Plantas at Elvas, involves strategies to select materials with high tillering, longer vegetative growth phase and short grain filling period. Introgression of winter genes onto spring pool has been successfully conducted. Behaviour of the new germplasm is presented.

Introduction

In Portugal , marked progress has been achieved with spring triticale varieties [1], with special reference to yield potential, improved test weight and yield stability [2] . Varieties based on CIMMYT germplasm have in particular shown a continuous growth cycle, thus making them suitable for late sowings.

Although farmers want to maintain triticale in their agricultural systems, some pressure is increasing for longer cycle material wich will allow earlier Autumn sowings.

Compared to wheat (Table 1) triticales show earlier heading and a longer grain filling period. These characteristics may, under the Mediterranean conditions of

285

H. Guedes-Pinto et al. (eds.), Triticale: Today and Tomorrow, 285–289.
© 1996 *Kluwer Academic Publishers. Printed in the Netherlands.*

South Portugal, result in lower yields. Late frosts that occur during flowering and moisture stress at the end of the cycle (Table 2) negatively affect grain set and weight.

At Estação Nacional de Melhoramento de Plantas - Elvas (ENMP) a flexible approach to various crosses and selection methods is being adopted to produce germplasm suitable for the new realities of agriculture and farmer demands. This strategy involves selection of materials with high tillering, a long vegetative growth phase and a short grain filling period.

Table 1 - Heading dates for the most cultivated spring triticale varieties compared to "Anza" and "Almansor" spring bread wheats.

| Varieties | 1990/91 | | 1991/92 | | 1992/93 | | 1993/94 | |
| | Date of | Sowing: 17/11/90 | Date of | Sowing: 21/11/92 | Date of | Sowing: 27/11/92 | Date of | Sowing: 24/11/93 |
	Heading Date	days after Jan. 1	Heading Date	days after Jan. 1	Heading Date	days after Jan. 1	Heading Date	days after Jan. 1
Bacum	26/3	85	3/4	94	28/3	87	17/3	76
Arabian	26/3	85	1/4	92	28/3	87	19/3	78
Beagle	31/3	90	31/3	91	31/3	90	26/3	85
Borba	31/3	90	6/4	97	1/4	91	25/3	84
Juanilho	29/3	88	6/4	97	29/3	88	21/3	80
Anza	8/4	94	11/4	102	10/4	100	30/3	89
Almansor	10/4	96	11/4	105	10/4	100	31/3	90

Table 2 - Annual last frost, total rainfall and April - May rainfall over a 10 years period.

Year	1984	1985	1986	1987	1988	1989	1990	1991	1992	1993	1994
Date of last frost	June 3-6	March 14-15	March 7	March 6, 9 ,10	April 7-10	April 7	April 7	Feb 20	April 10	April 18	April, 1 3-21
Total rainf- all (Sept- July)	642,1	652,2	364,7	498,8	712,4	433,9	806,9	526,3	386,3	384,4	431,8[*]
April- May rainf- all	95,7	140,6	55,1	98,6	97,0	167,3	151,7	32,6	87,5	124,7	82,5

*** Sept-May 1994 only**

Breeding Methods

Although CIMMYT materials are still the most important source of germplasm both for field evaluation and as parents in our crossing program, winter and facultatives genotypes are now being used extensively in order to 1) diversify the germplasm base and 2) to select genotypes with a longer growth cycle. Various strategies are being adopted to permit selection of materials with a longer vegetative growth phase where, flowering after the frost period. A shorter grain filling period to avoid the effects of high temperature and water stress during the terminal phase is also sought. To maintain the intergenomic genetic balance within triticale genotypes, the crossing program includes different types of crosses: three way crosses among triticales where first cross is spring x winter or facultative and the third parent is facultative; triticale x bread wheat x triticale and; triticale x durum wheat x triticale. The main goal is to transfer positive traits such as low vernalization requirement or daylength sensitivity, later flowering, shorter grain filling period and high tillering for grazing or forage production.

This breeding program follows a pedigree-single plant selection schem with some modifications. From the F2 to F4 generations, single plant selections are made on the basis of the proposed objectives. In the earlier generations a great number of plants are selected. In the F5, spikes are collected to produce F6 head-rows. This methodology allows better management of segregants, and selection of homozygous pure lines to meet the recomendations of the National Service responsible for variety evaluation to the National Catalogue. From F7, quantitative evaluation is performed and the selected advanced lines are included in a regional network. The results obtained from this network are of great importance once they indicate the genotype x environment interaction pattern. The resulting lines are well characterized for yield and yield stability.

Results

Hybridization between hexaploid triticale and tetraploid and hexaploid wheats has become a routine procedure in our triticale breeding program. Despite the problem of "recombination loss" pointed out by Lelley [3], we were able to select several interesting lines from these types of crosses. The resulting genotypes showed high fertility, later flowering and a shorter grain filling period, thus indicating possible translocations from the bread or durum wheats.
The most advanced lines have now been included in the regional network yield trials. They show good adaptation   to early sowings and perform better than "Lasko", the winter variety, which is too much late for the conditions of South Portugal (Table 3).

Table 3 - Mean data of four years (1990-1994) observations of two advancedtriticale lines compared with   "Clercal" , "Lasko" and "Crato". Trials were sown in early Autumn.

Genotype	Heading Time (Days from Jan 1st)	Yield (kg/ha)	Test weight (kg/hl)	TKW (g)
TTE 9302	104	6111	67.06	37.55
TTE 9303	102	5815	68.85	40.31
Clercal (facultative)	108	3690	65.93	34.39
Lasko (winter)	111	3332	62.76	31.89
Crato (spring)	89	4807	74.45	39.91

TTE 9302 and TTE 9303 are facultative genotypes that meet very well the demand of farmers to plant early in Autumn. In a trial conducted to evaluate daylength sensitivity and vernalization requirement, these two lines showed some response to photoperiod (Table 4).

Table 4 - Heading date of "TTE 9302" "TTE 9303", "Lasko" and "Arabian" over a three year period (1992-1994).

Varieties	Date	of	Sowing
	First Week of November	January	First week of March
TTE 9302	6/4	5/5	24/5
TTE 9303	26/3	26/4	22/5
Lasko (winter)	12/4	20/5	no heading
Arabian (spring)	17/3	25/4	13/5

References

1 - Bagulho, F.; Barradas, M.; Maçãs B.; & Coutinho, J. - Some remarks on triticale breeding evolution in Portugal. a) Role played by the National Plant Breeding Station. Proceedings of the 3rd Int. Triticale Symp.; 1994 Jun 13-17; Lisboa.

2 - Coutinho, J.; Maçãs, B.; Bagulho, F.; Gomes, C.; & Coco, J. - Registration of three new triticale varieties: Crato, Arruda and Alter. Proceedings of the 3rd Int. Triticale Symp.; 1994 Jun 13-17; Lisboa.

3 - Lelley, T. - Breeding triticale: an alternative approach. Proceedings of the 2nd Int. Triticale Symp. 282-285; 1990 Oct. 1-5; Passo Fundo. CIMMYT 1991.

VARIATION OF SOME PHYSIOLOGICAL AND SPIKE CHARACTERS AFFECTING THE REPRODUCTIVE BEHAVIOUR OF INTROGRESSIVE TRITICALE LINES WITH *T. monococcum* GENETIC INFORMATION

Wojciech SODKIEWICZ, Magdalena TOMCZAK

Institute of Plant Genetics, Polish Academy of Sciences, Poznań, Poland

Abstract

Introgression of genetic information from diploid wheat into hexaploid triticale gave wide genetic variability of the spike structure as measured by the number of spikelets per spike, size of spike and number of basal and distal flowers. Some secondarily hexaploid recombinants of triticale with A^mA^mRR amphiploid rebuild their genetical homozygosity very slowly, as in BC_1/F_6 generation about 20 percent of lines showed still morphological heterogeneity. The frequency of fertile flowers analyzed in spikes of BC_1/F_5 generation was high. Some of lines exhibited improved sprouting resistance.

Introduction

Diploid wheat *Triticum monococcum* L. is known to have many features which are of interest for hexaploid triticale and wheat breeding. Because of this the introduction of the genetic information of this primitive diploid wheat is valuable for enriching the hexaploid wheat and triticale germplasm. The possibilities of recombining these genomes in *Triticum aestivum/T. monococcum* hybrids are severely restricted because of the strong incompatibility barriers [1, 2, 3]. The same has been found in direct *Triticale* × *T. monococcum* crosses [4]. The successful synthesis of a *T. monococcum* / rye - amphiploid [5] offers the possibility to introduce the diploid wheat germplasm into hexaploid triticale by this amphiploid [6].

Triticale lines were crossed as females with a *T. monococcum* × *S. cereale* amphiploid, and by increasing the chromosome numbers of hypoploid BC_1 − plants

H. Guedes-Pinto et al. (eds.), Triticale: Today and Tomorrow, 291–297.
© 1996 *Kluwer Academic Publishers. Printed in the Netherlands.*

introgressive triticale lines have been developed [7]. This paper presents data on the variation of some characters affecting the reproductive behaviour of these lines.

Materials and Methods

The set of sixty four introgressive hexaploid triticale lines with *T. monococcum* genetic information (Tc/Tm) developed as a result of crossing two 6x-triticale stocks with the amphiploid *T. monococcum* × *S. cereale* (A^mA^mRR) were investigated.

The material consisted of:

1. 24 lines Tc/Tm resulted from initial crossing LT 176 "cv. Lasko" × A^mA^mRR
2. 40 lines Tc / Tm resulted from initial crossing LT 522 × A^mA^mRR

The material was grown on experimental field in 3-row plots consisting of 30 plants of BC_1/F_5 and 30 plants of BC_1/F_6 generations with parental triticale lines sown as a control for every ten plots.

On each plot the plants free of waxy (non-glaucous) spikes were recorded as having the morphological marker transmitted from *Triticum monococcum* as well as general visual plants homogeneity was ascertained. Ten primary spikes, typical for the line, were harvested from each plot of BC_1/F_5. The data assessed for each Tc/Tm line were: average length of the spike, number of spikelets per spike, number of basal and distal florets, and percentage seed set.

At harvest ripeness stage the sprouting resistance was ascertained based on artificial wetting of intact spikes in germination towels test. Number of seeds germinated per five spikes tested were scored.

Results and Discussion

Observations of the frequency of plants with waxy (glaucous) or without waxy (non-glaucous) spikes, made on the plants BC_1/F_5 and BC_1/F_6 generations are presented in Fig. 1 in relation to the earlier results from BC_1/F_4. The absence of waxy bloom on the glumes of spikes of introgressive *Triticale* lines with *T. monococcum* genetic information is one of the characteristic features inherited from diploid wheat. In F_1-hybrids, this character was expressed as dominant in respect to the bluish-green waxy spikes of maternal triticale lines. Percentage of lines which exhibited uniformity in respect of this character markedly increased in succeeding generations. Only 5.8 percent of BC_1/F_6 plots consisted of plants with and without non-glaucous marker. Opposite to this visual field homogeneity of other qualitative and quantitative characters increased slowly. In BC_1/F_6 generation about 20 percent of lines showed morphological instability and in respect to the number of homogenous lines, no marked progress was achieved as compared to BC_1/F_5 generation (Fig. 1). It leads to the conclusion, that some recombinants of triticale with diploid wheat, rebuild their genetic homozygosity particularly slowly. They still undergo genetic recombination, giving heterogeneous progenies.

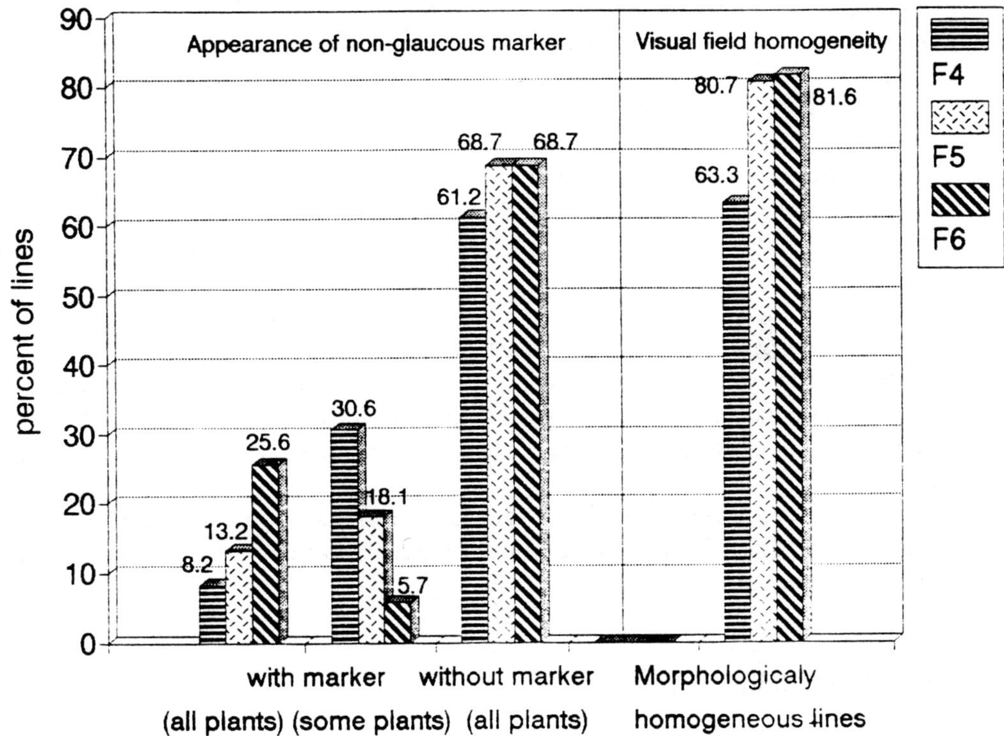

Fig. 1. Frequency of introgression of "non-glaucous" marker and morphologic uniformity of introgressive Tc/Tm lines in F_4, F_5 and F_6 generations

On the basis of the results of BC_1/F_6 progenies and the descent of particular lines, the number of such recombinants in BC_1/F_2 could be estimated as 6-10 percent.

The percentage of fertile flowers obtained in spikes of BC_1/F_5 generation of 64 introgressive Tc/Tm lines was high. General fertility level was comparable with that of parental triticale stocks (Fig. 2). About 50 percent of analyzed lines set seeds with similar frequency as the maternal triticale, but it is noteworthy that with 21.9 percent the seed setting was even greater.

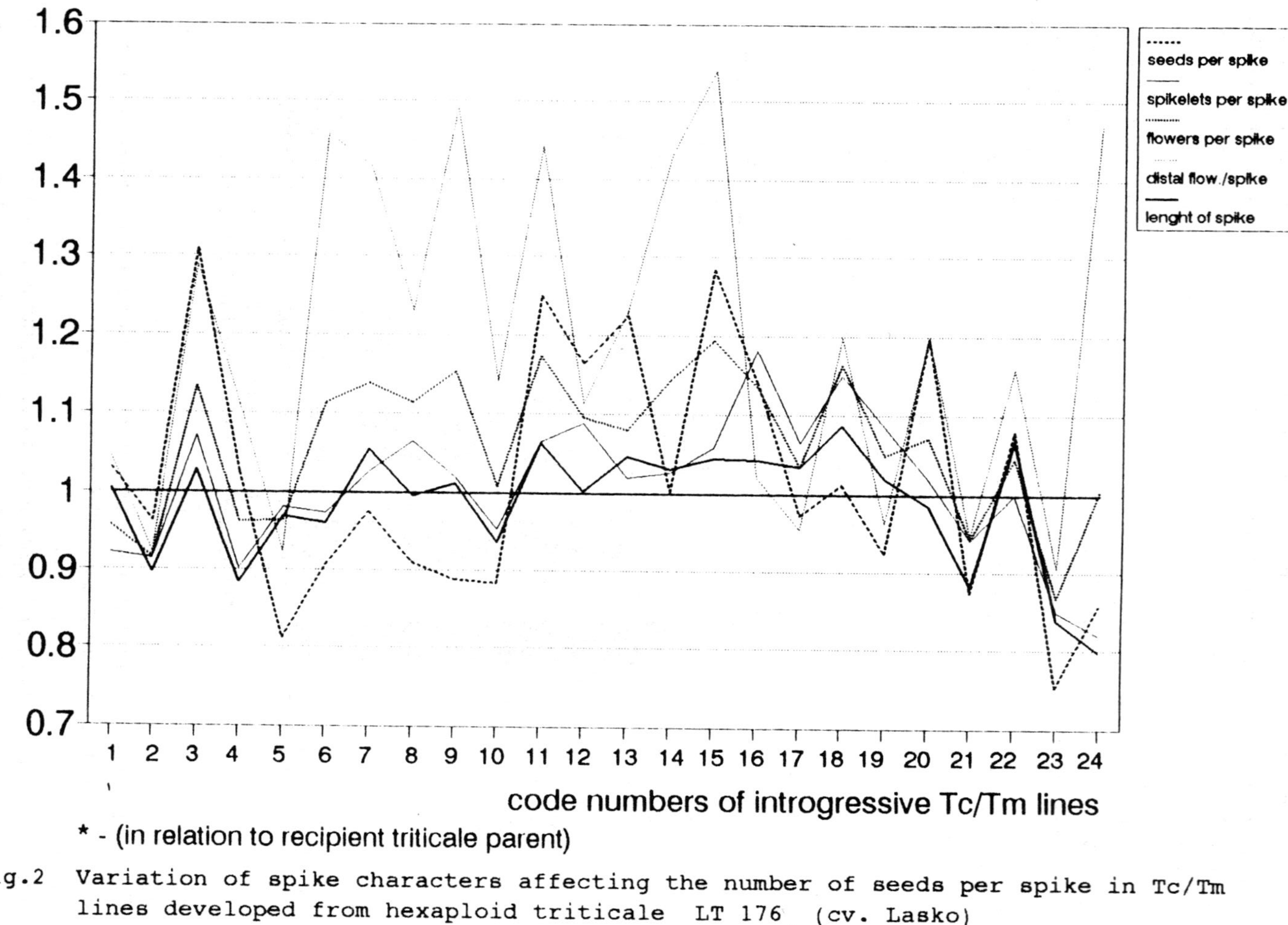

* - (in relation to recipient triticale parent)

Fig.2 Variation of spike characters affecting the number of seeds per spike in Tc/Tm lines developed from hexaploid triticale LT 176 (cv. Lasko)

Estimating these results from the point of view of influence of *T. monococcum* genes on fertility of recipient parent, it is necessary to remember that in earlier generations the restoration of the hexaploid chromosome number was done by checking spike fertility [7]. In that way, not only hypoploid plants but some triticale/*monococcum* recombinants with low fertility as well were eliminated.

The introgression of genetic information from diploid wheat into hexaploid triticale gave wide genetic variability of spike structure. The variation coefficient of the number of spikelets per spike was about 7.5 percent; size of spike (7.4%) and the number of flowers per spike (9.6%) shoved similar variation. The latter character was dependent mainly on great variation of the number of distal flowers.

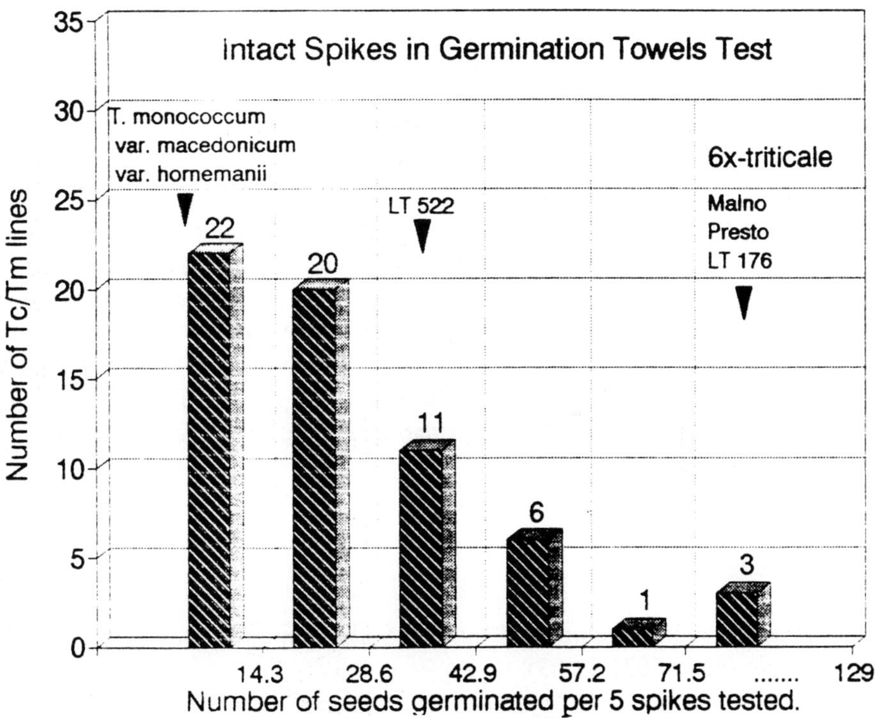

Fig. 3. Results of sprouting resistance test of triticale introgressive lines with *T. monococcum* genetic information

As the *T. monococcum* and *T. monococcum*/rye amphiploid had only two florets in their spikelets, it is interesting to note that there the lines with lower as well as higher number of distal flowers in relation to the parental form of 6x-triticale were found. Variation of spike characters and their relation to the number of seeds per spike in 24 introgressive lines developed from LT 176 ("cv. Lasko") are given in Fig. 2. All characters are presented in relative numbers to the average value of parental 6x-triticale.

The sprouting resistance of mature kernels was measured by artificial wetting of intact spikes. The objective of this study was to evaluate the expression of seed dormancy exhibited by kernels of *T. monococcum/S. cereale* amphiploid (Sodkiewicz unpubl. data). Preliminary results of testing of intact spikes in the germination towel test are presented in Fig. 3. An unexpectedly high number of lines showed improved sprouting resistance, some of them were comparable with the level of sprouting of wetted spikes of *T. monococcum*.

Conclusions

The stabilization phase of some secondarily hexaploid triticale/*T. monococcum* recombinants is prolonged, because they rebuild their genetic homozygosity particularly slowly.

That is not, however, a general rule because the homozygosity of the lines in respect of non-glaucous marker continues without special difficulties.

Most of the developed Tc/Tm lines of BC_1/F_6 generation displayed visual field homogeneity as well as good vigour and spike fertility.

The new genetic material introduced from diploid wheat genome did not seem to restrict spike productivity. Some of introgressive lines might have higher sprouting resistance as compared to parental triticale stocks and Polish standard triticale cultivars.

Acknowledgement:

This research work is supported by The State Committee for Scientific Research (project No. PB 5 5021 92 03).

References

1. The TT, Baker EP. Basic studies relating to the transference of genetic characters from *Triticum monococcum* L. to hexaploid wheat. Austral J Biol Sci 1975; 28: 189-199.

2. Bijral JS, Gupta BB, Singh B, Sharma TR, Karnwall KS. Studies relating to direct introgression from *Triticum monococcum* L. to hexaploid wheat. Cereal Res Comm 1990; 18:139.

3. Cox TS, Harrel LG, Chen P, Gill BS. Reproductive behaviour of hexaploid/diploid wheat hybrids. Plant Breeding 1991; 107: 105-118.

4. Neumann M, Sodkiewicz W, Kison H-U. On possibilities of genetic information transfer from *Triticum monococcum* to triticale. Genetica Pol 1985; 26 (2): 208-215.

5. Sodkiewicz W. Amphiploid *Triticum monococcum* L. × *Secale cereale* L. (AARR) – a new form of tetraploid triticale. Cereal Res Comm 1984; 12: 35-40.

6. Sodkiewicz W. Probleme und Möglichkeiten der Übertragung genetischer information von *Triticum monococcum* auf *Triticale* under Verwendung von Alloploiden aus diploidem Weizen und Roggen. Arch Züchtungsforsch 1985; 15: 79-82.

7. Sodkiewicz W. Introgression of Genetic Information from *Triticum monococcum* L. into Hexaploid *Triticale* by Hybridisation with a *T. monococcum* × *S. cereale* Amphiploid. I. Development of Introgressive Hexaploid Lines by Increasing the Chromosome Numbers of Hypoploid BC_1-Plants. Plant Breeding 1992; 109: 287-295.

BIOLOGICAL POTENTIAL OF TRITICALE GERMPLASM (xTRITICALE M.) AND ITS UTILIZATION IN BREEDING

Július Rychtárik
Research Institute of Plant Production, Piešťany,
Slovak Republik

Abstract

In 1989-1992 at RIPP in Piešťany the production potential of a selected collection of 347 triticale cultivars was studied with the aim of their selection and utilization in breeding programs. The cultivars originated mainly from European countries, where selection of triticale has reached a high level. The yield of the best triticale cultivars reached or exceeded 12 t.ha^{-1} and was significantly higher than the yield of the standard winter wheat variety "Viginta". The most important yield components were number of ears per m^2, number of grains in ear and thousand grain weight. The most productive cultivars were: Polish "CT 1/87", "MT 7226-54", "CHD 888", "LT 1/88", Bulgarian "No 23", "GT-AD 1/87-88", "GT-AD 2/87-88", German "TSW 3001-8-7", "TSW 2604", "TSW 301", "CIS 27466", British "CWT 1981/DB 327", "CWT 1982.7.27", French "TBT 55-24", "76245", Swiss "CH 50393" and USA "Tritical 177/736".

Introduction

In contrast to other species of cultivated plants, where natural genetic centres exist, triticale is a new manmade genus. Therefore each genetic resource was practically created by biologicaly and technicaly difficult processes. The number of genetic resources in comparison with other species is lower. Bethencourt and Konopka (1990) reported that triticale collections are maintained only in eight of 26 evaluated genebanks. The biggest collections are at CIMMYT - Mexico, in VIR, Sankt Peterburg - Russia and in Aberdeen - USA.

The triticale germplasm can be utilized in breeding programs in two ways: (1) in productivity improvement of

H. Guedes-Pinto et al. (eds.), Triticale: Today and Tomorrow, 299–308.

© 1996 *Kluwer Academic Publishers. Printed in the Netherlands.*

hexaploid triticale which is economicaly the most important and its best cultivars exceed the productivity potential of winter wheat and (2) partly in wheat breeding where substitution of wheat chromozomes for rye could be utilized. Rychtárik (1989) conducted a survey of high yielding cultivars which were developed during the last years in various states of Europe. Particularly in the 80s triticale breeding reached high levels. At present triticale is grown in approximately 2 mil. hectares.

The aims of our work were: to collect and study in field experiments domestic and foreign cultivars of triticale, and identify and select cultivars with high yield potential for utilization in breeding programs or directly in field production.

Materials & Methods

During the period of 1989-1992 at RIPP in Piešťany collection of 347 triticale cultivars was evaluated with the aim of selecting the best ones and utilizating them in breeding programs. In the collection were included cultivars from countries where triticale breeding reached high levels - Bulgaria, France, Canada, Germany, Poland, Romania, Switzerland, USA, Great Britain, CIS and the former Czecho-Slovakia.

Experiments were established at degraded chernozem soil with good sorption capacity and saturated colloid complex. The previous was silage maize. Fertilization: 100 kg N, 39 kg P, 108 kg K per ha.

Experiments were established as randomized blocks with four replications; the area of the plots was 2,5 or 10 m^2. (Experiments with 2,5 m^2 plots are indicated in the table as "A" and with 10 m^2 plots as "B"). Sowing rate was 4,5 million of germinated seeds. The control was the variety of winter wheat "Viginta". During the vegetation period the phenological observations were made and after harvest yield and yield components were evaluated. Five main characters were measured, namely: plant height (m), number of ears per m^2, thousand grain weight, yield ($t.ha^{-1}$) and protein content in dry matter (%). From the evaluated collections of particular countries the cultivars with the best yields were evaluated and compared to the standard winter wheat cultivar. The results of 2-3 years experiments were analysed by analysis of variance. Particular experiments are separated in the tables with broken lines.

Results & Duscussion

During the experimental years the annual rainfall was in the average about 140 mm lower than the longterm average (625 mm) and longterm average annual temperature 9.97°C. It is possible that meteorological conditions favored the drought resistant genotypes of southern origin.

Most of the varieties tested were of Polish and Bulgarian origin.

Table 1: The performance of selected Polish triticales under
Slovakian conditions

Kind of experiment and year	Cultivar	Plant height	Number of ears per m^2	Thousand grain weight	Yield	Protein content in dry mather
A	MT 6353-44	1,230	643	49,9	11,77	12,28
1987	CT 1/87	1,145	642	54,4	12,04[+]	11,66
1988	Viginta	1,099	726	48,2	11,10	11,80
A	MT 7226-54 CHD	1,017	601	54,4	12,53[+]	10,29
1988	70286	1,048	735	51,0	11,79[+]	9,84
1989	Viginta	0,963	780	47,5	10,45	11,19
A	CT 1 (CHD 888)	1,055	572	45,3	12,00[+]	10,01
1989	LT 1/89	1,035	644	49,3	12,00[+]	10,57
1990	Viginta	0,920	640	51,8	8,92	10,23
A	CTHD-774-85 CHD	1,300	510	43,9	10,38[+]	12,02
1990	888	1,222	473	35,1	10,05[+]	12,52
1991	Dagro	1,213	484	44,5	9,40	13,42
A	CT 2/91	1,149	386	35,5	8,70[+]	12,78
1991	MT 2/91	1,193	358	43,6	8,72[+]	11,78
1992	Viginta	0,994	335	45,4	7,68	13,41
B	CT 1/87	1,130	678	50,7	8,56	11,95
1987	Presto	1,110	610	46,7	9,76[+]	10,74
1988	Viginta	1,007	624	45,1	8,53	10,92
B	MAH 1086	1,014	451	49,5	7,43	15,53
1988	Ugo	1,001	483	50,2	7,34	12,74
1989	Viginta	0,904	668	44,2	4,11	13,95

The collection of Polish cultivars (Table 1) represented in our conditions the best biological material. The cultivars with the best yields "CT-1/87", "MT 7226-54", "CHD-888", "LT 1/88" significantly exceeded the cultivar of winter wheat Viginta and reached yields over 12 t.ha^{-1}. The high productivity of these cultivars due to relatively higher ear productivity which means higher number of florets, higher number and weight of grains and higher thousand grain weight. Better tillering of plants resulted in better density of canopy. The number of ears per m^2 reached the level of wheat. The high yield of the most important triticales was associated with high thousand grain weight and shorter stem, which only a little exceeds 1 m ("MT 7226-54", "MAH 1086", "Ugo"). Also in the past very high yields of Polish cultivars were achieved. Rychtárik (1989) reported on yields of some cultivars higher than 12 t.ha^{-1} (the highest being 13,31 t.ha^{-1}).

Table 2: The performance of selected Bulgarian triticales under Slovakian conditions

Kind of experiment and year	Cultivar	Plant height	Number of ears per m^2	Thousand grain weight	Yield	Protein content in dry mather
A	No 23	1,235	637	48,9	11,32[+]	12,50
1987	No 25	1,027	601	50,3	10,00	12,13
1988	Viginta	1,012	640	45,8	10,62	11,80
A	GT-AD 1/87-88	1,089	686	55,1	11,42[+]	10,97
1988	GT-AD 2/87-88	1,083	652	54,7	11,40[+]	11,91
1989	Viginta	0,963	780	47,5	10,45	11,19
A	GT 1/88/89	1,022	516	58,7	9,24	10,79
1989	GT 2/88/89	1,043	508	48,0	10,76[+]	11,13
1990	Viginta	0,920	640	51,8	8,92	10,23
A	GT-AD 1/91	1,157	397	48,3	8,12	14,31
1991	GT-AD 3/91	1,108	283	50,0	8,13	13,09
1992	Viginta	0,994	335	45,4	7,68	13,41
B	GT-AD 4/88	1,045	541	46,6	9,50[+]	10,97
1988	GT-AD 1/88	1,086	529	45,4	9,25	11,28
1989	Viginta	0,935	610	48,8	9,01	10,47
B	2333-22	1,239	664	37,7	9,36	13,63
1990	1593-1717	1,174	115	39,5	5,95	13,90
1991	Viginta	0,981	109	39,1	5,78	13,06
B	GT-AD 3/91	1,156	496	46,1	6,59	13,72
1991	GT-AD 1/91	1,095	620	46,9	7,06	13,90
1992	Viginta	0,934	560	41,7	7,64	14,22

Within the Bulgarian cultivars those with high yield were "No 23", "GT-AD 1/87-88", "GT-AD 2/87-88" (Table 2). Cultivars in this collection had shorter stems than the Polish cultivars and lower yields expressed as lower number of ears per m^2 compared to "Viginta" wheat. The highest thousand grain weights were observed in cultivars "GT 1/88/89" and "GT-AD 1/87-88". Protein content was at the level of wheat. If the above mentioned resources are to be used in selection it is necessary to take into account their lower winter hardiness which was found in field trials.

From the German collection (Table 3) eight cultivars were chosen with yields at the level of "Viginta" wheat or higher. Among the most productive were "TSW 3001-8-7", "TSW 2604", "TSW 301", which exceeded the yield of the control. Characteristic for these cultivars was their longer vegetative period and partial lodging. High thousand grain weight was shown by cultivars "TSW 3001-8-9" and "TSW 2602". Others cultivars compared to "Viginta" wheat had thousand grain weight lower. Their protein content did not exceed "Viginta" wheat.

Table 3: The performance of selected German triticales under Slovakian conditions

Kind of experiment and year	Cultivar	Plant height	Number of ears per m^2	Thousand grain weight	Yield	Protein content in dry mather
A	TSW 3001-8-7	1,053	550	51,8	11,25[+]	10,29
1988	TSW 2602	1,165	548	52,3	9,47	13,26
1989	Viginta	0,963	780	47,5	10,45	11,19
A	TSW 2605	1,140	584	46,7	9,36	11,35
1989	TSW 2604	1,085	612	45,7	10,76[+]	10,46
1990	Viginta	0,920	640	51,8	8,92	10,23
A	LO-16	1,149	389	39,5	8,51	13,46
1990	TSW 2742-85	1,314	346	45,2	9,40	12,52
1991	Dagro	1,213	335	45,4	9,40	13,42
A	TSW 301	1,127	545	37,2	8,22[+]	12,50
1991	TSW 403	1,259	480	40,4	8,19	12,82
1992	Viginta	0,994	484	44,5	7,68	13,41

From the cultivars of CIS (Table 4) only one strain 27466 outyielded "Viginta" with a yield of 11.04 t.ha^{-1}, with relatively high protein content. High protein content was found in the cultivars of Krasnodar origin "AD Zelenyj" (16.14 %) and "Krasnodarskij kormovoj" (16.77 %). The reason for the lower yields of the others cultivars were less tillering and non-standard canopy. In addition to the high protein content "AD Zelenyj" should be used as short-stem source.

Table 4: The performance of selected previous USSR triticales under Slovakian conditions

Kind of experiment and year	Cultivar	Plant height	Number of ears per m^2	Thousand grain weight	Yield	Protein content in dry mather
A	27 504	1,265	699	44,9	8,59	12,96
	27 466	1,424	474	52,6	11,04[+]	14,40
	Viginta	1,012	640	45,8	10,62	11,80
A	K-431 TRAT 104/1	1,081	462	48,1	7,24	13,81
	K-501 TRAT 46/4	0,996	656	50,8	7,61	13,59
	Viginta	0,936	642	44,7	8,48	15,03
A	AD Zelenyj	0,922	244	39,4	5,40	16,14
	Krasnod. kormovoj	1,063	392	53,2	5,50	16,77
	Viginta	0,950	236	46,6	6,53	14,00

In the last years the cultivars from Great Britain were performing well in our conditions due mainly to their well developed standard canopy, leaf disease resistance and often also to their short stature (Table 5). They are however late maturing with heat-shock grain shrinkage which lowers thousand grain weight and yield. At the level of winter wheat "Viginta" in plant height there were in our experiments the cultivars "CWT 1981/DB/327" and "CWT 1982.7.27", with higher yield than the standard. Some of these cultivars can be utilized in breeding programs for short stem ("CWT 1981") and productivity ("CWT 1982").

Table 5: The performance of selected Great Britain triticales under Slovakian conditions

Kind of experiment and year	Cultivar	Plant height	Number of ears per m^2	Thousand grain weight	Yield	Protein content in dry mather
A 1989 1990	SEMU 479	1,088	406	41,8	8,01	14,08
	SEMU 1977/290	1,073	460	43,2	7,81	14,94
	Viginta	0,936	642	44,7	8,48	15,03
A 1991 1992	Cumulus	1,133	356	43,2	6,37	15,30
	Viginta	0,950	236	46,6	6,53	14,00
A 1988 1989	CVT 1980/193/67	1,063	597	46,4	10,46	10,56
	Viginta	0,963	780	47,5	10,45	11,19
A 1989 1990	CVT 1981/DB/327	0,900	492	44,7	9,32	10,46
	Viginta	0,920	640	51,8	8,92	10,23
A 1990 1991	CVT 1982.7.27	1,314	480	42,6	10,06[+]	12,38
	Dagro	1,213	484	44,5	9,40	13,42

From the cultivars of former Czecho-Slovak Republic (Table 6) the cultivar "Ring" registered in 1991 showed very good yield parameters significantly outyielding the control. Yield components at high level were number of grains per ear and thousand grain weight. With yields at the level of the standard wheat cultivar were "KS 12", and "BR 14".

Table 6: The performance of selected Czecho-Slovak triticales under Slovakian conditions

Kind of experiment and year	Cultivar	Plant height	Number of ears per m^2	Thousand grain weight	Yield	Protein content in dry mather
A	UH-99	1,221	591	51,4	9,93	11,24
1987	KS-12	1,045	634	50,0	10,61	10,52
1988	Viginta	0,962	780	47,5	10,45	11,19
A	Ring	1,107	516	52,3	10,48[+]	10,57
1989	BR-14	1,015	480	49,7	9,28	11,13
1990	Viginta	0,920	640	51,8	8,92	10,23
B	Ring	1,032	444	47,3	7,66	13,33
1989	KS-21	1,162	143	50,5	6,86	13,51
1990	Viginta	0,904	668	44,2	7,11	13,95
B	UH-127	1,128	421	36,2	5,07	12,61
1990	KS-24	1,085	558	42,5	5,87	13,51
1991	Viginta	0,980	672	39,2	5,78	13,06
B	RA-Tc-10	1,085	530	42,2	5,07	13,93
1991	UH-113	1,228	488	48,4	6,80	13,84
1992	Viginta	0,933	560	41,7	6,43	14,22

From the others less numerous collections included in the experiments interesting from the breeding viewpoint could be the cultivars of French origin "TBT 55-24" and

Table 7: The performance of selected French triticales under Slovakian conditions

Kind of experiment and year	Cultivar	Plant height	Number of ears per m^2	Thousand grain weight	Yield	Protein content in dry mather
A	TBT 55-24	1,074	576	55,2	10,29	11,19
1988	76245	1,038	506	54,8	9,78	10,52
1989	Viginta	0,963	780	47,5	10,45	11,19
A	80 TT 362	1,205	508	58,3	9,36	11,46
1989	76245	0,965	516	59,3	10,28[+]	10,90
1990	Viginta	0,920	640	51,8	8,92	10,23

"76245" in which high productivity is combined with short stem and high thousand grain weight. High yields with high parameters of other evaluated characters were shown by the Swiss cultivar "CH 50393" (Table 8) and USA cultivar "Triticale 177/736" (Table 9). Cultivars of Romanian, Hungarian and Canadian origin were less productive and therefore are not mentioned in this paper.

Table 8: The performance of selected Swiss triticales under Slovakian conditions

Kind of experiment and year	Cultivar	Plant height	Number of ears per m^2	Thousand grain weight	Yield	Protein content in dry mather
A	CHD 50593	1,025	392	49,0	9,84[+]	11,35
1989	Viginta	0,920	640	51,8	8,92	10,23
1990						

Table 9: The performance of selected USA triticales under Slovakian conditions

Kind of experiment and year	Cultivar	Plant height	Number of ears per m^2	Thousand grain weight	Yield	Protein content in dry mather
A	Tritical 177/736	1,083	663	44,3	10,76[+]	11,28
1988	Tritical 333	1,090	540	41,9	9,20	10,25
1989	Viginta	0,963	780	47,5	10,45	11,19

[+] - significant differences
A - 2,5 m^2 of experimental area
B - 10 m^2 of experimental area

Cytogenetic analyses have shown that the best performing genotypes were hexaploid with a complete set of rye chromosomes.

The following conclusions can be drawn on the basis of the present work: 1. Productivity potential of the best triticale cultivars in our experiments reached or exceeded 12 $t.ha^{-1}$ and was significantly higher than the standard winter wheat cultivar "Viginta". 2. The most important yield components were number of ears per m^2, number of grains in ear and thousand grain weight. 3. The most productive cultivars were Polish cultivars "CT 1/87", "MT 7226-54", "CHD 888", "LT 1/89", Bulgarian "No 23", "GT-AD 1/87-88", "GT-AD 2/87- 88", German cultivars "TSW 3001-8-7", "TSW 2604", "TSW 301", British "CWT 1981/DB 327", "CWT 1982.7.27", "CIS 27466", French "TBT 55-24", "76245", Swiss "CH 50393" and USA "Triticale 177/736".

References

Bettencourt E.; Konopka J.: Directory of germplasm collections. 3 cereals, IBPGR, Rome, 1990, 264
Cvetkov S.: Triticale. Zemizdat Sofia, 1989, 124
Dotlačil L.; Rychtárik J.; Bareš I.: Genofond kultúrnych rastlín v ČSFR. Genetické zdroje rastlín, Nitra, 1991, 42-52
Rychtárik J.: Genetické zdroje tritikale. VÚRV Piešťany 1988, 36
Rychtárik J.: Produkčné schopnosti odrôd tritikale a reakcia na niektoré prvky agrotechniky Slovosivo IV., (8), 1989, 5-7

THE ANALYSIS OF HERITABILITY AND COMBINING ABILITY OF PROTEIN CONTENT IN TRITICALE

Sun Yuanshu[a] & Zhang Yuqing[b]
a. Institute of Crop Breeding and Cultivation, CAAS Beijing, 100081, China
b. Institute of Improvement and Utilization of Saline alkali Soil, HAAS, Anda 151400, China

Abstract
This paper describes the hereditary variation and heritability of grain protein content in triticale. Six triticale varieties having different protein contents were used as female parents and were hybridized with five stable triticale strains. Thirty combinations were made by an incomplete diallel cross. The protein content of the grain of offspring was estimated by the index of hereditary variation constitution and heritability. The experiment indicated that grain protein content was under the control of additive genetic effects. This was due to the polygene complement, and the interaction of genotype with environment. Superior parental materials with high protein contents can be selected by using the superiority of the character established in this study.

Introduction

Grain protein is important for human nutrition and health. Triticale is a new, man-made cereal crop. It has high grain yield and improved amino acid composition of the protein. Knowledge of the genetics of protein content is useful in developing a strategy to improve the protein content of triticale.

Materials and Methods

Six triticale strains of different protein contents, A(K800), B(80-20), C(80-177), D(K66), E(79-60), F(T-2), were used as female parents $P_1(I)$ and five stable triticale lines 1(WR822), 2(WR307), 3(WR326), 4(WR386), 5(WR242), were used as male parents $P_2(j)$. Thus, thirty combinations were made by an incomplete diallel cross. The hybrid seeds were planted in a field experiment from 1991 to 1993. Plots were arranged in a randomized complete block with four replicates. Each plot consisted of two rows, 5 m long, and 40 cm apart. Plants were spaced 5 cm within rows. Recommended crop management was used in the experiment.

H. Guedes-Pinto et al. (eds.), Triticale: Today and Tomorrow, 309–314.

© 1996 *Kluwer Academic Publishers. Printed in the Netherlands.*

Twelve gram samples of ripened grain were taken from each plot of parents and hybrid F_1 generation and were measured using a G-GQA31 instrument. The data were analysed by NCII method (Comstock and Robinson, 1952).

Results And Discussion

A large variation in protein content was detected among the different combinations (Table 1). The analysis of variance (Table 2) revealed that the protein content of the different combinations were significantly different ($P < 0.01$). The analysis also showed that the F values of the parent $P_1(I)$, $P_2(j)$ and their interaction were up to the level of highly significant in pattern 1. All but $P_2(j)$ in pattern 2 were also highly significant.

Selection of Parents and Combinations

Variance analysis 2 (Table 3) compared the general combining ability (GCA) variance of protein character ($MSP_1 + MSP_2$) with specific combining ability (SCA) variance ($MSP_{1,2}$). The former had an absolute dominance as compared with the latter. Comparison of GCA with SCA (Table 6) also indicated that the former had a dominant effect.

It was shown that protein characters possessed an absolutely dominant position with gene additive effect in the genetic constitution of the hybrid offspring. Variance was produced by additive interaction of intra-allelic and intra-nonallelic loci. This shows that the behavior of the character in hybrid offsprings is mainly influenced by the parental character value; hence, protein content of triticale grain is controlled by additive gene effects. The analysis of variation of grain protein contents with the environment factor showed that the constitution of additive and dominant was important to the character.

The estimation of $P_1(j)$ group parental heritability indicated that the protein content of offspring was > 77.83% which was transmitted from female parents. The estimate of $P_2(j)$ group parental heritability proves that the protein content of offspring was > 30.6% which was transmitted from male parents. The heritability of female parents is more important than that of male parents.

The results of the experiment showed the parent C(180-177) had the highest GCA (0.81), with the relative effect value being 34.62% (Table 4 and 5). This parent had good tolerance to humidity and high leaf withering, and had high protein content (16.16 %) in the grain. The effect of grain protein is marked with the genotype of additive gene effect. The parent A(K800) with the protein content of 15.32% was the second in the six strains. Its GCA was 0.67 and its relative effect value was 3.82%. It was inferior to C(80-177). This parent was high yielding and hardy. In contrast, D(K66) was short-strawed but was not good as a parent; it had a negative GCA with relative effect value of 10.10%. The triticale strain 5(WR242) was a good parent with the highest GCA of 0.56 and relative effect value of 3.20%. E3, F2, and G5 combinations had the highest values according to relative effect value or SCA; namely SE3 was equal to 5.71%, SE2, to 5.31% and SG5 to 5.17. SA1 value of 5.08% was inferior, while SF3 and SD4 with the values of -7.08% and -5.99%, respectively were the lowest. Therefore, selection of parents with high or low combining abilities and the relative effect values of SCA are useful in improving the potential protein contents of parental

Table 1. Protein contents of different combinations in Triticale.

Combination	Replicate				Total	Mean
	I	II	III	IV		
A(K800)1	19.80	19.70	19.16	18.15	76.81	19.20
2	18 92	18.90	18.80	18.00	74.62	18.66
3	18.05	18.15	18.00	18.00	72.20	18.05
4	17.36	17.26	17.00	17.00	68.62	17.16
5	17.13	18.50	17.90	18.00	71.57	17.88
B(80-20)1	18.20	18.21	18.22	18.10	72.73	18.18
2	18.24	18.20	18.22	18.20	72.86	18.22
3	17 98	17.90	17.80	17.81	71.55	17.89
4	18.26	18.20	18.02	18.05	72.53	18.18
5	18.10	18.10	18.15	18.10	72.35	18.69
C(80-177)1	18.71	18.60	18.69	18 69	74.69	18.67
2	18.00	18.01	17.90	18.00	71.91	17.98
3	17.36	17.36	17.40	17.39	69.51	17.38
4	18.82	18.80	18.79	17.80	74.21	18.55
5	19.16	19.80	19.09	19.10	76.35	19.09
D(K66) 1	15.79	15.70	15.70	15.78	62.97	15.74
2	15.80	15.79	15.71	15.79	63.09	15.77
3	16.00	16.10	16.05	16.11	64.26	16.07
4	14.00	14.10	14.05	14.11	56.26	14.07
5	17 00	17.10	17.20	17.19	68.49	17.12
E(79-60) 1	17.92	17.90	17.90	17.89	71.61	17.90
2	18.28	18.20	18.18	18.28	72.94	18.24
3	17.15	18.80	18.80	18.79	75.21	18.80
4	17.25	17.20	17.20	17.15	68.80	17.20
5	18.48	18.40	18.43	18.41	73.72	18.43
F(T2) 1	16.24	16.04	16.05	16.17	64.50	16.13
2	17.81	17.79	17.80	17.78	71.18	17.80
3	15.00	15.10	15.05	15.10	60.25	15.06
4	15.90	17.20	15.87	15.88	64.85	16.21
5	17.90	17.80	17.81	17.88	71.39	17.85
Total	526.18	528.11	524.94	522.76	2101.99	

Table 2. Covariance analysis of the protein contents

Source	df	SS	MS	F	0.05	0.01
Replicate	3	0.50315	0.168	2.85**	1.60	1.94
Combination	29	180.141	6.212	105.29**		
Error	89	5.149	0.059			
Total	119	185.7935				

Table 3. Covariance analyses of mating

Source	df	SS	MS	Model1	Model2	0.05	0.01
P_1j Female	5	115.022	3.004	389.99**	10.50**	2.33	3.25
P_2j male	4	21.30	5.325	90.25**	2.43	2.48	3.56
$P_1j \times P_2j$	20	43.821	2.191	37.14**	37.14**	1.70	2.11
Error	87	59.49	0.059				

Table 4. Specific combining ability of protein contents of Si, Sj and Si,j

S_1j	P_2j					
P_1	1	2	3	4	5	g_i
A	0.89	0.21	0.17	-0.4	-0.87	0.67
B	-0.04	-0.04	0.10	0.66	-0.57	0.58
C	0.22	-0.51	-0.64	0.85	0.20	0.81
D	-0.13	-0.24	0.63	-1.05	0.81	-1.71
E	-0.13	-0.13	1.00	-0.28	-0.24	0.59
F	-0.60	0.93	-1.24	0.23	0.68	-0.91
g_i	0.12	0.26	-0.31	-0.63	0.56	

Table 5. The relativity effect of parent GCA and SCA

P_1i	S_{ij} P_2j					
	1WR822	2WR207	3WR326	4WR836	5WR242	Yl
A(K 800)	5.08	1.20	0.91	-2.28	-4.97	3.82
B(80-20)	-0.23	-0.80	0.57	3.77	-3.25	3.21
C(80-177)	-1.26	3.48	-3.65	4.25	5.17	34.62
D(K 66)	-0.74	-1.37	3.60	-5.99	4.62	-10.10
E(79-60)	-1.88	-0.74	5.70	-1.60	-1.37	3.37
F(T-2)		-3.42	5.31	-7.08	1.31	3.88 5.19
gi	0.68	1.48	-1.77	-3.60	3.20	

Table 6. Estimate of covariance amount and heredity

$\sigma^2 e$	σ^2_1	σ^2_2	σ^2_{1+2}	σ^2_{12}	σ^2_g	S^2_p	h^2_g	h^2_l	Female			Male
									h^2_g	h^2_l	h^2_g	h^2_l
0059	104	013	117	0.53	170	176	9632	7178	9779	7793	9305	3062

and offspring. Selecting the strains with high protein contents in these combinations will be potentially valuable.

Offspring Selection

The analysis showed that GCA for protein contents was high. It is easy to stabilize in the offspring generations. Heritability of protein content was strong, particularly, since additive effects were dominant. It indicates that selecting this character in the early generations would be valuable as the strong parental heritability allows the character have more of a chance to be expressed in the offspring.

When both the male and female triticale parents had high protein content, the protein contents of the hybrids offspring were higher than those of their parents. Genotype complement and interaction between genotype and environment had effects on grain protein contents in triticale. The results of this experiment were consistent with those found from genetic studies of protein content in wheat. Hence crossing between triticale with high protein contents seems to be effective in obtaining new strains with protein content superior to their parents.

References

Amaya, A. and Pena, R. J. 1990, Triticale industrial quality improvement at CIMMYT, Proc. of 2nd ITS.Mexico DF, CIMMTY. 412421.

Avivi, L., 1978, High-grain protein content in wild tetraploid wheat triticale dicoccoides, Proc. 5th IWGS, 372-380.

Bhullar, B. S., 1978, Geneticesanalusis of protein in wheat, Proc. 5th IWGS 539-547.

Briggle, L. W. et al, 1987, wheat and wheat improvement, (second proc.)

Comstock, R. E. and Robinson, H. F., 1952, Estimation of average dominance of genes, *Heterosis*,(ed. J. W. Gowan), pp. 494-516, Iowa State College Press, Ames, Iowa.

Johnson, V. A. et al, 1978, Breeding progress for protein and Iysine in wheat, Proc. 5th IWGS. 825-830.

Law, C. N. 1983, Quantitative genetics studies in wheat, Proc. 6th IWGS, 539-547.

Li, Z. Z., 1985, The analysis of combining ability in wheat quality and agronomic character, ACTA Agronmica sinica Vol. NO2 121-130.

Proceddu, E., 1983, Genetics of grain protein in wheat seed proteins, Proc. 6th, IWGS 77-84.

Sun Y. S. 1991, Quality breeding of Triticale in Zhai, f. I(Edl. Quality breeding in crops, press by the publishing house of agri. (Beijing).

Zun, M. Y. et al, 1984, The analysis of heterosis and combining ability in six wheat cultivars for grain protein, Iysine content, ACTA AGR. Science 10. 4:237-224.

EVALUATION OF THE TRITICALE BREEDING MATERIALS WITH RESPECT TO THEIR REACTION TO FUNGAL DISEASES

Stanislaw Wegrzyn, Anna Strzembicka, Helena Grzesik and Zofia Gajda
Plant Breeding and Acclimatization Institute, Department of Cereals
Zawila, Kraków, Poland

Abstract

A total of 237 strains of Triticale were evaluated in the years 1991-93 for resistance to *Puccinia graminis* f.sp.tritici, *Puccinia recondita* and *Fusarium nivale* under field conditions using the method of artificial inoculations. The above mentioned strains were also tested at the seedling stage in a greenhouse for resistance to *Puccinia graminis* and *Puccinia recondita*. High resistance to all pathogens was found in over 50% of studied Triticale forms in both stages of development: seedling stage and adult plant.

Another experiment was carried out in some localities of Poland in the years 1991-93 on 140 strains of Triticale with a view to evaluating resistance to leaf rust and glume blotch in field under natural inoculation conditions. Based on the results to statistical analysis genetic variability of the breeding materials with respect to their reaction to the afore-mentioned pathogens was determined.

Introduction

For several years Triticale has been considered as one of the most disease resistant cereals. However, with a steadily increasing area of Triticale cultivation there is a growing threat of an outbreak of epidemic diseases which may result in considerable economic losses. Studies by some authors (Arseniuk *et al.* 1989, Arseniuk, Czembor 1991) and our personal investigation carried out in Plant Breeding and Acclimatization Institute, Department of Cereals, Kraków (unpublished) have shown that every year Triticale is more of less affected by *Septoria nodorum*. Some strains of Triticale have appeared to be a host responsive to leaf rust (*Puccinia recondita* f.sp.tritici) and stem rust (*Puccinia graminis* f.sp.tritici). In the years 1991 - 93 in the Plant Breeding and Acclimatization Institute studies were conducted on the appraisal of Triticale breeding strains response to more important fungal pathogenes.

Material and Methods

The winter Triticale reaction was estimated on 237 older (more advanced) breeding material from Plant Breeding and Acclimatization Institute and from Plant Breeding Company "Danko". A plant resistance to *P. recondita* f.sp.tritici and *P. graminis f.sp.tritici* was evaluated in the stage of seedlings under greenhouse conditions and in the dough phase in

H. Guedes-Pinto et al. (eds.), Triticale: Today and Tomorrow, 315–320.
© 1996 *Kluwer Academic Publishers. Printed in the Netherlands.*

field with the use of artificial inoculation. In a greenhouse 15 seedlings of each genotype were inoculated in the 2nd-leaf stage with a mixture of isolates of the mentioned pathogens. The infection was estimated after 14 days using a 0 (resistant) to 4 (susceptible) scoring scale. In field the plants were inoculated with a suspension of uredospores of leaf rust in the 8-9 phase of Feekeís scale and of stem rust in the 10.3 phase of Feekeís scale. All plant were inoculated with cultures mixture, its suspension being 10,000 spores/ml with mineral oil (Tween 20), using sprayer with fine nozzle. Inoculation was performed on a stripe of plants 50 cm wide in each 2-row plot. As a control were used uninoculated plant from the opposite side of the study plot. A degree of infestation was estimated at 2 and 4 weeks after infection based on, a 9-score scale (1-susceptible, 9-resistant).

An evaluation of the Triticale seedling resistance to *Fusarium nivale* was performed in field applying an artificial inoculation with fungus. For inoculation was used *Fusarium nivale* which was growing on medium composed of soil, sand and peat with rye meal. The plants were inoculated in the tillering stage. After inoculation the seedling were covered with a layer of leaves with a view to providing provocative conditions for disease development.

After 3 months the percentage of attacked plants was calculated. In the years 1991 - - 1993 a total of 237 Triticale strains were evaluated for their resistance to the mentioned pathogens.

During the study period the response of 140 Triticale strains were estimated with respect to *Puccinia recondita f.sp.tritici* and *Septoria nodorum,* under conditions of natural infection in a few localities in Poland. A scale of 1 (susceptible) to 9 (resistant) points was used.

The results were statistically computed with the analysis of variance considering the infestation with disease of individual localities. Thus relevant weight were computed for them according to the formula worked out by Wegrzyn (unpublished).

$$W_j = \frac{9 - \bar{x}_j}{\sum_{i=1}^{m}(9 - \bar{x}_i)}$$

where:

W_j = weight for j - locality

$\bar{x}_j$ = mean disease severity for j- locality

m = number of localities

$\sum_{j=1}^{m} W_j$ = being equal 1

Results and Discussion

The results of evaluating a total of 237 Triticale strains with respect to their resistance to fungal pathogens, performed in the years 1991-93 under conditions of artificial inoculation in field and in greenhouse, indicate that above 50 % of the tested strains were characterized by resistance to leaf rust and about 80% to stem rust, estimated at 7-9 scores in field and 0--2 in greenhouse (Table 1) of 75, 40 strains showed a joint resistance to both species of rust in 1991 and 42 strains in 1993 (Table 2). In the years 1991 and 1992 a considerable number of strains (about 75 %) was affected by snow mold (*Fusarium nivale*). In 1993 over 50 % of the studied strains displayed resistance to that pathogen (Table 1). A complex resistance to *P. recondita f.sp.tritici, P. graminis f.sp.tritici* and *Fusarium nivale* was shown by 17 strains in 1991 of which 7 strains showed resistance also under conditions of greenhouse. Of 78 studied strains in 1993, 30 strains showed joint resistance to all three pathogens including 20 which were also resistant under greenhouse conditions (Table 2).

In the years 1991-1993 under conditions of natural infection the reaction of 140 Triticale breeding strains to *P. recondita tritici* and *Septoria nodorum* was estimated. Evaluation was performed in some localities in various region of Poland. In 1991 a total of 45 strains (Table 3) were evaluated. Almost all strains (44) have shown resistance to *P. recondita f.sp.tritici* and only 12 strains displayed resistance to *Septoria nodorum* (over 7.0). It was found that 10 strains showed resistance to both pathogens. Of 50 strains studied in 1992, 40 were found resistant to leaf rust and 18 to glume blotch.

Resistance to both pathogens was found in 16 strains. In the last study year a total of 45 strains were tested of which 33 displayed resistance to leaf rust, where as almost all strains were affected by glume blotch. Only 2 strains combined resistance to leaf rust with to glume blotch.

Analysis of variance for leaf rust has shown significant differences in all studied strains in each year of study. In case of glume blotch significant differences were found only in younger strains. Interaction of entries x localities was non significant for both pathogens. Coefficients of variability (CV %) for entries (Table 3) are greater than those for interaction entries x localities indicates that genetic variability of tested Triticale strains is greater than influence of environmental conditions on infestation these strains by studied diseases.

While analysing the results of 3-year synthesis (Table 4) it can be seen that both pathogens can be threat to Triticale. Glume blotch (*Septoria nodorum*) attacks Triticale to a great degree - 92 strains were attacked being estimated at 5-7 scores and 22 strains at less than 5.0. Thus Triticale proved to be susceptible to the infestation by that pathogen which has been observed for several years (Arseniuk *et. al.* 1989, Arseniuk *et. al.* 1991, Wos 1992).

An incidence of leaf rust, although to a much lesser extent is becoming also a very serious problem. In spite of the fact that under conditions of natural infection majority of the studied strains have shown a high resistance to that pathogen, however, 30 strains were infested at a scale below 7.0 (Table 4). On the incidence of leaf rust in Triticale write more authors (Arseniuk, Czembor 1991). Under conditions of artificial infection a more diversified reaction of Triticale to leaf rust was observed. Of 237 studied strains as many as 94 have shown infestation to a noticeable degree (Table 1).

A number of the tested Triticale strains under conditions of artificial infection were affected to a high extent by fungus *Fusarium nivale* (Table 1). Similar results were obtained by other authors (Cichy 1990, Lacicowa 1986).

Summing - up

* Based on our own results and considering a possibility of an increased incidence of Triticale diseases it seems imperative to pay attention to a proper selection of sources of resistance to individual fungal pathogens in continuing resistance breeding of that cereal species

Table 1 - Number of Triticale strains resistant to fungal pathogens under conditions of artificial inoculation in field and greenhouse.

Years	No. of tested strains	Number of resistant strains				
		in field			in greenhouse	
		leaf rust	stem rust	snow mold	leaf rust	stem rust
1991	75	50	62	26	37	58
1992	84	48	–	23	46	68
1993	78	45	65	57	50	63
Total	237	143	127	106	133	189

Field studies: leaf rust and stem rust - estimation based on 9-score scale; 7-9 - resistant; 1 - susceptible
snow mold -% of seedling survival after inoculation-over 90%
Greenhose studies: 5-score scale; 0,1,2 - resistant; 3,4 - susceptible

Table 2 - Number of Triticale strains showing complex resistance to the studied fungal pathogens under conditions of artificial infection in field and greenhouse

Years	No. of tested strains	Number of resistant strains			In field and greenhouse	In greenhouse
		in field				
		leaf rust and stem rust	leaf rust, stem rust and snow mold	leaf rust and snow mold	leaf rust, stem rust and snow mold	leaf rust and snow mold
1991	75	40	17	17	7	–
1992	84	–	–	31	–	14
1993	78	42	30	30	20	–

Table 3 - Number of resistant strains (>7.0), mean and coefficient of variability (CV %) under conditions of natural infection.

	1991			1992			1993		
	a	b	Total	a	b	Total	a	b	Total
No. of tested strains	20	25	45	20	30	50	20	25	45
Leaf rust									
No. of localities	4	4		4	2		5	4	
No. of resistant strains	19	25	44	19	21	40	15	18	33
Mean	7.8	8.5		8.1	7.5		7.9	7.8	
CV % (entries)	17.7	7.7		16.7	25.5		34.5	35.6	
CV % (entries x localities)	12.9	5.7		15.5	25.2		22.5	27.8	
Glume blotch									
No. of localities	6	4		3	3		6	4	
No. of resistant strains	12	0	12	10	8	18	0	2	2
Mean	6.9	4.6		6.9	6.6		5.8	5.4	
CV % (entries)	18.3	42.5		12.3	15.6		22.6	27.3	
CV % (entries x localities)	17.5	37.5		11.2	10.1		19.3	20.4	
No. of resistant strains to leaf rust and glume blotch	10	0	10	10	6	16	0	2	2

a - older strains, more advanced
b - younger strains

Table 4 - Number of resistant Triticale strains (7-9), medium susceptible (5-7) and susceptible (< 5) - synthesis of 3-years studies under conditions of natural infection

	No. of tested strains	Leaf rust			Glume blotch		
		7 - 9	5 - 7	< 5	7 - 9	5 - 7	< 5
a	49	42	6	1	8	41	–
b	75	52	16	7	2	51	22
Total	124	94	22	8	10	92	22

a - older strains, more advanced
b - younger strains

- A relatively great number of resistant older strains, is indicative of selection made by breeders for resistance to those diseases.

References

Arseniuk E, Czembor H.J., Sowa W., Krysiak H., Preliminary studies of Septoria spp. on Triticale in Poland. Proc. 3rd International Workshop on Septoria Diseases of Cereals; 1989 July 4-7. Zurich; 16-18.

Arseniuk E., Czembor H.J., Triticale diseases in central Poland in 1991. Triticale Topics. International Edition 1991. No 7.

Cich H. Odpornosc na plesn sniegowa rodów i odmian pszenzyta ozimego wyhodowanych w ZD HAR Malyszyn. Biul. IHAR 1990; Nr 181-182; 83-87.

Lacicowa B. Wystepowanie *Fusarium nivale* (Fr.) Ces. na pszenzycie i podatnosc roznych rodów hodowlanych na porazenie. Rocznik Nauk Rolniczych 1986, Seria E. 16.

McInthos A.R., Luig N.H., Milne D.L., Cusick J. Vulnerability of triticales to wheat stem rust. Can. Journal of Plant Pathology 1983. 5; 61-69.

Wos H. Odpornosc pszenzyta ozimego na grzyb *Leptospheria nodorum*. Biul. IHAR 1992, Nr 181-182; 83-87.

FACULTATIVE TRITICALE: HYDROPONIC EVALUATION OF DOUBLED HAPLOIDS WITH SUPERIOR MID-GENERATION PROGENIES

Ahmad Arzani & Norman L. Darvey
The University of Sydney, Plant Breeding Institute, Cobbitty, NSW, Australia

Abstract

This study was conducted to compare the forage performance of doubled haploid (DH) lines derived from anther culture of superior individual plants with corresponding field-selected mid-generation progenies in a greenhouse hydroponic system. Although the differences between these two methods were not significant with respect to dry matter, total biomass and grain yield, the DH's were characterized both higher means and higher frequencies of high yielding lines than the corresponding mid-generation materials.

Introduction

Most inbreeding crops such as triticale are generally improved through the development of homozygous lines generated by self pollination. The doubled haploid (DH) method is the most rapid procedure for the extraction of fully homozygous lines from highly heterozygous populations.

A new breeding methodology for forage and dual-purpose triticale [1-3] using a greenhouse hydroponic system and anther culture was employed, and present study was aimed at comparing doubled haploid (DH) lines resulting from this methodology with corresponding backcross-F_3 (BC_1F_3), topcross-F_3 (TC_1F_3) and F_4 field-selected populations (traditional method).

Materials and Methods

DH lines derived from eight population of three types of crosses (BC_1F_1, TC_1F_1 and F_2) and corresponding field-selected lines of BC_1F_3, TC_1F_3 and F_4 populations (Table 1.) were grown in a greenhouse hydroponic system. Plant weight (PW) (roots + shoots), first

321

H. Guedes-Pinto et al. (eds.), Triticale: Today and Tomorrow, 321–324.
© 1996 *Kluwer Academic Publishers. Printed in the Netherlands.*

clipping dry matter yield (DM1), total biomass (TB) and grain yield (GY) were recorded. Data from lines derived from both methods of traditional field selection and doubled haploidy were combined-analysed.

Results and Discussion

For all characters (PW, DM1, TB and GY) the differences between the two methods were not significant (Table 2). However, differences among the three types of crosses (BCF$_3$, TCF$_3$, F$_4$/BCF$_\infty$, TCF$_\infty$, F$_\infty$) were significant (0.01>P<0.001) for PW and DM1. Populations within each type of cross were significant for PW, but not significant for DM1. The populations were ranked according to DM1, TB and GY in Table 1. Spearman rank correlation coefficients between the two methods were not significant for DM1 (R= -0.04), TB (R= 0.26) and GY (R=0.32).

Although the differences between the two methods of doubled haploidy and field-derived lines were not significant, the DH's possessed both higher means and higher frequencies of high yielding lines for PW, DM1, TB and GY compared to corresponding mid-generation materials (Table 3). This demonstrates that the rapid forage breeding methodology, using the hydroponic system and anther culture, is beneficial and highly recommended.

Table 1. Ranks of doubled haploid populations and field-derived BC$_1$F$_3$, TC$_1$F$_3$ and F$_4$ populations for DM1, TB and GY.

Populations	Doubled haploid			Field-selection		
Crosses*	DM1	TB	GY	DM1	TB	GY
1. M86-6068 /TW179//EP80	4	3	2	5	6	4
2. Flora /TW179//EP80	6	7	7	7	8	6
3. II85-15 /TW179//II83-237	7	5	4	2	3	2
4. II85-79/Key25 //II83-194						
////II83-194	1	1	1	3	2	3
5. 80424 (216) /II83-194	5	2	3	6	7	7
6. Plot 327 / TW179	8	4	8	4	1	5
7. Polony Q /TW179	3	8	6	1	4	8
8. II85-73 /TW179	2	6	5	8	5	1

* Crosses 1 to 3 are TC's, 4 is a BC and the remainder are selfs.

Table 2. Analysis of variance for plant weight (PW), first clipping dry matter yield (DM1), total biomass (TB) and grain yield (GY) of triticale lines developed by DH and field selection methods from four F_2, three TC and one BC populations.

Source of variation	Degrees of freedom	Mean square			
		PW^a	DM1	TB	GY
Method	1	0.393^{ns}	6.59^{ns}	276.3^{ns}	21.37^{ns}
Cross	2	1.296^{***}	62.65^{**}	5834.5^{ns}	5.15^{ns}
Population(Cross)	5	0.696^{**}	11.93^{ns}	1787.4^{*}	0.63^{ns}
Method x Cross	2	1.167^{**}	42.33^{**}	608.9^{ns}	0.92^{ns}
Population (Method x Cross)	5	0.937^{***}	71.60^{***}	558.4^{ns}	0.74^{ns}
Residual	504	0.138	8.98	629.4	0.50

a: Based on Log transformed data

$*$ P<0.05, $**$P<0.01, $***$P<0.001 and ns: not significant.

Table 3. Means ($\pm$S.E.) and the range of PW, DM1, TB and GY of triticale lines developed by doubled haploid (DH) and field selection (Sln.) methods derived from TC, BC and selfing crosses.

Populations (DH and corresponding selection lines)	Mean			Range		
	DM1	TB	GY	DM1	TB	GY
	------------(gm)------------			------------(gm)------------		
DH_1-C_1*	35.75 (1.79)	10.94 (0.22)	7.21 (0.30)	3-24	0-126	0-21
Sln_1-C_1	33.34 (1.60)	10.91 (1.24)	5.99 (0.22)	5-17	5-90	0-10
DH_2-C_1	31.61 (3.79)	10.52 (0.54)	5.38 (0.70)	5-15.2	0-79	0-15
Sln_2-C_1	26.11 (3.04)	10.30 (0.51)	4.38 (0.60)	7.6-17	10-60	1-8
DH_3-C_1	34.28 (4.50)	9.70 (0.45)	5.50 (0.73)	5-15.7	0-178	0-28
Sln_3-C_1	42.19 (8.80)	11.17 (0.62)	6.34 (1.44)	6.8-18	0-185	0-25
DH_4-C_2	51.72 (5.81)	12.78 (0.51)	10.10 (1.34)	6-18.4	5-133	1-30
Sln_4-C_2	43.53 (4.93)	11.15 (0.37)	6.24 (0.59)	6.9-13.5	0-90	0-10
DH_5-C_3	37.53 (3.30)	10.58 (0.64)	6.44 (0.66)	6.8-16.2	13-90	1-16
Sln_5-C_3	26.66 (4.70)	10.69 (0.48)	3.96 (0.74)	6-15.4	0-60	0-9
DH_6-C_3	35.43 (2.99)	9.41 (0.53)	4.95 (1.04)	5-16.9	3-67	1-13
Sln_6-C_3	44.55 (6.68)	11.04 (0.65)	5.40 (0.70)	7-14.9	0-105	0-19
DH_7-C_3	29.36 (4.10)	11.03 (0.48)	5.45 (1.04)	5.8-13.2	5-58	1-15
Sln_7-C_3	37.00 (12.5)	11.84 (0.81)	3.67 (1.30)	9.8-15.8	0-100	0-12
DH_8-C_3	31.67 (4.90)	12.14 (0.49)	5.50 (1.05)	7.7-23.5	3-90	1-18
Sln_8-C_3	34.44 (6.34)	7.94 (0.34)	6.49 (0.67)	6-10.6	0-85	0-16

*DH1 to DH8 populations were listed in Table 2. Sln1 to Sln8 are their corresponding field selection lines and C1, C2 and C3 refer to the TC, BC and selfing crosses respectively.

References

1. Arzani A. and Darvey N.L. New breeding methodology for forage cereals I. Early generation identification of superior triticale populations. Proceeding of Second International Symposium on New Genetical Approach in Crop Improvement,1992 Feb 16-20; Karachi. Pakistan (In press).
2. Arzani A. and Darvey N.L. New breeding methodology for forage cereals I. Early generation identification of superior triticale populations. Proceeding of Second International Symposium on New Genetical Approach in Crop Improvement,1992 Feb 16-20; Karachi, Pakistan (In press).
3. Arzani A. and Darvey N.L. Hydroponics and Anther culture: tools for rapid breeding of forage cereals. In: Imrie BC and Hacker JB, editors. Focused plant improvement: towards responsible and sustainable agriculture, Vol. 1. Proceeding Tenth Australian Plant Breeding Conference, 1993 Apr. 18-23.

IV - BIOTECHNOLOGY

BIOTECHNOLOGY FOR BASIC STUDIES AND BREEDING OF *TRITICALE*

Janusz Zimny[1] and Horst Lörz[2]

1. Plant Breeding and Acclimatization Institute - Radzików, Warszawa, Poland;
2. Institut für Allgemeine Botanik, AMP II, Universität Hamburg, Germany.

Abstract

Conventional breeding methods have, over the years, resulted in significant improvements in the yield, disease and quality of the cereal crops and in the future continued progress can be expected. I recent years, however, the emergence of plant biotechnology, based on recombinant DNA methodologies and on *in vitro* genetic manipulation of cells, has resulted in the development of several new techniques which will have application to triticale improvement.

We will report on the progress with cell biological and molecular approaches to improve cereal crops in general and specifically triticale. First transgenic triticale plants have been obtained by microprojectile bombardment of scutellar tissue with a plasmid carrying the gus- and bar-gene coding for glucuronidase activity and phosphinotricin (BASTA) resistance, respectively. The screenable and selectable genes are considered marker genes and will be used for further improvement of transformation efficiency in triticale. Subsequent experiments are aiming for the transfer of genes improving disease resistance, stress tolerance, and other quality characters.

Introduction

Plants carrying foreign genes have recently been obtained for many crops including wheat, rice, maize and barley and recently we have obtained transgenic *Triticale* plants. Genes important in the field of *Triticale* breeding, influence characters such as: yield potential, grain type and weight, disease resistance, tolerance to environmental stresses, preharvest sprouting resistance and fertility. They are to be localised and characterised by means of biochemical and molecular biological methods. These modern techniques are considered to be complementary to the classical plant breeding methods.

The ability of the conventional plant breeding programs to introduce new genetic information into plants islimited. Recent progress in plant and cell biology and molecular biology allow the use of newly developed techniques to complement the work of conventional plant breeding (1).

The new approaches of cereals biotechnology include: the production of complete or partial hybrids and cybrids by protoplast fusion - microinjection of DNA into cells - direct uptake of DNA into protoplasts - gene transfer using genetically modified bacterial or viral

H. Guedes-Pinto et al. (eds.), Triticale: Today and Tomorrow, 327–337.
© 1996 *Kluwer Academic Publishers. Printed in the Netherlands.*

vectors and, so far the most effective method, microbombardment of cells and tissue.

The most important facet for practical breeding is the eventual regeneration of whole plants possessing the desired agronomic alterations. For a number of the most important crops, usable techniques do exist but they are often not optimal. For a large number of crops especially for the most important ones cereals and legumes, efficient and reproducible methods for obtaining plants from single cells still need to be improved though there were several reports on successful regeneration.

Triticale and the other cereals are among the most important crops for human and animal nutrition, and therefore, their improvement by means of new techniques is a highly desirable objective. Genetic engineering of the major cereal crops including wheat, rye, *Triticale*, barley and maize has been limited for a long time due to the lack of an efficient transformation system and the difficulties to regenerate cereal protoplasts to mature plants. The prerequisite for all mentioned techniques for introduction and stable integration of desired genes is an efficient regeneration system.

In this paper we review the progress that has been made with *Triticale in vitro* culture and genetic engineering for the last twenty years, since the history of *Triticale* biotechnology is already 20 years old.

PLANT REGENERATION SYSTEMS IN *TRITICALE*

Regeneration of plants from *in vitro* cultured cells is an important step towards genetic manipulation of plants (2). Two pathways of *in vitro* plant regeneration exist: one is somatic embryogenesis the other - organogenesis; both are observed in the *Poaceae* family. Plant material used in the experiments should be morphogenic and should have the capacity of plant regeneration. For this reason researchers devote much attention to meristematically active tissues with high potency for cell division.

ANTHER CULTURE AND DOUBLE HAPLOID PRODUCTION

The development of homozygous lines through sexual crossing requires up to seven generations or even more. By using anther culture combined with spontaneous or induced chromosome doubling, homozygosity can be achieved within no more than two generations. The *Triticale* breeders are facing with the inequality of lines which causes a big problem especially when the registration of the new cultivar is concerned. It can be solved in a short time by the anther culture method.

The production of haploid *Triticale* plants by anther culture was first reported by Ya--Ying et al. (1973). Later investigations have shown the influence of many factors on the development of microspores into fertile plants. These include the plant genotype (3,4), the developmental stage of the microspore (5,6), cold pretreatment of the anthers (3,4,6), the influence of the growth conditions of the donor plants (7,5), culture medium and medium supplements (7,3) as well as seasonal effect (4).

In general N6 (8), B5 (9) and MB (7) media solidified with agar or liquid with Ficoll, supplemented with 2.4 D, for callus induction and the same media without 2.4 D or supplemented with IAA and/or kinetin for regeneration were used. Microspores in uninuclear stage from cold pretreated anthers were taken.

Among the regenerants different morphological variants were observed. It concerned mostly the head shape, size of awn, straw thickness and leaf blade width. Aneuploid and plants of higher ploidy level were also found (3). The phenomenon of albinism was noticed too by many authors (10,7,5,3). It seems to be connected with the growth temperature of the donor plant (10) and other culture conditions (5).

PLANT REGENERATION FROM SOMATIC TISSUE

Regeneration of cereal plants from different tissues and organs has been already described by several authors. Somatic embryogenesis is the most direct way for regeneration from tissue culture. Embryogenic cultures can be initiated from ontogenetically young explants containing meristematically active cells. *Triticale* plants have been regenerated via organogenesis or somatic embryogenesis from explants such as young inflorescences (11), leaf base (12) and microspores (5,3,4). Immature embryos proved optimal as a starting material for *in vitro* regeneration of *Triticale* (13,14,15), but some of the reports described somatic embryogenesis (15,16,17).

A number of factors influencing the regeneration of *Triticale* plants has been found. but the main elements of this procedure are as follows. Immature embryos isolated at the proper stage and cultured in optimal conditions show swelling on the surface of the scutellum after 5-8 days of culture. Callus proliferation is observed during the second week. The calli are different in morphology: soft, compact or watery. All these calli are capable of forming somatic embryos. Sometimes somatic embryos are growing directly on the periphery of explants. Globular structures are easy to recognize as young embryos since they have a small cavity - the place where coleoptile will grow. During the next two weeks fully developed embryos can be observed. They possess all organs typical for zygotic embryos such as: coleoptile, coleorhiza and scutellum. The embryos appear as a singular structure or in groups but frequently in the form of twin-embryos or polyembryos. The same process takes place when anther cultures are established.

Amongst the factors of importance for successful *in vitro* culture are the developmental stage of the explant, medium composition, and genotype.

The developmental stage of the immature embryo plays an important role for high efficiency of callus proliferation. Experimental data indicate that the late spherical coleoptile stage (LSCS) is the optimal one (18). The developmental stage of embryos taken for experiments was determined on the basis of their general condition and features during isolation. These criteria appeared to be more precise than measurement of the size of embryos (14,15) or time span from anthesis till the moment of embryo collection (19,20). The size of the embryo can be different within the same variety as well as in one plant or spike. Depending on the conditions, the time of reaching LSCS also varies - from 10 to 27 days after anthesis.

The efficiency and quality of regenerable callus depends on media composition. The most important component is the auxin. 2.4-D has been used most frequently (19,20,21,15) however Dicamba has proved to be very efficient too (22). 2.4.5-T has also been used (13,20). The most suitable basal medium for callus induction is MS (23) medium (19,13,21,15), but modified KAO (24) medium has been shown to be suitable for *Triticale in vitro* cultures (14,20,15).

The genotype has been recognized as further important factor in *Triticale* regeneration (19,14,15). Sharma et al. (1980) tested ten different genotypes for callus formation and plant regeneration capability. Only 7 of them regenerated plantlets. Nakamura & Keller (1982), Stolarz & Lörz (1986a) and Zimny (1989) also selected some cultivars or lines that were more regenerable than others. In our experiments over 500 lines were tested for the

330

embryogenic genotypes. All lines tested showed embryogenic response from immature embryos. The efficiency of somatic embryogenesis ranged from 10% to 100%. Generally, we can conclude that for every line, culture conditions and regeneration capacity can be improved.

To obtain organogenic regeneration, calluses were transferred to medium containing 1.1 μM 2.4-D (19), or 2.9-29 μM GA_3 (14).

In order to obtain germination of the somatic embryos as well as to prevent turning into the callus stage they were transferred onto a regenerating medium. A comparison of different concentrations showed that 4,5 μM kinetin stimulated embryo germination and growth of *Triticale* plantlets (15,16). Later plants were moved into glass vessels and when grown to 10-12 cm in size they were transferred to soil. Having been vernalized in the case of winter type *Triticale*, they matured in the greenhouse and set seeds.

The experiments ware carried out during 4 following years in order to study the genotypic dependence of somatic embryogenesis (22). Several sublines of the regenerants showed increasing ability for somatic embryo formation. More than 3000 plants have been regenerated, selfed and part of the progeny analyzed for somaclonal variation.

Somaclonal Variation

From the breeder's point of view both genetic instability and stability are important. Instability is desired in order to obtain genetic variation, but too high degree of variation resulting from multiple events may be difficult to handle in a breeding program and may require extensive cross breeding. On the other hand, stability of cultures and regenerants is essential for most stages in breeding programs, for commercial propagation of established cultivars, and for maintenance of germplasm.

Phenotypic variations have been observed for many years the in plants regenerated from tissue cultures. This phenomenon was called "somaclonal variation" (25). The mechanism of somaclonal variation is still not clear, but generally it is linked to the passage of disorganized growth of callus or cell suspension. Morphological, cytological and molecular variations can be observed. The differences among regenerated *Triticale* plants have been reported for such characters like plant height, time of flowering and morphology of whole plants and spike, leaf waxiness, hairy neck and fertility (21,26,15,16). Similar changes have been reported for anther culture derived plants (3). Variations in morphological and agronomic characters have also been assessed in the field for over 3000 somatic embryo derived lines (unpublished results). These plants showed wide range of diversity, e.g. reduced plant height, increased number of grains on the tiller, and variation in thousand grain weight. Bebeli et al. (1993) reported considerable variability in the R_2 generation. For many traits there were individual R_2 plants that were better then the best control. On the tissue culture level we have selected aluminium tolerant plants as well as deoxynivalenol tolerant plants (data not published). 26 somaclonal variants were screened for *Septoria* susceptibility and appeared significantly more resistant to *Septoria nodorum* blotch than the controls but were susceptible to *Septoria tritici* blotch (27).

Albinism has been frequently observed amongst regenerants (28) and seems to be strongly dependent on the genotype. This is particularly true for androgenic regenerants (data not published) and for subcultured callus (18). The question still exists - what causes the albinism? Day & Ellis (1985) reported deletions in plastid DNA and Schumann (1988) found degenerated plastids in albinotic regenerants of *Triticale*. On the contrary albino barley plants contained normal chloroplasts (29). According to Bernard (1980) plastids are blocked in their development and do not function. On the other hand cold conditions (5°C) allow to regenerate green plants, whereas warm conditions (27°C) inhibit chloroplast synthesis during their development.

Chromosome variation. Chromosome variation is known to occur naturally in *Triticale*. Cytological studies on *Triticale* regenerants showed different changes of chromosome number and chromosome breakages. The karyotype in roots was ranging from 2n=36+6t (t-telocentric) to 42+3t (20). The same authors observed modifications in the C-banding pattern of chromosomes. They found deletions and breakage's near the centromere with loss of breakage products. Hypoaneuploidy and hyperaneuploidy as well as acrocentric, telocentric and dicentric chromosomes were detected by Nakamura and Keller (1982). Jordan and Larter (1984) reported extreme chromosomal instability of one genotype resulting in univalents as observed in metaphase I. Aneuploidy and plants with higher ploidy were also found within regenerants from anther cultures (3). In contrary Stolarz and Lörz (1986a) reported normal number of chromosomes in root tips of regenerants and Brettel et al. (1986) found only occasional loss of chromosomes. Nakamura and Keller (1982) connected the chromosomal abnormality with the age of the callus, and in their opinion, that is the evidence that abnormality do not preexist in the explant. On the other hand Wang and Hu (1993) suggests that the chromosome variation can occur already prior to *in vitro* culture.

Molecular variation. There is a paucity of data concerning the changes at the molecular or biochemical level in *Triticale*. Jordan and Larter (1984) showed with electrophoresis studies that all regenerated plants had the same prolamin banding pattern as the donor *Triticale*. Variation existed in the intensity of the bands especially encoded by the rye genome. Brettel et al. (1986) analyzed the rRNA genes located at the *Nor* loci on chromosomes 1B, 6B and 1R. The rRNA gene loci were stable, when the chromosomes carrying these loci were present. The authors found reduction of a number of rDNA units in one plant. This reduction was heritable and correlated with reduced C-banding at the position of *Nor* - R1 on chromosome 1R. Other plants regenerated from the same callus and the parent did not carry the alteration. Kaltsikes and Bebeli (1992) tested the somaclonal variation in plant families coming from Triticale lines differing in telomeric heterochromatin on chromosome arms 7RL and 6RS and showed that rye telomeric heterochromatin influence the somaclonal variation and the presence of that heterochromatin leads to greater amounts of genetic variability in progeny of regenerants.

These studies indicate that somaclonal variation does occur in *Triticale* but in our opinion it can be sometimes mistaken with the instability of its genome.

CELL SUSPENSION AND PROTOPLAST CULTURE

Cell suspension culture. It is a general experience that the establishment of regenerable suspension cultures of cereals is difficult. On the other hand it is a prerequisite for *in vitro* selection and protoplast culture. Usually embryogenic callus was taken to initiate cell suspensions of *Triticale* (30,31). So far embryogenic callus from immature embryo has been used for cell suspension initiation (32,31), but lately regeneration potential of microspore derived cell suspensions of barley and wheat has been reported (33,34,35).
For *Triticale* immature embryo derived embryogenic lines have been used to establish the suspension cultures. Two suspensions have been established, one nonembryogenic and one embryogenic. Plants were regenerated from this suspension (31). Both suspensions

have been used for protoplasts isolation.

Protoplast culture. The new approaches of biotechnology include somatic hybridization to overcome sexual incompatibility barriers, the production of complete or partial hybrids and cybrids by protoplast fusion, introduction of defined genes mediated by chromosome transfer, microinjection of DNA into cells, direct uptake of DNA into protoplasts, and gene transfer using genetically modified bacterial or viral vectors (2). Plant protoplasts appear to be the most suitable recipient cells for DNA uptake experiments because they lack a cell wall and can be prepared in large quantities by enzyme digestion (36). Plants have been regenerated from protoplasts of many species, especially from those belonging to the *Solanaceae* and *Cruciferae*, but also from cereals e.g. *Oryza sativa* L. (37,38), *Zea mays* L.(39, 40, 41), *Triticum aestivum* (42), *Hordeum vulgare* (43,33). Plant regeneration from cereal protoplasts still is a very delicate process, depending on parameters which are often beyond the control of the investigator. *Triticale* protoplasts have shown the ability to form globular structures that developed to somatic embryos (44). Unfortunately these embryos did not develop to plants.

Despite the fact that protoplasts have been substituted by simpler techniques for transformation there still exists the need for development of efficient suspension and protoplast systems for in vitro selection, fusion and transient expression studies.

TRITICALE TRANSFORMATION

Recent progress in plant and cell biology as well as molecular biology enables the use of newly developed techniques to complement the work of conventional plant breeding (1) by direct gene transfer. The Ti plasmids of *Agrobacterium tumefaciens* have been converted into useful vectors for dicotyledonous plant transformation (45). The expression of genes mediated by T-DNA transfer in some monocotyledonous species has been reported (46,47). As far as cereals are concerned, a method independent of agrobacteria, for DNA mediated protoplast transformation of *Triticum monococcum* (48), *Oryza sativa* (49), *Zea mays* (50), and *Hordeum vulgare* (51) has been developed. Despite considerable efforts in the genetic engineering of plants, and notable achievements in some species, the world major cereal crops were proving remarkably recalcitrant to genetic transformation (52). For *Triticale* only transient gene expression in protoplasts has been reported so far (53,54). In this experiments cell suspension derived protoplasts were used. Plasmid DNA carrying a chimeric gene consisting of the *uidA* gene under the control of a CaMV 35S promoter were added to freshly isolated protoplasts in the presence of Ca^{++} and PEG. Enzyme activity could be detected in extracts of the treated protoplasts after 12h to 4 days. This indicates that a major problem for obtaining transgenic plants of *Triticale* is the establishment of long--term suspensions and plant regeneration from protoplats.

BIOLISTIC TREATMENT

The method of biolistic transformation (55) bypasses complications with regeneration from protoplasts by shooting DNA, which is precipitated on heavy metal particles, into plant cells and regeneration of plants from transformed cells. Following this method regeneration of transgenic plants was achieved with maize (56), rice (57), oat (58) and wheat (59).

Here we report the first transgenic plants obtained after particle bombardment of scutellar tissue of hexaploid *Triticale* (60).

The method of induction of somatic embryogenesis and plant regeneration from *Triticale* scutelar tissue has been developed previously (14,15). This tissue material was bombarded with the plasmid pDB1 (59) containing the ß - glucuronidase gene (*uidA*) under the control of actin-1 promoter (Act1) from rice and a selectable marker gene *bar* (phosphinothricin acetyl transferase) under the control of the CaMV 35S promoter. From 465 bombarded scutela about 4000 plants were regenerated. These regenerants were screened for enzyme activity by the histological GUS assay, and by spraying the plants with herbicide (Basta) solution. Twenty four regenerants showed GUS-activity and survived Basta spraying. PCR tests as well as Southern blot analysis showed the presence of both marker genes introduced in the genome of transgenic plants.

The plants were grown to maturity and set seeds. Pollen grains showed segregation of GUS-activity from 1.8:1.0 to 1:2 in different transformed plants. Within the progeny of these plants we found segregation of the introduced gene ranging from 2:1 to 15.6:1.0. To our knowledge we report here the first successful transformation of *Triticale* and we are confident, the reported system will be applicable to other *Triticale* genotypes too.

GENES OF INTEREST

All genes able to instil the plant with new traits of agronomic value ought to be identified. Genes can be isolated from other cereals, dicots, animals or microorganisms. At present much attention is focused on identifying genes and constructing chimeric genes influencing yield potential, pest resistance, herbicide resistance, product quality or stress tolerance. Of interest are also genes for antisense RNA that inhibit the normal flow of genetic information during translation reducing synthesis of a target protein.

As far as *Triticale* is concerned we expect that the biotechnology will help to overcome the problems with acid soil tolerance, disease resistance, herbicide resistance, sprouting of the grain in high humidity, and bread making properties. Herbicide resistance and some disease resistance can be regulated with one gene and these genes are already available. Bread making quality of *Triticale* could be improved by introduction of the *Glu--D1d* gene encoding high molecular weight glutenin. This gene is present in most high quality wheats. Recently a segment of chromosome 1D with *Glu-D1d* was transferred to 1R and 1A of *Triticale* through translocation (60,61) but it would be necessary to be able to introduce specifically this gene to any desired line or cultivar. Genes related to aluminium resistance hopefully will be isolated soon since isogenic lines already exist and specific gene isolation procedures can be applied (A. Aniol, personal communication). For other characters that are determined by several genes, the molecular breeding approach will be more difficult and time consuming.

Conclusions

Despite a rather slow development of *Triticale* in vitro culture methods, recent achievements are promising. Methods such as anther culture, embryo rescue, or *in vitro* selection are already routine procedures in the research related to breeding. Among transformation methods DNA uptake after PEG treatment of protoplasts and microprojectile bombardment appear the most effective for cereals and also for *Triticale*. Genetic and molecular analyses of the progeny of transformed plants are still needed to provide more information about the stability of the function of the "new" gene in the next generations. Numerous chimeric genes are available for research. Some of these genes could provide significant benefits also in future breeding of *Triticale*.

References

1. Davey MR, Rech EL, Mulligan BJ. Direct DNA transfer to plant cells. Plant Mol Biol 1989;13:273-285.
2. Lörz H, Göbel E, Brown P. Advances in tissue culture and progress towards genetic transformation of cereals. Plant Breeding 1988;100:1-25.
3. Sosinov A, Lukjanjuk S, Ignatova S. Anther cultivation and induction of haploid plants in *Triticale*. Z Pflanzenzüchtg 1981;86:272-285.
4. Charmet G, Bernard S. Diallel analysis of androgenic plant production in hexaploid *Triticale* (x.*triticosecale*, Wittmack). Theor Appl Genet 1984;69:55-61.
5. Bernard S. *In vitro* androgenesis in hexaploid *Triticale*: determination of physical conditions increasing embryoid formation and green plant production. Z. Pflanzenzüchtg 1980;85:308-321.
6. Kozdój J, Zimny J. Microspore development stages in chilled and unchilled anthers of *Triticale* (x *Triticosecale* Wittmack). Bul Pol Acad Sci 1993;2/93:108-116.
7. Bernard S. Etude de quelques facteurs contribuant a la reussite de l` androgenese par culture d`antheres *in vitro* chez le *Triticale* hexaploide. Ann Plantes 1977;27:639-635.
8. Chu CC. The N6 medium and its applications to anther culture of cereal crops. Proceedings of the Symposium on Plant Tissue Culture. Peking, May 25-30-1978. Science Press Peking 1978.
9. GamborgOL, Miller RA, Ojima K. Nutrient requirements of suspension cultures of soybean root.ExpCell Res 1968:50,151-158.
10. Schumann G. Untersuchungen zum Albinismus in Antherenkulturen von *Triticale*. Arch Züchtungsforsch Berlin 1988;18:2 115-122.
11. Eapen S, Rao PS. Plant regeneration from immature inflorescence callus cultures of wheat, rye and *Triticale*. Euphytica 1985;34:153-159.
12. Fedak G. Chromosome Irregularities in wheat and *Triticale* plants regenerated from leaf base callus. Plant Breeding 1987;99(2):151-154.
13. Eapen S, Rao PS. Callus induction and plant regeneration from immature embryos of rye and triticale. Plant Cell Tissue and Organ Culture 1982;1:221-227.
14. Nakamura Ch, Keller WA. Callus proliferation and Plant regeneration from immature embryos of hexaploid *Triticale*. Z Pflanzenzüchtg 1982;91:137-160.
15. Stolarz A, Lörz H. Somatic embryogenesis, *in vitro* manipulation and plant regeneration from immature embryos of hexaploid *Triticale* (x *Triticosecale* Wittmack). Z. Pflanzenzüchtg 1986a;96:353-362.
16. Zimny J, Rybczyñski JJ. Somatic embryogenesis of *Triticale*. In: Genetic Manipulations in Plant Breeding, 1986 Symposium Procceedings Berlin (West) 1985.
17. Bebeli P, Karp A, Kaltsikes PJ. Plant regeneration and somaclonal variation from cultured immature embryos of sister lines of rye and triticale differing in their content of heterohromatin. 1. Morphogenetik response. Theor Appl Genet 1988; 75:929-936
18. Zimny J, Lörz H. High frequency of somatic embryogenesis and plant regeneration of

rye (*Secale cereale* L.). Z Pflanzenzüchtng 1989;102:89-100.
19. Sharma GC, Bello LL, Sapra VT. Genotypic differences in organogenesis from callus of ten *Triticale* lines. Euphytica 1980;29:751-754.
20. Armstrong KC, Nakamura C, Keller WA. Karyotypic instability in tissue culture regenerants of *Triticale* (x *Triticosecale* Wittmack) cv. `Welsh' from 6-month-old callus cultures. Z Pflanzenzüchtng 1983;91:233-245.
21. Jordan MC, Larter EN. Somaclonal variation in *Triticale* (x *Triticosecale* Wittmack) cv. Carman. Can J Genet Cytol 1984;27:151-157.
22. Zimny J. Genotypic dependence of the somatic embryogenesis of *Triticale* (x *Triticosecale* Wittmack).In: Science for Plant Breeding. XII Eucarpia Congress;1989 Göttingen, Germany.
23. Murashige T, Skoog F. A revised medium for rapid growth and bioassay with tobacco tissue cultures. Physiol Plant 1962;15:473-497.
24. Kao KN. Chromosomal behaviour in somatic hybrids of soybean - *Nicotiana glauca*. Mol Gen Genet 1977;150:221-230.
25. Larkin PJ, Scowcroft:Somaclonal variation - a novel source of variability from cell cultures for plant improvement. TheorApplGenet 1981;60,197-214.
26. Brettell RIS, Denis ES, Scowcroft WR, Peacock WJ. Molecular analisis of somaclonal mutant of maize alcohol dehydrogenase. Molec Gen Genet 1986;202:235-239.
27. Scharen AL, Arseniuk E, Sowa W, Zimny J, Podyma W. Seedling resistance of *Triticale* and Triticum spp. germplasm to *Septoria nodorum* and *S.tritici.* Proceedings of the Second International *Triticale* Symposium, 1990; Passo Fundo, Brazil.
28. Ono H, Larter EN. Anther culture of *Triticale*. Crop Science 1976;16:120-122.
29. Clapham D. Haploid Hordeum plants from anthers in vitro. Z Pflanzenzüchtg 1973;69: 142-145.
30. Stolarz A, Lörz H. Somatic embryogenesis, cell and protoplast culture of *Triticale* (x *Triticosecale* Wittmack). In:Genetic Manipulation in Plant Breeding.Symposium Proceedings. 1986b Horn,Jensen,Odenbach,Schieder(eds) Walter de Gruyter and Co Berlin.
31. Zimny J. Somatic embryogenesis and plant regeneration of rye (*Secale cereale* L.) and *Triticale* (x *Triticosecale* Wittmack). In: Regulations of Plant Somatic Embryogenesis 1992 ed. by Griga , Tejklova, Sumperk-1992.
33. Jähne A, Lazzeri PA, Lörz H. Regeneration of fertile plants from protoplasts derived from embryogenic cell suspensions of barley (*Hordeum vulgare* L.). Plant Cell Rep 1991:10:1-6.
34. Lührs R, Nielsen K. Microspore cultures as donor tissue for the initiation of embryogenic cell suspensions in barley. Plant Cell Tissue and Organ Culture 1992;31: 169-178.
35. Schmitt M. In Vitro Kultur und Protoplastenregeneration von Weizen (*Triticum aestivum* L.) dissertation, University of Hamburg 1993.
36. Krens FA, Molemdijk L, Wullems GJ , Schilperoort RA. *In vitro* transformation of plant protoplasts with Ti-plasmid DNA. Nature: 1982;559-563.
37. Fujimura T, SakuraiM, Nagishi T, Hirose A. Regeneration of rice plants from protoplasts. Plant Tissue Culture Letters 1985;2:74-75.
38. Kyozuka J, Hayashi Y , Shimamoto Y. High frequency plant regeneration from rice protoplasts by novel nurse culture methods. Mol Gen Genet 1987;206:408-413.
39. Rhodes CA, Lowe KS, Ryby KL . Plat regeneration from protoplasts isolated from embryogenic maize cell cultures. Biotechnology 1988;6:56-61.

40. Prioli LM, Sondahl MR. Plant regeneration and recovery of fertile plants from protoplsts of *Zea mays* L. Bio/Technology 1989;7:589-584.
41. Shillito RD, Carswell GK, Johnsons CK, DiMaio JJ, Harms CT. Regeneration of fertile plants from protoplasts of elite inbred maize. Bio/Technology 1989;7:581-587.
42. Vasil V, Redwey F, Vasil IK. Regeneration of plants from embryogenic suspension culture protoplasts of wheat (*Triticum aestivum* L.) Bio/Technology 1990;8:429-434.
43. Yan Q, Zhang X, Shi J, Li J. Green plant regeneration from protoplsts of barley (*Hordeum vulgare* L.) Kexue Tongbao 1990;35:1581-1583.
44. Stolarz A. Cell and protoplast culture, Somatic embryogenesis and transformation studies in different formes of x *Triticosecale* Wittmack. In: Proceedings of the Second International *Triticale* Symposium 1990, Passo Fundo, Brazil.
45. Barton KA, Brill WJ. Prospects in plant genetic engineering. Science 1983;219:67:1-675.
46. Hooykaas-van Slogteren GMS, Hooykapas PJJ, Schilperoot RA. Expression of Ti plasmid genes in monocotyledonous plants infected with *Agrobacterium tumefaciens.* Nature 1984;311:763-764.
47. Hernalsteens JP, Thia-Toong L, Schell J, Van Montagu M. An *Agrobacterium* -transformed cell culture from the monocot *Asparagus officinalis.* EMBO Journal 1985;3 (13):3039-3041
48. Lörz H, Backer B, Schell J Gene transfer to cereal cells mediated by protoplast transformation. Mol Gen Genet 1985;199:178-182.
49. Ushimiya H, Fushmi T, Hashimoto H, Harada H, Syano K, Sugawara Y. Expression of a foreign gene in callus derived from DNA treated protoplasts of rice (*Oryza sativa* L.) Mol Gen Genet 1986;204:204-207.
50. Fromm ME, Taylor LP, Wallbot V. Stable transformation of maize after gene- -transformation by electroporation. Nature 1986;319:178-182.
51. Lazzeri PA, Bretschneider R, Lührs R, Lörz H. Stable transformation of barley via PEG - induced direct DNA uptake into protoplasts. Theor Appl Genet 1991;81:437-444.
52. Potrykus I. Gene transfer to plants :assessment and perspectives. Physiol Plant 1990;79:125-134.
53. Stolarz A, Lörz H. Protoplast culture and transformation studies of *Triticale* (x *Triticosecale* Wittmack). Plant Cell Tissue and Organ Culture 1988;2:227-230.
54. Zimny J, Rafalski A. Transformation study on *Triticale* protoplasts. Bull Inst Plant Breed Acclim 1993;197:127-132.
55. Sanford JC, Klein TM, Wolf ED, Allen N. Delivery of substances into cells and tissues using a particle bombardment process. J Part Sci Technol 1987;5:27-37.
56. Fromm ME, Morrish F, Armstrong A, Williams R, Thomas J , Klein TM. Inheritance and expression of himeric genes in the progeny of transgenic maize plants. Bio/Technology 1990;8:833-839.
57. Cao J, Duan X, McElroy D, Wu R. Regeneration of herbicide resistant transgenic rice plants following microprojectile mediated transformation of suspension culture cells. Plant Cell Rep 1992;11:586-591.
58. Somers DA, Rines HW, Gu W, Kaeppler HF, Bushnell WR. Fertile transgenic oat plants. Bio/Technol 1992;10:1589-1594.
59. Becker D, Bretschneider R, Lörz H. Fertile transgenic wheat from microprojectile bombardment scutellar tissue. Plant Journal 1994;5(2):299-307.
60. Lukaszewski AJ , Curtis Ch. Transfer of the *Glu-D1d* gene from chromosome 1D of bread wheat to chromosome 1R in hexaploid *Triticale*. Plant Breeding 1992;109:203-210.

60. Zimny J, Becker D, Brettschneider R, Lörz H. Fertile, transgenic *Triticale* (x *Triticosecale* Wittmack) Plant Mol Biol 1994;submitted.
61. Lukaszewski AJ , Curtis Ch,. Transfer of the *Glu-D1* gene from chromosome 1D to chromosome 1A in hexaploid *Triticale*. Plant Breeding 1994;112:177-182.
62. Bebeli P, Kaltsikes PJ, Karp A. Field evaluation of somaclonal variation in triticale lines differing in telomeric heterochromatin. J Genet Breed 1993;47:249-258.
63. Christou P. Genetic transformation of crop plants using microprojectile bombardment. The Plant Journal. 1992;2(3):275-281.
64. Day A, Ellis THN. Deleted forms of plastid DNA in albino plants from cereal anther culture. Current Genetics 1985;9:671-678.
65. Kaltsikes PJ, Babeli PJ. The effect of rye telomeric heterochromatin on the nature and size of variance in regenerated families of hexaploid triticale. J Genet Breed 1992;46: 359-362.
66. Klein T, Wolf M, Wu ED , Sanford R . High velocity microprojectiles for delivering nucleic acids into living cells. Nature 198;327:70-73.
67. Wang Y, Hu H. Gamete composition and chromosome variation in pollen-derived plants from octoploid triticale x common wheat hybrids. Theor Appl Genet 1993;85: 681-687.
68. Ya-Ying W, Ching-San S, Ching-Chu W, Nan-Fen C. The induction of pollen plantlets of *Triticale* and Capsicum annum from anther culture. Sci Sin 1973;16:147-151.

RFLP MARKERS AND THEIR APPLICATIONS IN CEREAL BREEDING

Katrien M. Devos & Mike D. Gale
Cambridge Laboratory, Colney Lane
Norwich NR4 7UJ, U.K.

Abstract

Individual RFLP-based maps are currently available for the Triticeae species wheat, barley and rte. However, an aspect of genetic mapping that is becoming increasingly important is the identification of intergenomic relationships, both within polyploids and between related species. In hexaploid bread wheat, a comparative genetic map of the A. B, and D genomes has been constructed, which, to date, includes some 1000 loci. The map reveals a strong collinearity of the homoeologous wheat genomes, that is disrupted only by a few chromosomal translocations. Comparative genetic maps of the wheat and rye genomes were constructed using a set of low copy homoeologous probes, previously characterized and mapped in wheat. The wheat and rye maps differ by a number of evolutionary translocations involving chromosome arms 2RS, 3RL, 4RL, 5RL, 6RS, 6RL, 7RS and 7RL. The presence of inter- and intrachromosomal rearrangements between 'donor' and 'recipient' genomes has important consequences for the development of gene introgression strategies.

Introduction

Over the last few years, a range of molecular markers, including both restriction fragment length polymorphism (RFLP) markers based on Southern blot technology and polymerase chain reaction (PCR)-based markers have been developed. Each of these marker types has specific characteristics which determine their applications. PCR-based markers are easy to apply, but require in the template DNA the presence of sequences that show nearly complete homology with the oligonucleotide primers. One base pair mismatch between the template sequence and 3'end of the primer may be sufficient to inhibit primer extension, and thus the generation of a PCR-product. Therefore, unless exons of highly conserved genes are targeted, it is unlikely that primers, designed to specific sequences in one species will

339

H. Guedes-Pinto et al. (eds.), Triticale: Today and Tomorrow, 339–348.
© 1996 *Kluwer Academic Publishers. Printed in the Netherlands.*

find their perfect complement in related species. This will restrict the use of PCR-based markers mainly to intraspecific applications. RFLP markers are more cumbersome but also more versatile. So, despite a number of inherent problems, RFLP markers have become valuable tools in genome analyses and plant breeding.

In Triticeae cereals, the importance of RFLP markers is illustrated by the existence of extensive RFLP-based maps in wheat (Gale et al. 1994), barley (Graner et al. 1991; Heun et al. 1991; Kleinhofs et al. 1993) and rye (Devos et al. 1993), and the generation of sequence tags for genes of agronomic importance is well under way. Furthermore, comparative mapping efforts within the Triticeae species have revealed the relationship between the wheat, barley and rye genomes (Devos et al. 1993a,b; Laurie et al. 1993). The genetic maps of wheat, the wheat.rye comparative maps and the consequences of this research for plant breeding will be further discussed.

PROBE CHARACTERIZATION

Bread wheat, *Triticum aestivum* (2n=6x=42), is an allohexaploid with three homoeologous genomes, A, B and D. Therefore, the hybridization behaviour of DNA probes to the three wheat genomes gives a strong indication about the degree of conservation of any particular sequence, and thus its potential utility as a molecular marker in other Triticeae species. Based on this characteristic, RFLP clones may be classified as 'homoeologous' or 'non--homoeologous' probes. Homoeologous probes detect homoeoloci on the three wheat genomes, and also hybridize to homoeologous locations in related triticeae species such as barley and rye (Fig. 1). However, homoeologous probes detect relatively lower levels of RFLP variation and tend to map into the centromeric regions of the chromosomes (Fig. 2). The non-homoeologous probes are usually genome-specific or detect loci in different genomes that are unrelated by homoeology (Fig. 3). In most cases, they fail to cross--hybridize to other Triticeae species. 'Non-homoeologous' probes generally detect relatively higher levels of polymorphism and their corresponding loci are more evenly distributed over the genome (Fig. 2).

The importance of probe characterization is clear when one considers that different breeding applications require specific marker types. Gene tags for marker-aided selection of agronomic traits will be useful only if they show high levels of variation, so that the selection can be carried out in any cross, and, therefore, 'non-homoeologous' probes would be most suitable for this type of application. On the other hand, when RFLP markers are used to follow the introgression of alien chromosome segments into wheat, homoeologous probes are preferred as they hybridize across species. Furthermore, the low level of polymorphism displayed by this type of probe is a benefit rather than a disadvantage as high levels of variation within the cultivated species will generally complicate a wheat/alien analysis.

THE WHEAT GENETIC MAP

To date, the genetic map of wheat consist od some 1000 loci. Tha main characteristics are the strong clustering of loci in the centromeric region which is a reflection of the concentration of recombination events in the distal chromosome regions, and the conserved collinearity of the three homoeologous genomes (Fig. 2). Within the wheat genome, three reciprocal translocations between chromosome arms 4AL and 5AL, 4AL and 7BS, and 2BS and 6BS have been identified and mapped. The 4AL/5AL translocation is likely to have taken place before the polyploidization of wheat, while the 4AL/7BS translocation dates back to the tetraploid wheat parent of modern bread wheat (King et al. 1994). The presence of intra - and interchromosomal translocations between wheat varieties and in wheat relative to related species is a factor that needs to be taken into account when considering gene transfer, as is illustrated by the relationship between the rye and wheat genomes.

THE COMPARATIVE WHEAT-RYE GENETIC MAP

The construction of comparative maps is achieved by mapping the genomes of different species with a common set of probes. Prerequisites for these probes are a low copy number and an ability to cross-hybridize to homoeoloci in the genomes of the species under investigation. Within the tribe Triticeae, hexaploid bread wheat is a logical choice as starting point for comparative mapping. Indeed, the genetic map of the allopolyploid wheat genome is itself a comparative map. Furthermore, in wheat, low copy homoeologous probes that are suitable candidates for the quest in genome relationships can be readily identified.

In rye, a genetic map was constructed using low copy wheat probes that had previously been characterized as 'homoeologous'. The result was quite surprising (Fig. 4). Previous marker studies using wheat-rye single chromosome addition lines had shown rearrangements in the rye genome relative to wheat involving chromosome arms 4RL, 5RL and 7RS (Naranjo et al. 1987; Liu et al. 1992). The construction of a comparative wheat-rye map not only fine-mapped the breakpoints of these translocations, but revealed the presence of additional rearrangements in the rye genome relative to that of wheat. In total, eight of the 14 chromosome arms had undergone translocations since the divergence of these two Triticeae species from their common ancestor. Despite this disruption of chromosomes, the marker order within each of the translocated segments was conserved.

EVOLUTION

Assuming that the ancestral genome from which the rye genome evolved had a basic composition similar to that of barley or the D-genome of wheat, it is possible to explain that the current structure of the rye genome evolved by at least seven translocations. Genetic mapping studies have shown that, within the limits of the experiments, the wheat 4AL/5AL and rye 4RL/5RL breakpoints are the same. The same or a similar 4/5 translocation is also present in the Triticeae species *Thinopyrum bessarabicum* (E^bE^b), *Aegilops umbellulata* (UU), *Dasypyrum villosum* (VV) and *Secale montanum* (R^mR^m), the progenitor of cultivated rye (Miller 1984; King et al. 1994). This would suggest either that this translocation occurred before the 'speciation' of the A, E^b, U, V and R^m genomes, but after the divergence of the B, D and H (barley) genomes, or that certain chromosome regions are hypersensitive to breakage. The presence of the grain esterase gene, *Est-5*, on the homoeologous group 3 chromosomes in wheat, on 6RL in *S. cereale* and $6R^m$ in *S. montanum* (Ainsworth et al. 1986), indicates that a translocations involving chromosomes 3 and 6 may also be present both in *S. montanum* and *S. cereale*. Similarly, the existence of a 4/7 translocation in both *S. cereale* and *S. montanum* was derived both from the similar plant morphology including the presence of the purple culm *(Pc)* gene, carried by 7BS in wheat (Law 1966), in the CS/4R and CS/4R^m addition lines, and from RFLP data (King et al. 1994). All other translocations are likely to characterize only *S. cereale.*

CONSEQUENCES FOR PLANT BREEDING

Rye carries a range of interesting genes such as genes conferring disease resistance and aluminium tolerance that are potentially useful in wheat. However, so far, most attempts to introgress rye chromosomes into wheat have failed to yield lines of good agronomic performance. The co-introgression of rye genes controlling unfavourable traits together with the selected gene(s) of interest, has frequently been quoted as the main cause for failure. Although this may be one of the causes, the consequences of the chromosomal rearrangements present in rye relative to wheat definitely cannot be overlooked. For example, introgression of a gene carried on the '7L segment' of rye chromosome 6R through induced recombination between 6R and either wheat chromosomes 6A, 6B or 6D would result in the replacement of a wheat homoeologous group 6 chromosome segment with a rye segment that is homoeologous to segments of the homoeologous group 7 chromosomes in wheat (Fig. 4). Consequently, the newly generated line would carry a duplication of a segment of the homoeologous group 7 segments, and a deletion of a homoeologous group 6 segment. The presence of chromosomal duplications and deletions, although viable in a hexaploid background, may affect the agronomic quality of the recombinant material. Similar negative effects would be expected in the production of wheat/rye substitution lines and substituted triticales. This problem, at least in the case of rye introgression, could be overcome by rigorous selection of recombinant lines with a balanced genome. A notable exception that reinforces the argument is the 1BS/1RS translocation present in many feed wheats. Examination of the rye-wheat comparative map (Fig.4) shows that the short arms of the group 1 chromosomes represent one of the few

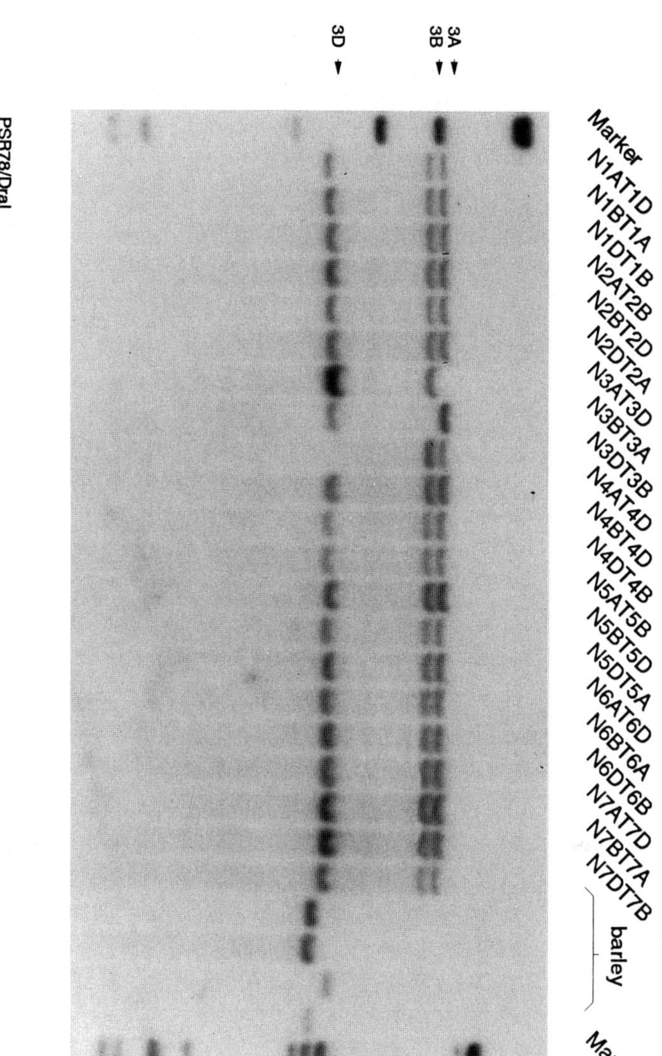

Fig. 1: Autoradiograph showing the hybridization pattern of a single copy homoeologous probe, PSR78, that detects copies on wheat chromosomes 3A, 3B and 3D.

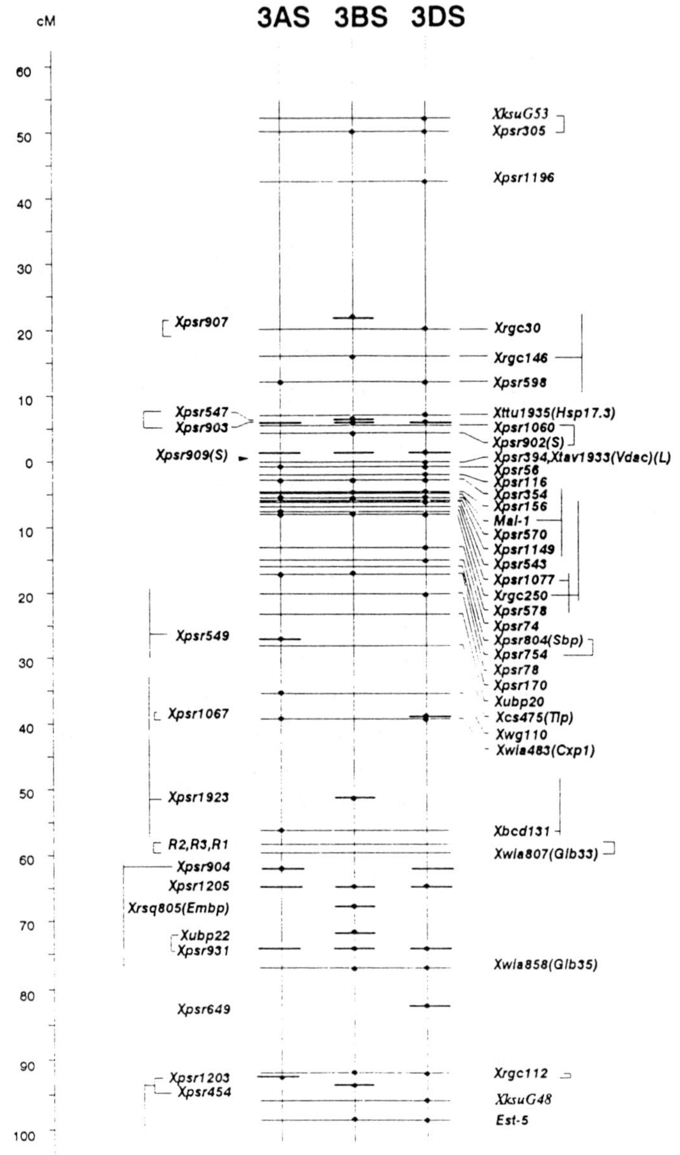

Fig. 2: A consensus map of the homoeologous group 3 chromosomes of wheat and rye.

● indicate mapped loci, *continuous lines over the A, B and D chromosomes* indicate homoeologous loci, *bold short lines* indicate map posotions of non-homoeologous loci..

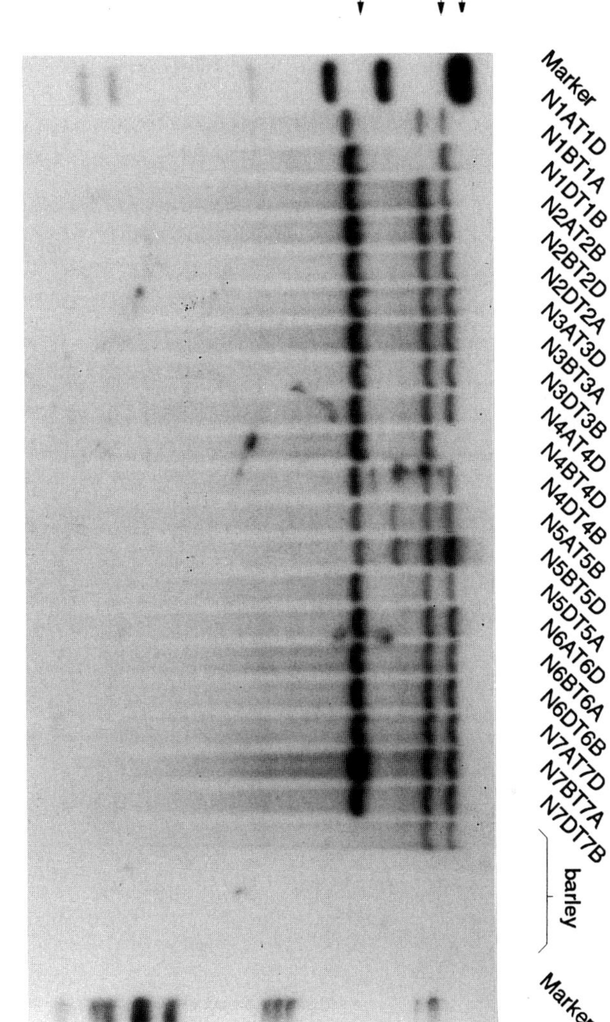

Fig. 3: Autoradiograph showing the hybridization pattern of a low copy non-homoeologous probe, PSR648, that detects copies on wheat chromosomes 1B, 4A and 7D.

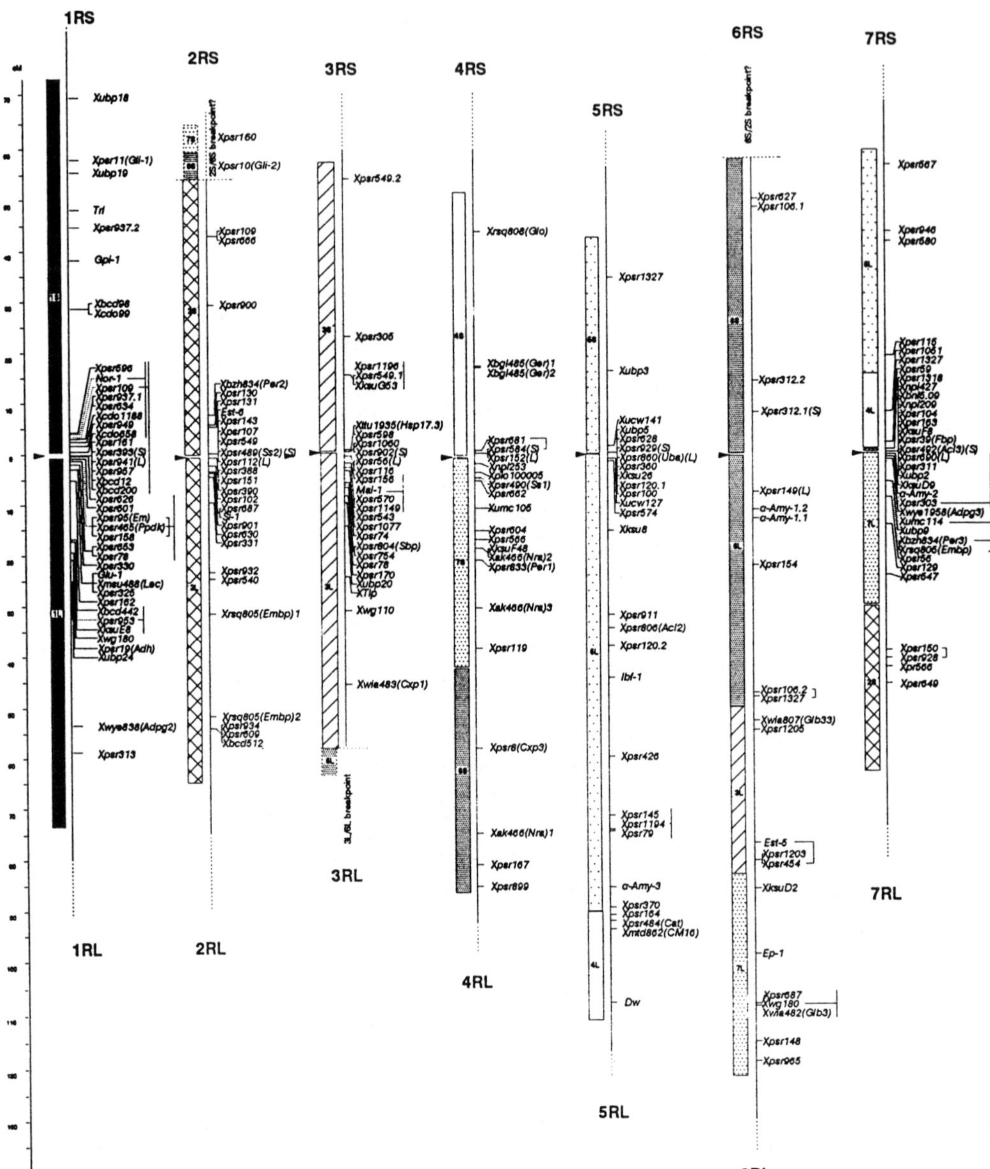

Fig. 4:   A genetic map of the rye genome, and its relationship to the wheat genome.

untranslocated regions. Thus these lines will be balanced. Furthermore, the many genes incorporated in the large introgressed rye segment give very few deleterious effects besides the negative effect on bread-making quality. Indeed, the evidence would indicate an enhancing effect on yield itself.

References

Ainsworth, C.C., T. E. Miller, M. D. Gale, 1986. The genetic control of grain esterases in hexaploid what. II. Homoeologous loci in related species. Theor Appl. Genet. 72: 219-225.

Devos, K. M., T. Millan and M. D. Gale, 1993a. Comparative RFLP maps of the homoeologous group-2 chromosomes of wheat, rye and barley. Theor. Appl. Genet. 85: 784-792.

Devos, K. M., M. D. Atkinson, C. N. Chinoy, R. L. Harcourt, R. M. D. Boebner, C. J. Liu, P. Masojc, D. X. Xie and M. D. Gale, 1993b. Chromosome rearrangements in the rye genome relative to that of wheat. Theor. Appl. Genet. 85: 673-680.

Gale, M.D., M. D. Atkinson, C. N. Chinoy, R. L.Harcourt, Q. Y. Li and K. M. Devos, 1994. Current status of the wheat genetic map. Theor. Appl. Genet. (In press).

Graner, A., A. Jahoor, J. Schondelmaier, H. Siedler, K. Pillen, G. Fischbeck, G. Wenzel and R. G. Herrmann, 1991. Construction of an RFLP map of barley. Theor. Appl. Genet. 83: 250-256.

Heun, M., A. E. Kennedy, J. A. Anderson, N.L. V. Lapitan, M. E. Sorrells and S. D. Tanksley, 1991. Construction of an RFLP map for barley (*Hordeum vulgare* L.). Genome 34: 437-447.

King, I.P., K. A. Purdie, C. J. Liu, S. M. Reader, S. E. Orford, T. S. Pittaway, T. E. Miller, 1994. Detection of interchromosomal translocations within the Triticeae by RFLP analysis. Genome (In press).

Kleinhofs, A., A. Kilian, M. A. Saghai Maroof, R. M. Biyashev, P. Hayes, F. Q. Chen, N. Lapitan, A. Fenwick, T. K. Blake, V. Kanazin, E. Ananiev, L. Dahleen, D. Kudrna, J. Bollinger, S. J. Knapp, B. Liu, M. Sorrells, M. Heun, J. D. Franckowiak, D. Hoffman, R. Skadsen and B. J. Steffenson, 1993. A molecular, isozyme and morphological map of the barley (*Hordeum vulgare*) genome. Theor. Appl. Genet. 86: 705-712.

Laurie, D. A., N. Pratchett, K. M. Devos, I. J. Leitch and M. D. Gale, 1993. The distribution of RFLP markers on chromosome 2(2H) of barley in relation to the physical and genetic location of 5S rDNA. Theor. Appl. Genet. 87: 177-183.

Law, C.N., 1966. The location of genetic factors affecting a quantitative character in wheat. Genetics 53: 487-493.

Liu, C. J., K. M. Devos, C. N. Chinoy, M. D. Atkinson and M. D. Gale, 1992. Non-homoeologous translocations between group 4,5 and 7 chromosomes in wheat and rye. Theor. Appl. Genet. 83: 305-312.

Miller, T. E., 1984. The homoeologous relationship between the chromosomes of rye and wheat. Current status. Can J. Gent. Cytol. 26: 578-589.

Naranjo, T., A. Roca, P. G. Goicoechea and R. Giraldez, 1987. Arm homoeology of wheat and rye chromosomes. Genome 29: 873-882.

SOMACLONAL VARIATION DETECTED ON THE MOLECULAR LEVEL IN THE MITOCHONDRIAL GENOME OF REGENERATED PRIMARY CMS TRITICALE PLANTS

Ralf Weigel, Markus Wolf, Micaela Stierle, Thomas Stoesser and Claus-Ulrich Hesemann

Institute of Genetics, University of Hohenheim, D-70593 Stuttgart, Germany

Abstract

The mitochondrial DNA of primary hexaploid triticale with and without *timopheevi* cytoplasm and the corresponding wheat cross parents were characterized by hybridization experiments using six different mitochondrial gene probes and four different restriction enzymes for Southern blot analyses with the novel non-radioactive digoxigenin technique. *Via* somatic embryogenesis 623 triticale plants were regenerated. The mitochondrial genome of 159 plants was investigated. Somaclonal variation of the chondriome was found to a high percentage in regenerated plants with *timopheevi*, but only to a low percentage in individuals with maintainer cytoplasm. In sixteen cases rearrangements in the mitochondrial DNA were detected. Furthermore, obvious alterations in the stoichiometry of specific hybridization bands have occurred. Some of these stoichiometric differences depend on the triticale-type.

Introduction

Using tissue or cell suspension culture somaclonal variation was found in many cultivated plant species. This type of variation has been detected as point, chromosomal and genomic mutations.

Plants are characterized by having two organellar genomes apart from the nuclear genome, the plastome and the chondriome. The organellar genomes can easily be isolated and purified. Thus and because of their limited size they represent an attractive model for molecular investigations. The plastome has been conserved during evolution and variability in regenerants is mainly known performing as albino mutants. Kemble and Shepard [1] have proved the stability of the plastome after *in vitro* cultivation. The chondriome however, is characterized by high heterogenity [2] resulting from homologous recombination of repeated sequences which occurs in almost all plant species. Falconet et al. [3] published a model for this mechanism. Many investigators have found mitochondrial DNA rearrangements in connection with somatic hybridization, cybridization, cytoplasmic male sterility and *in vitro* culture.

For example, the group of Rathbun [4] has established that rearrangements of mitochondrial DNA in the gene regions *cob*, *cox II* and *cox I* are not related to either cms or fertility-restored states in hexaploid wheat.

H. Guedes-Pinto et al. (eds.), Triticale: Today and Tomorrow, 349–355.

© *1996 Kluwer Academic Publishers. Printed in the Netherlands.*

There is an increasing number of papers about mitochondrial DNA variability in regenerants or in vitro cultivated cells of cereals. At least the results concerning wheat should be mentioned here. Rode *et al.* [5] found extensive changes in the restriction patterns of isolated mitochondrial DNA between cultivars and resulting callus cultures. By Southern blot analysis his group found a hypervariable region which included recombinationally active repeats responsible for these rearrangements. It has been proved recently, that nuclear genes control the mitochondrial rearrangement process in tissue cultures of wheat deriving from immature embryos [6]. These phenomena are connected with the appearance of new hybridization bands and result from changes in the relative amount of subgenomic components. Hartmann *et al.* [7] have demonstrated that the mitochondrial genome variability of long-term tissue culture in wheat is variety-specific. In the context of these results two further publications of Hartmann, Rode and co-workers have been published [8,9]. In the first paper, it was shown that in the wheat cultivar 'Chinese Spring' only few of the novel subgenomic configurations are organ or tissue specific. They have found a relation between culture duration and the extent of mitochondrial DNA rearrangements and propose a correlation between a specific organization of the mitochondrial genome and regeneration capability in tissue culture. In their second paper they detected subgenomic configurations in the donor plants, the tissue culture and also in the corresponding regenerants. In some cases subgenomic configurations were present in tissue culture and regenerated plants, but not in the donor plants. The occurrence of these configurations appeared to be time- and organ- or tissue-dependent. The most current publications about somaclonal variation in wheat are from Hartmann *et al.* [10] and from Chowdhury *et al.* [11]. The latter revealed quantitative differences in three mitochondrial genes only in cell suspension culture but not in embryogenic callus culture. Hartmann et al. found a rare recombination event along a short repeat. This fact gives proof to the current opinion distinguishing between high abundant recombination along longer repeats and low abundant recombination at short repeats [12].

Triticale is known to contain the cytoplasmic components of its maternal parent wheat. In so far, results concerning the chondriome in wheat should be transferable. Kück and his co-workers have found that in triticale and the corresponding wheat the cms trait is connected with changes in the structure of the mitochondrial genome [13]. These rearrangements are influenced by the nuclear genome of the parental rye. At the moment, there is only one publication about somaclonal variation detected on the molecular level in triticale. Brettel *et al.* [14] found variation at the Nor loci. Therefore molecular investigations of regenerated triticale plants are considered to be necessary to get more information about occurrence, extent and type of somaclonal variation in the structure of the chondriome of this intergeneric hybrid. At the same time this research represents important preliminary work about the connection between the phenotype of cytoplasmic male sterility and the corresponding molecular organization of the mitochondrial genome.

Materials and Methods

Several cytoplasmic male sterile triticale forms with *timopheevi* cytoplasm have been investigated. The triticale were classified as phenotypically sterile or fertile. Fertility was induced by restorer genes of the rye genome. As comparison the triticale with maintainer- (i.e. *durum-* or *turgidum-*) cytoplasm and the corresponding wheat parents including three different cytoplasms were analysed. Triticale plants were regenerated *via* somatic embryogenesis. For determination of the regeneration rates we used 450 regenerated plants in total. The regeneration rate depended on genotype and cytoplasm type.

We had 16 triticale forms at disposal, eight triticale with *timopheevi* cytoplasm, the rest with *durum* or *turgidum* cytoplasm. Entirely 3000 embryos were cultivated, about 45 petri dishes of each genotype, each petri dish containing four calli . 623 plants were regenerated, per form about 39 individuals on average. Up to now we have analysed 159 young regenerated plants on DNA variations of the mitochondrial genome. Six different probes and four different restriction enzymes were used. Total genomic DNA was isolated from green shoot material according to a simple method established at the institute of plant physiology at the University of Hohenheim. 7μg of DNA were cleaved overnight by 25 units of enzyme. DNA was transferred to a positively charged nylon membrane [15] and hybridized to a non-radioactive, digoxigenin labelled probe. Labelling, hybridization and colorimetric detection were carried out following the recommendations of "the DIG system user´s guide for filter hybridization" [16].

Results and Discussion

The non-radioactive labelling and detection method turned out to be sensitive enough to our purposes even though total DNA was used for hybridization and the mitochondrial inserts were not excised from the vectors. In relation to the total number of investigated triticale plants the extent of variation in regenerated plants was 2,5 percent. A small number of calli were used for Southern blot analyses, too. They showed divergent hybridization patterns to a much higher percentage. Although the number of calli investigated may not be representative, these findings suggest the occurrence of selection processes between callus phase and regeneration.

The mitochondrial hybridization patterns were analysed in triticale and the corresponding wheat parents with *timopheevi* and maintainer cytoplasm. In all cases, except the *coxII*- and *nad5*-probes, hybridization patterns are different between fertile and sterile triticales. Comparison of the patterns between wheat and triticale with the same cytoplasm reveals great similarity. In regenerants reliable mitochondrial variation was only detected using the probes pTae8 containing the *atp6, cox III, rps13* genes and LT1-173 containing the *nad1* exons b and c.

Two examples of somaclonal variation were chosen to demonstrate different possibilities in the realization of subgenomic configurations. Each shows the divergent hybridization patterns detected by comparison of regenerants having the same genotype (Figure 1a and b).

In the case of the pTae8 probed Southern blot (a) only the regenerant in lane 2 (from the left) shows the hybridization pattern found in the corresponding control plants. Lane 1 shows one additional, lanes 3 and 4 show two additional bands. The 10,0 kb band is supposed to represents a genecopy normally found in the paternal rye. In the source material this fragment is stoichiometrically underrepresented. What the somaclones are concerned amplification processes must have occurred. The 11,0 kb band could be the product of homologous recombination between the 12,2 kb and the 10,0 kb copy (B. Laser personal communication). With the LT1-173 gene probe the loss of a single hybridization signal has been detected (b). Hybridization patterns of the regenerants in lane 1-4 do not differ from those of the original triticale but there are obvious differences in the band stoichiometries. In lane 5 the usually distinct 7,0 kb signal is lost. This is supposed to be due to recombination or deletion processes. Adjacent to the 5´ flanking region of the *nad1* exon b there is a recombinational active repeated sequence [17]. The 1,9 kb and 2,0 kb bands represent the Hind III fragments containing the *nad1* exons. This fragments were totally conserved among all somaclones tested.

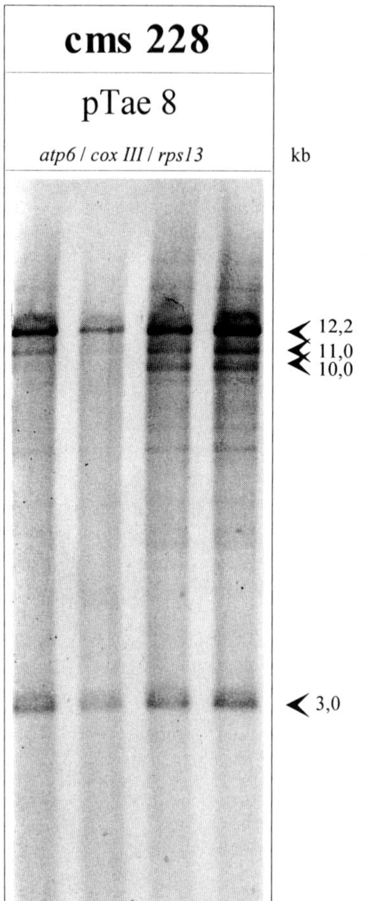

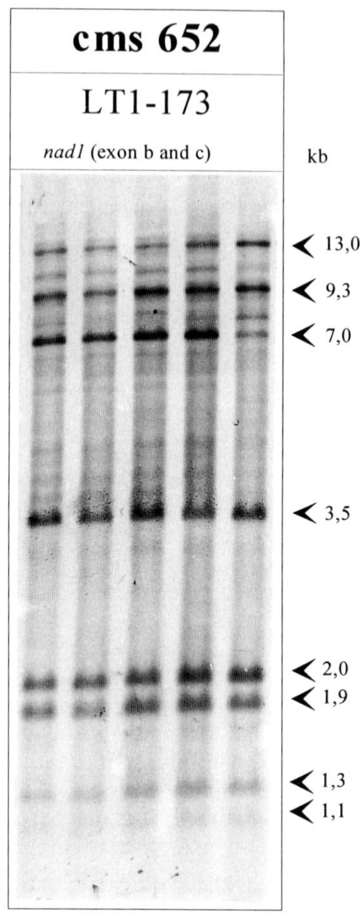

Figure 1 a (left) and b (right): Southern blot hybridization analyses of total genomic DNA of triticale regenerants with mtDNA probes. **(a)** Blot of Eco RI digests hybridized to the pTae8 probe consisting of the *atp6*, *cox III* and *rps13* genes. Only the regenerant in lane 2 shows the hybridization pattern found in the corresponding plants. Lane 1 shows one additional, lanes 3 and 4 show two additional bands.

(b) Blot of Hind III digests hybridized to the LT1-173 probe consisting of the *nad1* exons b and c and flanking regions. Hybridization patterns of the regenerants in lane 1-4 do not differ from those of the original triticale. There are obvious differences in the band stoichiometries. In lane 5 the usually distinct 7,0 kb signal is lost.

Somaclonal variation of the chondriome was found to a low percentage in regenerated plants with maintainer cytoplasm (stoichiometric differences only). We hypothesize that the structure of the mitochondrial genome is more stable in genotypes with fertile cytoplasm compared to plants with *timopheevi* cytoplasm. Furthermore the investigations have shown that type and extent of mitochondrial variation depends on the specific triticale genotype, the kind of probe and also on the restriction enzyme (Table 1). These results enhance those of the above mentioned publications [5-11].

Table 1: Summarized list of regenerated sterile plants showing somaclonal variation. Only those cases are indicated showing loss or gain of hybridization bands.

code number	number of investigated regenerants	applied restriction endonuclease	DNA probe
745	2	Hind III	LT 1-173
228	1	Hind III	LT 1-173
652	7	Hind III	LT 1-173
221	1	Hind III	LT 1-173
224	2	Hind III	LT 1-173
228	3	Eco RI	pTae 8

In future, the investigations should be continued using more triticale forms, probes with smaller inserts and new restriction enzymes. Thus, more detailed information about the occurrence and mechanisms of rearrangements in the chondriome due to somaclonal variation will be obtained.

Conclusions

The mitochondrial genome represents a miniature-model for the nuclear genome. At present, a limited number of results published with regard to wheat or triticale reveal enormous variability in the realization of mitochondrial subgenomic configurations in response to different in vitro culture conditions. In the chondriosome the variation in size and number of subgenomic configurations can be explained by homologous recombination events along repeated sequences but up to now, the mechanisms causing somaclonal variation are not well understood. In this context detailed molecular investigations may provide further information about this kind of variation. On the other hand, research work on new mitochondrial arrangements can also enlighten the relation between cms status and phenotype of the corresponding plant.

Acknowledgements

Financial support of this study by the DFG is gratefully acknowledged.

References

1. Kemble RJ, Shepard JF Cytoplasmic DNA variation in a potato protoclonal population. Theor Appl Genet 1984;69:211-216

2. Breimann A Mitochondrial DNA diversity in the genera of Triticum and Aegilops revealed by Southern blot hybridization. Theor Appl Genet 1987;73:563-570

3. Falconet D, Lejeune B, Quetier F, Gray MW Evidence for homologous recombination between repeated sequences containing 18S and 5S ribosomal RNA genes in wheat mitochondrial DNA. EMBO J 1984;3:297-302

4. Rathburn H, Song J, Hedgcoth C Cytoplasmic male sterility and fertility restoration in wheat are not associated with rearrangements of mitochondrialDNA in the gene regions for cob, coxII or coxI. Plant Mol Biol 1993;21:195-201

5. Rode A, Hartmann C, Falconet D, Lejeune B, Quetier F, Benslimane A, Henry Y, De Buyser J Extensive mitochondrial DNA variation in somatic tissue cultures initiated from wheat immature embryos. Curr Genet 1987;12:369-376

6. Hartmann C, De Buyser J, Henry Y, Morere-Le Paven MC. Dyer TA, Rode A Nuclear genes control chnages in the organozation of the mitochondrial genome in tissue cultures derived from immature embryos of wheat. Curr Genet 1992;21:515-520

7. Hartmann C., De Buyser J, Henry Y, Falconet D, Lejeune B, Benslimane A, Quetier F, Rode A Time-course of mitochondrial genome variation in wheat embryogenic somatic tissue cultures. Plant Science 1987;53:191-198.

8. Morere-Le Paven MC, Henry Y, Dr Buyser J, Corre F, Hartmann C, Rode A Organ/tissue - specific changes in the mitochondrial genome organization of in vitro cultures derived from different explants of a single wheat variety. Theor Appl Genet 1992a;85:1-8

9. Morere-Le Paven MC, De Buyser J, Henry Y, Corre F, Hartmann C, Rode A Multiple patterns of mtDNA reorganization in plants regenerated from different in vitro cultured explants of a single wheat variety. Theor Appl Genet 1992b;85:9-14

10. Hartmann C, Recipon H, Jubier M, Valon C, Delcher-Besin E, Henry Y, De Buyser J, Lejeune B, Rode A Mitochondrial DNA variability in a single wheat regenerant involves a rare recombination event across a short repeat. Curr Genet 1994;25:456-464

11. Chowdhury MKU., Vasil V, Vasil IK Molecular analysis of plants regenerated 1994;87:821-828

12. Andre C, Levy A, Walbot V Small repeated sequences and the structure of plant mitochondrial genomes. Trends Genet 1992;8:128-132

13. Mohr S, Schulte-Kappert E, Odenbach W, Oettler G, Kück U Mitochondrial DNA of cytoplasmic male - sterile Triticum timopheevi : rearrangement of upstream sequences of the atp 6 and orf 25 genes. Theor Appl Genet. 1993;86 :259-268

14. Brettell RIS, Palotta MA, Gustafson JP, Appels R Variation at the Nor loci in triticale derived from tissue culture. Theor Appl Genet 1986;71:637-643

15. Southern EM Gel electrophoresis of restriction fragments. J Mol Biol 1975;98:503-517

16. Biochemica Boehringer Mannheim GmbH The DIG system user's guide for filter hybridization 1993.

17 Bonen L The mitochondrial S13 ribosomal protein gene is silent in wheat embryos and seedlings. Nucl Acids Res 1987;24:10393-10404

INVESTIGATIONS ON MORPHOGENIC SUSPENSION CULTURES INITIATED FROM MATURE EMBRYOS OF TRITICALE

Frank J. Maier[1], Gitta Oettler[2], Claus-Ulrich Hesemann[3]

[1]Research Centre Biotechnology and Plant Breeding,
[2]State Plant Breeding Institute,
[3]Institute of Genetics; University of Hohenheim, Stuttgart, Germany

Abstract

Suspension cultures are a prerequisite for plant breeding selection methods at the micro-callus or cellular level. The ability of regeneration is essential in the case of *in vitro*-selection and transformation. In *Gramineae* the predominantly used explants to establish these cultures are immature embryos. Their donor plant management is a limiting factor, especially in winter cereals. The genetical and physiological state of the immature embryos is not defined. The use of mature embryos might increase efficiency to establish suspension cultures. The experiments presented here were conducted to (1) identify zones within mature triticale embryos which are competent for tissue culture and (2) establish (regenerable) fast growing cellular suspension cultures in triticale from these explants. In mature embryos cultured on solid medium the coleoptilar and hypocotylar zone was found to form nodular friable calli which showed shoot regeneration. Four carbon sources, two auxins, and four cytokinins, each in various concentrations, investigated in suspension cultures from the coleoptilar and hypocotylar zone did not result in any regeneration of shoots after 21 weeks. Nevertheless, many variants showed the potential for the establishment of fast growing, homogeneous cellular suspension cultures within a short period of two to three months. These cultures were in most cases able to form secondary macro-calli with morphogeneous capacity from single cells or micro-calli. They could be used as alternatives to cultures established from immature embryos in cases where regeneration is not of immanent importance.

Introduction

The establishment of suspension cultures can serve various purposes in plant breeding strategies: *in vitro*-selection, transformation, and research on the mechanisms of resistance or tolerance. With regard to the first two aims, the suspension cultures have to be regenerable in order to use the selected or transformed material in breeding cycles. To clarify mechanisms of resistance or tolerance, there is no absolute need for regenerable cultures [1,2].

The predominantly used explants to establish suspension cultures in *Gramineae* are immature embryos. Their genetical and physiological state, however, is not defined [3]. From such explants of a winter cereal fast growing homogeneous suspension cultures cannot be established in less than 5-6 months (including donor plant management

H. Guedes-Pinto et al. (eds.), Triticale: Today and Tomorrow, 357–364.

© 1996 *Kluwer Academic Publishers. Printed in the Netherlands.*

and initiation of culture). The application of mature embryos could increase efficiency.

The use of mature embryos to establish regenerable suspension cultures and even protoplast regeneration is described by several authors for *Lolium* [5,6]. In cereals, studies with mature embryos cultured on solid medium were, among others, performed by Lupotto [7]. The experiments presented here were conducted to identify possible regions within a mature triticale embryo which are able to produce (regenerable) cellular suspension cultures. Additionally, several medium components (carbon sources, auxins, and cytokinins), which in some cases promote regeneration ability [4,6,8], were tested in liquid media.

Materials and Methods

Triticale varieties Alamo (A), Lasko (L), Modus (M), Purdy (P), and one primary hexaploid triticale (HSE), all of winter habit, were used in this study. In earlier experiments, calli of the latter had shown the capacity of *h*igh *s*omatic *e*mbryogenesis [9] on a complex medium [10].

Seeds of triticale genotypes were surface sterilized by soaking them first in 70 % ethanol for one minute and then in a 100 % solution of sodium hypochlorite supplemented with 0.01 % Tween 20 for 30 minutes. Thereafter, the seeds were rinsed three times for 10 minutes in sterile bidistilled water. For all three experiments embryos with scutellum were excised and dissected as shown in Figure 1. All liquid media variants were cultured on a rotary shaker (90min^{-1}) at 26 °C and under diffuse light (24 h). Morphogenic capacity in all regeneration experiments was determined by evaluation of differentiations (roots, shoots) and formation of chlorophyll.

SOLID BASAL MEDIUM

Embryos of the four varieties were cut along their axis into 5-8 nearly equal sections. These sections were arranged in their natural sequence from coleoptile (CP) to coleorhiza (CR) on CC- [11] or MS- [12] medium supplemented with 20.4 µM 2,4-D (2,4-dichlor-phenoxyaceticacid, Serva) and solidified with 0.8 % agar (initiation condition) in such a manner that one of the cut surfaces was in contact with the medium (I. in Figure 1). Five in this way dissected embryos were cultivated in each petri dish (Greiner, 60/15 mm). After four weeks (Figure 2-A), the explants were transferred to regeneration medium (basal medium without phytohormones for six weeks).

LIQUID BASAL MEDIUM

From each of five embryos, two parts of every zone were cultivated in sixwell-plates (Greiner, 35x14 mm/ 7.5 ml) with 4 ml of liquid MS-medium supplemented with 20.4 µM 2,4-D (II. Figure 1). To assess the growth of the different zones, fresh weight of suspended calli and cell numbers (Fuchs-Rosenthal-counter) were determined after three weeks of initiation (Figure 2-B). Thereafter, the total biomass of each well was separated from the liquid medium by sieving through a filter paper under a soft vacuum and plated for regeneration on solid medium (0,8 % agar) supplemented with 0.45 µM KIN (6-furfurylaminopurine, Kinetin; Serva).

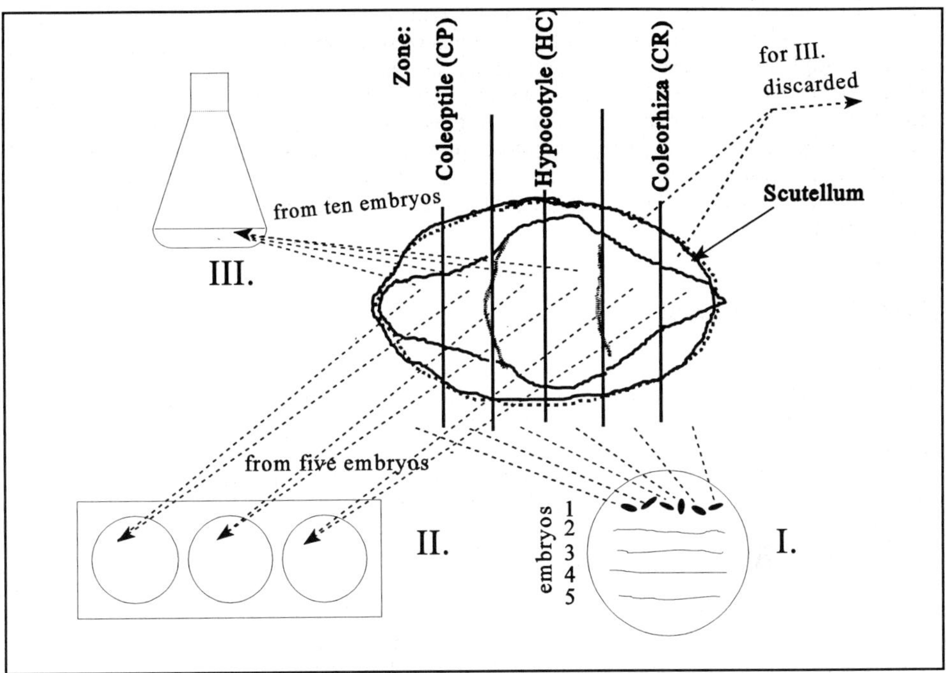

Figure 1: Preparation of mature triticale embryos for three experiments: I. Solid basal medium; II. Liquid basal medium; III. Medium compositions.

MEDIUM COMPOSITIONS

From the dissected embryos of HSE, the two coleorhizal pieces were not included in the experiment (III. Figure 1). The parts of the two other segments were added to 10 ml liquid MS-media supplemented with a number of carbon sources, cytokinins, and auxins in various concentrations (Table1).

Table 1: Medium components tested in a factorial design.

Carbon source[§]	Auxin	Cytokinin	
Sucrose	2,4-D (13.6 µM)	0	
Maltose	2.4-D (20.4 µM)	BAP[■]	
Glucose	DIC (27.2 µM)	BPA	(0.045, 0.45,
Sucrose+Maltose+Glucose	DIC (40.7 µM)	KIN	2.25 and 4.5 µM)
		PTU	

[§]90 mM each;[■]BAP: 6-benzylaminopurine (Sigma), BPA: N-benzyl-9-(2-tetrahydropyranyl)adenine (Sigma), 2,4-D: 2,4-dichlorphenoxyaceticacid, DIC: 3,6-dichloro-2-methoxybenzoicacid (Serva), KIN: 6-furfurylaminopurine (Serva), PTU: 1-phenyl-3-(1,2,3-thiadiazol-5yl)urea (Sigma)

Since the emphasis was on testing a large number of different media (altogether 272), the experiment was conducted with one replication only. Media were prepared under sterile conditions from stock solutions. After two weeks and then at intervals of one week, 5 ml of media were discarded and the same volume of new medium added. After 8 weeks, suspensions with calli were assessed for the proportion of necrotic aggregates. From the well suspended suspensions, 2 ml were filtered through a sieve (150 μm width of mesh), and the filtrate was given to 8 ml fresh medium. After further eight weeks, the number of newly formed suspended calli was counted. Thereafter, the calli were put on MS-medium without auxin, but supplemented with the same cytokinins as before and solidified with 0.8 % agar for regeneration.

Results

SOLID BASAL MEDIUM

After initiation, the different sections of the embryo on both media showed various sizes and types of calli. For both of these parameters there was, in general, a gradient between CP and CR. CP and HC formed large, compact nodular and yellow calli, whereas the calli from the CR were small, soft and watery (Figure 2-A). This applied to both media tested (CC and MS). The gradient was also found for the morphogenic capacity. CP and HC showed green rootlike morphogenesis, and small amounts of green shoots were formed by organogenesis. CR only rarely developed rootlike and green clusters. Since both basal media were equivalent with regard to the gradient, for further investigations the MS-salts were used.

LIQUID BASAL MEDIUM

The weight of suspended calli for zones after the initiation period showed the same gradient from CP to CR as described before for solid media (Figure 2-C). The weight of the explants was 100% for CP, 30% for HC, and less than 10% for CR. The range of formation of single cells and micro-calli was about 0.5 to 2 x 10^5 ml^{-1}. After the regeneration period, there was no development of green shoots, but the explants formed large amounts of green roots from the CP, many roots from the HC, and few white roots from the CR.

MEDIUM COMPOSITIONS

During the first two weeks of culture, both 2,4-D concentrations (13.6 and 20.4 μM) initiated some shoots, which, to force non-differentiated growth, were discarded. The influence of the carbon source on necrosis of the cultures was insignificant and ranged from 31.5 to 38.0%. On both 2,4-D concentrations, 30 % of the structures showed necrosis, whereas with both DIC concentrations necrosis was 40 %. The carbon sources had a noticeable influence on the formation of secondary calli. Averaged over all auxin-cytokinin combinations, the number of secondary calli formed from 2 ml of eight-weeks-old suspension (< 150 μm) after further eight weeks of cultivation was 22 for saccharose, 20 for maltose, 14 for the mixture, and 13 for glucose. Some variants with glucose, however, gave completely green calli. Within one auxin type, formation of secondary calli was induced with increasing concentrations. On average, 2,4-D gave 12 and DIC 23 secondary calli.

A development from high vacuolized cells to homogeneous and fast growing cellular suspension cultures, rich in cytoplasm was observed from initiation to 21 weeks on media with less than 2.25 μM of cytokinins (Figure 2-B).

Cytokinins had an obvious effect on the formation of necrotic structures. In general, the proportion of necrotic suspension particles increased with increasing cytokinin concentrations. Secondary callus formation was inhibited significantly by cytokinin concentrations higher than 4.5 µM (Figure 3).

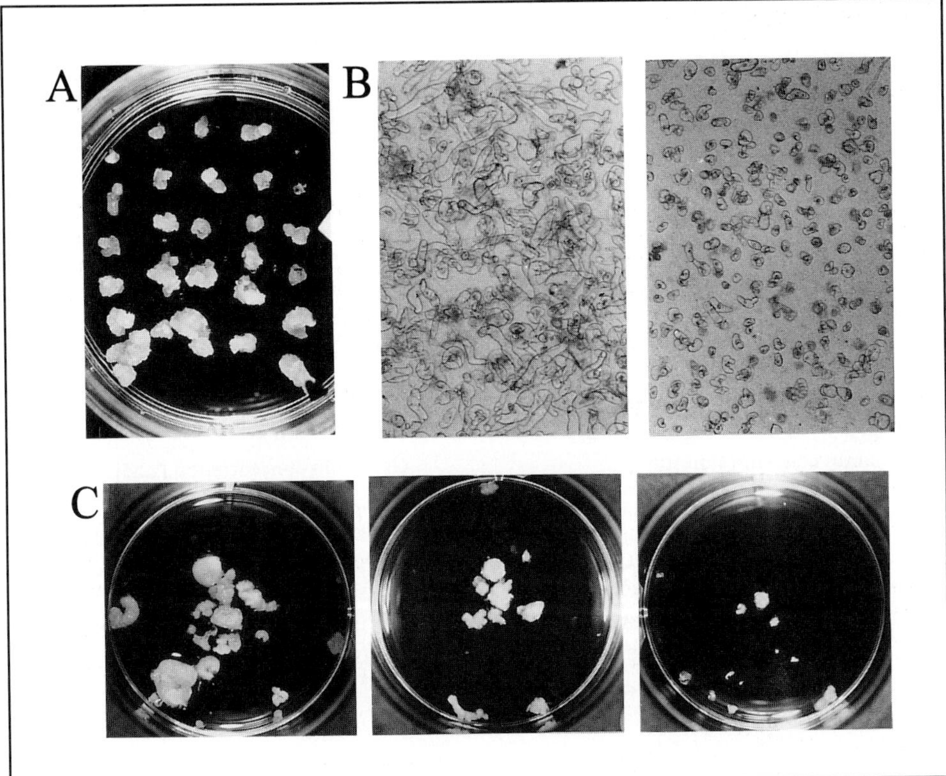

Figure 2: A) Pattern of callus development on solidified MS-medium in dependence on origin of explant within the mature embryo (top: CR, bottom: CP; left to right: embryo traces 1 to 5). B) Development from heterogeneous to homogeneous suspension culture (left: 6 weeks after initiation of suspension culture; right: about 5 month after initiation), magnification: 400x. C) Pattern of callus development in liquid MS-medium in dependence on origin of explant within the mature embryo (left: CP, middle: HC, right: CR; parts of ten embryos each).

Discussion

The apico-basal gradient with regard to callus formation observed in immature embryos of barley by Golds [4] was also found in mature triticale embryos in solidified and liquid medium. It can therefore be assumed that this result is also valid for immature embryos of triticale. The compact and nodular calli of CR and HC had particularly favourable appearances according to the four callus classes described by Redway [13], but gave less green shoots on the media tested. Carbon sources could be ranged for tissue culture growth: sucrose > maltose > glucose+maltose+sucrose > glucose. The presence of cytokinins does not seem to be beneficial, as only 0.45 µM BPA was superior to 0 µM. Measured by necrosis and formation of secondary calli, all cytokinins in liquid medium appeared to be toxic in concentrations up from 2.25 µM (Figure 3).

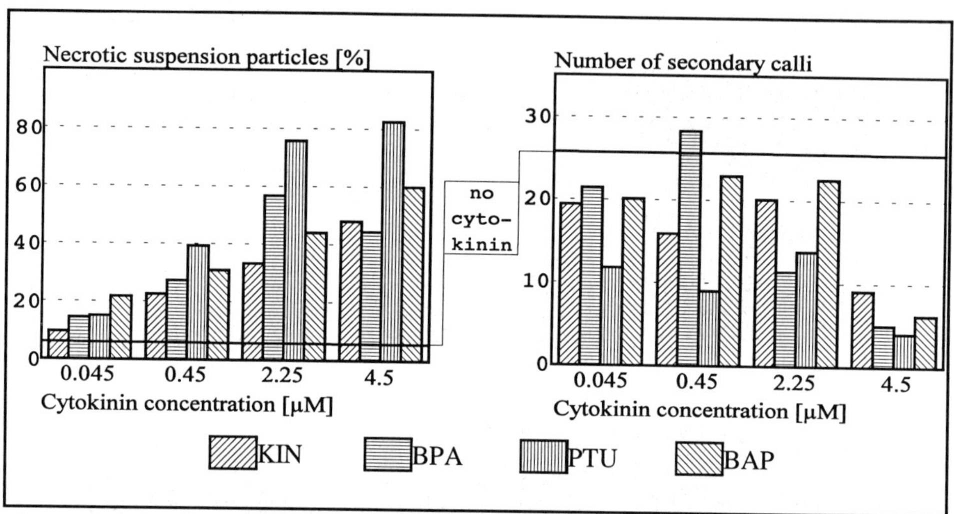

Figure 3: Proportion of necrotic structures after 8 weeks (left) and number of secondary formed calli of 2 ml of suspension (filtered < 150 µm) after further 8 weeks. Both variables induced with MS-salts, various carbon sources, auxins and cytokinins (mean over all carbon sources and auxins).

In preliminary experiments, 2,4-D appeared to be a more competent auxin than DIC. But the results of the study with various medium components suggests that both auxins are of the same strength.

A direct comparison of the two auxins was not possible, because different concentrations were used and some explants were taken from the flasks of the auxin variants. The capacity of DIC to induce formation of secondary calli could either depend on the strength of the two auxins or on the higher concentrations applied. To exclude plant development during initiation of tissue culture from mature embryos of triticale, in contrast to mature embryos of ryegrass [14], auxin concentrations should be higher than 20.4 µM. Sharma [15] used 2,4-D concentrations of 13.6 to 27.2 µM in callus cultures initiated from immature embryos of triticale. The mutagenic effect of 2,4-D and DIC should be inhibited by reducing the concentrations in maintenance media. Dalton [6], for example, described for mature embryos of *Lolium* a reduction of 2,4-D from 27.2 µM to 13.6 µM from initiation to maintenance medium of suspension cultures. In future, genetic investigations about mutagenic effects on medium components, particularly of the auxins, are of special interest.

Since the regeneration rate is very small and only via organogenesis (not of single cell origin), the use of mature embryos of triticale is not profitable to transform or select suspension cultures. But for the first time in cereals a method to establish suspension cultures from mature embryos is described. This method is suitable for some aspects of research (such as claryfying mechanisms of resistance or tolerance). The use of such cultures gives a time gain of about 50 % relative to suspension cultures initiated from immature embryos.

In further investigations without cytokinins, the best concentration and species of auxin in initiation and maintenance medium have to be found. The single media tested, should be observed to identify those which give the earliest homogeneous suspension culture. According to our experiences, such a culture could presumably be established in less than two months.

Amount and composition of nitrogen source is of decisive importance for embryogenesis and development of isolated barley microspores as found by Mordhorst and Lörz [16]. In further investigations with mature embryos of triticale this aspect should also be considered.

Acknowledgements

The project was financially supported by a grant from the Bundesministerium für Forschung und Technologie (BMFT-No. PBE0318897A). The responsibility for the contents of this publication rests with the authors. The authors thank Ms. Carmen Kolle for excellent technical assistance.

References

1. Miller JD, Arnison PG. Degradation of deoxynivalenol by suspension cultures of the fusarium head blight resistant wheat cultivar Frontana. Can J Plant Pathol 1986;8:147-50.
2. Ojima K, Ohira K. Characterization of aluminium and manganese tolerant cell lines selected from carrot cell cultures. Plant Cell Physiol 1983; 24:789-97.
3. Lörz H, Göbel E, Brown P. Advances in tissue culture and progress towards genetic transformation of cereals. Plant Breeding 1988;100:1-25.
4. Golds TJ, Babczinsky J, Koop H-U. Regeneration from intact and sectioned immature embryos of barley (Hordeum vulgare L.): the scutellum exhibits an apico-basal gradient of embryogenic capacity. Plant Sci 1993;90:211-18.
5. Creemers Molenaar J, van Eeuwijk FA, Krens FA. Culture optimization for perennial ryegrass protoplasts. J Plant Physiol 1992;139:303-08.
6. Dalton SJ, Thomas ID. A statistical comparison of various factors on embryogenic proliferation, morphogenesis and regeneration in Lolium temulentum cell suspension colonies. Plant Cell, Tissue and Organ Culture 1992;30:15-29.
7. Lupotto E. Callus induction and plant regeneration from barley mature embryos. Ann Bot 1984;54:523-29.
8. Rybczynski JJ. Plant tissue culture of Secale: A review. Euphytica 1990; 46:57-70.
9. Dormann M. Feld- und In-vitro-Untersuchungen zur Resistenz primärer Triticale (x Triticosecale Wittmack) gegenüber Fusarium graminearum und Microdochium nivale [dissertation]. Hohenheim: Univ. Hohenheim, 1992.
10. Stolarz A, Lörz H. Somatic embryogenesis, in vitro multiplication and plant regeneration from immature embryo explants of hexaploid triticale (x Triticosecale Wittmack). Z Pflanzenzüchtg 1986;96:353-62.

11. Potrykus I, Harms CT, Lörz H. Callus formation from cell culture protoplasts of corn (Zea mays L.). Theor Appl Genet 1979;54:209-14.
12. Murashige T, Skoog F. A revised medium for rapid growth and bio assays with tobacco cultures. Physiol Plant 1962;15:473-97.
13. Redway FA, Vasil V, Vasil IK. Characterization and regeneration of wheat (Triticum aestivum L.) embryogenic cell suspension cultures. Plant Cell Reports 1990;8:714-17.
14. Altpeter F. In vitro Selektion bei Lolium perenne (L.) unter Einsatz des Mykotoxins Deoxynivalenol zur Resistenzverbesserung gegenüber Microdochium nivale (Fr.) Sam. & Hall [dissertation]. Hohenheim: Univ. Hohenheim, 1994. Verlag Ulrich E. Grauer, Wendlingen. ISBN 3-86186-043-0.
15. Sharma GC, Bello LL, Sapra VT, Peterson CM. Callus initiation and plant regeneration from triticale embryos. Crop Sci 1981;21:113-118.
16. Mordhorst AP, Lörz H. Embryogenesis and development of isolated barley (Hordeum vulgare L.) microspores are influenced by the amount and composition of nitrogen sources in culture media. J Plant Physiol 1993;142:485-492.

RESPONSE OF MATURE TRITICALE EMBRYOS TO ALUMINIUM-TOXIC CALLUS INDUCTION MEDIA

Claudia S. Gaus[1], Gitta Oettler[2] and Claus-Ulrich Hesemann[3]
[1] Research Centre Biotechnology and Plant Breeding ,[2] State Plant Breeding Institute,[3] Institute of Genetics; University of Hohenheim, Stuttgart, Germany

Abstract

Callus induction of threefold cut mature embryos of one winter triticale (x *Triticosecale* Wittmack) variety ("Lasko") and three spring triticale genotypes (PFT 8512, Ocepar 1 and PFT 8710) was determined in a recently developed aluminium (Al)-toxic medium. Callus weight and pH alteration of spent medium were assessed at the end of the callus induction phase of 7 weeks in media containing different levels of $AlCl_3$ (0.1 - 10mM) and being sterilised either by filter or by autoclaving.

In filter-sterilised media with concentrations higher than 2mM none of the four examined triticale genotypes developed calli and the pH-level of the spent medium remained nearly constant. Lower concentrations of 0.5mM and 1mM Al yielded callus weights significantly higher than the pH 3.6-control without Al, but less than the pH 4.6-control, and pH increased on the same level as in either control. In all concentrations below 2mM Al, somatic embryos could be observed. In contrast, the toxic effect of autoclaved Al-toxic media was expressed by inferior callus quality - even the controls were poor - and did not result in any somatic embryo.

Consequently, filter-sterilised Al-toxic media should be preferred and could serve to select Al-tolerant cell lines, e.g. in suspension cell cultures, or may be useful to investigate Al-tolerance mechanisms at the celluar level.

Introduction

Acid soils with excess of soluble Al are main factors limiting plant growth in vast areas of the world, particularly in tropical and subtropical regions. Triticale shows a high level of tolerance and a considerable variation within and among genotypes. Since the toxic effects of Al and the mechanisms of its resistance are based on events of the cellular level, Al tolerance seems to be selectable in cell cultures.

Solutions with a constant amount of total Al and variable pH show a pH-dependent collinearity among Al species [1]. The toxicity of different mononuclear and polynuclear Al is discussed controversially. Whereas the mononuclear hydroxy-Al was found to be more toxic to many dicotyledons, the rhizotoxic species for wheat seems to be Al^{3+} [1,2]. With a maximum amount of this species below pH 4.0 [1], a further decrease of the pH might be suitable to create an Al-toxic medium, unless the growth of the control without Al is dramatically decreased at this pH. Due to the interacting effects of some components, such as calcium, phosphate or EDTA of normal culture media,

H. Guedes-Pinto et al. (eds.), Triticale: Today and Tomorrow, 365–371.
© 1996 *Kluwer Academic Publishers. Printed in the Netherlands.*

of some components, such as calcium, phosphate or EDTA of normal culture media, which could reduce the toxicity of Al, further changes of media composition should be made [3,4].

The reaction of plant cell cultures of dicotyledonous crop plants to Al-toxic media have been widely reported [3-5], but information on monocotyledons is very rare. The aim of this research with triticale was to (1) examine the toxic effect of Al in a modified CC-medium [6] with a pH of 3.6 and (2) test different sterilisation techniques of this medium with regard to its Al toxicity.

Materials and Methods

The study to investigate the influence of Al on callus induction of mature embryos was conducted with the winter triticale variety "Lasko" and the spring triticale genotype PFT 8710 (Experiment 1). In addition, for the test of different sterilisation techniques of Al-toxic media (Experiment 2), the spring triticale genotypes PFT 8512 and Ocepar 1 were used (seeds of all spring triticales were kindly made available by Dr. A.C. Baier, Passo Fundo, Brazil).

To obtain mature embryos, the seeds were surface-sterilised by immersing them in 70% ethyl alcohol for one minute, then in a 13% solution of sodium hypochloride for 30 minutes, followed by three rinses of 10 minutes in sterile destilled water. Thereafter they were soaked in sterile water for about 4 hours to facilitate preparation. With the help of a lancet-shaped dissecting needle embryos were excised from the seeds under sterile conditions. The seed coat was roughly unwrapped and embryos were dissected one time along the embryo axis and two times transversally to the first cut. The pieces of always five embryos were placed onto patches of cloth saturated with 7ml of liquid medium in plastic petri dishes (Greiner, 60/15mm) in three replicates and incubated at 26°C in the dark. After 7 weeks of callus induction, the fresh weight of calli and remaining embryo pieces of each petri dish was determined. The callus pieces were then moved on regeneration medium (MS [7]; with 1mg/l kinetin), and after another 4 weeks the number of plantlets was recorded.

The standard culture medium (CC [6]; without coconut milk and mannitol, with 1.5mg/l 2,4-D) used for callus induction was supplemented with unchelated iron. Of a freshly prepared $AlCl_3$ x $6H_2O$ stock solution (100mM), Al was slowly added as the last component at pH 4.6 in concentrations of 0.1-10 mmol/l medium. The pH of the medium decreased according to the supplemented Al concentration. In media, which were filter-sterilised, pH was readjusted with 0.1 or 1M NaOH to 3.6 ± 0.02. Those variants, which were sterilised by autoclaving for 5 minutes at 121°C, had a readjusted pH of 4.00 ± 0.02 before sterilisation. This was necessary because by heating the media became more acid, and after sterilisation no pH readjustment was made. Initial pH was recorded within 24 hours and final pH at the end of the experiments. In order to determine the pH change over time, pH of spent medium, which was squeezed out of the cloth, was ascertained over the three replicates.

Results and Discussion

AL-TOXIC MEDIUM

Studying Al toxicity in plant cell cultures, the occurrence of Al precipitation in the standard culture medium is a major problem. Preliminary experiments showed that the precipitations are mainly caused by the formation of Al hydroxide species when pH has to be readjusted after addition of Al. Because of a greater decline in pH when higher concentrations of Al are added (Table 1), the amount of alkaline solution rises accordingly. But the lower the requested initial pH of the Al-containing medium, the less NaOH has to be supplemented and consequently, less Al hydroxide species were formed. Al-containing CC-media (even with an Al concentration of only 0.1mM) with reduced pH of 4.6 instead of 5.6 could not be filter-sterilised (data not shown). Further decrease of pH to 3.6 in Al-containing and Al-toxic media led to no or almost no precipitation. Various authors report decreasing precipitation when using media with reduced phosphate and/or calcium [3,4]. This reduction had no perceptible effect in our standard medium. In CC-medium with pH of 3.6, phosphate concentration could be maintained, because decreasing the pH also diminishes Al phosphate precipitation [3]. Insoluble Al phosphate might slowly be solubilized by chelation of Al [8].

All media containing Al, even in a concentration of 20mM, could be filter-sterilised when pH was adjusted after the addition of Al at pH 4.6 to 3.6. In these variants the average pH change of 0.11 ± 0.02 (Table 1) over a period of 7 weeks (without cells) was negligible. The pH changes of autoclaved media were equally insignificant between initial and final values. The pH adjustment to 4.00 ± 0.02 prior to this sterilisation technique was found to be necessary to obtain the required initial pH of 3.6 ± 0.02, as a result of the calculated pH-decrease during autoclaving. After 7 weeks the pH of the control (without Al) changed to 0.3 units below the requested value of 4.6. The same could be observed for liquid media with a pH of 5.6 (data not shown).

Iron is reported to have toxic [9] and alleviating effects [3]. It was therefore added unchelated.

Table 1:   Changes of pH-values in liquid CC-media with different levels of Al at 24 hours (initial pH) and 7 weeks (final pH) after sterilisation (The pH was requested to be 3.6 before sterilisation if not stated otherwise.).

Al conc. [mM]	pH after Al addition at pH 4.6	Experiment 1 Filter-sterilised		Experiment 2 Filter-sterilised		Autoclaved *	
		Initial pH	Final pH	Initial pH	Final pH	Initial pH	Final pH
0 (control) #	-	4.64	4.26	◆	◆	◆	◆
0 (control)	-	3.62	3.66	3.68	3.58	3.68	3.64
0.1	4.00	3.59	3.61	◆	◆	◆	◆
0.5	3.44	3.61	3.49	◆	◆	◆	◆
1	3.32	3.61	3.50	◆	◆	◆	◆
2	3.16	3.60	3.47	◆	◆	◆	◆
3	3.13	◆	◆	3.56	3.46	3.45	3.48
4	3.10	◆	◆	3.56	3.43	3.44	3.49
5	3.07	3.61	3.47	3.54	3.43	3.49	3.58
6	3.05	◆	◆	3.56	3.46	3.54	3.59
10	2.96	3.60	3.47	◆	◆	◆	◆
20	2.96	3.61	3.47	◆	◆	◆	◆

* pH adjusted before autoclaving to 4.00 ± 0.02 # requested pH 4.6   ◆ not investigated

AL DOSE RESPONSE IN FILTER-STERILISED MEDIA

The pH measurements of spent media, which were squeezed out of the patches of cloth, reached for "Lasko" and PFT 8710 up to an Al concentration of 2mM significantly higher values compared to the initial and final pH without cells. The increase was in the same magnitude as in the Al-free media (for "Lasko" corresponding to the pH 4.6 control and for PFT 8710 to its pH 3.6 control). The means of pH-values were higher for "Lasko" than for PFT 8710, but showed a greater variability (Table 2). Marziah [10] observed similar changes of pH of Al-toxic media containing tepary bean cell culture. The small variability he observed, might be due to identical clones he used, whereas in our investigation embryos of triticale lines might possess a certain heterozygosity.

Table 2:   Callus fresh weights, changes of pH (means of 3 replicates and their standard deviation), and number of calli forming somatic embryos after 7 weeks of callus initiation on liquid CC-media with different levels of Al, and viable plantlets after regeneration.

Genotype	Al conc. [mM]	Fresh weight [mg]		pH of spent medium		No. of calli forming somatic embryos	No. of plantlets
Lasko	0 (control)	1203 ±	292	6.10 ±	0.67	4	5
	0 (control)	916 ±	38	5.91 ±	0.79	5	-
	0.1	772 ±	81	5.93 ±	1.06	3	10
	0.5	1067 ±	171	5.94 ±	0.39	2	36
	1	1061 ±	244	6.03 ±	0.64	1	10
	2	642 ±	171	5.27 ±	0.05	1	1
	5	32 ±	5	3.88 ±	0.05	-	-
	10	36 ±	4	3.63 ±	0.04	-	-
	20	28 ±	9	3.58 ±	0.00	-	-
PFT 8710	0 (control)	1289 ±	165	5.59 ±	0.30	2	-
	0 (control)	789 ±	106	5.20 ±	0.08	2	-
	0.1	1062 ±	201	5.30 ±	0.17	2	-
	0.5	1004 ±	106	5.27 ±	0.15	1	-
	1	824 ±	311	5.30 ±	0.14	2	-
	2	700 ±	98	5.32 ±	0.02	1	6
	5	58 ±	13	4.01 ±	0.07	-	-
	10	32 ±	8	3.62 ±	0.03	-	-
	20	38 ±	5	3.58 ±	0.02	-	-

Concentrations of 5mM and higher showed a declining increase with rising Al concentrations. The toxic effect of Al was also obvious in the inefficiency of the embryo pieces to induce calli at these concentrations. Embryos of "Lasko" did not change their appearance during 7 weeks of incubation, and those of PFT 8710 at 5mM showed merely a swelling of the scutellum. Concentrations of 2mM and less yielded callus fresh weights of >70 - 90% of the pH 3.6 control and >50% of the pH 4.6 control. For an Al concentration of 0.5mM a stimulating effect was observed (Table 2). These results were surprising, because most reported investigations of selection for Al tolerance in cell culture show a decreased growth of cells at Al concentrations of 0.1 - 0.5mM [3, 9-11] or a reduced callus induction rate at 0.15mM Al [4]. This confirms the relatively high Al tolerance of triticale in the field. The deviation of up to 25% from the mean (especially PFT 8710), which might be due to certain heterozygosity of the triticales, shows that there is adequate potential for selection.

Induced calli of both genotypes formed somatic embryos. Regeneration of plantlets for Al variants was successful for "Lasko" only (Table 2).

Table 3: Callus fresh weights, changes of pH and their standard deviation after 7 weeks of callus initiation on filter sterilised or autoclaved liquid CC-media with different levels of Al (means of 3 replicates).

Genotype	Al conc. [mM]	Sterilisation technique			
		Filter-sterilised		Autoclaved	
		Fresh weight [mg]	pH of spent medium	Fresh weight [mg]	pH of spent medium
Lasko	0 (control)	962 ± 72	6.08 ± 0.80	1025 ± 57	5.60 ± 0.28
	3	334 ± 147	5.38 ± 0.02	395 ± 145	5.35 ± 0.17
	4	62 ± 8	4.03 ± 0.06	71 ± 10	3.99 ± 0.10
	5	62 ± 3	3.82 ± 0.11	62 ± 6	3.84 ± 0.06
	6	49 ± 11	3.70 ± 0.04	52 ± 4	3.69 ± 0.02
PFT 8710	0 (control)	579 ± 614	4.53 ± 1.18	128 ± 48	3.62 ± 0.14
	3	398 ± 151	5.23 ± 0.10	227 ± 56	4.66 ± 0.80
	4	103 ± 29	4.29 ± 0.05	91 ± 12	4.13 ± 0.06
	5	88 ± 18	4.07 ± 0.06	58 ± 15	3.80 ± 0.07
	6	62 ± 8	3.85 ± 0.05	56 ± 12	3.79 ± 0.04
Ocepar 1	0 (control)	1529 ± 49	5.84 ± 0.25	1194 ± 325	5.47 ± 0.22
	3	779 ± 171	5.23 ± 0.11	506 ± 119	5.29 ± 0.07
	4	507 ± 235	5.14 ± 0.14	216 ± 122	4.88 ± 0.45
	5	199 ± 122	4.43 ± 0.39	75 ± 30	4.02 ± 0.14
	6	80 ± 10	3.91 ± 0.08	81 ± 1	3.91 ± 0.05
PFT 8512	0 (control)	1271 ± 176	5.49 ± 0.48	753 ± 100	5.29 ± 0.11
	3	618 ± 196	5.12 ± 0.09	134 ± 43	4.83 ± 0.19
	4	136 ± 32	4.45 ± 0.23	68 ± 22	4.19 ± 0.18
	5	97 ± 19	4.00 ± 0.06	63 ± 28	4.03 ± 0.22
	6	78 ± 7	3.80 ± 0.05	39 ± 2	3.72 ± 0.04

FILTER-STERILISED VS. AUTOCLAVED AL-TOXIC MEDIA

In Experiment 2, neither the autoclaved nor the filter-sterilised media induced embryogenic structures. Consequently, no plantlets could be regenerated.Callus fresh weights after incubation on autoclaved media were lower than in the filter-sterilised variants, except for "Lasko", and pH of the spent medium did not increase in the same way (Table 3). Obviously, even the short heating up to 121°C for 5 minutes changed the composition of the medium and affected the reaction of the dissected embryos.

Both experimentes of our investigation point to a great variability of the examined embryos of the winter and spring triticale to induce calli in Al-toxic media. Consequently, in a petri dish with five embryos of one genotype the once initiated division of a tolerant embryo could stimulate the physiological activity of cells, initiate the release of metal detoxificating substances such as citric acid [12], and result in further callus growth of possibly sensitive embryos. Bigger cell clumps could also have a higher capacity to complex Al and reduce its toxicity [11]. This might explain the great variability of callus fresh weight after 7 weeks of induction in media with an Al concertration of more than 5mM. Therefore, further studies should comprise more replicates and/or single embryos and a change of the medium every week to diminish these effects.

Decreasing the pH from normally 5.6 to 3.6, it is possible to produce Al-toxic media without or with very little Al hydroxide precipitation. Because phosphate concentration was not reduced, compared to other Al-toxic media reported so far, this medium might be suitable for induction and subculture of Al-tolerant cells. Using these media in suspension cultures, the study of resistance mechanisms seems to be possible in future investigations.

Acknowledgements

Financial support of this study by the "Vater und Sohn Eiselen-Stiftung", Ulm, is gratefully acknowledged. Thanks are due to Ms C. Arnold for her skillful and excellent technical assistance.

References

1. Kinraide TB. Identity of the rhizotoxic aluminium species. In: Wright RJ, Baligar VC, Murrmann RP, editors. Plant-soil interactions at low pH. Dordrecht: Kluwer Academic Publishers, 1991: 717-28.
2. Wagatsuma T, Ezoe Y. Effect of pH on ionic species of aluminum in medium and on aluminum toxicity under solution culture. Soil Sci Plant Nutr 1985; 31: 547-61.
3. Conner AJ, Meredith CP. Stimulating the mineral environment of aluminium toxic soils in plant cell culture. J Exp Bot 1985; 36: 870-80.
4. Kamp-Glass M, Powell D, Reddy GB, Baligar VC, Wright RJ. Biotechniques for improving acid aluminum tolerance in alfalfa. Plant Cell Reports 1993; 12: 590-92.
5. Wersuhn G, Kalettka T, Gienapp R, Reinke G, Schulz D. Problems posed by in-vitro selection for aluminium-tolerance when using cultivated plant cells. J Plant Physiol 1994; 143: 92-95.
6. Potrykus I, Harms CT, Lörz H. Callus formation from cell culture protoplasts of corn (Zea mays L.). Theor Appl Genet 1979; 54: 209-14.

7. Murashige T, Skoog F. A revised medium for rapid growth and bio assays with tobacco tissue cultures. Physiol Plant 1962; 15: 473-97.
8. Ojima K, Koyama H, Suzuki R, Yamaya T. Characterization of two tobacco cell lines selected to grow in the presence of either ionic Al or insoluble Al-phosphate. Soil Sci Plant Nutr 1985; 35: 545-51.
9. Wersuhn G, Nhi HH, Tellhelm E, Börner T. Response of potato cell cultures to aluminium and selection of tolerant cell lines. Potato Research 1986; 29: 399-02.
10. Marziah M. Changes of pH of legume cell suspension culture media containing aluminium. In: Wright RJ, Baligar VC, Murrmann RP, editors. Plant-soil interactions at low pH. Dordrecht: Kluwer Academic Publishers, 1991: 879-82.
11. Marziah M. Growth of peanut cells in suspension cultures treated with aluminium. In: Wright RJ, Baligar VC, Murrmann RP, editors. Plant-soil interactions at low pH. Dordrecht: Kluwer Academic Publishers, 1991: 875-77.
12. Ojima K, Abe H, Ohira K. Release of citric acid into the medium by aluminum-tolerant carrot cells. Plant & Cell Physiol 1984; 25: 855-58.

PLANT REGENERATION AND *IN VITRO* SELECTION OF TRITICALE

Gert Kleijer, N. Ndack Diop, Aldo Fossati
Station Fédérale de Recherches Agronomiques de Changins (RAC), Nyon, Switzerland

Abstract

Immature embryos have proven to be a good material for the production of cals and regeneration of plants in cereals. Fifty embryos of 15 to 18 days old of four winter triticale varieties were cultivated on two different media, N6 and MS with 7.95 mg/l 2.4D. After four weeks, the embryos were splitted in two parts, one part on a regeneration medium, the other on callus medium, which was repeated three times. The highest percentage cals from which plants could be regenerated was 30% for the variety Brio. Regeneration decreased with the duration of the culture of cals and varied for Brio from 985 plants regenerated after four weeks culture to 36 plants after 16 weeks. Important differences in regeneration capacity were observed between genotypes and the total number of plants regenerated using the N6 medium were 1584, 593, 657 and 215 for respectively Brio, Presto, Dagro and Meridal. N6 proved to be a better medium (3049 plants regenerated for four varieties) than MS (1114 plants regenerated). A preliminary trial was carried out to see whether in vitro selection of triticale for resistance to head blight (Fusarium) is possible. Six weeks old cals were placed on a selection medium prepared by the double layer technique. Fusarium culmorum and F. graminearum were cultivated on a solid medium for two weeks, autoclaved and after solidification, a second layer of the regeneration medium was added. Differences in number of regenerated plants were observed but were more dependent on the variety used than on this selective medium, on which for two varieties more plants could be regenerated on one of the selective media than on the standard medium. It could be that the two Fusarium isolates did not produce a toxin concentration high enough to influence callus growth and plant regeneration. This has to be repeated with other isolates with know toxin concentrations.

Introduction

Plant regeneration from tissue culture has been successful for all important species of the *Gramineae* family. The most suitable explant for the induction of somatic embryogenesis proved to be immature embryos [1]. In the case of triticale, plant regeneration through callus culture has been obtained for octoploids and hexaploids [2-6]. Genotypic differences in the ability to form callus were observed [7].

Head blight caused by *Fusarium* spp. can provoke severe damage on wheat and triticale, and

H. Guedes-Pinto et al. (eds.), Triticale: Today and Tomorrow, 373–377.
© 1996 *Kluwer Academic Publishers. Printed in the Netherlands.*

yield losses up to 50% have been observed in wheat [8]. The resistance to head blight is relative and differences between triticale varieties have been observed [9]. *Fusarium* spp. produce mycotoxins and the correlation of resistance to *Fusarium* and resistance to these toxins is very hight [10]. This opens an alternative way of screening for selecting resistance to *Fusarium* by *in vitro* selection.

In this paper, we describe plant regeneration of four triticale varieties after four to sixteen weeks callus culture and *in vitro* selection for resistance to head blight using the double layer technique [11].

Material and methods

The four varieties used in this study were Dagro and Presto from Poland, and the two Swiss varieties, Brio and Meridal.

Immature embryos were collected on the 15th to 18th day after anthesis on plants cultivated in a growth chamber. The immature seeds were surface sterilized during five minutes in 7% sodium-hypochlorite solution, washed once in sterile water, immersed twice for 15'' in 95% alcohol, then rinsed in three successive sterile water baths. The basal medium for callus proliferation consisted of solidified MS or N6 medium supplemented with 7.95 mg/l 2,4-D and 50g/l sucrose [4]. Fifty immature embryos per medium and per variety were placed, scutellum side up, at 28°C in darkness. After four weeks, calli were split into two parts One part was regenerated on a medium consisting of MS or N6 with 1 mg/l of kinetin and 30 g/l sucrose placed at 26°C with a photoperiod of sixteen hours during four weeks. Regenerated plantlets were transferred to glass tubes containing N6 medium with 30g/l sucrose without hormones before being transferred into pots and grown to maturity in the greenhouse. The other part was subcultured on N6 or MS with 0.5 mg/l 2,4-D and 50 g/l sucrose. After four weeks, calli were split again into two parts. This operation was repeated three times.

In vitro selection was carried out with 49 calli of Brio, Dagro and Meridal varieties differing in resistance to head blight [12]. The double layer technique used was the same as described by [11]. In our case, the second layer consisted of N6 medium supplemented with 1 mg/l kinetin and 30 g/l sucrose which allows regeneration on the double layer medium. Two strains of *Fusarium* were used, one from *F. graminearum* and one from *F. culmorum*. Calli remained on this selectif medium for six weeks. Regenerated plantlets were isolated and grew up to maturity. After these six weeks, the calli were transferred to a fresh regeneration medium without toxins for another four to eight weeks culture.

Results and discussion

Important differences were observed between the callus growth on N6 or MS medium. Calli on MS were significantly smaller (average 0.5 cm long and 0.4 cm large) than on N6 (0.58 / 0.45 cm) and showed much more necrosis. Twelve calli on MS were dead (no growth at all or completely necrotic) after four weeks culture, on N6 only one. This difference was accentuated during the first subculture. Genotypic differences in callus size were observed.

The number of calli which could be regenerated are presented in table 1. Brio showed the highest number and Meridal, the lowest number.

Table 1. Number of calli from a total of fifty which regenerated to plants four weeks after callus initiation

	N6	MS
Brio	14	15
Presto	9	1
Dagro	11	5
Meridal	3	6

The total number of plants regenerated during the different phases of regeneration and subculture are presented in table 2. Most of the plants could be regenerated during the first cycle of regeneration and their number decreased with increasing duration of callus growth for all four varieties. Genotypic differences were observed. The variety Brio showed the highest regeneration capacity and Meridal the lowest. Brio was the only variety which regenerated after sixteen weeks callus culture. Regeneration succeeded better on N6 medium than on MS with a total of plants regenerated of respectively 3049 and 1114. Some albinos were obtained. Earlier results [2,7] showed genotypic differences in callus formation and plant regeneration. The descendants of a part of regenerated plants are cultivated on the field to observe somaclonal variation.

The results of plant regeneration on the selectif medium are presented in table 3. On the double layer, only the control treatment of Brio showed a higher regeneration rate. For Dagro and Meridal, the highest rate was obtained on *F. graminearum* and *F. culmorum* respectively. For the three varieties, regeneration continued after transferring the calli from the double layer to normal regeneration medium, except for Dagro on *F. culmorum*. There was no influence of the toxins produced by *Fusarium* on plant regeneration. In wheat, important reductions in number of calli and embryoids have been observed [13], compared to the *Fusarium* toxin free medium using anther culture. An inhibition of callus growth on a double layer medium was observed in wheat [11]. In triticale, an increasing reduction of callus weight with increasing concentration of *Fusarium* toxins was observed [14]. This was also true for plant regeneration afterwards on a toxin free medium. In our study, no effect on regeneration was observed. This can be due to a very weak toxin production by the strains used. The trial has to be repeated with other *Fusarium* strains, if possible with known toxin production.

Table 2.   Number of regenerated plants of four varieties of triticale on two different medium during the different phases of regeneration (R) and subculture (SC) each subculture phase lasted four weeks. The calli on R II and SC II were coming from SC I

	BRIO		PRESTO		DAGRO		MERIDAL	
	N6	MS	N6	MS	N6	MS	N6	MS
R I	985*	530**	322	2	415*	113	112	108
SC I	2	0	0	0	0	0	6*	0
R II	396	52	249	76	178	69	97	82
SC II	32	0	11*	2	61	7	0	15
R III	112*	23	11	17***	3	7	0	1
SC III	21	9	0	0	0	0	0	0
R IV	36	0	0	0	0	0	0	0
SC IV	0	0	0	0	0	0	0	0
Total	1584	614	593	97	657	196	215	206

*   1 albinos
**   2 albinos
***   6 albinos

Table 3.   Number of plants regenerated on a double layer medium (R I) with *F. culmorum* (*Fc*), *F. graminearum* (*Fg*) or the control and during eight weeks following R I on a normal medium (R II)

	BRIO			DAGRO			MERIDAL		
	F.c.	C	*F.g.*	*F.c.*	C	*F.g.*	*F.c.*	C	*F.g.*
R I	635	807	486	22	115	215	202	170	64
R II	649	142	441	0	18	157	76	36	58
Total	1284	949	927	22	133	372	278	206	122

References

1.   Lörz H, Göbel E, Brown P. Advances in tissue culture and progress towards genetic transformation of cereals. Z Pflanzenzüchtung 1988; 100: 1-25.
2.   Immomen AST. Effect of karyotype on somatic embryogenesis from immature triticale (x *Triticosecale* Wittmack) embryos. Plant Breeding 1992; 109: 116-122.
3.   Stolarz A. Cell and protoplast culture, somatic embryogenesis and transformation studies in different forms of x *Triticosecale* Wittmack. Proc 2nd Int Triticale Symp Cimmyt Mexico 1991; 286-289.

4. Stolarz A, Lörz H. Somatic embryogenesis, *in vitro* multiplication and plant regeneration from immature embryo explants of hexaploid triticale (x *Triticosecale* Wittmack). Z Pflanzenzüchtung 1986; 96: 353-362.

5. Eapen S, Rao PS. Callus induction and plant regeneration from immature embryos of rye and triticale. Plant Cell Tissue Organ Culture 1982; 1: 221-227.

6. Nakamura C, Keller WA. Callus proliferation and plant regeneration from immature embryos of hexaploid triticale. Z Pflanzenzüchtung 1982; 88: 137-160.

7. Sharma GC, Bello LL, SapraVT. Genotypic differences in organogenesis from callus of ten triticale lines. Euphytica 1980; 29: 751-754.

8. Miedaner T, Walther H. Ermittlung der *Fusarium* Resistenz von Weizen im Aehrenstadium. Z Pflanzenkrankh Pflanzensch 1987; 94: 337-347.

9. Arseniuk E, Goral T, Czembor HJ. Reaction of triticale, wheat and rye accessions to graminaceous *Fusarium* spp. infection at the seedling and adult plant growth stages. Euphytica 1993; 70: 175-183.

10. Wang YZ, Miller JD. Effects of *Fusarium graminearum* metabolites on wheat tissue in relation to *Fusarium* head blight resistance. J Phytopathology 1988; 122: 118-125.

11. Ahmed KZ, Mesterhazy A, Sagi F. *In vitro* techniques for selecting wheat (*Triticum aestivum* L.) for *Fusarium* resistance. I. Double layer technique. Euphytica 1991; 57: 251-257.

12. Fossati A, Fossati D, Kleijer G. Triticale breeding in Switzerland. In : Proc III Int Triticale Symposium June 13-17 1994 Lisbon, Portugal.

13. Fadel F, Wenzel G. *In vitro* selection for tolerance to *Fusarium* in F_1 microspore populations of wheat. Plant Breeding 1993; 110: 89-95.

14. Maier FJ, Oettler G. Selection for the *Fusarium* toxin deoxynivalenol in callus cultures of triticale. Third European Seminar *Fusarium*-mycotoxins, Taxonomy, Pathogenicity and Host resistance. Radzikow, Poland Sept 22-24 1992.

INDUCTION OF HAPLOIDS IN TRITICALE [X *TRITICOSECALE* WITT.] BY CROSSING IT WITH MAIZE [*ZEA MAYS*]

Maria Rogalska Stanislawa[1], Wojciech Mikulski[2]
[1]Department of Plant Genetics and Breeding, Agricultural University, Poznan, Poland. [2]Plant Breeding Enterprise Szelejewo, Poland

Abstract

Four cultivars (Lasko, Presto, Vero, Prego) and four lines (MAH1590,SZD3801, SZD1740, SZD1834) of winter, hexaploid triticale have been pollinated by pollen of two hybrid forms of corn Hansa 200 and Scandia 180 in aim to obtain hybrid or polihaploid embryos. After pollination triticale plants were injected with water solution of 100 ppm of 2,4-D or with 1000 ppm of IAA. Total 3484 flowers were pollinated and 108 grains were obtained. Immature embryos were incubated on Murashige and Skoog' medium. Finnally 22 green, polihaploid plants were regenerated.

Introduction

The research project on haploid induction in triticale via crosses with maize has been undertaken with an emphasis on the improvement of triticale uniformity in breeding practice. In the literature, there is much information on wheat crosses with maize to induce wheat haploids [1,2,3,4,5,6].

Material and Methods

The method of induction of haploids is the same as that for crossing wheat with *Hordeum bulbosum*. This inspired us to apply the same method to induce haploids of triticale. The experiment was conducted on winter, hexaploid forms of triticale during 1993/1994. Four cultivars [Lasko, Presto, Prego and Vero] and four other lines [SZD1740, SZD1834, SZD3801 and MAH1590] were used as female parents. The pollen source was provided by two hybrid maize cultivars [Scandia 180 and Hansa 200]. Triticale spikes were

H. Guedes-Pinto et al. (eds.), Triticale: Today and Tomorrow, 379–382.
© 1996 *Kluwer Academic Publishers. Printed in the Netherlands.*

emasculated and isolated with paper bags. Two days later they were pollinated with maize pollen. A 100 ppm 2,4-D (2,4-dichlorophenoxyacetic acid) solution was applied twice: 1 hour and on the next day after pollination. The solution was injected into the uppermost internodes and sprayed over the spikelets. About 14 days after pollination, developed grains were collected, sterilized and embryos dissected under the microscope. The immature embryos were cultured following the procedures of Taira and Larter [7] and Gustafson [8]. Regenerated plants were exemined for chromosome number and at the tillering stage divided into single tiller-plants.The single tiller-plants were immersed in a 0,5% aqueous colchicine solution with 2% DMSO [dimethyl sulfoxide] and the wetting agent Tween-20 [15 drops/litre] for 5 hours at 20 C under light. Germination of maize pollen on triticale pistils was observed be means of a fluorescence microscope. During the first hour after pollination, pistils were collected every ten minutes and fixed in Carnoy's [6:3:1] solution and kept in the refrigerator. Then they were washed with water and transferred into 1N NaOH at + 60 C for 2 hours. Next, they were washed and stained with aniline blue fluorescent dye for two hours or overnight. The pistils were then placed on basic glass in a drop of glycerine, covered and observed under a microscope. The meiotic behaviour of PMC's in derived double-haploid plants [D1] was observed.

Results and Discussion

As a result of pollination of 3484 flowers of triticale with maize pollen, 108 grains were produced and 35 embryos were dissected. Most of the embryos were very small and in a globular stage. Only few embryos were well developed and vigorous. They developed into plants none of which was albino. As Table 1 indicates 23 doubled-haploid plants were obtained from this experiment after propagation and colchicine treatment. Only plants from the cross SZD1834 x maize survived all the treatments and they were named the D1 population. All D1 plants showed vigor, rapid growth and slight differentiation in spike morphology. The most striking feature of some plants of D1 was earliness. Seventeen plants from the D1 population were about 21 days earlier in development and ripening than their parent line SZD1834. As expected not all plants were fertile as they were not succesfuly doubled and were hence sterile or formed one or a few grains in spikes. This was the case for five plants, however the rest of the D1 plants produced 3667 grains. Observation of maize pollen germination on triticale pistils showed only a single maize pollen tube in one preparation made from a pistil collected 30 minutes after pollination and anothern in a pistil collected after 40 minutes. In this tudt we do not know whether the plants regenerated from immature hybrid embryos after maize chromosome elimination, or from apomictic development. The meiotic behaviour in the PMC's of D1 plants was very regular in fertile forms and was slightly disturbed in less fertile plants. In the five sterile plants at metaphase I, 16 to 19 univalents and 1 to 2 bivalents were seen. This is consistent with the undoubled chromosome numbers in these plants.

Table 1. Effectiveness of crossing triticale with maize

Genotype of triticale	Number of				
	flowers pollinated	grain produced	embryos	plants regenerated	doubled haploid plants
Lasko	641	30	4	1	0
Presto	390	1	0	0	0
Prego	251	2	1	0	0
Vero	272	13	4	3	0
SZD1740	428	28	8	2	0
SZD1834	386	15	9	9	23
SZD3801	496	19	9	1	0
MAH1590	650	0	0	0	0
Total	3484	108	35	16	23

Observation of maize pollen germination on triticale pistils showed only a single maize pollen tube in one preparation made from a pistil collected 30 minutes after pollination and anothern in a pistil collected after 40 minutes. In this tudt we do not know whether the plants regenerated from immature hybrid embryos after maize chromsome elimination, or from apomictic development. The meiotic behaviour in the PMC's of D1 plants was very regular in fertile forms and was slightly disturbed in less fertile plants. In the five sterile plants at metaphase I, 16 to 19 univalents and 1 to 2 bivalents were seen. This is consistent with the undoubled chromosome numbers in these plants.

Conclusions

On the basis of the obtained results there are some conclusions as follow:
1. Crosses of triticale with maize can be considered as an alternative method for the production of haploid plants. However, the number of dihaploids obtained was not high although they were strong and healthy.
2. A single maize pollen tube was observed in triticale pistils suggesting that in soem cases intergeneric fertilization could tooke place.
3. Outstanding vigor and earliness of dihaploid plants derived from triticale x maize crosses was observed.

References

1. Laurie, D. A., Bennett, M. D. The effect of the crossability loci Kr1 qnd Kr2 on fertilization frequency in hexaploid wheat x maize crosses. Theor. Appl. Genet. 1987; 73: 403-409.
2. Laurie, D. A., Bennett, M. D. The production of haploid wheat plants from wheat x maize crosses. Theor. Appl. Genet. 1988; 76: 393-397.
3. Inagaki, M., Tahir, M. Efficient production of wheat haploids through intergeneric crosses. Research Highlights, 1991; ISSN 0917-2378.
4. Ushiyama, T., Tobita, H., Shimizu, T., Kuwabara, T. High frequency haploid production by wheat x maize intergeneric cross. Japan. J. Breed. 1990; 40 (Suppl. 2): 310-311.
5. Suenaga, K. An effective method of production of dihaploid wheat [*Triticum aestivum*] plants by wheat x maize [*Zea mays*] crosses. Proc. of ICOBB in Miyazaki. 1991; 195-200.
6. Taira, T., Larter, E.N. Actor influencing development of wheat-rye hybrid embryos *in vitro*. Crop Sci. 1978; 8[2]: 27-32.
7. Gustafson, J. P. Production of Triticale germplasm. Triticale-Proceedings of Int. Symp. El Batan, 1973.

ENDOSPERM PROTEINS OF ANDROGENIC DOUBLE HAPLOID LINES OF 6X-TRITICALE

Juan M. Gonzalez[1], Ana Fuertes[1], Nicolas Jouve[1] & Sylvie Bernard[2]
[1]Department of Cell Biology and Genetics, University of Alcala de Henares,
Madrid, Spain
[2]Station d'Amélioration des Plantes, INRA, Domaine de Crouelle,
Clermont-Ferrand, France

Abstract

The analysis of prolamins by SDS-PAGE in F_5 progenies derived from the hybrid between the 6x-triticale lines 'Bolero' x 'Clercal' is reported. A total of 816 plants produced after three consecutive androgenetic progenies are studied. The electrophorectic profiles of 45.2% androgenetic plants showed missing groups of prolamins, which completely influenced specific regions of the proteinogram. The results are explained under different hypothesis: unreduced microspores, embryo fusion or gametoclonal variation consisting in either structural rearrangements of genes or chromosome regions, or aneuploidy affecting chromosomes carrying genes that encode or regulate the expression of prolamins.

Introduction

Androgenesis may be used to regenerate haploid plants from *in vitro* cultures of pollen or anthers. These plants may undergo spontaneous duplication of their chromosomes giving homozygous plant lines. The same may also be obtained by artificially inducing chromosome duplication.

However, these cultures do not always produce haploid specimens by androgenesis. Plants with different ploidies are regenerated from other cells present in the tissue cultured. Unlike double haploid androgenic plants, which show homozygosity, plants derived from partially or non-reduced microspores or from somatic cells, could show heterozygosity [1-3].

In vitro culture of plant cells and tissues frequently gives rise to variations in

H. Guedes-Pinto et al. (eds.), Triticale: Today and Tomorrow, 383–389.

© 1996 *Kluwer Academic Publishers. Printed in the Netherlands.*

regenerated plants. This variation is termed somatoclonal if plants are regenerated from somatic cells [4] and gametoclonal if they are derived from gametophytes [5]. These variations can be of use when trying to produce genetically diverse plants, but it is a serious problem when trying to maintain homogeneous material.

This investigation examines the gametoclonal variation arising from three consecutive androgenetic cycles of triticale lines from the F_5 of a cross between the triticales 'Bolero' and 'Clercal'. Analyses were made of their alcohol-soluble endosperm storage proteins (prolamins) and the degree of homogeneity of these plants, with respect to these proteins, was then established. The degree of gametoclonal variation was also determined.

Materials and Methods

Experimental plant material consisted of: 1) samples of two, 6x triticale cultivars ('Bolero' and 'Clercal'), 2) a series of F_5 pedigree-selected lines derived from the hybrid 'Bolero' x 'Clercal' (89-1, 90-1, 166-1, 263-1, and 279-1) and, 3) a series of androgenic plants from the F_5 lines, regenerated in three successive androgenetic cycles (Fig. 1). Haploid plants obtained in each androgenic generation were treated with colchicine in order to double their chromosome number. Throughout the three cycles, the plants of one generation were used to produce a new androgenetic generation.

Variability in prolamins was investigated using a minimum of ten seeds from each plant. Analysis was performed upon five plants of the first androgenetic cycle, eighteen of the second and eight of the third. Anther culture and colchicine treatments were carried out [6]. Acidic pH polyacrylamide gel electrophoresis (A-PAGE) was employed to analyse the prolamins [7].

Results and Discussion

F_5 plants from the cross between 'Bolero' and 'Clercal' varieties showed different electrophoretic patterns, with eight bands in the β and γ regions (Fig. 2a). This variation can be explained by the segregation of the different alleles received from the parental lines. These differences are still observed even after five generations of self-crossing.

Differences were also seen in the electrophoretic pattern of prolamins in plants of each androgenic generation; both between seeds of the same plant and between those of different plants (Fig. 2b). The analysis of plants regenerated in the second and third androgenetic cycles revealed variation in eight, and four plants, respectively. Further, in certain plants from these cycles, no prolamins were seen at all in the α region of the electrophorenogram.

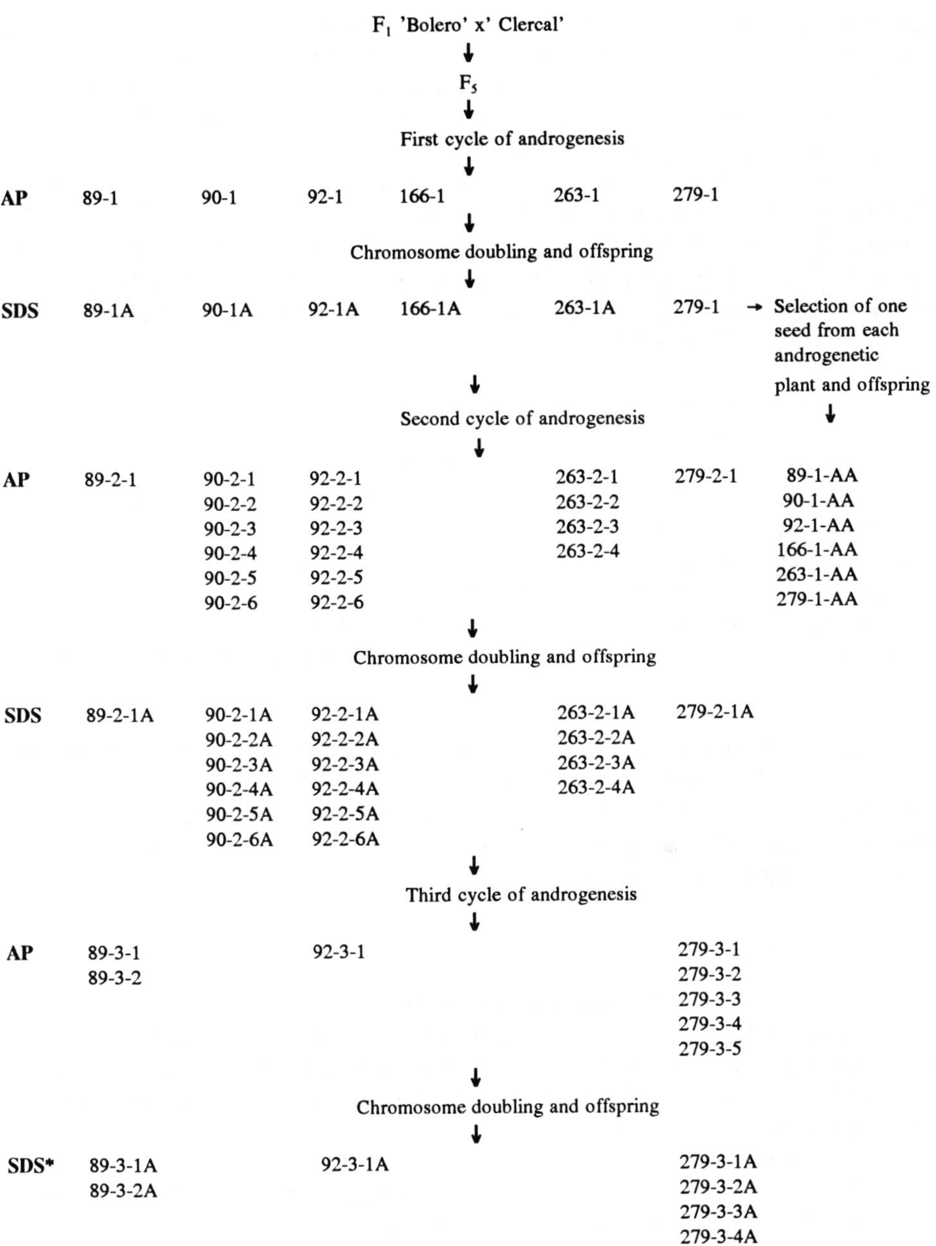

Fig 1.- Diagram showing the obtention of androgenetic material, from the cross between the 6x-triticale lines 'Bolero' and 'Clercal'. AP = androgenetic plants; SDS = Prolamin analysis; SDS* = prolamin analysis of seeds coming from individual spikes of each androgenetic plant.

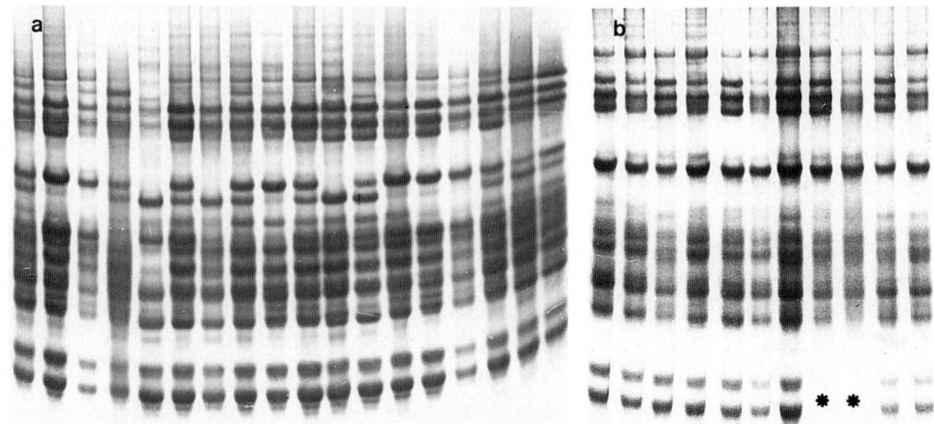

Fig2.- Example of variability for prolamin: a) in the F$_5$ after crossing 'Bolero'x'Clercal'; b) in the androgenetic plant 263-2-3A. * Samples lacking the alpha region.

Finally, a seed was chosen from every androgenic plant of the third cycle and these were allowed to produce descendants via normal, sexual reproduction. Analysis of these plants showed that variability existed in all but one plant. All results of prolamine analyses are shown in Table 1.

Table 1.- Results of the prolamin analysis in the androgenetic plants

	Number of androgenetic plants analysed	Number of seeds analysed	Number of variable androgenetic plants
First cycle of androgenesis	5	65	5
Second cycle of androgenesis	18	342	4
Third cycle of androgenesis	8	219	6
TOTAL	31	626	15

The appearance of segregated, and therefore heterozygous plants, regenerated from *in vitro* cultures of anthers, has also been reported in other species. Differences have been observed in molecular, biochemical, morphological and agronomical characteristics [2,3,8-11]. Possible explanations for these variations include: i) that plants were regenerated from partially or non-reduced microspores which would allow them to retain a degree of heterozygosity; ii) that fusion occurred between haploid embryos from different pollen grains early in development; and iii) through gametoclonal variation.

The first possible explanation can be drawn from the analysis of prolamins of the sexually produced descendants of seeds from plants of the third androgenetic cycle. Variation was seen between seeds of the same plant, in every plant but one. Similarly, four of the eighteen androgenic plants of the second cycle showed variation between seeds of the same plant. Further, different plants of the second cycle, all descendants of the same mother plant of the first cycle, showed different electrophoretic patterns. These results could indicate that some androgenic plants are heterozygous for prolamins and that they could have been regenerated from somatic tissues or, more probably, from partially or non-reduced microspores.

However, variation in prolamins in duplicated haploid plants could have arisen from early fusion between pro-embryos with different genetic constitution. Given that in the development of cereal crops, each ear proceeds from independent cellular lines originated very early in embryonic development, comparisons of seeds of the same and different ears can provide evidence for the above. In six of the eight androgenic plants of the third cycle, it was seen that seeds collected from different ears showed different electrophoretic patterns for prolamins, though no differences were seen between seeds of the same ear. This should be taken into account when seeds are collected from androgenic plants. A selection of seeds from different ears could give a mixture of genotypes.

Finally, gametoclonal variation (spontaneous genetic changes produced during cultivation of haploid plant cells [5]), is another possible source of variation. These changes usually affect the number and structure of chromosomes [12-14]. These changes also affect genes coding for molecular, biochemical and agronomical characteristics, both of the nuclear and cytoplasmic genomes [1-3,11,15-17].

The results also show that the α region of the electrophorenogram of some androgenic plants was completely absent. This was not observed in plants of the previous generation. This variation may therefore be gametoclonal, with the gene or chromosomal region coding for these prolamins lost. Given that α region prolamins are coded for by genes on chromosomes 6A and also 6B [18], the disappearance of the products they encode could be explained by alterations in the regulation of their expression. The most plausible explanations include aneuploidy, deletions, translocations, mobile elements or gene inactivation during *in vitro* culture.

The effects of genetic variation generated by *in vitro* culture is better tolerated by polyploid species such as wheat, potato [3,17,19], and triticale, since mutations can be compensated by duplicate genes on other chromosomes. Toleration is observed even in diploid species such as maize, rice and barley [2,20,21].

Acknowledgements.- The authors thank the financial assistance received from the Comision de Investigación Científica y Técnica (CICYT) of Spain, in the form of a research grant Nº AGR91-0191 (1992-1994), and a Spanish-French Concerted Action Nº HF-213-B (1993).

References

1. Zamir D, Tanksley SD, Jones RA Genetic analysis of the origin of plants regenerated from anther tissues of *Licopersicum esculentum* Mill. Plant Science Letters 1981;21: 223-27.

2. Guiderdoni E, Glaszmann JC, Coutois B. Segregation of 12 isoenzyme genes among doubled haploid lines derived from a japonica x indica cross of rice (*Oryza sativa* L.). Euphytica 1989;42: 45-53.

3. Rivard SR, Capadoccia M, Vincent G, Brisson N, Landry BS. Restriction fragment length polymorphism (RFLP) analyses of plants produced by *in vitro* anther culture of *Solanum chacoense* Bitt. Theor Appl Genet 1989;79:49-56.

4. Larkin PlJ, Scowcroft WR. Somaclonal variation. A novel source of variability from cells cultures from plant improvement. Theor Appl Genet 1981; 60:197-14.

5. Evans DA, Sharp WR, Medina-Filho HP. Somaclonal and gametoclonal variation. Ann. J Bot 1984;71:759-64.

6. Bernard S. *In vitro* androgenesis in hexaploid triticale: determination of physiological conditions increasing embryoid and green plant production. Z. Pflanzenzüchtg 1980;85: 208-21.

7. Lafiandra D, Kasarda DD. One and two dimensional (two-pH) polyacrylamide gel electrophoresis in a single gel: separation of wheat proteins. Cereal Chem. 1985; 62:314-19.

8. Wenzel G, Hoffmann F, Thomas E. Heterozygous microspore-derived plants in rye. Theor Appl Genet 1976; 48:205-08.

9. Orton TJ, Browers TA. Segregation of genetic markers among plants regenerated

from cultured anthers of broccoli (*Brassica oleracea* var. italica), Theor Appl Genet 1985;69: 637-43.

10. Colby LW, Peirce LC. Using and isozyme marker to identity doubled haploids from anther culture of *Asparagus*. Hort Science 1988;23: 761-63.

11. González JM, López LA, Bernard S, Jouve N. Prolamin analysis of progenies from androgenetic plants of triticale. Plant Breeding 1993;111:42-48.

12. Kathley DE, Scholl RL. Chromosomal heterogenety of *Arabidopsis thaliana* anther callus, regenerated shoots and plants. Z Planzenphysiol Bot 1983;112:247-55.

13. Charmet G. Bernard S, Bernard M. Origin of aneuploid plants obtained by anther culture in triticale. Can. J Genet. Cytol 1986; 28: 444-52.

14. Youssef SS, Morris R, Baenziger PS, Papa CM. Cytogenetic studies of progenies from corsses between "Centurk" wheat and its doubled haploids derived from anther culture. Genome 1989;32:622-28.

15. Powell W, Caligari PDS, Dunwell JM. Field performance of lines derived from haploid and diploid tissues of *Hordeum vulgare*. Theor Appl Genet 1986;72:458-65.

16. Baenziger PS, Wesenberg EM, Snail VM, AlexanderWL, Shaeffer GM. Agronomic performance of wheat doubled-haploid lines derived from cultivars by anther culture. Plant Breeding 1989;103:101-09.

17. Singsit C, Ozias-Akins P. Genetic variation in monoploids of diploid potatoes and detection of clone-specific random amplified polymorphic DNA markers. Plant Cell Reports 1993;12:144-48.

18. Shepherd KW. Chromosomal control of endosperm proteins in wheat and rye. Proc. Third, Inter. Wheat Genet. Symp; 1988 Camberra: 86-96.

19. Karp A, Maddock SE. Chromosome variation in wheat plants regenerated from cultured inmature embryos. Theor Appl Genet 1984;67:249-55.

20. McCoy TJ, Phillips RL. Chromosome stability in maize (*Zea mays*) tissue cultures and sectoring in some regenerated plants. Can J Genet Cytol 1982;24:559-65.

21. Finnie SJ, Foster BP, Chalmers KJ. Genetic stability of microspore-derived doubled haploids of barley: a cytological, biochemical and molecular study. Genome 1991;34: 923-28.

POSITIVE EFFECT OF TIMOPHEEVI CYTOPLASM ON ANTHER CULTURE RESPONSE OF TRITICALE

Ahmad Arzani & Norman L. Darvey
The University of Sydney, Plant Breeding Institute, Cobbitty, NSW, Australia

Abstract

The influence of *Triticum timopheevi* cytoplasm on the anther culture response in triticale was investigated using four reciprocal crosses. Significant differences were observed between and among reciprocal crosses for induction and regeneration, with the positive effect of timopheevi cytoplasm.

Introduction

Anther culture represents a potential tool of haploidization for both breeding and genetics research. However, low efficiencies of embryoid induction and green-plant regeneration in triticale have limited the application of this technology in breeding programs.
Cytoplasmic inheritance of androgenetic response indicates that the direction of the cross is also important in determining microspore development. A forage triticale population M86-6068 /TW179// EP80 (TCF$_1$) possessing the *Triticum timopheevi* cytoplasm was found to be the most amenable population to anther culture [1]. The present study was aimed at assessing the potential effect of the *T. timopheevi* cytoplasm on androgenetic response in triticale.

Materials and Methods

Two superior dual-purpose triticale plant selections from populations M86-6068/TW179//EP80 (TCF$_1$) and M86-6068/TW179 (F$_2$) in a *T. timopheevi* cytoplasm (from M86-6068) were crossed reciprocally to the following materials in normal wheat cytoplasm:
1) Sister accession of female parent (M86-6045).
2) 80465/II83-194 (both BC$_1$F$_1$ and F$_2$ plant selections)

H. Guedes-Pinto et al. (eds.), Triticale: Today and Tomorrow, 391–393.

© 1996 *Kluwer Academic Publishers. Printed in the Netherlands.*

3) The Polish topcross parent EP80.
The parents and crosses were grown in hydroponic units in a glasshouse in May, 1993 at Cobbitty. Semi-solid modified C17 medium [2] was used for induction.

Results

Differences among reciprocal crosses were significant for the proportion of responding anthers, embryoids per 100 anthers plated and green regenerants per 100 anthers plated but were not significant for albino regenerants (Table 1). Induction response was highest in those crosses in which M86-6068 was used as female parent.

Among reciprocal crosses, differences for green plants per 100 anthers plated were even more discrete than those obtained for induction. Furthermore, in the crosses involving M86-6068 as the female parent, green plant regeneration was markedly higher than where M86-6068 was used as the male parent. The differences between the reciprocal crosses involving sister lines (with and without the timopheevi cytoplasm) were also highly significant.

Table1. Means for anther culture response of reciprocal crosses of triticale with timopheevi (M86-6068) and normal wheat cytoplasm.

Crosses	(Cytoplasm)	Percentage of anther responding	Embryoids/ 100 anthers plated	Green /100 anthers plated	Albino /100 anthers plated
M86-6068-female/80465	T[a]	73.8 a*	79.6 ab	1.99 ab	4.30 a
M86-6068- female/		75.6 a	78.9 ab	1.66 bcd	4.83 a
M86-8045	T				
M86-6068-female/EP80	T	56.8 b	64.7 c	1.70 bcd	4.30 a
M86-6068	T	76.1 a	84.3 a	3.04 a	4.17 a
80465/M86-6068-male	N	46.5 de	50.8 d	0.69 e	4.08 a
M86-6045/M86-6068-		45.6 de	49.3 d	0.67 e	4.19 a
male	N				
EP80/M86-6068-male	N	40.2 e	40.1 de	0.70 e	4.23 a

* Numbers within columns not followed by the same letter differ at P< 0.01 with the LSD test.
(a) T, timopheevi and N normal wheat cytoplasm.

Discussion

Reciprocal differences between cytoplasms for anther culturability clearly demonstrate the beneficial effect of timopheevi cytoplasm in the M86-6068 population. This finding is in agreement with that of Charmet and Bernard (1984), who observed the positive effect of the timopheevi cytoplasm on anther culture of triticale. Desired combinations of genetic and cytoplasmic factors that govern androgenetic response in triticales could be used to markedly improve the overall response.

References

1. Arzani A. Breeding methodology in forage triticale [PhD dissertation]. Univ. of Sydney 1994.
2. Luckett DJ, Venkatanagappa S, Darvey NL, Smithard, RA. Anther culture of Australian wheat germplasm using modified C17 medium and membrane rafts. Aust J Plant Physiol 1991;18:357-367.
3. Charmet G, Bernard S. Diallel analysis of androgenetic plant production in hexaploid triticale (X *Triticosecale*, Wittmack). Theor Appl Genet 1984; 69:55-61.

GENETIC CONTROL OF ANTHER CULTURE INDUCTION IN TRITICALE

Larisa Nikolaevna Kaminskaya,
Lyubov Vladimirovna Khotyljova,
Svetlana Nikolaevna Matveenko

Institute of Genetics and Cytology,
Academy of Sciences, Minsk, Belarus

Abstract

For determining gene interaction in controlling induction of triticale anthers, Hayman's method of diallel analysis was applied by using dihaploid lines of spring hexaploid triticales and their hybrids obtained by us. Overdominance was shown to play a main role in controlling induction of triticale anthers, the character increase being determined by dominant genes. The triticale lines with the maximum number of dominant alleles controlling the character were selected. They can be used for raising the effeciency of callus and embryoidformation of triticale anthers. Heterosis effect in quantitative characters was revealed in most of hybrid combinations and a certain interrelation between the heterosis level in individual productivity components of forms and their new formation yield was observed. As a result of the research a new material: dihaploid triticale lines, as well as hybrids from crossing these lines was developed.

Introduction

A successful solution of the problem concerning mass production of triticale haploids required for application in genetic programs and breeding is inhibited by a number of reasons. Some of them lie in insufficient improvement of applied biotechnological methods, the others in poor study of genetic origin of callus and embryoid genesis and regeneration. For overcoming obstacles in more rapid development of line material by an *in vitro* anther culture, it is necessary to clarify genetic control of traits in new formation induction.

The available literature data in the control of embryoid formation show both an additive effect of dominant genes, and a nonadditive one in wheat and triticale [1-3]. The existence of a relation between embryoid formation and heterosis effect is reported. A more significant influence of the genotype and environmental factors in *in vitro* culture at the stage of callus and embryoidformation in comparison with the regeneration phase is supposed. The problems concerning heritability of trait variation

H. Guedes-Pinto et al. (eds.), Triticale: Today and Tomorrow, 395–399.
© 1996 *Kluwer Academic Publishers. Printed in the Netherlands.*

components of anther culture and the possibility of developing hybrids with a high efficiency of callusformation and regeneration on this basis are not quite clear [4].

The goal of our investigation was to study the structure of genetic trait variation of callus and embryoidformation and to reveal genes or genetic systems determining this trait as well as to clarify the role of the heterosis effect in productivity of donor plants for quantitative estimation of *in vitro* induction.

Material and methods

Homozygous lines of hexaploid spring triticales GL17, T441, Rosner, 6TA471 and Armadillo, produced by an *in vitro* anther culture, and hybrids from crossing this lines according to a complete diallel scheme were used as original material for investigations.

Triticale plants were grown in field experiment (three replications, 300 plants per plot). To study quantitative traits 50 plants of each form from each replication were randomly selected and used.

Heterosis effect of the hybrids was determined as ratio of average value of trait to one of best parent (%). For studying the genetic control of induction (sum of calli and embryoids) simultaneous investigations were carried out on anthers of parental lines and hybrids using conventional methods for preparation and sterilization of the material.

Culture was performed on the modified Gamborg medium B5. The total number of the explant anthers of each form was about 400 anthers taken from donor plants grown in field experiment. 10-15 plants in each of three replications were used. The efficiency of anther culture induction *in vitro* was determined as ratio of the number of new formations to the one of explant anthers (%).

The results of the investigations were processed according to the programme of Hayman's diallel analysis [5].

Results and discussion

Heterosis effect in quantitative traits particularly expressed bye estimating of such elements of spike and plant performance as grain number and weight was observed in most of the hybrid combinations as seen from table 1. The most considerable excesses in grain weight of plants in F_1-hybrids in comparison with parental forms were noted in 6TA471 x Rosner, T441 x Armadillo, T441 x Rosner and Armadillo x Rosner (200 - 400%). Such a high heterosis level may be accounted for by low values of grain weight of plants in dihaploid parental lines, which were produced by an *in vitro* anther culture and were less viable and partially sterile.

In table 2 are presented the data on the efficiency of callus and embryoid formation in all hybrids and initial forms. As seen, hybrids with a high heterosis level of grain weight per plant 6TA471 x Rosner, GL17 x Armadillo and T441 x Rosner are distinquished by a high per cent of new formations in comparison with other hybrids and parents. In a part of hybrids substantial reciprocal differences are expressed in the trait of induction. The significance of the difference between the hybrid forms studied

Table 1. Heterosis effect (%) in quantitative traits in triticale hybrids

Hybrid combination	Main spike		Grain weight per plant
	grain number	grain weight	
Rosner x Armadillo	5.1	30.0	116.7
Armadillo x Rosner	52.0	60.0	200.0
Rosner x 6TA471	62.3	61.5	161.1
6TA471 x Rosner	61.7	61.5	400.0
Rosner x T441	-7.5	0	58.8
T441 x Rosner	70.8	100.0	282.4
Rosner x GL17	-42.3	-50.0	0
GL17 x Rosner	-22.5	-22.2	57.1
Armadillo x 6TA471	60.4	53.8	150.0
6TA471 x Armadillo	59.4	46.2	144.4
Armadillo x T441	44.0	38.5	100.0
T441 x Armadillo	122.9	123.1	347.1
Armadillo x GL17	6.6	-5.6	66.7
GL17 x Armadillo	45.4	50.0	190.5
6TA471 x T441	47.6	61.5	166.7
T441 x 6TA471	57.5	76.9	172.2
6TA471 x GL17	19.6	11.1	109.5
GL17 x 6TA471	8.5	5.6	119.0
T441 x GL17	14.4	0	66.7
GL17 x T441	16.3	16.7	109.5

Table 2. Induction of new formations (%) from F_1 hybrid anthers of dihaploid triticale lines

Maternal form	Paternal form				
	Rosner	Armadillo	6TA471	T441	GL17
Rosner	2.6	2.2	5.6	7.2	5.5
Armadillo	2.4	1.1	3.8	2.6	5.2
6TA471	14.2	3.7	4.8	2.3	3.2
T441	8.3	4.0	1.6	5.5	0.6
GL17	2.9	11.8	2.9	7.0	6.6

was confirmed by the results of ANOVA. Such differences were caused by the fact that the forms included in the system of diallel crossings sharply differed in this trait.

The next stage of the research was study on gene interactions in control of the trait by using the programme the base of which was the method of Hayman's diallel analysis. The data obtained (Table 3) gave an idea of the heredity pattern of the trait "new formation induction" in triticale lines, namely: 1) overdominance is principal in the genetic control of the new formation induction trait in triticale; 2) the presence of overdominance in every locus is proved; 3) dominance in hybrid forms is directed towards parental forms with greater expression of the trait; 4) the ratio of the alleles with positive and negative effects is below a theoretical value 0.25 from which it results that dominant alleles with positive effects prevail; 5) the ratio of dominant and recessive genes in parental lines has shown considerable prevalence of the former;

Table 3. Genetic and environmental components of trait variation in induction of new formations from triticale anthers

$\hat{V}$	7.76
$\hat{V}_p$	4.99
$\hat{V}_m$	0.40
$\hat{W}$	0.68
$\hat{D}$	3.60 ± 1.00
$\hat{F}$	5.57 ± 1.02
$\hat{H}_1$	29.68 ± 5.42
$\hat{H}_2$	26.67 ± 4.92
$\hat{h}^2$	0.48 ± 3.32
$\hat{E}$	1.39 ± 0.82
$\dfrac{\hat{H}_1}{\hat{D}}$	8.24
$\overline{F}-\overline{P}$	0.58
$\dfrac{\sqrt{4\hat{D}\hat{H}_1}\ +\ \hat{F}}{\sqrt{4\hat{D}\hat{H}_1}\ -\ \hat{F}}$	1.47
$k=\dfrac{\hat{h}^2}{\hat{H}_2}$	0.02
$\overline{uv}=\dfrac{\hat{H}_2}{4\hat{H}_1}$	0.22

Abbreviations in formulae were made in according to Hayman [5]

6) the value h^2 was minor that's why it was impossible to reveal the number of genes or gene groups controlling the given trait.

We had determined the ratio of dominant and recessive genes in per cents. The lines included in the analysis were located according to the number of dominant alleles in the following way: GL17, T441, 6TA471, Armadillo, Rosner. Correlation between the trait expression and the number of dominant alleles was revealed, i.e. the line having a higher number of dominant alleles will have a higher mean value of the trait studied. The lines GL17 and T441 had the greatest number of dominant alleles (about 100%), Rosner - approximately equal number of dominant and recessive alleles. Among the forms studied there were not observed completely recessive ones in the analysed trait.

Thus, overdominance effects are of major importance in the genetic control of the trait studied. The increase in the trait was shown to take place due to the action of dominant genes, the lines with the maximal content of such genes being determined. Taking into account that the trait expression of new formation induction from triticale anthers is correlated with the heterosis level in the initial donor forms in the main performance elements, the results obtained can be used for working-out the most rational genetic and breeding programmes in triticale with an application of biotechnological methods.

References

1. Lasar MD, Baenziger PS, Schaeffer GW. Combining abilities and heritability of callus formation and plantlet regeneration in wheat (Triticum aestivum L.) anther culture. Theor and Appl Genet 1984;68:131-34
2. Deaton WR, Metz SG, Armstrong TA, Mascia PN. Genetic analysis of the anther culture response of three spring wheat crosses. Theor and Appl Genet 1987;74:334-38.
3. Charmet G, Bernard S. Diallel analysis of androgenetic plant in hexaploid Triticale (x triticosecale Wittmack). Theor and Appl Genet 1984;69:55-61.
4. Knudsen S, Due IK, Andersen SB. Components of response in barley anther culture. Plant Breed 1989;103:241-46.
5. Hayman BI. The theory and analysis of diallel crosses. Genetics 1954;39:789-809.

CALLUS FORMATION AND THE CULTURABILITY OF INFLORESCENCE PRIMORDIA OF MUTANT TRITICALES

Gallia Butnaru
University of Agricultural Sciences of Banat
County, Timisoara, Romania

Abstract

For a theoretical and practical reason the *in vitro* culture response of different Triticale genotyps "wild" and mutants was followed. The callusogenesis and culturability of spike explants were decisively influenced by the organogenetic stage.

The best growth medium was 66, MS supplemented with 2.4-D (1 mg/ l) and Magnetic Particles (MP θ = 0.15 x 10^{-3} g/cm³) pH 5.8. Callusogenesis ability was positive on mutant genotypes (T. Tim 501-17 - an average 45.2%). The variability was higher at the "wild" regenerants. The FMR_1 showed dwarfness (~9.7 cm; 40.4 %) with small but normal plant characters. There were nuclear and plastone mutations also (tubular leaf, variegation and lethal chlorophyll deficiency).

The plant characters of FMR_2 were more sharply differentiaded. The caulogenesis was strongly connected with the calli size (r = + 0.878).

Introduction

Cultures which undergo callus formation and plant regeneration have been initiated from young tissues, where the cell activity and division was continuing. The ability to establish tissue cultures varies widely with the donor tissue, donor plant genotype and its age (1). Organogenesis and plant regeneration have been initiated from a wide range of tissues (meristems, mesocotyl, hypocotyl, shoot tissues, roots, rachises mature and immature embryos). Currently, immature embryos appear to be the most generally acceptable donor tissue for initiation of cultures capable to regenerate plants (2,3).

This study focused on differences: in callus induction and plant regeneration of young inflorescences and their proper ontogenetic stage; in the ability of triticale magnetic fluids (MFs) mutants; in the level of characters expression of regenerated plants (R_1 and R_2) under field conditions.

Material and Methods

Growth media: There were 4 media used, (64-67) modified according to Murashing--Skoog (4, MS) and Gyulai (5) with 2.7 g/1; 22.8 mM level of ionic strength (NH_4NO_3 956; KH_2PO_4 1160; $MgSO_4x7H_2O$ 370; $CaCl_2x2H_2O$ 96). They were

H. Guedes-Pinto et al. (eds.), Triticale: Today and Tomorrow, 401–407.
© 1996 *Kluwer Academic Publishers. Printed in the Netherlands.*

supplemented with m-inositol 0.1 g/l; thiamine 2.0 mg/l; ac. nicotinic 1.0 mg/l; pyroxine 0.5 mg/l; glycine 2.0 mg/l. All media had a 130 Gs FM in $\theta = 0.15 \times 10^{-3}$ g/ cm^3 concentration. They were divided in some groups depending on 2.4 D 1 presence and pH :1) two of them, 66 and 67 contained 2.4 D mg/ l; 2) depending on pH they were 5.6 (64 and 66 media) and 5.8 (65 and 67 respectively).

The reason of dividing was to see if the Mfs can induce morphogenetic processes and if the FM mutants response could be different from initial triticale line ("wild").

Plant explants: three constant MF mutants of hexaploid triticale (T. Tim 501-5; T. Tim 501-17; T. Tim 501-9) were used to test their suitability *in vitro* culture. They were grown in the greenhouse until the spikes appeared in the flag leaf sheath. The culms were detached below the uppermost node, placed in MS liquid containing - 2, 4, 6 mg/l Ga$_3$, kept at cold temperature, 4°C, for 7 days to synchronize cell division. The inflorescence primordia (1, 2, 4, 8 cm length) were cut into 0.5 cm sections, and inoculated in Erlenmayer glasses containing 50 cm^3 medium. Explants for callus initiation were kept at 25°C, for 3 weeks in the dark. Calli were incubated 16 h long under dim orange light.

Data recovery and analysis for callusogenesis, callus traits and roots, shoots and plant morphological characters were observed both after darkness incubation, and under the light cycle respectively. There was determined the calli growth (cm^2), number of regenerated roots and shoots. For the proper data the Least Sygnificant Difference (LSD) was calculated too.

Results and Discutions

THE CALLUSOGENESIS AND THE CULTURBILITY OF SPIKE EXPLANTS

The explants differentiation, expressed in age, was decisive for the callus initiation processes as well as for plant regeneration (Table 1). Very young spikes (0.5-1.0 cm) or old spikes (5-8 cm) did not produce many calli for calusogenesis. The young explants survived a long time *in vitro*, without apparent loss of vigour. In the old spike explants, the vitrification process started after 25-30 days from planting, and the callus eventually induced on it soon stopped growing and turned necrotic. The middle-aged explants (3-4 cm length) estabilished callus and roots without shoot formation, even when transferred on fresh medium. The best explants of immature inflorescences were 1.2-2.5 cm length; they produced callus and revealed the morphogenetic capacity of callus tissues (6). A lot of aerial roots, on the calli surface, were developed but only a few cells in the callus mass were directly involved in shoot initiation processes. The calli measurements established higher growth dynamics for 2.4-D associated with magnetic particles (MPs). The mean values after 15 days of light were larger on T.Tim 501-9 and T.Tim 501-17 lines (1.4±0.6cm^2 and 1.36±0.4 cm^2) in comparison to T.Tim 501 (0.51±0.1 cm^2). The macroscopic calli aspects, as cell aggregation, pointed out a high diversity in colour and structure: white-silver or white-yellowish, and dark green, respectively, smooth-flask, friable and fibrogenous-firm. On the same surface, two or three different types of cell assembly were observed.

The fibrogenous-firm, white-yellowish calli developed embryo primordia and plantlets. On the remaining types of calli or cell aggregates non-embryogenetic initiations were observed. Plant regeneration occured only on the white calli. On the surface of the other calli, initiated under dim orange light incubation, several deep green spots appeared which developed aerial roots, but not shoots.

Seven to ten days after the inflorescence primordia, inoculation callus initiation was observed on all media with different intensity (Table 1). The media supplemented with 2.4-D (1 mg/l) and MPs (0-0.15x^{-3} g/cm^3) were effective in callus induction and plant regeneration of all triticales (7). In media supplemented with MPs, without auxins, only callus initiation with globular structure and roots, but not shoot regeneration was observed.

MUTANT CALLUSOGENESIS ABILITY

Callusogenesis differences were observed among the triticale genotypes studied (Table 2). Positive differences were found between the mutant genotypes and the wild line (T.Tim 501). T.Tim 501-17 line expressed a high speed of growth in both pH conditions (pH-5.6 and pH-5.8; Table 3) .T.Tim 501-9 grew better at lower pH (5.6). T-Tim 501-5 revealed a low growth rate and morphogenesis.

Table 1. The mutant percent of callusogenesis

Specification	Callusogenesis (%)															
Medium types	64				65				66				67			
Magnetic Particles (MP) content (g/cm^3)	θ - 0.15x10^{-3}								θ - 0.15x10^{-3}							
2.4-D (mg/l)	0								1							
pH	5.6				5.8				5.6				5.8			
Mutant lines	**Spike**								**length**				**(cm)**			
	1	**2**	**4**	**8**	**1**	**2**	**4**	**8**	**1**	**2**	**4**	**8**	**1**	**2**	**4**	**8**
T.Tim 501	0	75	18	0	0	67	63	0	0	98	90	12	0	83	70	6
T.Tim 501-5	0	71	36	10	0	68	51	10	0	100	100	23	0	90	90	14
T.Tim 501-9	0	81	48	10	0	64	58	0	0	100	94	34	0	90	88	19
T.Tim 501-17	0	80	43	19	0	73	61	13	0	100	100	30	0	90	93	20

A growth medium mixture, with 2.4-D plus MPs and pH 5.6, introduced on all explants a rapid and uniform callus growth, with significant differences on mutant genotypes. Under these conditions, the 2.0 cm T.Tim 501-17 spike explants yielded the highest percentage of calli, 99.5% on the average. T.Tim 501-17 showed the highest average culturability 45.2% and 57.5% on the 66 medium. Other mutant genotypes (T.Tim 501-9 and T.Tim 501-5) showed as well highly significant differences, not only between media, but also between genotypes (Table 2). T-Tim 501-17 callus growth rate was weak in the dark (0.20±0.01 cm^2) but proceeded a much high (6 fold increase) rate in the first 5 days of light (Table 3).

Table 2. The mean response of mutant genotypes and nutrition media to callus formation (%)

Media					
Genotype	**64**	**65**	**66**	**67**	**Average**
T.Tim 501	23.3	32.5	50.0	39.8	**36.4**
T.Tim 501-5	29.3	32.5	55.8	48.5	**41.5**
T.Tim 501-9	34.8	30.5	57.0	49.3	**42.9**
T.Tim 501-17	35.5	36.8	57.5	50.8	**45.2**
Average	**30.7**	**33.1**	**55.1**	**47.1**	

LSD 0.05 = 2.28
0.01 = 3.30

Table 3. The different calli growth and organogenesis rate on the 66 medium

No	Mutant lines		501	501-5	501-9	501-17
1	21 days after planting	C	0.16±0.01	**Darkness** 0.18±0.01	0.24±0.02	0.20±0.01
2	5 days	C	**Dim** 0.48±0.04	**orange** 0.64±0.02	**light** 0.77±0.05	1.20±0.7
3	15 days	C R Sh	0.51±0.1 2.1±0.7 -	0.78±0.1 2.5±1.4 -	1.40±0.6 7.3±0.9 1.2±0.9	1.36±0.4 12.4±3.1 2.7±5.5
4	25 days	C R Sh	stationary 4.2±0.6 1.2±2.6	stationary 3.3±0.4 -	stationary 11.9±1.2 1.2±1.0	stationary 21.8±1.7 3.1±3.4
5	35 days	C R Sh	stationary 6.1±0.7 1.4±2.3	stationary 6.7±1.1 -	stationary many 1.2±1.0	stationary many 3.1±3.4

C - calli surface (cm^2)
R - Root number
Sh-Shoot number

It should be emphasized that the embryogenic calli of mutant genotypes have regenerated several dark green (24%), albino (0.3%) and viridis (11%) plantlets, that are characterized by chloroplast mutations (7).

THE REGENERANTS CHARACTERS EXPRESSION

Regenerated plantlets (R), originated from wild and mutant types were transferred to the greenhouse after two or six weeks, depending on growth recovery after the *in vitro* period. Planting in stabilized peat and perlite (1:1) substrate was effective when the plantlets were MS-solution watered. The plantlets were established without problems: the recovery was quicker and better for the mutant genotypes, T.Tim 501-9 and T.Tim 501-17, after 3-4 days and 96.8 to 97.9% respectively.

The essential step in in vitro breeding triticale, as well as in the traditional one, is the screening process of novel potentially agronomically useful phenotypes from a large population of unwanted individuals. To identify the somatic variation of triticale it would be more efficient to screen the novel genotypes before plant regeneration. Regenerant (R) selection from different types of calli was a failure. Between Rs originated from white or green calli there were not morphological differences. Somaclones regenerated from the wild line have had a great coefficient of variability concerning the tiller number (values ranged between 4 to 87), and a middle one of the flag leaf length (values ranged between 5.9 to 48.3 cm). The imposition of MFs has already proved effective in selecting somaclones carrying novel characters (8).

The analyses of R_1 progeny proved that the regenerated plants were tiny; with little or no seed yield. Several R_1 progenies were observed to segregate for heading time, and morphological characters (number of leaves -3 to 12, tallness -5.5 cm to 16 cm). The magnetic fluid regenerants (MFRs) revealed small stature (Table 4; 9.7 cm) dwarfed (40.4%) with small, but with normal number of leaves. The physiological characters as: frost and heat tolerance were better on MFR mutants (T.Tim 501-9) due to the small cells, stuffed with endoplasmatic reticulum and a great amount of ribosomes (9). Among the first triticale regenerated plants there were discrete nuclear gene and plastone mutations: tubular leaf, variegation and lethal chlorophyll deficiency.

Among R_2 some morphological characteres were clearly differentiated: height from 83 cm to 123 cm; head length from 12.5 cm to 19.5 cm; spikelets number from 23 to 41 (Table 5).

Table 4. The R_1 T.Tim 501-17 explants grown on different media

Medium"Triticale" supplemented with: Characters	0	2.4-D 0.5 mg/l	MF θ - 0.075 g/cm^3
Height (cm)	9.3±2.0	24.0±1.0	9.7±0.9
Leaves: No green	2.6±0.3	5.0±1.0	3.2±0.4
No dry	4.0±0.0	13.5±2.5	5.0±0.7
Length (cm)	7.1±1.2	10.2±0.2	7.6±0.5
Width (cm)	0.4±0.07	0.4±0.04	0.4±0.02
Surface (cm^2)	14.05	56.61	18.69
Roots: Number	7.8±0.7	9.7±3.0	22.0±1.0
Length (cm)	4.5±0.3	6.1±0.9	1.2±0.1

Table 5. The R_2 - T.Tim 501-17 morphological character

The descendents	Height (cm)	No green leaves	No dry leaves	Head length	Spikelets number
1	83±1.4	4±0.01	3±1.0	17.6±0.5	35±1.8
2	101±0.6	5±0.01	2±0.9	17.4±0.4	25±0.8
3	103±1.1	4±0.06	1±1.1	16.7±0.4	27±0.7
4	110±1.8	4±0.04	2±1.0	12.5±0.7	23±1.1
5	116±1.9	4±0.07	2±1.0	17.6±0.3	27±0.9
6	122±1.1	4±0.06	1±0.7	19.5±0.4	29±0.9

Conclusions

Based on the experimental data, it is possible to draw the following conclusions:
Spike explants from mutant genotype T.Tim 501-17 had the highest capacity to form calli and regenerate plantlets (100% and 1.2/callus);
The most satisfactory introductive media were 66 and 67 with callus formation rates of 55.1 and 47.1%, respectively;
A period of about 28-45 days was nedeed for the regeneration of plantlets from spike calli;
The interaction between mutagenic treatment with MFs, and *in vitro* culture, containing MPs, may be a suitable method for triticale novel genotypes selection;
Shoot regeneration frequency seems not to be directly related to the callus-introduction frequency (r = +0.450), but very strongly connected with calli size (r = +0.878).

Aknowledgements

This work was supported by Ministry of Science and Technology - Romania. The author would like to thank D.Botau for suggestions and A. Andreica for the English proofreading.

References

1. Demarly Y., Biotechnologies du clonage des genotypes promotion de l'individu d'elite. In Demarly, Amelioration des plantes et biotechnologies. Ed.John Libbey. Paris, 1989: 45.
2. Babeli P., Karp A., Kaltiskes P.J., Plant regeneration and somaclonal variation from cultured immature embryos of sister lines of rye and triticale differing in their content of heterochromatin. Theor. Appl.Genet. 1988; 75:929.

3. Eapen S., Rao PS., Callus induction and plant regeneration from immature embryos of rye and triticale. Plant. Cell. Tissue Organ Culture. 1982;1:221.
4. Murashige T., Skoog F., A revised medium for a rapid growth and bioassasy with tabacco tissue cultures.Physiol. Plant. 15, 1962:, 473.
5. Gyulai G., Janovszky J., Kiss E., Lelik L., Csillag A., Heszky LE.,Callus initiation and plant regeneration from inflorescence primordia of intergeneric hybrid Agropyron repens (L) Beauv x Bromus inermis Leyss .CV. nanus on a modified nutritive medium. Plant Cell Reports 1992; 11 ; 266.
6. Balan M., Cachita-Cosma D., Grusea A., Callus production and plant regeneration from immature inflorescences of some forage grasses. Lucrarile celui de al V- lea Simpozion national de culturi de celule si tesuturi vegetale. 1993; 303.
7. Butnaru G., Corneanu M., Somatic embryogenesis and plant regeneration in tissue culture in medium with magnetic liquids. Sixth Int.Conf. Magnetic Fluids. Paris, 1992; 478.
8. Butnaru G., Rasa F., Badea M., Micropropagation and mutation in Triticale. *In Vitro* Plant Cultures, present and perspective; 1989: 69.
9. Butnaru G., Terteac D., Bujoreanu V., Nuclei ultrastructure under MLs influence. Seventh Int. Congr. Genet. Birmingham, 1993: 140.

ISOZYME AND ENDOSPERM PROTEIN MARKERS IN THE DETERMINATION OF CHROMOSOMAL CONSTITUTION IN X *Triticosecale* WITTMACK

Pilar Rubio[1], Consuelo Soler[2], Angeles Bernardo[1] & Nicolás Jouve[1]

[1]Department of Cell Biology and Genetics, University of Alcala de Henares, Madrid, Spain

[2]Department of Plant Breeding, C.I.T., I.N.I.A., La Canaleja, Alcala de Henares, Madrid, Spain

Abstract

The objective of this work is the evaluation of ability of a series of biochemical markers to verify chromosome composition of chromosome segregant materials or secondary lines obtained by crossing triticale and wheat. From all sytems essayed, isozymes of Glutamate Oxalacetate Transaminase, Malate Dehydrogenase and endosperm prolamins studied by SDS-PAGE, were the most effective. GOT, A-PAGE and 2D-PAGE offer the possibility to detect 6D(6A) and other chromosome substitutions.

Introduction

As crossing becomes more important as a means of obtaining triticale germplasm and of producing secondary lines with new chromosomal constitutions, ways to detect and study segregant chromosomes have become increasingly necessary. Electrophoresis has proved a useful tool in both the identification of individual varieties and the elucidation of chromosomal composition [1-4]. The abundance of biochemical markers that can be used in electrophoretic analysis, provide a convenient tool in the analysis of genetic variation in large and diverse samples.

Compared to the much slower and more difficult procedure of karyotyping, electrophoresis is a rapid alternative for chromosome analysis. Monodimensional electrophoresis offers the possibility of including material from several dozen plants on a single gel, allowing the analysis of several proteins and isozyme markers at the same time. Therefore, chromosomal composition and the differences between segregated materials can be examined simultaneously.

H. Guedes-Pinto et al. (eds.), Triticale: Today and Tomorrow, 409–415.

© 1996 *Kluwer Academic Publishers. Printed in the Netherlands.*

The purpose of this investigation was to determine the value of a series of storage proteins and isozymes as markers in the analysis of chromosomal composition using a range of triticale lines. The results seem to demonstrate that isozymes of Glutamate Oxaloacetate Trasaminase, Malate Dehydrogenase and gliadins are the most useful in understanding ploidy level and certain chromosome substitutions.

Materials and Methods

PLANT MATERIALS

Specimens of a total of forty six cultivated lines of x*Triticosecale* Whittmack, were used as experimental material. Samples included hexaploid, octoploid, primary, and secondary, and complete and substituted triticale lines. All experimental material was taken from the working collection of the authors.

ELECTROPHORESIS

Isozyme markers were studied using crude extracts obtained from young leaflets. Paper wicks were soaked in the extracts and inserted into horizontal starch (MDH, E.C.1.1.1.3.7) or polyacrylamide (GOT, E.C. 2.6.1.1.) gels [5-7].

Storage protein markers were investigated using sodium dodecyl sulfate polyacrylamide gel (10%) electrophoresis (SDS-PAGE) [8]. Two further protein separating techniques were tested when results obtained were inconclusive.: i) electrophoresis of the storage proteins in polyacrylamide gels with new characteristics (10%, with pH=3.1 aluminum lactate buffer, A-PAGE), and ii) two-dimensional (2-pH electrophoresis) [9,10].

Results and Discussion

The value of isozymes and endosperm proteins as markers differs depending on the goal of the investigation. In the analysis of variability within and between species, all variable and non-variable genes are of interest. Therefore, since allelic alternatives are required when measuring how closely different lines are related and when interpreting the residual segregation within each line, markers which vary are essential. However, to determine a definite association between a protein or isozyme band and the chromosome carrying the gene coding for the marker, monomorphic systems or systems that only vary with respect to chromosomal composition are more reliable.

The results of the isozymes and proteins studied, show that ten electrophoretic methods are of use, ACPH, CPX, α-EST, ß-EST, GOT-1, GOT-2, GOT-3, HMW-proteins, MDH-1 and MDH-2. The particular uses of each method are the following: i) for monomorphic systems: ß-EST, GOT-1, MDH-1; ii) for variable ystems, depending on the chromosome composition: GOT-2. GOT-3, MDH2, and iii) for polymorphic systems: ACPH, CPX, α-EST and HMW glutenins.

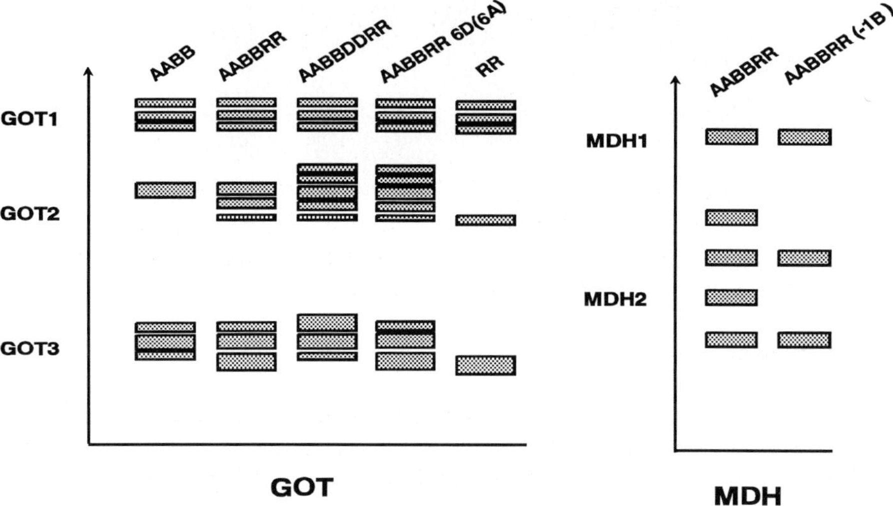

Fig 1.- Electrophoretical patterns of GOT and MDH useful as markers in the isozyme analysis of triticale.

All bands were considered as manifestations of phenotypic features for intravarietal and intervarietal comparisons. Only 8 of the 46 lines analyzed resulted monomorphic for the complete range of biochemical markers characterized. The results on the analysis of the variability for all isozyme systems and storage proteins in the triticale collection has been reported in a previous paper [11].

Glutamate oxalacetate transaminase, also called aspartate aminotransferase (AAT), appeared as three regions in the zymograms of all plant material examined (Fig 1a). These were previously reported as GOT-1, GOT-2 and GOT-3 [12]. The GOT system has been studied in several cereal species and their dimeric structure and monomorphism is well established [6,7,13-16]. The GOT-2 region presents a set of up to five bands. The pattern composed of the bands designated GOT-2.3, GOT-2.4 and GOT-2.5, occurred in 86.3% of the lines analyzed. GOT2.3 and GOT-2.4 showed a similar staining intensity. GOT-2.5 exhibited a less intense band (Fig 2). This result coincides well with the hypothesis of Diaz and Jouve [16], that GOT2 isozymes are dimers with approximately equal quantities of four subunits designated α^2, β^2, δ^2 and ρ^2. These sub-units are encoded respectively by the genes *Got-A2*, *Got-B2*, *Got-D2* and *Got-R2*, located in the long arms of chromosomes 6A, 6B, 6D and

6R [17]. The products of triplicate *Got-2* structural genes of wheat and their homologs in rye, could combine to produce different dimeric molecules.

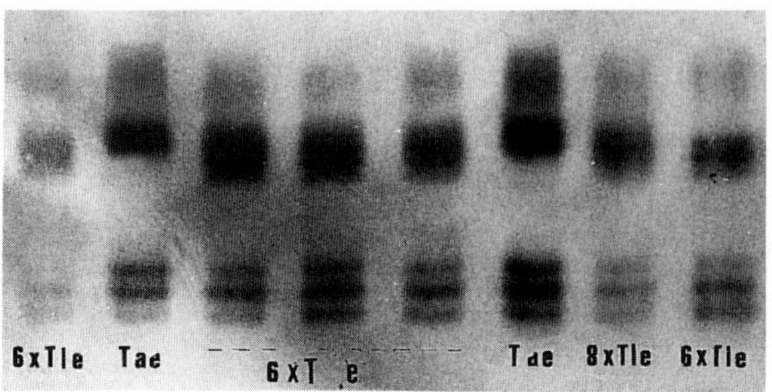

Fig 2.- Phenotypes of GOT in different triticale lines. Tle = Triticale; Tae = *T. aestivum*

The octoploid lines 'Forlani-8403', 'Forlani-8404' and 'Forlani-8408' and the hexaploids 'JK-F5/2' and 'Tegueff', were the only triticale lines that exhibited the bands GOT2.1 and GOT2.2. It can be suggested that bands GOT2.3, GOT2.4 and GOT2.5 are dimeric isozymes formed by the association of subunits α^2, β^2, and ρ^2. The GOT2.1 and GOT2.2 bands are probably the products of the dimeric association of subunits α^2, β^2, δ^2. The five isozyme bands present in the octoploid lines would then be caused by the random association of the four protein subunits into approximately equally active dimers. The 'Tegueff' line was the only hexaploid with a three band pattern (GOT2.1, GOT-2.2 and GOT-2.3).

This pattern led to our study of the line's chromosomal constitution. C-banding showed that this variety has $2n=42$ and the chromosome substitutions 2D/2R and 6D/6A. The presence of chromosomes 6R and 6D resulted in the appearance of bands for GOT2.1 ($\delta 2\delta 2$ homodimer), GOT2.2 ($\delta^2\beta^2$ heterodimer), GOT2.3 ($\beta^2\beta^2$ homodimer and $\delta^2\rho^2$ heterodimer), GOT2.4 ($\beta^2\rho^2$ heterodimer) and GOT2.5 ($\rho^2\rho^2$ homodimer).

Triticale malate-dehydrogenase (MDH) gave a pattern with two regions designated MDH-1 and MDH2 (Fig. 1b). MDH-2 presents at least two patterns which could be associated with different chromosomal compositions. The results demonstrate that the pattern, composed of four bands named Mdh-2a to Mdh-2d, was the most frequent (86.6%) in all the triticale lines. However, lines 'JK 147/1', 'Khobe-95' and 'Manolito-1630' lacked the bands Mdh-2a and Mdh-2c. These bands are associated with the presence of chromosome 1B and are encoded by gene *Mdh2B* [18-20]. The absence of chromosome 1B should be explained by homoeologous chromosome substitution 1R(1B) or 1D(1B).

Finally, two techniques were used to investigate the value of endosperm proteins as markers (Fig. 3). Monodimensional A-PAGE was carried out on all lines and was supplemented, in some lines, with two-dimensional electrophoretic analysis, firstly at an acidic pH and under the same conditions as for monodimensional A-PAGE and, finally at an alkaline pH.

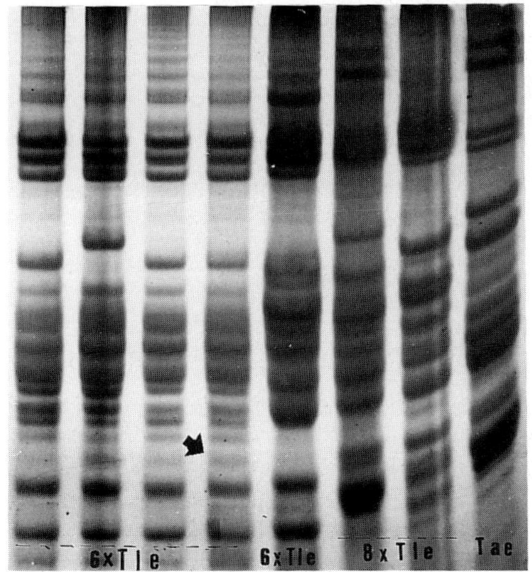
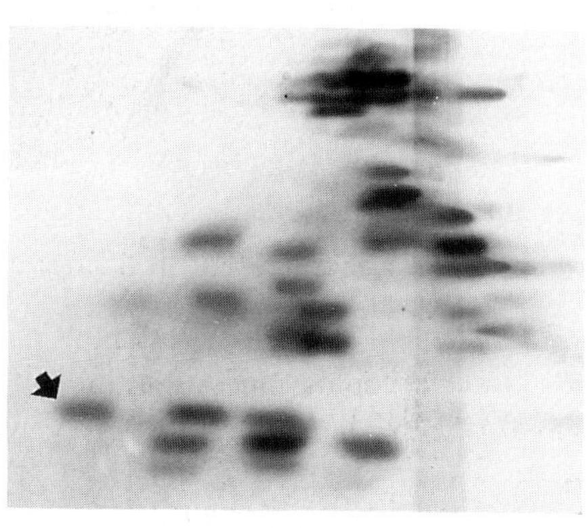

Fig 3.- Examples of the phenotypes obtained by A-PAGE and 2D-PAGE
showing gliadin markers for the presence of chromosome 6D.

These analyses were used to highlight gliadins encoded by genes on chromosomes belonging to homologous group 6. The presence of chromosome 6D was confirmed by the presence of an α-gliadin encoded by the gene *Gli-D2,* found in the short arm of the chromosome [21]. Using this marker as a reference, chromosome 6D was seen to be present in the 'Forlani 8403', 'Forlani 8404' and 'Forlani 8408' octoploid lines and in the 'JK-F5/2' and 'Tegueff' hexaploid triticale lines (confirmed by the appearance of GOT-2.1 and GOT-2.2 bands). Its presence in the octoploids is expected but its appearance in the hexaploids requires explanation.

The 'Forlani-8403', 'Forlani-8404' and 'Forlani-8408' lines, also showed an α-gliadin marker encoded by the gene *Gli-A2* on chromosome 6A, and a γ-gliadin encoded by *Gli-B2* on chromosome 6B. The hexaploid triticale 'Tegueff' showed the 6B marker but not that of 6A whilst the opposite was found for the 'JK-F5/2' variety. The most probable explanation is that both triticales conserve the 6R chromosome (confirmed by C-banding) with a 6D(6A) substitution in 'Tegueff' and a 6D(6B) substitution in 'JK-F5/2'. This is in agreement with the observation that about 20 percent of CIMMYT lines, and some European winter lines of triticale, have the 6D(6A) substitution, thus combining the breadmaking qualities of wheat with the robustness of rye.

Acknowledgements.- This work is included in a research program directed by Dr Consuelo Soler, investigating the multiplicaction, characterization and evaluation of winter cereal crops (Project Number 7689 of the INIA , of Spain).

References

1. Kahler AL, Hallauer AR, Gardner CO. Allozyme polymorphisms within and among open-pollinated and adapted exotic populations of maize. Theor Appl Genet 1986; 72: 592-601.

2. Marchylo BA. Identification of Canadian and American wheat cultivars by SDS gradient PAGE analysis of gliadin and glutenin subunits. Can J Plant Sci 1987; 67: 945-952.

3. McIntyre CL.. Variation at isozyme loci in *Triticeae*. Pl Syst Evol 1988; 160: 123-142.

4. Bernardo A, Jouve N, Soler C. Discriminación de variedades de cebada (*Hordeum vulgare* L.) mediante electroforesis. 2. Análisis de isoenzimas. Investigación agraria. Prod Prot Veget 1990; 5(1): 25-52.

5. Poulik HD. Starch gel eletrophoresis in a discontinuous system of buffers. Nature (London) 1957; 180: 47.

6. Hart GE. Glutamate oxaloacetate transaminase isozymes of *Triticum:* evidence for multiple systems of triplicat estructural genes in hexaploid wheat. In: "Isozymes: III Developmental Biology" (Markert, C. Ed.) New York: Academic Press, 1975: 637-657.

7. Salinas J, Benito C. Chromosomal location of malate dehydrogenase structural genes in rye *(Secale cereale* L.). Z. Pflanzenzchtg 1985; 94: 208-217.

8. Payne PI, Corfield KG, Holt LM, Blackman JA. Correlations between the inheritance of certain high-molecular-weight subunits of glutenin and breadmaking quality in progenie of six crosses of bread wheat. J Sci Food Agric 1981;32: 51-60.

9. Lafiandra D, Kasarda D. One- and two-dimensional (two-pH) polyacrilamide gel

electrophoresis in a single gel separation of wheat ptoteins. Cereal Chem 1985; 62: 314-319.

10. Tercero JA, Bernardo A, Jouve N. Encoding genes for endosperm proteins in *Hordeum chilense*. Theor Appl Genet 1990; 81: 127-132.

11. Rubio P, Jouve N, Soler C, Bernardo A. Genetic variation for isozymes and proteins in lines of the synthetic amphiploid xTriticosecale Wittmack. Genetics & Breeding 1994;(in press).

12. Tang KS, Hart GE. Use of isozymes as chromosome markers in wheatÄrye addition lines and in triticale. Genet Res Camb 1975; 26: 187-201.

13. Scandalios JG Genes, isozymes and evolution. Isozymes IV: Genetics and evolution. San Francisco: Academic press, Inc, 1975.

14. Hart GE, Langston PJ. Chromosome location and evolution of isozyme structural genes in hexaploid wheat. Heredity 1977; 39: 263-277.

15. Hart GE, Tuleen NA. Introduction and characterization of alien genetic material. Isozymes in Plant Genetics and Breeding, Part A. Amsterdam, Elsevier Science Publishers, 1983:: 339-362.

16. Diaz F, Jouve N. Structure of isozymes of the AAt-2 and AAT-3 systems of Aspartate Amino transaminase in Triticineae. Euphytica 91986; 35: 129-135.

17. Hart GE. Biochemical loci of hexaploid wheat (*Triticum aestivum*, 2n=42, genomes AABBDD). Genetics Maps 1984Vol 3: 485-490.

18. Bergman CW, Salinas J. Isozyme variants of esterase and malate dehydrogenase among wheat aneuploids. Agron Abstr 1972; 23.

19. Benito C, Salinas J. The chromosomal location of malate dehydrogenase isozyme in hexaploid wheat (*Triticum aestivum* L.). Theor Appl Genet 1983; 64: 255-258.

20. Diaz F, Fernandez JA, Jouve N. Structure and chromosomal location of malate dehydrogenase (Zone 2) isozymes in common and durum wheats. Euphytica 1986; 35: 509-513.

21. Sozinov A. 1984. Blocks of cereal storage proteins as genetic markers. Proc 2nd Int Workshop on Gluten Proteins, Wageningen: 121-127.

V - ABIOTIC STRESSES

STRESS TOLERANCE IN NEWER TRITICALES COMPARED TO OTHER CEREALS

Robin S. Jessop
Department of Agronomy and Soil Science, University of New England, Armidale NSW 2351, Australia

Abstract

The world status of triticale's level of adaptation to adverse environments is reviewed with particular emphasis on tolerance to acid soils (high in Al), low nutrient availability, drought stress and temperature stress. Tolerance to acid soils has been increased through screening for Al tolerance in nutrient culture systems; we still need better information both on the genetics of Al tolerance and its stability under field conditions. The ability of triticale to tolerate lower levels of phosphorus, nitrogen and the trace elements copper, manganese and zinc than wheat should be confirmed in a range of environments and utilised more widely. Drought tolerance (compared to bread wheats) appears to be improving, particularly with newer complete triticales; this again needs confirmation in differing environments with some more detailed measurements of the plant mechanisms involved. Both winter and spring triticales now have good tolerance of cold conditions compared to other cereals but further work is still needed to improve high temperature tolerance at flowering.

Introduction

The last ten years of triticale development throughout the world have seen the development of winter, spring and facultative cultivars with an ability to grow and yield well in a range of difficult soil and aerial environments. Triticale is now an accepted crop in regions where other cereals either have low yields or are economically less acceptable.

The constraints which the crop has been able to accept include low soil pH (and high aluminium levels), nutrient toxicities, low nutrient status (both macro and micro nutrients), drought/heat stress tolerance and generally low pesticide usage.

The current status of triticale in regard to these attributes will be examined with some comments where relevant concerning the need for continuing work.

419

H. Guedes-Pinto et al. (eds.), Triticale: Today and Tomorrow, 419–427.
© 1996 *Kluwer Academic Publishers. Printed in the Netherlands.*

LOW PH/ALUMINIUM STRESS

Early studies indicated the ability of hexaploid triticales to produce high grain yields on acid soils high in available aluminium [1]. One of the major aims of breeding work has been to produce triticales which are as well adapted to acid soils as rye [1] (W. Sowa, personal communication). Aluminium tolerance screening has been performed in many breeding groups by using modifications of the nutrient culture screening systems developed for wheat at Oregon Sate University; there appears to be good correlations with root growth after Al stress in nutrient culture and performance under acid soil field conditions [2]. The results from nutrient culture studies appear relative, however, since differing root growth may occur depending on root temperature (W. Sowa, personal communication) and a range of other parameters including pH and nutrient content of the solutions [3].

Cultivar differences in Al tolerance and therefore presumably acid soil tolerance have been widely detected [4,5] but in some work, such as recent studies reported by Baier (personal communication), relatively small differences between cultivars were recorded. It has been suggested that such small differences between cultivars may be related to a small sample of germplasm or to strong selection of most material under acid soil situations.

Since it has been shown that Al tolerance appears to have a high narrow sense heritability value, selection within early generation segregating populations should yield improved Al tolerant material. Results have also indicated that selection within apparently fixed line material can also result in improved Al tolerance (W. Sowa, personal communication); the longer term response stability of such material warrants further examination.

A new program in Germany (G. Oettler, see data at conference) is attempting to design an *in vitro* test using embros and calli on varying media which will predict Al plant tolerance; this test will be assessed under field conditions in Brazil (A. Baier).

The genetics of Al tolerance has undergone a considerable amount of study; Aniol and Gustafson [6] found that Al tolerance in triticale was associated with chromosomes 3R, 4R and 5R. It has also been suggested that the tolerance gene is located on chromosome 4D [7] but later evidence [8] suggests additive genetic control. In a review article, Manyowa and Miller [9] refer to a number of papers also indicating additive gene control; additionally, expression of Al tolerance appears related to the level of tolerance of both parents.

Recent data from Spain [10] and from CIMMYT (W. Pfeiffer, personal communication) indicate that the complete triticales are better adapted to acid soils. Since the species' ability to yield well under acid soils high in Al is one of its main *raison d'etre,* we still appear to need better information on:
1) the genetics of Al tolerance;
2) the relative yield stability of lines which are apparently higher in their tolerance to Al;
3) the influence of high levels of Al on quality factors in triticale grain (protein levels and baking quality);

4)   the range of Al tolerances in forage/dual cultivars;
5)   the growth reduction mechanisms within the plant of Al (and other toxicity/deficiency) problems.

Triticale is also often grown on high pH soils, particularly in some parts of north Africa, Spain [10] and in parts of Australia. In these situations, it appears that the substituted triticales may give better yields than the complete cultivars [10]; this finding agrees with earlier data from [11]. Such findings should be verified in other major alkaline high yielding soil environments. Assuming such findings are generally correct, this may require a change in the breeding and selection methods for countries such as Australia.

OTHER ELEMENT TOXICITIES

Under wet conditions in acidic soil, increased quantities of manganese (Mn) and iron (Fe) become soluble and may damage root growth in susceptible crops. Work with Mn tolerance in wheat has indicated that selection for tolerance in early generations after crossing should enable improvements to be made [12]. Similarly, Camargo *et al.* [2] have found similar tolerance to iron in nutrient solution compared to a tolerant wheat variety.

It would be interesting to know if varieties with improved tolerance to toxic levels of Mn, Fe and Bo have been found or released.

Triticale may have some advantages compared to other cereals under waterlogged conditions. Data in solution culture [13] indicated less reliance on aerated conditions (ie. greater tolerance of low O_2 levels) for Muir triticale than Gamenya wheat: after cycles of waterlogging/drainage, the seminal roots of the triticale were double the length of those in the wheat. Similar studies by [4] indicated a range of tolerances within selected wheat varieties.

Under field conditions in the cooler higher regions of Australia, Brent Scott (unpublished data) showed relative waterlogging tolerance was highest in rye and triticale and lowest with barley and wheat.

Results from hydroponic studies [14] also indicates that at high salt concentrations some triticale lines were more tolerant than rye; some lines had similar tolerance to barley.

LOW NUTRIENT AVAILABILITY

Triticale appears valuable in situations of low soil levels of Mn (and other trace elements including copper) and phosphorus (P). Highly calcareous soils low in available Mn occur in many Mediterranean type climates, including South Australia. Cooper *et al.* [15] showed that a range of complete triticales had intermediate tolerance to low Mn compared to rye (very good) and bread wheats (poor). Within the triticales there were also marked differences in tolerance.

Cooper's experiment was duplicated on a low zinc (Zn) soil to examine differences in Zn requirement. Again, differences in Zn tolerance between triticales were produced; commonly, those lines tolerant of low Mn were not also low Zn tolerant.

Copper (Cu) deficiency is common on light sandy soils in South and Western Australia. Early work [16] indicated that older triticales were intermediate between bread wheats and rye in their ability to produce grain yield on low Cu soils. Later results [17] showed that triticales which carry the 5R chromosome (and which often have a hairy peduncle) are tolerant of copper deficiency. The copper efficiency factor appeared to be carried on the 5R long arm close to the Hp locus. In his 1987 paper, Graham [17] comments that whilst triticale appears to be more tolerant of Zn and Mn deficiency than wheat, there is a wide range of tolerance in triticales to these two deficiencies.

In glasshouse experiments [18] triticale took up more Fe, Zn and Mn than wheat at all stages of growth; Cu uptake was greater during early growth.

Triticale's tolerance and its ability to yield well with low levels of soil phosphorus was widely discussed at Passo Fundo in 1990. This aspect of the crop is under study at a number of world locations; considerable work in this area has been completed with bread wheats [19]. Some published information on the comparative efficiency of P usage (with other cereals or between varieties) would be most useful.

Triticale's ability to tolerate reduced levels of many essential nutrients, although established in some locations, needs to be more widely confirmed and utilised to provide farmers and end-users with an economic alternative to other cereals.

DROUGHT STRESS

Carmargo *et al.* [3] suggested that triticale has great potential for high yields under both drought and high temperature stress. Results from Tunisia [20] and Spain support this view. Tunisian farmers have praised triticale's resistance to lodging combined with a modest increase in grain yield over wheat (5-6%) in average rainfall years . In drier years it has outyielded durum wheat by an average of 39% and bread wheat by 10%. Other authors [21] have shown that Antares triticale was able to maintain grain yield even under severe water stress; these authors suggested that its better drought tolerance was associated with earlier heading and greater water extraction by root uptake.

Comparisons of bread wheats and incomplete triticales in Australia [22] indicated that water use efficiencies for dry matter and grain yield were not greatly different between the species; in fact, the cultivar differences within species often overshadowed any marked species differences. Work in Western Australia [23] suggested that in moisture stressed situations on heavy clay soils in low rainfall regions (< 325 mm annual rainfall), wheat was higher yielding than triticale (Table 1). Thus, whilst the advantage of triticale compared with wheat on low rainfall acidic soils is clearly demonstrated in this Table, in other situations the other crops were higher yielding. Indian data [24,25] also indicate poorer triticale yields (compared to wheat) under drought conditions.

More recent work [26] has highlighted the advantages of the newer complete triticales; using a range of analyses of International Triticale Yield Nurseries data, Pfeiffer suggests that *'under drought conditions and in tropical highland and problem soil regions, complete triticales show distinct yield superiority and appear to have adaptive*

Table 1. Relative yields (kg/ha) for cereals in different cropping situations in Western Australia. (Yields expressed as a percentage of wheat yields in brackets).

Soil Type	Wheat	Al Sensitive Wheat	Barley	Triticale	Oats
Reliable wheatbelt (no Al)	2012 (100)	2245 (112)	2174 (108)	1974 (98)	1812 (90)
Low R/F[a] light soil (low Al)	2120 (100)	1742 (82)	1596 (75)	1805 (85)	2150 (101)
Low R/F[a] heavy soil (no Al)	1171 (100)	1193 (102)	1230 (105)	835 (71)	1307 (144)
Low R/F[a] acid soil (Al toxic)	790 (100)	466 (59)	348 (44)	1059 (134)	

Based on average of three years, oats on one
(Data from McLean and Barclay, 1990)
[a] R/F = Rainfall

advantages over wheats'. Such advantages are reduced in highly productive environments. With the advantages of the complete types in mind, the CIMMYT program has changed from 25:75 to 90:10 complete to substituted types between 1984 and 1991.

Whilst references to triticales' yield advantages compared to wheat under dry conditions have been found (even under British conditions!, Green, personal communication), detailed data on comparative water use/root pattern studies are hard to find. Fettell [27] also found higher comparative yields of triticales than wheat under dry conditions; this was associated with a greater percentage contribution of pre-anthesis assimilate to grain filling in the complete triticale Currency compared to bread wheats.

Using matched sets of similar development pattern wheats and triticales, more detailed information on comparative water use efficiencies for grain production is needed to evaluate both yield advantages and the physiological reasons (both shoot and root mechanisms) for such yield advantages.

The clear soil type interactions in Table 1 indicate that there may be other soil factors which influence relative triticale yields. Of interest here is the comment [10] that under Spanish conditions the complete and substituted triticales could be considered as two different crops.

Information supplied by Stan Nalepa (USA) has indicated that improved drought tolerance in their breeding program has been related to extensive/deep root systems in triticale/rye crosses; rye has generally been found to have the best drought tolerance.

Nalepa also suggests that drought during the shooting to heading stage of the crop can be critical for triticale. The differences between published reports in the relative tolerance of triticale to dry conditions may be related to varietal effects (phasic development patterns and type of triticale), timing and degree of drought conditions.

TEMPERATURE STRESS

Winter hardiness improvement has been a major achievement of the Danko breeding program [28]; this has also been an important aim of other winter triticale breeding programs (Nalepa, personal communication). For forage lines, the ability to recover after grazing under frosty conditions is one important attribute of triticale compared to oats [29]. Similarly, triticale has given high yields in cool environments in Brazil [30]. Under Italian conditions, triticale has yielded well in hilly areas and on both good and poorer soils [31].

Triticale also appears to have the capacity to produce high grain and forage yields in years of abnormal seasonal conditions. In Argentina, Acevedo and Lopez (personal communication) have released triticales tolerant to drought and cold stress; in 1992/93, unusually low radiation levels combined with waterlogging to produce an unusual season. A number of their lines were able to achieve very high forage and grain yields compared to other cereals in this adverse year.

Heat stress has been a serious concern with earlier triticale lines. Australian work in China (unpublished data, Australian Centre for International Agricultural Research), in which a range of winter and spring triticales were screened, suffered from severe head sterility; such findings are not unusual (Nalepa, personal communication). Nalepa has suggested that selection for improved pollen release and high viability under hot conditions is important in heat stressed environments.

OTHER TRITICALE ATTRIBUTES

Triticale, although only a small area crop in the U.K., has been promoted as a low cost cereal which is winter hardy, has a reduced N requirement and is the best cereal for whole crop silage (Semundo 1992). Data from CIMMYT [32] show that complete triticales were able to take up more nitrogen (N) than substitute types or bread wheats; this suggests that the complete triticales would be well suited to either low rates of N fertiliser application or low soil N fertility. Work in Kasmir [33] supports these findings and shows greater N yield response than in wheat. Data from central Germany [34] also indicate that compared to wheat, triticale has earlier and greater uptake of N combined with better utilisation of spring applied N.

Triticale is also promoted as being a crop which needs less pesticides. Semundo's Chris Green suggests that triticale's value lies in its ability to produce grain at a lower unit cost than other cereals. Survey data supplied from Semundo indicate that in 1990 triticale had only 3 pesticide applications compared to 8 for wheat.

Recent studies at Wagga Wagga Research Institute (Australia) by Deidre Lemerle and Kath Cooper (unpublished data) have compared the ability of 400 lines of different winter cereals to suppress the growth and seeding capacity of annual ryegrass (*Lolium rigidum*, a major cereal crop weed); the triticales within this study were in the top 10% of all the cereals tested for both weed suppression and ability to maintain grain yield in the presence of ryegrass. This work supports triticale's ability to be used with less herbicides than other winter cereal crops.

SOME OUTSTANDING PROBLEMS WITH TRITICALE

Reports from some locations (eg. Belgium, B. Bodson, personal communication) indicate that lodging, under high yielding conditions, still occurs due to high vegetative yields. It is suggested that this could be reduced using a lower sowing density (for European conditions 250 seeds/m^2), by applying fertilisers in three separate smaller treatments and by using growth regulators.

Bodson also indicates that low levels of grain set can occur under conditions of frequent rain (every day) and high relative humidity.

The use of growth regulators also appears relatively high in triticale; in data supplied by Chris Green (Semundo, U.K.) from a summary of pesticide usage on 761 farms, the percentage of crops treated with growth regulators was 84.6% for triticale and 68.8% for wheat. To be regarded as a low input crop this level of usage should be lower.

Conclusions

Based on the comments above, I would suggest that our crop does have the potential to increase in both area and value to end users. The crop also has potential to expand in newer areas (such as many African countries, [35]). We do still have some major problems to overcome; however the crop will continue to provide farmers in specific situations with good economic returns. Whilst we appear to now have well adapted cultivars available for many regions where triticale has agronomic potential, further market development is needed to assure its longterm future. We need to continue to improve the natural adaptability of triticale whilst also concentrating on making sure we satisfy the needs of end-users.

References

1) Aniol A. Breeding of triticale for aluminium tolerance. Genetics and breeding of triticale. Proceedings of Eucarpia meeting, Clermont Ferrand (France); 1984 July 2-5; Paris. Paris: INRA, 1985.

2) Camargo CEO, Dos Santos RR, Pettinelli Jn. A. Durum wheat: Tolerance to aluminium toxicity in nutrient solution and in the soil. Bragantia 1992;51(1): 69-76.

3) Camargo CEO, Felicio JC, Ferreira Filho AWP. Abiotic stresses; Mineral toxicities, low nutrient availability, drought and heat. Proceedings of the Second International Triticale Symposium (ITS); 1990 Oct 1-5; Passo Fundo, Brazil. Passo Fundo: CIMMYT, 1991.

4) Johnson JW, Cunfer BM, Manandhar J. Adaptation of triticale to soils in the southeastern USA. Proceedings of the Second International Triticale Symposium (ITS); 1990 Oct 1-5; Passo Fundo, Brazil. Passo Fundo: CIMMYT, 1991.

5) Jessop RS, McCallum RA, Wright R, Wright T. Selecting for improved low pH/Aluminium tolerance in Australian triticale varieties. Proceedings of the Second

International Triticale Symposium (ITS); 1990 Oct 1-5; Passo Fundo, Brazil. Passo Fundo: CIMMYT, 1991.

6) Aniol A, Gustafson PJ. Chromosome location of genes controlling aluminium tolerance in wheat, rye and triticale. Can J Genet Cytol 1984;26:701-705.

7) Lagos MB, Fernandes MIBM, Carvalho FIF, Camargo CEO. Localizacao do gene(s) de tolerancia ao crestamento em trigo cv. BH 1146 (Triticum aestivum L.). Trabalho apresentado na 13ª Reuniao Nacional de Pesquisa de Trigo; 1984 Cruz Alta, RS, Brazil, quoted by Carmargo et al 1990 (3).

8) Pinto Carnide O, Guedes-Pinto H, Vaz E. Aluminium tolerance behaviour of F2 6X--triticales and their progenies. Proceedings of the Second International Triticale Symposium (ITS); 1990 Oct 1-5; Passo Fundo, Brazil. Passo Fundo: CIMMYT, 1991.

9) Manyowa NM, Miller TE. The genetics of tolerance to high mineral concentrations in the tribe Triticeae - a review and update. Euphytica 1991;57:175-185.

10) Royo C, Rodriguez A, Romagosa I. Differential adaptation of complete and substituted triticale. Pl Breeding 1993;111:113-119.

11) Fox PN, Skovmand B, Thompson BK, Braun HJ, Cormier R. Yield and adaptation of hexaploid spring triticale. Euphytica 1990;47:57-64.

12) Camargo CE. Wheat Breeding. III Evidence of genetic control over tolerance of toxic manganese and aluminium in wheat. Bragantia 1983;42:91-103.

13) Thomson CJ, Colmer TD, Watkin ELJ, Greenway H. Tolerance of wheat (Triticum aestivum cvs Gamenya and Kite) and triticale (Triticosecale cv Muir) to waterlogging. New Phytol 1992;120(3):335-344.

14) Gorham J. Salt tolerance in the triticeae: ion discrimination in rye and triticale. J Exp Bot 1990;41(226):609-614.

15) Cooper KV, Graham RD, Longnecker NE. Triticale: a cereal for manganese-deficient soils. In: Graham RD, Hannam RJ, Uren NC, editors. International Symposium on Manganese in Soils and Plants. Dordrecht, Netherlands: Kluwer Academic Publishers, 1988: 113-115.

16) Harry SP, Graham RD. Tolerance of triticale, wheat and rye to copper deficiency and low and high soil pH. J Plant Nut 1981;391(4):721-730.

17) Graham RD. Triticale: a cereal for micronutrient deficient soils. Triticale Topics 1987;1:6-7.

18) Wojcieska U, Giza A, Wolska E. Growth, development, dry matter accumulation and nutrient uptake by spring triticale cv. MAH-183 and spring wheat cv. Kadett. II Changes in Fe, Mn, Zn and Cu contents during plant growth. Pamietnik-Pulawski 1989;94:77-97.

19) Jones GPD, Jessop RS, Blair GJ. Alternative methods for the selection of phosphorus efficiency in wheat. Field Crops Res 1992;30:29-40.

20) Anon. Triticale production and utilisation by Tunisian farmers; preliminary results of a farm survey. In: Farm Research Management Program, 1992 Annual Report. ICARDA, Aleppo, Syria. 1992.

21) Giunta F, Motzo R, Deidda M. Effect of drought on yield and yield components of durum wheat and triticale in a Mediterranean environment. Field Crops Res 1993;33(4):399-409.

22) Sweeney GC, Harris HC, Jessop RS. Water use by triticale and wheat. Proceedings of the First International Triticale Symposium (ITS); 1986 February 2-8; Sydney; Sydney: Australian Institute of Agricultural Science Publication, 1987.

23) McLean R, Barclay I. The role of triticale in Western Australian cereal cropping. Proceedings of the Second International Triticale Symposium (ITS); 1990 Oct 1-5; Passo Fundo, Brazil. Passo Fundo: CIMMYT, 1991.

24) Gill KS. Perspectives of two decades of research on triticale in India. Proceedings of the Second International Triticale Symposium (ITS); 1990 Oct 1-5; Passo Fundo, Brazil. Passo Fundo: CIMMYT, 1991.

25) Sinha SK, Aggarwal PK, Chaturvedi GS, Singh AK, Kailasnathan K. Performance of wheat and triticale cultivars in a variable soil-water environment. I. Grain yield stability. Field Crops Res 1986;13(4):289-299.

26) Pfeiffer WH. Triticale improvement strategies at CIMMYT: exploiting adaptive patterns and end-use orientation. Triticale Topics 1993;11:18-27.

27) Fettell NA. Yield physiology of triticale under water deficits: a comparison with wheat [dissertation]. Armidale: Univ. of New England, 1993.

28) Wolski T. New organisation and new winter triticale varieties in Poland. Triticale Topics 1992;8:21.

29) Andrews AC, Wright R, Simpson PG, Jessop R, Reeves S, Wheeler J. Evaluation of new cultivars of triticale as dual-purpose forage and grain crops. Aust J Exp Agric 1991;31:769-775.

30) Franco FA, MC Bassoi. Performance of triticale in Western Parana. Proceedings of the Second International Triticale Symposium (ITS); 1990 Oct 1-5; Passo Fundo, Brazil. Passo Fundo: CIMMYT, 1991.

31) Mosconi C, Rossi L. Triticale in Africa. In: Triticale 1987; Un Nuovo Cereale Da Considerare. Estratto da L'Informatore Agrario XLIII (39); 1987 October 8. ENEA.

32) Ortiz-Monasterio RJL, Sayre KD, Pfeiffer WH. Differences in nitrogen recovery among CIMMYT's bread wheats and complete and 2D(2R) substituted triticales. Triticale Topics 1993;11:6-9.

33) Bali AS, Shah MH, Lahari, Singh KN, Khanday BA, Koul RN. Studies on the production efficiency and economics of N fertilisation for wheat and triticale genotypes under Kashmir Valley conditions. Fert and Market. News 1991;22(3):11-13.

34) Karpenstein-Machan M, Heyn J. Yield structure of the winter cereals wheat and triticale at climatically marginal sites in north. Hesse Agribiol Res 1992;45(1):88-96.

35) Impiglia L. Triticale in Africa. In: Triticale 1987; Un Nuovo Cereale Da Considerare. Estratto da L'Informatore Agrario XLIII (39); 1987 October 8. ENEA.

HIGH LEVELS OF SALT TOLERANCE REVEALED IN TRITICALE

Robert M.D. Koebner, Paul K. Martin
Cereals Research Department, John Innes Centre, Norwich Research Park,
Colney NR4 7UJ, England

Abstract

We have demonstrated genetic variation for salt tolerance in triticales. The level of
tolerance, particularly in some of the Polish germplasm, exceeds that of wheat and is
comparable, if not superior, to that of tritipyrums (primary amphiploids between
hexaploid wheat and the highly tolerant *Thinopyrum* spp.). As far as we can tell, the
tolerant triticales have not been subjected to deliberate selection under saline conditions.
Since these triticales have been bred for productivity in a cropping environment, we
suggest that triticale should be considered as a primary cereal for saline-affected soils, in
preference to attempts to develop tritipyrum as a salt tolerant cereal.

Introduction

The ability of crop plants to tolerate salinity stress is becoming an important character as
the extent of arable area affected increases, both as a response to the spread of cropping
into more marginal areas and as a result of secondary salinization caused by poor
agronomic practices. In addition, the possibility of climatic change leading to less
favourable growing conditions in the temperate zones has increased the need for
adaptation of the major crop species to abiotic stresses such as salinity. Genetic variation
for response to salinity stress is known in wheat, although the level of tolerance that has
been achieved to date by breeding is not high. Higher levels of tolerance are present
among certain wild relatives of wheat, in particular those which are adapted to naturally
saline environments, and at least a part of this tolerance is expressed in amphiploids using
Thinopyrum species as the wild donor [1,2]. However, although triticale has acquired a
reputation for better performance than wheat on soils suffering from aluminium toxicity or
poor fertility, little attention has been paid to date to its response to salt stress in the field.
In limited genotypic comparisons (involving only two lines in each case) in field
experiments, it has been observed that it suffers both little foliar damage [3] and little
inhibition of tillering [4].

429

H. Guedes-Pinto et al. (eds.), Triticale: Today and Tomorrow, 429–436.
© 1996 *Kluwer Academic Publishers. Printed in the Netherlands.*

A major constraint in the assessment of new sources of salinity tolerance lies in the lack of efficient and effective screening systems. The ability to germinate at high salinity levels has been proposed as a selection criterion in wheat [5], but the correlation between this character and long-term plant growth in salinized culture is poor in many species, including triticale [2,6]. Field selection in saline-affected soils suffers from the lack of both spatial (horizontal and vertical) and temporal homogeneity with respect to salinity. Various hydroponic systems, which control both these factors, have therefore been employed to study variation in plant response to salt stress; however, even under these conditions, there is frequently a large non-genetic element of variation so that extensive genotypic replication is necessary to recognise genetic effects.

In this paper, we describe the use of a nutrient film hydroponic system (which permits good expression of the genetic potential to withstand salt stress, and which allows a substantial throughput of genotypes [7]) to screen a collection of wheat genotypes. During this screen it was noticed that a few triticales proved to be very tolerant of salinity stress. We have found that under these conditions, a number of triticales, particularly emerging from Polish breeding programmes, are very tolerant, and appear to be at least as well adapted to the stress as the very tolerant tritipyrums (wheat x *Thinopyrum* spp. amphiploids). Other triticales, in particular those coming from the CIMMYT breeding programme, do not show any appreciable levels of tolerance. In contrast to the tritipyrums, which have no history of breeding for agronomic worth, triticales have been bred for yield and adaptation over a long period in many environments. Thus, our conclusion is that triticale represents a readily exploitable resource for the breeding of a salt tolerant cereal.

Materials and Methods

HYDROPONICS SYSTEM

We used a nutrient film technique (NFT), as described more fully elsewhere [7]. The system involves growing plants in rockwool blocks (10 x 10 x 7.5cm) or plugs (2.5 x 2.5 x 4cm) under which a continuous flow of nutrient liquid is maintained. The nutrient medium is circulated from a holding tank by a submersible pump to the head of the tray holding the plants and flows back to the tank by gravity. The advantages of this system are: 1) that it provides an inert and absorbent, but not waterlogged solid support for root growth; 2) that the nutrient medium is continuously aerated and mixed by the flow; and 3) that individual plants can be conveniently removed from the hydroponicum and replanted into soil or another hydroponicum merely by lifting the block or plug. Plants were salinized at the two leaf stage by adding all the salt (NaCl/CaCl$_2$ 20:1, generally 200mM NaCl) in crystalline form, to the holding tank. Thereafter, the water level was maintained at a constant level to retain a constant salt concentration and the medium was changed every two weeks. In other experiments, plants were grown in vermiculite-containing pots standing in large saucers and were watered frequently from the top to replace evaporative and transpirative water loss, as measured by the level of standing water in the saucer. Every two weeks the pots were allowed to drain and a new culture solution was added to the saucer.

PLANT MATERIAL

The initial triticale studied was designated W4194 in the AFRC Institute of Plant Science Research & John Innes Institute Catalogue of Germplasm Collections 1989. This line is selection LT1033-79 from the Poznan (Poland) breeding programme. A further 21 triticales, representing accessions mainly from Poland and CIMMYT, were then studied (Table 1). In this experiment, each line was represented by five replicates and the plants were grown at 260mM NaCl for six weeks in small rockwool plugs. Under this level of salinity plant growth is severely retarded and the test is used to select genotypes able to survive successfully a level of salt stress that is lethal to all the wheats that have been tested to date. The tritipyrum CS(Ciano 2D) x *Th. bessarabicum* [8] was used as a tolerant control. The best nine triticale lines were then grown to flowering in large rockwool blocks as eight and two replicates in saline (200mM NaCl) and control (0 mM NaCl) solutions, respectively, to study the response of biomass accumulation to a more moderate level of salt stress, but one that still represents approximately 40% seawater (11.7 g/l salt). Plants were harvested (shoot only) at this stage. The material was oven dried to obtain dry weight of biomass.

Results

The initial observation that confirmed that triticale showed useful genetic variation with respect to salt tolerance was the survival of a plant of line W4194 under conditions where the salinity level was allowed to rise by evaporative and transpirative loss from 200mM to almost 500mM NaCl. Under this regime, W4194 was the only line among a number of triticales that survived beyond an early stage of development to reach flowering (Figure 1), although the plant was unable to set seed. In a follow-up experiment, using the 22 triticales shown in Table 1, a selection regime of 260mM NaCl was used to confirm the response of W4194 and to sample the extent of genetic variation for plant survival at salt concentrations lethal to wheat. From this experiment nine lines, all but one from Polish germplasm, were selected (Table 1). After six weeks in salt, plants of these lines could be removed from the hydroponicum and potted into soil, where they were able to resume normal growth and develop into fully fertile plants. Wheat seedlings rescued in the same way did not recover (Figure 2).

The response of W4194 was investigated in more detail. Under high salt conditions vegetative growth was only slightly inhibited, with only little chlorosis of leaf tissue at the tips of the leaves. The extension of the shoot at flowering was greatly reduced, with the ear only partially emerging from the boot (Figure 3). Interestingly, the time taken to flowering in salt and control was almost identical, in contrast to the almost universal pattern in cereals, whereby flowering is advanced by salt stress [4,9,10], sometimes by a substantial number of days. The response of triticale W4190 was similar, although this line is much earlier flowering than W4194.

Figure 1: Triticale W4194 grown in high salt NFT. Dead plants are CIMMYT triticales.

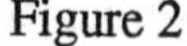

Figure 2

Figure 3

Figure 2: Triticale W4194 grown in 200mM NaCl / vermiculite (left) and in soil (right).
Figure 3: Rescued plants one week after transplantation into soil following six weeks at 260mM NaCl/NFT of (left) wheat (died) and (right two plants) triticale W4194.

The absolute and relative shoot dry weights and flowering times of the nine selected triticales grown at 0mM and 200mM NaCl are presented in Table 2. Biomass accumulation under salt was reduced, as expected, for all lines but the extent of this reduction varied from over 90% to under 70%. The highly tolerant tritipyrum suffered an 86% reduction and was one of the lowest yielders of all the lines in the test, reflecting the poor agronomic background of this primary amphiploid. Encouragingly, some of the selections performed better than W4194, both in absolute and relative terms. Advance in flowering times under salt stress was shown by most of the lines, although the extent of the advance, both in absolute and relative terms, was variable. One line (cv. Lasko) flowered a few days later in salt than in control.

Discussion

The indication from this preliminary work is that certain triticale breeding programmes have, perhaps serendipitously, been successful in selecting for good tolerance to salinity stress. Our data suggest that the levels of tolerance that can be achieved may be quite high. Some limited data from field experiments in saline soil have ranked the salt tolerance of triticale as good as, if not better than, that of wheat, which is itself one of the most tolerant of glycophytes [3,4]. It is of significance that the flowering time of some of these selections is not substantially advanced by the imposition of salt stress (unlike those of the two cultivars described in [4]), since this character has been correlated with salt tolerance [11] and is also expressed by barley cv. California Mariout, one of the most tolerant barley cultivars available (unpublished data). Nevertheless, it is clear that field experimentation is now needed to validate our conclusions.

As we have also found with wheat (unpublished data), tolerance to salt stress appears to be present in triticale germplasm which has never been subjected to selection under stress conditions. We suggest that the most likely explanation for this surprising observation is that the genetic basis of the tolerance lies in a pleiotropic effect exerted by genes which are important for other aspects of productivity and are therefore selected in breeding programmes aimed at, and conducted in, non-salt-stressed environments. There may indeed be a commonality between at least some of the genes responsible for tolerance to mineral stresses ('multitolerance' [12]) so that selection for aluminium tolerance, for example, which is important in Brazilian and other programmes, could have the effect of co-selection for salt tolerance. Rye chromosome 5R has been shown to be important in determining both the response to excess aluminium [13] in triticale and is also implicated in the improved copper efficiency character when present in a wheat background [14], suggesting that a common biochemical and/or physiological mechanism, perhaps mediated by phytosiderophores, may be involved in these responses. Although the pathways related to salt tolerance are likely to be complex, an aspect of it is probably related to an ability to exclude, compartmentalize or neutralize excess sodium, and the involvement of siderophores could be imagined in at least some of these processes.

A substantial body of research is being devoted to the chromosomal

Table 1. Triticales (and their provenance) screened for survival at 260mM NaCl. Those selected for further testing at 200mM NaCl are shown in bold type.

Line[a]	Provenance	Line[a]	Provenance
279	Rosner Canada	**4190**	Juanillo 207 CIMMYT
630	Armadillo E 1/68/B CIMMYT	**4194**	LT-1033-79 Poland
631	Armadillo T151 CIMMYT	4195	LT-1231-79 Poland
632	Armadillo T165 CIMMYT	**4196**	LT-1533-79 Poland
633	Armadillo T178 CIMMYT	4197	LT--79 Poland
634	Armadillo T179 CIMMYT	**4199**	LT-2211-79 Poland
636	Armadillo T200 CIMMYT	**4200**	LT-435-79 Poland
638	Armadillo T1333 CIMMYT	**4201**	LT-50-79 Poland
639	Armadillo T CIMMYT	**4202**	LT-676-79 Poland
641	Inia x Guarda T2 CIMMYT	**4206**	Loutoi Poland
4143	Caborca 79 CIMMYT	**5733**	Lasko Poland
Tritipyrum	ref [8]		

[a]Accession numbers of the AFRC Institute of Plant Science Research & John Innes Institute Catalogue of Germplasm Collections 1989

Table 2. Absolute and relative dry shoot biomass accumulation at flowering per plant, and flowering times of selected triticales grown under 0mM and 200mM NaCl NFT.

Line	B^c	FT^c	B^s	FT^s	B^s/B^c	FT^c-FT^s
4190	6.2	45	1.6	45	26	0
4194	26.4	54	4.9	49	19	+5
4196	32.4	55	4.0	51	12	+4
4199	27.0	59	5.4	36	20	+23
4200	29.3	60	3.6	49	12	+11
4201	90.0	94	4.1	67	5	+27
4202	32.4	54	4.0	36	12	+18
4206	29.6	70	9.2	44	31	+26
5733	36.8	39	3.1	43	9	-4
Tritipyrum	19.0	64	2.6	27	14	+37

B^c, B^s: Shoot dry weight (g) in control (0mM NaCl) and salt (200mM NaCl), respectively. B^s/B^c: % reduction in shoot dry weight in salt from control.
FT^c, FT^s: Flowering time (days) of main tiller in control and salt, respectively.
FT^c-FT^s: Difference in flowering times between control and salt treatment.

engineering of wheat for salt tolerance by accessing the genetic information of related wild species, such as *Th. elongatum* and *Th. bessarabicum*, which have evolved high levels of tolerance in nature. Although there is some dispute as to the genetic basis for the tolerance of these species, none of the lines carrying a single *Thinopyrum* sp. chromosome pair added to the wheat complement has a tolerance close to that of the parental wheat x *Thinopyrum* sp. amphiploid [1,15]. This has led to the suggestion that the breeding of tritipyrums may be an appropriate route to a salt tolerant cereal. While this strategy overcomes the problems of complex inheritance of the trait, it may be a simplistic one in view of the time scale and investment which has been required to bring triticale to a position where it is perceived to be viable as a crop. In this paper, we have demonstrated the possibility that triticale itself may be a good source of tolerance which appears to be as, if not more, effective than that contributed by the *Thinopyrum* spp. Thus, the attractiveness of tritipyrums is yet further diminished since advantage can be taken of the ready availability of adapted, high-yielding and good quality triticale germplasm, while the tritipyrums are still at the stage of being represented by a few primary amphiploids, mostly involving poorly adapted wheats as one parent and a handful of unselected *Thinopryum* spp. as the other parent.

While it is well recognised that the tolerance of tritipyrums has been contributed largely by the wild donor, the situation in triticale is not yet so clear. Although, in general, tetraploid wheats are less tolerant than hexaploid wheats, nevertheless some *Triticum durum*, and particularly some *T. dicoccoides*, lines are quite tolerant. Since there is very little information regarding genetic variation in the response of rye to salt stress, we are now initiating a germplasm screen of this species. We have also begun to cross tolerant triticales with susceptible tetraploid wheats to determine whether the rye genome has contributed the tolerance and, if so, how complex the inheritance of the trait may be.

Acknowledgement

The senior author thanks the British Council for the provision of a travel grant which enabled the presentation of this work at this meeting.

References

1. Dvorak J, Edge M, Ross K. On the evolution of the adaptation of *Lophopyrum elongatum* to growth in saline environments. Proc Nat Acad Sci USA 1988;85:3805-9.
2. Forster BP, Gorham J, Miller TE. Salt tolerance of an amphiploid between *Triticum aestivum* and *Agropyrum junceum*. Plant Breeding 1987;98:1-8.

3. Richards RA, Dennett CW, Qualset CO, Epstein E, Norlyn JD, Winslow MD. Variation in yield of grain and biomass in wheat, barley, and triticale in a salt-affected field. Field Crops Res 1987;15:277-87.

4. Francois LE, Donovan TJ, Maas EV, Rubenthaler GL. Effect of salinity on grain yield and quality, vegetative growth and germination of triticale. Agron J 1988;80:642-7.

5. Kingsbury RW, Epstein E. Selection for salt-resistant spring wheat. Crop Sci 1984;24:310-15.

6. Norlyn JD, Epstein E. Variability in salt tolerance of four triticales at germination and emergence. Crop Sci 1984;24:1090-2.

7. Martin PK, Humble J, Koebner RMD. Use of the nutrient film technique as a method for assessment of plant response to salt stress in the cereals. Acta Bot Soc Pol 1994; in press.

8. Martin PK, Koebner RMD. The effect of photoperiod insensitivity on the salt-tolerance of amphiploids between wheat (*Triticum aestivum*) and sand couch grass (*Thinopyrum bessarabicum*). Plant Breeding 1993;111:283-9.

9. Francois LE, Maas EV, Donovan TJ, Youngs VL. Effects of salinity on grain yield and quality, vegetative growth and germination of semi-dwarf and durum wheat. Agron J 1986;78:1053-8.

10. Maas EV, Grieve CM. Spike and leaf development in salt-stressed wheat. Crop Sci 1990;30:1309-13.

11. Taeb M. The genetics of salt and waterlogging tolerance in wheat (*Triticum aestivum*) [PhD dissertation]. Cambridge UK, University of Cambridge, 1991.

12. Manyowa NM, Miller TE. The genetics of tolerance to high mineral concentrations in the tribe *Triticeae* a review and update. Euphytica 1991;57:175-85.

13. Manyowa NM, Miller TE, Forster BP. Alien species as sources for aluminium tolerance genes for wheat, *Triticum aestivum*. Proceedings of the 7th Int Wheat Genet Symp (Miller TE, Koebner RMD, editors); Cambridge: Institute of Plant Science Research, 1988: 851-7.

14. Schlegel R, Werner T, Hulgenhof E. Conformation of a 4BL/5RL wheat-rye chromosome translocation in the wheat cultivar 'Viking' showing copper efficiency. Plant Breeding 1991;107:226-34.

15. Forster BP, Miller TE, Law CN. Salt tolerance of two wheat-*Agropyron junceum* disomic addition lines. Genome 1988;30:559-64.

ALUMINUM TOLERANCE IN TRITICALE, WHEAT AND RYE[*]

Augusto Carlos Baier[1], Daryl J. Somers[2] and J. Perry Gustafson[2,3].
[1]EMBRAPA-CNPT, C Postal, 569, Passo Fundo RS 99001-970, Brazil;
[2]Plant Science Unit, Dept of Agronomy; [3]USDA-ARS, PGRU, Curtis Hall,
University of Missouri, Columbia, MO 65211 USA.

Abstract

Acid soils and toxic aluminum (Al) limit the growth of plants on over 1.5
billion hectares worldwide. The use of lime can be reduced by using tolerant
cultivars making cultivation on acid soils more cost effective. This
presentation aims to report laboratory Al-tolerance evaluations of triticale,
wheat, and rye in Al containing hydroponic solutions. Rye showed the
longest and the Al-sensitive wheats showed the shortest roots in Al
containing solutions. Root elongation was impaired for all cultivars at high
Al concentrations. The Al-tolerant wheat and triticale cultivars had
intermediate root lengths in Al solutions. The newer complete triticale
cultivars and lines had longer roots than the 2D(2R) substituted triticale
types. Root dry weight, root length and relative root length gave similar Al-
tolerance rankings for the Al-tolerant wheat and triticales but not for rye.
The Al-tolerance genes of rye appeared to be expressed in triticale at levels
similar to rye alone. This simple test is useful for breeding programs in
regions with acid soils, although field performances should also be
considered, especially when comparing different species.

Introduction

Acid soil limits plant growth on over 1.5 billion ha of the arable land
worldwide. The reduction of liming and consequent lower soil pH and
increased aluminum (Al) toxicity, is limiting production in traditional wheat
producing areas [1]. Acid-soil tolerant cultivars provide an alternate,
inexpensive, ecologically clean and sustainable solution to liming or make
liming and fertilization more cost effective [1-3]. The value of using Al-
tolerant cultivars is evident from a study comparing near isogenic lines of
wheat that differed in Al tolerance; 31% more spikes, 66% more biomass,

---

[*]Names are necessary to report factually on available data; however, the
USDA neither guarantees nor warrants the standard of the product, and
the use of the name by USDA implies no approval of the product to the
exclusion of others that may also be suitable.

H. Guedes-Pinto et al. (eds.), Triticale: Today and Tomorrow, 437–444.
© 1996 Kluwer Academic Publishers. Printed in the Netherlands.

438

and 68% more grain yield was obtained from Al-tolerant lines grown on acid
soil [4].

Species and genotypes within species differ widely in acid-soil and Al
tolerance [5]. Acid-soil and/or Al tolerance was identified in wheat [2,3,6],
triticale and rye [7-9] genotypes from various origins. Triticale germplasm
with the full rye complement, possesses a better adaptation and yield
potential in marginal environments [10]. Camargo and Felicio [8] report
that the rye cultivars 'Goyerovo' and 'Branco' tolerate higher levels of Al in
nutrient solution, than tolerant wheat and triticale cultivars. In contrast,
Aniol et al. [7] concluded that rye inbred lines varied between the Al-tolerant
and the Al-sensitive wheat cultivars.

The Al-tolerant triticale, wheat, and rye cultivars from Brazil
represent an important resource for the evaluation of Al-tolerance
screening techniques. The objectives are to present information on the Al
tolerance of different triticale types (2D(2R)) substituted, recommended
cultivars, advanced lines), compared to wheat and rye genotypes using a
simple and rapid screening technique with Al containing hydroponic
solutions.

Materials and method

The germplasm included two rye cultivars, ten triticale genotypes and ten
wheat cultivars, all from Brazil, as well as one rye cultivar from Poland and
one rye landrace variety from Ecuador. Among the Brazilian triticale
genotypes there were three types: the 2D(2R) substituted types, developed
prior to 1980; the complete triticale cultivars, developed between 1980 and
1985; and the newer, advanced, complete triticale lines that were developed
between 1986 and 1990 and are being evaluated in field trials for
commercial release (Table 1).

Seeds of each genotype were placed on moist filter paper in petri
dishes, at $2°\pm1°C$ for 12 h and then were germinated for 24 h at room
temperature. Two replications of six seedlings of each genotype with similar
root lengths (3 to 10 mm) and endosperm size were then placed on plastic
mesh floating on 2 L of an aerated, low ionic strength hydroponics medium
[400 uM $CaCl_2$; 650 uM KNO_3; 250 uM $MgCl_2$; 10 uM $(NH_4))_2SO_4$;40 uM
NH_4NO_3 (pH 4.0)] containing Al, (modified from [7]). The Al was added as
$AlCl_3$ to the acidified hydroponics medium at concentrations of 0.5, 1.0, 2.0,
or 4.0 mg Al L^{-1} (1 mg L^{-1} = 37 uM of Al). The hydroponic tanks and
seedlings were placed in an environmental growth

Table 1. Germplasm types, cultivars, and pedigrees.

Germplasm type description Cultivars Pedigree (Line designation)	

Substituted triticale cultivars (Old):

Triticale BR1	M2A/CML (Panda)
CEP 18	TOB/8156//CC/3/Inia/4/SPY/5/M2A
(Teddy)	

Complete triticale cultivars:

Triticale BR4	BGL/CIN//MUS
CEP 22	BGL/CIN//IRA/BGL
IAPAR 38-Arauna	JLO/Panther
OCEPAR 3	CIN/CNO//BGL/3/Merino (Hare)

Complete triticale lines:

PFT 8922	MUS/Lynx//Yogui/3/MUS
PFT 107	Hare 263/Civet
TCEP 878	Hare 263/Civet
PR884	B6811-270-27Y-3Y-0M (Uron)

Al-sensitive wheat cultivars:

Anahuac 75	II 12300//LR 64/7C/3/NOR 67
IAPAR 30	ALD Sib//CNT 7/PF 70354/3/PAT
24/BB/KAL	

Al-tolerant wheat cultivars:

Trigo BR 23	CC/ALD/3/IAS 54-20/COP//CNT 8
Trigo BR 35	IAC 5*2/3/CNT 7*3/LD//IAC 5/HAD
Trigo BR 43	PF 833007/Jacui
EMBRAPA 15	CNT 10/BR 35//PF 75172/Tifton

Highly Al-tolerant wheat cultivars:

BH 1146	Ponta Grossa 1//Fronteira/Mentana
CNT 1	PF 11.1000.62/BH 1146
IAC 5-Maringa	Frontana//Kenya 58/Ponta Grossa 1
Trigo BR 37	Mazoe/F 13279//Pelado Marau

Rye:

Blanco	From Dr. Robert Metzger, originally from Brazil, selected for Al tolerance five cycles
Centeio BR 1	(Cultivar from Brazil)
Dankovskie Zlote	(Cultivar from Poland)
Ecuador	(Landrace from Ecuador)

chamber (26°±1°C 16 h day/8 h night). The seedling growth continued for 4 days with the solutions being changed each day to minimize changes in the pH and Al concentration.

The two longest roots from each seedling were measured and averaged for each genotype and Al concentration. The RTI was calculated by dividing the average root length in each Al concentration by the average root length in the solution with no Al. Roots of the six seedlings of each replication from selected genotypes were dried at 80°C for 5 hours.

Results and discussion

The evaluated germplasm represents a broad base of the Al tolerance present in triticale, wheat, and rye cultivars recommended for cultivation in Brazil. Root elongation was reduced in all genotypes in solutions with more than 1 mg Al L^{-1}. The Al-sensitive wheat cultivars had the shortest roots and the lowest RTI values in the solutions with Al. There were no Al-sensitive triticale and rye cultivars observed. Though rye had the longest roots, the rye RTI values were not significantly higher than the triticale and the highly tolerant wheat cultivars. The triticale cultivars showed little reduction in root length from 0.0 to 1.0 mg Al L^{-1}, but at higher Al concentrations, root growth was impaired (Fig. 1).

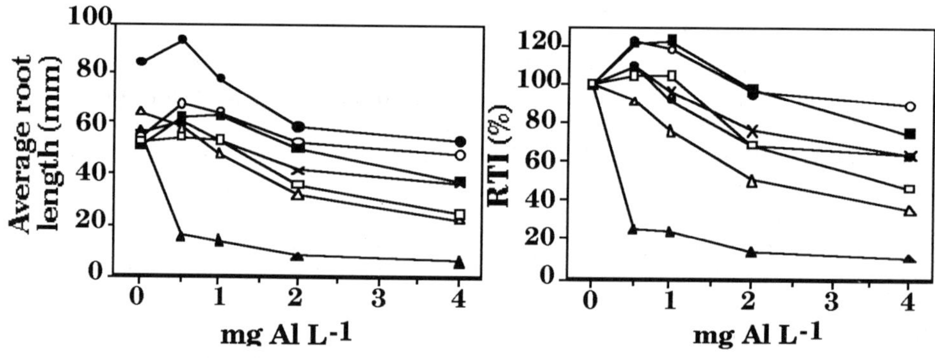

Figure 1. Root elongation and RTI of triticale, wheat and rye germplasm types in Al-containing hydroponic solutions. Two replications of six seedlings of each genotype were grown for 4 d in hydroponic solutions containing various Al concentrations. The average root length and RTI of each germplasm type was derived by averaging the two longest root measurements of all seedlings from each genotype for each germplasm type. Germplasm types: Al-sensitive wheat cultivars (▲), Al-tolerant wheat cultivars (△), highly Al-tolerant wheat cultivars (X), substituted triticale cultivars (□), complete triticale cultivars (■), complete, advanced triticale lines (O), rye (●).

The triticales were separated into 3 types in order to compare the cultivars/genotypes from different breeding periods. The first type, recommended prior to 1980, with the 2D(2R) substitution shows a relatively poor performance on very acid soils. The second type (recommended complete cultivars) and the third (complete advanced triticale lines) have a higher yield potential, an improved disease resistance, and better adaptation to acid soils.

In the solution with 1 mg Al L^{-1} (Table 2, Fig. 1) the Al-sensitive wheat cultivars had the shortest roots and the lowest RTI values. The Al-tolerant wheat cultivars and the substituted triticale cultivars 2D(2R) were intermediate between the sensitive wheat and the other two complete triticale types and rye. The highly Al-tolerant wheat cultivars and the complete triticales had similar root elongations and were shorter than the advance triticale lines and rye.

The longer roots in the advanced triticale lines indicates that in addition to the progress in yield and agronomy, Al tolerance is also being improved. This is consistent with Rajaram et al. [10] who considers that the triticale germplasm with the full rye complement possesses a better adaptation and yield potential in marginal environments. Further investigation is needed to evaluate if the longer and heavier roots, and the higher RTI values are also related to different genetic mechanisms of Al tolerance and/or to efficient phosphorus uptake mechanisms in triticale [5]. If so, it may be possible to make further improvements in the adaptation of triticale to acid, depleted and marginal soils.

Even though the long roots of the rye seedlings are indicative of its well known ability to grow on acid soils, they do not necessarily indicate better Al tolerance. Faster root growth is a constitutive growth habit for rye. In addition the lower root dry weights observed in rye in relation to triticale and wheat, reflect the fact that rye has long but thin roots. We propose the apparently greater root growth of rye over triticale is not related to greater Al tolerance and that rye Al tolerance is near fully expressed in triticale (Fig. 2). The differences in root elongation among and within rye cultivars, suggests that Al-tolerance in rye can be improved by hydroponic selection (Fig. 3). The improved Al-tolerance in rye could also be used in triticale and wheat.

Comparing root elongation, RTI and root dry weight, 'Anahuac 75', 'IAC 5-Maringa' and 'Triticale BR 4' had comparable Al-tolerance rankings. In contrast, rye had the longest roots, RTI values between the triticale and the Al-tolerant wheat and a root dry weight lower than the Al-tolerant wheat and 'Triticale BR 4' (Fig. 2). The different relative response of rye, suggests that the comparison of Al tolerance among species should be taken carefully. The improved Al tolerance in the newer, advanced triticale lines indicates that limitations on rye Al-tolerance gene expression are being progressively overcome by breeding in the newer germplasm (Table 2, Fig. 1).

Table 2. Average root elongation and RTI for the germplasm types and cultivars in the hydroponic solutions with 1 mg Al L^{-1}

Germplasm type Cultivar	Root elongation (mm)	RTI (%)
Sensitive wheat cultivar:	19.9 A*	34.9 A
Anahuac	18.1 a	34.6 a
IAPAR 30	21.7 a	35.2 a
Tolerant wheat cultivars:	45.2 B	71.0 B
Trigo BR 35	40.4 b c	65.7 b
Trigo BR43	42.9 b c d	76.3 b c d
EMBRAPA 15	48.6 c d e f	69.1 bb
Trigo BR 23	48.7 d e f	72.9 b c
Highly tolerant wheat cultivars:	49.5 C	89.0 C
CNT 1	44.4 c d	98.1 f g h i j
Trigo BR 37	45.2 c d	88.9 d e f g h
IAC 5- Maringa	51.0 d e f g	84.5 c d e f
BH 1146	57.7 g h	84.6 c d e f
Substituted triticale cultivars:	44.5 B	84.8 C
Triticale BR 1	35.4 b	85.2 c d e f
CEP 18	53.6 e f g	84.3 d e f
Complete triticale cultivars:	52.9 C	103.1 D
OCEPAR 3	47.9 c d e	87.9 d e f g h
IAPAR 38	48.5 d e f	98.9 g h i j
CEP 22	56.6 f g h	114.1 k
Triticale BR 4	58.6 g h	111.7 j k
Complete triticale lines:	57.7D	104.8 D
PR 884	50.7 d e f g	105.7 h i j k
PFT 8922	57.8 g h	109.2 j k
TCEP 878	58.5 g h	100.8 h i j k
PFT 107	63.9 h i	103.8 i j k
Rye Cultivars:	73.5 E	86.8 C
Landrace Ecuador	68.8 i j	78.7 b c d
Dankovskie Zlote	68.9 i j	86.2 c d e f g
Blanco	77.0 j k	88.8 d e f g h
Centeio BR 1	79.5 k	93.5 e f g h i

*Averages of germplasm types followed by same uppercase letter and of cultivars followed by same lower case letter do not differ by Fishers Protected LSD (p = 0.05).

The fact that triticale breeders are seemingly selecting for additional Al tolerance indicates that Al-tolerance levels can be increased along with improving other agronomic parameters. This increased level of Al tolerance would allow for the improved varieties to be grown on even more marginal soils.

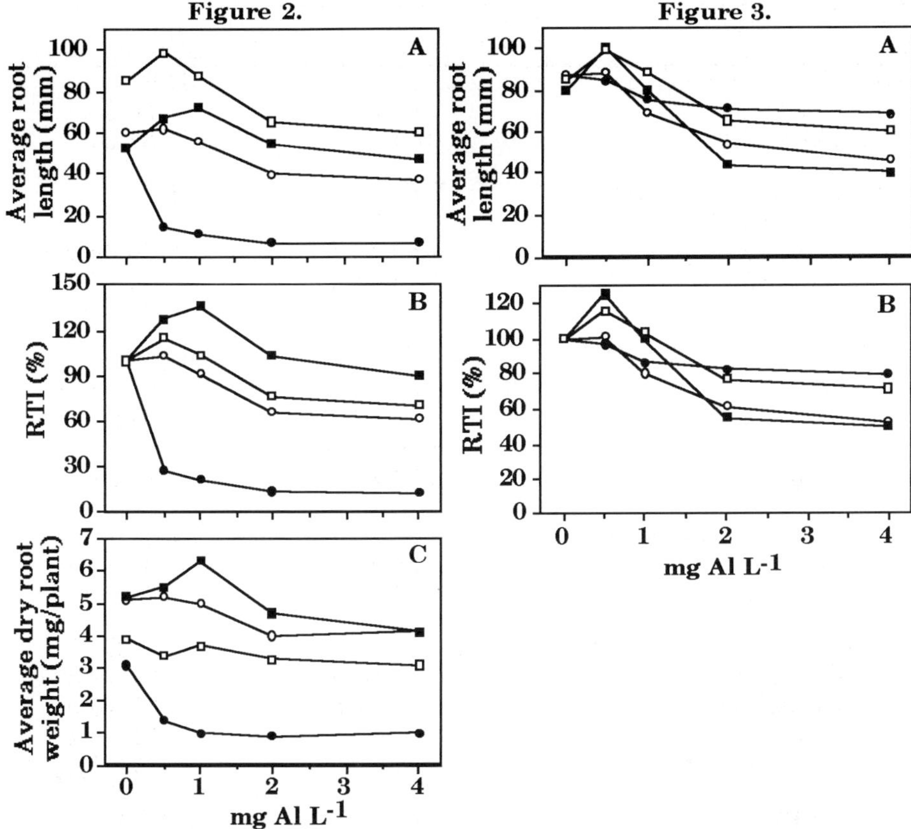

Figure 2. Root growth responses of rye 'Centeio BR 1' (□), 'Triticale BR 4' (■), 'IAC 5 Maringá' (Al-tolerant wheat) (○), and 'Anahuac 75' (Al-sensitive wheat) (●) in Al-containing hydroponic solutions. Six seedlings of each genotype were grown for 4 d in various concentrations of Al. The root length(A) was an average of the two longest roots from each of the six seedlings in two replications; RTI (B) was root length relative to the zero Al solution; and root dry weight (C) was from the combined roots of all six seedlings of each treatment in two replications.

Figure 3. Root growth responses of four ryes 'Centeio BR 1' (□), 'Dankovskie Zlote' (■), 'Ecuador landrace' (○), and 'Blanco' (●) in Al-containing hydroponic solutions. Six seedlings of each genotype were grown for 4 d in various concentrations of Al. The root length (A) was an average of the two longest roots from each of the six seedlings in two replications, RTI (B) was root length relative to the zero Al solution.

444

References

1. Briggs KG, Taylor GJ. Success in wheat improvement for poor soils: Experience with the aluminum tolerance system in NW Canada. Proceedings, workshop on adaptation of plants to soil stresses. University of Nebraska, (INTSORMIL/USAID) 1994; in press.

2. Foy CD, Armiger WH, Briggle LW, Reid DA. Differential aluminum tolerance of wheat and barley varieties in acid soils. Agron J 1965;57:413-417.

3. Foy CD, da Silva AR. Tolerance of wheat germplasm to acid subsoil. J Plant Nutr 1991;14:1277-1295.

4. Carver BF, Whitmore WE, Smith EL, Bona L. Registration of four aluminum-tolerant winter wheat germplasms and two susceptible near-isolines. Crop Sci 1993;33:1113-1114.

5. Rajaram S, Kohli MM, Lopez-Cesati J. Breeding for tolerance to aluminum toxicity in wheat. Wright et al. RJ editors, Plant-soil interactions at low pH. Kluwer Academic Publishers, The Netherlands. 1991;1019-1028.

6. Mesdag J, Slootmaker LAJ. Classifying wheat varieties for tolerance to high soil acidity. Euphytica 1969;18:36-42.

7. Aniol A, 1991. Genetics of acid tolerant plant. Wright RJ et al. editors, Plant-soil interactions at low pH, Kluwer Academic Publishers, The Netherlands. 1992; 1007-1017.

8. Camargo CEO, Felicio JC. Tolerance of wheat, triticale and rye cultivars to different levels of aluminum in nutrient solution (English summary). Bragantia, Campinas. 1984;43:9-16.

9. Camargo CEO, Felicio JC, Ferreira Filho AWP. Triticale: Aluminum tolerance in nutrient solution (English summary). Bragantia, Campinas. 1991;50:323-330.

10. Rajaram S, Varughese G, Abdalla O, Pfeiffer WH, van Ginkel M. Accomplishments and challenges in wheat and triticale breeding at CIMMYT. Plant Breeding Abstracts 1993;63:131-139.

SCREENING CULTIVARS FOR ALUMINIUM TOLERANCE

Ana Maria Antunes[1] , José Pereira [2] and Maria Antonieta Nunes[1]
1) Instituto de Investigação Científica Tropical, Lisboa, Portugal
2) Instituto Nacional de Investigação Agrária, Oeiras, Portugal

Abstract

Triticales are recognized among cereals as particularly tolerant to excess of Al and adequate for cultivation in acid soils. Because there are great diferences in tolerance among genotypes, the present work was done to evaluate the tolerance of advanced genetic lines produced in the breeding program of the Estação Nacional de Melhoramento de Plantas at Elvas.
A root growth technique using seedlings in nutrient solutions having different doses of Al (0, 5, 10, 15 and 20 ppm) was used for 17 genetic lines of triticale and 3 of wheat. The length of the longest (primary) root was measured at 3 times during growth after two days in the nutrient solution without Al; after inhibition in the nutrient solutions containing Al; and after recovery for 2 days in the initial solution.
The solution with 5 ppm Al concentration was the best, as a single one, to discriminate among the different genotypes. All triticale lines were clearly more tolerantes than the wheat ones. These had only 11.3 % growth in the Al solution with 5 ppm Al relative to the growth in the solution without Al, whereas the average for triticales was 46.2%. Among the triticale lines, Arabian ranked higher in tolerance with only 30% reduction in the root growth in contrast with Beagle which presented the strongest inhibition (75%).

Introduction

Soil acidity is a major growth limiting factor for plants in many parts of the world [1]. In Portugal, the soils used for agriculture are mainly acid, and it represents about 80% of the acreage [2]. Al toxicity is probably the most important yield limiting factor for crops in most acid soils [3, 4]. The agronomic solutions for acidity problems (and A1 toxicity) are: a) soil liming , b) use of tolerant cultivars and c) both techniques simultaneously. The triticale (artificial breed between *Secale*

445

H. Guedes-Pinto et al. (eds.), Triticale: Today and Tomorrow, 445–451.
© 1996 *Kluwer Academic Publishers. Printed in the Netherlands.*

cereale L and *Triticum sp.)* is very rustic and particularly tolerant to acid soils and aluminum. The National Plant Breeding Station (ENMP) located in a cereal producing area, retook in 1969 a program for triticale improvement to the center and southern regions of the country. Triticale was used by the Portuguese farmers for the first time in 1979 and it showed promising adaptability to adverse agroclimatic conditions [5]. In the present work advanced genetic lines and cultivars of triticale produced by ENMP have been evaluated for Al tolerance. This will identify genetic sources to be later used for improving low pH/Al tolerance similarly to what is being done e.g. in Australia and North of Portugal .

The aim of the present work was to evaluate the tolerance to Al for advanced genetic lines and cultivars produced by ENMP. A root growth technique using seedlings in nutrient solutions having different doses of Al (0, 5, 10, 15 and 20 ppm) was used in 17 genetic lines of triticale and 3 of wheat. The length of the longest primary root was measured 3 times during the growth period: after 2 days in the nutrient solution without Al; after inhibition in the nutrient solutions containing Al and after recovery in the initial solution.

The solution containing 5 ppm Al was the best, as a single one, to discriminate among the different genotypes. All the triticale lines were clearly more tolerant than the wheat ones. These showed only 11.3% growth in the 5 ppm Al solution relative to the growth in the solution without Al, whereas the corresponding average for triticale was 46.9°/o. Among the triticale lines, Arabian ranked higher in tolerance showing only 30% reduction in the root growth in contrast with Beagle which suffered the strongest inhibition (75%).

Material and Methods

Ten advanced genetic lines, seven cultivars of triticale and three cultivars of wheat were used in a test of root growth in nutrient solutions having different doses of Al. The cultivars of wheat *(Triticum aestivum)* were the following: Anza - obtained in Estação Nacional de Melhoramento de Plantas, Elvas, Portugal (ENMP) and included in the National Catalogue of Varieties (CNV) in 1982; BH-1146 - obtained in Instituto Agronómico de Minas Gerais, Brazil, worldwide known as tolerant to Al among wheat and IAS - obtained in Secretaria de Agricultura, Rio Grande do Sul, Brazil, also refereed as tolerant to Al. The triticale *(Triticosecale* Wittm.): Arabian, Arruda, Bacum, Beagle, Borba, Crato and Juanilho have been obtained in ENMP and included in the CNV from 1982 to 1991. The breeds TTE 9101, 9102, 9103, 9104, 9105, 9201, 9202, 9203, Borba X Faísca and Rat X Borba are advanced genetic lines of triticale which undergo studies of adaptation in several regional sites supervised by ENMP.

The seeds were germinated over cotton covered with filter paper soaked with water, at 23°/26° C. Ten homogeneous seedlings of each cultivar having 10-15 mm root length were selected, placed into plastic buckets containing three liters of a basic nutrient solution and grown for two days with continuous aeration. Afterwards the treatment solutions consisting of the diluted (1:10) basic nutrient solution free from phosphate with $FeCl_3.6H_2O$ substituting iron citrate, and added by different doses of Al, was given to the plants. The composition of the basic nutrient solution [8] is the following: 4 mM $Ca(NO_3)_2$; 4 mM KNO_3 ; 2 mM $MgSO_4.7H_2O$; 0.5 mM $KH_2 PO_4$; 0.435 mM $(NH_4)_2SO_4$; and micronutrients 0.8 μM $ZnSO_4$; 30 μM NaCl; 0.1 μM $NaMoO_4.2H_2O$; 10 μM H_3BO_3; 10 μM $C_6H_5O_7Fe.5H_2O$; 2 μM $MnSO_4.4H_2O$; 0.3 μM $CuSO_4$. The pH was set at 4 and adjusted daily with 1N H_2SO_4. The Al concentrations were 0, 5, 10, 15 and 20 ppm applied as $Al_2(SO_4).18H_2O$.

The length of the longest (primary) root was measured with the precision of 0.5 mm, at 3 times during growth: after 2 days in the basic nutrient solution without Al (C1); after 2 days inhibition in the nutrient solutions containing Al (C2); and after recovery for 3 days in the initial basic solution (C3). The experiment was performed with root temperature controlled at $25 \pm 1°C$ by a waterbath, in a cabinet with 12 hours of light from Phillips HPLRG lamps, irradiance of 150 μmol m^{-2} s^{-1}; air relative humidity about 60% and temperature $22 \pm 4°C$ during the day.

Usually a set of 10 cultivars were tested each time with five Al doses, replicated 3 times independently, in a split - plot scheme. Analysis of variance was applied to the whole set of cultivars using the relative growth [(C2-C1)in Al sol./(C2-C1)in the 0ppm Al sol.] to express the inhibitory Al effect. A more detailed description is given else where [9].

Results and Conclusions

Two experiments have been performed, with a set of 10 cultivars in each one. The analysis of variance for the increments in root length showed significant differences among cultivars, treatments and the interaction, in both experiments. The cultivars exhibited different values for absolute growth in the solution without Al ranging between 30.8 to 40.9 mm, respectively Arabian and TTE 9104 , which suggests that the genetic potential is not the same. The absolute growth of the wheat Anza was one of the highest. Concentrations higher than 5 ppm Al have been assayed but, this one has been recognized as the best to show good evidence of the inhibition for the group of 20 cultivars under study. However, if wheat germplasm was the major object of screening, a range of lower concentrations should be preferable.

Table 1. Root growth in the Al solutions (5, 10, 15 and 20 ppm) as a percentage of growth in the solution without Al. Each value corresponds to a mean of 30 seedlings. Cultivars are listed in decreasing order for the mean value in all the four of treatments.   LSD (P≤0.01): Cultivars = 5.34; Treatments = 5.91; Interaction Treat.xCv. = 10.7 (Values with the same letter are not statistically different).

N°	Cultivars	5 ppm	10 ppm	15 ppm	20 ppm	Mean
1	Arabian	69.9	27.6	22.5	12.1	32.8 a
2	9203	55.9	23.4	14.2	9.8	25.8 b
3	9103	58.4	15.1	12.3	10.7	24.2 bc
4	9104	50.5	20.7	15.1	9.9	24.1 bc
5	9102	48.1	19.4	14.0	13.7	23.8 bc
6	Rat x Borba	62.4	13.4	7.3	9.0	23.0 b-d
7	9101	46.0	15.8	12.4	9.4	20.9 b-e
8	Crato	43.7	17.1	12.9	9.5	20.8 b-e
9	9202	43.3	17.8	11.9	9.7	20.7 b-e
10	Borba	40.1	17.3	12.8	10.2	20.1 c-f
11	Borba x Faísca	50.4	13.7	9.4	5.7	19.8 c-g
12	Arruda	41.6	13.7	9.0	8.8	18.3 d-g
13	9105	43.0	11.0	6.7	4.9	16.6 e-h
14	Bacum	39.7	11.3	7.0	6.3	16.1 e-h
15	9201	32.8	13.4	6.9	7.1	15.0 f-i
16	Juanilho	34.8	11.2	5.7	6.3	14.5 g-j
17	IAS-	14.6	9.8	9.4	12.3	11.5 h-k
18	Beagle	25.2	4.1	5.5	5.2	10.0 i-k;
19	BH - 1146	12.2	10.4	7.0	7.1	9.2 jk
20	Anza	7.1	7.9	6.6	6.9	7.1 k
Mean		40.9 a	14.7 b	10.4 bc	8.7 c	

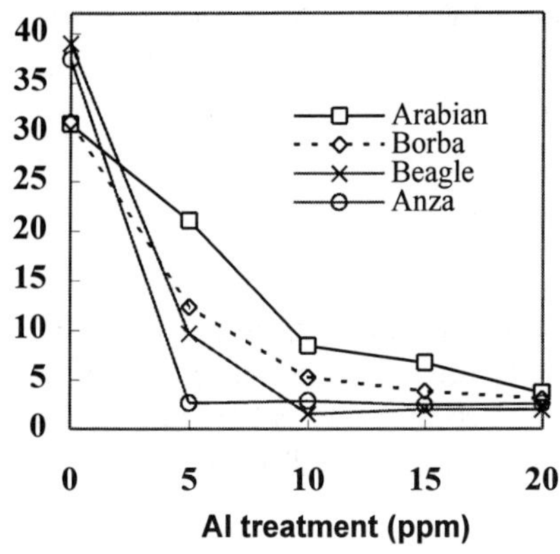

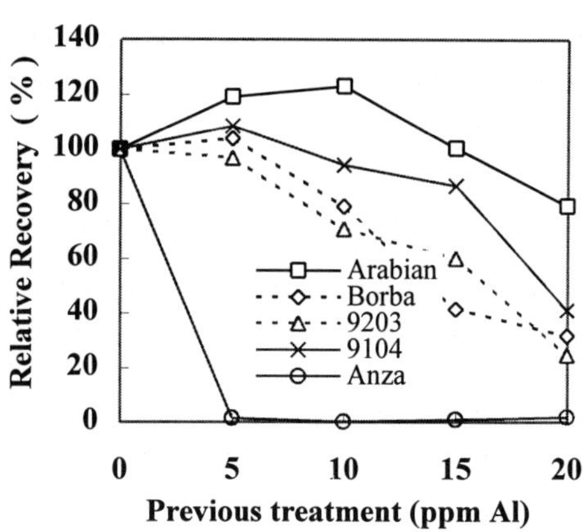

Fig.1 Absolute growth of primary root (C2-C1) during two days, in the treatment solutions containing from 0 to 20 ppm Al, for 3 representative cv. of triticale and one of the wheat Anza. Means of 30 seedlings. LSD (0.01): Treat. = 5.15; Cv. = 2.07.

Fig.2 Root growth during the recovery from the treatments with several Al concentrations, as % of the growth in the nutrient solution without Al, for 4 cvs. of triticale and for wheat Anza. LSD(0.01): Treat.=22;Cv.=14.7.

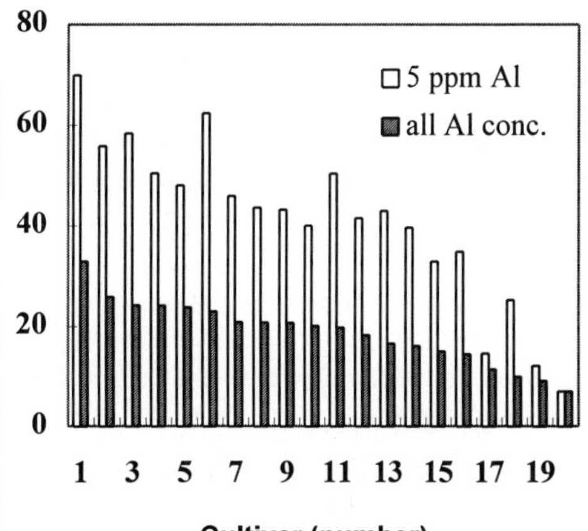

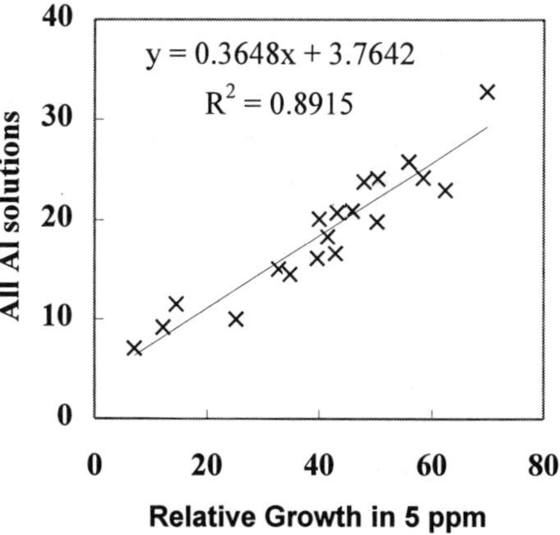

Fig.3 Average root growth in all Al solutions and in the single solution containing 5 ppm Al concentration, relative to the growth in the solution without Al. Code for cv. and statistical significance data in table 1.

Fig.4 Correlation between the relative growth (%) in the solution containing 5 ppm Al and the mean relative growth for all the set of Al treatment solutions.

Figure 1, presenting four typical examples of the growth response to the Al dose, shows the prominent tolerance of Arabian. Further reduction in growth due to Al concentrations higher than 5 ppm, was small. Most cultivars behaved in an intermediate way. However, the significance of the differences can only be appreciated through the relative growth, as will be presented later on.

The cultivars also differed in the extent of the Al after effect as appreciated during the recovery (C3-C2) in the nutrient solution (Fig. 2). Seedlings of most cultivars agree with Borba and TTE 9203 that recovered completely after the 5 ppm treatment and recovered to about 75% after the 10 ppm solution. Anza was irreversibly affected by 5 ppm Al, whereas the triticale Arabian was very tolerant showing even some growth stimulation in the nutrient solutions containing 5 and 10 ppm Al. Slightly affected was TTE 9104 that recovered completely after the 10 ppm Al treatment and showed strong problems only in the 20 ppm Al concentration. Some authors have used (C3-C2) to classify wheat cultivars [8]. We consider this criterion less interesting for several reasons: 1) the possibility of recovery is not realistic in the field; 2) the test will require more manipulations and time; 3) the primary root is sometimes difficult to identify due to stimulated growth of secondary roots, which contributes to large error. So, this criterion was rejected in favor of the root growth during the treatments.

The two sets of experiments could be combined and interpreted together, as shown in Table 1. The ANOVA applied to cultivars, treatments and interactions showed significative differences. Through the application of the LSD (P≤1%) test, the significance of differences was evaluated and groups of cultivars with different tolerance organized. The average for cultivars in all Al treatments clearly shows that Arabian had significantly higher growth than any of the others. On the other extreme of sensitivity to Al were the triticale Beagle and the three cultivars of wheat. Anza presented the lowest growth compared with the Brazilian wheat. Other authors [10] also found great sensitivity of this cultivar to acid soils in the region of Évora.

As a simplification of the method, the response of the root growth to the single 5 ppm Al solution can he used instead of the average response to the whole set of solutions. Figure 3 shows that the solution 5 ppm Al alone was responsible for about 40% of the average root inhibition indicating that such concentration induces a representative response. In effect a good linear correlation ($R^2 = 0.89$) was found between the relatives growths in both solutions (Fig. 4). With a few exceptions the ranking of cultivars by this criterion is similar. Beagle (cv. n° 18) at 5 ppm concentration showed better performance than the wheat, which is in good agreement with field observations in acid soils. Also the cultivars Rat x Borba and Borba x Faísca (cvs. n° 6 and 11) may be more tolerant than was appreciated by the overall test.

Some conclusions can be drawn from the results: 1) the germplasm of all the triticale presented superior tolerance than wheat; 2) there are great differences among genetic lines of triticale which can be appreciated either by the average relative growth of the primary root in the whole range of Al concentrations or, by the relative growth in the single solution with 5 ppm Al; 3) Arabian, a Portuguese cultivar already recognized as having high productivity and quality, showed the highest tolerance.

Acknowledgements

This work has been sponsered by Junta Nacional de Investigação Científica e Tecnológica.

References

1. Foy CD . Plant adaptation to acid , aluminum toxic soil. Commun Soil Sci Plant Anal 1988, 19: 959-87.
2. Almeida LAV . A matéria orgânica e a calagem na fertilização da terra. Memórias da Ordem Eng. 1955,89: 1-16.
3. Kamprath EJ and Foy CD . Lime-fertilizer-plant interactions in acid soils. In: Fertilizer technology and use. 3rd ed.: Soil Sci Soc Am 1985, 49: 91-116.
4. Foy CD and Silva AR . Tolerance of wheat germplasm to acid subsoil. J Plant Nutr 1991, 14: 1277-95.
5. Bagulho F and Carvalho M . Results and perspectives in triticale breeding. Tag-Ber Akad Landwirtsch-Wiss DDR 1988, 266: 407-13.
6. Jessop RS, MacCallum RA, Wright R and Wrigth T . Selecting for improved low pH/Aluminum tolerance in Australian triticale varieties. Proceedings of the Second International Triticale Symposium, Passo Fundo (Brazil), CIMMYT (Mexico DF), 1991: 164-170.
7. Carnide OP, Guedes-Pinto H and Vaz E . Aluminum tolerance behavior of F_2 6x-triticales and their progenitors. Proceedings of the Second International Triticale Symposium, Passo Fundo (Brazil), CIMMYT (México DF), 1991: 268-273.
8. Camargo CEO and Oliveira OF . Tolerância de cultivares de trigo a diferentes níveis de alumínio em solução nutritiva e no solo. Bragantia 1981, 43: 21-31.
9. Pereira JAV . Comparação da tolerância ao alumínio em triticale usando um teste de crescimento radicular. Univ.Évora, 1994 (Relatório de Licenciatura).
10.Carvalho M and Bagulho F . Produção de cereais - solos ácidos: toxicidade de manganês.Melhoramento1992,32:301-15.

SCREENING METHODS FOR ALUMINIUM TOLERANCE IN SEEDLINGS OF TRITICALE

Dorothee Morath[1], Gitta Oettler[2], Albrecht E. Melchinger[1]

[1]Institute of Plant Breeding, Seed Science and Population Genetics,
[2]State Plant Breeding Institute; University of Hohenheim, Stuttgart,
Germany

Abstract

The main reason for inhibited plant growth in many parts of the world are toxic concentrations of Al^{3+} in combination with low mineral availabilty of P and reduced uptake of Ca^{2+}. For developing an effective screening method, seedlings of varieties and strains of winter triticale (×*Triticosecale* Wittmack) were tested in nutrient solution at pH 4.0 with several concentrations of aluminium (Al), calcium (Ca), and phosphorus (P). They were exposed to P and/or Ca for 7 days and during this period to Al for 48h ("pulse method"). Regrowth of Al tolerant roots was assessed by staining them in 0.1% aqueous solution of Eriochrome cyanine R. Characterization of genotypes was based on the percentage of tolerant seedlings. Ca was the main factor for increasing the tolerance in the four varieties used. These results indicated a protecting effect of divalent cations such as Ca^{2+} to Al toxicity in triticale. With Al addition, P stimulated root regrowth of the sensitive variety up to 1.8mM, but had no effect in tolerant varieties. A decreasing tolerance at 3mM P indicated a toxic effect to sensitive genotypes. Concentrations of 1.48mM Al and 2mM Ca maximized differences between genotypes and, thus, were used for screening 11 strains in three replications. Correlations of these results with symptom scores of field tests in different Al-toxic soils were nonsignificant at pH 5.5 (r = -0.19) but moderate (r = -0.47) at pH 5.0. The results indicated that choosing the same pH level in the nutrient solution as under field conditions could result in a higher correlation. Such a test would then be suitable for breeders to screen a large number of genotypes for Al tolerance.

Introduction

Approximately 40% of the world's arable soils are characterized by pH values below 5.0. Under acid conditions, Al is mobilized and regarded as the main toxic factor for plant growth. These conditions cause a deficiency of Ca, Mg, and K and a decrease in P and Mo solubility, which can also inhibit plant growth [1]. Al toxicity can be ameliorated by Ca or Mg ions in the soil, because these are able to displace Al ions from the external binding sites of the roots [2]. Complexing or chelation with phosphate and organic acids, respectively, also reduce Al toxicity in the soil [3].

Liming can neutralize most exchangeable Al, but is often uneconomic and offers no permanent corrective measure for deeper soil layers. Alternatively, the tremen-

H. Guedes-Pinto et al. (eds.), Triticale: Today and Tomorrow, 453–459.

© 1996 *Kluwer Academic Publishers. Printed in the Netherlands.*

dous genetic variability among and within cereal species can be exploited for selecting Al-tolerant genotypes [4].

The mechanisms of Al tolerance are still under investigation [3]. It is known that Al causes disturbances of mitotic division in the meristematic cells and this leads to a reduction in root growth and inhibition of lateral root formation [5]. But Al treated roots are able to regain growth [6].

Seedling tests in nutrient solution measuring the inhibition of root growth are suitable to screen genotypes for Al tolerance in breeding programs. Most authors tested only small numbers of seedlings per genotype, and even the reaction of triticale to Ca and P in Al-toxic solutions has not yet been investigated.

The objectives of the present study were in Experiment 1 to (1) clarify the influence of Ca and P on Al tolerance of triticale seedlings of different genotypes after a short-term Al treatment ("pulse test" after Moore *et al.* [7]) in a nutrient solution and (2) find the optimum concentration of these components for maximizing differences among the four varieties tested. Additionally, in Experiment 2 the results of the screening method based on the optimum nutrient solution were compared with symptom scores under acid field conditions.

Materials and Methods

PLANT MATERIALS

Four registered winter triticale varieties Alamo, Boreas, Clercal, and Lasko were employed (Experiment 1). Eleven advanced strains from the Hohenheim winter triticale breeding program were investigated in Experiment 2. They had been grown and scored (after flowering) in a two-year observation trial in Brazil on two acid soils with pH 5.5 and 5.0.

METHODS

About 450 seeds per genotype were sterilized in 10% NaOCl solution for 10min and rinsed three times in bidistilled water for 10min. Sterilized seeds were soaked for 24h in petri dishes with 1mM $CaSO_4$. Subsequently, three samples of approximately 130 germinated seeds per genotype were transferred to polyethylene fly nets covering 1-liter plastic pots completely filled with aerated nutrient solution. Three pots were used for each of the four varieties. The solution [8] with the lowest Ca concentration contained (mM): 0.4 $CaCl_2{\times}2H_2O$, 6.5 KNO_3, 2.5 $MgCl_2{\times}6H_2O$, 0.1 $(NH_4)SO_4$, 0.4 NH_4NO_3. Al was added as $AlCl_3{\times}6H_2O$ for 48h from a freshly prepared solution. P was added in the required concentrations as KH_2PO_4, and Ca concentrations were varied. There were four pots for one concentration at three different levels. During the experimental period, solutions were exchanged twice and each time the pH was adjusted with HCl to 4.0. The plants were grown in a growth chamber at 22°C, 16h daylength, 2000lx light intensity, and 50% relative humidity.

The Al pulse test [7] was combined with the staining technique of Polle *et al.* [9]. After growing for four days in a solution without Al, the seedlings were transferred to the nutrient medium supplied with Al at various concentrations. After 48h of incubation in this medium, seedlings were thoroughly dabbed and transferred for root regrowth to a nutrient solution without Al for 24h. They were then removed, washed with bidistilled water and stained with 0.1% aqueous solution of Eriochrome cyanine R for 10min. After staining, the excess dye was washed with deionized water.

Experiment 1: In order to investigate the influence of Al, Ca, and P, the four varieties were first tested at 10 different Al concentrations (0, 0.74, 1.48, 1.85, 2.22, 2.59, 2.97, 3.34, 3.71, 5.93mM). Due to technical limitations, only three concentrations could be tested in parallel. At concentrations of 1.48mM Al, Ca was varied between 0.4, 1, 2, 4, and 8mM. At 2mM Ca, which guaranteed good differentiation among the varieties tested, 0.6, 1.8, and 3mM P was added. The experiment was conducted in a completely randomized block design with four pots in one block. Blocks represented the different concentrations of the above mentioned components and were replicated temporally staggered.

Experiment 2: According to the results of Experiment 1, addition of 1.48mM Al and 2mM Ca to the nutrient solution was chosen for maximizing differences among varieties. This composition was used for a test with 11 strains and variety Lasko in three replications. For comparison of the results obtained from the nutrient solution with those of the field symptom scores, the percentage of tolerant seedlings of each strain was correlated with the scores of plants grown on two different acid soils. The experiment was conducted in a completely randomized block design. Each pot represented one genotype and three temporally staggered replications were represented by blocks.

RECORDING OF RESULTS

Root regrowth after Al shock was assessed by staining the roots with Eriochrome cyanine R. Seedlings with apical meristems damaged by a given Al pulse had intensively blue stained root tips, while those not damaged had a stained tip followed by a white section. Connected to this zone was a stained region again. This ring formation of roots was used as a criterion for Al tolerance. A plant with more than 50% tolerant roots was characterized as tolerant. Thus, each genotype could be divided into sensitive and tolerant plants.

STATISTICAL ANALYSIS

An analysis of variance was conducted with the data on percentage of tolerant seedlings per genotype and concentration (block). Effects of genotypes and concentrations of Al and P were considered fixed.

Results

EXPERIMENT 1

Al: Increasing Al levels reduced root length after 48h exposure. Roots of the control without Al were normally developed. Al caused brown discolorations at the tips and the sensitive ones appeared stunted and thick. The percentage of tolerant plants of all varieties decreased with increasing Al concentrations (Figure 1).

456

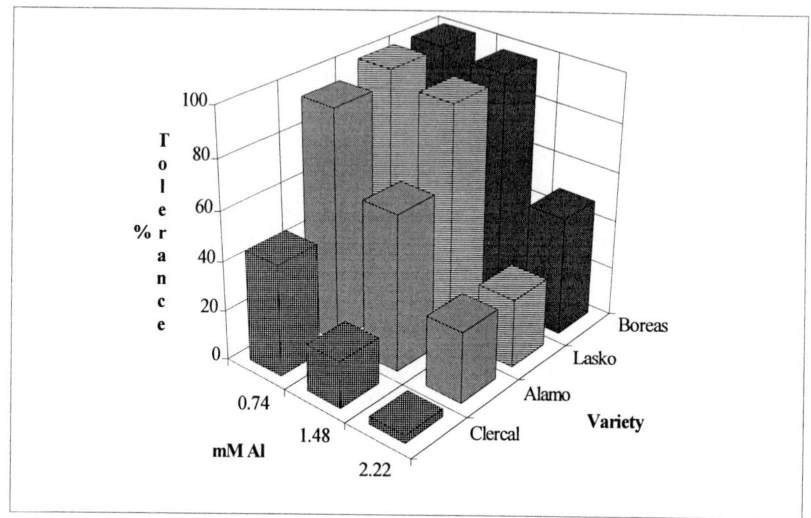

Figure 1: Mean percentage of tolerant seedlings of four triticale varieties
tested in nutrient solution at three Al concentrations across four replications

Highly significant differences between varieties were found (Table 1). The best differentiation between varieties was observed at 1.48mM Al. Therefore, this concentration was
used for varying the Ca levels.

Table 1: Mean squares (MQ) obtained from analysis of variance for
percentage of tolerant seedlings after staining roots with Eriochrome cyanine R in four triticale varieties tested at different Al
and P concentrations

| | | Al | | P |
Source	df	MQ	df	MQ
Concentration	2	14238 **	1	269
Variety	3	8349 **	3	14576 **
Replication	3	2785 **	1	536
Pooled error	39	448	10	2609

** Significant at probability level $\alpha=0.01$

Ca: With increasing Ca concentration, an increase in Al tolerance was observed, but the
effect of Ca on improved Al tolerance was smaller for tolerant varieties (Boreas and
Lasko) than for Alamo and the sensitive variety Clercal. For maximizing differentiation,
2mM Ca proved to be most suitable.

P: Increasing P addition to the nutrient solution increased root growth *per se*. P had no
effect on improving root regrowth for tolerant genotypes after Al treatment. It stimulated, however, the sensitive Clercal up to 1.8mM, and at 3mM P a decrease of tolerance
was recorded. Additionally, as differences between P concentrations 0 and 3mM were
not significant (Table 1), P was not added in further tests.

EXPERIMENT 2

Field: Symptom scores (1 = tolerant, 9 = sensitive) on soils containing high levels of Al showed a wide range of 1 to 7 at pH 5.0 as compared to 1 to 4 at pH 5.5. A correlation of r = 0.48 between these symptom scores suggested that these differences are due to soil effects.

Nutrient solution: By testing 11 strains and Lasko with 2mM Ca and 1.48mM Al, extremely sensitive genotypes like 22 or 70 were characterized very well, but other sensitive triticales like 69 or 85 showed more than 90% tolerance (Figure 2). Thus, a differentiation was not possible.

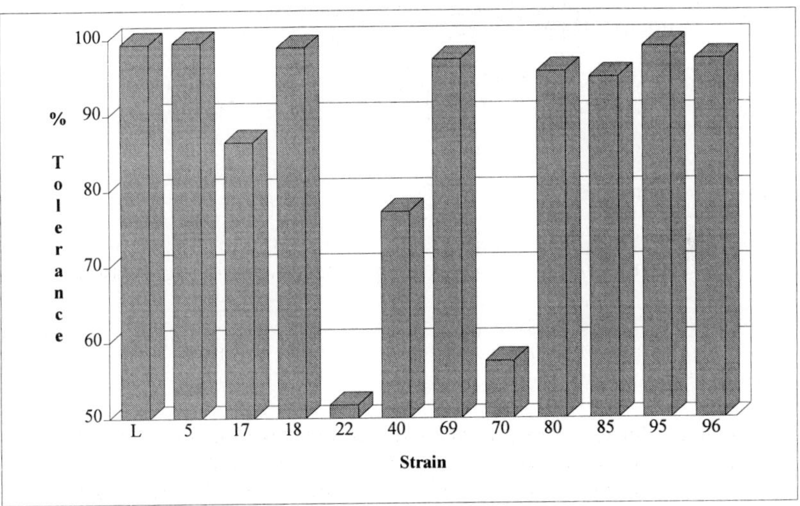

Figure 2: Mean Al tolerance across three replications of 11 triticale strains and variety Lasko (L) in nutrient solution with 2mM Ca and 1.48mM Al

Comparison: Table 2 compares the results from the nutrient solution and the field scores. In both tests, strain 22 was found to be the most sensitive genotype. Likewise, for strain 70 the 57% tolerance corresponded well with the field score 6 at pH 5.0, but was not consistent with the score at pH 5.5. No correlation (r = -0.19) was found between percentage of tolerant seedlings of each genotype with the respective value of mean scores on a less acid soil (pH 5.5). A comparison with field scores on soil of pH 5.0 yielded a moderate correlation (r = -0.47).

Table 2: Percentage of tolerant plants in nutrient solution and mean field scores (1= tolerant, 9 = sensitive) at pH 5.0 and 5.5 for 11 strains and variety Lasko (L)

Designation		L	5	17	18	22	40	69	70	80	85	95	96
Tolerance [%]			99	99	86	99	52	77	97	57	95	95	99	97
Field scores:	pH 5.0	1.0	3.5	5.0	3.0	7.0	2.0	6.0	6.0	4.5	7.5	1.0	3.5
	pH 5.5	1.0	3.5	3.5	3.5	4.0	1.5	3.0	1.5	1.0	3.0	1.0	1.0

Discussion

EXPERIMENT 1

Al: The great variability for Al tolerance among and within varieties in triticale are consistent with similar findings in wheat [3]. The genotype-specific reduction in root growth with increasing Al concentration could be attributed to different mechanisms of Al tolerance.

Ca: Improvement of Al tolerance (less root inhibition) by increased Ca concentrations indicates that Ca^{2+} and other divalent cations can protect triticale roots from the toxic effects of Al. This is also known from wheat [10]. Presumably, increasing Ca concentration reduces Al uptake into the roots or Al binding to the rhizodermis.

P: Enhanced formation of lateral roots and regrowth of root tips by addition of P to the nutrient solution could be explained by a better P supply. However, P addition did not improve Al tolerance in Alamo, Boreas and Lasko. Foy [4] concluded from investigations with wheat that Al tolerance is associated with the ability to tolerate low lelvels of P. Thus, varieties may react more tolerant to surplus of Al in acid soils. Nevertheless, the tolerance of Clercal depended on P supply. Ben [11] described similar reactions of triticale tested in the field under different P levels. The initial increase in Al tolerance of Clercal was possibly due to a better P supply. The decrease at 3mM P could be explained with a toxic effect of P to Al sensitive varieties. To avoid precipitation of $Al(PO_4)$, Moore *et al.* [7] suggested to omit P from the Al treatment in solutions.

EXPERIMENT 2

A comparison of Hematoxylin-stained roots from nutrient solution tests with plant scores in soils with different acidity showed no differences between the tolerant genotypes on less acid plots for wheat [12]. The authors attributed this to a sufficiently strong tolerance of the better genotypes. Acid stress of the soils was insufficient to differentiate between wheat genotypes. In our experiment, phenotypic differences at pH 5.5 were too minor for an efficient selection of Al tolerance. The low selection pressure became apparent in the range of symptom scores (Table 2). Consequently, acid soils with pH below 5.0 appear to be more suitable for selection.

Conclusions

Soil solution tests could be a suitable method to clarify the mechanisms of Al tolerance and to investigate factors like root releases of chelators for Al, nutrient uptake and nutrient availability of the soil [13].

Staining of seedling roots is a simple method and, therefore, suitable for screening of Al tolerance in breeding programs. If the selection pressure (pH) in the nutrient solution is adapted to field conditions, a stronger correlation could be expected. Symptom scores of vegetative plant growth alone are not sufficient for testing the correlation to nutrient solution tests, because yield and its components are of equal importance to breeders.

Acknowledgements

The Vater und Sohn Eiselen-Stiftung, Ulm, is gratefully acknowledged for supporting this research. We also thank Dr. A.C. Baier, Passo Fundo, Brazil, for conducting the field observations.

References

1. Haug A. Molecular aspects of aluminium toxicity. CRC Crit Rev Plant Sci 1984; 1:345-73.
2. Kinraide TB, Parker DR. Cation amelioration of aluminum toxicity in wheat. Plant Physiol 1987; 83:546-51.
3. Taylor GJ. Mechanisms of aluminium tolerance in Triticum aestivum (wheat) V. Nitrogen nutrition, plant induced pH, and tolerance to aluminium; correlation without causality? Can J Bot 1988; 66:694-99.
4. Foy CD. Plant adaption to acid, aluminum-toxic soils. Comm in Soil Sci Plant Anal 1988; 19:959-87.
5. Marschner H. Mechanisms of adaption of plants to acid soils. In: Wright RJ *et al.*, editors. Plant-soil interactions at low pH. Dordrecht: Kluwer Academic Publishers, 1991: 683-702.
6. Bennet RJ, Breen CM, Bandur VH. A role for Ca^{2+} in the cellular differentiation of root cap cells: A re-examination of root growth control mechanisms. Environ Exp Bot 1990; 30:515-23.
7. Moore DP, Kronstad WE, Metzger RJ. Screening wheat for aluminium tolerance. In: Klatt AR, editor. Wheat production constraints in tropical environments. Mexico D.F.: CIMMYT. 1988:287-95.
8. Polle E, Konzak CF, Kittrick JA. Visual detection of aluminium tolerance levels in wheat by hematoxylin staining of seedling roots. Crop Sci 1978; 18:823-27.
9. Lopez Cesati J, Villegas E, Rajaram R. CIMMYT's laboratory method for screening wheat seedling tolerance to aluminium. In: Kohl MM, Rajaram S, editors. Wheat breeding for acid soils: Review of Brazilian/CIMMYT collaboration 1974-1986. Mexico, D.F.: CIMMYT, 1988: 59-61.
10. Aniol A. Aluminium uptake by roots of two winter wheat varieties of different tolerance to aluminium. Biochem Physiol Pflanzen 1983; 178:1-20.
11. Ben JR. Response of triticale, wheat, rapeseed and lupine to phosphorus in soil. CIMMYT, Proceedings 2nd Int Triticale Symposium; 1990 Oct 1-5; Passo Fundo. Mexico DF: CIMMYT, 1991. 207-11.
12. Ruiz-Torres NA, Carver BF, Westermann RL. Agronomic performance in acid soils of wheat lines selected for hematoxylin staining pattern. Crop Sci 1992; 32:104-07.
13. Wright RJ, Baligar VC, Ritchey KD, Wright SF. Influence of soil solution aluminum on root elongation of wheat seedlings. Plant and Soil 1989; 113:294-98.

THE VARIABILITY OF ALUMINIUM TOLERANCE AMONG TRITICALE STRAINS AND CULTIVARS BRED IN POLAND

Andrzej Aniol
Plant Breeding and Acclimatization Institute, Radzików, Poland

Abstract

In order to assess the aluminium tolerance level in breeding materials of triticale approximately 200 strains and cultivars were tested in 1983 and 1993. Tested strains originated from the main triticale breeding centers in Poland and were both winter and spring types. The resistance of root apical meristems to aluminium shock applied in nutrient solution to 4-days old seedlings was used as a measure of aluminium tolerance of tested genotypes [5]. Triticale strains were tested at least twice in the independent experiments using approx. 50 seedlings of each cultivar in single experiment. Aluminium tolerance of screened genotypes was compared to two checks: Al- tolerant wheat cultivar BH 1146 and rye variety Dankowskie Zlote. Triticale strains of Al-tolerance below the level found in tolerant wheat were classified as very sensitive to aluminium, while those of Al-tolerance approaching the level of rye were regarded as very tolerant to aluminium. Two classes of intermediate tolerance were also established.

The variability of aluminium tolerance found in tested genotypes was compared to that found ten years ago among analogous material of advanced strains and new cultivars. The ranking of Triticale cultivars with respect to aluminium tolerance was compared to their performance in preliminary and official field trials.

It is concluded that very large variability exists among advanced Triticale strains with respect to aluminium tolerance: from very sensitive ones at the level of wheat cultivars up to very tolerant similar to rye. The proportion of very sensitive strains and cultivars increased during the last decade, especially among winter forms. The reasons and consequences of this shift are discussed.

Introduction

Aluminium tolerance is regarded as an important prerequisite for plant adaptation to soil acidity [1]. Natural soil acidification is in recent decades accelerated bu use of mineral fertilizers and industrial pollution.

The differential reaction to aluminium ions was found between cereal species as well as among cultivars of the same species. Rye (*Secale cereale* L.) was found to be the most tolerant cereal but also in this crop wide variability was found among inbred lines [2]. However, rye grain has a poor feeding value because of the presence of some antinutritional compounds. Triticale as a wheat-rye hybrid theoretically should combine good feeding value of wheat grain and tolerance to soil acidity of rye parent. For this reasons breeding programs of triticale were developed and executed in Poland in recent

461

H. Guedes-Pinto et al. (eds.), Triticale: Today and Tomorrow, 461–465.

© 1996 *Kluwer Academic Publishers. Printed in the Netherlands.*

decades. Developed triticale cultivars were successfully implemented into the Polish agriculture. This new cereal crop is now cultivated on approx. 750 000 ha. Feeding value of triticale grain is much better than that of rye and is comparable with the feeding value of wheat or barley grain [3]. However the adaptation of triticale cultivars to less fertile often acid soils occupied by rye is not sufficient. It is an important problem in Poland where 60% of arable land is classified as acid or very acid soils [4].

Materials and Methods

Triticale cultivars and strains tested for tolerance to aluminium originated from inter-station and preliminary field trials located in Radzików in 1982/93 and 1992/93 seasons. The most advanced strains and top cultivars from the main Triticale breeding centers were included in those trials, therefore this material is representative for the Triticale germplasm utilized in the breeding programs.

ALUMINIUM TOLERANCE SCREENING TEST

Seeds were sterilized with 0.1% $HgCl_2$ aqueous solution for 10 min., rinsed excesively with water, and germinated overnight on filter paper in Petri dishes. Sprouted kernels were sown next day on polyethylene net fixed in lucite frames. Styrofoam blocks were attached to the framers with rubber bands and floated on the surface of vigorously aerated nutrient solution. Containers with nutrient solution were placed in water bath at 20°C under 16 h per day illumination with low intensity incandescent light. Standard nutrient solution was used with aluminium added as indicated in the experiments [5]. Four days old seedlings were used for incubation with aluminium. After Al-shock, seedlings were removed, thoroughly washed for 1-2 min. in running tap water and transferred to Al-free nutrient solution for the next 48 h. The root regrowth or additional root growth after Al-shock was easily assessed by staining roots with 0.1% aqueous solution of Eriochrome cyanine R dye for 10 min. After staining, the excess of dye was removed by washing under tap water. During all stages of growth, and particularly during the Al-treatment the nutrient solution was maintained at pH 4.5.

Triticale strains and cultivars were tested at 16 ppm (0.593mM) aluminium added as aluminium chloride during 24 h. Approx. 50 seedlings of each strain were tested al least twice. Per cent of seedlings with root regrowth was calculated and compared with BH 1146 Al-tolerant cultivar used as a standard. Triticale strains with less than 50% tolerant seedlings were classified as very sensitive to aluminium (class 1), those showing more than 50% of tolerant seedlings were classified as sensitive (class 2, level of tolerance observed in Al- tolerant wheat). All strains classified in group 2 were further tested at the same Al-conc. but for 48 h with rye cultivar Dankowskie Zlote used as a check. At this conditions apical meristems of Al-tolerant wheat standard were completely destroyed while those of rye survived in more than 75% seedlings. Triticale strains with more than 25% but less than 50%

of tolerant seedlings were classified as Al-tolerant (class 3) and strains with more than 50% of tolerant seedlings were grouped in class 4 of very tolerant genotypes.

Results and Discussion

VARIABILITY BETWEEN STRAINS.

The distribution of tested Triticale strains from 1982/83 and 1992/93 in different Al-tolerance classes is presented in Figures 1.

The important changes occurred during the last decade in the distribution of winter Triticale strains in different Al-tolerance classes (Fig.1); the number of very sensitive strains increased markedly and is accompanied by the decrease of the number of Al-tolerant strains (class 3). At the same time some tendency to increase in number of very tolerant strains (class 4) can be observed. The observed decrease in Al-tolerance level among the most advanced winter Triticale strains is a undesirable development, which, if continued, could adversely affect the future of this crop in Polish agriculture. In my opinion, Triticale has its place in our farming system mainly as an substitute for rye in production of good quality feeding grain on rye soils. It means that the level of aluminium tolerance, similar to that found in rye populations, is very important character in this cereal.

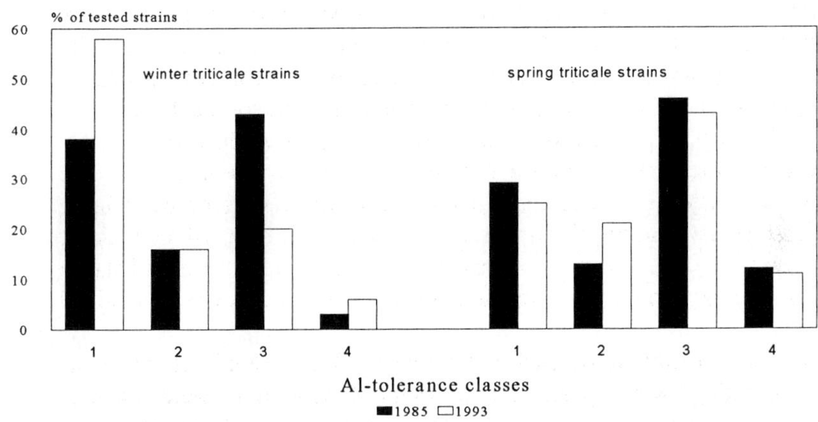

Figure 1. The distribution of winter Triticale strains in different Al-tolerance classes in 1992 / 93 as compared to the distribution of analogous breeding material one decade ago. The criteria of tolerance classes are described in the "Material and Methods"; class 1- very sensitive; class 2 - sensitive; class 3 - tolerant and class 4 - very tolerant.

One can speculate on many reasons for observed decrease in Al-tolerance level among the winter Triticale strains. Triticale was developed and introduced into the Polish farming in the environment of central-planned economy; in the situation of permanent shortages of feed grain when agricultural production was geared at maximum yield with little attention paid to the costs of inputs, often heavily subsidized. Under these circumstances, breeders were pressed to develop high yielding cultivars at possible short time. This tendency in Triticale breeding resulted in selection and introduction of wheat-type Triticale cultivars outyielding rye and even wheat on good rye and wheat soils. This tendency is illustrated by the data on soil pH at locations where the official variety testing of Triticale and rye were performed (Fig.2).

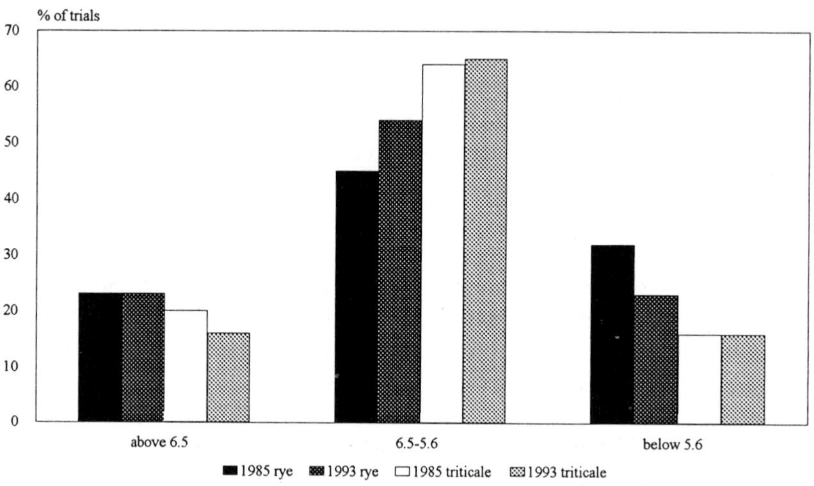

Figure 2. Soil pH at locations of official variety testing for winter Triticale and rye.

Only 16% of Triticale trials were performed on soils of pH below 5.6 in which the potential aluminium toxicity might occur. This is in sharp contrast with the situation on farms where 60% of arable land is classified as acid (soil pH_{KCl} 4.6-5.5) or very acid (soil pH below 4.5).
It means that Al-tolerance is practically eliminated from the official variety evaluation system. It is not surprising that winter Triticale production is restricted to the best rye and wheat soils. The described situation is also the reason for observed decline and stagnation of Triticale acreage when the economical system was changed into the market oriented system; when grain shortage was suddenly changed into grain surplus wheat grain with established industrial market is much more attractive than Triticale.

The changes in the distribution pattern in different Al-tolerance classes among spring forms of Triticale are much smaller (Fig.1). However some trend can be seen; here also the increase in number of Al-sensitive strains in class 2 is accompanied by a tendency to decrease in tolerant and very tolerant classes 3 and 4. The smaller changes in distribution pattern in spring Triticale as compared to winter forms might be explained by the fact that spring Triticale is a very recent crop in Poland with only a few cultivars registered so far and might be less affected by changes in farming system and variety testing criteria.

VARIABILITY WITHIN STRAINS.

The variability in aluminium tolerance between Triticale strains described above makes the selection of more tolerant cultivars possible. Situation of the breeder is even better because of variability found within strains. A good example is winter cultivar Bogo (Tab.1).

Table 1. Heterogeneity of Triticale strains with respect to aluminium tolerance.

Strain	root regrowth after Al - treatment					
	16 ppm / 24 h			16 ppm / 48 h		
	# seedl.	%	mm	# seedl.	%	mm
OZH 252 / 2 / 88	95	64	11	60	40	7
Gabo	62	92	10	69	0	0
Bogo	114	36	5			
Bogo (hairy)	101	52	10			

Selection within this cultivar, based on a morphological character led to the development of two sublines: normal stem and hairy neck stem and this sublines also differ markedly in Al-tolerance. One can conclude that in some Triticale cultivars and advanced lines Al-tolerance character can be significantly improved by subline selection procedure.

Data presented in Table 1 indicate also, that bulk screening procedure resulting in allocation of tested strains to the established Al-tolerance classes should be treated only as an approximate assessment of this character. In some strains allocation to Al-tolerance class based on the percentage of tolerant seedlings might be misleading; e.g. seedlings of Gabo cultivar when tested at 16 ppm Al for 24h survived in more than 90% and might be allocated to tolerant class while seedlings of strain OZH 252/2/88 survived this Al-shock only in approx. 60% and might be regarded as sensitive. However when tested at the same Al-conc. one day longer even a single seedling of Gabo did not survive while approx. 40% of OZH 252 / 2 / 88 seedlings survived this Al-shock. The above results suggest that before a particular Triticale strain is used as Al-tolerance source in breeding program its Al- tolerance should be assessed more precisely than in routine screening

References

1. Foy CD, Chaney RL, White MC. The physiology of metal toxicity in plants. Ann. Rev. Plant Physiol. 1978; 29:511
2. Anioł A, Hill RD, Larter EN. Aluminum tolerance of spring rye inbred lines. Crop Sci. 1980; 20: 205-208.
3. Rakowska M, Neuman M. Nutritive value of Triticale as food and feed. Hod. Ros. Nas. 1980;.24: 615-624.
4. Boguszewski W. Wapnowanie Gleb. 1980. PWRiL Warszawa.
5. Anioł A. Aluminium uptake by roots of two winter wheat varieties of different tolerance to aluminium. Biochem. Physiol. Pflanzen 1983; 178: 11-20.

TOLERANCE TO ALUMINIUM IN SPRING TRITICALES

Grzegorz Budzianowski[1], Walenty Maćkowiak[1],
Kazimierz Paizert[1], Barbara Apolinarska[2]

[1]Plant Breeding and Acclimatization Institute, Experiment
 Station "Małyszyn", 66-400 Gorzów Wlkp., Poland
[2]Institute of Plant Genetics of Polish Academy of Sciences
 60-479 Poznań, Poland

Abstract

The results from many experiments conducted over 4
years to determine the tolerance of 169 entries of Polish
spring triticale and 63 entries of spring triticale from
CIMMYT to aluminium are summarized. All experiments were
conducted in nutrient solutions at varying aluminium concentrations (5, 15 and 25 ppm). The LD50 concentration of Al
ions (ppm) was determined for each entry. All cultivars,
strains and lines were classifed as very sensitive (LD50
<5ppm), moderately sensitive (LD50 5-15ppm), moderately
tolerant (LD50 15-25ppm) or tolerant (LD50 >25ppm).
Most of Polish and CIMMYT's spring triticales was
ranked in the moderately sensitive and moderately tolerant
categories. A greater number of Al-tolerant lines was found
among CIMMYT's lines (18 genotypes) than Polish strains
(14). The pedigrees of the most Al-tolerant Polish strains
could be traced to closely related lines; however, a wide
range of tolerance levels to aluminium was found among
strains of similar origin. Advanced lines with 6D/6A substitution from CIMMYT often exhibitied a higher percentage of
seedlings with root regrowth than lines with the unmodified
(AABBRR) chromosome constitution at different concentrations
of Al ions in nutrient solutions.

H. Guedes-Pinto et al. (eds.), Triticale: Today and Tomorrow, 467–474.
© 1996 *Kluwer Academic Publishers. Printed in the Netherlands.*

Introduction

Acid and very acid soils predominate in Poland and occupy about 61% of agricultural areas [1]. In such soil conditions aluminium toxicity is regarded as the most important yield-limiting factor [2]. Since aluminium toxicity cannot always be corrected by changing soil conditions, breeding for Al tolerance is an alternative solution [3].

The large-scale screening for Al tolerance of Polish winter and spring triticale genotypes was performed in 1984 and a wide spectrum of tolerance was found in the breeding material [4]. In the last decade triticale has became widely cultivated in Poland. To date four spring cultivars of triticale have been released for production in Poland [5].

Progress of breeding spring triticale in Poland and at the International Maize and Wheat Improvement Center (CIMMYT) has created a wide range of new genotypes. The increased levels of exchange of breeding materials between breeding programmes has increased the possibility for changes in Al tolerance. Because of this potential the authors decided to screen recently relesed triticale genotypes for Al tolerance. Data for both Polish spring lines and CIMMYT releases are summarized.

Materials and methods

Cultivars for assessment came from the Plant Breeding and Acclimatization Experiment Station Małyszyn (ZDHAR Małyszyn,159), Plant Breeding and Acclimatization Institut Radzików (IHAR, 8), and Plant Breeding Company "Danko" (2). Seed samples of lines from CIMMYT were kindly supplied by Dr. G. Varughese in 1990 (26 lines) and Dr.W.H. Pfeiffer and Dr. P.N. Fox in 1992 (37 lines) from the 23rd International Triticale Yield Nurseries (ITYN).

Data on the chromosome constitution of the 14 CIMMYT lines from 1990 were kindly supplied by Prof. A.J. Łukaszewski, University of California, USA. The remaining 12 lines from 1990 were analysed for their chromosome constitution using the C-banding technique [6].

The modified-pulse method as described by Anioł [7,4] was used with only minor modifications - $AlCl_3x6H_2O$ was substituted for $AlKSO_4x12H_2O$ as the source of aluminium. All the genotypes were subjected to three test solutions containing 5, 15 and 25ppm aluminium. The number of seedlings with root regrowth after aluminium stress was determined for each entry. Genotypes were tested at least twice in independent experments using approximately 50 seedlings each entry in a single test. Standard cultivars of known Al tolerance were used in each test; there were tolerant and sensitive cultivars of spring wheat (BH 1146 and Jara),spring triticale c.v. Jago (bred in ZDHAR Małyszyn) and spring rye c.v. Strzekęcińskie.

The LD50 (ppm), i.e. the concentration of aluminium ions at which 50% of seedlings were irreversibly damaged, was calculated for each entry using the probit method [8].

Results and discussion

In order to rank all genotypes for relative tolerance to Al toxicity we used four categories based on the LD50 [Table 1]. These categories allowed comparisons of Al tolerance level between cultivars of spring cereals [Table 2], and among strains and lines from different breeding centries [Table 3].

Table 1. Categories for tolerance to solution aluminium based on the aluminium concentration (ppm) required to reduce number of seedlings with root regrowth by 50% (LD50)

Category of Al tolerance	Range in LD50 (ppm)
Sensitive	LD50 < 5
Moderately sensitive (MS)	5 < LD50 < 15
Moderately tolerant (MT)	15 < LD50 < 25
Tolerant (T)	LD50 > 25

Table 2. The tolerance of selected cultivars of spring cereals to increasing Al concentration in the medium

Species and cultivar	% of tolerant seedling			Category of Al tolerance
	Al concentration (ppm)			
	5	15	25	
Spring triticale				
JAGO	86	59	37	MT
Spring wheat				
Jara	0	0	0	S
BH 1146	100	37	0	MS
Spring rye				
Strzekęcińskie	100	100	63	T

Most examined entries from Poland were either moderately tolerant or moderately sensitive (39 and 41% of all forms tested, respectively). Similar distribution of relative tolerance to Al was observed among lines from CIMMYT. However, more triticale genotypes were found in the tolerant category (Al tolerance level similar to rye) among the lines from CIMMYT, compared with the Polish strains.

Table 3. Aluminium tolerance of spring triticale strains bred in Poland and advanced lines from CIMMYT, Mexico

Location	Percentage of forms with Al tolerance category			
	sensitive	moderately sensitive	moderately tolerant	tolerant
Poland (n-169)	11.2	41.4	39.1	8.3
CIMMYT (n-63)	14.2	33.3	23.8	28.6

Pedigrees of the most AL tolerant Polish strains could be traced to a closely related lines [Table 4]. There were also clear influences of germplasm from CIMMYT on Al tolerance in many lines.

Table 4. Pedigrees of the most aluminium tolerant strains of spring triticale bred in Poland

Strain	Pedigree
MAH*7695-9/1	238M2A Beagle x A-8-1<French form>
MAH 7746-50	[(Armadillo x sw Bastion) x Beagle] x A-8-2
MAH 7746-80	- " -
MAH 13313-20	7695-R-1 x 7746-10 <7746-10=JAGO>
MAH 14269-5	[(6T6T x sw Kolibri) x Beagle] x 7695-502
MAH 14271-2	4226-7 x 7746-9
MAH 17582-1	(4226-7 x 7746-10) x 4226-17
MAH 18341-1	7746-10/7 x EDA"S"/PND"S"RM"S"-51
MAH 19150-1	[(7746-9-23 x 7746-9)x 7746-10/7]x 13313-10
MAH 19172-1	(Caborca 79 x Jago) x 13313-3-20
MAH 19182-1	(7746-10/3 x Tapir"S") x 13313-10-2
MAH 19183-1	(7746-10/3 x Hare"S"-164) x 13313-10-2
RAH**1024	MAH 7746-3/9 - reselection
RAH 1025	N.9-Cin"R" - reselection

*; ** - strains from ZDHAR Małyszyn and IHAR, respectively
sw - spring wheat

However, a wide range of Al tolerance was found among strains from the same origin. For example, LD50 values for 12 strains selected from hybrid 7746 ranged from 4.5 to >25 ppm (unpublished data). The present results and data obtained during screening of breeding material from 1984 [4] showed a similar distribution of both sets of genotypes tested. The percentage of recently released strains which could be considered as desirable for breeding and the percentage of Al tolerant ones from 1984 were 47 and 40%, respectively.

Table 5.Aluminium tolerant lines of spring triticale from CIMMYT, Mexico

Entry number	Name or cross
S*-60	STIER 25
S-128	NIMIR
S-154	GNU 8
S-155	STIER 29
S-156	GAUR 3
S-168	RAM"S"2//PF70354/BOW"S"/3/RAM"S"2
23 ITYN**1	CANANEA 79
9	BAGAL 3
10	BAT/RHINO 3
14	BGL/CIN//MUS/4/DLF99/3/M2A/F3SNP//BGL
15	EMS 6TA.313A HEAD7/3/BCM/GPR//PANDA/4/TAPIR
20	GAUR 2
21	HARE 132/TESMO 3//ZEBRA 79
22	KATZE//JUP/BJY/3/TESMO 3
26	MERINO/JLO//REH/3/HARE 267
28	NIMIR 4-1
30	REH/HARE 212-7
36	ZEBRA 32/3/BGL/CIN//MUS/4/LYNX

*; ** entry number from year 1990 and 1992, respectively.

Table 6. Differential reaction of lines with 6D/6A substitution and lines with AABBRR karyotype to aluminium

Chromosomal constitution	Number of lines	Mean % of tolerant seedlings		
		Al concentration (ppm)		
		5	15	25
6D/6A	12	81.9	62.9	47.4
AABBRR	14	66.0	29.7	12.0

The names and numbers of crosses of the most Al tolerant lines from CIMMYT are presented in Table 5. All advanced lines from 1990 which were classifed to Al tolerant category had a 6D/6A substitution. This new chromosome constitution was identified in both spring and winter triticale, and within a few years spread to about one-fourth of CIMMMYT's spring materials [9]. It is likely that the 6D/6A substitution arise from crosses between octoploid and hexaploid triticale at CIMMYT.

Table 7. Aluminium tolerance and chromosomal constitution of advanced lines of spring triticale bred in CIMMYT, Mexico

Name or cross	Chromosomal constitution	% of tolerant seedlings Al concentration (ppm)		
		5	15	25
GRF"S"/YOGUI 1	AABBRR	74.2	8.4	0.0
YOGUI 1/GRF"S"	AABBRR	89.0	27.2	11.3
CIVET"S"	AABBRR	74.1	28.2	0.0
LAMB 2	AABBRR	62.9	6.7	0.0
FAHAD 3	AABBRR	51.8	4.1	0.0
DAHBI 1	AABBRR	89.3	37.5	5.1
ILO"S"/PTR"S"//YOGUI 1	AABBRR	77.0	51.5	32.2
KER 4	AABBRR	100.0	54.0	12.6
SIKA 26	AABBRR	100.0	79.1	31.0
TATU 3	AABBRR	100.0	37.2	31.0
PTR"S"CASTOR"S"//BTA"S"	AABBRR	41.2	34.3	28.5
IGUANA 4-3	AABBRR	63.9	48.2	16.9
ANOAS 3	AABBRR	0.0	0.0	0.0
ANOAS 5	AABBRR	0.0	0.0	0.0
REH"S"/HARE 212-12	6D/6A	90.2	67.1	44.7
STIER 25	6D/6A	100.0	90.8	82.7
MUS"S"*2/ZEBRA 79	6D/6A	24.8	0.0	0.0
FARAS 1	6D/6A	68.9	0.0	0.0
ARD 4	6D/6A	88.8	77.0	41.4
NIMIR 4-1	6D/6A	76.7	70.6	62.1
HARE 7265/YOGUI 1	6D/6A	87.4	56.5	42.0
GNU 8	6D/6A	92.8	87.2	57.8
STIER 29	6D/6A	86.4	80.1	71.9
GAUR 3	6D/6A	100.0	97.2	69.7
RAM"S"2//PF70354/BOW"S"/3/ RAM"S"2	6D/6A	100.0	88.3	71.9
RONDO//REH"S"/HARE 212-11/ /3/REH"S"/HARE 212-11	6D/6A	66.9	40.4	24.1

Among 26 lines from 1990 twelve had 6D/6A substitution and the remaining 14 lines had an unmodified AABBRR chromosome constitution. A comparison between the average level of Al tolerance of lines with 6D/6A substitution and average level of Al tolerance of the unmodified ones showed that the occurance of tolerant seedlings was much more frequent among the former than among the latter. Such differences were especially evident at the aluminium concentration of 25 ppm [Table 6]. This phenomenon, probably in part, explains the expansion of genotypes with 6D/6A substitution among advaned lines at CIMMYT (A.J. Łukaszewski, personal communication). However, a wide range of categories of Al tolerance was found both among lines with 6D/6A and among unmodified ones [Table 7]. Neverthless, the presence of a new chromosomal constitution suggests that the optimal karyotype has not been determined and, probably, a single optimal chromosomal configuration for all conditions is unlikely [9, 10].

The germplasm from CIMMYT has been used as a source of a new variability by breeders from many triticale breeding centries. The results given in the Tables indicate that incorporation of the CIMMYT's germplasm into Polish triticale remais a suitable way to improve Al tolerance.

Acknowledgments

The authors wish to thank Professor Adam Łukaszwski, University of California for providing the data achromosomal constitution and Mrs. Danuta Biała for providing technical assistance.

References

1. Boguszewski W. Wapnowanie gleb. PWRiL Warszawa,1980.
2. Foy CD, Chaney RL, White MC. The physiology of metal toxicity in plants. Ann Rev Plant Physiol 1978;29:511-566.
3. Foy CD. Plant adaptation to acid, aluminum toxic soils. Commun Soil Sci Plant Anal 1988;19:959-987.
4. Anioł A. Breeding of triticale for aluminum tolerance. Genetics and breeding of Triticale, Eucarpia meeting, Clermont-Ferrand 2-5 July 1984. - INRA, Paris, 1985.
5. Maćkowiak W, Paizert K, Mazurkiewicz L, Woś H. Osiągnięcia i problemy hodowli pszenżyta w Polsce. Biul IHAR 1993; 187:143-166.
6. Łukaszewski AJ, Gustafson JP. Translocations and modyfications of chromosomes in triticale x wheat hybrids. Theor Appl Genet 1983;64:239-248.
7. Anioł A. Aluminum uptake by roots of two winter wheat

of different tolerance to aluminum. Biochem Physiol Pflanzen 1983;178:11-20.

8. Finney D. Probit Analisys. University Press,Cambridge, 1952.

9. Łukaszewski AJ. Chromosome constitution of hexaploid triticale lines in the recent International Yield Trials. Plant Breeding 1988;100:268-272.

10. Pfeiffer WH, Fox PN. Adaptation of Triticale. Proceedings of the Second International Triticale Symposium; 1990 October 1-5; Passo Fundo, Rio Grande do Sul, Brazil. CIMMYT, Mexico DF, 1990.

CHANGES IN pH OF THE ROOT MEDIUM PRODUCED BY CULTIVARS OF TRITICALE

Ana Maria Antunes[1], José Pereira [2] and Maria Antonieta Nunes[1]
1) Instituto de Investigação Científica Tropical, Lisboa, Portugal
2) Instituto Nacional de Investigação Agrária, Oeiras, Portugal

Abstract

The induction of an increase in the pH of the root medium has been indicated for several plants as part of a strategy for living in acid soils. The possibility of such occurrence has been evaluated for some genetic lines of triticale and wheat for which there are also other indications of tolerance to acidity and high Al concentration. After germination, ten seedlings of each strain were grown in a convenient volume of nutrient solution with or without 5 ppm of Al for five days. The pH, ion conductivity of the solution and the length of the primary root were also measured after 2 and 5 days of growth. The change in pH was small in the presence of Al, but in the standard or diluted (1/5) nutrient solutions there were significant differences among genotypes. In standard solution the triticale Borba increased the pH significantly whereas the wheat Anza as well as the triticale Arabian induced minor changes. Since Arabian has been recognised as tolerant to Al this indicates that other mechanisms (not modification of pH) must be present in such plants.

Introduction

Under acid soil conditions the major constraints to plant growth and production are the following: (I) H^+ toxicity; (II) Al toxicity; (III) Mn toxicity; (IV) Mg^{2+} , Ca^{2+} and K^+ deficiency due to decrease in the concentration of these cations in soil solution and to the competition in cation uptake by Al^{3+} and Mn^{2+}; (V) P and Mo deficiencies on account of insolubility problems; (VI) general nutrient deficiency due to the frequent increase in leaching (e.g. B deficiency) [1]. The relative contribution of each constraint to plant stress depends on environmental and genetic factors. Al and Mn toxicities are probably the most important growth limiting factors for plants in many acid soils, specially where the pH is below 5.0 [2,3].

Inhibition of root growth is the main effect of Al toxicity. Al reduces both plant rooting depth and branching, which affects the capacity for water and mineral nutrient uptake and increases the risk of drought stress. Young seedlings are generally more

H. Guedes-Pinto et al. (eds.), Triticale: Today and Tomorrow, 475–480.

© 1996 Kluwer Academic Publishers. Printed in the Netherlands.

susceptible to Al toxicity than older plants [2,3]. Strategies of plant adaptation to excess Al vary widely with species and within varieties of a single species and may be regulated separately or concerted [1,4].

Some dicots and specially monocots have the ability to increase the pH of the rhizosphere or nutrient solution what may contribute directly to decrease the solubility and avoid toxicity of Al by precipitation [5]. Some Al-tolerant cultivars or genetic lines of snapbean, *Phaseolus vulgaris* L.[6], rice, *Oryza sativa* L.[7], wheat, *Triticum aestivum* L., rye, *Secale cereale* L., triticale, x*Triticosecale* Witt., and barley, *Hordeum vulgare* L.[4,8,9,10] increase the pH of their rhizospheres or the pH of nutrient solutions. In these experiments, some Al-sensitive cultivars showed smaller decrease or no significant effect in the pH and in consequence were probably exposed to higher Al concentrations in the root medium. The pH changes in root zones are often associated with other mechanisms of Al tolerance and affect mineral nutrition of plants in various ways [4,7,8,9,10,11].

The purpose of the present study was to evaluate the potential adaptation to Al among varieties of triticale and wheat, through an eventual mechanism of avoidance of Al stress by early plant-induced pH changes in nutrient solutions.

Materials and methods

Three advanced genetic lines of triticale, two cultivars of triticale and one of wheat included in the National Catalogue of Varieties, which follow studies of adaptation under the supervision of ENMP, Elvas, Portugal, have been used namely : TTE 9201, TTE 9202, TTE 9203, Arabian, Borba, and Anza.

All seeds were disinfected with a 5%active Cl sodium hypochlorite solution for 8 minutes, well washed and germinated between sheets of filter paper, over cotton soaked with deionized water for 2 or 3 days at 23°-26° C in the dark. Ten seedlings of each cultivar having 10-15 mm root length were selected and transferred for growth during 48 hours into plastic jars containing 150 ml of nutrient solution continuously aerated. Then they were placed for 72 hours in diluted nutrient solution (1/5), free from phosphate, with ferric chloride in instead of Fe-citrate and added, or not, 5 ppm Al applied as $Al_2(SO_4)_3.18H_2O$. The composition of the standard nutrient solution was the following: 4mM $Ca(NO_3)_2$; 4mM KNO_3; 2mM $MgSO_4.7H_2O$; 0.5mM KH_2PO_4; 0.435mM $(NH_4)_2SO_4$ and micronutrients 0.8 μM $ZnSO_4$; 30 μM NaCl; 0.1 μM $NaMoO_4$ and 10 μM H_3BO_4; 10 μM $C_6H_5O_7Fe.5H_2O$; 2 μM $MnSO_4.4H_2O$; 0.3 μM $CuSO_4$ [Moore *et al.* ref. *in* 12]. The initial pH of standard nutrient and treatment solutions was adjusted at 4.00 with 1N H_2SO_4. The experiments were performed having root temperature controlled at $25\pm1°C$ by a water bath, under irradiance of 150 μM Q m^{-2} s^{-1}, and 12 hours light.

The length of the longest primary root was measured twice with ruler after two days in standard nutrient solution (C1) and after three days in treatment solutions with or without Al (C2). The differences in root length were used to calculate the absolute extension increment AE=C2-C1 ; relative extension RE=(C2-C1) in 5 ppm Al divided by (C2-C1) in 0 ppm ; and relative extension rate RER=(C2-C1).(C1.days)$^{-1}$ The pH and ion conductivity of the final solutions were measured after 2 and 5 days of growth.

A spit-plot scheme with five replications (blocks), six cultivars as main plot, and the two treatments as sub-plot was used. ANOVA for a factorial experiment and LSD-test ($P \le 0.05$) for the significant factors or interactions was applied.

Results and discussion

The ANOVA for the AE showed significant differences among cultivars, treatments (mean values 42.1 mm in 0 ppm and 11.6 mm in 5 ppm, with LSD = 4.2 mm) and interaction cultivar x treatment. The cultivar Anza had the longest absolute extension without Al, followed by 9203, Arabian, 9201, 9202 and Borba. In treatment solution with Al, Anza showed the shortest extension, and was significantly different from Arabian and 9203 (Table 1).

Table 1. Mean values (n=50) of absolute extension (AE, mm) and relative extension rate (RER, mm mm^{-1} day^{-1}), for 0 ppm and 5 ppm Al, and relative extension (RE= AE in 5 ppm . AE in 0 ppm $^{-1}$).

Cultivar	A E		RER		RE
	0 ppm	**5 ppm**	**0 ppm**	**5 ppm**	
9201	39.0 (bc)	12.3 (de)	0.313 (ab)	0.128 (c)	0.315 (bc)
9202	30.9 (c)	10.4 (de)	0.284 (ab)	0.113 (c)	0.337 (ab)
9203	48.8 (b)	16.5 (d)	0.320 (a)	0.162 (c)	0.338 (ab)
Arabian	41.6 (b)	17.2 (d)	0.325 (a)	0.154 (c)	0.413 (a)
Borba	29.8 (c)	9.6 (de)	0.256 (b)	0.111 (c)	0.322 (ab)
Anza	62.7 (a)	3.2 (e)	0.268 (ab)	0.019 (d)	0.051 (c)
	LSD=10.29 ($P \le 0.01$)		LSD=0.058 ($P \le 0.05$)		LDS=0.205 ($P \le 0.01$)

(Values with the same letter are not statistically different)

Without Al the initial RER for Arabian and 9203 was significantly different from the rate for Borba (Table 1).

The most interesting variable to compare cultivars for Al tolerance seems to be RE, since it can eliminate differences in the initial root length of cultivars [13]. By this criterion Arabian was the most Al-tolerant cultivar, followed in order by the group of three advanced genetic lines, Borba and finally the wheat Anza (Table 1). This classification agrees with previous results obtained by the authors with the same varieties in a group of 20 submitted to a test for Al tolerance [14].

All the cultivars were able to decrease the concentration of H^+ in standard nutrient and treatment (1/5 diluted) solutions, because nitrogen was supplied mainly as NO_3^+.

The pH increase in solutions without Al was always higher than in the solutions containing Al (Table 2). The most interesting pH changes were observed in standard nutrient solution, which has probably the composition more close to the general plant nutrient needs. In the diluted treatment solution containing Al the pH changes induced by the different cvs. were small but followed the same trend as in the standard solution. In standard solution cv. Borba induced the biggest pH change and Arabian, similarly to the wheat Anza, induced the smallest, whereas the RE for Borba was the smallest and for Arabian the biggest (Fig.1). These results suggest that in such cvs. the capacity for pH changes is not related to the capacity for root growth in a medium with Al. However the ionic strength of the root medium may somewhat alter the relative changes in the pH induced by the varieties.

The conductivity of the solutions changed in proportion to the root extension of cultivars. Therefore the decrease observed was more important in the diluted solutions without Al (Table 2).

Table 2. Mean values of pH and conductivity ($\mu S\ cm^{-1}$) after two days in the standard nutrient solution and after the next three days in the treatment solutions (0 ppm and 5 ppm Al).

Cultivar	pH			Conductivity		
	Standard	**0 ppm**	**5 ppm**	**Standard**	**0 ppm**	**5 ppm**
9201	4.79 (ab)	4.66 (c)	4.33 (d)	1686.9 (a)	295	436
9202	4.63 (b)	4.78 (c)	4.20 (de)	1678.7 (abc)	352	460
9203	4.77 (ab)	5.27 (a)	4.32 (d)	1716.9 (a)	283	401
Arabian	4.47 (b)	5.00 (b)	4.26 (de)	1690.5 (ab)	294	403
Borba	5.13 (a)	5.24 (a)	4.36 (d)	1654.1 (bc)	330	457
Anza	4.53 (b)	5.39 (a)	4.14 (e)	1637.9 (c)	251	360
	LSD=0.44	LSD=0.18		LSD=41		
	(P≤0.01)	(P≤0.01)		(P≤0.05)		

(Values with the same letter are not statistically different)

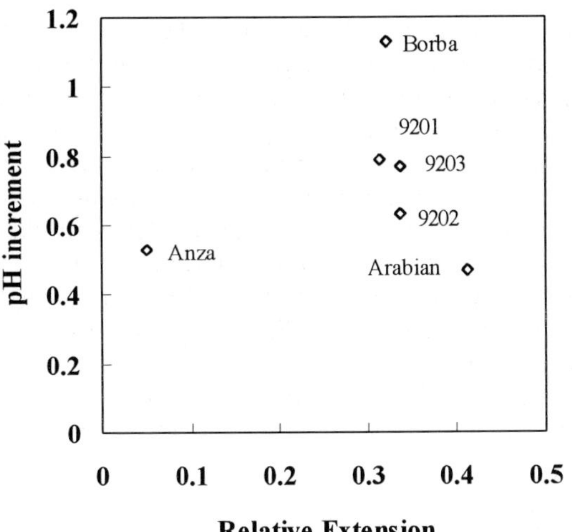

FIG.1 Relation between the relative extension of roots (RE=AE in5ppm/AE in0ppm) and the pH increment of root medium after two days in standard solution.

Conclusions

Early potential root growth in the solution without Al was different among cultivars either expressed by absolute extension or by relative extension rate. Therefore, to evaluate the effect of Al, the growth in the solution with Al has to be corrected by the potential growth of the same variety in the solution without Al. The highest tolerance to Al excess, expressed by the criterion of relative extension (RE), was shown by Arabian, Borba and the advanced lines of triticale 9203 and 9202, and the lowest by TTE 9201 and Anza. Significant differences between them were smaller than those reached in previous trial [13], because the plants were allowed to induce changes in pH media, which would contribute to better performances of the lines less adapted to Al excess.

It is expected that Borba, 9201 and 9203 in acid soils would increase the pH of the rhizosphere which would reduce Al-toxicity to plants, as happened in nutrient solutions.

Since Arabian and 9202 behaved as tolerant to Al this indicates that other mechanisms of adaptation than pH change shall be present in these cultivars.

Acknowledgements

This work has been sponsored by Junta de Investigação Científica e Tecnológica.

480

References

1. Marschner H. Mechanisms of adaptation of plants to acid soils. Plant and Soil 1991; 134: 1-20.
2. Foy C D. Plant adaptation to mineral stress in problem soils. Iowa State J Research 1983; 57: 339-54.
3. Foy CD. Physiological effects of hydrogen, aluminum and manganese toxicities in acid soil. In: Adams F (ed.) Soil acidity and liming. Agronomy 1984; 12: 57-97.
4. Mugwira LM, Elgawhary SM, Patel SU. Aluminium tolerance in triticale, wheat and rye as measured by root growth characteristics and aluminium concentration. Plant and Soil 1978; 50: 681-90.
5. Foy CD, Chaney RL, White MC. The physiology of metal toxicity in plants. Ann Rev Plant Physiol 1978; 29: 511-66.
6. Miyasaka SC, Buta JC, Howell RK, Foy CD. Mechanism of aluminum tolerance in snapbeans. Plant Physiol 1991; 96: 737-743.
7. Sivaguru M, Paliwal K. Differential aluminum tolerance in some tropical rice cultivars.II Mechanism of aluminum tolerance. J Plant Nutr 1993; 16: 1717-1732.
8. Foy CD, Burns GR, Brown JC, Fleming AL. Differential aluminium tolerance of two wheat varieties associated with plant-induced pH changes around their roots. Soil Sci Soc Am Proc 1965; 29: 64-67.
9. Mugwira LM, Patel SU. Root zone pH changes and ion uptake imbalances by triticale, wheat and rye. Agron J 1977; 69: 719-22.
10. Mugwira LM, Elgawhary SM, Patel KI. Differential tolerances of triticale, wheat, rye and barley to aluminum in nutrient solution. Agron J 1976; 68: 782-87.
11. Marschner H, Römheld V, Horst WJ, Martin P. Root-induced changes in the rhizosphere importance for the mineral nutrition of plants. Z Pflanzenernaehr Bodenkd 1986; 149: 441-56.
12. Camargo CEO, Oliveira OF. Tolerância de cultivares de trigo a diferentes níveis de alumínio em solução nutritiva e no solo. Bragantia 1981; 40: 21-31.
13. Pereira JAV. Comparação da tolerância ao alumínio em triticales usando um teste de crescimento radicular. Évora: Univ.de Évora, 1994 (Relat. Final Curso).

ALUMINUM TOXICITY AND PROTEIN SYNTHESIS IN NEAR ISOGENIC LINES OF *TRITICUM AESTIVUM* L. DIFFERING IN ALUMINUM TOLERANCE*

Daryl J. Somers[1], Keith G. Briggs[3] and J. Perry Gustafson[1,2],
[1]Plant Science Unit, Department of Agronomy, [2]USDA-ARS
PGRU, University of Missouri, Columbia, USA, [3]Department of
Plant Science, Faculty of Agriculture, Forestry and Home
Economics, University of Alberta, Edmonton, Canada

Abstract

To further understand the genetics and physiology of Al stress, a pair of
near isogenic lines of wheat (*Triticum aestivum* L. em Thell.) and several
sister lines differing in Al tolerance were examined for root growth and
levels of protein synthesis following Al stress in a hydroponic system. The
Al-sensitive recurrent parent (Katepwa) showed marginal root elongation in
low Al concentrations. In contrast, the Al-tolerant near isogenic line
(Alikat) and Al-tolerant donor (Maringa) had much greater root elongation.
All of the genotypes tested showed an increase (approximately 2 fold) in the
level of protein synthesis after 3 d in Al, based on incorporation of L-[^{35}S]-
Met. This increase in protein synthesis occurred at lower Al concentrations
in Katepwa than in Alikat and Maringa and was directed entirely toward
the microsomal fraction of the cell. The redistribution of L-[^{35}S]-Met
labeled proteins between the cytoplasmic and microsomal fraction was
directly correlated (r=0.75) to Al tolerance and root elongation.

Introduction

A major factor that limits crop production worldwide is the presence of acid
soils. When plants are subjected to an artificial Al stress, the primary effect
is the inhibition of root tip growth and subsequently, stunted root
development [for review see 1,,2].

One approach is to utilize Al-tolerant cultivars selected by screening
germplasm collections or to create Al-tolerant cultivars by conventional
breeding practices. For use in these approaches, hydroponic and other
methods are available to assess germplasm or controlled crosses, for

---

*Names are necessary to report factually on available data; however, the
USDA neither guarantees nor warrants the standard of the product, and
the use of the name by USDA implies no approval of the product to the
exclusion of others that may also be suitable.

481

H. Guedes-Pinto et al. (eds.), Triticale: Today and Tomorrow, 481–488.
© 1996 *Kluwer Academic Publishers. Printed in the Netherlands.*

evidence of Al tolerance [3,4]. It is our opinion that a detailed understanding of the genetics of Al tolerance will lead to more efficient development of improved Al-tolerant cultivars.

Much attention has been devoted to the analysis of protein synthesis during Al stress with the suggestion that proteins may be exuded to chelate Al or are membrane bound to block Al entry. The expression of Al-induced polypeptides appears limited to only a few polypeptides located in both the cytoplasm and microsomal fractions [5,6,7,8].

Research that makes use of near isogenic pairs is valuable for analyzing traits controlled by a few genes. A near isogenic line (Alikat) that is agronomically identical to the recurrent parent (Katepwa), differing only in Al tolerance, was acquired from the University of Alberta breeding program (Edmonton, AB, Canada) for this study. Our objectives were to use these wheat lines to assess their level of Al-tolerance in a low ionic strength hydroponic system and to look for differences in the level of protein synthesis between the pair of near isogenic lines, as well as for sister near isogenic lines and other cultivars.

Materials And Methods

PLANT MATERIAL.

The germplasm consisted of *Triticum aestivum* L. (cv Maringa, Al tolerant) (cv Katepwa, Al sensitive) and Katepwa*3/Maringa (cv Alikat, Al tolerant). Also included were *T. aestivum* (cvs PT741, Cutler, Al tolerant) and several BC_3 and BC_4 Alikat sister lines. All the genetic stocks and Al-tolerance designations of each line were determined at the University of Alberta in laboratory soil tests. One cultivar of *Secale cereale* L. (cv Blanco, Al tolerant) was acquired from the USDA-Sears collection, University of Missouri.

GROWTH AND AL STRESS.

Seeds of each cultivar were germinated for 36 h. The seedlings were then grown on plastic mesh floating over 2 L of an aerated, low ionic strength hydroponics medium [400 µM $CaCl_2$; 650 µM KNO_3; 250 µM $MgCl_2$; 10 µM $(NH_4)_2SO_4$; 40 µM NH_4NO_3 (pH 4.0)] containing Al at 26°C ±1°C with a 14 h day for 3 d with the solutions being changed each day.

The root tolernce index (RTI) was calculated as a ratio of the average root length from Al-stressed plants : the average root length of plants grown without Al.

RADIOLABELLING AND INCORPORATION INDEX.

Five seedlings from each treatment were placed in 3 mL of the final hydroponics solution containing L-[^{35}S]-Met (100 µCi mL^{-1}) and the roots were radiolabelled for 2 h. The seedling roots were then washed, excised and divided (where indicated) into whole roots, or root tips (2 to 3 mm) for separate homogenization. The homogenates were centrifuged and the supernatants were frozen. The TCA precipitable and total intracellular radioactivity of each sample were determined. The level of protein synthesis (incorporation index) was expressed as the ratio of the average TCA precipitable radiolabel : total intracellular radiolabel.

SOLUBLE *VERSUS* INSOLUBLE PROTEIN SYNTHESIS.

Following radiolabelling, root tips were collectively homogenized. The homogenates were separated into CRUDE, SOLUBLE (cytoplasmic) and INSOLUBLE (microsomal) fractions by ultracentrifugation. The TCA precipitable radiolabel content of the cellular fractions was expressed as a ratio between soluble : insoluble subcellular fractions.

Results

The effects of Al on root growth, expressed as a root tolerance index (RTI), showed some degree of root growth inhibition, over the range of Al concentrations assayed (Fig. 1). Katepwa (Al sensitive) showed a marked decrease in RTI at 0.5 μg mL^{-1} Al. Maringa and Alikat (both Al tolerant) had higher RTI values throughout the range of Al concentrations with Alikat showing intermediate RTI values at higher Al concentrations (Fig. 1). The inset graph (Fig. 1) showed Maringa and Alikat RTI values were unaffected at 0 to 0.5 μg mL^{-1} Al in contrast to Katepwa which showed a strong decrease in RTI at 0.1 μg mL^{-1} Al. The toxic dose of Al differed between these three lines. A significant reduction in RTI was observed at 0 to 0.1 μg mL^{-1} Al for Katepwa, 0.5 to 1.0 μg mL^{-1} Al for Alikat and 1 to 2 μg mL^{-1} Al for Maringa (Fig. 1).

The changes in the level of protein synthesis of whole roots following Al stress were expressed as an L-[^{35}S]-Met incorporation index. Over the range of 0 to 5.0 μg mL^{-1} Al, all three cultivars showed an increased L-[^{35}S]-Met incorporation. Katepwa (Al sensitive) had the greatest incorporation index of 1.68 at 2.0 μg mL^{-1} Al (Fig. 2). Figure 2 depicts the incorporation index at low Al concentrations. Alikat and Maringa showed no change in the incorporation index over the range of 0 to 0.5 μg mL^{-1} Al, whereas the incorporation index of Katepwa increased at 0.3 μg mL^{-1} Al and was similar to Alikat and Maringa at lower Al concentrations (Fig. 2).

The large increase in the incorporation index of all three cultivars was evaluated in subcellular fractions comprising soluble (cytoplasmic) and insoluble (microsomal) proteins. Katepwa (Al sensitive) had the highest increase (2.19 fold) in the incorporation index of root tips at 1.0 μg mL^{-1} Al

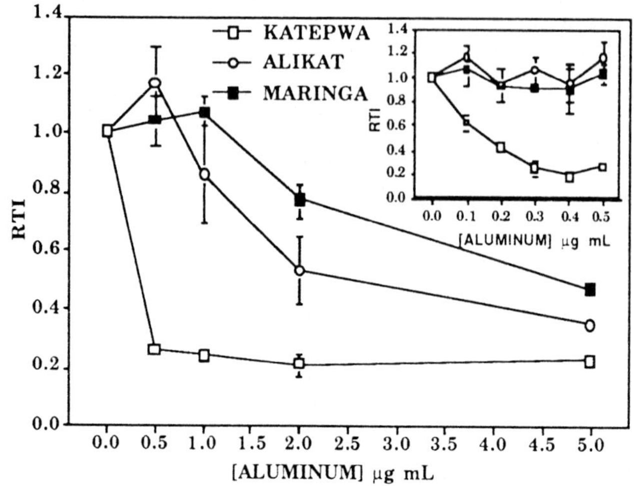

FIGURE 1.

Figure 1. Root tolerance index (RTI) of wheat lines treated with various Al concentrations . The two longest roots on each seedling were measured and the growth was expressed relative to seedlings grown without Al. Each data point is the average of 3 replications of 6 seedlings including standard error bars.

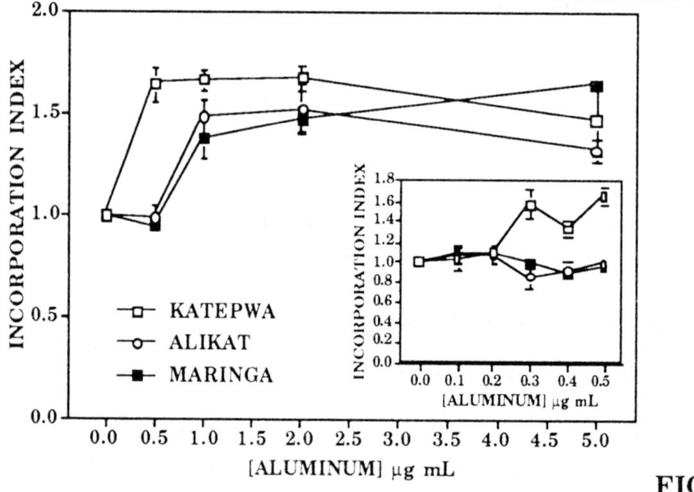

FIGURE 2.

Figure 2. The relative L-[^{35}S]-Met incorporation index of whole roots of three wheat lines treated with various Al concentrations. The level of incorporation of radiolabel into TCA precipitable polypeptides was determined and then expressed relative to incorporation values of plants grown without Al. Each data point is the average of 3 replications of 5 seedlings including standard error bars.

Table I. The effect of 1 µg mL^{-1}Al for 3 d on L-[^{35}S]-Met incorporation into the soluble and insoluble protein fractions of root tip cells in the three wheat lines.

Line	[Al] (µg mL^{-1})	Incorporation Index	Distribution (% of crude)		Ratio (soluble - insoluble)
			Soluble	Insoluble	
Katepwa	0	1.0±0.06	65.4±3.3	30.2±1.8	2.17
	1	2.19±0.30	36.1±3.6	53.4±2.2	0.68
Alikat	0	1.0±0.04	65.8±3.7	31.9±1.4	2.06
	1	1.67±0.19	36.9±2.2	47.4±3.1	0.78
Maringa	0	1.0±0.06	62.8±4.0	30.7±1.7	2.05
	1	1.98±0.40	42.2±4.3	50.5±9.5	0.84

The quantity of TCA pecipitable radiolabel in each cell fraction was determined. Data from three replications of n=4 seedlings.

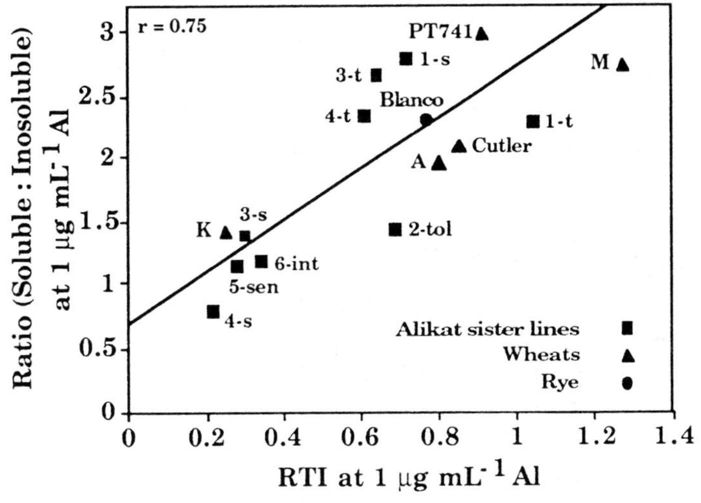

FIGURE 3.

Figure 3. A correlation of the ratio of soluble - insoluble TCA precipitable radiolabel between 0 and 1 µg mL^{-1} Al *versus* the RTI at 1 µg mL^{-1} Al of various lines. The amount of incorporated radiolabel into TCA precipitable polypeptides was determined for each fraction and was expressed as a ratio of soluble - insoluble between 0 and 1.0 µg mL^{-1} Al. There were three Alikat sister lines (1,3,4) that segregated (S-sensitive; T-tolerant) and three Alikat sister lines (2,5,6)that did not segregate for Al tolerance (Sen,Tol,Int). Each data point is the average of 4 seedlings.

in comparison to 1.67 and 1.98 fold increases for Alikat and Maringa (both Al tolerant) respectively (Table I). The distribution of TCA precipitable radiolabelled protein was similar for all three lines at 0 µg mL^{-1} Al; the majority of the radiolabel (62.8 - 65.8%) was in the soluble fraction and the remainder (30.2 - 31.9%) was in the insoluble fraction (Table I). Following an Al stress of 1 µg mL^{-1}, the distribution was reversed, and the majority of the radiolabelled protein (41.4 - 53.4%) was found in the insoluble fraction with the remainder (36.1 - 42.2%) in the soluble fraction (Table I).

The redistribution of TCA precipitable protein in the subcellular fractions following Al stress was also expressed as a ratio between soluble : insoluble fractions. The data showed no significant differences between Katepwa, Alikat and Maringa at 0 µg mL^{-1} Al. The altered ratio at 1 µg mL^{-1} Al was highest for Maringa lowest for Katepwa and intermediate for Alikat (Table I).

The above analysis was extended to several Alikat sister lines, two tolerant wheats and one rye. A correlation of the RTI values at 1 µg mL^{-1} Al *versus* the soluble : insoluble TCA precipitable radiolabel ratio between 0 and 1.0 µg mL^{-1} Al showed a strong, direct relationship (r=0.75) (Fig. 3). The best evidence to support this relationship came from two Alikat sister lines which showed both a segregated RTI value and segregated change in ratio following the Al stress (Fig. 3).

Discussion

The three lines Katepwa, Alikat and Maringa all showed a degree of Al sensitivity, the Brazilian wheat (Maringa) was the most Al tolerant, based on its ability to maintain greater root elongation over the 3 d experiment (Fig. 1). Katepwa leveled off at a 0.2 RTI value (Fig. 1), which suggested the plants were Al sensitive yet continued to grow marginally in Al. Hematoxylin staining experiments clearly show the root tips of Al-sensitive wheat stop growing in the presence of Al [9,10]. We observed that Alikat and Maringa RTI values were not significantly different over the range of 0 to 0.5 µg mL^{-1} Al suggesting that the Maringa gene(s) were fully expressed in Alikat within these Al concentrations.

Alikat showed an intermediate Al-tolerance level, based on RTI, at Al concentrations between 0.5 to 5 µg mL^{-1}, suggesting that Alikat lacks the Al-tolerance gene expression capabilities of Maringa at these Al concentrations. A similar pattern of root growth was observed in an alternate BC$_3$ near isogenic system [11].

Protein synthesis has been implicated as a possible means of defense, so we examined whether the level of protein synthesis was altered following Al stress in the whole cell and subcellular fractions. In this system, protein synthesis increased dramatically based on the levels of incorporation of L-[^{35}S]-Met. This response appeared to be related to Al sensitivity as Katepwa responded at lower Al concentrations whereas Alikat and Maringa responded at similar and higher Al concentrations.

We were interested if the increase in incorporation of L-[^{35}S]-Met was directed toward the membrane fraction of the cell which could indirectly support the Al-tolerance theories of extracellular chelation or blockage of Al entry. An Al-stress concentration of 1 μg mL^{-1} both optimized the induced levels of protein synthesis and provided differences in the L-[^{35}S]-Met incorporation index values between the three lines (Fig. 3). Following Al stress, there was a clear redistribution of TCA precipitable radiolabel that increased the microsomal proportion by approximately 1.63 fold (Table I).

The analysis of the redistribution of TCA precipitable radiolabel in the subcellular fractions was extended to ascertain whether the ratio of soluble - insoluble TCA precipitable radiolabel between 0 and 1.0 μg mL^{-1} Al were related to Al tolerance. This ratio was lower in Al-sensitive genotypes such as Katepwa, and Al-sensitive segregants of the Alikat sister lines. The greater the ratio, the more Al tolerant the genotype (Fig. 3; r=0.75). These results would suggest that Al-tolerant genotypes may be constitutively Al tolerant and require less *de novo* expression of proteins, to be directed toward the microsomal fraction.

Based on the Al-dose/response curves (RTI data), we can speculate that not all of the Al-tolerance has been transferred from Maringa to Alikat. In addition, a large increase in protein synthesis was observed following Al stress; this increase in protein synthesis was largely directed toward the microsomal fraction of the cell and was correlated to the Al tolerance of the genotype. Further experimentation will be conducted to examine the changes in microsomal proteins, including polypeptide profiles in conjunction with segregating material, in order to associate new microsomal proteins with Al tolerance.

References

1. Wright RJ. Soil aluminum toxicity and plant growth. Commun in Soil Sci Plant Anal 1989;20(15/16):1479-1497.

2. Bennet RJ, Breen CM. The aluminum signal: New dimensions to mechanisms of aluminum tolerance. Wright RJ et al. editors, Plant-soil interactions at low pH, Kluwer Academic Publishers 1991; 703-716.

3. Briggs KG, Taylor GJ, Moroni JS. Optimizing techniques for selecting acid soil tolerant wheat - case studies with aluminum and manganese tolerance. Tanner DG, Mwangi W. editors, Proc seventh regional wheat workshop for eastern, central and southern Africa. Nakuru, Kenya: CIMMYT 1992;37-48.

4. Wheeler DM, Edmeades DC, Christie RA, Gardner R. Comparison of techniques for determining the affect of aluminum on the growth of, and the inheritance of aluminum tolerance in wheat. Plant and Soil 1992;146:1-8.

5. Delhaize E, Higgins TJV, Randall PJ. Aluminum tolerance in wheat: Analysis of polypeptides in the roots apices of tolerant and sensitive

genotypes. Wright RJ et al. editors, Plant-soil interactions at low pH, Kluwer Academic Publishers 1991;1071-1079.

6. Ownby JD, Hruschka WR. Quantitative changes in cytoplasmic and microsomal proteins associated with aluminum toxicity in two cultivars of winter wheat. Plant Cell Environ 1991;14:303-309.

7. Picton SJ, Richards KD, Gardner RC. Protein profiles in root-tips of two wheat (*Triticum aestivum* L.) cultivars with differential tolerance to aluminum. Wright RJ et al. editors, Plant-soil interactions at low pH, Kluwer Academic Publishers 1991;1063-1070.

8. Rincon M, Gonzales RA. Aluminum partitioning in intact roots of aluminum-tolerant and aluminum-sensitive wheat (*Triticum aestivum* L.) cultivars. Plant Physiol 1992;99:1021-1028.

9. Aniol A, Gustafson JP. Chromosome location of genes controlling aluminum tolerance in wheat, rye and triticale. Can J Genet Cytol 1984;26:701-705.

10. Ownby JD. Mechanisms of reaction of hematoxylin with aluminum-treated wheat roots. Physiol Plant 1993;87:371-380.

11. Delhaize E, Ryan PR, Randall PJ. Aluminum tolerance in wheat (*Triticum aestivum* L.) II. Aluminum-stimulated excretion of malic acid from root apices. Plant Physiol 1993;103:695-702.

RESPONSES OF TRITICALE AND WHEAT TO HYPOXIA

Jerry W. Johnson and Bingru Huang
Department of Crop and Soil Sciences, University
of Georgia, Georgia Sta., Griffin, USA

Abstract

The responses of triticale (cv. Florico) and wheat (cvs. Bayles and Savannah) to hypoxia were evaluated by growing plants in nutrient solutions flushed with air (aerated control), or a mixture of 5% O_2 and 95% N_2 (hypoxia). Hypoxia for 21 d reduced the number of crown roots for Bayles and Florico, while the number of crown roots of Savannah was increased. Seminal and crown root lengths were reduced significantly for all genotypes. Hypoxia reduced lateral root length significantly for Bayles and Savannah, but not for Florico. Reduction in root dry weight after 21 d of hypoxia was less for Florico than Bayles and Savannah. Shoot growth was inhibited by hypoxia, to a less extent for Florico and Savannah than Bayles. Hypoxia stimulated formation of aerenchyma in crown roots, to a greater extent for Savannah and Florico than for Bayles. Stomatal conductance was reduced for Bayles, but not for Florico and Savannah.

Introduction

Low oxygen supply in the rhizosphere due to waterlogging reduces plant growth and yield (1,2). Productivities in waterlogged soils may be increased by improving soil drainage and by the introduction of waterlogging tolerant genotypes. Crops vary in waterlogging tolerance or susceptibility. Among the small-grains, winter wheat is more tolerant than winter barley or rye (3). Little is known about the tolerance of triticale (X <u>Triticosecale</u>) to hypoxia or waterlogging. Comparing one waterlogging-tolerant triticale genotype with two wheat genotypes (4) reported that shoot fresh weight decreased more for one of the wheat genotypes than the triticale genotype. Our present study was conducted to compare the growth and anatomy of one genotype of triticale with two wheat genotypes known differ in waterlogging tolerance.

489

H. Guedes-Pinto et al. (eds.), Triticale: Today and Tomorrow, 489–495.
© 1996 *Kluwer Academic Publishers. Printed in the Netherlands.*

Materials and Methods

Two wheat cultivars, `Bayles', sensitive to waterlogging, and `Savannah', tolerant to waterlogging, and one triticale cultivar 'Florico' were grown in full-strength Hoagland's solution (5) contained in a 1000 ml polyethylene culture jar. Plants were grown in a walk--in growth chamber, providing daily maximal/minimal air temperatures of 20/15°C, a 12-h photoperiod, and a PAR of 800 μmol m^{-2} sec^{-1}. When crown roots were just starting to emerge from the shoot base at 14 d after planting, the nutrient solution was flushed with a gas mixture containing 5% O_2 and 95% N_2 at 50 ml min^{-1} (hypoxia) or bubbled with air at 150 ml min^{-1} (aerated control).

The experiment was a completely randomized design with four replicates. Treatment effects were determined by analysis of variance. Differences among treatment means were separated by the LSD at the 0.05 level of probability.

Stomatal conductance (g_s) was determined on attached leaves with a steady state porometer (Li-Cor 1600). Leaf area was determined with a leaf area meter (Li-Cor 3100). Number of tillers and of seminal and crown root axes were recorded. Total length of seminal and crown root axes, and total lateral root length were determined with a pseudo--color image analysis system (AgVision). Shoots and roots were dried at 80°C until no further weight loss occurred. Root to shoot dry weight ratio was then calculated.

To examine the formation of aerenchyma (intercellular air-filled space), free-hand cross sections were cut at 5 cm from the tip of crown roots. The percentage of the cortex that comprised aerenchyma was determined with project drawings and a pseudo-color image analysis system.

Results and Discussion

ROOT GROWTH AND FORMATION OF AERENCHYMA

Crown root production was significantly inhibited for Florico and Bayles after 21 d of hypoxia (Fig. 1). For Savannah, however, the number of crown roots increased significantly under hypoxic conditions. The increase in the number of crown roots in waterlogged plants is generally considered to be an adaptive response of plants to excessive water. The greater the crown root formation, the greater the waterlogging tolerance (6).

Crown root length was reduced significantly under hypoxia for all genotypes, with a 45% reduction for Bayles and Savannah, and 30% reduction for Florico (Fig. 2). Seminal root length was also decreased significantly for the three genotypes after 21 d of hypoxia. Lateral root growth was significantly retarded for Bayles and Savannah, but not for Florico.

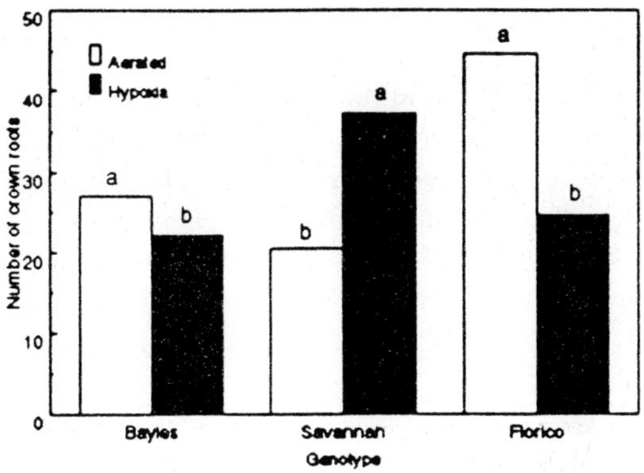

Figure 1. Crown root production as affected by hypoxia. Columns marked with the same letters are not significantly (P>0.05) different within genotypes.

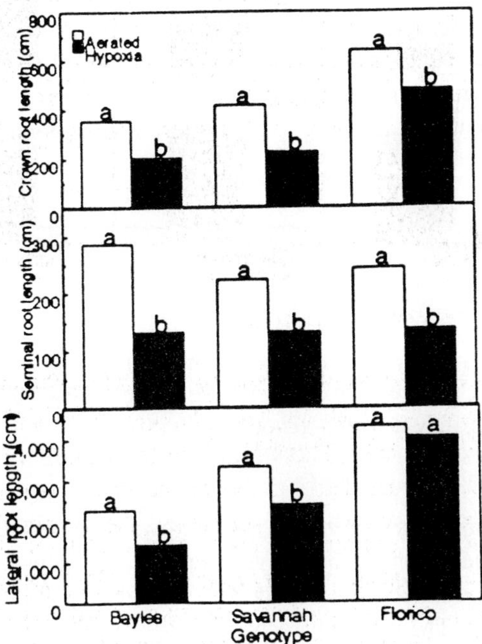

Figure 2. Total length of crown roots, seminal roots and lateral roots as affected by hypoxia. Columns marked with the same letters are not significantly (P>0.05) different within genotypes.

Reduction in root dry weight under hypoxia was 65% for Bayles, 33% for Savannah, and 15% for Florico (Fig. 3).   Root:shoot dry weight ratio decreased significantly only for Bayles, while the ratio remained at the control value for Savannah and Florico after 21 d of hypoxia. The reduction in root:shoot ratio for Bayles indicated that the root system was more sensitive to hypoxia in the rooting medium than the shoot for waterlogging-sensitive genotypes.

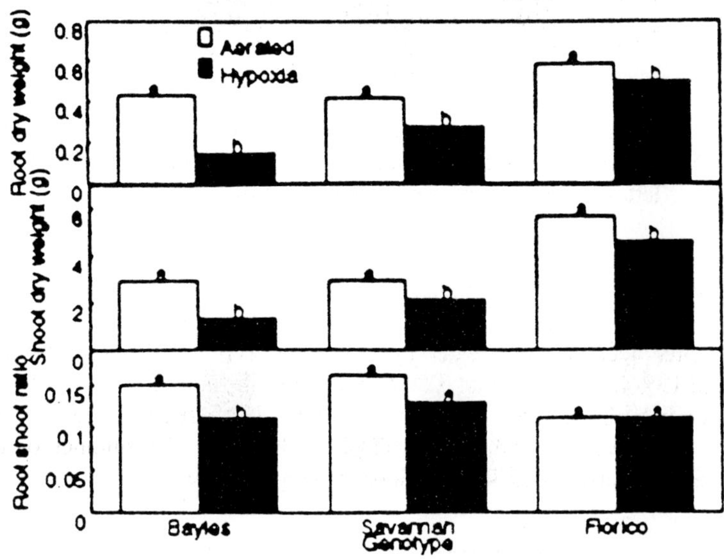

Figure 3.   Root and shoot dry matter production and root:shoot ratio as affected by hypoxia. Columns marked with the same letters are not significantly (P>0.05) different within genotypes.

Hypoxia enhanced formation of aerenchyma in crown roots for all genotypes, to a greater extent for Florico and Savannah than Bayles (Fig. 4). Development of aerenchyma in roots may help offset the reduction in shoot and root growth caused by the lack of oxygen in the root zone (7). The more advanced development of aerenchyma in roots of Florico and Savannah could account for their better root growth than for Bayles under hypoxia.

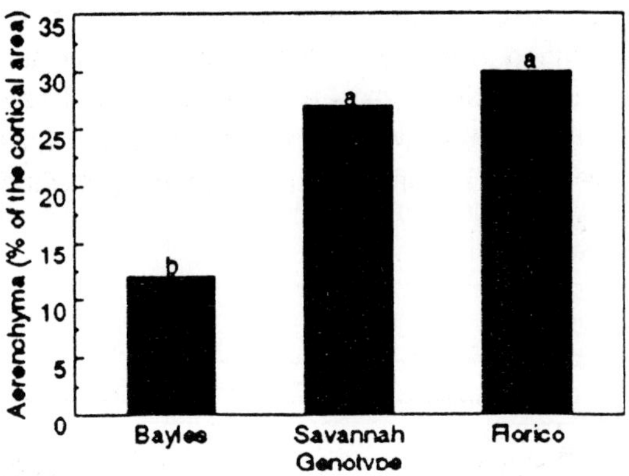

Figure 4. Formation of aerenchyma under hypoxia. Columns marked with the same letters are not significantly (P>0.05) different.

SHOOT GROWTH

Shoot dry matter accumulation under hypoxia declined by 54% for Bayles, 33% for Savannah, and 19% for Florico (Fig. 3). The reduction in shoot dry weight was attributed to the reduction in both leaf area and number of tillers. Leaf area was reduced by 60% for Bayles, 16% for Savannah, and 40% for Florico (Fig. 5). The number of tillers was reduced by 56% for Bayles and 30% for Savannah and Florico.

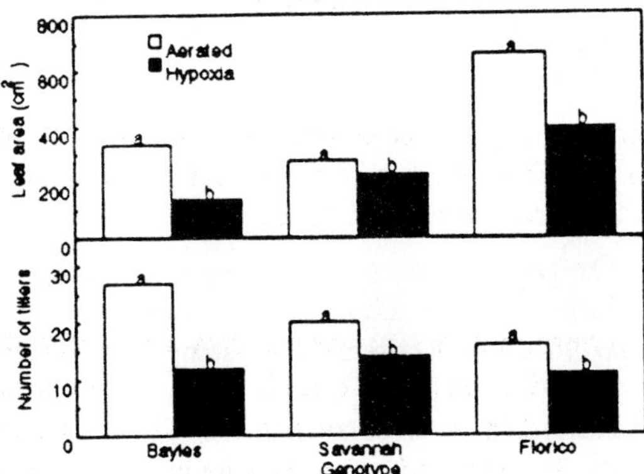

Figure 5. Leaf area and number of tillers as affected by hypoxia. Columns marked with the same letters are not significantly (P>0.05) different within genotypes.

STOMATAL CONDUCTANCE

Stomatal closure is one of the earliest and most sensitive indicators of plant responses to waterlogging (8). Stomatal conductance declined significantly for Bayles under hypoxic conditions, while the conductances for Savannah and Florico under hypoxia were not significantly different from the control plants (Fig. 6). The maintenance of stomatal opening could facilitate photosynthesis.

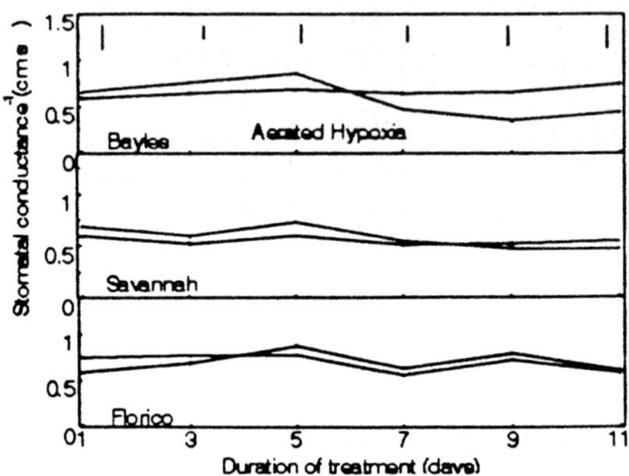

Figure 6.   Response of stomatal conductance to hypoxia. Bars indicate LSD (P=0.05) for treatment comparison.

In summary, the tolerance of hypoxia for triticale cv. Florico was greater than the waterlogging-sensitive cultivar, Bayles, and similar to the waterlogging-tolerant cultivar, Savannah. The better tolerance of triticale cv. Florico and wheat cv. Savannah could be attributed to the more advanced formation of aerenchyma, the development of a more extensive crown root system, and maintenance of stomatal opening.

References

1. Drew, M. C., 1991. Oxygen deficiency in the root environment and plant mineral nutrition. In: M. B. Jackson, D. D. Davis & H. Lambers (Ed.), Plant Life Under Oxygen Deprivation, pp. 301-316. Academic Publishing, The Netherlands.

2. Huang, B., J. W. Johnson, D. S. NeSmith & D. C. Bridges, 1994. Growth, physiological and anatomical responses of two wheat genotypes to waterlogging and nutrient supply. J. Exp. Bot. 45:193-202.

3. Bourget, S. J., B. J. Finn & B. K. Dow, 1966. Effect of different soil moisture tensions on flax and cereals. Can. J. Soil Sci. 46:213-216.

4. Thomson, C. J., T. D. Colmer, E. L. J. Watkin & H. Greenway. 1992. Tolerance of wheat (Triticum aestivum cvs. Gamenya and Kite) and triticale (Triticosecale cs. Muir) to waterlogging. New Phytol. 120:335-344.

5. Hoagland, C. R. & D. I. Arnon, 1950. The water-culture method for growing plants without soil. California Agricultural Experimental Circular 347:1-32.

6. Krizek, D. T., 1982. Plant response to atmospheric stress caused by waterlogging. In: M. N. Christiansen & C. F. Lewis (Ed.). Breeding Plants for Less Favorable Environments, pp 293-335. John Wiley and Sons, Inc.

7. Jackson, M. B. & M. C. Drew, 1984. Effects of flooding on growth and metabolism of herbaceous plants. In: T. T. Kozlowski (Ed.). Flooding and Plant Growths, pp 47--128. Academic Press, New York.

8. Bradford, K. J. & T. Hsiao, 1982. Stomatal behaviors and water relations of waterlogged tomato plants. Plant Physiol. 70:1508-1513.

VI - DISEASES

TRITICALE DISEASES - A REVIEW

Edward Arseniuk
Plant Breeding and Acclimatization Institute, Radzików
Blonie, Poland

Abstract

Triticale in some early research concepts was doomed to serve as a bridge to transfer resistance to leaf diseases from rye into wheat. Over the recent several decades, owing to a hard, inventive and consistent work of researchers and breeders it has become one of more important cereal crops in several countries around the world. In addition to triticale commercial and agronomical values it still is and it will continue to be bridging transfers of resistance genes to various pathogens and pests mainly from rye to wheat.

On the other hand, triticale is a crop on which meet pathogens of wheat and rye. In the latter respect triticale is also a bridge facilitating a direct contact between the pathogens, e.g. between physiological forms of the most important cereal rusts. Such contacts stimulate somatic hybridization and may finally result in new hybrid pathotypes carrying virulence genes (factors) to all three hosts, i.e. triticale, wheat and rye. In addition to the above, passaging of wheat or rye physiological forms of pathogens through the hybrid triticale tissue may adapt them to cross infections of triticale progenitors. Although still scarce, there is some evidence illustrating the above facts and it is to be discussed in the paper.

Triticale over a long time has been considered as resistant to diseases. Although, this opinion is no longer true, in comparison to wheat and rye it still looks as a healthy crop. The first triticale disease which occurred in epidemic proportions was stem rust (*Pucinia graminis* f. sp. *tritici*) in Australia. Leaf and stripe rusts (*P. recondita* f. sp. *tritici* and *P. striiformis*) are also gaining in importance everywhere triticale is grown. The same concerns several facultative pathogens, such as the most damaging to triticale *Septoria nodorum, Fusarium* spp., *Bipolaris sorokiniana*,

H. Guedes-Pinto et al. (eds.), Triticale: Today and Tomorrow, 499–525.
© 1996 *Kluwer Academic Publishers. Printed in the Netherlands.*

500

Pseudocercosporella herpotrichoides, and *Gaeumannomyces graminis*.
Triticale diseases are being controlled by various means. One of such highly recommended means is to increase genetic diversity for resistance in triticale germplasm. This could be achieved through conventional and inconventional breeding methods, e.g. use of somatic embryogenesis and genetic engineering. In the paper those methods will be discussed and examples given.

Introduction

Despite of controversial and even adverse opinions about triticale, the crop undergoes a steady develpoment. This is clearly illustrated by the figures on the world triticale hectareage reported in the literature. They consistently climb up with the elapsing time, e.g. 1,075,800 ha in 1986 versus 2,367,800 in 1992 (Triticale Topics No. 11, Dec. 1993). This means, that triticale is gaining new "followers" and is able to defend itself among the other small grains as a competitive, profitable and reliable crop. The vast literatural data show, that the combination of positive agronomic traits of *Triticum* (productivity, uniformity, quality) and *Secale* (hardiness, disease resistance, nutritional value) in triticale is still possible to improve.

In the world literature on triticale the positive characters of its progenitors are mentioned more frequently than the negative ones. While, it is quite obvious that a hybrid combines all types of traits of its parents from the positive to the negative ones. One of the most prominent advantages of a hybrid is to show a transgressive segregation of, at least, some parental characters. All the above seems to hold true concerning triticale - a man-made hybrid between tetraploid and/or hexaploid wheats (*T. durum* Desf; 2n=28, *T. aestivum* L.; 2n=42) and the diploid rye (*Secale cereale* L.; 2n=14).

Resistance to diseases in triticale has been considered as one of its most important and durable advantages. Although, this advantage disappears with the expanding hectareage and the cultivation time, in comparison to wheat and rye triticale still looks as a healthy crop. The above opinion is documented by the tabulated data (Table 1). The table has been prepared on the basis of papers abstracted in the Wheat, Barley and Triticale Abstracts (WBTA). It gives approximate figures since papers on triticale diseases in many local or regional periodicals are not widely distributed and surveyed by international bibliographic databases. Also, proceedings of many conferences are not covered by the bibliographic surveys, e.g. ones published after the triticale symposia held to date. Nevertheless, Table 1 and the present review is meant as a supplementation of reviews of Saari et al. (1986) and Singh & Saari (1991).

Table 1	List and approximate reporting frequency of pathogens affecting triticale over 1984 - 1993 and *Triticum aestivum, T. durum* and *Secale cereale* over 1991 - 1993 according to papers abstracted in the Wheat, Barley and Triticale Abstracts.																	
No.	Pathogen	Disease	19 84	19 85	19 86	19 87	19 88	19 89	19 90	19 91	19 92	19 93	To-tal	Inciden-ce in Poland	*T. aes-ti-vum*	*T. du-rum*	*S. ce-re-ale*
1	2	3	4	5	6	7	8	9	10	11	12	13	14	15	16	17	18
		Fungal pathogens observed on all plant organs															
1.	*Bipolaris sorokiniana* (Teleomorph: *Cochliobolus sativus*) & *Bipolaris* spp.	Kernel Smudge, Common foot and root rot, Spot blotch	4	-	-	1	1	1	2	3	2	2	16	Rare	71	2	2
2.	*F. graminearum, F. avenaceum,* (Teleo-morph: *Gibberella* spp.) *F. culmorum, Microdochium* (=*Fusarium) nivale* (Teleom: *Micro-nectriella nivalis*)	Head blight (Scab), Leaf spot, Pre-emergence blight, Seedling blight (Pink snow mold), Foot and root rots	2	2	3	1	4	3	9	5	8	3	40	Often	111	5	6
		Fungal pathogens affecting heads, seed and leaves															
3.	*Alternaria alternata, A. triticina*	Sooty mold, Leaf blight, Black point	-	-	1	-	-	-	1	1	1	1	5	Rare	16	-	1
4.	*Claviceps purpurea*	Ergot	2	-	1	1	-	2	-	-	-	-	6	Rare	1	-	1
5.	*Phaeosphaeria (Lepto-sphaeria) nodorum,* (Anamorph: *Stagonospora* (*Septoria) nodorum*), *Ph. avenaria* f.sp. *triticea*	Glume and leaf blotches	2	-	2	2	3	3	3	3	4	4	26	Rare	82	3	7
6.	*Pyrenophora (Drechslera)* spp.	Leaf spots and blotches	-	-	-	-	-	1	1	1	-	1	4	Often	2	-	-

Table 1 cont.d

1	2	3	4	5	6	7	8	9	10	11	12	13	14	15	16	17	18
7.	*Tilletia tritici* (=T. *caries*), *T. foetida*	Stinking smut	-	-	-	1	-	2	1	1	1	1	7	-	32	3	-
8.	*T.* (=*Neovosia*) *indica*	Karnal bunt	-	-	1	1	-	1	1	1	2	-	7	-	36	3	-
9.	*Ustilago segetum* var. *tritici* (=*U. tritici*)	Loose smut	1	-	1	1	-	-	-	-	-	1	4	Trace	4	3	-
	Fungal pathogens affecting primarily leaves, leaf sheaths, stems and occasionally heads																	
10.	*Erysiphe graminis* f.sp. *tritici*	Powdery mildew	-	1	1	-	5	1	1	5	-	4	18	Rare	220	-	6
11.	*E. g.* f.sp. *secalis*	Powdery mildew	-	-	-	-	-	-	2	1	-	-	3	Rare	-	-	6
12.	*Puccinia graminis* f.sp. *tritici (secalis)*	Stem rust	2	3	1	-	2	6	3	3	-	5	25	Rare	136	14	5
13	*Puccinia recondita* f.sp. *tritici /secalis)*	Leaf rust	1	2	2	3	6	6	9	4	2	8	43	Often	342	16	12
14.	*Puccinia striiformis* f.sp. *tritici*	Stripe (yellow) rust	1	-	1	-	2	1	1	1	-	-	7	Rare	150	2	6
	Fungal pathogens affecting foliage																	
15.	*Ramularia collo-cygni*	Leaf spot	-	-	-	-	-	1	-	-	-	-	1	-	-	-	-
16.	*Rhynchosporium secalis*	Scald	-	-	1	-	2	-	-	-	-	-	3	Rare	3	-	1
17.	*Mycosphaerella graminicola (S. tritici)*	Leaf blotch	-	-	2	1	-	2	-	-	-	-	5	Trace	60	2	-
	Fungal pathogens affecting lower stems and roots																	
18.	*Gaeumannomyces graminis*	Take-all	2	1	3	1	2	4	2	-	1	2	18	Often	95	2	6
19.	*T. ishikariensis, T. incarnata*	Speckled snow mold	-	-	1	-	-	1	2	-	-	-	4	-	10	-	3
20.	*Myriosclerotinia (Scleortinia) borealis*	Snow scald	-	-	-	-	1	1	1	-	-	-	3	-	3	-	2
21.	*Pseudocercosporella herpotrichoides*	Eye-spot	-	-	-	-	1	2	4	1	-	1	9	Often	78	-	9
22.	*Rhizoctonia cerealis*	Sharp eyespot	2	-	-	-	1	-	-	-	-	-	3	Rare	6	-	1
23.	*Rhizoctonia solani*	Root rot	-	-	1	-	1	-	1	-	-	-	3	Rare	23	-	-

Table 1 cont.d

1	2	3	4	5	6	7	8	9	10	11	12	13	14	15	16	17	18
						Bacterial pathogens											
24.	*Pseudomonas syringae* f.sp. *atrofaciens*	Basal glume rot	-	-	-	-	1	-	-	1	-	2	4	-	13	-	5
25.	*Xanthomonas campestris* pv. *undulosa, X. c.* pv. *translucens*	Bacterial stripe (streak) Black chaff	-	-	1	-	1	2	3	1	3	1	12	Trace	13	2	3
					Viruses, viruslike agents and nematodes												
26.	Brome Mosaic Virus		-	-	1	-	-	-	-	-	-	-	1	-	1	-	-
27.	Flame chlorosis		-	-	-	-	-	-	-	-	-	1	1	-	3	-	-
28.	Wheat Streak Mosaic Virus		-	-	-	1	-	-	-	-	-	1	2	-	19	-	1
29.	Barley Yellow Dwarf Virus		-	-	4	-	-	-	1	1	-	3	9	Trace	26	-	-
30.	Barley Yellow Mosaic Virus		-	-	1	-	-	-	-	-	2	-	3	-	1	-	-
31.	Root knot/lesion nematodes: *M. incognita, M. javanica, M. arenaria, P. thornei*		-	-	1	-	1	-	-	-	-	1	3	-	23	1	2
32.	Cereal cyst nematodes: *Heterodera avenae, H. glycines, H. latipons*		-	-	-	-	-	-	2	-	3	-	5	-	51	1	14
Total			19	9	30	14	34	40	50	33	30	41	300 1	-	163	59	99

- = not reported yet

Fungal pathogens affecting all plant organs

This group of fungi comprise species of *Bipolaris, Fusarium* and *Microdochium (=Fusarium) nivale* (ascigerous stages: *Cochliobolus* spp., *Gibberella* spp.- if known and *Micronectriella nivalis,* respectively). Common features of the above pathogens are following: nonspecificity, ability to attack all plant parts at all growth stages and a synthesis of potent phyto- and zootoxic mycotoxins.

In table 1 are listed diseases they cause. Seedling blights very often result from a seedborne inoculum of the fungi. Roots, crowns and stem bases are attacked by the pathogens under a broad range of temperatures and moisture conditions. Leaf and head infections frequently coincide with wet weather and temperatures of 20° - $25^{\circ}C$. Leaf symptoms are most obvious on lower leaves after heading.

Bipolaris sorokiniana (=Helminthosporium sativum) [ascigerous stage: *Cochliobolus sativus*] is the most often reported species from the genus *Bipolaris* in the literature (Chattannavar 1991a,b,c, Chauhan & Singh 1987, Lawn &, Sayre 1992, De Oliveira &, Baier 1993, Raghuchander 1991, Kuwite 1989, Reis 1991, Reis & Ambrosi 1991, Sisterna 1989). The other species, these are *B. specifera* and *B. ravenelli* (Sisterna 1989). Most of the above reports on the pathogens comes from India and Brazil, but ones are also from Mexico, Argentina and Tanzania. This would mean that the seed-borne pathogens are especially severe in warm, subtropical triticale-growing areas where they incite seedling and leaf blights in triticale (Raghuchander et al. 1988, 1991a,b, Reis & Ambrosi 1991) and also heavily infect triticale seed (Reis 1991, Satvinder et al., Vargas et al. 1991). Reports on a severe disease of basal stems in drought-stressed plants, called "dryland root rot", was not possible to identify in the WBTA. However, Singh & Saari (1991) cite reports on its damaging effect in triticale. The disease, incited by *B. sorokiniana,* is widely distributed and of great economic importance in wheat and other cereals (Wiese 1987).

Not much has been published about control of the diseases incited by *B. sorokiniana*. Resistant germplasm in triticale is rather scarce (Wildermuth & McNamara 1987, Chattannavar et al. 1991) and chemical control, as the most effective, is recommended (Lawn & Sayre 1992, Reis 1991, Reis & Ambrosi 1991).

Occurrence of diseases incited by *Fusarium* spp. and *Microdochium nivale* is reported in almost all geographical regions where triticale is grown (Arseniuk & Czembor 1990a,b,1991 - Poland, Cristani 1992 - Italy, De Oliveira & Baier 1993 - Brazil, Haesert et al. 1987 - Belgium, Ittu et al. 1990 - Romania, Klein et al. 1989 - Australia, Lawn and Sayre 1992 - Mexico, Łacicowa & Kiecana 1986 - PL, Martin et al.1992 - Canada, Pavlov & Karajova 1991 - Bulgaria, Sisterna 1984 - Argentina).

The analysis of the literature data shows that in addition to *M. nivale* about 20 other *Fusarium* spp. damage the crop in various ways. The most common are *F.*

graminearum, F. culmorum, F. avenaceum, F. sambucinum var. *coeruleum, F. crookwellense, F. sporotrichioides* (Arseniuk et al. 1991b,d,1993a,b, Klein et al. 1989, Łacicowa & Kiecana 1987, Pavlov & Karajova 1991, Wakuliński & Chełkowski 1993). The above species are able to attack triticale and other cereals over an extremely broad range of environments and temperatures. *F. graminearum* seems to prevail in warmer regions, *F. avenaceum* rather in cooler growing areas, while *F. culmorum* strives well in environments with an intermediate temperature. On the other hand, *M. nivale* incites snow molds at temperatures close to $0^{\circ}C$ or attacks any parts of growing plants during cool, wet periods. All the pathogens have simple nutritional requirements, are seedborne and omnipresent as saprophytes (Cristani 1993, Grzelak 1993, Wakuliń ski & Chełkowski 1993, Wiese 1987).

Yield losses afflicated by the pathogens are of dual nature, quantitative and qualitative. Data on reductions of grain yield, at least in triticale, are rather scarce, though authors stress that they might be substantial, e.g. Klein et al. (1989) reports that yield losses of grain at two experimental sites amounted to 35% and 18%, respectively. Arseniuk et al. (1993a,b) report that under experimental conditions the thousand kernel weight, number of kernels per head and head weight in 4 commercially grown triticale varieties were reduced by about 15%, 18% and 22%, respectively.

The other, even more important aspect is the grain quality diminished by its contamination with non-host specific mycotoxins of the pathogens (Perkowski et al. 1988, Raghuchander et al. 1988, Satvinder et al. 1993). The problem of mycotoxins is much better understood in the *Fusarium* - triticale than in the *Bipolaris* - triticale pathosystems. There are significant intervarietal differences in grain contamination with mycotoxins, e.g. Miller et al. (1985) report that deoxynivalenol (DON) concentration in kernels of susceptible varieties of triticale, wheat and rye was up to 8-fold higher than in resistant ones. Perkowski et al. (1991) reported that DON content in *Fusarium* Damaged Kernels (FDK) of, either triticale or wheat, was increasing by 0.30 - 0.46 mg/kg of 1% FDK. It is to add, that DON belongs to trichothecenes   potent toxins causing hemorrhagic syndrome, food/feed refusal, vomiting, pseudo-AIDS syndrome and infertility in humans and animals (Kiecana et al. 1987, Mirocha et al. 1993). The compound seems to be involved in pathogenesis and is also used in an *in vitro* screening for resistance to *Fusarium* spp. of triticale and wheat germplasms at the cell and callus level (Fadel & Wenzel 1993, Maier & Oettler 1993, Simmonds et al. 1993).

Diseases caused by *Fusarium* spp. and *M. nivale* could be controlled chemically but also breeding for resistance might be effective to some extent. Some of the authors recommend sources with higher resistance identified among tested triticale genotypes, e.g. Polish winter variety 'Malno' and 37TK (Ittu et al. 1992), 'Munstertaler' (Kleijer 1988), BR-2 (De Oliveira & Baier 1993). The variety 'Malno' in our experiments (Arseniuk et al. 1993a,b) showed to be the most resistant among 4 triticale, 4 rye and 5 wheat varieties at the seedling stage after inoculation with isolates of 8 *Fusarium* spp.. However, under field conditions the resistance to head blight was not so high, though still at a good level. It is also to point out that resistance to *Fusarium* spp. showed by seedlings not always is shown by heads at the adult growth stage (Arseniuk et al. 1993a,b). On average, in comparison to wheat and rye triticale shows intermediate resistance in seedlings and heads to *Fusarium* spp. (Arseniuk et al. 1993a,b). In some reports triticale is considered to be more susceptible on heads to *Fusarium* spp. than wheat (Martin et al. 1991). On the other hand, triticale seedlings may also be more resistant to infection by *M. nivale* than the wheat ones. Rye is usually rated as the most *Fusarium* resistant among the three cereal species in question. A general comment on the above reports is that results of comparative studies depend mainly upon the cereal genotypes used.

Resistance to snow mold (*M. nivale*) could also be improved through the use of *Fusarium* spp. resistant parental genotypes for synthesis of primary triticale (Dormann & Oettler 1991,1993) and intravarietal selection of sublines with superior resistance to the disease (Cichy & Maćkowiak 1993). Also somaclonal variation may offer a possibility to improve triticale resistance to *Fusarium* diseases (Kaltsikes & Bebeli 1993, Maier & Oettler 1993).

Fungal pathogens affecting leaves, heads and seed

Alternaria alternata and *A. triticina* are saprophytes or weak parasites and attack plants that are nutritionally deficient, lodged or damaged by other diseases. Most often these fungi are associated with heads and seed (Chattannavar 1991b,c, Lupashku et al. 1993, Chauhan 1986, Vargas et al. 1991), but also they may cause leaf blights (Chattannavar et al. 1991b, Shukla 1990).

On ergot (*Claviceps purpurea*) resistance in triticale not much information have been published since the 2nd Triticale Symposium and the reader is referred to the reviews by Saari et al. (1986) and Singh and Saari (1991). The disease occurs also extensively in Poland (Arseniuk & Czembor 1991, Małuszyńska 1992, Zamorski 1993a) and losses caused by it are rather qualitative than quantitative in nature. Małuszyńska (1992) showed, that on 3-years average, 36.2% of winter triticale seed samples were contaminated with sclerotia of *C. purpurea*. One third of the latter samples did not comply with the quality standard requirements concerning the content of the fungal resting bodies. Considerable intervarietal differences were reported by the author. The seed of 'Largo' triticale was the most heavily contaminated with sclerotia.

Considerable body of literature has been accummulated on stagonospora (septoria, phaeosphaeria) leaf and glume blotch of triticale (Table 1). In triticale the disease is incited by the same pathogens as in wheat and rye, i.e. *Phaeosphaeria nodorum* (=*Leptosphaeria nodorum*) [anamorph: *Stagonospora* (=*Septoria*) *nodorum*] and *Ph.* (*L*) *avenaria* f.sp. *triticea* [anamorph: *St.* (*S*) *avenae* f.sp. *triticea*] (Abreu & Marques 1989, Abreu et al. 1991, Arseniuk et al. 1990a, Arseniuk 1993, Pokacka 1985, Zamorski et al. 1993a). The literature data show that *Phaeosphaeria* (= *Stagonospora*, syn. *Septoria*) *nodorum* blotch on triticale occurs with various intensity in temperate (Abreu & Marques 1989, Arseniuk et al. 1990a,b,c, 1991c,1993, Baturo & Grib - personal communication, Haesert et al. 1987, Scott and Benedikz 1986, Skajennikoff & Rapilly 1983, Sowa 1986, Wo 1993) and warm triticale-growing areas (Reis 1991, Vargas et al. 1991, Cunfer & Youmans 1983, Kokhar & Pacumbaba 1987, Sapra et al. 1976). Author of this paper is not aware of reports on occurrence of this disease on triticale in Australia. It is very unlikely that it does not occur on the Australian continent. Occurrence of (stagonospora) phaeosphaeria avenae blotch on triticale so far has been reported only in Poland (Arseniuk et al. 1989, Arseniuk 1993).

In the literature it is difficult to find comparable data on triticale grain yield losses afflicted by the pathogens. Under natural mild infection of triticale plants the total grain yield per hectare could be reduced by 10% (Arseniuk et al. 1991a). Heavier seed infections/infestations with *S. nodorum* may increase the yield loss up to 15%. However, the disease severity is strongly influenced by weather and if it is not conducive for the disease development, even a heavy seed infection may not result in a grain yield loss (Arseniuk 1993). Under experimental conditions where plants were artificially inoculated with a spore suspension of *S. nodorum* the total grain yield was reduced up to 37.3%, 44.3% and 19.7% in the most susceptible genotypes of winter triticale (n=141), winter wheat (n=18) and winter rye (n=5) entries, respectively (Arseniuk 1993). On average, triticale response to stagonospora nodorum blotch appeared to be intermediate between wheat and rye, but closer to wheat than to rye (Arseniuk et al. 1991c Arseniuk 1993). The latter crop showed to be the most resistant to the disease (see Table 3 for a closer analysis). Similarly, for the spring triticale entries (n=160, mainly of the CIMMYT origin) under experimental conditions the average reduction in the total grain yield amounted to 22.0% and for spring wheat ones (n=18) the figure was 28.1% (Arseniuk 1993). No close association between resistances to stagonospora nodorum blotch and septoria tritici blotch in both winter and spring triticale types at the seedling stages were found (Scharen et al. 1991).

Although, *Phaeosphaeria* spp. are not specialized pathogens and major genes for resistance to the pathogens in triticale and other cereals are not known some progress through breeding for the disease resistance seems to be achieveable. Though, complete resistance in triticale and wheat seems to be non-existent, the crop germplasm showing both yield tolerance and higher levels of leaf and head resistance to the pathogens has been identified (Arseniuk 1993, Wo 1992). Also, similarly to *Fusarium* diseases some improvement in tritcale resistance to stagonospora (septoria) nodorum blotch may come from somatic embryogenesis. In our studies (Arseniuk 1993,Table 3) somaclones with a relatively higher resistance, than their parental or donor varieties, to the disease have been found.

There are only few papers abstracted in the WBTA on *Pyrenophora* spp. (anamorph: *Drechselera* spp.) (Kuwite & Braun 1989, Satvinder 1990). Triticale was affected by *P. teres, P. lolii* (anamorph: *D. siccans*), *P.(D.) graminea, Setosphaeria turcica* (anamorph: *Exserohilum turcicum*) and *S.(D) rostrata* (Satvinder et al. 1990, Sisterna 1989). Saari et al (1986) and Singh and Saari (1991) cite reports describing a frequent occurrence of *P.(D) tritici-repentis* in tropical and temperate regions. In the WBTA this pathogen is not mentioned, though, Wiese (1987) considers it to be present on wheat the world over. In Poland this is a quite common pathogen of triticale and wheat observed in the field every year (Arseniuk & Czembor 1991, Pokacka 1985, Zamorski - personal communication). The author observed *P. teres* (net blotch) on triticale, too (unpublished data). The reported *Drechslera* spp., *Alternaria alternata*, as well as, *D. avenacea, D. secalis, A. tennuissima, A. triticina* (seed of mainly CIMMYT origin) and described in the previous section *Bipolaris* (*Cochliobolus*) spp. are more or less frequently isolated from triticale seed at IHAR Radzików, PL (Grzelak-personal communication). The above mentioned fungi are known to produce biologically active, very potent and dangerous to humans and animals AAL and fumonisin toxins. The compounds have carcinogenic properties and cause leukoencephalomalacia - softening of the brain matter - (Mirocha et al. 1993) in animals (horses). It is to add, that triticale response to the pathogens described in this passage, as well as the economic importance of diseases incited by the pathogens await for more comprehensive studies.

There are several reports in the literature (Table 1) on incidence of *Tilletia* spp. and *Ustilago* spp. in triticale, but most of the reports prove that the fungi are of minor importance (Aujla et al. 1990, Bekele et al. 1985, Schevchenko & Karpachev 1985). Winter et al. (1992) found that after an artificial inoculation Lasko triticale appeared highly susceptible to *T. tritici*. Sethi et al (1988) showed limited evidence that the rye chromatin may play a role in triticale resistance to infection with *N. indica* and *U. segetum* var. *tritici*. According to Warham (1988) the 1B/1R translocations may be a source of morphological resistance in triticale to *N. indica*. Grewal et al. (1990) indicated that resistance to *U. tritici* may be transferred from triticale to *T. aestivum*.

Fungal pathogens affecting basically leaves, stems and occasionally heads

In comparison to wheat the incidence of powdery mildew on triticale is of minor importance. As it is in the case of bunts and smuts, the crop often shows complete freedoom from the disease (Arseniuk & Czembor-own observations, Iliev et al. 1990, Zamorski 1993a). This is especially true for the incidence of *E. graminis* f.sp. *secalis* (Rigin & Lebedeva 1990). The latter authors suggest that resistance to *E.g.* f.sp. *tritici* in addition and substiution lines of wheat was associated with the rye chromosomes 2R and 6R.

In contrast, the other genus of the obligatory and highly mutable pathogens, *Puccinia* spp., occurs on triticale the world over (Bekele et al. 1985, Haesert et al. 1987, Iliev et al. 1990, Wilson and Shaner 1989a,b, Zamorski 1994, Zwer et al. 1992) and among the triticale diseases is most often reported (Table 1). Cereal rusts, when occur in epidemic proportions afflict heavy yield losses in wheat (Wiese 1987) and triticale (Singh & McIntosh 1988, Singh & Saari 1991). According to Udachin and Tarankov (from Schevchenko 1985) every 10% of leaf infection of triticale with leaf rust results in 3% of grain yield loss.

It is well known that triticale is attacked by wheat and rye physiological forms of three rusts *P. graminis*, *P. recondita* and *P. striiformis* (McIntosh & Singh 1986, Saari et al. 1986, Singh & Saari 1991). Leaf rust is spread over a broad range of environments, but prefers temperate climate with temperatures of 15 - 22°C and not limiting moisture. Author's own as well other observations (Zamorski - personal communication) show that the rust in addition to infection of leaves and leaf sheaths is able to attack triticale glumes and awns. Stem rust prefers temperatures near 26°C and it is one that created the most serious problems for triticale growing, to date (Zwer et al. 1992). Stripe rust prefers humid and cool (3°C - 15°C) environments and it is the most rarely reported rust on triticale, to date (Table 1).

For information on genetics of stem and leaf rusts resistance in triticale, origin and distribution of resistance genes and pathogenic changes in populations of the rusts the reader is referred to reviews and papers presented at the previous triticale symposia (McIntosh & Singh 1986, Saari et al. 1986, Singh & Saari 1991) as well as the one of Zwer et al. (1992). The problems are disussed by the cited authors very comprehensively. Information on the above questions for stripe rust is still lacking.

Other authors already noticed that triticale is more easily attacked by the wheat physiological forms of the rusts than by the rye ones (NcIntosh & Singh 1986, Saari et al 1986, Singh & Saari 1991, Zwer et al. 1992). Since the time the reports were published not much has changed concerning the physiological forms attacking triticale. This phenomenon is undoubtedly related to the amount of wheat chromatin in triticale and the hectareage of wheat (ca. 221,316,000 ha in 1993, Rocznik Statystyczny za 1993 r.) grown worldwide and very often neighbouring with triticale. Due to triticale genome composition it is very unlikely that rust forms specific only to triticale will ever evolve. Since rye is usually resistant to the special forms of rust attacking wheat and wheat resistant to such forms attacking rye it was reasonable to expect that triticale, a complex genome species, will be cross-resistant (or cross-protected) to rusts and other pathogens. The reality shows to be different. As it was already stated in the introduction, the cross-bred species inherited from its progenitors both types of traits, the positive and the negative ones, e.g. in addition to rust resistance also rust vulnerability. Moreover, as stated by McIntosh et al. (1983) and McIntosh and Singh (1986), for triticale protection against rusts it is important to know what resistance genes rye and wheat gametes carry before their union and to what extent these genes are subsequently expressed in the amphiploid (McIntosh & Singh 1983). The new strains of primary triticales arise usually from deliberate crossings but whether triticale breeders are in the total control of traits the wheat and rye gametes carry it is another matter. On the other hand rust resistance is only one of many agronomic traits under consideration. Also, new triticale varieties can be created in many other ways, than just by straight crossing of rye × wheat. This way, to combine all the positive traits in one triticale genotype becomes a challenging task that requires extensive evaluations of germplasms and knowledgeable decisions.

Table 2	Reaction types of small grain species to inoculation with *Puccinia* spp.							
Puccinia spp.	Isolate origin	Winter triticale		Winter wheat	Winter rye	Barley 'Haisa'	Oats	*Agropyron caninum*
		'Lasko'	'Grado'	'Grana'	'Motto'			
P. graminis avenae	oats	-	-	-	-	-	++	-
P. graminis hordei	barley	-	-	-	-	++	-	-
P. graminis secalis	rye	n	n	-	++	-	-	+
P. graminis tritici	wheat	+	+	++	-	-	-	-
P. graminis (?)	triticale	++	++	+	+	-	-	-
P. graminis (?)	*A. caninum*	+-	+-	-	+	-	-	+
P. recondita secalis	rye	c	c	-	++	-	-	-
P. recondita tritici	wheat	++	++	++	-	-	-	-
P. recondita (?)	triticale	++	++	++	+	-	-	-
P. recondita (?)	*A. caninum*	-	+-	-	c	c	-	+
P. coronata	oats	-	-	-	-	-	++	-
P. hordei	barley	-	-	-	-	++	-	-

Explanations: (?) - forma specialis not known; "++" - very positive reaction, sporulation abundant, uredia large; "+" - positive reaction, sporulation intermediate, variable to medium size uredia, chlorotic areas present; "+-" - reaction not uniform, sporulation scarce, uredia of small size, chlorotic and necrotic areas abundant; "-" - no reaction; "c" - only chloroses; "n" - only necroses.

The occurrence of putative hybrids in populations of rusts of wheat, rye and possibly other cereals is an other important question. Information on the subject is still scanty and comes from a few papers (Arseniuk & Czembor 1991, Luig & Watson

1972, McIntosh et al. 1983). It is known that the rust species can hybridize through spontaneous crossings on their alternate hosts, e.g. *P. g. tritici* and *P. g. secalis* on *Berberis vulgaris*, as well as somatically on graminaceous hosts (Luig & Watson 1972, McIntosh et al. 1983). Oku and Tsuchizaki (1993) proved that hybrid cultures of *Erysiphe graminis* ff. sp. *secalis* × *tritici* segregated on wheat without resistance genes, with wheat resistance genes *Pm*1, 3b and 3c and rye resistance genes *Pm*7 and *Pm*8. Similarly, Arseniuk and Czembor (1991 and Table 2) showed that urediospores of *P. graminis* and *P. recondita* coming from triticale incited rusting of both wheat and rye. Since a deliberate hybridization of the rusts before the cross-inoculation tests was not made it is presumed that those rusts urediospores could be the putative (somatic) hybrids. Other typical examples of hybrids are given by McIntosh et. al. (1983). The authors showed that virulence alleles were carried by both pathogens *P. g. secalis* for wheat and *P. g. tritici* for rye. The reported data indicate that hybridizations between formae speciales *secalis* and *tritici* of either rusts or powdery mildew fungi are a source of new virulences compatible with wheat and rye resistances under laboratory and field conditions. However, more research is needed to show the role of hybrids in cross-infections of wheat and rye with "triticale rusts".

In table 3 results of our own studies are reported, which show that resistance in triticale to yellow (stripe) rust could be improved through somatic embryogenesis. The reported data are supported by findings of Oberthur et al. (1993). The latter authors reported that two Coker 916 wheat somaclones had increased field resistance to leaf rust. The resistance the somaclones carried appeared to be heritable and controlled by following genes: *Lr* 3, 10, 14a, 15, 20, and either 27 or 31.

Fungal pathogens affecting foliage

There are only a few pathogens affecting triticale leaves (Table 1). They occurr sporadically and have not caused serious diseases. Rhynchosporium scald was found at low levels also in Poland (Arseniuk & Czembor 1991, Zamorski et al. 1993a,b).

Mycosphaerella speckled leaf blotch is a minor disease worldwide. It is infrequently found on triticale grown in temperate (Haesert et al. 1987, Pokacka 1985, Sowa 1986) as well as in warmer (Eyal & Talpaz 1990) regions. It is characteristic that Scott & Benedikz (1986) in the cold wet weather of the Great Britain providing ideal conditions for severe *M. graminicola* infection could not identify the disesease symptoms on triticale plantations. May (1983) reported a successful transfer of resistance to speckled leaf blotch from triticale to wheat, however, whether the resistance in the selected triticale × wheat hybrids was derived from rye or durum wheat was not established.

| Table 3 | Reaction of somaclones[1], donor winter triticale varieties and conventional varieties/lines of winter triticale, wheat and rye to artificial inoculation with *Septoria (=Stagonospora) nodorum, S. tritici* and *Puccinia striiformis* at the seedling and adult plant stages. |

	1st-leaf seedlings*						Adult plants**																	
Entry	Septoria nodorum blotch				Septoria tritici blotch		Height (cm)		Heading in Julian days		Mean % disease (SNB)on:				% reductions in the total yield and test weights due to						*Puccinia striiformis* % disease/			
	Radzików		Bozeman								leaves		heads		SNB infection						Reaction type			
	% dis.	Ran-ge	% dis.	Ran-ge	% dis.	Ran-ge	Me-an	Ran-ge	Me-an	Ran-ge	% dis.	Ra-n-ge	% dis.	Ra-n-ge	Yie-ld	Ran-ge	TKW	Ran-ge	Hclt-wht	Ran-ge	Lea-ves	Ran-ge	Hea-ds	Ran-ge
Bolero	48.0	-	81.0	-	58.0	-	128	-	150	-	41.0	-	10.0	-	9.7	-	17.3	-	4.7	-	5.3,R	-	0.0	-
Bolero, Soma-clones, n=5	41.0	32-48	53.0	41-70	64.0	55-72	124	111-129	149	148-150	32.0	12-45	29.4	27-33	30.1	15.0-40.6	12.9	8.3-18.5	4.0	3.0-6.7	33.8,R MR,MS,S	6-80	3.1	0-5
RAH-101	50.0	-	75.0	-	77.0	-	126	-	152	-	50.0	-	11.0	-	17.0	-	11.3	-	2.3	-	14.0,R	-	0.0	-
RAH-101, Soma-clones, n=5	45.0	36-51	49.0	25-71	86.0	81-89	123	117-131	149	148-152	47.8	42-53	15.2	12-20	20.8	14.3-29.7	9.3	6.5-13.3	3.1	1.7-4.3	10.8,R MR	1-50	0.5	0-3
GWT-88-16	43.0	-	65.0	-	68.0	-	123	-	146	-	54.0	-	31.0	-	18.7	-	17.7	-	4.3	-	16.3,R	-	-	-
GWT-88-16, Somaclones, n=5	43.0	-	52.0	-	56.0	-	122	-	149	-	46.0	-	29.0	-	32.3	-	13.3	-	6.7	-	7.5,R	-	-	-
Lasko	46.0	-	53.0	-	61.0	-	125	-	148	-	41.0	-	17.0	-	15.7	-	10.7	-	2.3	-	23.7,R	-	0.5	-
Lasko, Somaclo-ne, n=1	45.0	-	58.0	-	62.0	-	124	-	150	-	46.0	-	20.0	-	13.6	-	13.0	-	2.7	-	7.5,R	-	0.5	-
Malno	41.0	-	65.0	-	81.0	-	118	-	148	-	43.0	-	19.0	-	19.3	-	11.3	-	3.7	-	-	-	0.5	-
Malno, Somaclo-ne, n=1	45.0	-	68.0	-	83.0	-	124	-	149	-	48.0	-	11.0	-	15.7	-	7.7	-	2.3	-	-	-	0.0	-
Triticale conven-tional varieties/ lines, n=141	45.0	31-50	60.0	15-86	65.0	28-87	124	96-138	148	141-153	47.3	32-60	20.0	7-53	16.6	6.3-37.3	11.7	4-28.7	3.1	0.3-7.3	-	-	-	-
Wheat conven-tional varieties/ lines, n=18	41.0	23-55	48.0	13-66	72.0	49-87	92	71-106	155	143-160	48.6	31-60	30.0	13-45	21.2	9.7-44.3	14.4	8.3-28.0	4.1	0.7-8.3	-	-	-	-
Rye conventional varieties, n=4	60.0	56-65	68.0	55-73	76.0	69-82	125	138-149	135	133-136	7.3	-	2.0	0.0-4.0	12.5	5.3-19.7	6.1	4.0-7.3	0.9	0.3-1.7	-	-	-	-
LSD	4.0	-	5.0	-	5.0	-	13.8	-	2.7	-	17.3	-	19.8	-	n.s	-	15.0	-	4.5	-	-	-	-	-

Explanations: 1 - mean of R_1 and R_2; * - means of 3 experiments, ** - means of 2 experiments over 2 years, TKW = thousand kernel wieght, Hclt-wht = hectoliter weight

Fungal pathogens affecting lower stems and roots

Gaeumannomyces graminis inciting take-all disease most often was reported as an important disease in Europe (Arseniuk & Czembor 1991, Haesert et al. 1987, Hollins et al. 1986, Gutteridge et al. 1993), USA and specifically in Georgia (Rothrock 1988), Mexico (Lawn & Sayre 1992) and Australia (Cooper 1991, Murray et al. 1987). Singh and Saari (1991) on behalf of their cooperators considered this disease as important also in Brazil and the Iberian Peninsula. Nothing is known on genetics of resistance to this disease in triticale. All the small grains except oats are susceptible to take-all. In literature reports rye is rated as the least susceptible, wheat the most and barley and triticale as intermediate. However barley was still significantly more resistant than triticale (Gutteridge et al. 1993, Rothrock 1988). Hollins et al. (1986) showed that triticale resistance is not reliably expressed in the greenhouse tests and should be evaluated in the field. The latter authors reported that in seven addition lines none of the rye chromosome pairs conferred resistance in the presence of the hexaploid wheat genome. If the rye chromosomes were substituted for the wheat ones the resistance was improved. Wallwork (1989) by increasing rye chromosome numbers also increased take-all resistance in triticale and wheat × triticale hybrids. Since the results are conflicting a further research is needed to clarify the effect of rye chromatin on take-all resistance in triticale. Gutteridge et al. (1993) found that under the disease stress triticale may be a good substitute for wheat on less fertile soil with low inputs, but not where the inputs are high.

Reports on eye-spot (*Pseudocercosporella herpotrichoides*, anamorph of *Tapesia yallundae*) and sharp eye-spot (*Rhizoctonia solani*) come mainly from European countries with temperate climate (Arseniuk & Czembor 1991, Bojarczuk & Bojarczuk 1988, Cichy 1992, Gindrat et al. 1988, Haesert et al. 1987, Scott et al. 1989). Upon the host attacked by populations of *P. herpotrichoides* several types of isolates are distinguished (Scott et al. 1989). Triticale was more severly attacked by rye (R) than by wheat (W) type of isolates (Bojarczuk & Bojarczuk 1988a). On average, the commercially grown triticale varieties and some strains appeared significantly more susceptible than the standard ones of rye and wheat (Bojarczuk & Bojarczuk 1988b). Among the standard wheat varieties was a French variety 'Capelle' with a durable resistance to the pathogen. As an other source of resistance for wheat is used *Aegilops ventricosa* (Bojarczuk & Bojarczuk 1988b, Scott et al. 1989). Eye-spot poses a serious threat for bread wheat and most probably the same is with triticale but the problem is still not recognized sufficiently and requires further studies.

The other root and stem base attacking pathogens *Typhula ishikariensis, T. incarnata, Myriosclerotinia (Scelrotinia) borealis* and *Rhizoctonia solani* on triticale are reported sporadically (Neate 1989). They occur in regions with long lasting snow cover and are of polyphagous type (Wiese 1987) attacking graminaceous and other crops.

Bacterial pathogens

Among the bacterial pathogens the most widely spread on triticale are two pathovars of *Xanthomonas campestris,* i.e. pv. *undulosa* (Duveiller et al. 1993, Duveiller & Bragard 1992, Mehta 1990) and pv. *translucens* (Akhtar 1990, Bekele et al. 1985, Cunfer & Scolari 1982, Duveiller 1989, Kotlyarov 1991). The pathogens were reported to occur primarily in warm and subtropical regions, but they also seem to be present in temperate triticale-growing areas (Arseniuk & Czembor 1991, Kotlyarov 1991, Sowa - personal communication, Wolski - personal communication). The disease caused by the bacteria is known most commonly on wheat as black chaff. The black chaff phase, is preceded by water-soaked streaking or striping of leaves - most common symptom on triticale (Johnson et al. 1987). Duveiller (1989) described in *X. campestris* 6 pathovars with overlapping host ranges. The definitions of the pathovars based on the host range seem to be unclear, e.g. Cunfer and Scolari (1982) showed rather a low specificity for the pv. *translucens* that strains from triticale were equally virulent to triticale, wheat and rye, much less virulent to barley, and nonpathogenic to oats and grasses. Strains from barley, in comparison to such ones from rye, wheat and triticale were the most restricted in host range. As it is noticed above, in the literature some authors write about the pv. *translucens,* the others about pv. *undulosa.* On the orher hand, El-Sadek (1992) reported occurrence of *X. c.* pv. *vignicola* on triticale roots.

Resistance in triticale to bacterial leaf stripe was shown to be controlled by one dominant gene (Johnson et al. 1987) and in bread wheat by five genes which differed in strength of expression (Duveiller 1993).

Viruses, virus-like agents and nematodes

Several diseases incited by viruses and nematodes have been reported to occur on triticale. Among viral diseases barley yellow dwarf (BYDV) is of the major importance. BYDV and wheat streak mosaic virus (WSMV) are also observed in Poland, mainly in triticale nurseries (Sowa, personal communication). Collin et al (1990) reported a considerable genetic variability for tolerance to BYDV in triticale. The tolerance was observed to be heritable. Nevertheless, the disease caused highly significant reduction in tillering, plant height and biomass in all triticale and wheat lines studied irrespective of their level of tolerance (Nkongolo et al. 1993).

Root knot (*Meloidogyne* spp.) (Sasser et al. 1988), root lesion (*Pratylenchus thornei*) (Lawn & Sayre 1992) and cereal cyst nematodes (*Heterodera* spp.) (Cooper 1991, Romero et al. 1991) have been found on triticale, but all at a low frequency (Table 1), though of considerable importance (Cooper 1991). Moderate resistance and tolerance to attack of *M. javanica* among triticale genotypes was reported (Sharma 1986). Triticale was ranked second in resistance to attack of *H. avenae*, after oats but before barley and wheat cultivars. Yield loss was mainly due to the reduction of head number/m^2 and the number of grains/head (Romero et al. 1992). Attempts to transfer resistance to the nematode from triticale to wheat are reported by Cooper (1991).

Other diseases

In the Wheat, Barley and Triticale Abstracts (WBTA) was not possible to identify reports on ascochyta leaf spot (*Ascochyta* spp.), phoma leaf spot (*Phoma glomerata*) and cephalosporium stripe (*C. gramineum*). All this diseases are occurring in Poland (Arseniuk & Czembor 1991, unpublished data, Sowa-personal communication, Martyniuk 1993) and elsewhere (Saari et al. 1986). In studies conducted by the author the first two pathogens have frequently been isolated from leaves of seedlings as well as of adult plants. *P. glomerata* was also observed on triticale in the USA (Hosford 1975). Since *Ascochyta* spp. and *P. glomerata* require prolonged wet postinoculation periods to cause leaf spotting they might be of importance in wetter seasons and wetter triticale growing areas.

The first thorough study and description of *C. gramineum* on triticale was done by Martyniuk (1993). The fungus infects triticale mainly through roots with broken stele. Once inside the plant conidia of the pathogen spread systemically in xylem vessels and colonize whole plant. A considerable variability occurs among triticale genotypes in response to cephalosporium stripe. On average, the crop is more susceptible to the disease than bread wheat and rye (Martyniuk 1993).

Discussion and conclusions

Table 1 shows pathogens and diseases on triticale, with a reference to bread and durum wheats and rye, reported in the world literature over the decade 1984 - 1993. This was a decade of successful triticale introductions into the commercial production of cereals in many countries. An interesting feature of the tabulated data is that they do not show a steep increase in the numbers of pathogens attacking triticale over the years. But the stem rust epidemic described in Australia (McIntosh et al. 1983, McIntosh and Singh 1986, Singh and McIntosh 1988, Zwer et al. 1992)) no other severe epidemics have been reported to occurr in triticale, to date.

Average reporting frequency of triticale pathogens in the papers published over the decade stays 30 per year. This is by 20-fold less than for bread wheat, 1.5-fold more than for durum wheat and comparable to the annual reporting frequency of pathogens on rye (Table 1). While comparing the literatural data on pathogens of the cereals in question, one should take into consideration the world hectareage (221,316,000 ha of wheat versus 2,367,800 ha of triticale, cited from Polski Rocznik Statystyczny za 1993 r. and Triticale Topics 11, 1993, respectively), the economic importance and the time span of cultivation by the man of the respective cereals. It is obvious that all these factors are closely interrelated.

This as well the former (Saari et al. 1986, Singh & Saari 1991) reviews indicate, that triticale inherited resistance and susceptibility to all diseases from its progenitors - *T. aestivum*, *T. durum* and *S. cereale*. The literature data show that triticale actually is a bridge species between its progenitors for both traits - resistance and susceptibility to diseases. All pathogens attacking triticale are, either, of wheat or rye origin. Physiological forms of the pathogens specific only for triticale are not known to evolve and in the literature have not been reported, yet. The putative hybrids between special wheat and rye forms of rust and powdery mildew fungi evolving sexually on alternate hosts and somatically on triticale are compatible with all three hosts - triticale, wheat and rye (Arseniuk & Czembor 1991, Luig & Watson 1972, McIntosh et al. 1983, Oberthur et al. 1993, Oku & Tsuchizaki 1993).

Numerous comparisons with wheat, rye and triticale on disease reactions show, with a few exceptions, that triticale usually reacts intermediately between wheat and rye, more often closer to wheat than to rye. The exceptions are powdery mildew and cereal cyst nematodes. In comparison to wheat and rye triticale resistance to the two disease agents is superior. Since rye is generally considered to be susceptible to powdery mildew (Miedaner et al. 1993) it is rather difficult to justify that the resistance to the disease in triticale is of rye origin. The crop could achieve the superior resistances, in part from rye germplasm and in part by a new combination of genes.

Triticale resistance to bunts, smuts, ergot and rhynchosporium scald can easily be explained by the phenomenon of genetical cross-protection inherited from rye and wheat. Highly positive effects of the rye chromatin incorporated into triticale on its resistance to bunts and smuts and the wheat chromatin on triticale resistance to ergot and rhychosporium scald are evident. A positive effect of rye chromatin is also implicated in triticale resistance to other diseases, e.g. take-all (Hollins et al. 1986, Wallwork 1989). Although, the direct studies are lacking, the effect of rye chromatin seems to be positive on triticale response to necrotrophic fungi, like *Ph. nodorum* and *Fusarium* spp.. Something opposite appeared to be in the case of eye-spot (Bojarczuk & Bojarczuk 1988a) and cephalosporium stripe (Martyniuk 1993) where the effect of rye chromatin seems to be negative or at least dubious, since triticale looks to be more susceptible to the diseases than its progenitors.

The analysis of presently available literature data allows on a closing generalization, that the rye chromatin exerts a positive effect on resistance of triticale to diseases of wheat origin, and conversely, the wheat chromatin prevents triticale from disastrous damages caused by diseases of rye origin. Thus, the objective to synthesize triticale for its enhanced disease resistance is still valid, but it is to be challenged and verified over both the time and space of triticale expansion.

Acknowledgements

The contributions of prof. H. J. Czembor, dr. W. Sowa and dr. J. Zimny into the preparation of this review are gratefully acknowledged.

Bibliography

Abreu C.A. & C.P. Marques, Septoria nodorum blotch in Tras-os-Montes: yield losses. Proc. 3rd International Workshop on *Septoria* diseases of Cereals. Zürich (Switzerland). 1989, July 4-7: 121-123.

Abreu C.A., H. Guedes-Pinto & J.S. Bento, 1991. Septoria nodorum blotch in Northern Portugal. Proceedings of the 2nd International Triticale Symposium. Mexico D.F., CIMMYT: 274-277.

Akhtar M.A. Occurrence of bacterial blight on triticale in Pakistan. In: Wheat, Barley and Triticale Abstracts (WBTA)(1990) 7 (5): 615.

Arseniuk E., H.J. Czembor, W. Sowa & H. Krysiak, 1989. Preliminary studies of *Septoria* spp. on triticale in Poland. Proc. 3rd International Workshop on *Septoria* Diseases of Cereals, Zürich, (CH). July 4-7: 16-18.

Arseniuk E., H.J. Czembor, W. Sowa & H. Krysiak, 1990a. *Septoria* diseases on triticale in Poland. Proc. International Wheat Symposium, Albena (Bulgaria). June 1989.

Arseniuk E. & H. J. Czembor, 1990b. Breeding aspects of the studies on triticale, wheat and rye diseases. Zeszyty Problemowe: Hodowla Zbóż , Prace Badawcze Grup Problemowych, Część II - Hodowla Pszenżyta: 65-76. (Published in 1992 in Polish).

Arseniuk E. & H.J. Czembor, 1990c. Major diseases of triticale, wheat and rye in Poland. In: Triticale in Eastern Europe. Proceedings of Triticale Conference organized within frames of the Council for Mutual Economic Assistance, June 1990, Ma yszyn, Poland. pp.117-132. (In Russian).

Arseniuk E. & H.J. Czembor, 1991. Triticale diseases in Central Poland in 1991. Triticale Topics No. 7.

Arseniuk E., A.L. Scharen, H.J. Czembor, W. Sowa & H. Krysiak., 1991a. Reaction of triticales from different geographic regions to septoria nodorum blotch under climatic conditions of Central Poland. Proceedings of the 2nd International Triticale Symposium. Mexico D.F., CIMMYT: 252-255.

Arseniuk E., A.L. Scharen, W.E. Grey & H.J. Czembor, 1991b. Transmission of *Septoria nodorum* and *Fusarium* spp. on triticale seed. Proceedings of the 2nd International Triticale Symposium. Mexico D.F., CIMMYT: 260-263.

Arseniuk E., P.M. Fried, H. Winzeler & H.J. Czembor, 1991c. Comparison of resistance of triticale, wheat and spelt, to septoria nodorum blotch at the seedling and adult plant stage. Euphytica 55: 43-48.

Arseniuk E., A.L. Scharen & H.J. Czembor, 1991d. Pathogenicity of seed transmitted *Fusarium* spp. to triticale seedlings. Mycotoxin Research 7A: 121-127.

Arseniuk E., T. Góral & H.J. Czembor, 1993a. Reactions of triticale, wheat and rye accessions to grami naceous *Fusarium* spp. infection at the seedling and adult plant growth stages. Euphytica 70: 175-183.

Arseniuk E., T. Góral & H.J. Czembor, 1993b. Pathogenicity and differential response in *Fusarium* spp. X*Triticosecale* Wittmack/*Triticum aestivum* L./*Secale cereale* L. pathosystems. Hod. Ro l. Aklim. Nasien. 37 (3): 17-23.

Arseniuk E., 1993. Septoria glume and leaf blotch of triticale. Final Technical Report of Project No.: PL-ARS -138, pp. 26.

Aujla S.S., Sharma I. & S.K. Gill. 1990. Sources of resistance in wheat to Karnal bunt. Indian Phytopathology 43: 91-93.

Bekele G.T., B. Skovmand, L.I. Gilchrist & E.J. Warham, 1985. Screening of triticale to certain diseases occurring in Mexico. Genetics and Breeding of Triticale, Eucarpia Meeting, Clermont Ferrand (France). July 2-5, 1984: 559-564.

Bojarczuk M. & J. Bojarczuk, 1988a. Problem of biological specialization of the fungus *Pseudocercosporella herpotrichoides* (Fron.) Deighton). Hod. Rośl. Aklim. Nasiennictwo [1988, publ. 1990] 32: 95-99.

Bojarczuk M. & J. Bojarczuk, 1988b. Problem of triticale resistance to *Pseudocercosporella herpotrichoides* (Fron.) Deighton). Acta Acad. Agricult. Techn. Olst. Agricultura 47: 123-131.

Chattannavar N.S., T. Raghuchander & S. Kulkarni, 1991a. Effect of *Bipolaris sorokiniana* on biochemical constituents of triticale. Proceedings of the 2nd International Triticale Symposium. Mexico D.F., CIMMYT: 238-240.

Chattannavar N.S., T. Raghuchander & S. Kulkarni, 1991b. Triticale varietal screening against *Bipolaris sorokiniana*. Proceed. of the 2nd International Triticale Symposium. Mexico D.F., CIMMYT: 229-230.

Chattannavar N.S., S. Kulkarni, Patil K.N. & Y. Hedge, 1991c. Sowing dates on the intensity of alternaria blight in triticale. In: WBTA (1992) 9 (5): 625.

Chauhan S.S., 1986. Seed mycoflora of triticale and their control. In: WBTA (1986) 3

(1): 79.

Cichy H., 1992. Resistance of winter triticale to *Pseudocercosporella herpotrichoides*. Biuletyn IHAR 181-182: 89-92.

Cichy H. & W. Maćkowiak, 1993. Intravarietal differences in winter triticale resistance to snoe mould (*Fusarium nivale*). Hod. Rośl. Aklim. Nasien. 37 (3): 115-120.

Collin J., A. Comeau & C.A. St-Pierre, 1990. Tolerance to Barley Yellow Dwarf Virus in triticale. Crop Science 30: 1008-1014.

Cooper V.K., 1991. Breeding triticale for Australian problem soils. Proceedings of the 2nd International Triticale Symposium. Mexico D.F., CIMMYT: 188-195.

Cristani C., 1992. Seed-borne *Microdochium nivale* [Ces. ex Sacc.) Samuels (=*Fusarium nivale* (Fr.) Ces.] in naturally infected seeds of wheat and triticale in Italy. Seed Science and Technology 20: 603-617.

Cunfer M. B. & L.B. Scolari, 1982. *Xanthomonas campestris* pv. *translucens* on triticale and other small grains. Phytopathology 72: 683-686.

Cunfer M.B. & J. Youmans, 1983. *Septoria* on barley and relationships among isolates from several hosts. Phytopathology 73: 911-914.

De Oliveira R.M.A. & A.C. Baier, 1993. Industrial quality and resistance to spot blotch and scab intriticale, wheat and rye. Pesquisa Agropecuaria Brasileira 28: 603-608.

Dormann M. & G. Oettler, 1991. Effect of parental genotype on resistance to *Fusarium* spp. in primary triticales. Proceedings of the 2nd International Triticale Symposium. Mexico D.F., CIMMYT: 225-228.

Dormann M. & G. Oettler, 1993. Genetic variation of resistance to *Fusarium graminearum* (head blight) in primary hexaploid triticale. Hod. Rośl. Aklim. Nasien. 37 (3): 121-127.

Duveiller E., 1989. Research on "*Xanthomonas translucens*" of wheat and triticale at CIMMYT. EPPO Bulletin 19: 997-103.

Duveiller E. & C. Bragard 1992. Comparison of immunoflorescence and two assays for detection of *Xanthomonas campestris* pv. *undulosa* in seeds of small grains. Plant Disease 76: 999-1003.

Duveiller E., C. Bragard, & H. Maraite. 1993a. Bacterial diseases of wheat in the warmer areas   reality or myth? In: WBTA (1993) 10 (4): 490.

Duveiller E., M. Van Ginkel & M. Thijssen, 1993b. Genetic analysis of bacterial leaf streak caused by *Xanthomonas campestris* pv. *undulosa* in bread wheat. Euphytica 66: 35-43.

El-Sadek S.A.M., 1992. Bacterial blight of cowpea: 3 - survival of the causal organism in the rhizosphere of host and non-host plants. In: WBTA (1992) 9 (4): 496.

Eyal Z. & H. Talpaz, 1990. The combined effect of plant stature and maturity on the response of wheat and triticale accessions to *Septoria tritici*. Euphytica 46: 133-141.

Fadel F. & G. Wenzel, 1993. In vitro selection for tolerance to *Fusarium* in F_1

microspore populations of wheat. Plant Breeding 110: 89-95.

Gindrat D., C. Cavegn & L. Zufferey, 1988. Fungicide resistance of cereal eyespot: the situation in Switzerland. Revue Suisse d'Agriculture 20: 141-147.

Grewal A.S., S.S. Aujla & K.S. Gill, 1990. Triticale - a source of resistance to loose smut. Abstracts of the 2nd International Triticale Symposium, October 1-5, 1990, Passo Fundo, RS, Brazil.

Grzelak K., 1993. Occurrence of seed-borne *Fusarium* spp. on triticale seed in relation to seed decay and seedling root rot. Hod. Rośl. Aklim. Nasien 37 (4): 75-79.

Gutteridge J.R, D. Hornby, T.W. Hollins & R.D. Prew, 1993. Take-all in autumn sown wheat, barley, triticale and rye grown with high and low inputs. Plant Pathology 42: 425-431.

Haesert G., A. Baets, de & A. Daneels, 1987. Diseases of triticale and their control. Med. Fac. Landbouw. Rijksuniv. Gent 52: 797-806.

Hollins T.W., P.R. Scott & R.S. Gregory, 1986. The relative resistance of wheat, rye and triticale to take-all caused by *Gaeumannomyces graminis*. Plant Pathology 35: 93-100.

Hosford R.M., 1975. *Phoma glomerata*, a new pathogen of wheat and triticales. Phytopath. 65: 1236-1239.

Iliev V.I., S. Mikhova, I. Stoyanov & S. Tsvetkov, 1990. A study on the resistance of new triticale lines to rusts and powdery mildew. Plant Science Vol. XXVII/5: 36-42. [Bu] (Also in: WBTA 1991/8/1:125).

Ittu G., Ittu M. & N.N. Saulescu, 1990. Preliminary results on *Fusarium* infection of the ears in a collection of varieties and lines of triticale. In: Wheat, Barley and Triticale Abstracts (1992) 9 (2): 236.

Johnson J.W., B.M. Cunfer & D.D. Morey, 1987. Inheritance of resistance to *Xanthomonas campestris* pv. *translucens* in triticale. Euphytica 36: 603-607.

Kaltsikes J.P. & P.J. Bebeli, 1993. Somaclonal variation causes changes in the inter-relationships between traits in hexaploid triticale. Japanese J. Breeding 43: 45-51.

Kiecana I., J. Perkowski, J. Chełkowski & A. Visconti, 1987. Trichotecene mycotoxins in kernels and head fusariosis susceptibility in winter triticale. European Seminar "*Fusarium* - Mycotoxins, Taxonomy, Pathogenicity", Warsaw (Poland), 8-10 September: 53-56.

Kleijer G., 1988. Resistance of Swiss wheat cultivars to snow mould. Revue Suisse d'Agriculture 20:65-67.

Klein T.A., L. Burgess & W.F. Ellison, 1989. The incidence of crown rot in wheat, barley and triticale when sown on two dates. Australian J. Experimental Agriculture 29: 559-563.

Koczowska I. & M. Wiwart, 1990. Reaction of rye and winter triticale cultivars to different isolates of *Fusarium nivale* (Fr.) Ces.. Biuletyn IHAR 173-174: 71-73.

Khokhar L.K. & P.R. Pacumbaba, 1987. Alternative gramineous hosts of *Leptosphaeria nodorum* including two new records for the USA. J. Phytopathology 120: 75-80.

Kotlyarov V.V, 1991. Methods of evaluating resistance to bacterial diseases in wheat and triticale. In: WBTA (1991) 8 (6): 779.

Kuwite C. & L. Braun, 1989. 1988 Tanzania wheat disease survey. In: WBTA (1991) 8 (2): 261.

Lawn D.A. & D.K. Sayre, 1992. Soilborne pathogens on cereals in a highland location of Mexico. Plant Disease 76: 149-154.

Luig N.H. & I.A. Watson, 1972. The role of wild and cultivated grasses in the hybridization of formae speciales of *Puccinia graminis*. Aust. J. Biol. Sci. 25: 335-342.

Lupashku G.A., N.N. Balashova, I.V. Morar, S.L. Koretskaya & I. P. Byukli, 1993. Root rots of triticale. WBTA 1993 (3): 363.

Łacicowa B. & I. Kiecana, 1986. Występowanie *Fusarium nivale* (Fr.)Ces. na pszenżycie i podatność różnych rodów hodowlanych na porażenie. Rocz. Nauk Rol., Seria E, 16: 139-150 (In Polish).

Łacicowa B., & I. Kiecana, 1987. Występowanie *Fusarium culmorum* (W.G. Smith) Sacc. i *Fusarium avenaceum* (Fries.) Sacc. na pszenżycie oraz podatność rodów hodowlanych na porażenie. Rocz. Nauk Rol., Seria E (Ochrona Roślin) 17: 161-178 (In Polish).

Maier F. J. & G. Oettler, 1993. Selection for the *Fusarium* toxin deoxynivalenol in callus cultures of triticale. Hod. Rośl. Aklim. Nasien. 37 (3): 43-49.

Ma uszy ska E., 1992. Occurrence of ergot [*Claviceps purpurea* (Fr.) Tul.] in the seed material of winter triticale (X *Triticosecale* Wittmack). Biuletyn IHAR 184: 3-10.

Martin R.A., A. J. Mac Leod & C. Caldwell, 1991. Influences of production inputs on incidence of infections by *Fusarium* species on cereal seed. Plant Dis. 75: 784-788.

May C.E., 1983. Triticale × wheat hybrids and the introduction of speckled leaf blotch resistance to wheat. Proc. 6th International Wheat Genetics Symposium, Kyoto, Japan: 175-179.

Martyniuk S., 1993. Studies on cephalosporium stripe disease of cereals. D. Sc. Dissertation, pp. 66.

McIntosh A.R., N.H. Luig, D.L. Milne & J. Cusick, 1983. Vulnerability of triticales to wheat stem rust. Can. J. Plant Pathol. 5: 61-69.

McIntosh A.R. & S.J. Singh, 1986. Rusts - Real and potential problems for triticale. Proceedings of International Triticale Symposium, Sydney, Occasional Publication no. 24, pp. 199-207.

Mehta Y.R., 1990. Management of *Xanthomonas campestris* pv. *undulosa* and hordei through cereal seed testing. Seed Science and Technology 18: 467-476.

Miedaner T., H.K. Schmidt & H.H. Geiger, 1993. Components of variation for quantitative adult-plant to powdery mildew in winter rye. Phytopathology 83: 1071-1075.

Mirocha J. Ch., Junping Chen, Weiping Xie & Yechun Xu, 1993. Biology and chemistry of fumonisin and AAL toxins. Hod. Rośl. Aklim. Nasien. 37 (1):13-21.

Miller J.D., C.J. Young & R.D. Sampson, 1985. Deoxynivalenol and *Fusarium* head blight resistance in spring cereals. Phytpath. Z. 113: 359-367.

Murray G.M., J.B. Scott, Z. Hochman & J.B. Butler, 1987. Failure of liming to increase grain yield of wheat and triticale in acid soils may be due to the associated increase in incidence of take-all (*Gaeumannomyces graminis* var. *tritici*). Austral. J. Exp. Agiculture 27: 411-417.

Neate S.M., 1989. A comparison of controlled environment and field trials for detection of resistance in cereal cultivars to root rot caused by *Rhizoctonia solani*. Plant Pathology 38: 494-501.

Nkongolo K.K., Comeau A. & C. A. Saintpierre, 1993. Relationships between leaf chlorosis and certain agronomic traits in wheat, triticale and wheat-triticale lines infected with the barley yellow dwarf virus (BYDV). Can J. Plant Sci. 73: 1225-1231.

Oberthur E.L., S. A. Harrison, T.P. Croughan & D.L. Long, 1993. Inheritance of improved leaf rust resistance in somaclones of wheat. Crop Science 33: 444-448.

Oku T. & T. Tsuchizaki, 1993. Compatibility of *Erysiphe graminis* hybrids f. sp. *secalis* × f. sp. *tritici* with wheat lines involving resistance genes from rye. J. Phytopathology 138: 77-83.

Pavlov P. Y. Karayova, 1991. Species composition of *Fusarium* fungi on triticale in Bulgaria. In: WBTA (1991) 8 (3): 397.

Perkowski J., J. Chełkowski, P. Błażczak, C. Snijders & W. Wakuliński, 1991. A study of the correlations between the amount of deoxynivalenol in grain of wheat and triticale and percentage of *Fusarium* damaged kernels. Mycotoxin Research 7 [A (Suppl.) Part II): 102-114.

Perkowski J., J. Chełkowski & W. Wakuliński, 1988. Deoxynivalenol and 3-acetyldeoxynivalenol and *Fusarium* species in winter triticale. Mycotoxin Research 4: 97-100.

Pokacka Z., 1985. Badania nad plamistościami liści pszenicy ze szczególnym uwzględnieniem roli *Septoria nodorum* Berk. Prace Naukowe IOR, Tom XXVII, Zeszyt 2: 5-31 (In Polish).

Raghuchander T., S. Kulkarni & K.R. Hedge, 1988. Studies on leaf blight of triticale caused by *Bipolaris sorokiniana* (Sacc.) Shoem. anamorph of *Cochliobolus sativus* (Ito & Kurib.) Drechsler ex Dastur. Plant Pathol. Newsletter (1988) 6 (1-2): 43-44.

Raghuchander T., S. Kulkarni & K.R. Hedge, 1991a. An unrecorded pathogen on triticale in India. Indian Phytopathology 44: 262.

Raghuchander T., S. N. Chattannavar & S. Kulkarni, 1991b. Bioassay of nonsystemic fungicides against *Bipolaris sorokiniana*, causal agent of leaf blight of triticale. Proceedings of the 2nd International Triticale Symposium. Mexico D.F., CIMMYT: 231-233.

Reis M.E., 1991. Control of *Bipolaris sorokiniana* in triticale seeds. Proceedings of the 2nd International Triticale Symposium. Mexico D.F., CIMMYT: 212-214.

Reis M. E. & I. Ambrosi, 1991. Economical threshold to control leaf blights of triticale by spraying fungicides. Proceedings of the 2nd International Triticale Symposium. Mexico D.F., CIMMYT: 215.

Rigin V.B. & V. T. Lebiedeva, 1990 (publ.). Interactiopn of the wheat and rye genomes in the control of resistance to powdery mildew. In: WBTA (1991) 8 (5): 667.

Romero M.D., A. Valdeolivas & C. Lacasta, 1991. Incidence of *Heterodera avenae* on the growth and yield of cereals in Spain.: In: WBTA (1992) 9 (3): 367.

Saari E., G. Varughese & O.A. Abdalla, 1986. Triticale diseases: Distribution and importance. Proceedings of International Triticale Symposium, Sydney, Occasional Publication no. 24, pp. 208-231.

Sapra V.T., L.J. Hughes & A.L. Scharen, 1976. Preliminary observations on the incidence of *Septoria nodorum* on wheat, rye, and triticale in Alabama. Proc. *Septoria* diseases of Wheat Workshop, University of Georgia, Experiment, Georgia, USA: 57-60.

Satvinder K., S.P.A. Mann & S.K. Aulakh, 1990. Pathogenic potential of seven *Drechslera* species prevalent in cereal seeds. Indian Phytopathology 43: 179-185.

Scharen A.L., E. Arseniuk, W. Sowa, J. Zimny & W. Podyma, 1991. Seedling resistance of triticale and *Triticum* spp. germplasm to *Septoria nodorum* and *S. tritici*. Proceedings of the 2nd International Triticale Symposium. Mexico D.F., CIMMYT: 256-259.

Schevchenko V.E. & V.V. Karpachev, 1985. Genetic resistance to fungus diseases in all triticale varieties. Genetics and Breeding of Triticale. Eucarpia Meeting, Clermont - Ferrand, 2-5 July 1984:565-571.

Scott P.R. W.P. Benedikz, 1986. *Septoria, Fusarium* ear blight. In: Annual Report of the Plant Breeding Institute 1985. Cambridge, UK (1986): 97-99.

Scott R. P., T.H. Hollins & R.W. Summers, 1989. Breeding for resistance to two soilborne diseases of cereals. Vort. Pflanzenzüchtg. 16: 217-230.

Sethi G.S., Plaha P., & S.K. Gill, 1988. Transfer of Karnal bunt [*Neovosia indica* (Mitra) Mundkur] resistance from rye (*Secale cereale* L.) to wheat (*Triticum aestivum* L. em. Thell.). In: Proc. 7th International Wheat Genetics Symposium, Cambridge, UK, 13-19 July 1988 (Eds. Miller T.E. & D.M.R. Koebner), pp.439-

442.

Sharma R.D., 1986. Reaction of some tritcale genotypes to *Meloidogyne javanica*. In: WBTA (1986) 3(1):79.

Shukla R., 1990. Effect of atmospheric temperature, relative humidity and age of host plant on the development of *Alternaria* leaf blight of triticale. In:WBTA (1990)7(2): 326.

Simmonds J., J. Fregeaureid, R. Pandeya, D. Sampson & G. Fedak, 1993. Potential of anther culture in breeding for *Fusarium* head blight resistance in wheat. Cereal Res. Com. 21: 141-147.

Singh S.J. & R.A. McIntosh, 1988. Allelism of two genes for stem rust resistance in triticale. Euphytica 38: 185-189.

Singh R.P. & E.E. Saari, 1991. Biotic stresses in triticale. Proceedings of the 2nd International Triticale Symposium. Mexico D.F., CIMMYT: 171-181.

Sisterna M.N., 1989. Graminicolous species of *Drechslera, Bipolaris* and *Exserohilum* in Argentina. In: WBTA 6 (1): 105.

Sisterna M.N., 1984. *Gibberella zeae* (Schw.) Petch. (*Fusarium graminearum* Schw.) on triticale in Argentina. In: WBTA 1 (2): 219.

Skajennikoff M. & F. Rapilly, 1983. Aggressivness of *Septoria nodorum* on wheat and triticale. Effects of the host and infected organs. Agronomie 3: 131-140.

Sowa W., 1986. Studies of lines of the 42-chromosome triticale (X *Triticosecale* Wittmack). Part I & II. Hod. Roślin, Aklim. Nasien., Vol. 30, No. 1/2.

Vargas R.P., E.C. Picinini, E.M. Reis & A. Baier, 1991. Seed health of triticale and wheat produced in different regions in Brazil. Proceedings of the 2nd International Triticale Symposium. Mexico D.F., CIMMYT: 216-220.

Wakuliński W. & J. Che kowski, 1993. *Fusarium* species causing scab of wheat, rye and triticale in Poland. Hod. Rośl. Aklim. Nasien. 37 (3): 137-142.

Wallwork H., 1989. Screening for resistance to take-all in wheat, triticale and wheat-triticale hybrid lines. Euphytica 40: 103-109.

Warham E.J., 1988. Screening for Karnal bunt (*Tilletia indica*) resistance in wheat, triticale and barley. Can. J. Plant Path. 10: 57-60.

Wiese V.M., 1987. Compendium of wheat diseases. APS Press, Second edition, 112pp.

Wildermuth G.B. & B.R. McNamara 1987. Susceptibility of winter and summer crops to root and crown rot infection by *Bipolaris sorokiniana*. Plant Pathology 36: 481-491.

Wilson J. & G. Shaner 1989. Individual and cumulative effects of long latent period and low infection type reactions to *Puccinia recondita* in triticale. Phytopathology 79: 101-108.

Wilson J. & G. Shaner 1989. Inheritance of leaf rust resistance of four triticale cultivars. Phytopathology 79: 731- 736.

Winter W., H. Krebs & I. Banziger, 1992. Susceptibility of cereal cultivars to some smut diseases. Landwirtschaft Schweiz 5: 293-297.

Woś H., 1992. Winter triticale resistance to *Leptosphaeria nodorum* Mül. Biuletyn IHAR 181-182: 83-87.

Woś H., 1993. Occurrence of *Phaeosphaeria nodorum* in Gorzów Wlkp. region. Resistance of winter triticale cultivars to septoriosis. Biuletyn IHAR 187: 113-119.

Zamorski C., M. Schollenberger & M. Sieklucka, 1993a. Occurrence of diseases on triticale, wheat and rye in 1992. Hodowla Roślin i Nasiennictwo 5: 14-16. (In Polish).

Zamorski C., M. Schollenberger & M. Sieklucka, 1993b. Susceptibility differentiation of triticale cultivars and germplasm lines to brown rust and scald. Materiały Sympozjum: " Biotyczne środowisko uprawne a zagrożenie chorobowe roślin." Olsztyn, 7-9.09.1993 r., pp. 435-441. (In Polish, English summary).

Zamorski C., M. Schollenberger & M. Sieklucka, 1994. The role of winter triticale crop in overwintering of *Puccinia recondita* f.sp. *tritici* and *P. striiformis*. Zeszyty Naukowe A. Rol. w Szczecinie. (In print).

Zwer P.K., F.R. Park & A. R. Mc Intosh, 1992. Winter stem rust in Australia 1969-1985. Austral. J. Agic. Res. 43: 399-431.

SCREENING FOR FUSARIUM SCAB RESISTANCE IN TRITICALE

Mariana Ittu, Gheorghe Ittu & Nicolae N. Săulescu
Research Institute for Cereals and Industrial Crops
8264, Fundulea, Călărași - Romania

Abstract

The incidence of Fusarium scab in the areas intended for triticale in Romania is relatively high. Breeding for resistance to this pathogen is the main way to reduce yield losses. All breeding materials beginning with the F4 - F5 generations are tested by artificial inoculations in the field. The continuous selection pressure based on the visual score of Fusarium attack in heads and the reduction of head weight (as % of control) allows the identification of genotypes with high and stable levels of resistance according to both criteria (maximum 5 - 20% intensity of attack in heads and over 50% from the weight of control). Such genotypes are 37 TK1-105 111 R, 219 TR2 - 2, 6135 TW5 1101 and 1102, 5780 TW0 - 21, 318 TR8-1201, etc. Most of these genotypes have in their pedigrees the Romanian line 37 TK1 and the Polish cultivar Malno.

Introduction

Triticale is one of the crops recommended to be cultivated in Romania mainly in the hilly regions where poor and acid soils are frequent.
The new cultivars Plai and Colina, recently released at Fundulea, due to their high yield potential, improved level of general resistance to pathogens and a good quality, offer a better alternative to exploit ecological conditions rather restrictive for other small grains crops.
Fusarium scab could be very damaging to triticale when climatic conditions during anthesis are favorable.
Genetical resistance is accepted as the most efficient way to prevent both the yield and quality losses produced by this pathogen [1-3].

H. Guedes-Pinto et al. (eds.), Triticale: Today and Tomorrow, 527–533.
© 1996 *Kluwer Academic Publishers. Printed in the Netherlands.*

Developing new varieties with improved level of resistance to Fusarium, became an objective of major concern in our triticale breeding program. As a consequence of multiannual screening trials under artificial inoculations with Fusarium a continuous progress in this direction has been registered [4].

The objectives of the present study were to evaluate triticale germplasm with the aim to identify new sources of resistance to Fusarium scab for use in breeding programs.

Material and methods

20 triticale genotypes formerly selected as very resistant have been assessed in a multiannual screening test carried on between 1991 - 1993, in order to evaluate the reaction of resistance in the variable environmental conditions of different years.

204 new triticale genotypes were tested one year (1993) with respect to their resistance against Fusarium scab.

All the trials were performed in a disease nursery under artificial inoculations by injecting a suspension of Fusarium mycelia and spores directly in four florets from the middle of 10 spikes/genotype at anthesis. Inoculum was obtained in liquid Czapek - Dox medium with continuous aeration for 7 days.

Reaction to scab was assessed by visual score of infection intensity (extension of infection in the spike) on a 1-9 scale (where 1=very resistant; 9=very susceptible) and by the relative weight of inoculated heads as compared with noninoculated ones.

The two criteria proved in previous researches a good reliability for assessment of resistance to Fusarium scab in wheat and triticale [5].

Results and discussions

Evaluation for three successive years of twenty selected triticale genotypes generally confirmed the high level of resistance previously identified.

The data presented in figure 1 demonstrate a very significant, negative correlation (r=-0,92) between the two criteria used (extension of Fusarium infection and the relative weight of inoculated heads).

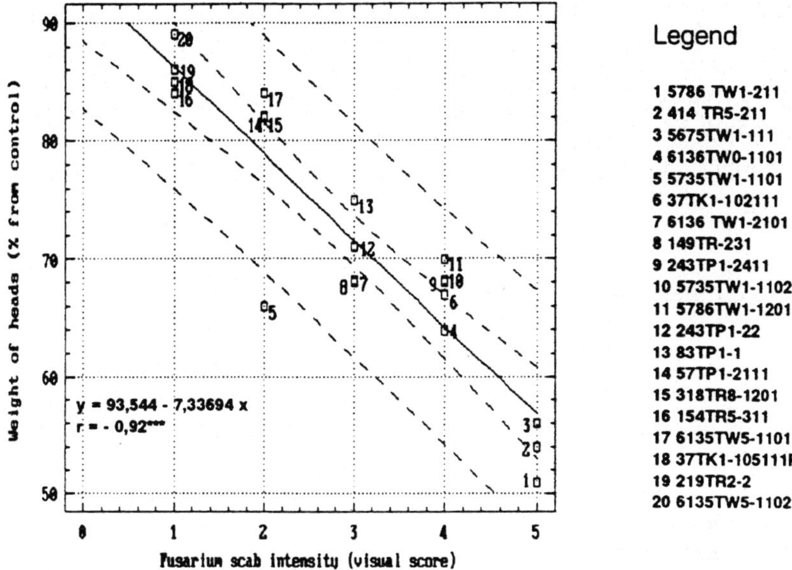

Fig.1 Genotypes of triticale with high level of resistance to Fusarium scab
(average 1991 - 1993)

These results indicate the triticale genotypes 37 TK1 - 105 111R, 219 TR2 - 2, 6135 TW5 1101 and 1102 as the most resistant to Fusarium scab according to the small reductions in the weight of heads following inoculation correlated with a low degree of Fusarium extension in spike (intensity of attack) in average on all the three experimental years.

The weight reduction of inoculated versus noninoculated spikes was higher in the other genotypes included in this experiment. However even the more affected genotypes such as 5786 TW1 - 211, 414 TR5 - 211, 5675 TW1 - 111, etc. had less than 50% spike weight reduction under inoculation, which correspond to an acceptable level of resistance.

Considering the strong influence of the environment among the factors affecting the Fusarium pathosystems, the good agreement of data obtained in the three years of testing is remarkable and the classification of the genotypes can be regarded with a reasonable degree of certainty.

In the one year trial conducted in 1993 with 204 new triticale genotypes scab reaction varied from highly resistant to very susceptible, irrespective of which from the two assessment criteria have been utilized.

According to the visual score of Fusarium extension (intensity of attack) presented in figure 2, the highest number of genotypes were recorded in the classes corresponding to susceptibility (6-9), most of them being moderately susceptible and only 11% very susceptible. Only 6% of the triticale genotypes were classified as very resistant and resistant.

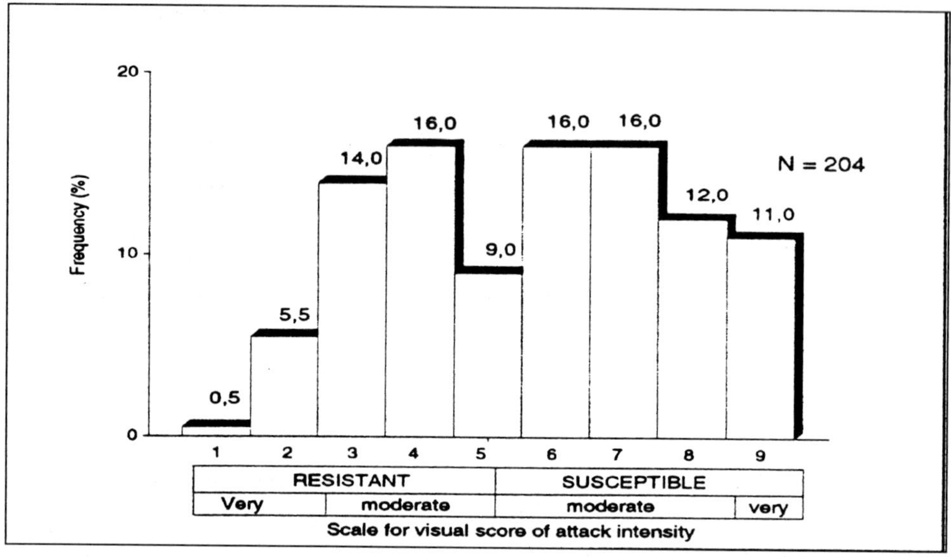

Fig. 2. Distribution of triticale genotypes according to the visual score of attack intensity (Fusarium scab)

The weight of inoculated heads varied from 19 to 91% from the weight of noninoculated ones (figure 3). At half of these triticale genotypes, the weight of inoculated heads varied between 61 and 91% from the control. Among them, triticale genotypes 5780 TWO - 21, 7999 TW1 - 1, 8025 TW1 - 1, 8041 TW1 - 1, 5780 TW0 - 211, 8000 TW1 - 1, 93 TU - 2, 191 TR1 - 01 and 484 TS2 - 111 were classified as very resistant according both to low intensity of Fusarium scab attack and reduction of head weight (figure 4).

Lower reduction of the relative weight of inoculated heads compared to the noninoculated ones recorded in these genotypes suggests that a new step in the breeding of resistance to Fusarium scab in triticale has been accomplished.

The results of the present screening test carried out under artificial inoculation are in good agreement with the data observed in the conditions of the natural strong epidemic of Fusarium scab recorded in our area in 1991 (unpublished data).

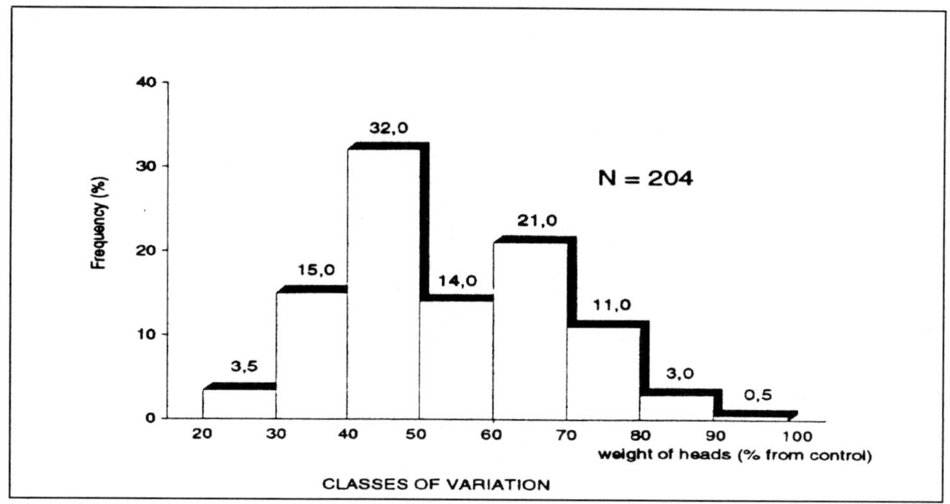

Fig. 3. Distribution of triticale genotypes according to the weight of heads (% from control)

A mention has to be made on the origin of the most resistant triticale genotypes which were identified. Most of the new triticale genotypes classified now as resistant are derivatives of the line 37 TK - 1 selected from the cross 1684 TG 3- 2/AD206 // TF2. As both AD206 and TF2 are susceptible to Fusarium, the presumable source of resistance is the line 1684 T6 3 - 2 (Chinese wheat 974- 1253/Rye Snoopy - Danae/Tcl Py 152 - 74 // Tcl Mayo Armadillo).

At present we cannot speculate if resistance comes from the Chinese wheat or from the rye parent. The highest level of resistance was found, in progenies of combination 5780 TW, which combines the Romanian triticale line 37 TK - 1 and the Polish cultivar Malno, previously identified as resistant [4]. Therefore it seems that the two sources of resistance might be complementary.

Acknowledgements

- Presentation of this paper in the 3rd International triticale Symposium was financially supported by CIMMYT.
We thank Dr. G. Varughese - CIMMYT (Mexico) and Dr. H. Braun - CIMMYT (Turkey) for their interest in our work.

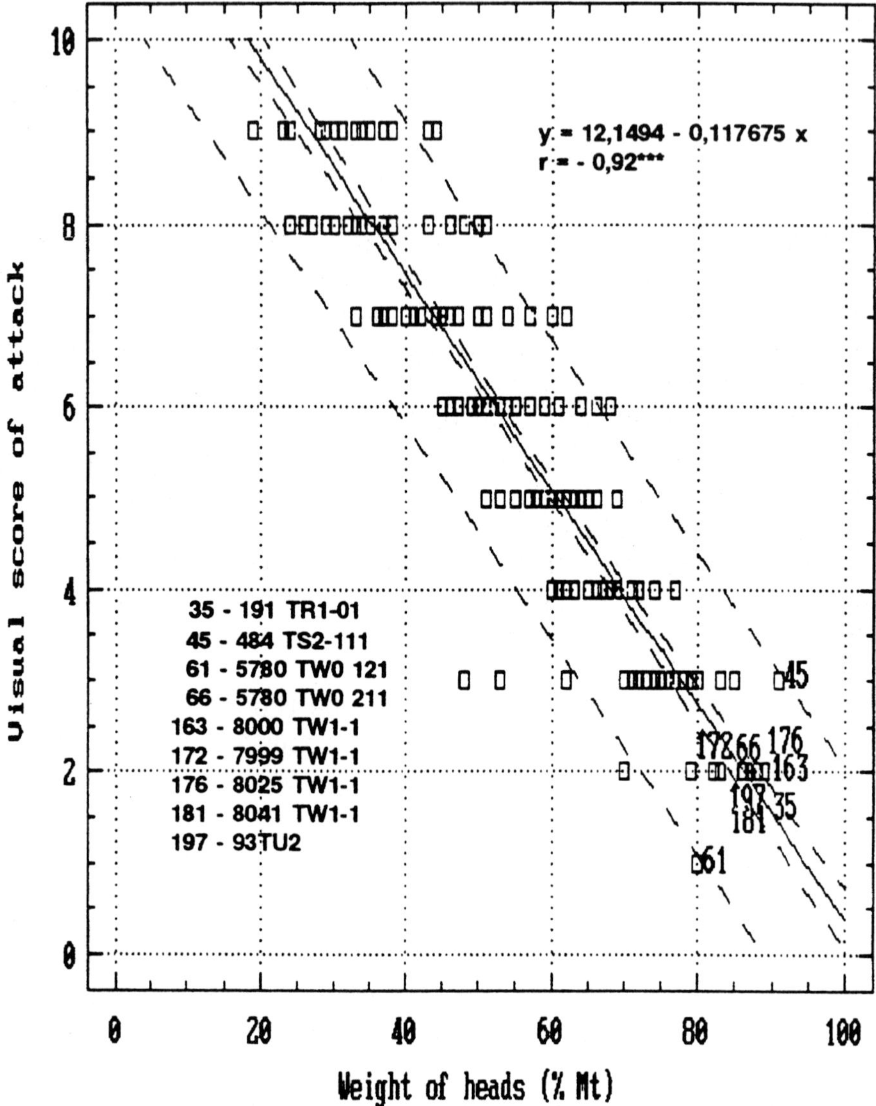

Fig. 4. Correlation between weight of spikes (% of control) and visual score for head attack of Fusarium scab, following artificial inoculation in triticale (Fundulea 1993)

References

1.	Arseniuk E., Goral T., Czembor H.J. Pathogenicity and differential response in Fusarium spp x Triticosecale Witt./Triticum aestivum L./Secale cereale pathosystems. Plant Breeding Acclimatization and Seed Production 1993; 37:17-25.
2.	Grzelak K. Occurence of Fusarium spp. on triticale seed in relation to seed decay and seedling root rot. Plant Breeding Acclimatization and Seed Production 1993; 37:75-81.
3.	Dormann M., Oettler G. Genetic variation of resistance to Fusarium graminearum (head blight) in primary hexaploid triticale. Plant Breeding Acclimatization and Seed Production 1993; 121-129.
4.	Ittu G., Ittu Mariana, Săulescu N.N. Preliminary results concerning resistance to Fusarium scab in triticale (Triticosecale Witt.).Rom. Probl. genet. teor. aplic. 1990; 22:91-98.
5.	Ittu Mariana, Săulescu N.N., Ittu G. Improved method of assessment of resistance to Fusarium scab in wheat.Rom. Probl. genet. teor. aplic. ;22:1-8.

INHERITANCE OF HEAD RESISTANCE OF WINTER TRITICALE TO *PHAEOSPHAERIA NODORUM*

Woś Henryk
Plant Breeding and Acclimatization Institute
Station Małyszyn, Myśliborska 81, 66-400 Gorzów Wlkp., Poland

Abstract

The leaf and head septoriosis in winter triticale is caused by the fugus *Stagonospora nodorum* which belongs to the most dangerous diseases of this species. This report tries to define the genes responsible for winter triticale head resistance to *S. nodorum* and the segregation ratios. Two winter triticale cultivars were used to crosses: cultivar Bogo, was characterized by susceptible heads and cultivar Almo, has significantly high resistance of heads. From the data of the F-2 generation hybrids the Mendelian segregation ratio was 3 susceptible head to 1 „resistant".

A recessive gene was responsible for the head resistance to the pathogen. There appears to be some cytoplasmic effects on head resistance but differences were not statistically significant. In triticale species, a linkage block of rye occurs with the governing leaf and head resistance to *Stagonospora nodorum*.

Introduction

The glume blotch disease caused mainly by the fungus *Stagonospora nodorum* (Berk.) Castelani et Germano [=*Septoria nodorum* Berk.] [teleomorph *Phaeosphaeria nodorum* (Muller) Hedjaroude] (=*Leptosphaeria nodorum* Muller) is one of the more dangerous diseases of leaves and heads of triticale in Poland [1,5,6, 8,9].

Many reports discuss the inheritance of resistance to *Stagonospora nodorum* in wheat cultivars, but there are few investigation concerning triticale. Finding the genes associated with resistance to *S. nodorum* in triticale and the ratio of segregation makes the breeding for resistant cultivars more efficient. This problem is the subject of this report.

Materials and methods

During the summer of 1991, two crosses were made under field conditions: winter triticale Almo (as the mother) x winter triticale Bogo, and the reciprocal cross. The heads of cultivar Bogo are susceptible to infestation by the fungus *S. nodorum* but cultivar Almo has significantly high resistance to glume blotch. The cultivar Bogo has hairs on its glumes, but cultivar Almo none. One head of the well-tillering plant of Almo

H. Guedes-Pinto et al. (eds.), Triticale: Today and Tomorrow, 535–539.

© 1996 *Kluwer Academic Publishers. Printed in the Netherlands.*

was pollinated with Bogo, the second head of the same plant was isolated against cross-pollination. A similarly reciprocal cross was done on a plant of Bogo and one head was isolated. From the cross between Almo and Bogo obtained 29 seed, and from the reciprocal cross 24 seeds were obtained. In September of the same year, seeds derived from the hybrids and isolated heads of the parents were sown by hand in spacing of 20 x 10 cm. In 1992, before flowering, the heads of each plant were isolated against cross pollination and harvested. The separate hybrids and parents were sown in rows with spacing of 20 x 15 cm in September 1992. In 1992 the F-2 generation plants, along with the two parental cultivars were twice inoculated with S. nodorum spores after heading and during a little rain. Spore concentrations were adjusted to 5 x 10^6 spores per milliliter of suspension. Then the estimation of infected parent and hybrid heads was carried out with a 1 - 9 scale (1-susceptible and 9-resistant). The fully resistant plants were not observed, and highest estimate was 6. Investigated material was divided into two groups:1) plants with scores 1-3 were found as group of susceptible plants and 2) plants with scores of 4-6, grouped as „resistant" plants. The Mendelian segregation was 3 : 1 (3 susceptible plants and 1 resistant plant) and was verified by χ^2 test. In order to find the significant difference of mean resistance in the simple cross Almo x Bogo and reciprocal cross the C test (Cochrana and Coxa) [4] was used.

Results and discussion

So far no resistant lines to glume blotch have been found in triticale, in spite of screening hundreds of triticale lines during a few years at Małyszyn Station. As it has been shown in figure 1, and in the literature [4, 9] the cultivars Bogo and Almo differ significantly with regard to head resistance to S. nodorum.

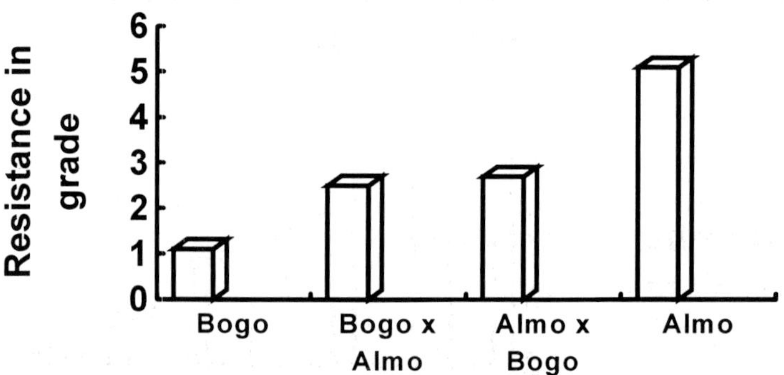

Fig1. Mean values of head resistance of parent Bogo, Almo and F-2 generation of crosses Bogo x Almo and Almo x Bogo

In Fig.2-1 some frequency distribution diagrams of crosses Bogo x Almo and Almo x Bogo (Fig.2-2.), are presented depending on the level of resistance. The frequency diagram made on the infection results for the cross Almo x Bogo is close to the normal distribution. The diagram for the cross Bogo x Almo clearly shows an adventage of susceptible plants (162) in relation to the reciprocal cross (87 plants).

It was observed that susceptible plants were characterized by the presence of glume hairness, but the resistant plants had no hairness on its glumes. The hairness might cause moisture to remain longer on the heads, and consequently create better conditions for development of the fungus. Besides, it has been suggested that the fungus can penetrate into plant through the gentle hairs easier.

Fig. 2-1. Frequency distribution of F-2 generations in the winter triticale crosses Bogo x Almo

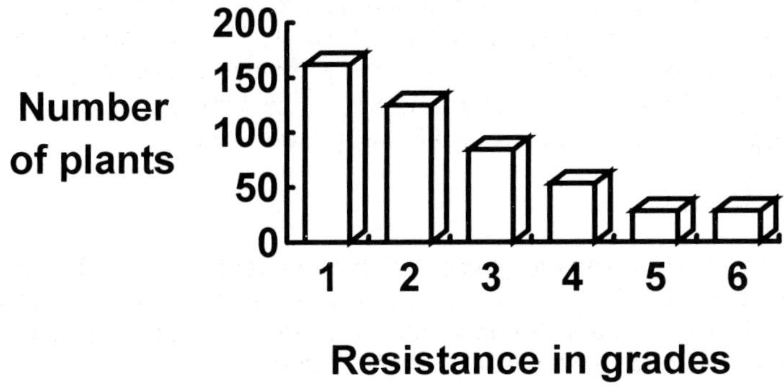

Fig. 2-2. Frequency distribution of F-2 generations in the winter triticale crosses Almo x Bogo

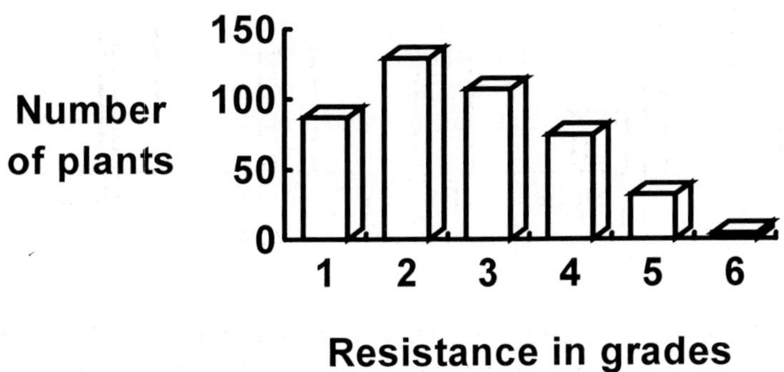

The means for resistance of the heads for the cross Bogo x Almo was 2.5, and for

Almo x Bogo 2.7 [Fig. 1]. The cytoplasmic effect on infection of spike was analyzed by C Test comparing of the means of F-2 hybrid heads for the two crosses. The difference was not significant at α=0.05 level. The use of a larger number of cultivars in diallel crosses may give a better estimation of the maternal effects on head resistance to *S. nodorum*. In both crosses of F-2 generation, the Mendelian segregation ratio was 3:1 (3 susceptible plants and 1 resistant) [Table 1].

The means for resistance of the heads for the cross Bogo x Almo was 2.5, and for Almo x Bogo 2.7 [Fig. 1]. The cytoplasmic effect on infection of spike was analyzed by C Test comparing of the means of F-2 hybrid heads for the two crosses. The difference was not significant at α=0.05 level. The use of a larger number of cultivars in diallel crosses may give a better estimation of the maternal effects on head resistance to *S. nodorum*. In both crosses of F-2 generation, the Mendelian segregation ratio was 3:1 (3 susceptible plants and 1 resistant) [Table 1].

Table 1
Winter triticale hybrids segregation for reaction to *Stagonospora nodorum* in F-2 generations under field conditions with artificial inoculation.

Cross	Number of seeds	Number of plants F-2		Segregation ratio	P value
		Suscceptible	Resistant		
Bogo x Almo	24	371	109	3 : 1	0.246
Almo x Bogo	29	323	111	3 : 1	0.783

A similar segregation was obtained by Wong et al. [7] by estimating the resistance of seedling leaves in wheat. On the basis of F-2 generation data, it appears that the dominant gene is responsible for susceptibility of heads and, respectively, the recessive gene is for resistance. The above results are a confirmation of Fried's and Meister's [2] work conducted on adult plants in wheat. They considered the dominant gene as responsible for susceptibility of the leaves. Moreover, the cited authors have indicated that different genes are responsible for leaf and head resistance. The above example also appears to apply to triticale, with the cultivar Bogo heaving good leaf resistance and high head susceptibility. There is the hope that hybrids derived from crossing good yielding, susceptible to the head infection cultivars, like Bogo, with lines with high head resistance will result in segregates resistant with favourable agronomic traits. Triticale contains the rye genom, which shows high resistance to leaf and head septoriosis, but the expression of the rye gene(s) resistance to *S. nodorum* is not revealed. In my opinion, the introduction of resistance genes from wheat allows for grater expression of resistance than the introduction of resistance genes from rye. This is suggested from the experience from the expression of resistance of strains derived from hybrids (triticale x resistant wheat) x triticale. Triticale has inherited a number of unfavourable traites from wheat and rye.

For lack of effective genetic sources within the winter triticale the improvement of unfovourable traits, might come about through the introduction of strong sources from species where this trait is generally worse.

Conclusions

1. Hybrids of F-2 generation have segregated in the ratio of 3 plants with susceptible head to 1 resistant plant to *Stagonospora nodorum.*
2. The recessive gene was responsible for the spike resistance to the pathogen.

3. No cytoplasmic effect on the spike resistance was statistical proven.
4. Crossing good yelding forms characterized by susceptible to *S. nodorum* head blight with the resistance lines can all result in highyelding resistant plants.
5. The heads of F-2 generation plants with hairs on its glumes were more infected by fungus than the plants withcut hairs.
6. In triticale, suppression of rye gene(s) governrning leaf and head resistance to *Stagonospora nodorum* occurrs.

References

1. Arseniuk E, Czembor H J, Sowa W, Krysiak H. Wstępne badania nad septoriozą pszenżyta. Biuletyn IHAR 1990; 173-174:65-69.
2. Fried PM, Meister E. Inheritance of leaf and head resistance of winter wheat to *eptoria nodorum* in diallel cross.Phytopathology 1987:vol. 77; N10:1371-1375.
3. Maćkowiak W, Budzianowski G, Łukańko U. Charakterystyka odmian pszenżyta ozimego i jarego hodowli ZD HAR Małyszyn oraz ich reakcja na niektóre czynniki środowiska. Zeszyty Naukowe Akademii Rolniczej w Szczecinie 1994; 162.
4. Oktaba W. Elementy statystyki matematycznej i metodyka doświadczalnictwa. Warszawa; PWN, 1966.
5. Pokacka Z, Korbas M, Staniszewska H. Choroby występujące na pszenżycie i ich zwalczanie. Ochrona Roślin 1987; 2 :7-9.
6. Pokacka Z. Choroby liści pszenżyta. Ochrona Roślin 1991; 5-6:11-13.
7. Wong LSL, Hughes GR. Genetic control of seedling resistance to *Leptosphaeria nodorum* in wheat. Proceedings Third International Workshop on *Septoria* Diseases of Cereals; 1989 July 4-7; Zurich, Switzerland; 136-138.
8. Woś H. Odporność pszenżyta ozimego na septoriozę liści i kłosów. Praca doktorska. Gorzów 1992.
9. Woś H, Maćkowiak W, Paizert K. Występowanie *Phaeosphaeria nodorum* w rejonie Gorzowa Wlkp. Odporność pszenżyta ozimego na septoriozę. Biuletyn IHAR 1993; 187:113-119.

EVALUATION OF NEW WINTER OCTOPLOID TRITICALES FOR THEIR REACTION TO SINGLE AND COMBINED INFECTIONS WITH BYDV AND *TYPHULA ISHIKARIENSIS*.

Benoit Bizimungu[1], J. Collin[1], C.-A. St. Pierre[1] and A. Comeau[2]
[1]Département de Phytologie, Université Laval, Québec, Canada
[2]Station de Recherches, Agriculture et Agro-alimentaire Canada, Sainte-Foy, Québec, Canada

Abstract

Six octoploid triticales were compared to hexaploid triticales, wheats and ryes for their reaction to single and combined infection with barley yellow dwarf virus (BYDV) and *Typhula ishikariensis* Imai. Agronomic characters like winter survival and biomass yield indicated that ryes and wheats are respectively tolerant and susceptible to both diseases. Hexaploid triticales showed a good tolerance to BYDV, but not to *T. ishikariensis*. Among octoploid lines, reactions to the two pathogens varied from tolerant to susceptible. These results suggest that synthesis of octoploid triticales may provide variability for reaction to both diseases.

Introduction

Barley yellow dwarf virus infection occurring in the fall was shown to reduce cold tolerance and winter survival of winter cereals [1]. Prolonged and persistent snow cover in late spring favors development of snow mold fungi, which may represent an additional major winterkill factor [2]. Four species of snow mold fungi have been detected in Québec (Canada), including *Typhula ishikariensis* Imai which causes the speckled snow mold [3]. Mold infection results in inconsistent winter survival and makes winter cereal production an uncertain venture.

Developing resistant varieties is a promising approach to control these diseases. However, the task is complicated by the lack of sources of resistance in wheat [4, 5, 6]. Among related species, rye (*Secale cereale* L.) is a potential source of resistance to both diseases [5,6]. Some hexaploid triticales (X *Triticosecale* Wittmack) show better tolerance to BYDV than wheats [7], but are as susceptible to snow molds [5,6]. This study was conducted to evaluate the reaction of a few primary octoploid triticales as potential sources of tolerance to single and combined infections with BYDV and T. *ishikariensis*, in comparison with wheats, ryes and hexaploid triticales.

H. Guedes-Pinto et al. (eds.), Triticale: Today and Tomorrow, 541–547.
© 1996 *Kluwer Academic Publishers. Printed in the Netherlands.*

Materials and Methods

PLANT MATERIAL

Six primary octoploid triticale populations (Caldwell/Kustro (Ca/Ku), Caldwell/Puma (Ca/Pu), Frankenmuth/Cougar (Fra/Co), Frankenmuth/Kustro (Fra/Ku), Frankenmuth/ Puma (Fra/Pu) and Fukuho/ Cougar (Fu/Co)) were compared to five wheats (Augusta, Borden, Ruby, Norstar and Frankenmuth), three ryes (Cougar, Gauthier and Kustro), and two hexaploid triticales (K9-6 and OAC Wintri). Triticale cv K9-6 and OAC Wintri are BYDV tolerant checks and rye cv Gauthier is known to have outstanding winter hardiness.

DESIGN OF EXPERIMENT

One trial was conducted on campus at Sainte-Foy, in a wide plastic tunnel (an unheated plastic house) to protect the plants from winter damage. In mid-december, the plants were covered with glass wool blankets to simulate snow cover, until mid-april, when the outdoor temperature warms up. The tunnel proved to be suitable for evaluating winter cereals independently of other abiotic factors, such as flooding and ice stress, which are usually confounded with disease effects in the field [6]. Two other trials were conducted in the field at the St-Augustin experimental farm in 1991-92 and 1992-93.

Field experiments were arranged in a split-split-plot design, with BYDV inoculation or uninoculated control in main plots, genotypes in subplots, and *T. ishikariensis* inoculation or protection by PCNB fungicide spray in sub-subplots. In the tunnel, blocks were nested within BYDV inoculation treatments to ensure a better control of viruliferous aphid movements. The trials had four blocks (replications) in the field and two in the tunnel. The experimental unit consisted of four 65 cm rows in the field and two in the tunnel. BYDV inoculation was carried out at the 2-leaf stage, according to the procedure decribed by Comeau [8], using the severe "Cloutier" PAV isolate. Control plants were sprayed with pirimicarb to prevent accidental infection. The same operation was carried out four days later in the tunnel to kill viruliferous aphids, but was not necessary in the field since low fall temperatures restrain aphid activity. The *T. ishikariensis* inoculum consisted of autoclaved grains of rye infested with sclerotia. In the field, the inoculum was applied in the fall just before the first persistent snow, at a rate of at least 5000 sclerotia per plot. Inoculation was carried out 1 month later in the tunnel.

Winter survival was assessed in the spring, after snow melt, and plant height was measured at maturity. Grain and biomass yields were recorded at harvest. Statistical analyses were carried out with the GLM procedure of SAS statistical software (SAS Institute Inc.).

Results and Discussion

The effectiveness of artificial inoculations varied across experiments. The most interesting results were obtained in the tunnel in 1991-92, where both BYDV and *Typhula* inoculations gave good results. In the field, BYDV inoculation also gave good results, but *Typhula* inoculation effects were negligible. Thus, the field experiment was repeated in 1992-93, where both diseases were observed at a good level of infection.

The differential reaction of genotypes to speckled snow mold infection was most obvious in the tunnel experiment (Figures 1 & 2) where a significant "genotype x *Typhula*" interaction was observed (P<0.0001). BYDV inoculation alone had no apparent effect on winter survival, but decreased biomass yield of the wheat genotypes. Octoploid triticales showed a reaction to BYDV similar to that of the hexaploid triticales, which are considered tolerant to BYDV. *Typhula* inoculation affected winter survival of most of the genotypes, which resulted in a reduction of biomass yield. Rye genotypes were the most tolerant to *Typhula*, with cv. Gauthier being the most tolerant across environments. The Caldwell/Puma

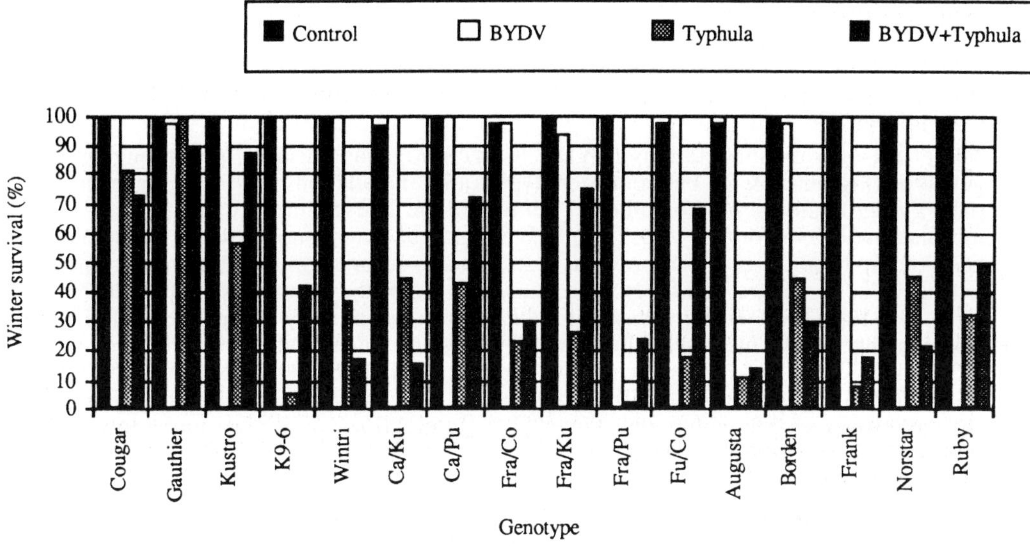

Figure 1. Winter survival following single and combined inoculations with BYDV and *T. ishikariensis* in the tunnel in 1991-92

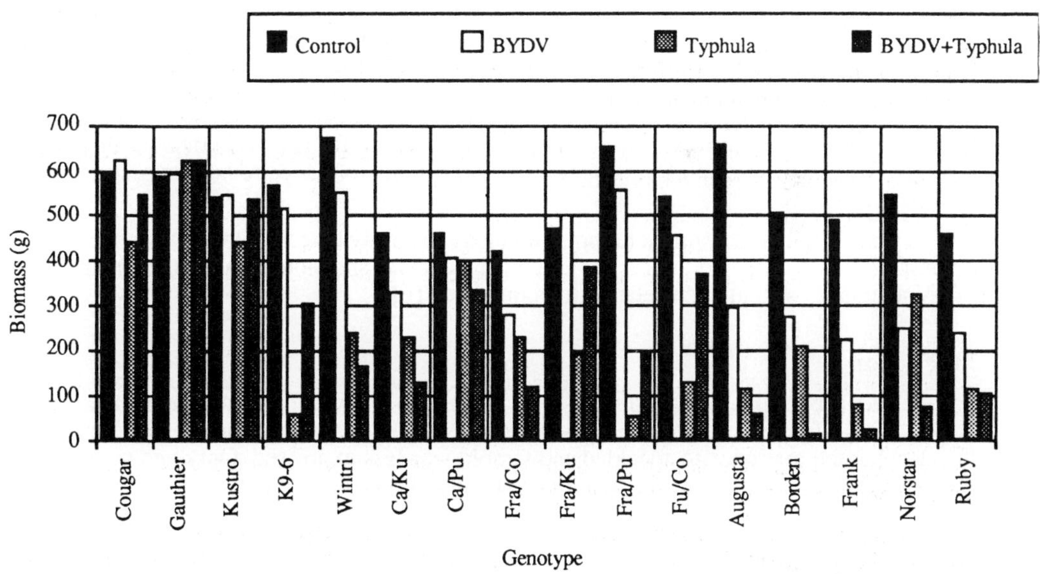

Figure 2. Biomass yield following single and combined inoculations with BYDV and *T. ishikariensis* in the tunnel in 1991-92

octoploid showed the highest tolerance to Typhula and Frankenmuth/Puma was the most susceptible. Both hexaploid cultivars were also sensitive, with K9-6 being more sensitive than Wintri. Wheat cvs Frankenmuth, Augusta and hexaploid triticale K9-6, as well as octoploid triticale Frankenmuth/Puma had the lowest winter survival under speckled snow mold inoculation.

The effect of joint inoculations (BYDV + *Typhula*) was variable, depending on the genotype (Figures 1 & 2). For some genotypes, joint inoculation resulted in higher losses than *Typhula* alone, but for others *Typhula* alone caused more damage than joint inoculation (P=0.06 for the "virus x *Typhula* x genotype" interaction). Such a phenomenon had already been observed in previous tests. This is surprising since the most common interaction between viral and fungal infections seems to be the increased susceptibility to the fungus [9, 10]. However, some cases of reduction of susceptibility are reported [11, 12]. It is known that BYDV increases soluble carbohydrate and reduces nitrogen in infected leaves [12], while snow mold resistance in winter wheat has been associated with the rapid accumulation of high levels of carbohydrate in the crowns [15, 4] and slower depletion of the carbohydrate throughout the winter [14]. Interactions of BYDV and snow mold infection effects are therefore both possible.

The 1991-92 field experiment (data not shown) confirmed that the ryes and the two hexaploid triticales are tolerant to BYDV. The wheat genotypes showed different levels of sensitivity. None of the octoploid triticales was as BYDV tolerant as the hexaploid in the field, and Frankenmuth/Cougar looked as sensitive as the wheat genotypes. No significant *Typhula* effect was observed and the experiment was repeated in 1992-93.

In 1992-93, both BYDV and *Typhula* inoculations gave satisfactory results in the field, although the *Typhula* inoculation was not as severe as in the tunnel. Again, ryes were tolerant to both diseases. *Typhula* or BYDV alone generally had small effects on winter survival, but joint inoculations often resulted in serious winter damage for many cultivars (Figure 3). This damage can probably be explained by the fact that BYDV infected plants became more vulnerable to *Typhula* infection. BYDV effect was visible on biomass yield (Figure 4). Biomass yield of hexaploid cv. K9-6 was reduced by double inoculation. Frankenmuth/Cougar showed the greatest sensitivity to BYDV infection among the octoploid triticales. Caldwell/Puma showed an acceptable tolerance to both stresses, as in the 1991-92 trial.

Data on plant height and grain yield (not presented) generally agreed with the above results. Biomass yield is considered the best criterium, since it was shown to be a good estimator of BYDV tolerance, with high correlation to grain yield [16].

Conclusion

As expected, rye cultivars proved to be very tolerant to both BYDV and *T. ishikariensis*. Useful levels of tolerance to BYDV were expressed in some unrecombined octoploid triticales, but the tolerance to speckled snow mold was less expressed. Only one octoploid (Caldwell/Puma) was tolerant to both diseases and showed a stable winter survival and biomass yield under single and combined infections. These results suggest that synthesis of primary octoploid triticale may provide new genetic variability for reaction to both diseases. Other octoploid triticales are now being synthesised with rye cv. Gauthier. Head selection has also been initiated to isolate the more desirable types among the octoploid populations.

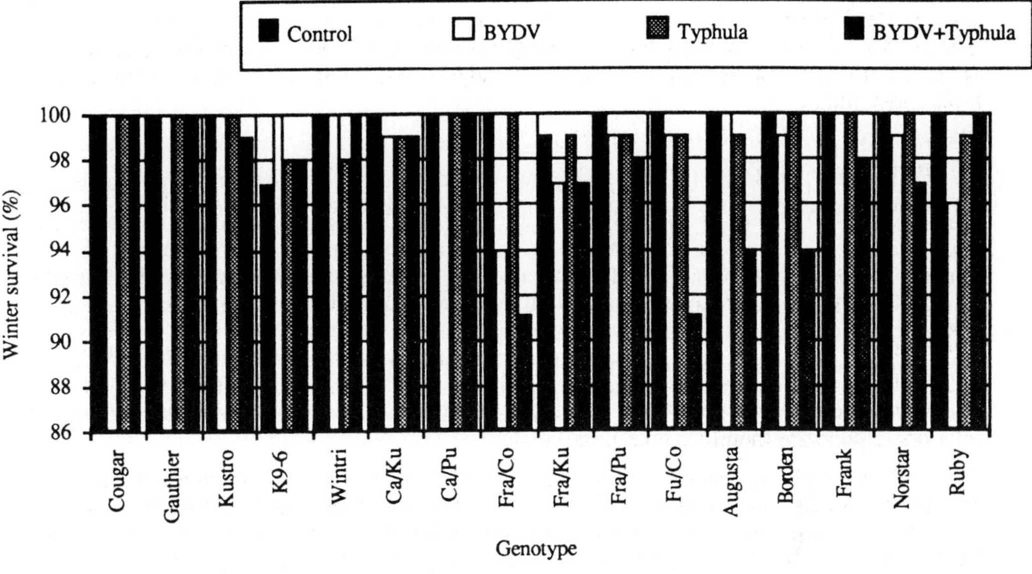

Figure 3. Winter survival following single and combined inoculations with BYDV and *T. ishikariensis* in the field in 1992-93

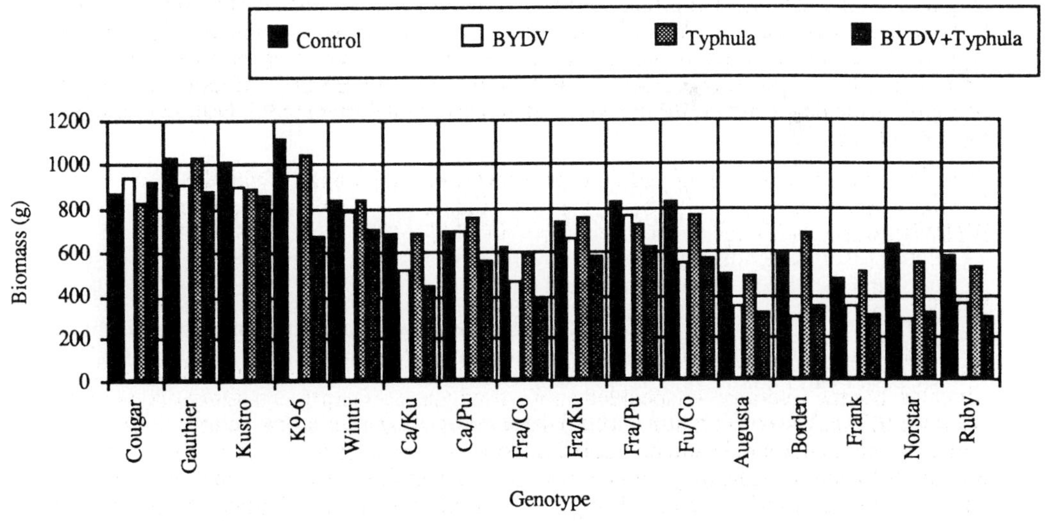

Figure 4. Biomass yield following single and combined inoculations with BYDV and *T. ishikariensis* in the field in 1992-93

Acknowledgements

Thanks are due to the UNR-CIDA-Laval Project and CORPAQ for the financial support provided and to D. Marois for his excellent technical assistance.

References

1. Paliwal YC, Andrews CJ. Barley Yellow Dwarf Virus-Host plant interactions affecting winter stress tolerance in cereal. In: Burnett PA, editor. World perspectives on Barley Yellow Dwarf. CIMMYT, Mexico. 1990: 313-320.

2. Andrews CJ, Pomeroy MK, Seaman WL. The response of fall-sown cereals to winter stresses in Eastern Ontario. Can J Plant Sci 1986; 66:25-37.

3. Pouleur S, Couture L. Présence et distribution des moisissures nivéales des céréales d'automne au Québec. Can J Plant Pathol 1988; 10:183-187.

4. Bruehl GW. Developing wheats resistant to snow mold in Washington State. Plant Dis 1982; 66:1090-1095.

5. Litschko LD, Burpee LL, Goulty LG, Hunt LA, McKersie BD. An evaluation of winter wheat for resistance to the snow mold fungi Microdochium nivale (Fr.) Samu & Hall and Typhula ishikariensis Imai. Can Plant Dis Surv 1988; 68:161-168.

6. Rioux S, Couture L, St-Pierre CA. Evaluation of resistance to speckled snow mold in winter wheat. Can J Plant Pathol 1993; 15:284-292.

7. Collin J, Comeau A, St-Pierre CA. Tolerance to barley yellow dwarf virus in triticale. Crop Sci 1990; 30:1008-1014.

8. Comeau A. Aphid rearing and screening methods for resistance to barley yellow dwarf virus in cereals. In: Barley Yellow Dwarf- A Proceedings of the Workshop, CIMMYT, Mexico, D.F, Mexico. 1984:60-71.

9. Cavelier M, Maroquin C. Sensibilité variétale de l'escourgeon à Typhula incarnata Lasch et au virus de la jaunisse nanisante de l'orge. Bull Rech Agron Gembloux, 1980; 15:309-326.

10. Sward RJ, Kollmorgen, JF. The effects of the interaction of barley yellow dwarf virus and take-all fungus on the growth and yield of wheat. In: Burnett PA, editor. World perspectives on Barley Yellow Dwarf, CIMMYT, Mexico, 1990; 305-312.

11. Potter LR. Interaction between barley yellow dwarf virus and rust in wheat, barley and oats, and the effects on grain yield and quality. Ann Appl Biol 1982; 100:321-329.

12. Varughese J, Griffiths E. Effects of barley yellow dwarf virus on susceptibility of barley cultivars to net blotch (Pyrenophora teres) and leaf blotch (Rhynchosporium secalis). Plant Pathol 1983; 32:435-440.

13. Jensen SG. Metabolism and carbohydrate composition in barley yellow dwarf virus-infected wheat. Phytopathology 1972; 62:587-592.

14. Kiyomoto RK. Carbon dioxide exchange and total nonstructural carbohydrate in soft white winter wheat cultivars and snow mold resistant introductions. Crop Sci 1987; 27:746-752.

15. Kiyomoto RK, Bruehl GW. Carbohydrate accumulation and depletion by winter cereals differing in resistance to Typhula idahoensis. Phytopathology 1977; 67:206-211.

16. Collin J, St-Pierre CA, Comeau A. Analysis of genetic resistance to barley yellow dwarf virus in triticale and evaluation of various estimators of resistance. In: Burnett PA, editor. World perspectives on Barley Yellow Dwarf. CIMMYT, Mexico. 1990: 404-414.

AN EVALUATION OF SELECTION CRITERIA FOR TOLERANCE TO *FUSARIUM GRAMINEARUM* IN TRITICALE

Rebeca Gonzalez, Castrejon S.A. and Eulalio Venegas Gonzalez
Campo Exoperimental Morelia, INIFAP-SARH
Morelia. Mich. Mexico

Abstract

Scab caused by *Fusarium graminearum* Sschw. (Giberella zea Pech) is a major production constraint of triticale grown under high rainfall situations in the high valleys of Sierra Tarasca in Central Mexico. Twenty-one spring triticale lines were evaluated during 1992 to assess associations between 13 different characters and tolerance to scab, estimated as coefficient of infection (CI). Data were analyzed using phenotypic correlation analysis, principal component analysis, and cluster analysis.
Days to heading were negatively associated with CI and explained 80% of the scab infection in the spike. Four indexes based on four principal components were defined. Genotypes were grouped into six clusters. Genotypes combining high levels of resistance to *F. graminearum* with high grain yield and high test weight could be identified.

Introduction

In Mexico, triticale is currently grown on about 4,000 ha. The State of Michoacan in southcentral Mexico is a distinct production areas where triticale is cultivated under rainfed, high rainfall growing conditions on highly organic, but acidic soils. Hence, scab caused by *Fusarium graminearum, F. avenaceum, F. culmarum,* and *F. nivale* constitutes a major production constraint [1,2]. Environmental conditions in the Tarascan Highlands, such as free moisture and high humidity, are ideal for the propagation of the scab-causing pathogens and favor the release of ascosporas since the spikes are wet for a extended period around anthesis stage. Early anthesis and a prolonged anthesis period favor infection [3,4] and results in a negative correlation between the relative coefficient of infection of head scab and the number of days to heading. Infection decreases by 15 to 30% for every 10 days of delay in anthesis [5].
The importance of *Fusarium* and its actual potential to cause crop failures in triticale production warrant in-depth analysis of relevant variables associated with the disease. This analysis is facilitated by the use of multivariate methods, such as principal component analysis and grouping of the genotypes according to resistance as reported in peanuts, cassava, and beans [6,7]. The objective of this study was the identification of

H. Guedes-Pinto et al. (eds.), Triticale: Today and Tomorrow, 549–556.
© 1996 *Kluwer Academic Publishers. Printed in the Netherlands.*

selection criteria for tolerance of Triticale to *F. graminearum* using the principal component and cluster analyses.

Materials and Methods

During 1992, the following materials were examined: 18 triticale advanced lines developed at CIMMYT and selected for adaptation and yield-tested over two to five years at the Morelia Experimental Station of the Mexican National Institute for Forestry and Agriculture (INIFAP-CEFAP Morelia), the check triticale varieties Eronga-83 and Tarasca-86, and the bread wheat check variety Curinda 87 (Table 1). The experiment was planted as a randomized complete block with three replications in 6-row, 5-m plots with 30-cm row spacing. The center 3 m2/plot were harvested for grain yield estimation. Ten spikes from each plot were used to determine disease components.

Table 1. Cross and selection history of the genetic material evaluated triticale during 1992 at Potzumaran, Mich., Méx.

Variety number	Cross and selection history
1	Faro 13
2	Merino/Jlo 170/Tesmo CTM-16227-OM-OY-17Y-2B-OY
3	Faro/Zebra 79//Bta CTM-245411-1M-OY-OM-OY-1M-OY
4	Jlo/Ptr//Yogui CTM-10233-028Y-OM-OY-16M-1Y-2B-OY
5	Tesmo/Lira//Bgl/Jlo CIT-3049-OY-5M-3Y-3M-1Y-OB
6	Hare 7265/Yogui SWT-1697-05Y-OM-OY-4M-4Y-OB
7	Cml/Pato//Bgl/3/Ita Bulk/4/Drira/Fas 204/5/ Hare 57/6/asad CTM-19603-3M-OY-OM-OY-7M-4Y-7B-OY
8	Hare 263/Civet CTM-10189-064Y-OM-OY-4M-2Y-3B-OY
9	Merino/Jlo//Zebra 32 CTM-15605-OM-OY-OH-OY-OH-2OY-2B-OY
10	Zebra 32//Bgl/Cin//Mus CTM-16005-OM-OY-OH-OY-OH-6Y-2B-OY
11	Bgl/Cin//Mus/3/Bgl/Jlo CTM-13801-043Y-OM-OY-15M-1Y-1B-4Y-2B-OY
12	Yav79/3/Snp/PI 275357//p.5D/4/Spt/5/Bglt/ Inc/6/Yogui 1 CIT-3507-1B-1Y-1B-2Y-2M-OY
13	Zebra 79/Lynx

Table 1. Continued.

Variety number	Cross and selection history
	CTM-20767-OM-OY-OM-12Y-6B-1Y-OB
14	Tatu 1*2/China 7
	CTM.86B.328-1Y-1M-1Y-2M-3RES-OB
15	Lt 1117.82//Mus/Lynx/3/Uron 5
	CTM 86.1327-1OM-1Y-1B-3Y-1B-5RES-OB
16	Asad*2//Prl/Toni
	CITM 868.323-1Y-2M-2Y-1M-4RES-OB-OPZ
17	Nimir 1/Gnu 1
	CTM 86.414-OY-OM-OY-23M-4RES-OB-OPZ
18	Ram#2*2//Pf 70354/Bow
	CITM 86.221-2B-1Y-9M-2Y-2M-4RES-OB-2PZ
19	TARASCA TCL-86
20	CURINDA M-87
21	ERONGA TCL-83

The experiment was planted after maize in June 1992 at Potzumaran in the state of Michoacan, which is a test site with consistently high natural scab epidemics. Long-time average precipitation at Potzumaran is 1300 mm, relative is humidity 80%; the weather is generally cloudy. The materials were sown in the 1992 spring-summer cycle. Nitrogen and P_2O_5 were applied at rates of 120 and 80 kg/ha.

Variables measured were:

Spike length (cm)
Number of spikelets/spike
Number of grains/spike
Number of uninfected grains/spike
Number of infected grains/spike
Weight of uninfected grains/spike
Weight of infected grains/spike
Total weight of grains/spike
Days to heading
Days to maturity
Plant height (cm)
Hectoliter weight (kg/hl)
Grain yield (kg/ha)
Coefficient of infection (CI):

The spike infection coefficient (%CI) was determined by using the double-digit scale of Saari and Prescott [8] and, following van Beuningen [5], the two digits were multiplied and the results divided by the highest value multiplied by 100.

The 14 variables measured were analyzed using principal component analysis (PCA) to identify, based on the correlation matrix, underlying relationships to the principal components, and to select suitable variables for subsequent cluster analysis. Each group was described and characterized with emphasis on CI, grain yield, days to heading, and hectoliter weight. Further, multiple regression analysis was conducted to determine associations between days to heading, days to maturity, hectoliter weight, grain yield, number of infected grains, weight of infected grains and CI.

Results and Discussion

CI was negatively correlated with number of days to heading and reflects the coincidence of peak inoculum around the anthesis of the early maturing genotypes. Hence, number of days to heading was positively correlated with grain yield and plant height.

PCA showed that spike length, number of uninfected grains/spike, days of maturity, and plant height did not contribute to the variability of CI and that the 10 remaining variables were sufficient. Since the above-mentioned four traits are of agronomic importance and their influence on CI may not have been detected, probably due to a lack of variability in the germplasm studied or masking or dominating effects of other variables, they should be further investigated in future studies.

The first four principal components (PC) explained 89% of the total variability (Table 2). Each selected PC was used in an index that was constructed from a combination of variables representing >90% of the highest PC value. Table 3 shows that the first PC was associated with an index comprising variables measured on the spike: number of spikelets/spike, number of grains/spike, weight of grains/spike, and weight of uninfected grains/spike. The second PC was related to an index of grain damage since number of infected grains/spike and their weight of infected grains/spike was negatively correlated with hectoliter weight. The third PC was associated with an index related to days to heading and CI the time in which *F. graminearum* develops. Results are in agreement with van Beuningen [5], who found in studies with wheat that days to heading and plant height determine *F. graminearum* infection, if these variables are correlated with CI. The fourth PC corresponds to the relationship between the severity of *F. graminearum* infection and grain yield.

Table 2. Variance, proportion, and accumulated proportion of variability explained by the first four principal components.

Principal components	Variance (ß)	Proportion explained	Accumulated Proportion
1	7.0	50.5	50.2
2	3.1	22.5	72.6
3	1.7	12.2	84.9
4	0.6	4.3	89.2

Table 3. Characteristics of vectors associated with the first four principal components.

Characteristics	Associated vectors			
	1	2	3	4
Spike length	-0.30	0.13	0.30	0.18
Number of spikelets/spike	-0.35	0.18	0.10	-0.14
Number of grains/spike	-0.34	0.06	0.09	0.11
Number of uninfected grains/s	-0.27	-0.31	0.24	0.02
Number of infected grains/s	-0.16	0.48	-0.18	-0.11
Weight of uninfected grains/s	-0.33	-0.18	-0.10	0.05
Weight of infected grains/s	-0.16	0.43	-0.28	0.04
Total weight of grains/spike	-0.36	0.01	0.02	0.08
Days to heading	-0.20	-0.17	-0.54	-0.34
Days to maturity	-0.24	0.28	0.30	-0.22
Plant height (cm)	-0.31	-0.14	-0.13	-0.26
Hectoliter weight (kg/hl)	-0.14	-0.49	0.13	-0.13
Grain yield (kg/ha)	-0.30	-0.07	-0.21	-0.45
Coefficient of infection (CI)	0.14	0.19	0.50	-0.68

Based on the PCA, Figure 1 depicts the dendrogram of the 21 genotypes studied. With the four first PCs and using a value of 5.7 for the Euclidian distance, the genotypes separate into six groups. Average days to heading was associated with cluster groups and varied between 46, 52, 57, 64, 68, and 69 days (Table 4). Group 4, a single-member group consisted of the wheat check and is not discussed here. The genotypes tolerant to *F. graminearum* are in the first group (Table 4): genotypes 1, 19, 2, 14, 15. Genotypes 1, 19, and 2 showed low hectoliter weight compared to genotypes 14 and 15 and had about the same level of scab tolerance. For genotypes 14 and 15, grain yield was not reduced by scab infection and was similar to the genotypes of groups two and three. However, groups two and three had similar CIs between 30 and 100%. With the exception of genotype 10, grain yields of the genotypes of group 5 were affected by scab, but hectoliter weights were above those of group 1. The late maturing and tall genotype 7, the advanced line CML/PATO//BGL/3/ITA_BULK/ FAS204/5/HARE57/6/ASAD, formed group 6 and combined high yield potential in an acid soil environment with tolerance to *F. graminearum*.

The regression analysis (Figure 2) was calculated with group means with CI as the dependent variable and days to heading as the independent variable. Days to heading was closely associated with CI ($P < 0.05$). Based on the regression equation [$Y = 221 - 3.14(heading)$], each day delay in days to heading reduces the CI values by three units in the observed range.

Conclusions

Seven of 20 triticale genotypes studied combined tolerance to *F. graminearum* with high grain yields. Of those, TATU1*2/CHINA_7 (entry 14), LT1117.82//MUS/LYNX/3/

554

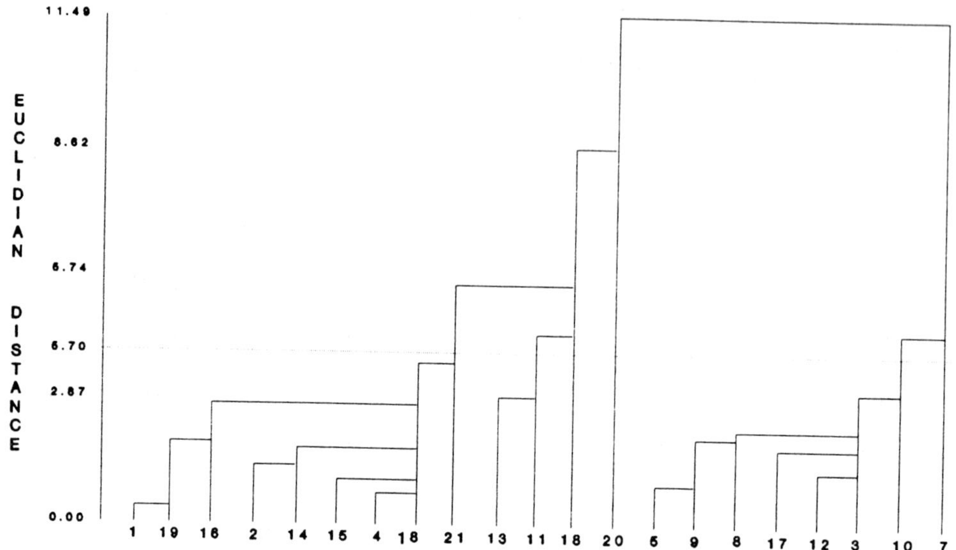

FIGURE 1. DENDROGRAM OF 21 TRITICALE GENOTYPES GROUPED FOR SELECTION BY *F. graminearum*

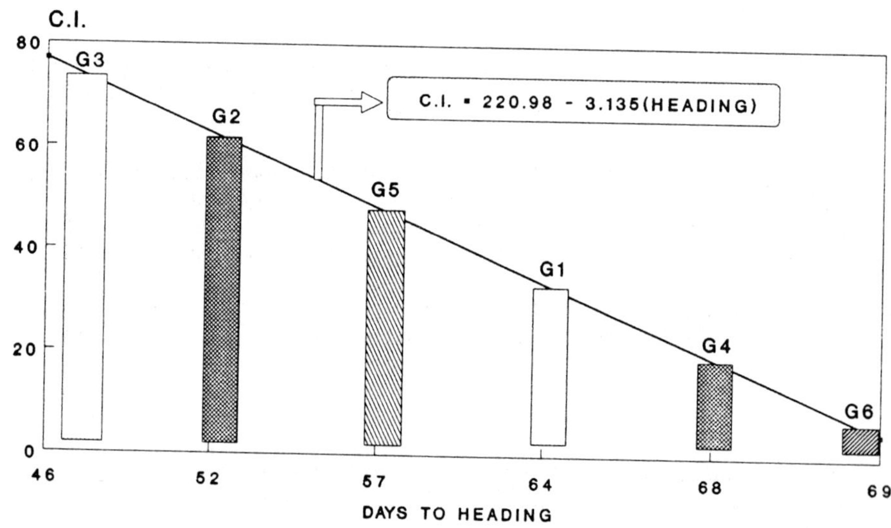

FIGURE 2. DAYS HEADING AND FUSARIUM INFECTION COEFFICIENT LINEAR REGRESSION FOR SIX TRITICALE GENOTYPE GROUPS

Table 4. Agronomic characterization of cluster groups.

Group number	Genotypes	Days to heading	Days to maturity	Plant height (cm)	HW	Grain yield (kg/ha)	CI %
1	1,19,16,2,14, 15,4,6	64	170	120	59	3760	25
2	21,13,11	52	168	107	54	2078	36
3	18	46	168	95	57	2672	100
4	20	68	146	106	62	2422	13
5	5,9,8,17,12,3,10	57	170	121	61	3173	29
6	7	69	171	130	62	4344	9

URON5 (entry 15), ZEBRA32/3/BGL/CIN//MUS (entry 10), and CML/PATO// BGL/3/ITABULK/FAS204/5/HARE57//6/ASAD (entry 7) had acceptable test weights, while FARO 13 (entry 1), TARASCA 86 (entry 19) and MERINO/JLO170//TESMO (entry 2) showed had low hectoliter weights.

The most important variable measured was days to heading, which was highly correlated with CI ($R2 = 0.80$); each additional day in days to heading reduced the Scab CI disease score by three units.

Principal component analysis revealed the association of number of spikelets/spike, grains/spike, weight of grains/spike, and weight of uninfected grains/spike with an index of spike variables. Infected grains/spike and weight of infected grains/spike were correlated with hectoliter weight and were related to a grain damage index.

Cluster groups were associated with differences in days to heading, grain yield, and CI. Given infection levels had genotype-specific effects on grain yield.

References

1. Wiese MV. Head blight or scab. In Compendium of Wheat Diseases. The American Phytopathological Society. St. Paul, Minn. USA. 1977: 16-17.

2. Zillinsky FJ 1984. "Enfermedades causadas por *Fusarium spp*. In Guía Para la Identificación de Enfermedades en Cereales de Grano Pequeño. CIMMYT, México. 1984: 59-69.

3. Melo RE. Fusariosis: Biología y Epidemiologia de *Gibberella Zeae* en trigo. In Taller sobre la Fusariosis de la espiga en América del Sur. 1987 Sept 7-11, Encarnación, Paraguay, 1989.

4. Kohli MM. Análisis de la fusariosis del trigo en el Cono Sur": Taller Sobre la Fusariosis de la Espiga en América del sur; 1987 Sept 7-11, Encarnación, Paraguay, 1989.

5. van Beuningen LT. "Tres años de evaluación de germoplasma del Cono Sur para Resistencia a la fusariosis": Taller sobre la fusariosis de la espiga en América del Sur. 1987 Sept 7-11, Encarnación, Paraguay, 1989.

6. Galindo RGA. "Caracterización morfológica de germoplasma de yuca (*Manihot esculenta* Crantz)". Tesis profesional. Univ. Aut. de Chapingo, 73 p. México, 1982.

7. Venegas GE. Uso de análisis multivariado en la agrupación y caracterización de 200 tipos de frijol. Tesis profesional. IIMAS,UNAM, México, 1986.

8. Saari EE, Prescott JM. A scale for appraising the foliar intensity of wheat diseases. Plant Dis. Rep. 1975;59:377-380.

RESISTANCE OF TRITICALE TO ROOT LESION NEMATODE IN SOUTH AUSTRALIA

Vivien Vanstone, Mohammed Farsi, Tony Rathjen and Kath Cooper
University of Adelaide, Waite Campus, Glen Osmond, 5064, AUSTRALIA

Abstract

The root lesion nematode (*Pratylenchus neglectus*) occurs throughout the cereal growing regions of South Australia. *P. neglectus* has a wide host range, infecting all cereals as well as crops grown in rotation with cereals (grain legumes, pasture legumes and oilseeds). However, nematode multiplication differs both between and within host species. Damage caused by *P. neglectus* impairs root function, limiting water and nutrient uptake, leading to poor growth and yied decline. In preliminary experiments, a yield loss of 20% has been recorded for an intolerant wheat variety.

Trials were conducted at infested sites to determine the the ability of cereal species and varieties to multiply the nematode. Further tests were conducted in the glasshouse to compare nematode multiplication on roots of triticale, wheat and rye varieties. Roots of triticale were found to contain fewer nematodes than the other cereals. Triticale is thus a useful rotational crop for areas infested with the root lesion nematode.

Introduction

The root lesion nematode (*Pratylenchus neglectus*) occurs throughout the cereal growing regions of South Australia. *P. neglectus* has a wide host range, infecting all cereals as well as crops grown in rotation with cereals (grain legumes, pasture legumes and oilseeds). However, nematode multiplication differs both between and within host species. This has implications for crop rotation strategies, as growth of different crops will result in varying nematode populations in the soil to infect subsequent crops.

Damage caused by *P. neglectus* impairs root function, limiting water and nutrient uptake, leading to poor growth and yield decline. For cereals, symptoms of infection include:

    disintegration of primary and secondary root cortices;
    reduction in numbers of root hairs;
    a lack of or stunting of lateral roots along main root axes;
    brown lesioning and discolouration of roots.

Above-ground symptoms are indistinct, but plants may show symptoms of nitrogen deficiency and lack vigour. In preliminary experiments, a yield loss of 20% has been recorded for an intolerant wheat variety.

Trials were conducted at infested field sites to determine ability of cereal species and varieties to multiply the nematode. Further tests were conducted in the glasshouse to compare nematode multiplication on roots of triticale, rye and wheat varieties. Roots of triticale were found to contain fewer nematodes than the other cereals. Triticale is thus a useful rotational crop for nematode-infested areas.

H. Guedes-Pinto et al. (eds.), Triticale: Today and Tomorrow, 557–560.
© 1996 *Kluwer Academic Publishers. Printed in the Netherlands.*

Materials and Methods

FIELD EXPERIMENTS

Plots (four rows wide x 4.2m long) of cereal varieties were sown at sites infested with *P. neglectus* in 1992 (five sites) and in 1993 (six sites). The 1992 unreplicated trial consisted of two triticale varieties, one rye, four durums, nine barleys and 44 wheats. In 1993, three replicates of two triticale, one rye, one durum, one oat, five barley and 20 wheat varieties were tested. Roots were sampled in September - October, three to four months after sowing. Plants contained in three 15cm sections of plot row were removed from each plot and bulked. Soil was washed from roots, and roots chopped into 1cm lengths. Nematodes were extracted from roots by misting for five days, and counted microscopically.

GLASSHOUSE EXPERIMENT

Tahara, Muir, Abacus and Currency triticale, King II rye, Spear and Molineux wheat and a selection of breeders' wheat lines were grown at 22°C in 600ml cups of pasteurised soil. Each cup contained one seedling, inoculated at emergence with 250 *P. neglectus* (mixed stages + 150 eggs) reared on aseptic carrot cultures in the laboratory. All cereal varieties were replicated eight times.   After seven weeks, soil was washed from roots, and nematodes extracted for counting.

Results

FIELD EXPERIMENTS

In 1992, roots of Currency and Tahara triticale and SA rye contained fewer nematodes than all other cereals tested (Figure 1). Barley and durum also multiplied fewer nematodes than the majority of wheats.

Similarly, in 1993 (Figure 2), Currency and Tahara triticale were infected with fewer nematodes than all but Galleon barley.

Overall, nematode numbers were ten times lower in 1993 than in 1992. However, ranking of varieties was similar for both years, confirming species and variety differences for *P. neglectus* multiplication on the roots of cereals in the field.

GLASSHOUSE EXPERIMENT

Roots of Tahara, Muir, Abacus and Currency triticale contained fewer *P. neglectus* than did the majority of wheats (Figure 3). Abacus allowed significantly lower nematode multiplication than did the commercial wheat varieties, Spear and Molineux. This has been confirmed by several tests.

Conclusions

No currently grown, commercial varieties of wheat have sufficient resistance to *P. neglectus*. Wheat with resistance comparable to that of Abacus triticale is being sought for incorporation into our breeding program.

Growth of suitable rotational crops will reduce nematode numbers in the soil, limiting damage incurred by subsequent, susceptible crops. Triticale is more resistant to *P. neglectus* than are the majority of varieties of other cereals, and is thus a useful rotational crop for the South Australian wheat belt.

Acknowledgments

The financial support of the Grains Research and Development Corporation is gratefully acknowledged, as is the assistance of Paul Lonergan.

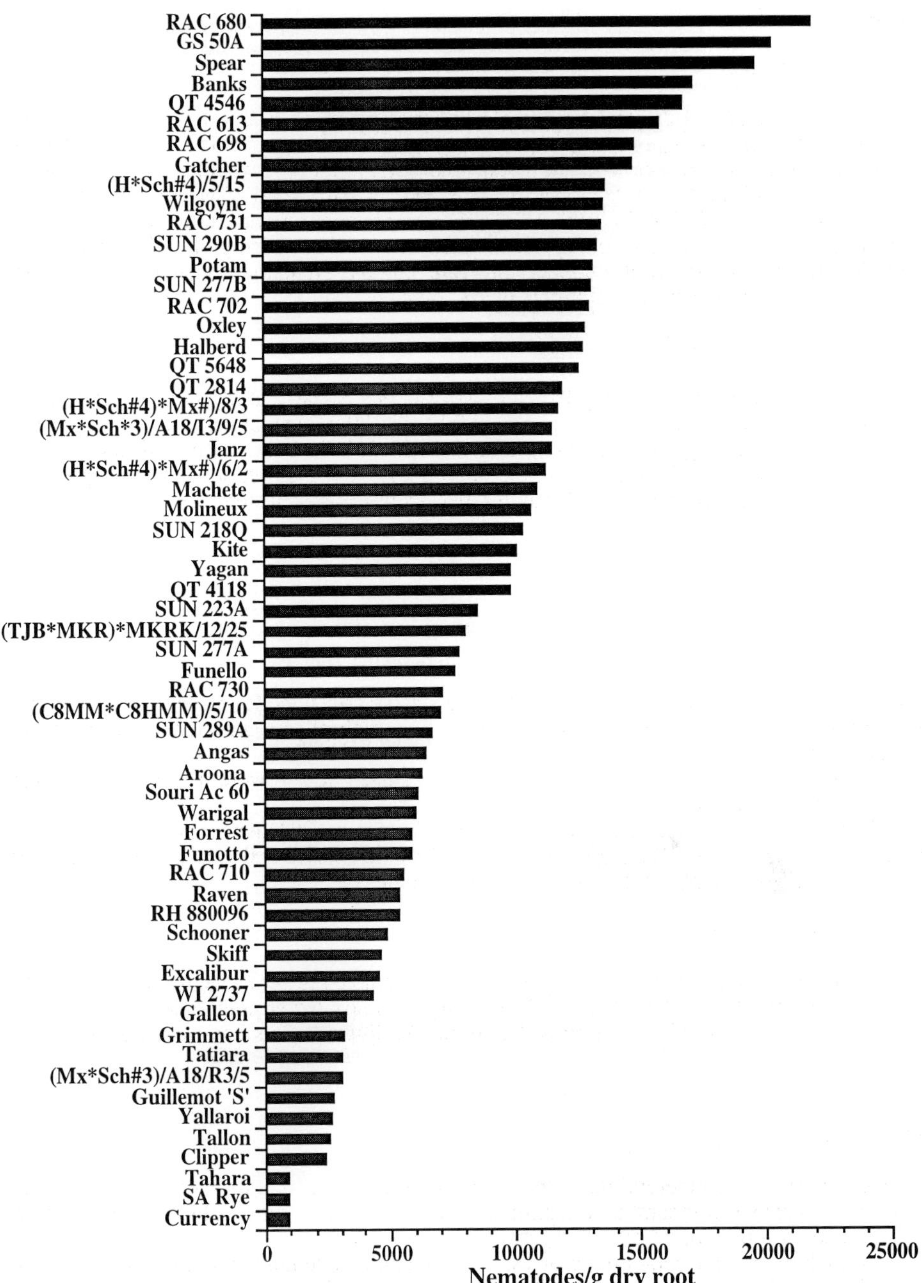

Figure 1: Number of *Pratylenchus neglectus* extracted from the roots of field-grown cereal varieties in 1992. Values are means taken over five sites.

560

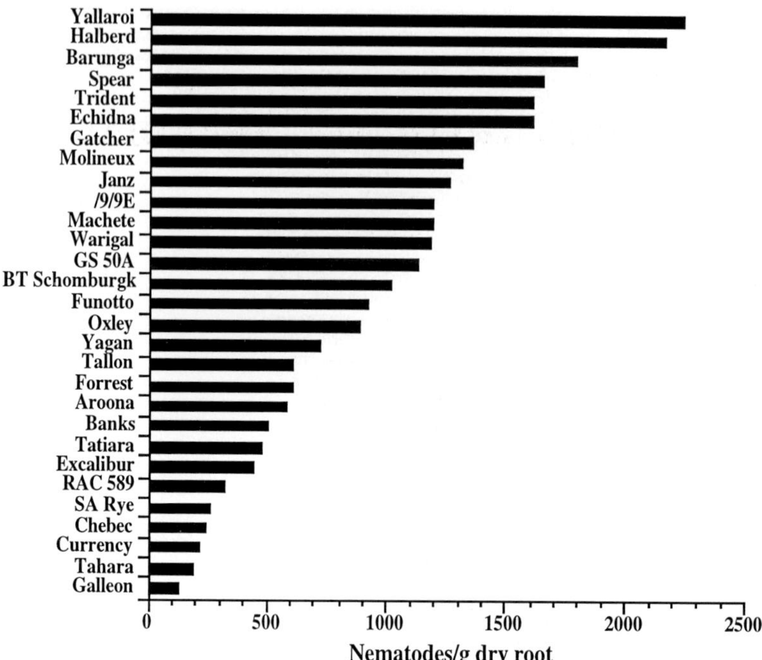

Figure 2: Number of *Pratylenchus neglectus* extracted from the roots of field-grown cereal varieties in 1993. Values are means taken over six sites.

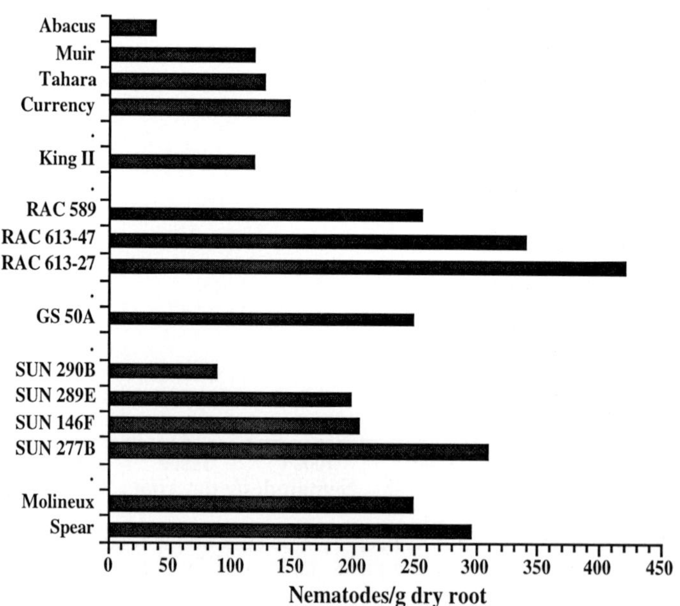

Figure 3: Number of *Pratylenchus neglectus* extracted from the roots of triticale, rye and wheat varieties inoculated in the glasshouse. Values are the means of eight replicates (LSD 95% = 176.8).

VII - TRITICALE BREEDING

BREEDING STRATEGIES FOR TRITICALE[*]

Augusto C. Baier[1] and J. Perry Gustafson[2]
[1]EMBRAPA-CNPT, C. Postal, 569, Passo Fundo RS, 99001-970, Brazil;
[2]USDA-ARS, PGRU, Curtis Hall, Un. of Missouri, Columbia, MO 65211
USA.

Abstract

A long-standing scientific vision -- the dream of combining in one and the
same plant, the yield potential and grain quality of wheat with the
rusticity of rye -- is materializing as triticale becomes an established
economic crop. The ability of scientists world-wide to shape food supply
strategies is being reduced, as the interactions between farmers and
consumers are increasingly being regulated by world-wide marketing
strategies. Thus, the success of a triticale breeding program depends
more than ever on a precise evaluation of the consumer's requirements
(they are the ultimate clients), and on a broad evaluation of the
environment and the farmers' limitations and potentialities (they are the
first clients). High yield potential, harvest index, protein and lysine
content, adaptation to acid and sandy soils, nutrient uptake efficiency,
resistance to diseases, tolerance to moisture and cold stresses are all
desirable features of triticale. Using triticale as a feed is an immediate
option, because of the high yield and good protein and amino acid balance.
Improved milling and bread making will increase its potential for human
nutrition. Sprouting resistance, broader adaptation, smoother grains in
stressed conditions, frost damage at flowering, milling and baking quality,
and poor primary triticales, are all challenges facing breeders.
Intergenomic barriers limiting gene expression between wheat and rye,
are still posing considerable difficulties to breeders. Basic research
involving potential tools (anther culture, molecular markers, etc.) for
breeders to manipulate this man-made intergeneric hybrid will need to be
developed hand-in-hand with multi-environment field evaluations. Rye
with better amino acid content or balance, plant type, biotic or abiotic
stress resistance or tolerance, nutrient uptake efficiency, broader
adaptation or lower secalin content, is critical to the improvement of
triticale. The type of rye which will give the best genome balance when

---

[*]Names are necessary to report factually on available data; however, the
USDA neither guarantees nor warrants the standard of the product, and
the use of the name by USDA implies no approval of the product to the
exclusion of others that may also be suitable.

H. Guedes-Pinto et al. (eds.), Triticale: Today and Tomorrow, 563–569.
© 1996 *Kluwer Academic Publishers. Printed in the Netherlands.*

placed in a wheat background has yet to be established. Finally, the study of agronomic practices for each environment will be important to fully explore triticale's potential.

Introduction

The first self fertile triticale was obtained from a cross of hexaploid wheat and rye more than a century ago. Intensive breeding started in the early 1950s at the hexaploid level in Hungary, in the former USSR, in Spain, and in Canada. A decade later the International Maize and Wheat Improvement Center (CIMMYT) started a breeding effort in cooperation with Canadian programs [1] which resulted in an important impact on the development of this crop. The idea of combining the qualities of wheat with the hardiness of rye on marginal soils was stressed. The breeding program at CIMMYT, with its two full selection cycles per year and its specific evaluation under many environments, created a broad and improved germplasm base which was used around the world.

Alongside specific breeding strategies, it is necessary to evaluate the socio-economic environment, to detect possible bottlenecks and potential growth areas for triticale production.

As cereal production costs increase and the prices of grains decrease, the potential for triticale production on marginal soils increases. The cost to re-supply depleted soils with appropriate amounts of fertilizers and correctives is becoming too expensive, and the intensive use of ammonia based fertilizers is increasing the acidification process of many soils, thus opening space for more hardy crops such as triticale.

Plant breeders that devote their energy to triticale are not to blame for the slow commercialization. Global marketing and the world surpluses of wheat sold at bargain prices take away many potential markets for triticale. The increased quality requirements are also fueled by the wheat surplus. As soon as world market prices for wheat and other feed grains again reflect production costs, triticale will regain attention.

There was a long standing socio-economic vision in Brazil: to reduce the dependency on wheat imports. But, as the country acquired the potential to near self-sufficiency in wheat, and as triticale matured as a crop, the quality issue was elevated to exclude triticale and to drastically reduce the internal demand for wheat.

The ultimate "clients"

The success of a triticale breeding program depends more than ever on a precise evaluation of the consumer's requirements; they are the ultimate "clients". The ability of scientists to shape food supply strategies is being reduced, as the interaction between farmers and consumers are increasingly regulated by world-wide marketing strategies.

Feed is the most immediate and important option, because of the high protein yield and good amino acid balance [2], especially in regions where triticale can be cultivated in rotation with summer grown grains. Grazing, hay, plant and grain silage, or the dry grain, all have equal potential. In Brazil it has been evaluated in mixtures for broilers and for swine [3], and as wet grain silage for milk cows and swine (Sadia and Batavo, personal comm.). The conservation process for the grain silage is simple and the acceptance by animals has been excellent. In all experiments, there were no significant differences when compared to corn.

The historic price evolution during the year for corn shows an increase of 20 to 25% in the months prior to corn harvest from October to January (Triticale is harvested in South Brazil during October and November). This price increase results from insufficient production and inadequate storage facilities in the poultry and swine production regions, and high transportation costs from the seaports or from the corn production regions to the animal production regions. This relatively small price difference of corn is crucial for a further expansion of triticale as a feed grain in South Brazil.

There is also a potential for human consumption where milling and baking is operated by small processing units and where consumers and producers are close. Improved milling and bread making quality of the newer breeding lines, has increased the potential for human nutrition. Extensive tests have been performed in many places, including in Brazil to show that milling and baking operations can be adapted to process triticale grain into acceptable bread products.

The farmers, the first "clients"

The evaluation of the farmers' needs and potentialities is extremely important. They are the potential buyers and growers of the released cultivars. Triticale has the best adaptation to environments with acid and aluminum-toxic soils and to marginally dry regions world-wide.

The small poultry and swine producing farmers of South Brazil are ideal "clients". Triticale enables them to double crop their fields (corn or soybeans in summer and triticale in winter). A second crop improves soil protection and the amount of available feed and, therefore the number of finished animals per area. Acceptance is restricted where harvest with combines is not feasible.

Interaction of agronomy and breeding

A Swedish example outlines the importance of the interaction of breeders and agronomists as well as the relevance of high yield. Rye was the predominant crop 100 years ago. Since then, rye cultivation has decreased markedly and winter wheat has constantly gained in area because of the ever increasing yield differences. Yield has risen from 1 t/ha for both crops, at the beginning of the century, to more that 6 t/ha for winter wheat, but only to 4.5 t/ha for rye in 1990 [4].

The experiments over the last 20 years, in Brazil, indicate that triticale has the potential for production between the 24° to 30°S parallels, and at altitudes from 400 to 1200 m. The soil is acid in these regions, has medium to high levels of aluminum and is low in phosphorus availability. At lower altitudes or latitudes the cultivation has not spread and evaluation was discontinued. It is speculated that the higher temperatures observed at lower altitudes during tillering prevent the expression of triticales full yield potential.

In this region, corn and soybeans are cultivated on approximately 10 million ha during the warmer season and only part of this area is second-crop cultivated with wheat, oats, barley, and grasses during the second colder season. The interaction with farmers has shown that in a large part of this region, triticale is an ideal crop for double cropping. In this manner, triticale is sown from April to June and is harvested in October or November, while corn and soybeans are sown in November and December and harvested in March and April.

Rajaram [5] stated that better resistance to biotic and abiotic stresses due to the full rye complement resulted in adaptive advantages of complete, over substituted types 2D(2R) triticales in most locations. This observation implies that the future of triticale as a commercial crop lies with the complete types on marginal environments in South Brazil. The yield of the direct drilled trials in Passo Fundo have improved from around 2 t/ha in 1976 to 6 t/ha in 1989 (Table 1). This increment is attributable in almost equal parts to improved cultivars, as well as, to improved cultivation practices. The yield increase of 'Triticale BR 1' from 2.3 t/ha to 4.5 t/ha can be attributed to improved soil fertility and cultivation practices. On the other hand the yield advantage in 1987/89 of 'Triticale BR 4' over 'Triticale BR 1', can in part be attributed to genetic improvement. The lower average for 1990/92 was in part caused by climatic changes, but also because of poor agronomic management resulting in reduced soil fertility. Not surprisingly, this strongly suggests at least two very important messages: breeding exhibits more potential when agronomic practices are improved simultaneously, and that gains in soil fertility can be lost in a very short time.

Table 1. Average yield (Kg/ha) of old and new recommended triticale cultivars and of a wheat check in trials at Passo Fundo, RS, Brazil.

years	Wheat	Triticale BR1	Triticale BR2	Triticale BR4
1976-78	1800	2300	-	-
1981-83	2600	3200	3700	-
1984-86	3200	3800	5200	-
1987-89	4100	4100	5700	6300
1990-92	2800	2500	3100	3500

The comparison of trials yields on acid and amended soils indicates that beside the well known aluminum tolerance of the newer triticale cultivars, the high yield potential is equally important. Table 2 indicates that the newer triticale cultivars perform just as well on amended soil as on acid soils despite their excellent aluminum tolerance.

Table 2. Relative yield (average 1989/92) of triticales and a wheat check in trials on acid and amended soils in Passo Fundo, RS, Brazil.

Soil type	Wheat	Trit. BR1	Trit. BR2	Trit. BR4	Embrapa 17
			(%)		
Acid	100 (2.6 t/ha)	96	108	119	123
Amended	100 (3.2 t/ha)	94	125	138	141

Comparing the yield results of the cultivars released during the last 20 years indicates that the substituted triticales (2D/2R), released in the late 1970s and early 1980s, tended to show higher yields on amended soils, whereas more recently released complete triticale cultivars containing the complete rye genome and even newer advanced breeding lines show a significantly better performance on acid soil. Even more promising is the indication that the new advanced complete lines show yet another advance in increasing triticale's tolerance to aluminum.

Root elongation in hydroponic solutions containing aluminum indicates that progress in selecting for improved Al tolerance of triticale is being made. The newer complete cultivars and advanced lines show longer roots in the solutions containing Al in comparison to the older substituted cultivars. This indicates that the Al tolerance in triticale is being improved steadily alongside other agronomic characteristics, which does not seem to be the case with wheat in Brazil.

The resistance to scab (*Fusarium graminearum*) and leaf spotting diseases (*Bipolaris sorokiniana*) have also improved. Resistance to leaf spotting diseases and the adoption of crop rotation are contributing to reduce their importance. Interestingly *Septoria* spp. diseases have lost importance in Brazil due to the adoption of crop rotation.

568

Selection and breeding

A valuable resource for any breeding program to start with is the genetic
diversity coming from the germplasm distributed by CIMMYT. The
triticale breeding programs in Brazil, as well as many others around the
world, have received strong and vital support from CIMMYT. The training
provided by CIMMYT in Obregon, Mexico as well as the possibilities for
participation on symposiums and congresses is of upper-most importance
to breeding programs in developing countries. Joint projects to evaluate or
to solve specific problems, could improve the flow of knowledge still more.

The germplasm distributed by CIMMYT during the last 25 years
has improved markedly, in yield potential, grain filling, harvest index,
disease resistance, and aluminum tolerance. The intensive breeding work
is complemented by the ability to grow two full selection cycles each year
and the selection for specific agronomic problems under many different
environments. No program in Brazil, with the presently available
resources, could have made comparable progress.

Breeding for marginal areas with acidic, sandy or alkaline soils;
trace element deficiency or toxicity; and moisture stress constitutes a
major effort in triticale improvement [6]. Precise descriptions of target
environments will allow breeders to focus on specific traits and
karyotypes to establish trait-based selection and breeding procedures in
multi-disciplinary approaches [5].

The use of specific selection procedures to evaluate simultaneously
in stressed soils and environments, and on high yield potential sites in
South Brazil where a large potential exists for triticale, is important.
Further, the interaction with farmers and meat processing industries that
have well developed extension services, is also indispensable.

Genomic imbalance

The genomic imbalance in primary triticales has been and still is a major
cause for the slow rate of advancement. For example, the expression of
aluminum tolerance genes located on rye chromosomes, when
incorporated into aluminum sensitive wheats can be suppressed by the
wheat background, depending on which wheat chromosomes are replaced
[7]. The control of aluminum tolerance expression from rye was observed
by Gustafson and Ross [8] to be located on a number of wheat
chromosome arms, with a few arms tending to enhance expression and
many tending to reduce it.

Aneuploidy in primary triticales is still attributed to many of the
negative interactions between the wheat and rye genomes. Lelley [1]
suggests that breeders concentrate on crosses of triticale with wheat or
rye, followed by backcrosses to the parental triticale. Evaluating large
numbers of crosses, Lelley [1] further suggested that this strategy would
reduce the cytological instability. In addition, he suggests that selection
tarts in large F3 or F4 populations in order to improve efficiency. Crosses

to rye are also utilized to improve aluminum or sprouting tolerance. However in our breeding program in Passo Fundo, the introduction of variability for aluminum tolerance, resistance to scab, and resistance to leaf spotting diseases, from local ryes and wheats, have not yet produced the expected results.

Improved selection strategies such as anther culture and double haploid techniques are also extremely important to accelerate the creation and adaptation of a new balance between the genomes. Since triticale is the result of the cross of an allogamous and an autogamous species, breeding should not be restricted to methods used for autogamous species. It remains to be seen whether new insights into the genetic mechanisms determining the function of an allopolyploid species, and new breeding strategies, will be able to accelerate breeding progress and help triticale ultimately to live up to its promise [1].

References

1. Lelley T. Triticale, still a promise? Plant Breeding 1992;109:1-17.

2. Hill GM. Quality: Triticale in animal nutrition. Proceedings of the Second International Triticale Symposium. EMBRAPA/CNPT, CIMMYT and ITA; 1990 Oct 1-5; Passo Fundo, RS. Brazil 1991; pp 422-427.

3. Ferreira AS, de Lima GJMM, Zanotto DL, Bassi LJ. Triticale como alimento alternativo para suínos em crescimento e terminação (Triticale as an alternative feed for pigs during growing and finishing - English Summary). Revista da Sociedade Brasileira de Zootecnia 1992;21:300-308.

4. MacKey J. Demonstration of genetic gain from Swedish breeding of small grains (English summary). (Belagg for gjorda framsteg inom svensk strasadesforadling). Sveriges Utsadesforenings Tidskrift 1993;103:33-43.

5. Rajaram S, Varughese G, Abdalla O, Pfeiffer WH, Van Ginkel M. Accomplishments and challenges in wheat and triticale breeding at CIMMYT. Plant Breeding Abstracts 1993;63:131-139.

6. National Research Council. Triticale: a promising addition to the world's cereal grains. Washington, D.C. 19989; pp 103.

7. Aniol A, Gustafson JP. Chromosome location of genes controlling aluminum tolerance in wheat, rye and triticale. Can J Genet Cytol 1984;26:701-705.

8. Gustafson JP, Ross K. Control of alien gene expression for aluminum tolerance in wheat. Genome 1989;33:9-12.

TRITICALE: POTENTIAL AND RESEARCH STATUS OF A MAN-MADE CEREAL CROP

Wolfgang H. Pfeiffer
International Maize and Wheat Improvement Center (CIMMYT)

Abstract

Triticale (Tcl) improvement concentrates on generating various crop and utilization options. These are associated with crop adaptive patterns, economic comparative advantages, issues of environmental sustainability, as well as issues of consumption, marketing, and end uses. By focusing on Tcl's comparative advantages over other crops regarding these options, CIMMYT has shifted its Tcl breeding objectives. International yield data and past experience indicate high genetic progress in Tcl grain yield under marginal and high production conditions (> 1.5%/year). Since grain yield in marginal environments reflects the genotypic response to the total environment, progress has to be associated with advances for all biotic and abiotic factors involved. Under near optimal conditions, comparison in maximum yield trials at Cd. Obregon over three location years of Tcls developed in the 1980s and 1990s reveals overall yield progress to be at +17% due to increases in harvest index (+16%), grain biomass production rate (+21%), grains/m^2 (+17%), spikes/m^2 (+12%), and test weight (+12%) and a decrease in height (-11%). Genetic variabilities for these traits suggest that gains in genetic grain yield potential can be maintained in the future. High rates of progress in germplasm developed from crosses involving interspecific Tcl x wheat, spring x winter Tcl, and 2D(2R) x complete R type suggest that these gene pools are promising sources with which to achieve future genetic gains. The shift in breeding emphasis to complete R genome and 6D(6A) types in the 1980s was accompanied by improved test weights, but negative effects on baking quality. Subsequent research with wheat-Tcl grain and flour blends on milling and baking properties suggested blends for commercial use. In 1990, specialized requirements and markets for products for human consumption, feed grain, and growing interest in forage and dual purpose triticales were prompted by end-use oriented and expanded breeding objectives. By 1994, germplasm products for the range of utilizations are available. Current quality improvement hinges on introgressing high-molecular-weight glutenin subunits, particularly allelic variants at loci Glu-D1, Glu-B1 and Glu-A1. Spectacular increases in SDS sedimentation in 1D(1A), 1D(1B), and 1Rs.1Dl Tcls that carry D genome non-enzymatic storage proteins suggest a future breakthrough in improving Tcl's industrial quality, while additional gains can be expected from exploiting gliadins and secalins.

H. Guedes-Pinto et al. (eds.), Triticale: Today and Tomorrow, 571–580.

© 1996 *Kluwer Academic Publishers. Printed in the Netherlands.*

Focus on Comparative Advantages and Utilization Options

Over less than one century, Tcl has been transformed from a botanical curiosity to a commercial crop. After overcoming Tcl's initial problems, modern Tcl improvement has to concentrate on two objectives: 1) generating crop options and 2) generating utilization options. Crop options are associated with crop adaptive patterns, economic comparative advantages, and issues of environmental sustainability. Utilization options refer to consumption, marketing, and end-uses. By focusing on Tcl's comparative advantages regarding these options, CIMMYT has shifted and expanded its Tcl breeding objectives.

Crop Options

Regression techniques and multivariate analysis with International Triticale Yield Nursery (ITYN) data and empirical data have been used to define Tcl's adaptive pattern, compare Tcls with other commodities, and direct breeding strategies [1,2,3,4,5]. Results indicated that Tcls, particularly those with the complete R genome, show distinct adaptive advantages over wheat and 2D(2R) substituted types:

> • Where marginal lands are afflicted with abiotic stresses, such as low and erratic rainfall, soil acidity, phosphorus deficiency, and trace element toxicity (e.g., boron) or deficiency (e.g., manganese); and

> • Where there is a high risk of biotic stresses, such as BYDV, leaf rust, powdery mildew, bunts, smuts, insect pests (e.g., Russian wheat aphid, Hessian fly), and weeds.

Confirmed by recent research [6], these observations imply, that Tcl is an ideal low-input crop for non-extractive, sustainable agriculture and organic farming. Future breeding agendas should exploit these areas and develop germplasm products tailored for sustainable and organic farming systems and their commodity markets.

Further, results revealed that both Tcl types can rival wheat production under high production conditions. These findings have prompted a shift in CIMMYT's emphasis towards complete R genome types for marginal environments. The production ratio of completes to 2D(2R) Tcls has shifted gradually from 25:75 in 1985 to 95:5 in 1992.

Area and production figures reflect increasing acceptance of Tcl, but the crop is still underutilized. Hence, advances in crop enhancement have to be maintained at current high rates to establish and expand Tcl area and production.

Progress in Tcl grain yield under marginal and high production conditions have been high (> 1.5%/year). Grain yields across locations (Figure 1), particularly if marginal environments are involved, reflect the genotypic response to the total environment, hence, progress has to be associated with advances for all biotic and abiotic factors involved [3]. The constant increase in average grain yields from ITYNs 19 to 23 compared to the checks (Figure 1) was accompanied by improvements in test weight (Figure 2). Although current rates of progress in test weight are modest when compared with genetic gains in the 1980s, Figure 2 depicts that Tcl top test weights have reached the level of the bread wheat check. Projected rates of genetic advance from the diverse Mexican test sites correspond to figures on the international level and suggest higher

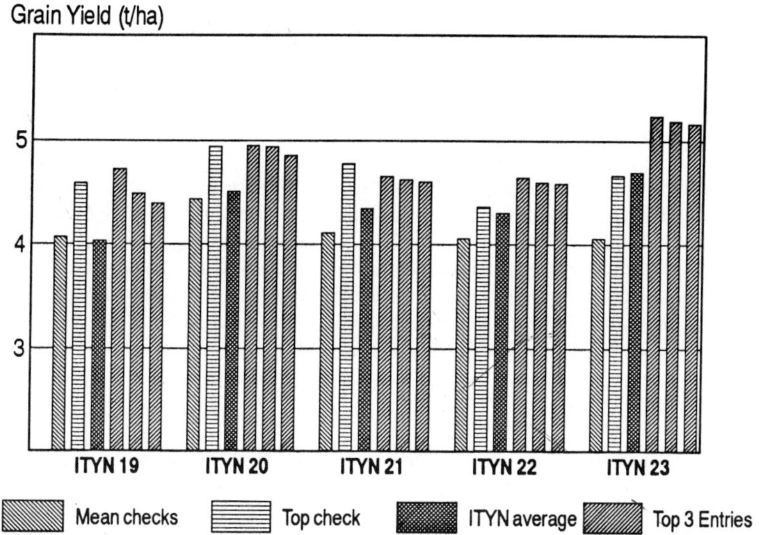

Figure 1. Triticale grain yield performance 19th to 23rd ITYN.

Mean yield across sites. Checks: Can 79, Alm 83, Gen 81 B-Wheat, Beagle, Eronga 83.

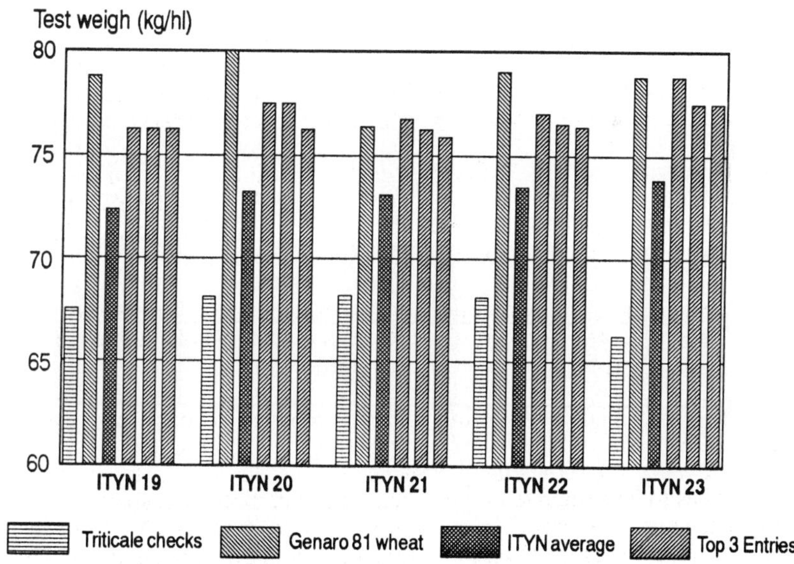

Figure 2. Triticale test weight performance 19th to 23rd ITYN.

Mean test weight across sites. Triticale checks: Can 79, Alm 83, Beagle, Eronga 83.

rates of progress for grain yield under near optimal and disease-prone high rainfall situations when compared with moisture stress environments. For test weight, advances were equivalent. These results are reflected in variety releases. During 1991-94, more than 30 CIMMYT-derived Tcls have been registered by National Agricultural Research Systems.

Variety release data and results from recent ITYNs imply adaptive advantages of complete Tcls that carry a 6D(6A) chromosome substitution. The 6D(6A) substitution rapidly spread to about 50% of the CIMMYT advanced spring germplasm and probably to winter germplasm through introgression of CIMMYT genotypes. These observations suggest the exploitation of karyotypic variability as a promising strategy to increase yield potential and adaptation since similar effects can be expected from other D(A) and D(B) substitutions and translocations which are now available. The optimal Tcl karyotype has to be determined and a single optimal chromosomal constitution for all situations is unlikely to emerge.

A prerequisite of CIMMYT's breeding efforts aimed at protecting high genetic yield potential is incorporating resistance to abiotic and biotic stresses as buffering mechanisms. According to this hypothesis, genetic progress under high productive and stress-prone growing conditions was based on improvements of Tcl's genetic yield potential. Grain yields from yield trials conducted at Cd. Obregon under irrigated, near optimal conditions indicate spectacular increases in genetic yield potential in Tcl (Figure 3). Grain yields of the highest yielding Tcl increased from 2.5 t/ha in 1968 to 9.7 t/ha in 1991.

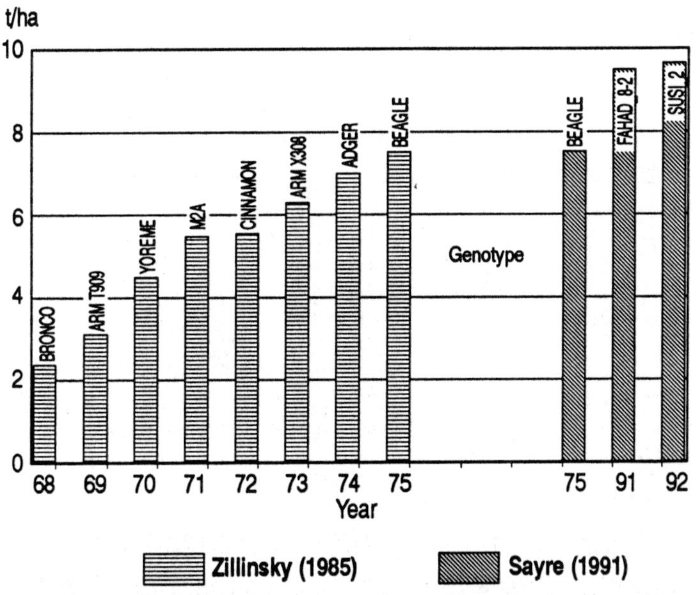

Figure 3. Triticale grain yield potential 1968-91.
Cd. Obregon data.

Figure 4 displays agronomic components associated with this advances in Tcl grain yields. Comparison over three years in maximum yield trials at Cd. Obregon of complete Tcls developed in the 1980s and 1990s reveals overall yield progress to be at +17%. Increases in biomass, if at all, were modest, but a major contribution resulted from increases in harvest index (+16%) and spikes/m^2 (+12%). With approximately the same number grains/spike, number of grains/m^2 (+17%) increased considerably, while plant height has decreased, on average, from 140 to 125 cm (-11%). At the same time, test weight increased from 68 to 76 kg/hl (+12%). Modest reduction in days to maturity and grain-fill duration by 4 and 3 days, respectively, were accompanied by a 10% reduction in vegetative growth rate and straw yields, but a 21% increase in the grain biomass production rate.

A comparison among the top three yielding Tcls and bread wheats may reflect agronomic components with potential pay-offs in raising Tcl's genetic yield potential (Figure 5). Average grain yields in Tcl over the last 3 years have been slightly above wheat. However, exploitation of higher biomass of Tcl (+4%) via increases in harvest index may result in further gains in yield. The very low tillering capacity (spikes/m^2; -40%) and

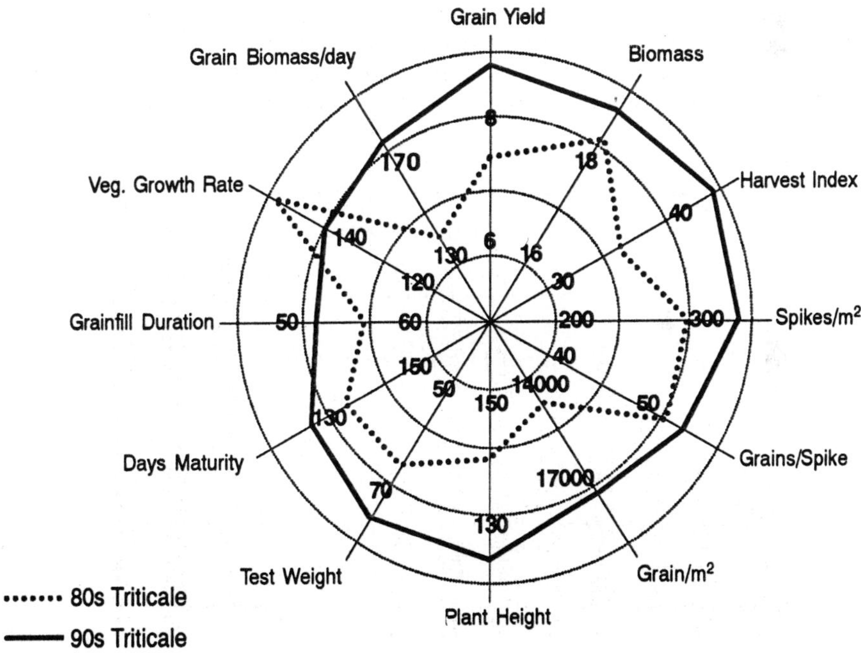

Figure 4. Comparison of agronomic components of the three highest yielding 80s and 90s triticales measured in 1991, 1992, and 1993 at Cd. Obregon.

Data source: Ken Sayre, CIMMYT Agronomy Section.

576

associated number of grains/m^2 (-16%) of triticale compared with wheat require special
attention in future crop improvement work. Past experience indicates that, in years with a
high number spikes/m^2, wheat yields surpass Tcl yields. This deficiency cannot be
compensated by a 40% higher number of grains/spike and 21% higher 1000 grain weight
in Tcl. Tcl exceeds the average plant height of wheat (+39%) by 35 cm (90 cm).
Although differences in maturity (Tcl, +5%) and grain-fill duration (Tcl, +16%) between
Tcl and wheat have decreased, the long grain-fill duration of Tcl compared with wheat is
a major deficiency in Tcl and special attention in breeding efforts is warranted. Long
grain-fill duration provokes vulnerability in terminal stress and late or early frost
situations with drastic effects on grain yield.

Progress for grain yield and agronomic traits in future breeding builds on existing genetic
variabilities, trait heritabilities, and associations with other traits. These parameters
suggest that gains in genetic grain yield potential and a number of value-added traits can
be maintained in the future (Table 1), but breeding for such traits should be guided by
monitoring population parameters to estimate projected genetic gains and identify traits
for which low genetic variability requires the development of special trait populations
[7,8]. Average, minimum and maximum values for the top 5% in grain yield when
compared with the base population (Table 1) indicate associations among traits; grain
yield, 1000 grain weight, plant height, and maturity days are positively correlated.
Average correlation coefficients in different years vary from 0.4 to 0.5 (P < 0.01) with

**Table 1. Mean, minimum (Min), maximum (Max), and phenotypic
standard deviation (Psd) for agronomic components and
industrial quality parameters measured on 1500 triticale
advanced lines (Total) and the 5% highest yielding
triticales (Top 5%) at Cd. Obregon during Y92-93 growing
cycle.**

	Total n=1500				Top 5% n=150			
	Mean	Min	Max	Psd	Mean	Min	Max	Psd
Grain yield (t/ha)	6.9	5.0	8.4	0.5	8.0	7.8	8.4	0.2
Test weight (kg/hl)	76	70	81	1.7	77	72	80	1.6
1000 g weight (g)	45	32	62	4.8	48	39	58	4.6
Plant height (cm)	119	75	145	6.8	123	110	145	7.3
Days heading (days)	80	68	99	4.8	82	75	96	4.4
Days anthesis (days)	87	75	106	4.4	89	82	103	3.9
Days maturity (days)	133	119	146	4.9	135	128	144	3.5
Grain-fill (days)	46	36	58	2.9	46	41	52	2.8
Protein % (%)	12	9	16	1.0	12	10	15	1.1
Grain hardness	46	36	65	4.1	46	39	55	3.8
SDS sediment. (ml)	5.4	3	8	0.9	5.3	4	7	0.8

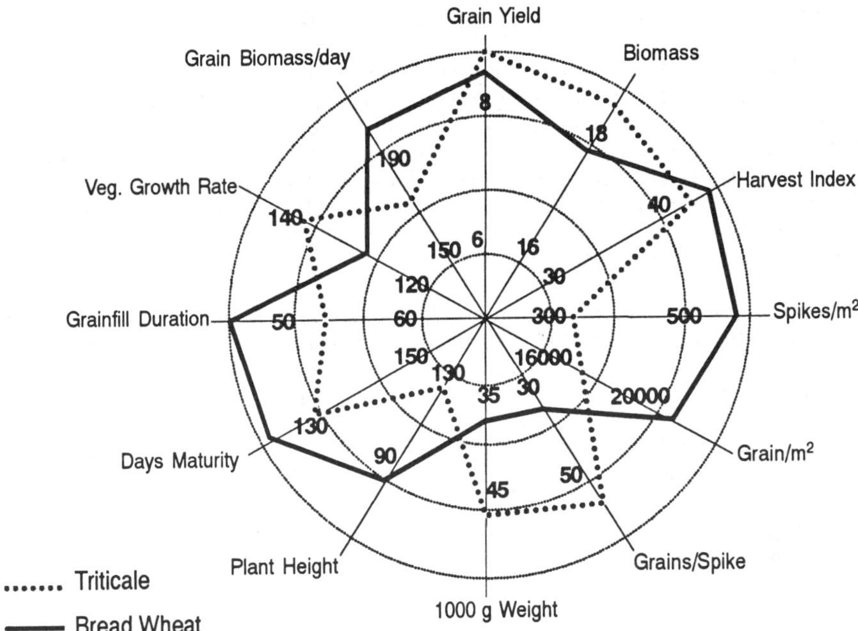

Figure 5. Comparison of agronomic components of modern bread wheats and triticales measured in 1991, 1992, and 1993 at Cd. Obregon.

Data source: Ken Sayre, CIMMYT Agronomy Section.

closer associations in early maturing Tcls, particularly for correlations among grain yield, plant height, and 1000 grain weight. Since, for example, plant height and 1000 grain weight are positively correlated across and within maturity groups, further increases in spikes/m^2 via reduced plant height have to be balanced against a probable decline in 1000 grain weight. For example, grain-fill duration, trait heritabilities in repeatable, and particularly stress-prone environments are low, and genetic variability is moderate (Table 1). Hence, parent building via the development of a special trait population are employed to enhance current incremental steps of progress. Population parameters for a large number of disease scores are published in recent International Triticale Screening Nursery (ITSN) reports [2,4,5] and indicate good variability for resistance to wide spectrum of rust and foliar pathogens to ensure future progress. High rates of progress in germplasm developed from crosses involving interspecific Tcl x wheat, spring x winter Tcl, and 2D(2R) x complete R type suggest that these gene pools are promising sources with which to achieve future genetic gains.

Utilization Options

The shift in breeding emphasis from 2D(2R) to complete R genome, particularly 6D(6A) types in the 1980s was accompanied by improved test weights, but negative effects on baking quality [9]. Research with wheat/Tcl grain and flour blends on milling and baking properties suggested blends for commercial use. Specialized requirements and markets for products for human consumption, high protein or high energy feed grain, and growing interest in forage and dual purpose triticales were prompted in 1990 by end-use oriented and expanded breeding objectives (Table 2). Further, projected genetic gains for quality traits based on existing genetic variability and heritability estimates, and phenotypic correlations among these traits, suggested that breeding should be targeted to human and animal consumption. Results indicate that, in general, genetic variability rather than heritabilities are limiting genetic progress and suggest that the development of special trait populations for such traits (e.g., protein concentration) is warranted [7,8,9]. Tcls with trait combinations required for specialist markets, such as malting, can be derived from the classes given in Table 2. By 1993, the whole range of products for the different groups given in the utilization scheme was available.

Table 2. End-use oriented structure in triticale breeding.

Human consumption	Animal consumption

· Baking quality
· Cookie quality
· Semolina quality
	· Feed grain
	   Energy
	   Protein
	· Forage/grain dual purpose
	· Forage (grain/cut forage)

Current Tcl production is nearly exclusively used as animal feed grain. Hence, the advantages of Tcl grain in animal nutrition, particularly in the monogastrics and poultry sector, are well documented and will not be addressed here. The possibilities to tailor Tcl regarding specific nutritional requirements for feed grain have not been exploited in crop enhancement.

Current quality improvement hinges on introgressing high-molecular-weight glutenin subunits (HMW) particularly allelic variants at loci Glu-D1 and Glu-B1. These subunits are closely associated with industrial quality parameters, and can easily be identified with SDS-PAGE. Evaluation of the baking properties of Tcl, including octoploid primary Tcls, revealed that the relative importance of the HMW loci Glu-A1, Glu-B1, and Glu-D1 approximated a ratio of 20:30:50, the effects of allelic variants were the same as in BW, and a strong negative effects of 6D(6A) on baking quality [9]. Consequently, favorable HMW loci on Glu-B1 such as 7+8 and 17+18 were transferred from BW to TCL and are combined in targeted crosses. However, results revealed that exploiting the existing variability for baking quality will not solve inherent limitations in industrial quality since Tcls not only lack the Glu-D1 but carry the major endosperm proteins Sec-1, Sec-2, and Sec-3 on the rye genome with negative effects on baking quality.

A new era in Tcl quality came in 1991 with the availability of Glu-D1 in 1D(1A) and 1RS.1DL in the Tcl Rhino substitution series developed by A.J. Lukaszewski at the University of California at Riverside (Figure 6). Drastic increases in SDS sedimentation values in a poor quality genetic background (Rhino) and a good quality genetic background (Passi) suggest a future breakthrough may be forthcoming. Since 1992, the 1D(1A)--which carries band 5+10--1D(1B), and 1D(1R) substitutions are available. Presently, Tcls with two, three, and four doses of Glu-D1 are being developed. Once Glu-D1 has been exploited, future research should be extended to gliadins and secalins.

FORAGE AND FORAGE/GRAIN DUAL PURPOSE TCLS

The development of forage and forage/grain dual purpose Tcls brings a new area of Tcl enhancement to CIMMYT. This demand-driven effort will complement crop and livestock enterprises in developing countries. The requirements, in terms of growth habit and trait combinations, are highly specific to the target environment and management. Resources are being allocated to develop a range of germplasm products using the STcl, FTcl, and WTcl gene pools, which are suitable for dual purpose and multiple forage situations. Evaluation of forage nutritional parameters and testing under livestock grazing pressure is being conducted by collaborating National Agricultural Research Centers and capitalizes on expertise and facilities outside of CIMMYT. The germplasm we are distributing includes next to advanced lines germplasm products such as F2s and F4s, which allow different levels of fine-tuning by national program colleagues under relevant situations.

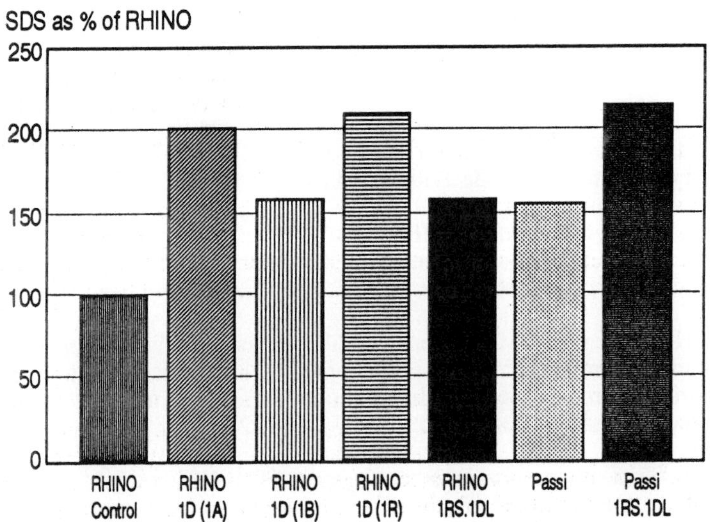

Figure 6. Effect of D-Genome High-Molecular-Weight glutenin subunits on SDS-Sedimentation.

A. Lukaszewski Rhino substitution series HMW subunit 2 + 12 on 1D, 5 + 10 on 1RS.DL

Early forage research in 1990 concentrated on determining the forage potential of STcl/FTcl compared with oats and barley as alternate forage commodities in one-cut situations. Results suggested that existing STcls were suitable for early forage, hay, and whole-crop silage. From 1991 onwards, a representative sample of the late STcl, FTcl, and WTcl gene pool was evaluated for forage potential in multiple forage harvests and grain-recovery potential [9]. Guided by results and implications, a new international nursery (FWTcl) was assembled including advanced and segregating FTcl and WTcl grain and forage materials and distributed to 75 cooperators. Parallel to the dissemination effort, traits such as reduced awns from different sources were identified, evaluated, and incorporated in the target gene pool.

References

1. Fox PN, Skovmand B, Thompson BK, Braun HJ, Cormier R. Yield and adaptation of hexaploid spring triticale. Euphytica 1990;47:57-64.

2. Pfeiffer WH, Trethowan RM, Fox PN, Amaya A, Peña J, Cardenas F, Magana RI. The nineteenth international triticale yield nursery (ITYN). Results of the 1987-88 triticale nurseries. CIMMYT, Mexico, D.F., 1990.

3. Pfeiffer WH, Fox PN. Adaptation of triticale. In Proceedings, 2nd International Triticale Symposium, Passo Fundo, Rio Grande do Sul, Brazil, 1-5 Oct. 1990. 1991: 54-59.

4. Pfeiffer WH, Alcalá M, Crossa J, Vega EH, Magaña RI. The Twentieth International Triticale Yield Nursery (ITYN). El 20o Ensayo Internacional de Rendimiento de Triticale. The Twentieth International Triticale Screening Nursery (ITSN) El 20o Vivero Internacional de Selección de Triticale. CIMMYT, Mexico, D.F., 1991.

5. Pfeiffer WH, Fox PN, Corbett J, Rodriguez MT, Magaña RI. The twenty-first and twenty-second International Triticale Yield Nurseries (ITYN). El 21o y 22o Ensayos Internacionales de Rendimiento de Triticale. (ITYN). CIMMYT, Mexico, D.F., 1993.

6. Ortiz-Monasterio I, Sayre KD, Pfeiffer WH. Differences in nitrogen recovery among CIMMYT's bread wheats and complete and 2D(2R) substituted triticales. Triticale Topics 1993;11:6-9.

7. Pfeiffer WH, Trethowan RM, Immonen AST, Amaya A, Peña RJ. Population parameters and their implications for applied breeding and projection of expected genetic advance in triticale. Proceedings, 2nd International Triticale Symposium, Passo Fundo, Rio Grande do Sul, Brazil, 1 Oct.-5 Oct. 1990. 1991: 121-124.

8. Pfeiffer WH. Triticale improvement strategies at CIMMYT: existing genetic variability and its implication to projected genetic advance. In Proceedings, 5th Portuguese Triticale Conference. Elvas, Portugal, 22 May-24 May, 1990 1994: in press.

9. Pfeiffer WH. Triticale improvement strategies at CIMMYT: exploiting adaptive patterns and end-use orientation. In Tanner DG, Mwangi W, editors. Proceedings, 7th Regional Wheat Workshop for Eastern, Central and Southern Africa, Nakuru, Kenya, 1992: 73-85.

SEMIDWARF WINTER TRITICALE

Tadeusz Wolski, Jaroslaw Gryka
Danko Hodowla Roslin
Ltd. Wspolma, Warsaw, Poland

Abstract

Data of breeding of semi-dwarf winter triticale with the use of the Hl rye dominant dwarfing gene, of the programme of Danko Ltd. are presented.

Two semi-dawrf winter triticale varieties have been presented now for registration in several European countries. They are around 20 cm shorter then conventional cultivars. The breeding and testing methods are presented and the future of semi-dwarf triticale is briefly discussed. Promising results of field trials from several locations are given, showing the progress in lodging resistance and grain yield. The competition of short triticale with wheat is outlined.

Introduction

The general trend to reduce inputs into agriculture is one of the reason of growing interest in triticale, which is regarded as a low cost crop. Still most triticale cultivars are tall and more or less susceptible to lodging. The use of growth regulators increases the costs of produtcion and in some countries is forbidden. This reduces the changes of varieties of the type of Lasko and Alamo [Presto].

Judging from the example of wheat, it seems that the productivity of short triticale plants should be transmitted from the stem to the ear and the grain yield could be increased.

In the DANKO [former Poznan Plant Breeders] programme the Hl rye dominant dwarfing gene has been used. The cultivar Prego [CHD 888] introduced into Poland, Germany, Swedenand Denmark inherited some of the advantages of Lasko, combined with much improved lodging resistance and winterhardiness.

The present paper deals with the second phase of this programme concerning the breeding of semidwarf winter triticale. Previous results were presented by the senior author at the Second International Triticale Symposium [1]. Two semidwarf winter triticale varieties have been presented now for registration in several Euroepan countries. They are about 20 cm shorter then convential cultivars. The breeding and testing methods are presented and the future of semidwarf triticale is briefly discussed.

581

H. Guedes-Pinto et al. (eds.), Triticale: Today and Tomorrow, 581–587.

© 1996 *Kluwer Academic Publishers. Printed in the Netherlands.*

Table 1. Results of a 2 - years trial with semidwarf triticale strains at 6 locations in 1992-93

VARIETY	HEADING IN MAY [days]	HEIGHT [cm]	HEIGHT HOMOGENITY [9-1]	LODGING RESISTANCE [9-1]	DENSITY * [heads/1m2]	TKW [g]	TEST WEIGHT [kg/hl]	YIELD [dt/ha]
DEBO [DED194]	30	95	7.3	9.0	395	50.6	69.0	80.0
DALO [LAD794]	29	98	6.6	8.4	432	45.2	72.1	78.8
ALAMO - ST.	22	120	7.5	6.2	481	42.8	74.9	76.7
MALNO - ST.	24	120	6.4	5.8	397	42.3	73.3	72.5
AVERAGE ST.	23	120	7.0	6.0	439	42.6	74.1	74.6

* ONE YEAR RESULTS

9 POINT IS THE BEST

Promising results of field trials from several locations are given, showing the progress in lodging resistance and grain yield. The competition of short triticale with wheat is outlined. Some data on grain quality are given.

Material and Methods

The test material consisted of F9-F11 lines obtained from crosses of an octoploid triticale [Lanca wheat * L506/79 rye] with the hexaploid triticale line LAD 627/80, crossed with hexaploid - Dagro in the case of Debo [DED194] and Alamo sib in the case of Dalo [LAD794]. The semidwarf character of this material is based on Hl rye dwarfing gene [2].

Continous single plant or ear selection was practiced. From the F4 generation onwards progenies were tested in unreplicated microtrials and separately multiplied. They were selected for yield, lodging resistance and short straw. Semidwarf lines reached uniformity relatively late, as tall segregants appeared in several generations.

When uniform enough, strains were bulked and tested in trials at 2-3 locations in 4 replications on 5m2 plots. The separately planted multiplications of tested lines underwent rogueing.

The two best and most uniform semidwarf varieties were tested in trials in 4 replications on 10 m2 plots in 6 locations in Poland. They were compared with tall standards: Alamo [Presto] and Malno with border short and tall plots in order to eliminate neighbour effects. They were compared also at 4 locations with two wheat standards Emika and Almari, the latter being the widest grown winter wheat cultivar in the country. In this case the comparison was not direct, as wheats were grown in separate trials.

In 1992/93 the two semidwarf varieties and two more advanced lines were tested in a number of foreign trials. LAD 176/90 is a Dalo sib line and DED 52/90 is our only successful line derived from a triticale * triticale cross with wheat CIMMYT germplasm in the background: Lasko * URSS Tcl * 3310-Bgl S * Grado.

Although triticale grain is used in Poland mainly as animal feed, the possibility of producing triticale bread seems of interest. The semidwarf progenies tested in microtrials and selected for agronomic trials were tested for two quality characters: protein content and SDS.

Results and Discussion

Table 1 presents a comparison of Debo and Dalo semidwarf triticale with tall standards Alamo [Presto] and Malno based on mean resultats of a 2-years trial at 6 locations in Poland. Both short varieties are later heading, similar in height homogenity and density of stand, superior in lodging resistance, yield and 1000 kernel weight, inferior in test weight. Judging from new material earliness and test weight may be improved.

The same short and tall triticales are compared with two winter cultivars at 4 locations in a 2 years trial [Figure 1]. The semidwarfs showed a clear superiority in Dankow on sandy soil. Possibly the root system of semidwarfs with the rye Hl gene is superior to this of wheat derived semidwarfs.

In Choryn in two dry seasons Debo outyielded the tall triticales and yields of wheat were inferior to those of all triticales. In Laski on medium soil short triticales were superior to tall ones, but Debo only outyielded wheat. In Debina on very heavy soil wheat was better and short triticales - superior to tall ones.

Probably the relatively lower rate of nitrogen was here the cause of the inferiority of semidwarfs as compared to wheat [Table 2]. Results of some semidwarfs tested in 18 European trials are presented in Table 3. Debo and Dalo are superior to older varieties but two new semidwarfs DED 52/90 and LAD 176/90 are still better.

Figure 2 presents the protein content and SDS values of 32 semidwarf and 33 tall single plant progenies tested in a microtrial in Laski and preselected for agronomic value. They are compared to the short and tall standards. Semidwarfs seem inferior in protein content, but superior in SDS. The same difference is visible when comparing both standards. This speaks in favour of the use of semidwarf triticale grain for baking.

Conclusions

Presented results relate to the first step of breeding semidwarf winter varieties based on the Hl gene. A marked progress in lodging resistance and yield has been obtained and it may be presumed that with further recombination breeding an improvement in such characters as test weight, earliness and grain quality may be expected.

It seems that the presented advanced semidwarf material may be competitive to tall triticales, while the competition to wheat should be tested with adequate fertilization. Such testes are under way.

References

1. Wolski T. Impact of semidwarf rye germplasm on triticale improvement. Proceedings of the Second International Triticale Symposium; 1990 Oct. 1-5; Passo Fundo. Mexico DF, CIMMYT 1991, 144-149.

2. Wolski T., Tymieniecka F. Effect of the rye mutant EM1 dwarfing gene on octo- and hexaploid triticale. Semidwarf mutants and their use in cross-breeding III. 1985 Dec 16-20; Rome. Viena: IAEA, 1988, 247-249.

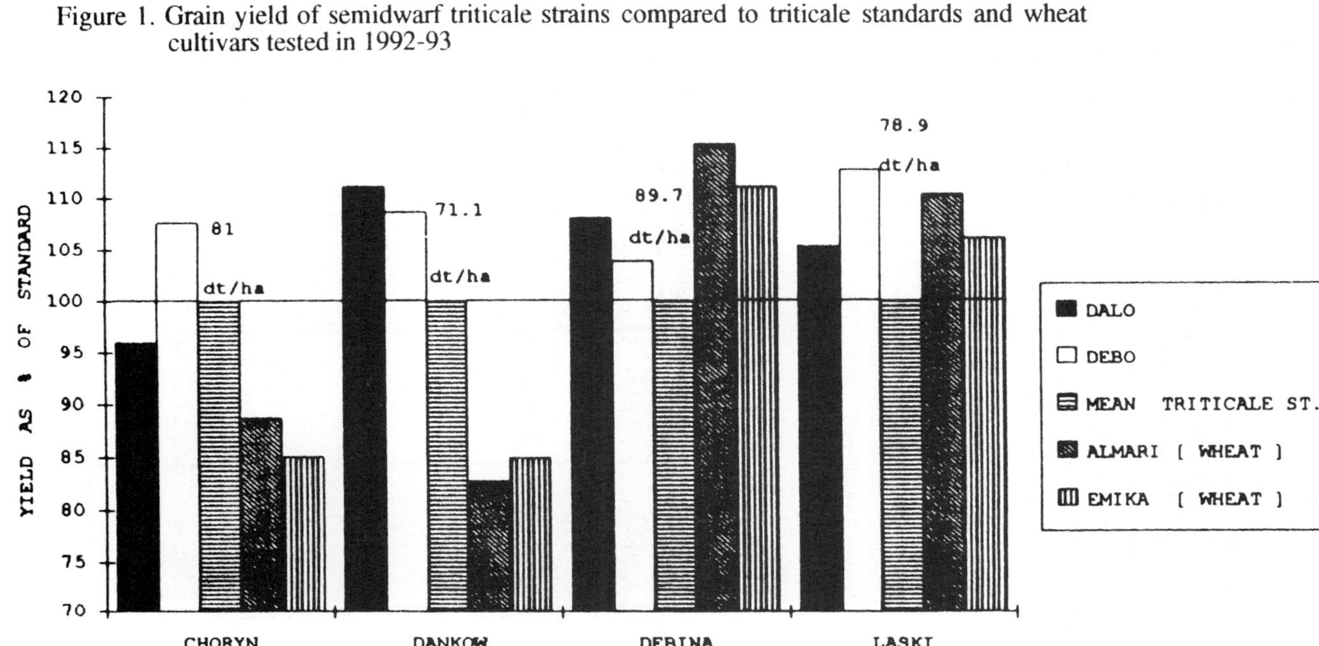

Figure 1. Grain yield of semidwarf triticale strains compared to triticale standards and wheat cultivars tested in 1992-93

585

Table 2. Nitrogen application in trials with wheat and
triticale, 1993

STATION	NITROGEN [kg / ha]		
	SEMIDWARF TRITICALE	TALL TRITICALE	WHEAT
CHORYN	90	80	100
DANKOW	97	97	101
DEBINA	97	97	122
LASKI	80	56	89

Table 3. Yields of advanced semidwarf triticale strains in some European
countries compared to standards in 1993

STRAIN	YIELD AS % OF STANDARDS				
	GERMANY	SWEDEN	BELGIUM	DENMARK	POLAND
DEBO [DED 194]	103	113		110	117
DALO [LAD 794]	111	111	110	108	112
DED 52/90	115			115	120
LAD176/90	113	117		115	119
AVERAGE STANDARD [dt/ha	78.45	74.52	85.6	87.61	76.46
STANDARDS :					
ALAMO [PRESTO]	96		102		101
LASKO	101		98		
PURDY	102				
LOCAL	92				
PREGO		100		103	
DAGRO				97	
MALNO					99
NUMBER OF TRIALS	5	3	2	2	6

Figure 2. The protein content and SDS value in single plant triticale progenies tested in microtrials in Laski in 1992

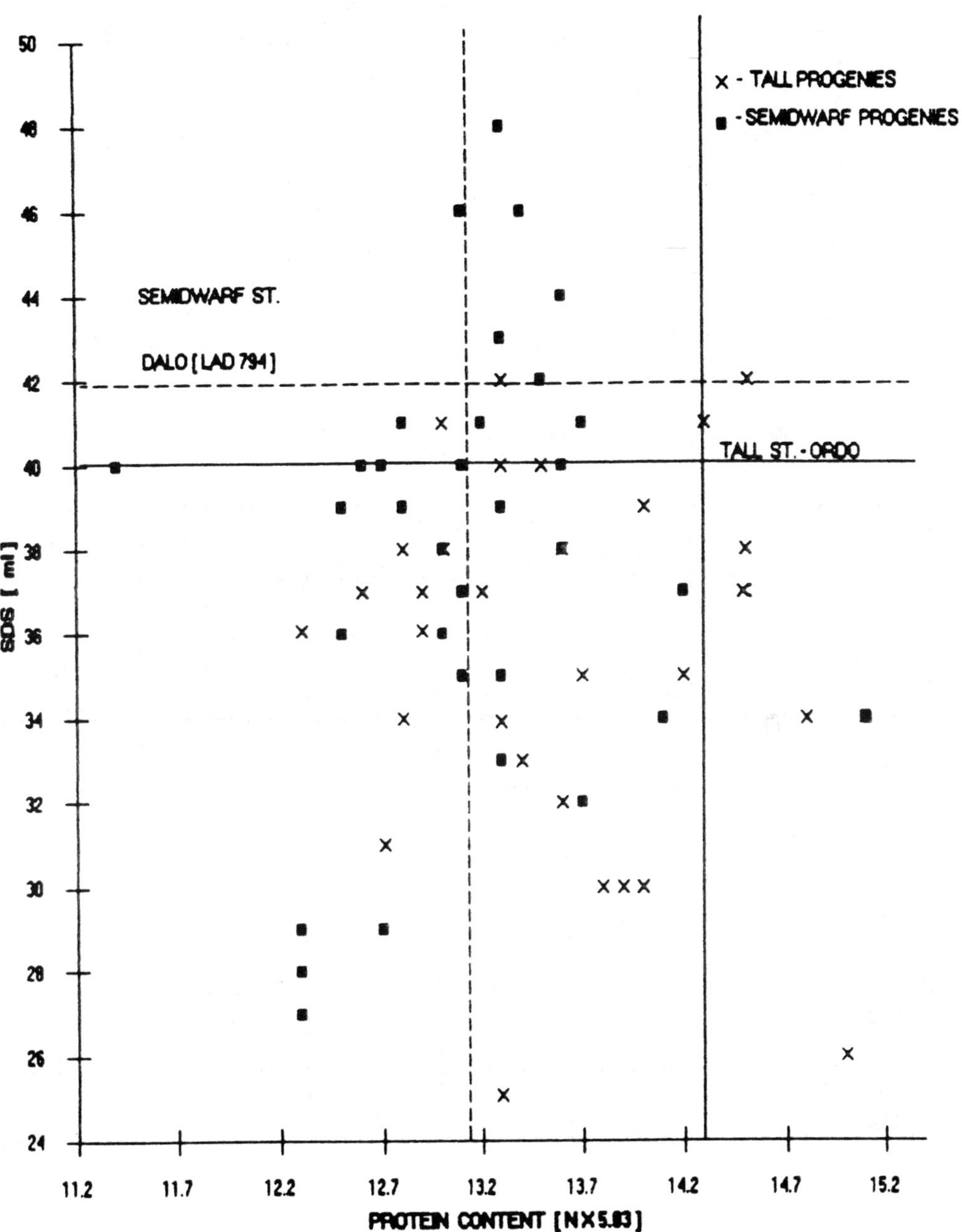

DEVELOPING SPRING AND WINTER TRITICALE WITH REDUCED-
-AWN EXPRESSION

Donald F. Salmon, James H. Helm and Patricia E. Jedel
Field Crop Development Centre, Alberta Agriculture, Food and Rural
Development, Bag #47, Lacombe, Canada, T0C 1S0

Abstract

Work on incorporating reduced-awn expression into triticale was initiated at the Field Crop
Development Centre in 1983 using spring triticale lines with low falling number and the
sprouting resistant spring wheat germplasm line RL4137. In both the spring and winter
triticale, selections from crosses with improved triticale parents produced since 1988 appear
to have potential in the production of future cultivars and as parental lines. To date
evaluation of advanced lines has been relatively limited to either biomass yield or tolerance
to sprouting conditions. However, few of the advanced lines appear to demonstrate improved
levels of sprouting resistance expressed in the form of falling number or actual dormancy
testing. Future work will concentrate on a detailed evaluation of selections with the reduced-
-awn characteristic for sprouting resistance as well as biomass yield and quality.

Introduction

New varieties of spring triticale (X *Triticosecale* Wittmack L.) such as Wapiti, Frank, Banjo, and
AC Copia have demonstrated superior biomass yield and similar feeding quality to spring barley
[1, 2] under a range of environmental conditions in western Canada. In a similar fashion, the
winter triticale varities Wintri and Pika provide a very high quality long-season grazing crop [3]
when spring seeded alone or in mixtures with spring barley (*Hordeum vulgare* L.) as well as an
early silage/conserved fodder crop when conventionally planted in the fall. In Alberta, nearly
250,000 ha of annual forage is harvested as a whole plant fodder. Approximately 50-60% of
the crop is harvested during the early dough stage as a dried and baled hay crop (greenfeed).

The rough awns in triticale and some barley varieties could pose a problem for feeding as a
dried fodder. Consequently, the program at the Field Crop Development Centre initiated an
effort to develop triticale cultivars with a reduced level of awn expression. The sprouting
resistant wheat germplasm line RL4137 was utilized as a potential source for incorporating both
the reduced-awn characteristic and sprouting resistance into spring and winter triticale.

589

H. Guedes-Pinto et al. (eds.), Triticale: Today and Tomorrow, 589–592.
© 1996 *Kluwer Academic Publishers. Printed in the Netherlands.*

Methodology

SPRING TRITICALE

In 1983 the spring triticale 7631-ED4B (high falling number) line was used as the female parent in a cross to the sprouting resistant wheat germplasm line RL4137. The F_1 was backcrossed with pollen for the triticale line 7431A-68E4 (high falling number).

The F_1BC was subsequently crossed with the triticale lines Panther "S" [4] and Otter "S" to create primary parental populations. Subsequent segregating generations were subjected to bulk modification and line selection based on the reduced-awn characteristic. Reduced-awn selections were first entered into the yield trial system in 1990 and pre-registration trials in 1993.

WINTER TRITICALE

In 1988 spring triticale lines with the reduced-awn characteristic were used in crosses to the cold tolerant winter triticale composite 87DE01. The F_1 was grown under artificial conditions to produce the F_2. The F_2 and subsequent generations were subjected to natural selection for cold tolerance as well as visual selection for the reduced-awn characteristic grown under field conditions. In 1993, ten F_6 lines were identified with sufficient cold tolerance to be used in the crossing program and advanced to the yield trial system in 1994 . Lines in the 1994 hardiness and yield trials are also being evaluated for potential as a spring seeded forage crop in combination with barley as well as being evaluated as a fall planted early silage/conserved fodder crop.

Results

SPRING TRITICALE

Selections from the early series of parental populations (7631-ED4B/RL4137//7431A-68E4/3/Otter "S" .. 85L012 and 7631-ED4B/RL4137/7431A-68E4/3/Panther "s"..85L013) have not performed well in comparison to conventional triticales in the program (Table 1). The main problems have been small seed size, poor seed development, weak straw and only fair seed and biomass yields. Few of these lines have demonstrated any improvement in sprouting resistance either measured by FN or actual sprouting trials.

Crosses made after the original parental populations had been subjected to several generations of modification range from near-awnless to completely awned and are agronomically similar in other characteristics to the newer conventional triticale varieties (Table 2). The cross ID 88L012 (7631-ED4B/RL4137// 7431A-68E4/3/Otter "S" /4/Wapiti/Panda R 203 Sel) has demonstrated high yield potential, good seed type and acceptable agronomic type.

Table 1. Performance of reduced-awn types from the original parent populations in comparison to the check cultivars Wapiti, Frank, Banjo, and AC Copia plus two conventional lines in 1993

Variety/ Line	Yield[*] %	Biomass[*] %	Twt kg hl^{-1}	1000k g	FN sec
Wapiti	100	100	65	44	66
Frank	103	102	66	40	63
Banjo	94	98	66	44	75
AC Copia	108	-	69	46	66
84L012012[t]	102	98	67	44	63
85L012002	91	90	65	39	74
85L012012	96	95	64	37	76
85L013016	84	95	62	37	66
85L013021	88	90	66	37	78
86L024001[t]	95	95	62	39	**149**

* Yield expressed as a % of Wapiti. [t] Conventional awned lines.

Table 2. The 1993 performance of new generation reduced-awn spring triticale lines compared to the check cultivars Wapiti, Frank, and Banjo

Variety/ Line	Yield[*] %	Mat days	Ht cm	Twt kg/hl^{-1}	1000k gm
Wapiti	100	126	118	66	47
Frank	100	125	114	68	41
Banjo	102	127	130	67	45
88L012003	102	125	116	66	44
88L012013	112	126	113	65	46
88L012014	109	125	118	65	43
88L012015	109	126	113	66	42
88L012018	95	126	108	66	42

* Yield represented as % of Wapiti.

WINTER TRITICALE

Work on the winter triticale populations has been limited to the selection of agronomically acceptable plant type combined with winter hardiness levels at least equal to the winter wheat cultivar Norstar (Table 3). The line 88DL01077 appears to have winter hardiness levels intermediate between Norstar and the winter triticale variety Pika.

In contrast to the spring triticale, severe winter conditions in the field places extreme selection pressure on populations. In general, selection of types with acceptable seed quality has not been as difficult. This may be due to the use of more advanced germplasm in the initial winter triticale cross than in the spring program initiated in 1983.

Preliminary evaluations of spring seeded lines in a binary combination with barley for silage and fall grazing has indicated that the lines such as 88DL01077 perform similar to the cultivars Wintri and Pika for regrowth (data not presented).

592

Table 3. Percentage winter survival potential of early generation reduced-awn winter triticale lines in 1993.

Variety/ Line	Black soil[*]	Brown soil
Norstar (wheat)	83	82
Musketeer (rye)	100	100
Pika	88	92
88DL01010	79	72
88DL01031	68	75
88DLO1036	76	68
88DL01037	75	75
88DL01076	74	68
88DL01077	88	80
88DL01078	74	81

* The black soil zone is a high snowfall area with good ground insulation during the winter whereas the snow cover is more limited in the brown soil zone.

Acknowledgements

The authors wish to acknowledge the work of T.R. Duggan the senior breeding technician in the triticale program as well as other technical staff at the Field Crop Development Centre.

Literature

Zobell, D.R., L.A. Goonewardense, and D. F. Engstrom. 1992. Use of triticale silage in diets for growing steers. Can. J. Anim. Sci. 72:181-184.

Korasani, G.R., E.K. Okine, J.J. Kennelly, and J.H. Helm. 1993. Effects of whole crop cereal grain silage substituted for alfalfa silage on performance of lactating dairy cows. J. Dairy Sci. 76:3536-3546.

Baron, V.S., H.G. Nadja, D.F. Salmon, and A.C. Dick. 1993. Cropping systems for spring and winter cereals under simulated pasture: Yield and yield distribution. Can. J. Plant Sci. 73:703-712.

Salmon, D.F., and J.H. Helm. 1985. Preharvest and postharvest dormancy in spring triticale. Agron. J. 77:649-652.

OUTCROSSING IN WINTER TRITICALE, MEASURED BY OCCURENCE OF TALL PLANTS

Wladyslaw Sowa and Halina Krysiak
Plant Breeding and Acclimatization Institute, Radzików, Poland

Abstract

Studies were undertaken to determine outocrossing rate for 11 varieties and 4 strains of Polish winter triticale and French variety Clercal. Experiment were carried out in 1988-1991. The outcrossing rate was defined as percent of hybrid plants in populations of about 4000 individuals. Long stemmed, 32-40 cm taller than tested varieties, homozygotic triticale line RAH-108/84 was used as outcrossing test. Long stem of tester line is controlled by a single dominant gene. In order to mach flowering period of tester line, tester was spaciously grown. Results of the experiments revealed differences between tester genotypes; the outcrossing rate ranged from 1.1% to 10.5% depending on variety and weather conditions. Less likely to outocross varieties did reacted to the weather conditions (not significant differences between the years). Only high outcrossing rate of Almo and Lasko varieties was increased by warm and dry weather. Outcrossing rate of the tested varieties was strongly reduced by space isolation, 3 m wide space isolation reduced outcrossing rate by nearly 4 times. We suggest applying space isolation starting from early breeding stages of triticale, when material is multiplicated in small plots, and using only inside plot plants for seed multiplication. This method would reduce time needed for achieving uniformity of triticale strains.

Introduction

In Poland today triticale takes close to 1/3 of it's world acreage. Up to 1993, 16 Polish varieties of winter triticale as well as 3 varieties of spring triticale have been registered. This paper reflects on new problem with methodology of reproduction of seed material. Occurance of phenotypically different plants on the seed plantations may be a reason for disqualification of the plantation. Outcrossing is one of the biological causes of the occurance of the nontypical individuals. This phenomenon was frequently observed in breeding of half-dwarf wheats. The tendency to outcrossing in wheat is well documented (Strebejko, 1976, Griffin 1987, Leighty and Taylor 1927, Martin 1990). However, very litle information in the literature are available about outcrossing in varieties of triticale. Generally it is accepted that the tendency to outcrossing in triticale is higher than in wheat, because of rye contribution to the triticale genome. (Rye is a typical crosspollinating specie).

The studies of Rossi (1985) showed high outcrossing in triticale, on the other hand Mogileva (1988) report that outpollination in triticale is rather rare. The knowledge of crosspollination rates will result in improvements on the methodology of seed production, specially in determining of necessary spatial isolation. Outpollination can

H. Guedes-Pinto et al. (eds.), Triticale: Today and Tomorrow, 593–596.
© 1996 *Kluwer Academic Publishers. Printed in the Netherlands.*

594

also be a serious problem in breeding, specially in the early generations, where different lines and strains are grown close to each other.

Our experiments were conducted to determine the tendency to outcrossing of polish varieties of winter triticale.

Materials and Methods

The trials were conducted on experimental fields in Plant Breeding and Acclimatization Institute, Radzików, Poland in 1988-1991. The subject of study were: 12 polish varieties of winter triticale, 3 strains and the french variety Clercal. Material were sown in the optimal dates (24.09.1988; 25.09.89;3.10.90) with the Øyjord sowing drill; 400 seeds/m^2 were sown on plots size of 1m x 10 m. Plots were layed with their longer axis on the east-west line. The outpollinator strain (RAH-101-8/84) was sown with half density in the distance of 45 cm from the short axis of the plots of tested varieties.

Strain RAH-101-8/84 is characterized by long (170 cm) straw, early and long flowering period which was extended by strong tillering (the tillers were flowering later than the main shoots). Straw lenght of tested varieties was 138 cm in Lasko, 134 cm in Presto, 132 cm in Malno, and about 130 in all other varieties. All F_1 hybrids were tall like the pollinator strain, 30-35 cm taller than the tested variety.

The feature of long straw in pollinator strain (RAH-101-8/84) is confered by a single dominant gene.

Determination of the crosspollination rates was performed by counting of the tall plants in the plots of next generation of individually harvested heads. 80 heads were harvested from each plots at two distances from tester: 45 cm and inside plot - 3 m from the pollinator. The spikes were thrashed and sown in plots of 10 m^2 (4000 kernels per plot). The experiment were carried out in two cycles 1988-1990 and 1989-1991. The results were calculated as the percent of hybrid plants.

Results and Discussion

The results of the experiments are presented in table 1. The frequency of outcrossing was decreasing with the distance to the pollinator-strain. The number of hybrid plants in the offspring of the spikes collected at 45 cm from the tester was many times higher than in the offspring of spikes collected at 3 m from the tester.

The weather conditions had a strong impact on the outpollination rates, specially in varieties showing higher tendency to outcrossing. Warm and dry weather of 1990 resulted not only in early flowering but also greatly increased outpollination rates of Lasko, Almo and Bolero, when compared to the results of summer of 1989.

Varieties which are less likely to outpollinate (Malno, Grado, Dagro) did not show significant differences in outpollination rates between the two years.

The tendency to crosspollination varies among tested varieties of triticale. It is quite probable that the higher level of outpollination in some varieties results from a higher male sterility in flowers. Outpollination may also occur in male fertile flowers when stigma are already mature enough to accept pollen and the chaff is open but anthers have not released pollen yet. This unique stage in flower development lasts for 22 minutes according to Mogileva (1985) and according to Kociuba (1974) for 15 to 60 minutes. Unfolded chaff let the allien pollen grains to fall on the stigma.

Table 1. The outcrossing rate of winter triticale cultivars, depending on the distance of the pollinator tester and years.

Cultivars	Outcrossing in %			
	Distance			
	45 cm		3 m	
	1990	1991	1990	1991
Malno	1.60	1.80*	0.50	0.62
Ugo	1.70	4.82	0.35	0.59
Almo	3.75*	10.06*	0.45	4.21*
Lasko	2.75*	10.47*	0.45	4.76*
Grado	2.15	2.44	0.45	0.61
Dagro	1.85	2.56	0.35	2.00
Bolero	2.15	5.71	0.40	-
Largo	1.70	-	0.50	0.53
Presto	1.10*	4.00	0.20*	1.35
Moniko	2.45	3.65	0.60	0.05
Purdy	-	4.06	-	-
LAD-187	1.10	-	0.25	-
LAD-285	2.00	-	0.30	-
Tewo	1.70	-	0.85	-
CHD-1024	1.55	-	0.55	-
Clercal	2.11	-	0.65	-
Average	1.95	4.96	0.47	1.55
LSD (P≤0,05)	0.57	2.87	0.16	1.56

In wheat, anthers burst during the opening of chaff or even earlier. According to Kociuba (1974) in most of triticale forms, anthers burst 3 to 30 minutes after emerging from the flowers. However, large varietal differences in duration of the open flower stage were observed. The degree of chasmogamy depends on the duration of this stage. Leighty and Taylor (1927) reported that degree of outpollination in wheat may reach 34% depending on the variety and the weather conditions. Experiments of Martin (1990) with dwarf varieties revealed outpollination rates of 0.3% to 5.6%.

In our experiment the outpollination rates in triticale ranged from 1.1% to 10.5%. Mogileva (1988) argues that triticale is fully self pollinating plant and that possibility of outpollination is very low.

The results our data similar of described those of Rossi (1985), Tarkowski and Wolski (1985), and confirm the occurence of partial chazmogamy in winter varieties of triticale.

Even the less likely to outpollinate varieties - Malno, Dagro, Grado reached high outpollination rates of 1.6% to 2.5%.

This still high rates show possible difficulties with breeding of triticale, specially in it's early stages. However, even short distance between plots can reduce outpollination. In our experiments 3 meter distance between plots decreased the outpollination by several times (Tab. 1). Old varieties (Lasko and Almo) showed high up to 10.5% tendency to outpollination. Presto and Bolero are outpollinating at the medium level of 4% and 5.7% respectively. The newest varieties showed a little oupollination and there was no significant differences between of them. Tarkowski and Wolski (1985) observed that outpollination in winter triticale may occur in up to 10% of the flowers, depending on the variety and weather conditions. Spring forms of triticale are even more likely to outpollinate. Tarkowski and Kociuba (1989) inform that even though forms of triticale express tendency to outpollination, most of the breeders consider it as a self pollinating plant.

In our experiment, the new strains were less likely to outpollinate than the old varieties. This may be a resulted of higher cytogenetical stability of the forms used for crossing in the new breeding programs. Strains showing high ununiformity are rejected from breeding programs in early stages of the breeding process.

300 meters isolation distance used in seed production requested in OECD countries seems to be too high for reproduction of the new varieties. The use of protection coat of the same variety around the plot, may greatly reduce the number of hybrid plants in the next generation. Results of our experiment show that even small isolation distance can reduce outpollination rates by many times and therefore using isolation in early steps of breeding procedure could save time for achieving uniform variety.

References

1. Griffin W.B. 1987. Outcrossing in New Zealand wheats measured by occurence of purple grain N.Z.J. Agric. Res. 30; 287-290.
2. Kociuba W. 1974. Biologia kwitnienia Triticale. Prace grupy problemowej ds. Triticale. IHAR-Radzików, 102-107.
3. Leighty C.E. and J.W. Taylor 1927. Studies in natural hybridization of wheat. J. Am. Soc. Agron. 10; 865-887.
4. Martin T. J. 1990. Outcrossing in twelve hard red winter wheat cultivars. Crop Science. V.30, nr 1; 59-62.
5. Mogileva V.I. 1988. Biology of triticale flowering. EUCARPIA - Triticale. Tag-Ber. Akad., Landwirtsch. - Wiss. DDR, Berlin 266; 237-245.
6. Rossi L., C. Mosconi. 1985. Achievements and problems of triticale in the Italian Agriculture. Genetics and Breeding of Triticale. EUCARPIA meeting, Clermont-Ferrand. 2-5 July 1984; 473-485.
7. Strebeyko P. (red.) 1976. Biologia pszenicy, PKWN.
8. Tarkowski C., Kociuba W. 1989. Biologia kwitnienia. Biologia pszenzyta. PWN; 99-110.
9. Tarkowski C., Wolski T. 1985. Podstawy hodowli, odmianoznawstwa i kwalifikacji polowej roslin uprawnych. Zeszyt 2 Zboza. IHAR; 155-164.

ESTIMATES OF COMBINING ABILITY FOR SOME TRAITS IN TRITICALE

Stanislaw Wegrzyn & Helena Grzesik

Plant Breeding and Acclimatization Institute, Department of Cereals Kraków, Poland

(in collaboration with Dr. W. Mackowiak, Dr. E. Witkowski and M. Kawecka,

M. Sc of Plant Breeding Stations at Malyszyn, Smolice and Ozansk)

Abstract

Combining ability analysis of some traits of winter triticale in crosses according to topcross system of four maternal and five paternal forms over generation F_1 - F_3 is reported. The F_1 generation was tested in one locality while the F_2 and F_3 generations were tested in three localities. Five traits were analyzed: plant height, spike length, grain number and yield per spike, as well as 1000 grain weight. Based on the results of statistical analysis of the parents, F_1, F_2 and F_3 generations, a highly significant general combining ability was found for all traits under study except 1000 grain weight in the F_2 as well as grain yield per ear in the F_3 generation.

Specific combining ability in F_1 generation was significant for plant height, spike length, grain number per spike and grain yield per spike. In the F_2 generation it was significant only for weight of 1000 grains, while in F_3 generation specific combining ability was non significant.

Introduction

It is commonly know that certain varieties combine well producing superior offspring whereas others varieties produce poor progeny. It is possible that one reason for this is that combining ability often depends upon complex interaction systems among genes. In our country studies on estimating the combining ability of triticale have been undertaken only recently by Wegrzyn and Grzesik (1992), Wegrzyn *et al.* (in press).

The aim of this study was to evaluate the combining abilities of some traits in several varieties and strains of winter triticale over the F_1-F_3 generations.

597

H. Guedes-Pinto et al. (eds.), Triticale: Today and Tomorrow, 597–601.
© 1996 *Kluwer Academic Publishers. Printed in the Netherlands.*

Materials and Methods

The experiments used two strains and seven varieties of winter triticale as well as the F_1, F_2 and F_3 generations. Crossing was performed according to a topcross system: four maternal forms (Clercal, Salvo, Dagro, MAH-384,) x five paternal forms (LAD-183, Bolero, Lasko, Malno, Newton).

Field studies including 20 crosses and 9 paternal forms were carried out for three years based on the method of randomized blocks. The F_1 generation was tested in Cracow in three replications while the F_2 and F_3 generations were tested at three localities (Malyszyn, Ozansk and Smolice) in four replications. Seed were sown in 6-row plots, the width of the plots being 1.5 m at 5x2.5 cm spacings.

After harvest, biometrical measurements were carried out of all collected plants except border plants. Analysis was carried out on five traits: plant height, spike length, grain number and yield per spike, as well as 1000 grain weight. The results were analysed using a computer progam which was developed in the Department of Cereals, at the Plant Breeding and Acclimatization Institute in Cracow.

In the analysis of variance (Table1), the sums of squares for general combining ability of both the maternal and paternal forms were combined; however, estimation of the effects of general combining ability was done separately for maternal and paternal forms (Table2).

Results and Discussion

Based on the analysis of variance (Table1) general combining ability of the studied varieties and strains of triticale was highly significant with respect to all tested traits in each generation except 1000 grain weight in the F_2 generation as well as grain yield per spike in the F_3 generation. Specific combining ability in F_1 generation was significant for plant height, spike length as well as number and yield of grains per spike. In the F_2 generation specific combining ability was significant only for 1000 grain weight. In the F_3 generation no trait under study showed a significant specific combining ability. As previously mentioned, the experiment with the F_1 was carried out at one locality. The crossing program was designed to provide information on behavior of the genotypes under soil-and-micro climatic conditions typical of a given environment where the experiment was conducted. Interaction of genotype and environment is of great significance in plant breeding, particularity with respect to such quantitative traits as yield, since these interactions will decrease the efficiency of breeding process expected with respect to selection. With this in mind, experiments with F_2 and F_3 generations as well as their parents involved three localities.

Interactions of general specific combining abilities x environment both F_2 and F_3 generations were statistically non significant. This means that environmental conditions had no significant effect on the results of gene action. Although there are few papers on the combining ability of the triticale, the results obtained from diallelic crosses by such authors as kaltsikes and Lee (1973), Reddy (1976), Gill *et al.* (1979), Rao and Joshi (1979), Carillo *et al.* (1983), Brar *et al.* (1985), Dhindsa *et al.* (1985) and Behl *et al.* (1985,1988) indicate significant general and specific abilities for almost all yield traits. GCA always higher than SCA.

Based on estimated effects of the general combining ability for maternal and paternal forms (Table 2) it can be noted that varieties, which display significant positive effects of GCA for specific traits, will increase the values of those traits in offsprinng. However, those exhibiting significant negative effects of GCA will decrease them in offspring. Varieties: Bolero, Clercal and Newton display significant positive effects of GCA for grain number per spike. These varieties will increase the value of this trait in the offspring.

Significant effects of specific combining ability which occurred in F_1 generation for such traits as plant heigth, spike lenght, grain number and yield per spike are indicative of considerable involvement of non-additive gene action. Thus selection of improved genotypes with respect to these traits in the early generations will be difficult. However, the significant positive effects of specific combining ability, particulary with respect to yield components, indicates a real possibility for utilizing such hybrids in heterotic breeding.

Heterosis for culm length and ear length in the hybrids was negligible; however, it increased considerably for grain yield per ear. Similar results were obtained by Spiss and Goral (1992) and Grzesik (1991). Among the hybrids studied above, significant positive effects of specific combining ability (table 3) for grain yield per spike were shown by two hybrids, Dagro x LAD-183 and Dagro x Bolero. For grain number per spike, significant positive effects of specific combining ability occurred in crosses Clercal x Bolero, Clercal x Malno, Salvo x LAD-183, Salvo x Newton, Dagro x Bolero and MAH-384 x Newton. These hybrids are thus recommended for heterosis breeding of triticale.

A significant specific combining ability was also observed for 1000 grain weight in the F_2 generation. Statistically significant effects of specific combining ability were found in three hybrids, Salvo x LAD-183 (2.82[x]), Dagro x LAD-183 (3.25[xx]) and MAH-384 x x Newton (2.50[x]).
The selection of improved genotypes with respect to 1000 grain weight in the F_2 generation of the above hybrids will be difficult. In the remaining hybrids these effects were non-significant.

Conclusions

- Based on the statistical analysis of the parents and of F_1, F_2 and F_3 generations of twenty triticale crosses, a highly significant combining ability was found for all traits under study except grain weight per spike in the F_3 generation.

- Specific combining ability in F_1 generation was significant for plant height, spike length, grain number per spike and grain yield per spike. In the F_2 generation it was significant only for 1000 grain weight while in the F_3, specific combining ability was non-significant.

- Varieties Bolero, Clercal and Newton showed highly significant effects of general combining ability for grain number per spike. These varieties will increase the value of this trait in their offspring.

References

1. Behl RK, Singh VP. Genetics of grain relation to total biological yield in Triticale. Wheat Information Service 1988; 67: 25-29.

2. Behl RK, Singh VP, Spalony L. Genetics of some grain characters in hexaploid triticale. Genetica Polonica 1985; 26(4): 479-485.
3. Brar GS, Sandha GS, Vira DS. Multienvironmental diallel analysis for combining ability in triticale. Crop Improvement 1985; 12/2: 106-110.

4. Carillo JM, Monteagudo A, Sanchez-Monge E. Inheritance of yield components and their relationship to plant height in hexaploid triticale. Zeitschrift fur Pflanzenzuchtung 1983; 90: 153-165.

5. Dhindsa GS, Sandha GS, Gill KS. Genetics of combining ability for yield and its components characters in triticale. Journal of Recherach. Punjab Agricultural University, 1983; 22/2: 199-205.

6. Gill KS, Bhardwaj HL, Dhindsa GS. Heterosis and combining ability in triticale. Cereal Research Communications, 1979; 7: 303-309.

7. Grzesik H. The results of studies carried out on dwarf mutants of winter hexaploid triticale. Cereal Research Communications, 1991; 19: 1-2, 91-99.

8. Kaltsikes PJ, Lee J. The mode of inheritance in hexaploid triticale. Zeitschrift fur Pflanzenzuchtung, 1973; 69: 135-141.

9. Rao VR, Joshi MG. A study of inheritance of yield components in hexaploid triticale. Zeitschrift fur Pflanzenzuchtung, 1979; 82: 230-236.

10. Reddy LV. Combining ability analysis of some quantitative characters in hexaploid triticale. Theoretical and Applied Genetics, 1976; 47: 227-230.

11. Spiss L, Góral H. Efekt heterozji u pszenzyta. Zeszyty Problemowe IHAR. Hodowla Zbóz. Radzików, 1992; 41-48.

12. Wegrzyn S, Gzresik H. General and specific combining of F_1 generation hybrids in some varieties and strains of hexaploid winter triticale. Annual Wheat Newsletter, 1992; 38: 164-165.

13. Wegrzyn S, Spiss L, Góral H. Oszacowanie zdolnosci kombinacyjnej kilku cech pszenzyta ozimego. Zeszyty Naukowe Akademii Rolniczej w Szczecinie (in press).

EFFECT OF LOW NITROGEN INPUT ON AGRONOMIC TRAITS IN TRITICALE

Gitta Oettler
Universität Hohenheim, Landessaatzuchtanstalt,
Stuttgart, Germany

Abstract

The development of genotypes for areas with restricted nitrogen (N) application requires information on genetic variation for agronomic traits in low-productivity environments. Thirty-six winter triticales (x *Triticosecale* Wittmack) were grown under two levels of N to (1) investigate the effect of N on important agronomic traits and (2) estimate genetic parameters. The genotypes were tested at two sites with zero and a site-specific N treatment in two-replicate balanced lattices with 5m^2 plots. A severe stress condition at zero N supply was revealed by an average reduction of grain yield of almost 40%. This was primarily caused by a nearly 40% lower tiller density, whereas the number of grains per spike was reduced by 17% only. Heading time did not respond to N. Thousand-grain and test weight increased slightly under low N. In general, genotypes, N and their interaction were significant sources of variation. Genotype x N interaction variances, however, were mainly small. Heritabilities for both N regimes were moderate to high. Correlations between the N rates for all traits were moderate to high and significant. It was concluded, therefore, that selection in a high N environment will result in a correlated response for most traits under low input conditions also. Nevertheless, the incorporation of a low-productivity environment in the selection process might be advisable.

Introduction

Agricultural production in Central Europe has increased during the past decades as a result of plant breeding and intensification of cultural practices with the excessive use of agrochemicals. Overproduction and environmental problems, such as contamination of the underground water by chemical residues, jeopardize modern agricultural systems. With a view to a more environment-friendly crop production, some European countries have imposed various restrictions on management practices. Amongst these limitations, the application of reduced N inputs has high priority.

H. Guedes-Pinto et al. (eds.), Triticale: Today and Tomorrow, 603–608.

© 1996 *Kluwer Academic Publishers. Printed in the Netherlands.*

Testing and selection in plant breeding programmes have mainly been performed in high-productivity environments. In future, the improvement and development of crops with high N efficiency will increasingly become of importance. Efficient genotypes are those that have the ability to give high performance in many important traits under less than optimum N (or other nutrient) supply [1]. A prerequisite of breeding for low N supply is the existence of genetic variation in N efficiency.

In contrast to wheat [2,3], rye [4], and oats [5], little information on the effect of N stress on the performance in low-productivity environments is available for triticale. Graham *et al.* [6] and Wiethölter *et al.* [7], investigating only one and two triticale genotypes, respectively, gave some indication of the effect of N on several agronomic traits. The present study with two N regimes and 36 genotypes was undertaken, therefore, to (1) determine the effect of N on important agronomic traits and (2) estimate genetic parameters.

Materials and Methods

Three winter triticale cultivars and 33 advanced breeding lines of diverse pedigrees were grown in 1993 at two ecologically different sites in southern Germany, Stuttgart-Hohenheim (HOH) and Eckartsweier (Rhine valley; EWE). The designs were 6x6 balanced lattices with two replicates for each of the two N level experiments. Nitrogen regimes were: NO = no nitrogen; N1 = site-specific 70-90 kg/ha N in three applications at early tillering (EC 22/24), jointing (EC 30/32), and pre-boot (EC 39) growth stages (according to Zadoks *et al.* [8]). Triticale followed oats. Genotypes were grown in $5m^2$ plots at a seeding rate of 300 grains per m^2. Data were recorded on a plot basis for days from May 1st to heading, plant height (cm), grain yield (q/ha), 1000-grain weight (g; TGW), and test weight (kg/hl). Tiller density (spike bearing tillers per m^2) was counted on 1m in a row, and the number of grains per spike was assessed on five randomly chosen spikes per plot.

Analyses of variance were conducted separately for the two lattice experiments (NO, N1) at each location to obtain lattice-adjusted mean values and repeatabilities. Variance components and heritabilities were estimated in combined analyses for each site, using the lattice-adjusted means, and the latter parameter was calculated for the two N rates also. Genotypes and sites were considered random and N levels fixed. The computer programme PLABSTAT [9] was used throughout.

Results

Mean performance at both N levels for nearly all traits was higher at Hohenheim than at Eckartsweier (Table 1). The zero N input resulted in a mean reduction of grain yield and tiller density of nearly 40%, a reduction of almost 20% in the number of grains per spike and plant height, and no change in heading time. Thousand-grain and test weight increased slightly.

Mean heritability estimates for both N levels across sites were moderate to high (0.52 to 0.96) and very similar (Table 1). Tiller density had consistently low or moderate repeatabilities/heritabilities due to the high error in measuring this trait.

Table 1. Mean values ($\bar{x}$) and repeatability(R)/ heritability(h^2) estimates for several agronomic traits of 36 triticales tested under two nitrogen levels (NO, N1) at two sites (HOH, EWE) in 1993

Trait	Par-ameter	HOH		EWE		Series		NO as
		NO	N1	NO	N1	NO	N1	% N1
Days to	$\bar{x}$	21.0	21.0	14.3	14.6	17.6	17.8	99.3
heading	R/h^2	0.96	0.98	0.90	0.88	0.87	0.96	
Plant	$\bar{x}$	117	126	77	111	97	119	81.5
ht.(cm)	R/h^2	0.69	0.86	0.42	0.78	0.62	0.80	
Yield	$\bar{x}$	60.9	84.5	23.2	53.3	42.1	68.9	61.0
(q/ha)	R/h^2	0.75	0.68	0.40	0.46	0.70	0.52	
Tillers	$\bar{x}$	325	441	202	394	264	417	63.2
per m^2	R/h^2	0.13	0.18	0.21	0.16	0.61	0.60	
Grains	$\bar{x}$	48.2	48.2	38.4	56.3	43.3	52.3	82.8
per spike	R/h^2	0.50	0.33	0.69	0.69	0.57	0.83	
TGW	$\bar{x}$	47.4	46.5	41.9	40.4	44.7	43.5	102.8
(g)	R/h^2	0.87	0.74	0.88	0.60	0.87	0.84	
Test wt.	$\bar{x}$	71.8	70.7	70.1	66.3[§]	71.0	68.5	103.6
(kg/hl)	R/h^2	0.95	0.85	0.95	-	0.85	0.84	

[§] One replication only

Combined analyses over N rates at the two sites showed that, in general, genotypes, N, and their interaction were significant sources of variation. The genotype x N interaction effect, however, was mainly small, as revealed by the variance components (Table 2).

The correlation coefficient for yield between N regimes was moderate (r = 0.51, significant at P $\leq$ 0.01). A scatter plot for this trait demonstrates that some genotypes had high or low performance at N1 and NO, whereas others experienced a shift in ranking (Fig. 1). For the other agronomic traits investigated, correlation between N levels was moderate to high and significant (results not shown), except for tiller density with a value of 0.32 only.

The genotype x N supply interaction is exemplified by a sample of nine genotypes that fell into three groups as suggested by Schinkel and Mechelke [2]: group 1 = relatively better performance under low N supply, group 2 = relatively better performance under high N supply, group 3 = intermediate performance (Fig. 2).

Table 2. Estimates of variance components and heritabilities (h^2) for several agronomic traits of 36 triticale genotypes (G) tested under two nitrogen (N) levels at two sites (HOH, EWE) in 1993

Trait	HOH				EWE			
	G	N	G x N	h^2	G	N	G x N	h^2
Heading	7.7**	-0.1	0.2**	0.98	4.5**	0.1	0.4**	0.92
Plant ht.	24.3**	47.8**	1.9	0.90	11.2**	587.0**	4.7*	0.70
Yield	23.2**	278.3**	2.8	0.85	5.2	451.7**	30.2**	0.17
Tillers[§]	6.9**	66.4**	-0.1	0.57	2.6	183.0**	1.1	0.30
Grains	21.7**	-0.5	3.4	0.70	24.6**	160.1**	11.3**	0.72
TGW	8.1**	0.3*	0.9*	0.88	4.8**	1.1**	1.6**	0.79
Test wt.	3.7**	0.6**	0.3**	0.93	3.3**	7.1**	1.8**	0.78

*,** Significant at P = 0,05 and 0.01, respectively; [§] Variance components x 10^{-2}

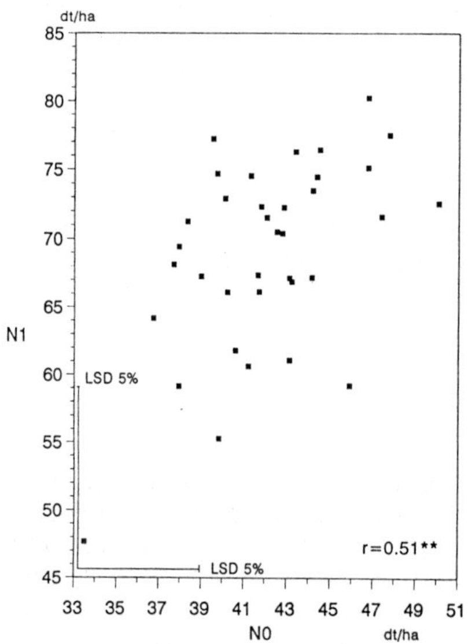

Fig.1 Scatter plot of grain yield for 36 triticales tested under 2 nitrogen regimes (N0, N1) at two sites in 1993.

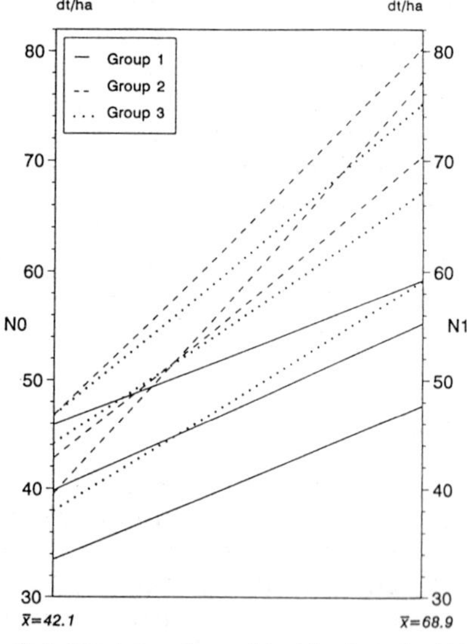

Fig.2 Genotype x nitrogen interaction for grain yield of 9 triticales classified into three groups.

Discussion

With regard to yield components, general responses of triticale to reduced N level were, in decreasing order of importance, lower tiller density, less grains per spike, and increased 1000-grain weight. This is in good agreement with studies in wheat [10] and rye [4]. In the triticale genotype investigated by Graham *et al.* [6], however, the number of grains per spike was not affected by N level.

A decrease in grain yield of nearly 40% indicates severe N stress at N0, compared to normal supply. The stress was considerably higher on the light sandy soil at the site EWE, as manifested in the severely reduced performance. But obviously, triticale can exploit a lower N level better than the traditional cereals wheat and rye, where a reduction in grain yield of more than 50% is reported [4, 10]. This might be due to a different root distribution and/or a more efficient root system of triticale.

Increased test weight, a lower number of grains per spike in response to N stress, and a consistently weak negative correlation between them at both sites and N rates, indicates a compensating effect for these traits. Grain shrivelling in triticale is a source-sink problem. Apparently, for plumper and bigger kernels part of the sink has to be sacrificed under normal and stress conditions also. Sahan and Singh [11] drew attention to this source-sink relationship earlier.

Although genotype x N interaction was mainly significant, it was of minor importance. Accordingly, correlations between N0 and N1 for all traits were moderate to high. Schinkel [12] in a similar study with wheat, also found significant, but small genotype x N interaction. Graham *et al.* [6] even reported the lack of a significant interaction. In maize, however, the genotype x N interaction amounted to nearly half the genotypic variance [13]. Desirable genotypes for low input conditions would be those that exhibit high performance, combined with a relatively low response to increased N level and therefore perform significantly better under stress (group 1 in Fig. 2).

Greater heritabilities in high as compared to low N environments were reported for wheat [12] and oat [5]. In the present experiment, however, heritabilities were not systematically higher at N1 and thus triticale seems to follow rye [4].

Conclusions

Genetic variation of response to N level is high for most agronomic traits. Due to the moderate to high correlations between N0 and N1 for nearly all traits, similar heritabilities for both N regimes, and minor importance of genotype x N interaction, a selection for N efficient genotypes appears feasible at the conventional N level. For most traits, this will result in a correlated selection response at the lower intensity level also. Nevertheless, because interaction effects are present, the breeder may be well advised to include a low-productivity environment at some stage in the selection process also.

References

1. Fox RH. Selection for phosphorus efficiency in corn. Comm Soil Sci Plant Anal 1978; 9: 13-37.
2. Schinkel B, Mechelke W. A method to estimate the prospect of specific breeding for nutrient efficiency. In: El Bassam N, Dambroth M, Loughman BC, editors. Genetic aspects of plant mineral nutrition. Wageningen: Kluwer Academic Publ., 1990: 449-56.
3. Spanakakis A, Viedt A. Performance of winter wheat cultivars under reduced nitrogen conditions. In: El Bassam N, Dambroth M, Loughman BC, editors. Genetic aspects of plant mineral nutrition. Wageningen: Kluwer Academic Publ., 1990: 465-73.
4. Miedaner T, Hartmann A, Geiger HH, Wortmann H, Stölken B. Vergleich der genetischen Variation für Ertrag und Qualität bei ortsüblicher und minimaler N-Düngung des Hybridroggens. Proceedings 9.Int Tagung Getreideverarbeitung u. Getreidechemie; Bergholz-Rehbrücke, 1993: 106-12.
5. Atlin GN, Frey KJ. Selecting oat lines for yield in low-productivity environments. Crop Sci 1990; 30: 556-61.
6. Graham RD, Geytenbeek PE, Radcliffe BC. Responses of triticale, wheat, rye and barley to nitrogen fertilizer.Aust J Exp Agric Anim Husb 1983; 23: 73-79.
7. Wiethölter S, Peruzzo G, Baier AC. Triticale response to nitrogen in two oxisols in Southern Brazil. Proceedings 2nd Int Triticale Symposium; 1990 Oct 1-5; Passo Fundo, Brazil. Mexico DF: CIMMYT, 1991: 221-22.
8. Zadoks JC, Chang TT, Konzak CF. A decimal code for the growth stages of cereals. Weed Res 1974; 14: 415-21.
9. Utz HF. PLABSTAT. Ein Computerprogramm zur statistischen Analyse von pflanzenzüchterischen Experimenten. Version 2F. Institut für Pflanzenzüchtung, Saatgutforschung und Populationsgenetik, Univ. Hohenheim, 1991.
10. Bruckner PL, Morey DD. Nitrogen effects on soft red winter wheat yield, agronomic characteristics, and quality. Crop Sci 1988; 28: 152-57.
11. Saharan RP, Singh VP. Source-sink relationship as a factor for grain shrivelling in triticale. Indian J Genet 1982; 42: 1-4.
12. Schinkel B. Untersuchungen zur nutzbaren genetischen Variation für Stickstoff-Effizienz bei Winterweizen [dissertation]. Hohenheim: Univ. Hohenheim, 1991.
13. Landbeck M, Seitz G. Grundlagen der Züchtung auf Stickstoffeffizienz bei Mais. Vortr Pflanzenzüchtg 1992; 22: 57-66.

GENETIC VARIATION, G*E INTERACTIONS AND SELECTION RESPONSE FOR HAGBERG FALLING NUMBER IN TRITICALE

Louis Jestin & Hélène Bonhomme.
INRA - Génétique et Amélioration des Plantes
Domaine de Crouelle, Clermont-Ferrand, France .

Abstract

A three-year experiment was carried out on triticale, using 6x adapted genotypes, to assess the genetic variation and G x E interaction of the Hagberg Falling Number trait . Significant interactions of genotype with year and site appeared in this study . Broad sense heritability appeared relatively moderate for that trait, and preliminary results on F3 segregating progenies showed mostly additive effects .

Introduction

More attention has been given relatively recently, in the INRA triticale breeding programme, to the utilization quality of the harvested grain, for the food and feed industries.

Among the important traits currently considered, the Hagberg Falling Number (HFN) plays a prominent role , as it is both an indicator of the presprouting process and of a high pre-germination α–amylase activity potentially interacting with dough properties [1].

Material and methods

The material studied included:

1- A "Hagberg" series comprising 25 breeding lines selected for high Hagberg FN in 1991

2- Advanced lines present in multisite trials (plains + mountains) from 1991 to 1993;

3-Partially inbred F3 progenies from a diallel cross in which series of crosses with BOLERO (high HFN) vs. MAGISTRAL (low HFN usually) were compared.

Hagberg Falling Number data were recorded in seconds according to the standard procedure , with 2 measurements per 150g-samples of whole-kernel meal. Additional information on site data (e.g. rainfall) and on plant covariables (e.g . earliness, height, presprouting ,TKW and grain shrivelling) were collected whenever possible.

H. Guedes-Pinto et al. (eds.), Triticale: Today and Tomorrow, 609–613.
© 1996 *Kluwer Academic Publishers. Printed in the Netherlands.*

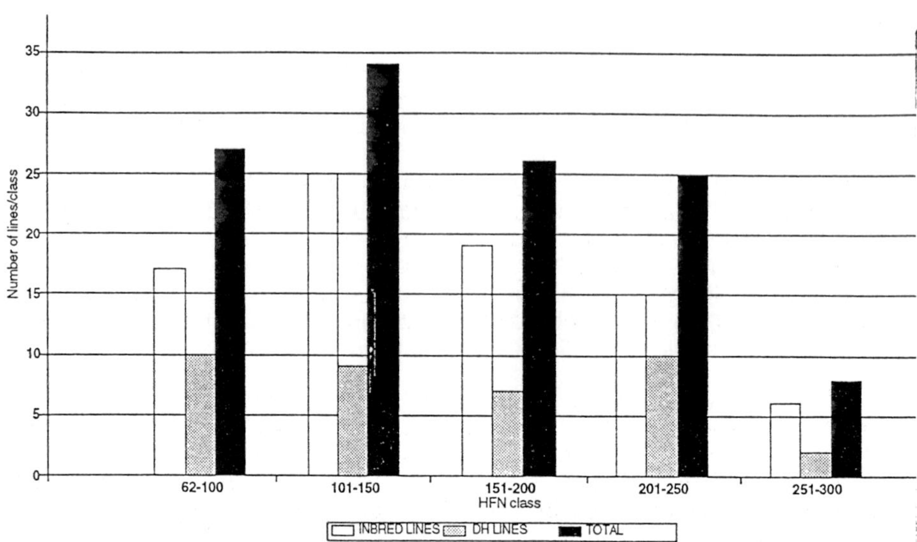

Figure 1- *Diversity of Hagberg Falling number values in triticale breeding material: 1990-91 INRA lines; Clermont-Ferrand nursery.*

Table 1- *Hagberg test response of 25 triticale genotypes : year effect according to rainfall. (*) INRA clermont-Ferrand nursery samples.*

| | | YEAR | 1991 | 1993 | | |
| | JUNE +JULY | rainfall | 94mm | 173mm | | |
HFN STABILITY CLASS (*)	VARIETY /LINE NAME	ORIGIN	HFN (s)	HFN (s)	EARLINESS 9=v.early	TKW class
I =highly stable	OS88012	SERASEM (F)	268	220	4	3
	SALVO	LASKI (PL)	267	246	6	4
II =some stability	83TBT91-2	INRA (F)	285	124	6	5
	OS87909	SERASEM (F)	275	170	4	4
	CI648	INRA (F)	223	120	5	3
	87HD13-1	dt	221	131	8	7
	88HD118-12	dt	280	135	2	6
III =some resistance to HFN collapse	82TT48-31	INRA (F)	263	98	6	4
	BOLERO	LASKI (PL)	234	99	6	1
	85TT14-3	INRA (F)	219	88	8	7
	87HD49-14	dt	248	107	8	5
IV =very low HFN in moist conditions	83TT149-2	INRA (F)	238	66	2	3
	83TT149-5	INRA (F)	266	70	4	6
	REH/HAR212	CIMMYT	232	62	2	3
	83TBT82-1	INRA (F)	231	63	6	4
	CWT83-78	PBI (GB)	216	62	6	5
	T88026	BENOIST (F)	216	67	6	3
	83TT161-1	INRA (F)	212	62	4	7
	89HD118-14	dt	262	74	4	6
	88HD118-155	dt	240	62	4	5
	89HD113-14	dt	236	62	5	5
	88HD115-946	dt	228	62	3	7
	88HD118-720	dt	221	62	4	7
	88HD115-473	dt	219	68	2	4
	88HD113-466	dt	213	62	4	7

RESULTS

A survey of the variation encountered in the bulk of 1991 advanced material was carried out: the HFN range observed was 62-400 s (Figure 1). The top 25 entries of 1991 were further studied in 1992 and 1993.

Years with strongly contrasting climate , such as 1991 (very dry) and 1992 (very wet in June) showed contrasting results, 1992 giving low figures for genotypes prone to pre-sprouting, although having high values in 1991 dry conditions:in fact, all entries were more or less presprouted in 1992, except SALVO and , to a lesser extent CI648 . In 1992, June + July rainfall reached 314 mm, a figure which should be compared to the 1961-90 thirty-year average of 118 mm.

Site interaction for HFN values was also considered (Table 2). The distribution observed with DH breeding material derived from anther culture appeared to be roughly the same as that of classical inbred lines of similar genetic background. In addition , the earliness/development pattern seemed also to interact significantly, so that no clear-cut and simple response model can be proposed for the moment.

Table 2- *Genotype x environment (GxE) interaction typical for triticale Hagberg Falling Number data: samples of 7 genotypes in 3 sites of INRA trials ,1992 harvest.*

VARIETY /LINE	HAGBERG F N (s) in 3		SITES	MEAN
	ST-GENES (63)	DIJON (21)	COSTAROS (43)	OVER SITES
MAGISTRAL	64	78	64	69
LASKO	142	115	-	(129)
CF82TT24	127	64	164	118
CF82TT48-3	71	62	112	82
CF82TT48-33	65	64	155	95
CF-CI2-10	62	62	62	62
DI86TT34-1	71	67	116	85

The overall data gave indications of moderate broad sense heritability. On the other hand, the average values of F3 (still segregating) progenies involving contrasting parents, e.g .BOLERO and MAGISTRAL (Table 3), suggested mostly additive effects, as previously reported [2].

A study of more numerous and homozygous progenies is now required to obtain more reliable results on the genetic control of the HFN.

Table 3- *Effect of the parent on the average Hagberg Falling Number value of triticale progenies (breeding material from 8 INRA crosses; 1993 harvest.)*

MALE PARENT -> HFN (s)→	CF77022 92	CENTRAL 74	CF 76010 62	DOMITAL 148
FEMALE PARENT HFN (s) ↓	F3 PROGENIES			
BOLERO 232	112	116	62	114
MAGISTRAL 75	73	71	64	65

Conclusion

Empirical selection pressure for the character, exerted on the 1988-1991 crops has led to an appreciable increase of the percentage of lines showing HFN figures superior to 150 s. (12 to 55 %) in dry growth conditions (Figure 2).

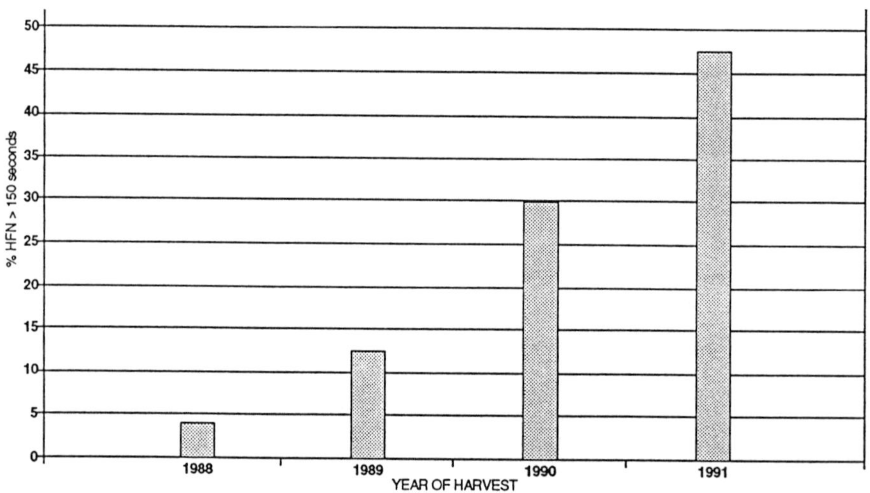

Figure- 2 *Evolution of % INRA triticale material with higher HFN values (1988-1991).*

It is expected that the HFN character will continue to be routinely used in selection. Other characters, such as % dietary fiber and ability to produce breakfast cereals might also be worth considering in the future.

Acknowledgments : The skillfull technical assistance of J.M. Legendre and R. Gaze is gratefully acknowledged.

References

1. Branlard G, Bernard M, Antraygue C, Barloy D. Shrivelling and alpha amylase activity in triticale and its parental species: preliminary studies. Proceedings   EUCARPIA Meeting: Genetics and breeding of triticale;1984 July 2-5; Clermont-ferrand. Paris: INRA Publ.: 617-631.

2. Oettler G . Analysis of a diallel cross among eight secondary hexaploid triticales for falling number. Tag.-Ber. Akad. Landwirtsch.-Wiss.DDR, Berlin 1988; 266:643-652

PREHARVEST SPROUTING RESISTANCE IN TRITICALE: PRELIMINARY RESULTS

Geert Haesaert and Antoon E.G. De Baets
Industriële Hogeschool, C.T.L.
Voskenslaan, Gent
Belgium

Abstract

Humid weather conditions near harvest ripeness often cause preharvest sprouting problems in triticale. The susceptibility of triticale to sprouting is under the climatological circumstances of the cool and wet regions of West-Europe, one of the most limiting factors for the further development of triticale.

Data from the different experiments reveal distinct differences between triticale genotypes in their level of preharvest sprouting. In general it could be concluded that 'Alamo', 'Pablo', 'Lasko' and 'Lukas' are very susceptible to sprouting while 'Purdy', 'Uno' and 'Ego' showed a good preharvest sprouting resistance.

A good correlation between the results of seed and spike germination tests was found. Data of preharvest sprouting under field conditions were also very similar to data of the germination tests. On the contrary the correlation between germination data and falling number were low and statistically not significant.

Introduction

Humid weather conditions close to harvest ripeness often cause preharvest sprouting problems in triticale. The susceptibility of triticale to sprouting is under the climatological circumstances of the cool and wet regions of West-Europe, one of the most limiting factors for the further development of triticale.

Preharvest sprouting reduces seed quality and vigor, so that its reduction should be an important breeding objective. In order to breed succesfully for preharvest sprouting resistance, it is necessary to dispose of a workable method to determine sprouting and to have information about the genetic variability and the inheritance of the character. The objectives of the present study were:

(1) to evaluate three different methods of measuring preharvest sprouting

(2) to investigate the genetic variation of preharvest sprouting in the present triticale germplasm

(3) to measure the influence of preharvest sprouting susceptibility or resistance of varieties on their progenies.

H. Guedes-Pinto et al. (eds.), Triticale: Today and Tomorrow, 615–622.
© 1996 *Kluwer Academic Publishers. Printed in the Netherlands.*

Materials and Methods

EXPERIMENT 1: In 1991 the sprouting resistance of 14 varieties was investigated using three different methods: seed germination test, intact spike germination test and falling number (1). The varieties used in the comparative study were grown in $12m^2$ drilled plots on the experimental farm of the I.H.- C.T.L. at Merelbeke. The experimental design was a completely randomized block with 4 replications. The harvest period of 1991 was without precipitation and no visible sprouting symptoms were detected.

At harvest ripeness 20 primary spikes from each plot were randomly selected. In this way 80 spikes of each variety were collected. Fourty of them were used to carry out the seed germination test while the other part was taken for conducting the spike germination test.

The spikes for the seed germination test were hand threshed. The half of the material was dried tot determine the seed moisture content. From the remaining half, 8 x 20 seeds of each variety were placed in individual petri diskes, containing a double filterpaper (Schleicher and Schüll, n° 595) moistened with 4 ml of distilled H_2O. The germination test took place at 20°C and the germination counts were conducted after 6 days.

For carrying out the spike germination test, random samples of 20 spikes were arranged on a Jacobsen germination table according to an uncompletely latin square with 4 replications. The spikes were placed between two humid filter papers. A paper strip in contact with the waterbad of the germination table moistened continuously the filterpapers around the spikes. The germination temperature was 20°C. After 6 days the germination was assessed visually using a 1-10 scale (1: no germination and 10: 100 % germination).

EXPERIMENT 2: The genetic variation of preharvest sprouting in the present triticale germplasm was determined from 71 selected genotypes of the I.H.-C.T.L. variety collection and the entries of the annual yield nurseries.

The 71 selected genotypes (Table 2) were grown on 2 rowed plots, 2m length with 25cm row spacings during the growing season 1990-1991. Twenty spikes from each variety were randomly collected at harvest ripeness. The sprouting susceptibility was measured using the method described above for the spike germination test.

High rainfall during the harvest period of the growing seasons 1987-88, 1988-89 and 1992-93 permitted us to evaluate the preharvest sprouting susceptibility of the entries from the annual yield nurseries. All yield trials were arranged in randomized blocs with 4 replications. Preharvest sprouting was assessed visually by controlling 80 spikes per variety by using a 1-10 scale (1: no germination and 10: 100 % germination).

EXPERIMENT 3: The objective of this experiment was to explore the heritability and genetic variation of preharvest sprouting. For that purpose 35 F_2 progenies of triticale x triticale crosses and 12 F_4 progenies of triticale x wheat (*Triticale aestivum* L.) crosses were included in the experiment. Bulks of F_1-plants, F_2-plants and in case of triticale x wheat populations F_3-plants were grown separately during the subsequent seasons. Progenies and as much as possible their respective triticale parents were grown in $24m^2$ drilled plots on the experimental farm of the I.H.-C.T.L. at Merelbeke during the growing season 1991-92. A random sample of 100 spikes was taken from each plot. A mean sprouting index was determined for each progeny and included parent using the method described for the spike germination test.

For these progenies of which at least one of the parents was included the narrow-sense heritability (h^2n) was calculated according to the method of Falconer (1981): $h^2n = 2b$ (b is the regression coefficient of the linear regression model between progenies and parent).

Arcsin square root transformation was applied directly to the germination results and to sprouting score. All significance tests are based on transformed data, however results have been presented in untransformed units. Kendall (1959) rank-correlation was used to compare the two germination tests with the falling number test.

Results

EXPERIMENT 1 (Table 1): 'Alamo' and 'Pablo' and to a less extent '5A 85.21-52', 'Lasko' and 'Bolero' showed a rapid an high germination rate in the seed germination test. On the other hand 'Ego', 'Purdy', '5B 85.7-20', 'CHD 888' and 'CTL 84-61' were significantly lower in germination then the other varieties tested. In general the results of the spike wetting test were very similar with the results of the seed germination test. Once again 'Alamo' and 'Pablo' appeared to be most susceptible to preharvest sprouting. Also 'Lasko' and 'Lukas' had a relatively high sprouting score. The high level of grain dormancy of 'Ego', 'Purdy', '5B 87.7-20', 'CHD 888' and 'CTL 84-61' from the seed germination test were confirmed by the results of the spike wetting experiment. Only 'Bolero' reacted different for the two germination tests: the seed germination rate was high while a very low spike sprouting index was noted.

Table 1: Preharvest susceptibilty of several triticale varieties

Variety Rank	Germination (1)	(2) kernels	Falling spike	numbers
Pablo	5.0	51.3	7.3	62.0
Alamo	10.5	80.0	4.6	65.0
5A 85.21-52	16.0	31.3	3.5	62.3
Lasko	21.5	31.3	4.8	160.3
CHD 1089	24.5	17.5	3.3	65.3
Lukas	25.0	20.0	4.2	119.0
6A 84.6-6	27.0	18.8	2.6	92.8
CHD 989	31.0	6.3	3.6	212.0
Bolero	33.0	30.0	1.5	123.5
CHD 888	34.5	13.8	1.1	71.3
5B 85.7-20	36.0	11.3	1.9	107.8
Ego	37.0	6.3	2.2	114.5
CTL 84-61	37.5	13.8	2.1	127.0
Purdy	40.0	9.0	2.5	166.5

(1) Sum of ranknumbers according to KENDALL (1957)
(2) germination of seeds in %; germination of spike according to a 1-10
　　scale with 1= no germination and 10=100 % germination

'Pablo', 'Alamo', '5A 85.21-52' and 'CHD 1089' had a significantly lower falling number than the other varieties while 'CHD 989', 'Purdy', 'Lasko', 'C.T.L. 84-61' and 'Bolero' noted the best falling numbers. From the '5A 85.21-52',

Table 2: Classification of triticale varieties based on their susceptibility to preharvest sprouting

Class 1: 1-2.5(1)	Class 2: 2.5-5	Class 3: 5-7.5	Class 4: 7.5-10
Bokolo: 1.3	Ego: 2.6	LT 464-77: 5.1	CHD989: 7.5
CHD 702: 1.3	Purdy: 2.9	CHD 312: 5.1	Mical: 7.5
CTL 84-61: 2.4	Uno: 2.9	CHD 738: 5.2	LT 2148: 7.8
	Clervix: 2.9	Rahum: 5.2	Largo: 7.9
	A 84.6.28: 3.0	CHD 130 : 5.4	CHD 521: 8.3
	Dino: 3.0	Oct ch(2): 5.4	LT 1714: 8.3
	CHD 1832: 3.2	JubxR1: 5.4	CHD 395: 8.3
	R 1980: 3.3	LT 217-19: 5.4	Lukas: 8.5
	LT 2211: 3.3	Beagle: 5.7	Panola: 8.9
	Bison: 3.4	CHD 1089 : 5.9	Pablo: 8.9
	Impala: 3.6	SES 79-83: 6.1	Alamo: 8.9
	CHD 431: 3.9	C 85-9(2): 6.1	
	SES 77-21: 4.0	Serval: 6.2	
	Grace: 4.2	CHD 897: 6.3	
	Civit: 4.2	LT 1439: 6.3	
	JubxR3(2): 4.3	CHD 1025: 6.3	
	CTL 86-1: 4.4	CHD 782: 6.3	
	SES 78-9: 4.8	CHD 1024: 6.4	
		Salvo: 6.4	
		CHD 928: 6.4	
		Triticor: 6.5	
		Rosko: 6.5	
		Bolero: 6.6	
		CTL 86-1: 6.6	
		LT 2129: 6.7	
		LT 419-74: 6.7	
		Mizar: 6.9	
		SES 79-4: 7.0	
		CT 312-83: 7.1	
		LT 464-77: 7.1	
		A 691178: 7.1	
		CHD 333: 7.2	
		Lasko: 7.2	
		LAD 766: 7.4	

(1) Sprouting index according to a 0-10 scale with 0: no germination and 10: 100 % germination
(2) octaploid

'Purdy' and 'C.T.L. 84-61' a close correspondence was found between falling numbers and germination data. For the other varieties the results were not so similar.

As already mentioned a good correlation between the data of the two germination tests was found (r = 0.748**). However, the correlation between the results of both germination tests and falling number values was statistically not significant (resp. r = -0.33 NS and r = 0.180 NS respectively for seed germination and falling number and for spike germination and falling number). Nevertheless the coefficient of concordance (3) was significant what means that statistically a rank-correlation exist between the 3 methods studied and a calculated order based on the rank-correlation of the 3 methods could be made. This rank-correlation order illustrated that 'Pablo', 'Alamo', '5A 85.21-52' and 'Lasko' seemed to be most susceptible to preharvest sprouting while 'Purdy', 'Ego', 'C.T.L. 84-61' and '5B 85.7-20' showed a rather good level of sprouting resistance.

EXPERIMENT 2: Based on their average sprouting index determined by a spike wetting test, the varieties studied were classified into 4 classes (Table 2).

A wide range of sprouting values were obtained going from 1.3 to 8.9. However, most of the varieties exhibited a sprouting index higher than 5. Only 5.6 % of the varieties showed a sprouting index between 0-2.5 (Class 1) while 15.5 % could be classified in class 4 (7.5 -10). Of the varieties studied 56.65 % had a value between 5 and 7.5. 'Bokolo', 'CHD 702', 'CHD 888' and 'CTL 84-61' and to a less extent 'Ego', 'Uno' and 'Purdy' performed well concerning sprouting resistance. On the other hand 'Alamo', 'Pablo', 'Lukas', 'CHD 395-84', 'LT 174-83' and 'CHD 521' showed heavy sprouting symptoms.

It can be concluded that most of the triticale varieties tested are susceptible to preharvest sprouting and the genetic base in the present triticale germplasm for preharvest sprouting resistance is very small.

Data of the preharvest evaluation under field conditions during the growing seasons 1987-88, 1988-89 and 1992-93 were very similar with these of the seed and spike germination tests (Table 3). 'Alamo', 'Pablo', 'Lasko' and 'Lukas' were also under field conditions characterized by serious sprouting problems while 'Uno', 'Purdy' and 'Ego' were less susceptible to sprouting in the spike. Remarkable was the sprouting susceptibility in 1993 of the rather new varieties 'Origo', 'Prego', 'Trimaran' and 'J47'.

Table 3: Preharvest sprouting of triticale varieties under field
condition

Variety	Growing season		
	1987-88	1989-90	1992-93
Lasko	4.0(1)	3.5	4.0
Bolero	3.0	2.0	-
Alamo	7.0	7.0	5.0
Pablo	7.0	5.0	4.0
Lukas	5.0	4.0	4.0
Purdy	2.0	2.0	3.0
Ego	1.0	-	3.0
Falko	-	-	4.0
Galtjo	-	-	4.0
Origo	-	-	5.0
Prego	-	-	7.0
J47	-	-	9.0
Trimaran	-	-	6.0

(1) Preharvest sprouting according to a 1-10 scale with 1: no
germination and 10: 100% germination

620

EXPERIMENT 3: Means of sprouting values of the progenies studied are presented in figure 1. Following remarkable conclusions can be made:
- all progenies derived from the variety 'Alamo' were characterized by a high preharvest sprouting susceptibillty. The progenies of 'Bolero', 'LT 2129-83' and 'Pablo' exhibited also low levels of sprouting resistance.
- F_3-populations deriving from 'Uno', 'Salvo' and 'Purdy' showed improvement to preharvest sprouting.
- F_4-populations of triticale x wheat crosses showed in general better sprouting tolerance than the triticale parent.
- narrow-sense heritability estimate was 0.49 (Figure 2). Only these progenies of which at least one of the parents was included were used for estimating the heritability.

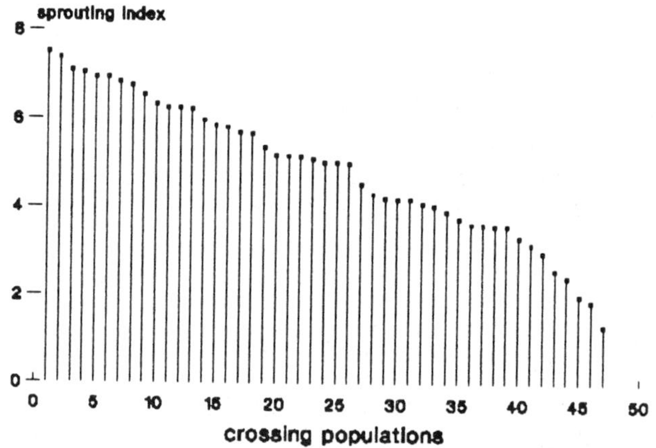

Figure 1: Preharvest sprouting susceptibility and tolerance of 35 F_3 populations triticale x triticale en 12 F_4 triticale x wheat populations (preharvest sprouting according a scale 1-10: 1: no germination and 10: 100 % germination) with:
1, 2, 5 and 7: crossing population of 'Alamo'
2, 8, 15 and 18: crossing population of 'Pablo'
3, 6, 11, 12 and 14: crossing population of 'Bolero'
23, 26 and 32: crossing population of 'Purdy'
23, 34 and 43: crossing population of 'Uno'
28, 29, 30, 35, 37, 39, 40, 42, 44, 45, 46 and 47:
Triticale x wheat progenies

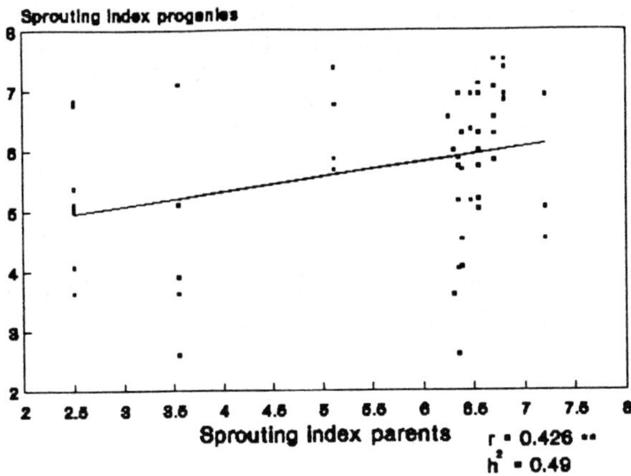

Figure 2: Linear regression model between sprouting index of triticale parents and their progenies

Conclusion

Data from the different experiments reveal distinct differences between triticale genotypes in their level of preharvest sprouting. Most part of the 71 genotypes studied showed a high preharvest sprouting susceptibility. Less than 10 % of the genotypes could be classified as more or less preharvest sprouting resistant and are a potential genetic source for breeding to more preharvest resistance.

In general it could be concluded that 'Alamo', 'Pablo' 'Lasko' and 'Lukas' are very susceptible to sprouting while 'Purdy', 'Uno' and 'Ego' showed a good preharvest sprouting resistance. This was confirmed by the results of seed and spike germination tests, falling number and field observations.

A good correlation between the results of seed and spike germination tests was found. Data of preharvest sprouting under field conditions were also very similar to data of the germination tests. On the contrary the correlation between germination data and falling number were low and statistically not significant.

Intact spike germination tests permit rapid assay of many seeds and should detect not only dormancy but also other properties as for example germination inhibiting compounds in the bracts. Seed germination are a direct assay of seed dormancy. Falling number is probably not the method of preference because of the low correlation with visual sprouting.

Sprouting susceptible or sprouting resistant varieties increased or decreased clearly the level of preharvest tolerance of their progenies. Crosses deriving from 'Alamo', 'Bolero', 'LT 2129' and 'Pablo' increased the susceptibility while populations deriving from 'Uno' and 'Purdy' showed improvement to preharvest sprouting. In general F_4-populations of triticale x wheat crosses exhibited better sprouting tolerance levels than the triticale parents. A narrow-sense heratibility of 0.49 was calculated.

References

1. Paterson, A.H., Sorrels, M.E. & Obendorf, R.L. (1989). Methods of evaluation for preharvest sprouting resistance in wheat breeding programs. Can. J. Plant Sci. 69: 681-689.
2. Falconer, D.S. (1981). Introduction to quantitative genetics. Longman, London and New York, 340 pp.
3. Kendall, M. (1957). Rankcorrelation methods. Griffin, London, 160 p.

FACULTATIVE TRITICALE: COMPARISON OF LOCAL AND CIMMYT GERMPLASM IN HYDROPONICS AND UNDER FIELD CONDITIONS

Ahmad Arzani & Norman L. Darvey
University of Sydney, Plant Breeding Institute, Cobbitty, NSW, Australia

Abstract

Superior lines were identified from a hydroponicaly grown CIMMYT nursery, on the basis of dry matter yield and recovery from three sequential clippings. These lines were then compared with superior mid-generation triticale derived lines from local breeding program under field conditions. The locally adapted materials were superior to the CIMMYT nursery with respect to forage yield and recovery from clipping.

Introduction

Triticale has generally been utilized as a grain crop. However, in view of the relatively high biomass of many primary triticales, recent interest has been focused on its potential for grazing as a viable alternative to forage oats. Preliminary evaluation for forage potential in facultative triticales from a CIMMYT nursery was conducted and further comparisons were made to local germplasm under field conditions.

Materials and Methods

Initially two hundred and twenty lines from the CIMMYT nursery (origin MI 90-91) were grown in a quarantine greenhouse for six weeks and then transplanted to both a custom-made greenhouse hydroponic system and the field (Cobbitty-1992). The resistance of these materials to prevalent stem, leaf and stripe rusts were assessed in the field. The pathotypes (pt.) used were stem rust: pt. 343-1,2,3,4,5,6,7,8,9 (PBI accession number), leaf rust: pts. 104-2,3,6,(7) (76694), 10-1,2,3,4 (72469) and stripe rust: pt. 110 E143 A+ (861725). Three sequential clippings were applied to these materials in the glasshouse and twelve superior lines identified on the basis of plant weight (PW), dry matter yield from clippings (DM1, DM2 and DM3) and tiller recovery from clippings (TR1, TR2 and TR3). The superior CIMMYT lines were

H. Guedes-Pinto et al. (eds.), Triticale: Today and Tomorrow, 623–625.
© 1996 *Kluwer Academic Publishers. Printed in the Netherlands.*

then grown with selected mid-generation field derived selections (BC$_1$F$_3$, TC$_1$F$_3$ and F$_4$) from 14 populations in the local breeding program under field condition at Cobbitty in May 1993 (Table 1). These were sown in hill plots with three replicates and subjected to a single clipping.

Results and Discussion

The correlation between field data and greenhouse data for total biomass and grain yield was not significant in the preliminarily evaluation the materials in the CIMMYT nursery . This was attributed to the reduced vigor of the majority of field grown CIMMYT nursery materials which were susceptible to the three rusts, especially stem rust under field conditions.

Means of 14 superior local populations and 12 superior CIMMYT nursery lines for DM1, TB and GY are shown in Table 2. The locally adapted germplasm was far superior to the materials in the CIMMYT nursery for forage yield and in particular for recovery from clipping, thus leading to a huge differences in total biomass and grain yield after single clipping.

Table 1. Ranking twelve superior CIMMYT triticale lines.

Local populations		CIMMYT lines
1. 80465	F$_4$	1. 79 B077003
2. 337 (P)	F$_4$	-0M1-0M1-1YI-2MI-0MI
3. UH 71	F$_4$	2. MS B219/A876
4. II85-79/Key25	BCF$_3$	-M86-6068-0MI-1YI-3MI-0MI
5. 337 (U)	BCF$_3$	3. Flora//EMS M83/6039
6. II80-20-11	BCF$_3$	M11 T88-8,9,10 (9 Espigas)-1MI-0MI
7. 80424 (216)	F$_4$	4. ANOAS 3/S TIER6
8. Polony Q	F$_4$	CTM86.119-1MI-1BI-0MF
9. Plot 327	F$_4$	5. 78W008001/DZDP.4
10.II85-3	TCF$_3$	M86-6172-Bulk0MI-1YI-1MI-0MI
11.M86-6068	TCF$_3$	6. SPN/DZDP-4//2 * LT933.82
12.II85-15	TCF$_3$	M86-6281-1MI-1MI-1MI-0MI
13.II85-73	F$_4$	7. M86-6278-1MI-1MI-2MI-0MI
14.Flora	TCF$_3$	8. DAWS/SNP//B164/3/A876/Yoco
		M84-6710-1-11MI-2MI-2YI-1MI-0MI
		9. MS B219/A876
		M85-6070-1-18MI-2MI-1MI-3MI-0MI
		10. K88-421-430 (7Espigas)-1MI-0MI
		11. M86-6278-1MI-1MI-3MI-0MI
		12. MS B219/A876
		M85-6070-1-18MI-2MI-1MI-1MI-0MI

Table 2. Means of 14 superior local populations and 12 superior CIMMYT lines for DM1, total biomass (TB) and grain yield (GY) (g/plot).

	Local*			CIMMYT*	
DM1	**TB**	**GY**	**DM1**	**TB**	**GY**
1. 41.2	109.5	38.8	1. 18.0	72.3	16.7
2. 45.8	123.6	39.3	2. 31.3	58.4	13.4
3. 41.8	110.6	38.0	3. 17.4	43.4	13.0
4. 41.4	99.6	33.6	4. 32.0	70.0	18.7
5. 43.6	108.9	33.7	5. 25.3	63.4	11.7
6. 36.3	122.8	42.3	6. 18.7	60.0	21.0
7. 36.3	96.5	33.3	7. 30.5	73.4	16.7
8. 42.8	119.2	40.4	8. 27.2	66.7	15.0
9. 49.3	127.5	43.4	9. 35.1	75.0	21.7
10. 44.5	102.9	31.9	10. 22.6	51.7	16.0
11. 48.6	113.1	38.4	11. 27.2	60.0	15.0
12. 45.4	109.7	38.9	12. 16.2	43.4	8.7
13. 30.5	88.4	29.8	-	-	-
14. 34.3	82.7	26.0	-	-	-

* Populations listed in order in Table 1.

**Numbers within column of each population not followed by same letter differ at P=0.01 with the LSD test

AN ATTEMPT AT TETRAPLOID TRITICALE IMPROVEMENT

Boguslaw Lapinski[1], Barbara Apolinarska[2], Grzegorz
Budzianowski[3], Malgorzata Cyran[1], Maria Rakowska[1]

[1]Plant Breeding and Acclimatization Institute - Radzików;
[2]Institute of Plant Genetics, Polish Academy of Sciences - Poznan;
[3]PBAI Experiment Station-Malyszyn. Gorzów Wlkp. Poland

Abstract

Since 1982 a small 4x triticale breeding programme has been carried out, aiming at creation
of new lines with fertility, vigour and feeding value close to the 6x triticale cultivars and
with the rye level of adaptation to marginal soil and weather conditions.

In order to keep the same chromosome composition, only one line of 4x triticale was
used in crossing with 6x triticale. F_1 hybrids with 6x triticale were back-crossed to 4x
triticale one or two times. The resulting populations with 50 %, 25 % and 12.5 % of the
new genes were bred separately. The breeding method was based on gene circulation
between outcrossing bulks and individual selection plots.

Tetraploid triticales kept the triticale level of resistance to diseases and nutritional
properties of the grain. Kernel filling in the advanced lines was not worse than in 6x
triticale. On an average, survival of seedlings on the acid media with high Al ions
concentrations was higher for 4x lines than for 6x triticale cultivars. Yield potential has
improved much in comparison to the initial primary 4x triticale line, however it is still
unsatisfactory (maximum 60 % of 6x triticale MALNO check in 1989, on 5m^2 plots).
Sterility is still the main problem.

Introduction

Poor vigour and low fertility of the first tetraploid triticales restricted their practical
application to a bridge component for chromosome manipulations in related crop plants.
However, there are reasons to expect, that on the 4x ploidy level triticale could be an
interesting new crop. Its main advantage over hexaploid forms is the possibility of
eliminating the aneuploidy problem [1,2]. There are also better conditions for the more
distinct expression of rye characters because of higher proportion of rye genomes in the
chromosome complement. Thus, 4x triticale could develop better tolerance to marginal
weather and soil conditions. This genome constitution seems also more convenient for

H. Guedes-Pinto et al. (eds.), Triticale: Today and Tomorrow, 627–634.
© 1996 *Kluwer Academic Publishers. Printed in the Netherlands.*

maintaining heterochromatin of rye chromosomes, which has been dramatically reduced while subjected to the domination of wheat chromosomes in 6x triticale [3,4]. Therefore, 4x triticale has a chance for a different way of evolutionary development and new range of breeding products. However, it is not an easy task, as not much is known about the influence of inbreeding depression, self-incompatibility, cytoplasmic differences and incoherence of a wheat mixogenome. The role of the last factor has been confirmed in some publications on the chromosome constitution of 4x triticale [5-7].

The challenge for scientists and breeders seems particularly provocative as tetraploids with one pivotal and one mixed genome exist as successful wild or cultivated plants in the *Triticum-Aegilops* group [8].

The presented attempt at 4x triticale improvement is based on a breeding procedure which seems stimulating or at least unlikely to destabilize a new genetic system.

Materials and Methods

In order to avoid cytogenetic disturbances only one line, P1N (= 653-50) was selected at the start of the breeding programme as the initial 4x component. The line seemed interesting because of shorter straw, relatively good fertility, grain plumpness and large anthers (possibly valuable in restoration an outcrossing propagation system). Lukaszewski's and Apolinarska's cytogenetic investigation revealed the presence of chromosomes 1A, 2B, 3A, 4A, 6A and 7B of the wheat genome and rye chromosome 5R in tetrasomic condition [9]. The line was derived from the cross of 6x triticale P11 with 2x rye No. 751.

Numerous 6x triticale lines and cultivars were the source of new variation. Advanced coadaptation of their wheat and rye chromosomes was expected to exert a positive influence on the poor 4x triticale homeostasis. The new genes were introduced into 4x breeding population in three doses: A - 50% (simple cross), B - 25% (first back-cross) and C - 12.5% (second back-cross). Back crosses to P1N line were applied in order to facilitate back-regulation to the 4x level and to restrict karyotype variation within breeding population.

The segregating material was the result of crosses from years 1982-1990. During this period 36 controlled simple 6x x 4x triticale crosses had been made. Besides, 40 spontaneous 5x F_1 hybrids were included, which appeared frequently in the neighbourhood of 6x triticale nurseries in Malyszyn. The number of first back-crosses was equal to the number of simple crosses' F_1-s and about 70 % of them were used in the second back-cross.

The breeding scheme was a composition of the progeny selection and bulk populations with conditions for spontaneous intercrossing (see Figure 1). There has been an exchange of materials between these two parts. Bulks delivered interesting plants to progeny selection and majority of lines returned to the bulks as about 5% (of seed grain mass) admixture.

The recycling of genes proceeded in three separated groups: A, B and C, created according to the number of back-crosses. The separation of groups was not complete because of possibility of intercrossing through the 2m wide isolation belts.

The bulk populations were founded in 1986. The number of breeding units on the individual selection plots surpassed at that time 200 and varied between 200 and 400 in the following years.

The main selection criteria were those connected with fertility: number of grains per stalk - until 1991 - and spike index, expressed as grain mass percent of total spike mass -since 1992. The former criterion was rejected as inducing strong selection pressure towards big, overdeveloped or even branched spikes with large numbers of flowers. For spike index, the associated selection pressure for small-size, delicate glumes and better developed kernels seemed more profitable. A ten spikes random sample was considered sufficient for evaluation of a line and 60 spikes for a bulk.

The other agronomic criteria were not different than in 6x triticale breeding.

Screening of the aluminium ions tolerance was made on the liquid media according to the procedure described by Aniol [10].

Evaluation of grain nutritive value was conducted with the use of the same methods as described in publication of Boros [11].

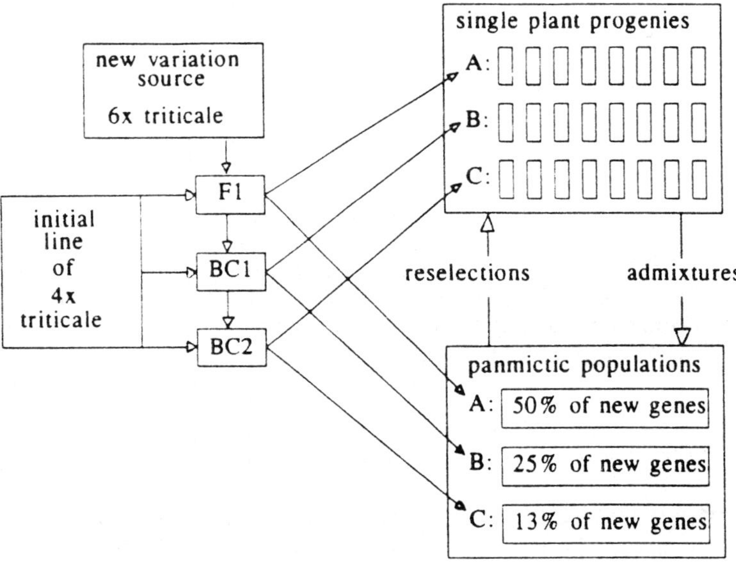

Figure 1. Breeding scheme of 4x triticale.

Results and Discussion

The crossing caused wide variation in the segregating populations, which enabled selection of improved forms.

In 1989 five first crossing-selection cycle derived secondary lines were compared with the initial 4x line and with 6x triticale on single 5m^2 plots. The yield of P1N parent was 1.82 t/ha, that of secondary lines varied between 2.46 and 3.54 t/ha, and of 6x triticale check (MALNO) amounted 5.96 t/ha. The results were paralell to those of spike fertility [12].

In the present stage of breeding, 4x triticale lines resemble inbred lines of rye showing inbreeding depression and admixture of vigorous F$_1$ hybrids.

630

Table 1. Comparison of average spike index for bulks, bulk reselection lines and individual selection lines derived materials carrying different proportions of new genes from 6x triticale. Radzików, 1993.

	individual sel.	bulk	bulk reselection
A: simple cross	2.5	2.2	3.7
B: 1st back-cross	3.3	7.2	8.1
C: 2nd back-cross	6.3	6.4	8.3

The lines derived from the first crossing-selection cycle showed distinct improvement of fertility when compared to the initial 4x-line. The number of grains per stalk has changed from 20 to 25-30 in the best yielding lines. However, in the same conditions, 6x triticale MALNO set above 48 grains per spike. At the same time the bulk populations yielded 25-28 grains per stalk [12].

Spike index was, on average, lower for the lines derived from the first breeding cycle controlled crosses than for the lines reselected from the respective bulks, as show the results of 1993 season presented in Table 1. It supports the idea of bulks presence in the breeding scheme. Bulks themselves showed fertility on the same or better level than the average of the individual selection lines. Probably bulks worked as spontaneous crossing blocks and the reselected lines represented the second (or more advanced) cycle of crossing and selection.

The differences between A-,B- and C-type materials indicate the necessity of uniform genetic background for a successful breeding population. The Apolinarska's investigation on chromosome constitution revealed diversity of karyotypes (unpublished). Among five C-bulks derived lines none retained tetrasomic status of chromosome 5R. In other homoeologous groups two lines had the same chromosome composition as P1N standard, two had one chromosome difference and one showed difference of two chromosomes. The other authors' publications contain evidence on almost regular pairing of homoeologous wheat chromosomes and on frequent chromosome substitutions in 4x triticale [7,13,14]. In such situation back-crossing to the 4x triticale parent seems really important for stabilization of karyotype.

The recently selected lines, mainly those from the C-bulk, showed spike index above 74%, close to that of 6x triticale (78-80% for MALNO). The only request at this fertility level would be stability of the character. Table 2 presents it for the best individuals and the best progenies. The highest indices were recorded for 1-3 spike plants reselected from bulks, however high results were seldom repeated in the next year. The best progenies were derived from better branched plants with not extreme spike index.

Since the segregating materials were produced, every year several plants appeared with extraordinarily high fertility exceeding 80 grains per spike. However in the next generation it dropped to the average or lower level. Such plants, when found on line selection plots, were usually off-types. It suggested their hybrid origin. The plants like 6A.1 and 6A.7 could not be considered a success, at least in line selection of a predominantly self--pollinating plant. On the other hand these high fertility plants may show importance of heterosis and inbreeding depression for the fertility problem.

P1N founder line showed relatively good grain filling when compared to other 4x lines. Its test weight was about 58 kg/hl. Numerous secondary lines reached 6x triticale level with results close to 70. Even bulks showed test weight above 67. The results for MALNO 6x check varied close to 71. Thus, grain shrivelling is not expected to be a difficult problem on 4x ploidy level.

Table 2. Spike index for the selected plants of 1992 season and for their progenies in 1993.

| Plant & line | 1992 | | 1993 | |
	No. of spikes	spike index (%)	spike index (%)	cross type
1A.8	3	79.2	70.5	A
6A.1	1	78.1	53.7	A
6A.7	2	76.1	51.9	A
3C.8	3	76.0	67.5	C
3C.7	3	75.9	66.9	C
3C.4	2	75.3	67.7	C
5C.1	1	74.4	73.6	C
C49..2	5	66.0	78.4	C
3C143.1	8	70.9	77.1	C
3C.3	3	70.6	76.6	C
4C.A	4	64.0	74.6	C

[*]) plant from the controlled cross (F7 generation); all other plants reselected from bulks.

Tetraploid triticale in grain shape resembles more rye than 6x triticale. It is elongated and the germ end is more acute. The difference in proportions is distinct enough to serve as a diagnostic character for 4x forms.

In spite of good plumpness germination of grain was low and unstable in the majority of 4x lines. It requires attention, because high germinability is necessary on weak soils to which 4x triticale is destined.

The rejection of seven pairs of wheat chromosomes could result in the loss of resistance genes in 4x triticale. However no increase of susceptibility to any disease was noticed, with the exception of ergot (*Claviceps purpurea* (Fr.) Tul.), which attacked semisterile materials in some years. Improvement of fertility should result in better resistance. Differences independent on fertility were noticed in ergot infection intensity.

According to the results of Budzianowski's investigation the expression of low pH and Al^{+++} ions tolerance is proportional to the rye genome share in triticale total genome complement. 4x triticale survival at high Al^{+++} concentrations (20 ppm at pH=4) was, on average, higher than for 6x cultivars (39.2% and 25.5%, respectively). Thus, possibilities of selection on 4x ploidy level seem better, especially with consideration of differences in breeding population size. However, the top results for 6x triticale (89.5% survival for SALVO and 83.3% for GRADO) show that ploidy level is not necessarily the limiting factor for Aluminium ions tolerance [15].

The initial P1N line was the weak one in this respect (only 5.3% of survival, in comparison to 12% for MALNO check in 1989) [11]. The secondary lines results, reaching 60%, were developed by breeding on average quality soils, without any screening and specific artificial selection pressure.

The reduced proportion of wheat chromosomes created the danger of more intense expression of rye genome controlled antinutritive compounds of grain. Rye-like properties of product could foil the idea of 4x triticale breeding. Thus the investigation of feed characters by Rakowska and Cyran on several 4x lines was particularly impotrant. Results are presented in Table 3, together with those for both parental species.

Body mass gain of chickens and feed efficiency ratio were, on average, lower for 4x triticale than for the 6x one. However they were much closer to 6x triticale than to diploid rye. The same rule was valid for protein digestibility in rats, content of alkyloresorcinols (AR), suspension viscosity and for soluble dietary fibre content (SDF). The last two indices were investigated as the ones most tightly correlated with results on animals, according to findings of Boros [11].

There is no reason to be too optimistic about higher protein content. Probably it will decrease, like in 6x triticale, as yield will grow. However, the increased content of lysine was an advantage of the rye genome. In this character domination of the wheat genome over the rye one was relatively weak.

The extremal values for particular compounds and indices show high plasticity of the tested 4x triticale in suspension viscosity, SDF and AR. It means, that better results may be reached without great breeding effort. Thus, tetraploid triticale could be an interesting alternative of other crops, if fertility and yield are improved.

Table 3. Comparison of grain nutritive value indices of 6x triticale, 4x triticale and 2x rye. (Rakowska, Cyran, 1990-1993). Different species' results came from different experiments. hy - hybrid rye forms, po - population of Dankowskie Zlote rye cultivar, ob. - number of forms investigated.

Indices	6x triticale			4x triticale			2x rye		
	ob.	average	range	ob.	average	range	ob.	average	range
body mass gain/10 days chicken (g)	8	53.3	46.9-57.6	7	49.3	44.4-53.0	13 hy	37.5	31.9-46.0
feed efficiency ratio	8	2.39	2.21-2.55	7	2.45	2.32-2.54	13 hy	3.03	2.75-3.34
protein content (%)	8	13.4	9.4-14.7	7	14.7	13.4-17.6	13 hy	9.6	9.0-11.2
lysine content (g/16g N)	8	3.58	2.94-4.09	7	3.65	3.54-3-81	13 hy	3.73	3.54-3.91
protein digestibility,rat (%)	8	87.4	85.3-89.8	10	87.1	84.3-89.1	13 hy	77.0	73.2-80.0
soluble dietary fibre (%)	4	2.11	1.78-2.50	6	2.78	2.18-3.35	1 po	3.94	
suspension viscosity (cP)	8	41	20-60	11	202	82-423	13 hy	2111	1461-3007
alkyloresorcinols (mg/kg)	8	533	430-673	11	664	494-839	1 po	1244	

Conclusions

During twelve years of 4x triticale breeding it attained 6x triticale level of grain plumpness and surpassed its average indices for low pH and aluminium tolerance.

Grain nutritional properties have not changed much with the decrease of ploidy level, in spite of removal of seven pairs of wheat chromosomes. The same is valid for resistance to diseases, with exception of ergot, provoked by sterility.

Yield potential of the initial form was almost doubled in some secondary lines, however it is still much below the 6x triticale standards. The main limiting factor is partial sterility, which is being slowly eliminated through natural selection supported by artificial one for spike index.

Hexaploid triticale proved to be a valuable source of new variation for all considered characters.

Bulk aided with reselected plants progenies is a successful breeding method, at least in the preliminary phase of 4x triticale improvement.

Back-crossing to one initial founder parent was important for stabilization of fertility.

Heterosis seems necessary to provide 4x triticale with full fertility and vigour.

References

1. Krolow K-D. Research Work with 4x Triticale in Germany (Berlin). In: Triticale: Proceedings of international symposium. El Batan (Mexico). 1-3 October 1973. IDRC-024e. 1974: 51-60.

2. Lukaszewski AJ, Apolinarska B, Gustafson JP, Krolow K-D. Chromosome pairing and aneuploidy in tetraploid triticale. I. Stabilized karyotypes. Genome 1987;29;4: 554-561.

3. Gustafson JP, Bennett MD. The effect of telomeric heterochromatin from *Secale cereale* on Triticale (x *Triticosecale*). I. The influence of the loss of several blocks of telomeric heterochromatin on early endosperm development and kernel characteristics at maturity. Can J Genet Cytol 1982; 24: 83-92.

4. Pilch J. Analysis of the chromosome constitution and the amount of telomeric heterochromatin in the widely and narrowly adapted hexaploid triticales. Theor Appl Genet 1981;60: 145-149.

5. Lukaszewski AJ, Apolinarska B, Gustafson JP, Krolow K-D. Chromosome Constitution of Tetraploid Triticale. Z Pflanzenzucht 1984;93: 222-236.

6. Badaev NS, Badaeva ED, Dubovets NI, Bolsheva NI, Bormotov VE, Zelenin AV. Formation of a synthetic karyotype of tetraploid triticale. Genome 1992;35: 311-317.

7. Hohmann U. Stabilization of tetraploid triticale with chromosomes from *triticum aestivum* (ABD) (ABD)RR (2n=28). Theor Appl Genet 1993;86: 356-364.

8. Zohary D, Feldman M. Hybridization between amphiploids and the evolution of polyploids in the wheat (*Aegilops-Triticum*) group. Evolution 1962;16: 44-61.

9. Lapinski B, Apolinarska B. Polish work upon 4x-triticale. In: Proceedings of EUCARPIA meeting on triticale. Clermont-Ferrand (France).2-5 July 1984. INRA. Paris. 1985: 261-265.

10. Aniol A. Aluminium uptake by roots of two winter wheat varieties of different tolerance to aluminium. Biochem Physiol Pflanzen 1983;178: 11-20.

11. Boros D. Comparison of antinutritional factors in control rye varieties and triticale breeding materials. In: Proceedings of the EUCARPIA meeting of the Cereal Section on Triticale. Schwerin (GDR).22-26 June 1987. ISSN 0138-2659. Tagungsbericht 1988; 206: 585-590.

12. Lapinski B, Budzianowski G. Tetraploid triticale breeding in the PBAES Malyszyn. Bull IHAR 1992;183: 119-124.

13. Lukaszewski AJ, Apolinarska B, Gustafson JP, Krolow K-D. Chromosome pairingand aneuploidy in tetraploid triticale. II. Unstabilized karyotypes. Genome 1987;29;4: 562-569.

14. Krolow K-D, Lukaszewski AJ, Gustafson JP. Preliminary results on the incorporation of D- and E- genome chromosomes into 4x triticale. In: Proceedings of EUCARPIA meeting on triticale. Clermont-Ferrand (France), 2-5 July 1984. INRA. Paris.1985: 290-295.

15. Budzianowski G. Tolerance of winter triticale seedlings of different ploidy level (4x,6x,8x) to low pH and high concentrations of aluminium ions. Zeszyty Problemowe IHAR. Hodowla zbóz.1990:131-137.

SOME REMARKS ON TRITICALE BREEDING EVOLUTION IN PORTUGAL
A) Role played by the National Plant Breeding Station

Francisco Bagulho, Manuel T. Barradas, Benvindo Maçãs and
José Coutinho
Estação Nacional de Melhoramento de Plantas, Elvas, Portugal

Abstract

The main steps of the triticale breeding evolution at NPBS are referred, showing its impact on the triticale utilization by the portuguese farmers.
Yield potential and test weight of the main varieties representing this evolution are presented.
New breeding strategies, consequence of the cereals price policy, are conducting to a different plant type, better adapted to a low input cropping systems. This new model of variety, based on longer growth cycle, can be decisive to the future of the triticale cultivation in Portugal.

Introduction

The first attempt to include wheat-rye crosses in a breeding program in Portugal was reported by Victoria Pires (1939). The objective was the selection of plants joining the morphological wheat characteristics and some rye physiologic traits, such as cold resistance, adaptation to poor soils or earliness [1].
In this sense, in the 50's, were realized crosses involving wheat and rye varieties. In the Cytogenetics Lab of the NPBS were obtained primary octoploid triticale plants [2,3]. These lines permitted some interesting theoretical studies but they did not show progress from agronomic point of view.
Since 1967 a fruitful cooperation with CIMMYT and National Plant Breeding Station gave us the possibility to observe at Elvas a large amount of germplasm. In 1969 we were considered as participants of the ITYN network. A continuous expansion of CIMMYT relationship gave the Elvas breeders a large amount of improved triticale germplasm and thus led to the establishment of a breeding program for the new cereal [4].
Based on this material severe selection criteria were applied in order to select varieties showing good adaptability to the agricultural prevalent system of Southern Portugal.

H. Guedes-Pinto et al. (eds.), Triticale: Today and Tomorrow, 635–641.
© 1996 *Kluwer Academic Publishers. Printed in the Netherlands.*

Alentejo, our main wheat region, is also the most important triticale region. According to an official information (INGA), in 1993 the triticale area in Alentejo was 98% of the total area occupied by triticale in Portugal.
The Portuguese farmers planted for the first time a triticale variety, Armadillo, in the growing season 1979.
Since then, triticale became a major crop in this Region, occupying the marginal soils normally released from wheat crop.

Breeding evolution

Following Armadillo, a second generation of varieties, showing very interesting characteristics, was distributed to the farmers in the beginning of the 80's. Successful varieties were Arabian, Bacum and Beagle, presenting very good behaviour in our difficult ecological conditions [5]. These varieties expressed excellent results in many farmers fields and stimulated decisively triticale cultivation in all the Region. They were included in the official list for certified seed production in 1982.
In fact, the high yielding capacity of these varieties under poor conditions was the main reason of the quick spread out of a new major crop for the traditional wheat areas.
Borba, registered in the National Catalogue in 1986, was selected by NPBS as an advanced line in a CIMMYT Screening Nursery and has been in the last years the most commercialized variety in Portugal. Another complete triticale, Juanillo, was also registered in 1988, with good results.
All this material has poor seed, high grain shrivelling, and test weight strongly influenced by environmental conditions [6].
To improve the grain characteristics, we are doing severe selection not only in CIMMYT germplasm but also in different segregating populations. In the 90's, new CIMMYT material favoured the selection of varieties expressing better grain characteristics, such as Alter, Arruda and Crato, which were recently included in the National Catalogue [7].
At present ten triticale varieties are registered in the Portuguese National Catalogue. Eight of these varieties were obtained by NPBS. However at least five varieties included in EC catalogues of other members are also commercialized.
In 1992/93, certified seed was 40% of the total planted area and our varieties represented more than 50% of the seed sold [8]. Borba 17.0%, Arabian 16.1% and Juanillo with 9.7%, were the top varieties.
In order to develop varieties with better adaptability to particular conditions of our agriculture, we are selecting promising lines in the segregating populations produced by NPBS. One of the most important traits to this kind of agriculture, with reduced costs and low inputs is the growth cycle. In brief terms we point out the need to select varieties adapted to early sowing and showing themselves both long vegetative period and short grain filling stage. However a considerable amount of breeding germplasm we are manipulating shows pronounced earliness at the flowering time. We may take into consideration that under agro-ecological conditions of the Southern Portugal late frost,

occurring in second fortnight of April, is an important constraint. Consequently, a cross program is being carried out by using adequate parents, aiming the induced genic systems fitted to those purposes.

Insensitivity to the photoperiod and non vernalisation requirements are characteristics of our traditional germplasm. So, we are doing crosses involving winter or alternative types, triticales and wheats, to increase genetic variability. Promising populations have been obtained, favouring selection of interesting lines, showing higher tillering and more adequate growth cycles. On the reverse, grain is not so attractive as the last registered varieties.

Summarizing up, we may say that one of the main objectives of NPBS program consists of changing the growth model of the actual triticale spring type [7,9].

Results

Triticale cultivation had a very quick development in Southern Portugal. Starting with an experimental area of 10 ha, in 1979, attained around 60 000 ha, ten years later. In fact, since the end of 80's triticale was transformed in a major crop, representing more than 10% of the wheat area (Table 1).

Table 1. Wheat and triticale in Portugal

	Área (ha)			Production (x10³ kg)			Yield (kg/ha)		
Year	Triticale	Wheat	Triticale Vs Wheat(%)	Triticale	Wheat	Triticale Vs Wheat(%)	Triticale	Wheat	Triticale Vs Wheat(%)
1988	16 479	293 681	5,6	24 718	411 153	6,0	1 500	1 400	107
1989	56 667	330 441	17,1	85 000	627 838	13,5	1 500	1 900	79
1990	38 000	259 000	14,7	34 200	284 900	12,0	900	1 100	82
1991	38 520	260 000	14,8	69 336	556 400	12,5	1 800	2 140	84
1992	47 970	233 100	20,6	52 767	349 500	15,1	1 100	1 500	73
1993	36 390	180 510	20,1	76 419	397 122	19,2	2 100	2 200	95

Source: ANPOC; INGA

The promising experience of Armadillo on farmer fields and the excellent performance of the second generation of triticales, Arabian, Bacum and Beagle, favoured a good acceptance of the new crop. In fact, results obtained at Elvas first trials (Table 2) permitted a good knowledge of this cereal and contributed decisively to the successful utilization of an unknown plant species by many farmers.

Table 2. Behaviour of the basic germplasm Elvas trials (1975, 1976, 1978)

Variety	Yield			Test Weight		
	Mean		Range	Mean		Range
	Kg/ha	%	Kg/ha	Kg/ha	%	Kg/ha
Armadillo	2957	100	2327-3880	73	100	72-75
Arabian	3543	120	2827-5193	72	99	70-72
Bacum	4631	157	3860-5211	72	99	70-74
Beagle	3953	134	2876-5730	70	96	69-71

Triticale is today very important to a lot of poor soils, specially acidic soils. It permits to maintain the traditional system involving cereals and extensive livestock, where triticale straw has been successfully used for feeding, with better results than wheat straw.

At the same time, grain market for animal feeding has been a good option to the growers, because the price was often quite stimulant. Triticale grains are frequently used for pig and poultry by feeding industry.

To the favourable evolution of the triticale cultivation it was decisive the dynamics of the new varieties distribution. The organization of the National Catalogue, in 1980, allowed a continuous spread of improved varieties. In this sense NPBS is developing an aggressive breeding program which permitted to select the varieties: Arabian, Bacum and Beagle, in a first group; Borba and Juanillo, the second group; and Alter, Arruda and Crato the last one.

In Fig. 1 we can observe yield data of this germplasm at Elvas trials during four years. All these varieties were selected in CIMMYT germplasm in different breeding stages and no significant differences were observed on the yield potential.

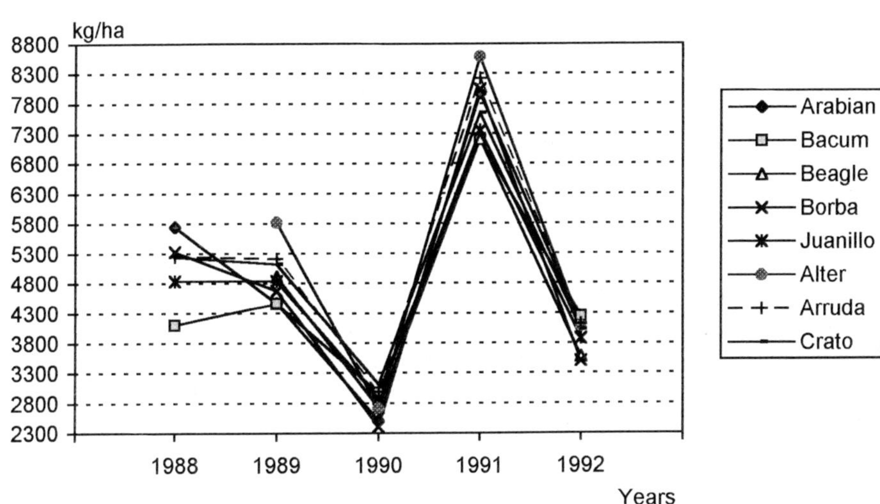

Fig.1 - Yield potential of the portuguese varieties Elvas Trials

Fig. 2 shows some progress on the newer wave of varieties, representing higher yielding capacity along the seasons.

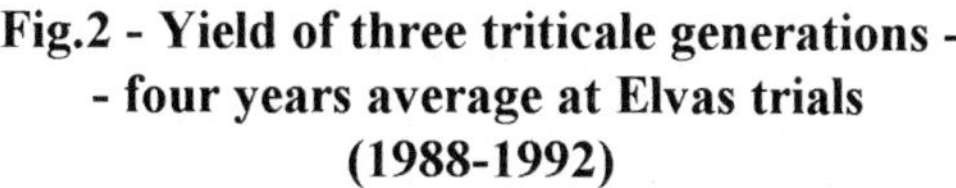

**Fig.2 - Yield of three triticale generations -
- four years average at Elvas trials
(1988-1992)**

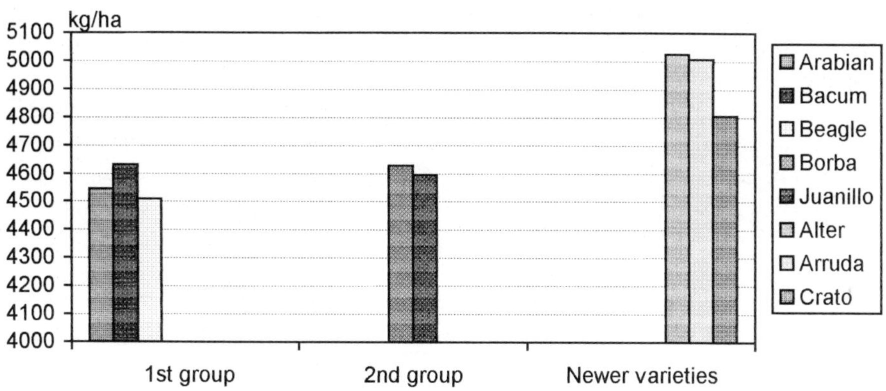

Arabian, Bacum and Beagle have high yielding capacity, good adaptability to different conditions, but they show low hectoliter weight; the substituted lines are phenotipically unstable what is not permitted in the certified seed fields.

Borba and Juanillo have similar behaviour, showing a slight progress comparatively to the first group. They are much more stable under the morphological point of view and they have better potential to the weeds competition.

Alter, Arruda and Crato are more advanced types, representing a superior breeding step. They have better yielding potential; the grain is much more attractive and less affected by the environment (Fig. 3).

Of course, all this material has some common weak points. They have early flowering time and a large grain filling period. In the last growing seasons, under severe drought and frost conditions during the flowering stage, triticale spike fertility was quite disappointing. Lodging problems can be frequent.

On the other side, until now this material expresses normally good tolerance to the main diseases, rusts and *Septoria tritici*. But, in the dry seasons they express high susceptibility to some insects, like *Calamobius filum* [10,11,12], responsible for high percentage of neck break.

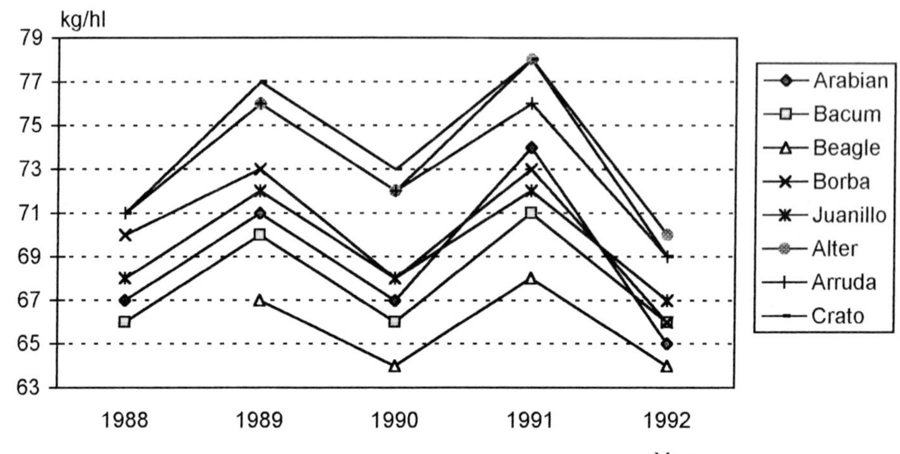

Fig.3 - Test weight of the Portuguese varieties
Elvas trials

Conclusions

After very few years of cultivation, triticale has significant possibilities to be an important crop for many regions of Portugal. Breeding efforts must be done to improve some of the main triticale constraints, to allow increase significantly yield potential and to select varieties for particular end uses.

Another strong constraint is a consequence of the cereals price policy defined at EU. Lower prices than competing cereals will have a serious negative impact in the future of this crop.

Data of the NPBS adaptation trials show the excellent possibilities of the triticale in relation to wheat, in different locations [13]. As triticale is a very new crop, we can hope that a strong potential can be exploited by breeding work.

Lower prices can be compensated with better performance, particularly in marginal areas, and, at the same time, triticale can be selected to a low input cropping system able to decrease production costs.

Success of triticale breeding will depend on the capacity to increase the variability of the germplasm base, in order to obtain new varieties well adapted to the needs of the tomorrow agriculture.

References

1. Victoria Pires, DR. - Linha geral de um plano para melhoramento de cereais e forragens. Oeiras. 1939.

2. Villax,E., Mota,M. & Ponce-Dentinho,A. - Dois novos triticales: Melhoramento 1954; 7: 29-56.

3. Villax,E. - Le comportement de nos triticales et la production des triticales neufs en masse: Melhoramento 1957; 10: 67-80.

4. Bagulho,F. & Barradas,M.T. - O melhoramento de triticales na EMP. I Reunião Portuguesa Sobre Triticales; 1977 June 2; Vila Real. Mimeo.

5. Barradas,M.T. - Perspectivas do triticale na problemática da cerealicultura portuguesa. Melhoramento 1982; 27: 29-48.

6. Bagulho,F. & Varughese,G. - Adaptation of triticale in the south of Portugal. Gen. and breed. of Triticale, EUCARPIA Meet., Clermont-Ferrand (France), 1984 July 2-5 INRA, Paris, 1985.

7. Bagulho,F., Maçãs,B. & Coutinho,J. - Metodologia e objectivos do melhoramento de triticale na ENMP. Melhoramento 1992; 32: 109-20.

8. ANSEME - Cereais de Sequeiro. Produção e Comercialização de Semente Certificada. Campanha de 1992/93.1993.

9. Maçãs,B., Silva,J.C., Bagulho,F. & Coutinho,J. - Produtividade em variedades de triticale e de trigo mole. Melhoramento 1992; 32: 141-46.

10. Amaro,F. - Cálculo de prejuízos causados por *Calamobius filum* (Rossi) [Coleoptera: Cerambicidae] em cultivares de Triticale. Diss. INIA Conc. Inv. Aux. 1989.

11. Amaro,F.S. - Prejuízos causados por *Calamobius filum* (Rossi) [Coleoptera: Cerambicidae] em Triticale (var. "Bacum") em Évora. Melhoramento 1992; 32: 243-57.

12. Mexia,A.M.M. - A cultura do trigo e os prejuízos causados por algumas pragas. Subsídios para o seu cálculo. Dissert. Mestr. Prod. Veg. Lisboa. UTL-ISA, 1985.13. Coutinho,J., Maçãs,B., Bagulho,F. & Coco,J. - Estudo comparativo do comportamento do triticale e do trigo mole em seis anos de ensaios. Melhoramento 1992; 32: 133-40.

TRITICALE RESEARCH AND BREEDING PROGRAMMES IN FRANCE : RECENT DEVELOPMENTS

Michel Bernard, Sylvie Bernard, Hélène Bonhomme, Corinne Faurie, Georges Gay, Louis Jestin

*INRA- Génétique et Amélioration des Plantes
Domaine de Crouelle, Clermont-Ferrand, France .*

Abstract

After outlining the recent development of triticale crop and breeding efforts on this species in France - 160 000 ha in 1993 -, the main aspects of genetics and breeding programmes are described . This includes the selection schemes involving 3 generations of bulks, broadening the diversity of 6x triticale germplasm pool via primary 8x - with kr compatibility genes - and use of anther culture to produce DH lines . Information is also given on tetraploid triticales and concerning genetic resources conservation .

Introduction : importance of triticale in France

During the past 10 years, the acreage of Triticale cultivated in France has increased regularly to reach 160 000 ha nearly in 1993, being until recently the sole cereal increasing its area (Table 1) . Cultivation areas are concentrated in cattle breeding regions, replacing rye crops in mid-altitude areas of the Massif Central and in Brittany. The position of triticale in the future is not easy to predict, with the changed deal related to the European agricultural policy ; however the overall disease resistance and "low input" characteristics of this species are often considered as serious assets.

Triticale breeding efforts involve both public -INRA- and private organizations-at least 5 companies - resulting in a moderate but continuous flow of new cultivars, e.g. recently : TRICK, TRIMARAN, AUBRAC, COLOSSAL...

Table 1- *Recent evolution of triticale cultivation in France.*

HARVEST YEAR	WINTER TRITICALE ACREAGES	
	CROP (ha)	SEED INCREASE (ha)
1984	80 000	2440
1986	100 000	4160
1988	125 000	3380
1990	145 000	4730
1991	159 000	4840
1992	174 000	5558
1993	159 000	4121

H. Guedes-Pinto et al. (eds.), Triticale: Today and Tomorrow, 643–647.
© 1996 *Kluwer Academic Publishers. Printed in the Netherlands.*

Triticale genetics and breeding

BREEDING FOR ADAPTED CULTIVARS

The main breeding objectives still associate a high regular yield without lodging, resistance to important diseases (brown rust, Septoria, eyespot) together with winter adaptation, low presprouting and easy threshing [1].
Breeding methods among reasonably adapted gene pools involve first a conventional approach consisting in crossing together fairly acceptable 6x x 6x genotypes - or 6x by bread wheats - . The bulk method is generally applied to early generations in selective environments since instability and/or sterility problems are often encountered at this stage, whereas the pedigree approach is delayed to the F5 onwards .

The adaptation to mid-altitude is ensured by selecting the segregating material in a station at about 800m above sea-level, where root system installation, winter hardiness, and appropriate growth cycle can be conveniently scored: current critical features of genotypes in hill areas consist in too early stem elongation and floral development, and the ensuing risk of ear damage by frost - sterility - or conversely a too late grain filling and maturation period in summer, leading to a risk of water stress - shrivelling and yield losses - or, later, moist weather causing presprouting during harvest time. Efforts are therefore carried out to breed "ideotypes" combining a relatively delated stem elongation together with a sufficiently early maturation.

The first yield trials are initiated on sufficiently homozygous, stable and fertile F6 material. Frequent proneness to outcrossing leads us to bag the material from F4 onwards to secure a better homozygosity.

BROADENING THE GENETIC DIVERSITY

Primary triticales : improving wheat x rye cross compatibility. Since it is assumed that the presently available variability among 6x adapted Triticales is quite limited, significant efforts are made towards increasing the genetic diversity through new primary triticale combinations. After a preliminary assessment of their potential interest vs. shortcomings, selected 8x genotypes are crossed with 6x improved lines to produce more adaptable types.

Table 2- *Crossability with rye of F1 [Courtot 5B and 5A monosomic x compatible wheats]*

GENOTYPES	Allelic constitution	% Seed set
COURTOT	Kr1Kr1 Kr2Kr2	7 %
NORIN 29	kr1kr1 kr2kr2	95 %
FUKUHOKOMUGI	kr1kr1 kr2kr2	86 %
[mono5B]COURTOT	Kr1 - Kr2Kr2	18 %
F1(mono5B*N.29) x rye	kr1 - Kr2kr2	70 %
F1(mono5B*FUKUOK.) x rye	kr1 - Kr2kr2	47 %
F1(mono5A*N.2 9) x rye	Kr1kr1 kr2 -	3 %
F1(mono5A*FUKUOK.) x rye	Kr1kr1 kr2 -	0 %

To facilitate the introgression of selected soft wheat as well as rye germplasm, the cross compatibility genes *kr1* and *kr2* have been introduced into various wheat genotypes (Table 2). Use of appropriate markers to assist these conversions would be helpful in the future [2].

Developing tetraploid triticales. Primary and secondary 4x Triticales (A/B A/B RR) were also obtained, the latter - those derived from primary 4x intercrossing - being much more vigorous and fertile. Various methods to develop substituted R/D primary Triticales have also been investigated. Series of intergeneric hybrids, involving Aegilops squarrosa as D-genome donor, have been obtained, with amphiploids showing acceptable levels of vigour and fertility in field nursery conditions. This material has been used to develop secondary populations (Table 3). In particular, hexaploid Triticales with complete R and D genomes have been developed [3], [4] .

Table 3- *Results of crosses of Aegilops squarrosa (D genome) with rye (R genome) vs. with tetraploid triticale ([A/B] and R genomes).*

	Ae.squarrosa x rye	Ae.squarrosa x T4x
F1 generation (greenhouse)		
Number of pollinated ears	13	4
%viable embryos/poll.florets	22.9	19.0
Number of F1 plants raised	30	11
% plants /100 embryos	20.1	28.9
Amphiploidized plants (greenhouse)		
Number of colchicinated plants(Co)	30	11
Number of fertile Co plants	24	11
Number of fertile ears	86	8
% fertile ears	7.6	2.9
Number of grains/plant (range)	14.8(0->74)	3.6(0->22)
C1 selfed amphiploids (field)		
Number of plants	36	9
Mean plant height (range) cm	131.6(120->180)	95(30->130)
Nr tillers/plant (range)	35.0 (16->55)	10.4(2->15)
Mean number of grains/plant(range)	33.7(0->143)	32.7(0->70)

Triticale genetic resources conservation. Broadening of the variability is also ensured via exchanges within various cooperation networks (CIMMYT, EUCARPIA, French Triticale Breeders) . As regards older germplasm, the management of the collection and completion of a computerized database according to IPGRI recommendations (for wheat) is in progress. Presently, the triticale database is installed on a microcomputer. The status of the related collection is that of an active one, with medium-term cold storage.

Use of anther culture to produce DH lines

Anther culture has been successfully applied to Triticale : doubled haploids are routinely produced from the most promising crosses (usually in F3) and for other research purposes. This technique enables a rapid fixation and helps to reveal at an early stage both outstanding lines and also those likely to be constitutively unstable [5] .

The method (media, inflorescence pretreatments, genetic control..) has been thoroughly investigated. Although unpredictable genotypic differences exist among genotypes, green haploid plants can be produced from most genotypes in this species : DH lines produced in 1992 from F6 and F3 lines, grown in 1993 in greenhouse after colchicine treatment , and in the field nursery in 1993-94 (Table 4). Still, the cost remains somewhat high, therefore androgenesis is used in priority on more attractive progenies. Care has to be taken to maintain homozygosity in further generations, as Triticale is relatively prone to outcrossing. The present best DH lines compare favourably with those derived from the bulk or pedigree schemes, as exemplified by the presence of DH lines in trial networks.

Table 4- *Typical response in DH production using F3 or F6 material from an unsorted series of of 17 hexaploid triticale crosses.*

T6x x T6x Cross (a few F3 or F6 from each)	Anthers in culture	Embryos /100 anthers	Green plants	Green plants /100 anthers	%albino plants	Fertile DH followed
(BR xSALVO)	4800	7.6	18	0.37	66	11
(UH x 78TT114)	3960	28.7	256	6.46	18	37
(CHD265 x SALVO)	4320	4.4	23	0.53	8	12
(CHD265 x 80T26-5)	3600	12.6	8	0.22	84	2
(LT1369 x 80TBT43-2)	4240	13.3	30	0.71	30	16
(LT1369 x MAGISTRAL)	4960	18.5	97	1.96	37	58
(BOLERO x SALVO)	4440	35.1	128	2.88	62	50
(BOLERO x 80T86-14)	11640	23.2	296	2.54	40	168
(SALVO x CI539)	3400	2.9	7	0.21	61	5
(DAGRO x 80TBT43-2)	4700	2.4	21	0.45	47	6
(LASKO x 80TBT43-2)	5500	11.6	78	1.42	39	39
(CLERVIX x 80TBT43-2)	5600	3.8	5	0.09	90	3
(CI648 x CENTRAL)	2160	26.0	30	1.39	64	7
(CI2-10 x CENTRAL)	2160	8.0	11	1.15	56	3
(CI2-7 xBOLERO)	2040	5.5	3	0.15	83	1
(CI2-7 x CENTRAL)	2040	29.3	79	3.87	52	27
(CI41-3 x PRESTO)	2160	08.8	29	1.34	38	15

Androgenesis on cultivars is also used to produce aneuploid plants, preferably with 40 or 40 + 2t chromosomes. Identification of missing chromosomes in a set of 200 DH lines is in progress. Possessing such an aneuploid set should facilitate the location of genes of interest, or the mapping of the Triticale genome .

Conclusion: future prospects.

Other research directions in the future should include more attention focused on grain quality and utilization possibilities, especially by the food processing industry. For this purpose, a better knowledge of triticale grain components : proteins and carbohydrates essentially (including non-starchy polysaccharids) is required, in particular regarding average composition, genetic vs. environmental variation, and potential improvement through breeding practices.

References

1   Bernard M, Jestin L, Bernard S, Gay G, Bonhomme H. Le Triticale en France: historique, évolution et perspectives.C R Acad Agr France 1992 ; n°4 : 23-36.

2   Gay G, Bernard M. Production of intervarietal substitution lines with improved interspecific crossability in the wheat cv. Courtot. Agronomie 1994 ; 14 : 27-32.

3   Bernard M, Bernard S, Joudrier P, Autran JC. Biochemical characterization of lines descended from 8x triticale x 4x triticale cross. Theor Appl Genet 1990 ; 79 : 646-653.

4   Bernard S, Bernard M . Creating new forms of 4x, 6x and primary triticale associating both complete R and D genomes. Theor Appl Genet 1987 ; 74 : 55-59.

5   Faurie C. Androgénèse in vitro chez le triticale : recherche de conditions environnementales optimales. Influence du niveau de génération des plantes- mères sur la réponse à la méthode et sur les caractéristiques cytologiques et agronomiques des plantes androgénétiques ; 1992 ; Thèse de doctorat n° 427 - Université de Clermont-Ferrand II, France. 125p.

TRITICALE BREEDING IN SWITZERLAND

Aldo Fossati, Dario Fossati & Gert Kleijer

Station Fédérale de Recherches Agronomiques de Changins (RAC), Rte de Duillier, case postale 254, CH-1260 Nyon, Switzerland

Abstract

Triticale research in Switzerland started in the beginning of the fifties and was focused on the enhancement of the fertility of octoploid triticales. From 1976 onwards, research on triticale was gradually abandoned to make place for the development of the more promising hexaploid triticale. Today 11.000 ha are cultivated in Switzerland. Triticale will get increasing importance in marginal agricultural zones of Switzerland and Europe. This is dependent on the enhancement of yield stability. The aim of our breeding programme is the creation of short straw triticales with good grain formation and resistance to diseases and preharvest sprouting. The first step in shortening straw lenght by recurrent selection resulted in two new varieties about 10 to 15% shorter than the variety Laskowithout reduction in yield potential. A very rigorous selection is carried out for grain formation from the F2 generation. This objectif is especially difficult to achieve for short straw triticales. The resistance level of our material to yellow rust, brown rust and Septoria nodorum is good. An artificial infection method is being studied at our institute for the screening of resistance to head blight (Fusarium). The cold tolerance level of our material is sufficient for our climatic conditions. However triticale is rather sensitive to snow mould. Attemps are undertaken to transfer resistance genes from landraces of wheat into triticale. Five varieties of our programme have so far been registrated : Gaetan and Formulin in France, Brio, Meridal and Tridel in Switzerland. Other lines are being tested in official trials in several European countries.

H. Guedes-Pinto et al. (eds.), Triticale: Today and Tomorrow, 649–653.
© 1996 *Kluwer Academic Publishers. Printed in the Netherlands.*

Introduction

About 11.000 ha of triticale are actually cultivated in Switzerland; this represents 7% of the cereal growing area. Triticale is cultivated between 600 m and 900 m elevation on farms specialized in animal production. It outyields other cereals in these regions and the forage value of the grain is superior to that of barley. Furthermore the rusticity of triticale avoids the farmer an important supervision of the culture.
Triticale culture may spread to cereal growing regions during the next years for different reasons. A triticale market is developing to meet demand by manufactures of feed for livestock. The cultivation of bread wheat will become less interesting after the opening of the borders as a consequence of GATT agreements. Finally the development of rustic species like triticale will be favoured by the obligation for farmers to use more ecological production techniques.
One of the aims of our breeding programme is to secure cultivation of triticale in high altitudes by increasing its snow mould resistance [1]. Our programme focuses also on the extension of triticale cultivation to other regions by selecting productive triticales with good grain formation, resistant to sprouting, to lodging (short straw) and to leaf and ear diseases [2, 3].

Material and methods

Two hundred and fifty to three hundred crosses between hexaploid triticales are carried out each year. This programme is completed by the creation of primary triticales [4] and a number of crosses between triticale and wheat. Artificial inoculation with *Septoria nodorum* and brown and yellow rust, is carried out from F_2 onwards. Field selection of short strawed, disease resistant plants is completed in the laboratory by a severe screening for grain formation. Anther culture is applied on some promising F_2 populations [5]. Disease resistance is continuously monitored in specific artificially inoculated nurseries parallel to the yield and official trials. New breeding lines are cultivated at different locations, three for the yield trials (three years) and ten for the official trials. These are located between 400 m and 900 m elevation.
Breeding lines are sown in a nursery at 1200 m elevation with a snow covering period of three to four months. A good natural infection with *Fusarium nivale*, *Typhula ishikariensis* and *T. incarnata* is obtained in these conditions. The same lines are also inoculated artificially with *F. graminearum* and *F. culmorum* to select for head blight resistance.

Results

The varieties Brio, Meridal and Tridel have been released in Switzerland, Gaetan and Formulin in France. Their most important characters are presented in table 1. Other lines are tested in official trials in several European countries.

Table 1. **Some agronomic characteristics of three Swiss varieties compared to Dagro** (data are averages of the official trials in ten locations over three years)

	Dagro	Brio	Meridal	Tridel
year of release	*1988*	*1991*	*1992*	*1994*
Grain yield (kg/ha)	7821	7836	7635	8196
Test weight (kg/hl)	68.5	69.8	67.1	68.7
Plant height (cm)	129	118	109	110
Lodging (score)	1.8	1.5	1.2	1.2
Falling number (second)	113	77	66	108
Protein (in % DM)	12.1	12.1	11.7	11.7

Straw length has been decreased by 15% to 20% in our breeding material in comparison to traditional varieties. At the same time, yield was increased by 5% to 10%. Disease resistance is good and sensitivity to mildew has never been observed. A high resistance level for brown rust, yellow rust and *Septoria nodorum* has been maintained thanks to an appropriate choice of parents and selection following artificial inoculation of the nurseries (table 2).

An important number of triticale lines are tolerant to head blight as shown by the first artificial inoculations. On the other hand, we did not obtain any progress in grain formation. Visual sorting out and measuring of test weight allowed the maintenance of an intermediate level for this trait. Progress for resistance to snow mould and sprouting have not yet been made either.

652

Table 2.   Disease resistance of three Swiss varieties compared to Dagro

Data are averages over three years, except *Fusarium* (one year). The scale for yellow and brown rust is 1 = resistant, 9 = very susceptible. For *Septoria*, the index under the disease development curve is given as percent of the trial average. For *Fusarium*, the percent thousand kernel weight infected/non-infected is given.

	Dagro	Brio	Meridal	Tridel
Yellow rust	2.0	2.2	2.0	1.7
Brown rust	1.2	1.2	1.0	1.0
Septoria nodorum **on leaf** **on spike**	102 114	104 105	86 91	102 100
Fusarium	85.5	78.9	75.1	75.6

Discussion and perspectives

The varieties studied have good agronomic characteristics but are not resistant to snow mould. The natural infection occurring in our nursery at 1200 m elevation is extremely severe and allows only absolute resistance to snow mould. We thus tried to sow our material in a nursery at 900 m elevation, but the snow covering period was very variable from one year to another. Actually we can only assert that the resistance level of triticale is low and that it is impossible to detect genetic variability. This is the first indispensable step for recurrent selection. The severity of this test has allowed to identify five wheat landraces with absolute snow mould resistance. Research is carried on in two directions : one has the objective to transfer resistance of wheat into triticale (crosses of triticale X wheat) and the creation of primary triticales, the other focused on developing methods for laboratory tests, independent of the climate, by the use of *in vitro* selection.

The new varieties satisfy to a large way to the requirements of cereal growers. A good yield potential with lodging resistance and resistance to the most important diseases has been realized. Yield stability is not yet satisfactory. However we could show that one of the reasons is the variation in test weight. Research is carried on to determine the causes of these variations and to improve the selection techniques. Measures of falling number proved not to be very useful for triticale as grain forage crop. Values are generally very low and genotypic variation weak. Furthermore no relationship between falling number and grain formation or germination rate could be demonstrated. Grains with a falling number of 70, germinated for more than 90%.

New programmes

HYBRID TRITICALE

This programme is implemented in collaboration with the Swiss Seed Producers Federation. We produce fifty to one hundred hybrids each year with a gametocide with the aim to control the seed production techniques and to define the characteristics of male and female lines. The most interesting F_1 hybrids were obtained by crosses between triticales with dwarfing genes and tall triticales. Increasing the agronomic value of female lines is necessary before producing a commercial hybrid. We have several lines with the Rht3 gene. These are characterized by poor tillering and bad grain formation. Other sources of dwarfing genes have been introduced into our material [3].

SPRING TRITICALE

A series of crosses is carried out between CIMMYT triticale lines and Swiss spring wheats and winter triticale in order to adapt spring triticale to our climatic conditions. The spring triticale variety Sandro was released in 1992 and is a selection of the line Stier"S" from CIMMYT. Sowed early (end February, beginning of March), this variety profits by a long tillering period. Yield is much higher than barley or spring wheats. Late sowing (end March, beginning of April) however, resulted in low spike densities and yield.

TRITICALE FOR GREEN FORAGE

A mixture of rye and vetch is sown by the farmers in autumn to assure green forage production in spring. The cutting is carried out before heading of rye to avoid a loss of the forage value. Triticale having a later heading would allow more flexible utilization of this type of production. New varieties must have higher dry matter production than rye. A specific selection programme to improve this character is under way.

References

1. Fossati A, Kleijer G. Selection of triticale for marginal areas. Hodowla Roslin Aklimatyzacja Nasienmictwo 1980; 24: 513-516.
2. Fossati A. Sélection du triticale en zone marginal. CR Acad Agri France 1992; 78: 15-21.
3. Fossati A, Fossati D. Le triticale : du laboratoire au champ. Rev Suisse Agri 1992; 24: 31-37.
4. Kleijer G, Fossati A. Production and fertility of hexaploid primary triticales. In Proc III Int Triticale Symposium June 13-17 1994; Lisbon, Portugal.
5. Kleijer G. L'androgénèse chez le triticale : premiers résultats. Rev Suisse Agri 1991; 23: 299-303.

CROP MANAGEMENT OF TRITICALE IN BELGIUM

Geert J.W. Haesaert & Antoon E.G. De Baets
Industiële Hogeschool, C.T.L.
Voskenslaan 270
9000 Gent - Belgium

ABSTRACT

Since about 15 years triticale is introduced in practical farming in Belgium. Triticale is mainly grown on sandy soils as a substitute for rye and winter barley.

Triticale yielded in comparison with the established cereals very well on the loamy sand soil at Merelbeke. 'Alamo' exhibited during severals years a high and regular yield. In general the level of disease resistance was satisfactory. Results of fungicide experiments showed that triticale had a low yield response to fungicide treatments. Most herbicides recommend for wheat can be used for triticale. However, varietal differences in reaction to methabenzthiazuron, isoproturon and mecoprop was found. Growth regulator sprays kept lodging and leaning to acceptable levels and affected the yield positively. On the loamy sand soil of Merelbeke the optimum nitrogen level was likely to be in the region of 130-160 kg/ha.

INTRODUCTION

Since about 15 years triticale is introduced in practical farming in Belgium. The area sown rose from 200 ha in the early 1980's to 8853 ha in 1993. Triticale is mainly grown on sandy soils as a substitute for rye and winter barley. Much of the interest in triticale is due to its wide adaptation to contrasting enviromnents, its higher level of resistance to the majority of diseases that attacks wheat and barley and its in general simple crop management. In recent years there has also been considerable interest in triticale from the typical wheat farmers because it is seen as a low cost alternative to the established cereals.

Due to the growing interest and the lack of comprenhensive information about the possibilities and the crop management practices of triticale, an extensive research project was started up in 1980 at the technical university, C.T.L.

This article deals with the most recent results of this research programme.

MATERIAL AND METHODS

The experiments were carried out on a loamy sand soil (3.3 % clay (0-2 μm), 3.6 % fine loam (2-6 μm), 9.6 % coarse loam (6-20 μm) en 83.5 % sand (>60 μm); 2.4 % organic matter, C.E.C.: 6.1 meq/100 g; pH KCl: 5.5)) at the experimental farm of the technical university,

H. Guedes-Pinto et al. (eds.), Triticale: Today and Tomorrow, 655–661.
© 1996 *Kluwer Academic Publishers. Printed in the Netherlands.*

C.T.L. All experiments were set up according to completely randomized block designs with 4 replications. Plot size was always 12 m² with 15 cm row spacings. All herbicide and fungicide applications were done with a knapsack sprayer under low pressure and with 300 l/ha water. The triticale was harvested mechanically and the grain yield was calculated to a 15 % moisture content.

RESULTS

YIELD POTENTIAL OF TRITICALE

During several growing seasons the yield potential of triticale wa compared with that of wheat, rye and winter barley (table 1). In each trial at least 2 varieties of each species was included. The most optimal crop husbandry measures were taken for the 4 species studied. To minimize the effect of shading and competition between genotypes of different height buffer plots were included in the experimental design.

Tabel 1: Average yield of triticale, wheat, rye and winter
barley during several growing seasons

Year	Triticale (kg/ha)	Wheat	Rye	Winter Barley
1984	7180	95.9(1)	96.2	98.6
1985	8797	74.3--(2)	79.0--	93.8
1986	6728	84.3--	96.8	-
1987	6832	73.2--	99.3	-
1988	6610	79.1--	93.9-	-
1989	7977	75.7--	89.7(3)	93.9
1990	7290	97.6	89.0 (3)	-
1991	6686	107.9	81.4-- (3)	-
1992	6684	90.4-	89.3- (3)	-

(1) Relative yield in % of yield of triticale for the same
    year
(2) -;--;+;++: yield significantly different (lower or
    higher) from that of 'Alamo' at P0.05 and P0.01
    respectively according to LDS multiple range test
(3) Hybride rye was included

Under the soil and climatological circumstances of Merelbeke, triticale was in general more productive than wheat, rye and winter barley (86.5 %, 90.4 % and 95.4 % respectively). Even the more intensive crop management measures taken on the wheat plots and the introducing of hybrid rye, led not to higher yields compared to triticale.

CHOICE OF VARIETY

Until today 5 varieties of winter triticale are liscensed in Belgium namely 'Alamo', 'Lasko', 'Lukas', 'Trimaran' en 'Salvo'. 'Alamo' takes more than 80 % of the total surface. Since 1980 a screening programme was started up at our institute to assess the material available commercially and in the earlier stages of development (1). Yield results of the last 5 years and of the frequently tested varieties are presented in table 2.

Table 2: Average yield (relative yield as % of Alamo) of
         triticale varieties studied on the experimental
         fields of the I.H., C.T.L.

Variety	1988	1989	1991	1992	1993
Alamo (1)	10328	7691	7229	6813	7480
Lasko	85.7--	83.3--	90.4-	95.0	93.0-(2)
Purdy	90.0-	90.4	101.6	95.4	96.7
Lukas	92.9-	88.1	88.6-	98.9	90.4
Pablo	79.9--	96.7	72.6--	97.1	78.8--
Bolero	96.1	90.2	100.1	-	-
Domino	90.6-	98.2	92.4-	-	-
Ego	-	-	99.8	-	108.8+
Trimaran	-	-	-	100.1	110.5++
Falko	-	-	-	-	101.5
Origo	-	-	-	-	102.9
Pegro	-	-	-	-	96.9
Galtjo	-	-	-	-	96.4

(1) yield of 'Alamo' in kg/ha
(2) -;--;+;++: yield significantly different (lower or
    higher) from that of 'Alamo' at PO.05 and PO.01
    respectively to according LDS multiple range test

The triticale varieties and genotypes tested in these trials
showed a wide range of yield potential. Especially 'Alamo' exhibited
a high and regular yield. 'Lasko' during the 1980's the most sown
and best yielded variety seemed to be a distinct lower yield
potential than 'Alamo'. 'Lukas', 'Domino' and 'Pablo' were also not
yielded very well. The new generation of varieties such as 'Ego',
'Falko', 'Origo' and especially 'Trimaran' demonstrated a same yield
level of 'Alamo' and are very promised to the future. Test weight of
the varieties studied fluctuated between 64.4 and 71. 'Alamo',
'Lukas', 'Lasko', 'Ego', 'Falko', 'Galtjo', 'Origo' and 'Pegro'
performed better than the other varieies with an average test weight
of more than 70.

Since some of the triticales tested were rather tall and weak it
was decided to applied a moderate rate of chlormequat to the
varieties in order to avoid excess lodging. Nevertheless, lodging
was frequently observed on the plots of 'Lasko' and 'Lukas'. In 1989
and 1993 lodging was also substantial on the 'Alamo' plots. 'Purdy',
'Tramaran' and 'Pablo' exhibited a good strawstiffness and lodging
resistance.

In general 'Alamo' and 'Trimaran' matured about 8 till 10 days
earlier than 'Purdy', 'Domino' and 'Pablo'. The other varieties
showed a moderate earliness. Remarkable is that of the varieties
with a late maturity only 'Purdy' yielded rather good.

DISEASE RESISTANCE

Until today diseases have not appeared as a serious limiting
factor for the yield of triticale under the Belgian growing
circumstances. Nevertheless, triticale is susceptible to most of the
diseases which attack wheat and rye, though it normally shows higher
levels of resistance than eihter of these crops.

In general the level of disease resistance of the varieties studied on the experimental fields of the technical university, C.T.L. was satisfactory (table 3). All varieties included in the trials were immune to powdery mildey, *Erisiphe graminis*. Leaf blotch, *Septoria tritici*, was the most widespread occuring disease. Every year a more or less important infection level of *Septoria* could be noted. The varietal differences in susceptibility to leaf blotch were rather small, although 'Purdy' and 'Ego' seemed to be less susceptible. On the other hand a little more septoria spots were observed on the leaves of 'Trimaran' during the last 2 growing seasons. Leaf rust, *Puccinia recondita*, was presented every other year. 'Lasko', 'Lukas' and 'Bolero' were recorded as the most susceptible to leaf rust. It is also clear that the susceptibility of 'Lasko' and 'Lukas' is increased during the last growing seasons due to the higher selection pressure of the pathogen. Glume blotch, *Septoria nodorum*, and head blight, *Fusarium ssp.*, were occasionally observed. Head blight was a problem by extended rainfall near harvest ripeness. Under these weather conditions especially 'Alamo' showed a high infection level. Glume bloth occured particulary on spring triticales.

Table 3: Susceptibility of several triticale varieties to leaf blotch, *Septoria tritici* and leaf rust, *Puccinia recondita*

Variety	Leaf blotch (1)						Leaf blight (1)		
	1988	1989	1990	1991	1992	1993	1988	1989	1990
Lasko	5.5	4.5	2.5	3.0	6.0	3.0	1.0	8.0	6.5
Bolero	3.5	3.5	3.0	2.0	–	–	2.0	3.0	5.0
Alamo	4.0	3.5	3.0	2.0	6.0	2.0	1.0	2.0	1.0
Mical	4.5	3.5	3.5	–	–	–	1.0	1.5	1.0
Purdy	4.0	3.5	2.5	4.0	5.0	–	1.0	1.0	1.0
Lukas	–	3.5	2.5	2.0	5.0	3.0	–	6.5	6.5
Salvo	4.0	3.5	–	–	–	3.0	1.0	1.0	–
Uno	6.0	3.5	–	–	–	–	1.0	3.0	–
Clercal	5.0	–	4.5	–	–	–	3.5	9.0	8.0
Pablo	–	3.5	–	4.0	5.0	3.0	–	2.5	–
Domino	–	3.4	–	2.0	–	–	–	1.0	–
Ego	–	–	–	2.0	–	2.0	–	–	–
Trimaran	–	–	–	–	6.0	3.0	–	–	–

(1) 1-9 scale with 1 no infection and 9: 100 % infection

During the last 10 years fungicide experiments were carried out to obtain information on the yield response of triticale. This aspect was particularly interesting in light of apparent high level of disease resistance shown by triticale varieties. In the fungicide trials the varieties 'Lasko' (5x), 'Bolero' (3x) and 'Alamo'(2x) were used. Most of the fungicide treatments were done in growth stage 59 (all spikes visible)(2). The results from the fungicide trials showed that the triticales had generally low yield response to the fungicide treatments. The best results were obtained with fungicide mixtures on base of chlorthalonil, applied in GS 59. Treatments with tebuconazol and epoxyconazol influenced also the yield positively.

CHEMICAL WEED CONTROL

Most herbicides recommend for wheat can be used for triticale. Of the pre-emergence applications studied chlortoluron, methabenzthiazuron, neburon, neburon + isoproturon, chortoluron + isoxaben, isoproturon + diflufenican and prosulfocarb appeared to be very safe for triticale (cv. 'Lasko' and 'Alamo'). The trifluralin mixtures with chlortoluron and especially with linuron were less selective.

Of the post-emergence treatments studied chlortoluron was more selective than isoproturon. Isoproturon at 1250 g/ha (the normal rate for wheat) always caused an inhibition and a decrease in yield. The rather new herbicides fenoxaprop-p-ethyl (+safener) and clodinafop (+safener) and the sulfonylurea derivations metsulfuron-methyl and thiameturon were caused no crop injury by application before GS 31. It must be mentioned that for the substituted urea herbicides isoproturon and methabenzthiazuron and for mecoprop clearly varietal differences were found (3,4). The similar reaction of closely related varieties on the one hand and the physiological nature of susceptibility on the other hand suggest that the varietal differences are under genetic control.

EFFECT OF GROWTH REGULATORS

The use of growth regulators led to a serious reduction of the plant height. Ethephon + mepiquat-Cl, applied in GS 37-39 shortened more than ethephon and chlormequat, applied in GS 30-31 and 37-39 respectively. In general the two spray programmes of chlormequat in GS 31 and ethephon and ethephon + mepiquat-Cl showed more inhibition than the single applications. The spread treatment of chlormequat (in GS 30 and GS 31 respectively) not always gave a stronger reduction than the single application. The results of chlormequat + imazaquin were similar to those of chlormequat. The lenght of the spike was not influenced seriously by the treatments of growth regulators (5).

Growth regulator sprays kept lodging and leaning to acceptable levels even when the nitrogen input was high. Ethephon + mepiquat-Cl showed the best anti-lodging effect. Yield was mostly affected positively by the applications of growth regulators (table 4). Yield increases were most important after ethephon + mepiquat-Cl in GS 37-39, certainly at high nitrogen levels. The spreaded application of chlormequat led in general not to higher yield increases compared with the single applications of chlormequat.

Some varietal differences in reaction were noted following treatments with regulators.

Table 4: Mean effect of growth regulators on the relative yield of triticale

Growth regulator	1984	1985	1986	1987(1)	1988(1)
chlormequat (GS 31)	105.5	105.4	99.3	96.1	101.1
ethefon (GS 37-39)	102.2	102.9	-	-	-
ethefon + mepiquat(GS 37-39)	103.2	103.9	103.3	101.1	110.2

(1) Average of several varieties and N-levels

RESPONSE TO NITROGEN

The mean results of the investigation to the optimal nitrogen
level for triticale are presented in table 5. On the loamy sand soil
of Merelbeke the optimum was likely to be in the region of 130-160
kg/ha. Only in case of an extremely wet spring higher rates can be
recommended. The different nitrogen fractions must be done in
fuction of the condition of the crop. In general it can be concluded
that too much early nitrogen promotes more vegetative growth and
increase the risk of lodging.

Table 5: Relative yield (as % of the trial average) of
         triticale at 6 rates of nitrogen in combination with
         growth regulator sprays

| Growing | Nitrogen level in kg/ha | | | | | |
season	100	130 +(1)	130 +(2)	160 +(1)	160 +(2)	200 +(1)+(2)
1987 (3)	–	99.2	102.5	100.0	100.3	92.4
1988	92.2	94.1	101.3	102.3	108.1	–
1989	85.0	109.0	112.2	103.1	108.6	97.2
1990	94.3	108.2	110.3	102.0	96.3	88.6
1991	89.3	103.5	105.7	105.8	90.4	93.1
1992	97.7	99.6	–	107.6	–	88.2
average	91.7	102.3	106.3	103.5	100.7	91.9

(1) chlormequat: 850 g/ha in GS 31
(2) ethephon + mepiquat-Cl: 388 g + 763 g/ha in GS 39
(3) average yield of 2 varieties in 1991 and 1992, of 3
    varieties in 1987, 1989 and 1990 and 5 varieties in 1988

REFERENCES

1. DE BAETS A. & HAESAERT G. (1989). Triticale: rassenkeuze en teelttechniek. Landbouwtijdschrift, 1989, 42(2): 233-241
2. ZADOKS, J.C., CHANG, T.T. & KONZAK, C.F. (1974). A decimal code for the growth stage of cereals. Weed research, 14, 415-421.
3. HAESAERT G. (1991). Reaktieverschillen voor herbiciden bij triticalecultivars. Doctoraatswerk, Rijksuniversiteit Gent, Fakulteit van de landbouwwetenschappen, 259 p.
4. HAESAERT G., A. DE BAETS, M. VAN HIMME & R. BULCKE (1992). Varietal differences in reaction of triticale to substituted urea herbicides. Med. Fac. Landbouww. Univ. Gent, 57/3b, 1053-1065.
5. HAESAERT G., A. DE BAETS & A. DANNEELS (1989). Effect of growth regulators on triticale. Med. fac. Landbouww. Univ. Gent 54/2a, 419-434.

TRITICALE BREEDING IN FUNDULEA-ROMANIA

Gheorghe Ittu and Nicolae N. Săulescu
Research Institute for Cereals and Industrial Crops
8264, Fundulea, Călăraşi-ROMANIA

Abstract

The triticale breeding program started in Romania at Fundulea in 1971. The first cultivar TF 2 was released in 1985 and the largest area grown reached 25.000 ha in 1987.
In 1992 and 1993 two other commercial cultivars Plai and Colina were registered. Both of them show significant improvement in yield capacity, test weight, tillering ability, spike fertility, winter hardiness and resistance to lodging in comparison with TF 2.
Progress in resistance to lodging was achieved by transfer of Rht 1 and H 1 genes for reduced plant height from wheat and rye respectively. In preliminary trials some of the short straw triticale lines had similar yields with the normal height type, while the level of winter hardiness and Septoria nodorum resistance is still lower.

Introduction

Triticale in Romania is a new small grain crop, which just started spreading.The main area where triticale is intented to be grown is the cooler hilly region around Carpatian mountains, with poor fertility and low pH soils.
The short history of this new small grain crop in our country began in 1927, when the first winter wheat/rye hybrids were produced [1]. A second step is represented by the synthetical octoploid released in 1939 [2].
A regular breeding program in hexaploid triticale, started later at Fundulea in 1971, based mainly on the spring forms from CIMMYT and winter ones originated from Russia, Canada, Poland and Germany [3].
Later on, many primary octoploid and hexaploid, as well as, secundary triticale forms have obtained [4,5,6].
The first result of Fundulea breeding was the TF 2 cultivar, officialy registered in 1985 [7].It is a selection from the cross between TCL 3, a CIMMYT spring triticale and a winter triticale from Canada.TF 2 reached the largest area (25000 ha) in

H. Guedes-Pinto et al. (eds.), Triticale: Today and Tomorrow, 663–668.
© 1996 *Kluwer Academic Publishers. Printed in the Netherlands.*

1987. Other two cultivar were registered in 1992 (Plai) and 1993 (Colina) after an extensively years of testing in official trials (1989-1993). Detailed description of their agronomical performances in different environmental conditions from Romania is given below.

Materials and Methods

Plai is a F5 selection from the cross 1684 TG 3-2/AD 206//TF 2 made in 1980. 1684 TG 3-2 is a Romanian winter hexaploid restitution triticale line, derived from the cross Chinese wheat 974-1253/Snoopy-Danae (rye)/3/Py 152-74/Mayo-Armadillo of which the main characteristics are good winter hardiness, lodging resistance, tillering ability and stand establishment. AD 206, a well known Ukrainean substitution winter triticale, is an early type with good test weight and a high level of winter hardiness.TF 2, a winter midlate Romanian triticale cultivar, with large spikes,good tillering ability, aluminium tolerance and vigurous seedling but susceptible to lodging and not enough winter hardy.
Colina is a F4 progeny from the cross CT 3/121 TJ 2-3 made in 1985.CT 3 is a Polish midlate winter triticale line with large, fertile spikes, resistant to Septoria nodorum and with acceptable level of sprouting resistance. 121 TJ 2-3 is an early Romanian winter hexaploid substitution strain, selected from the cross Sadovo 1/Lc 79-90//AD 206, with high test weight and winter hardiness.
In both crosses selection started in F2 generation and continued in F3-F4 generations as headrows selection.

Results

The yields of the two new cultivars Plai and Colina in 52 and 62 trials, respectively, during 1989-1993, were significantly superior to those of the control TF 2.In average, Plai and Colina yielded with 0,87 t/ha and 1,07 t/ha more than TF 2 (Fig.1 and 2).
Regression line was used in difining areas of adaptation of the Plai and Colina cultivars when compared with TF 2. Figures 1 and 2 show the regression of yields of Plai and Colina against the yields of the commercial cultivar TF 2. The cultivars Plai and Colina gave lower yields that TF 2 in only 6 of 52 trials, and 1 of 69 trials respectively. On the other hand the slope of the two regression lines are close to one suggesting that Plai and Colina performed better than TF 2 in poor as well as in good growing conditions. Lower coefficients of variation recorded in Plai and Colina as compared to those of TF 2, show their better yield stability.
The yield potential of Plai and Colina is due to their high fertility of the ears as well as the high number of uniform and well developed ears per unit area. (Fig. 3)
Both cultivars Plai and Colina also show a significant progress in comparison with

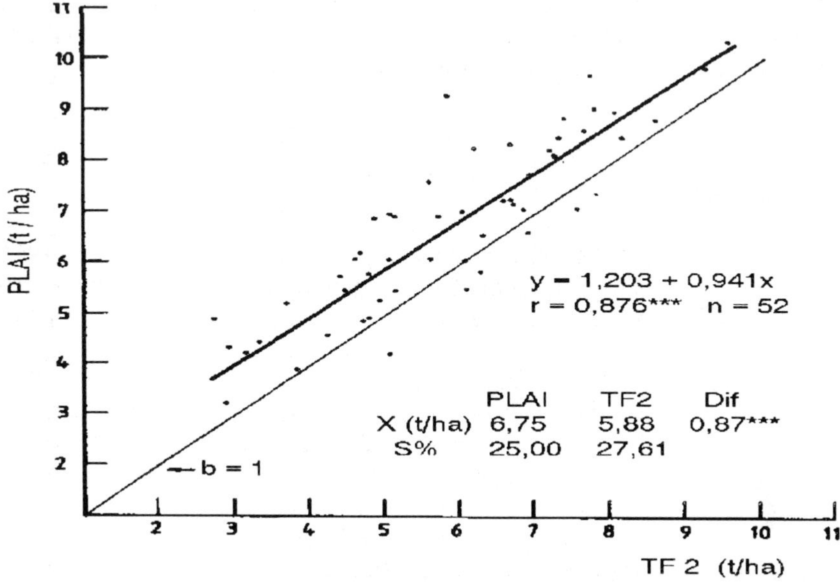

Figure 1. Comparision between yield of triticale cultivars Plai and TF 2 in 52 trials (1989 - 1992)

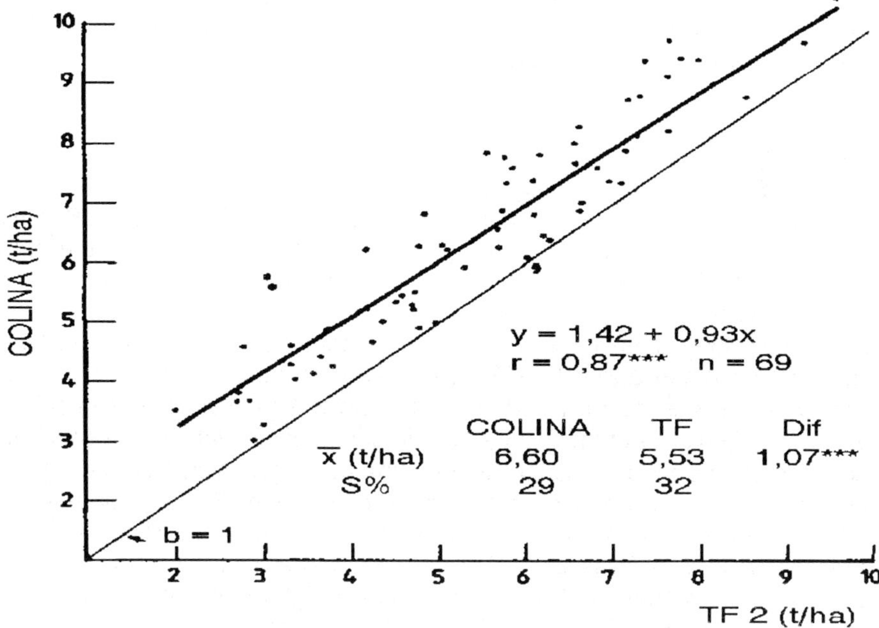

Figure 2. Comparision between yield of triticale cultivars Colina and TF 2 in 69 trials (1990 - 1993)

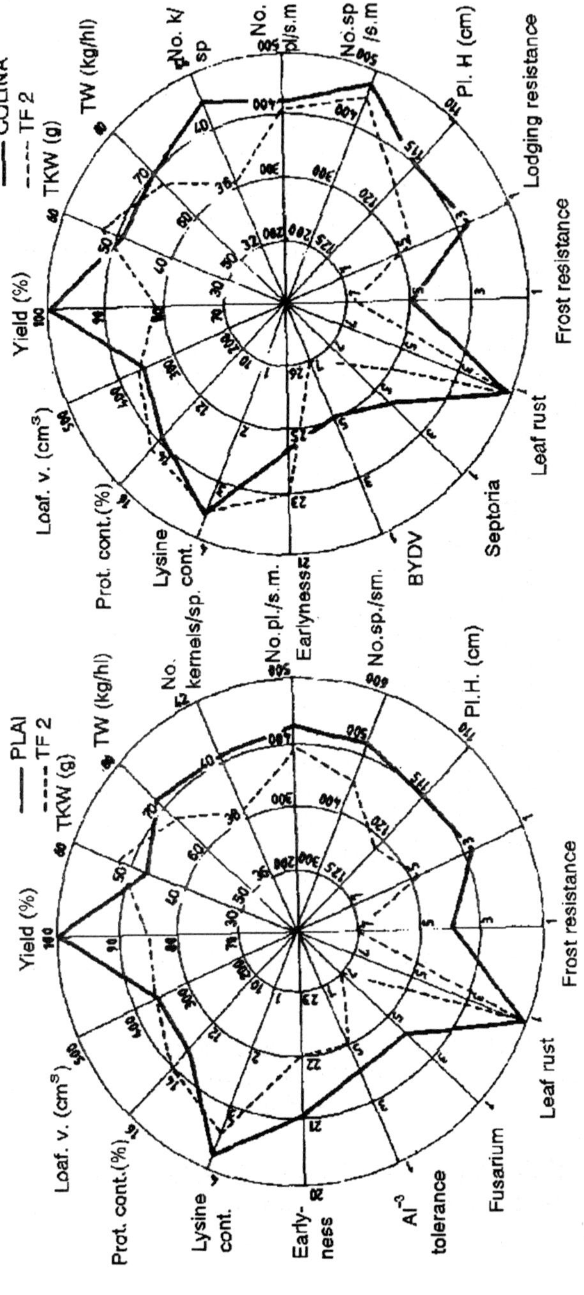

Figure 3. Polar diagrams of the main agronomic characteristics of the new triticale cultivars Plai and Colina compared with TF-2.

TF 2 in agronomical traits such as tillering ability, winter hardiness, lodging resistance.

Plai also represented a real improvement for test weight, earliness, resistance to Fusarium scab and aluminium tolerance, while Colina is superior for sprouting, Septoria nodorum and BYDV resistance.

Protein content of the grain is similar to TF 2 and to the average level of the present commercial winter wheat cultivars. The breadmaking quality is rather poor in all three cultivars. An acceptable loaf volume was obtained only by blending trIticale and bread wheat flour in a proprtion of 50:50.

During the last period in our breeding program more efforts were allocated to improve lodging resistance.In this context the first short straw lines were obtained with Rht 1 and H 1 gene transfered from the winter wheat Fundulea 133 and from Russian rye mutant EM 1 respectively.

Several lines with Rht 1 gene, selected from the cross 5735 TW, were already tested in 1991 in three locations in preliminary yield trials and their yield performances was similar with the normal height type (Fig.4). However, none of them had acceptable levels of winter hardiness and Septoria nodorum resistance.

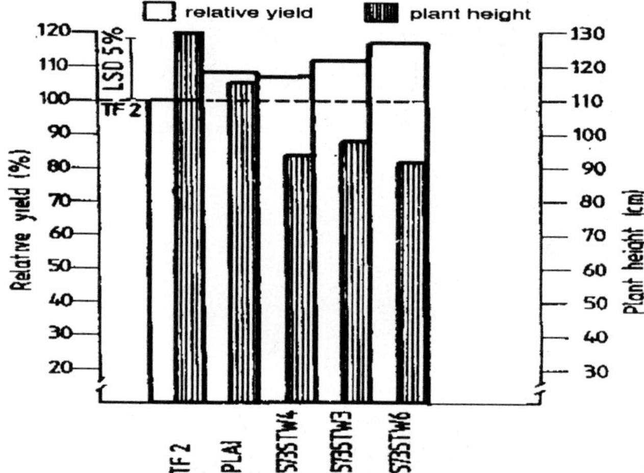

Figure 4. Relative yield and plant height of 3 short straw lines in comparison with TF 2 and Plai cultivars

More promising are the short straw lines carriers of the dominant H 1 gene selected from the hybrid combination 431 TU. The genotypes with this short straw gene have improved stiff straw and winter hardiness in comparison with the lines carriers of the Rht 1 gene.

For the moment none of the short straw lines firstly selected, with Rht 1 or H 1 genes, has a perspective to be registered. However, both types are extensively used in the breeding program to recombine short straw with other favorable agronomical characteristics.

Acknowledgements

- Presentation of this paper in the 3 rd International triticale Symposium was financially supported by CIMMYT. We thank Dr. G. Varughese - CIMMYT (Mexico) and Dr. H. Braun - CIMMYT (Turkey) for their interest in our work.

References

1.Saulescu N. Hibridul griu-secara. Revista Adamache-Iasi 1927;
2.Priadcencu Al. Contributii la studiul hibrizilor indepartati.AN. ICAR, XX 1939;
3.Saulescu N.N., Eustatiu N. Triticale-progress and prospects in breeding. Probl. genet. teor. aplic. 1972; XV: 314-346.
4. Ittu G., Mariana Ittu, Saulescu N.N., Preliminary results of resistance in triticale to Fusarium head blight. Probl. genet.teor.aplic. 1990; XXIII: 91-98.
5. Ittu G., Aurelia Jilaveanu, Saulescu N.N. Preliminary results of genetical resistance to BYDV in winter hexaploid triticale. Probl. genet. teor. aplic. 1990; XXIII: 147-150.
6. Verzea M. Crossability of tetraploid and hexaploid wheat with diploid rye. Proceedings of the Second International Triticale Symposium; 1991; Mexico DF. CIMMYT 157-160.
7. Ittu G., Saulescu N.N., Tapu C., Ceapoiu N. Triticale cultivar TF 2 (X Triticosecale Witt.). An. ICCPT,1986; LVI: 43-55.

BREEDING TRITICALE FOR THE SOUTH WESTERN CAPE

Herman Roux & Frans Marais
Department of Genetics, University of Stellenbosch, 7600 South Africa.

Abstract

It is expected that triticale will in future be planted on large tracts of marginal land in the south western Cape Province where wheat production is becoming uneconomical. As triticale is to be used primarily as a fodder crop and to replace a percentage of the feed grains imported into the area, the breeding programme wil be focused on the development of feed grain and dual purpose varieties. Breeding material via the CIMMYT nurseries has made a very valuable contribution towards reaching this goal. One year trial data indicated that some triticale genotypes significantly outyielded wheat (grain and fodder) if planted on marginal soils and grazed. The high levels of disease-, aluminium- and shattering resistance further adds to the profitability of growing triticale. A primary concern is, however, the very low grain protein contents attainable under these low input growing conditions.

Introduction

The south western Cape Province is climatically suited to the production of spring type small grains. It also includes large stretches of marginal, shallow and acid soils where wheat yields are often very low and unprofitable (Marais, 1985). The winters are wet with mild temperatures and are inducive to the development of a wide variety of plant diseases (Scott 1990). Compared to other small grain cereals triticale appears to be best suited for production on the marginal soils (unpublished data).

The utilization of triticale produced in the South Western Cape Province is roughly as follows: human consumption 2%, cover crop 8%, forage crop 30% and feed grain 60% (unpublished data). A small amount of triticale is used in the manufacture of breakfast cereals and specialty baked products. As a cover crop between vineyards and orchards, it has largely replaced rye. As a forage crop it is favoured for its ability to be grazed early in the winter when no other green fodder is available and to be cut later in the season as a silage, hay or grain crop. The grain yield is mostly used as a substitute for maize in animal feed. In 1993 approximately 60,000 ha were planted to triticale in South Africa of which 40,000 ha were in the south western Cape Province (unpublished data). The area under triticale cultivation is expected to grow rapidly because of the diversity in its utilization and primarily because of the growing demand for triticale as a partial substitute for maize in animal feeding. The price of triticale grain is currently based on the cost of delivered maize, with a premium being paid for high protein content grain. In 1993, bread

H. Guedes-Pinto et al. (eds.), Triticale: Today and Tomorrow, 669–674.
© 1996 *Kluwer Academic Publishers. Printed in the Netherlands.*

wheat production in the south western Cape Province ($\pm$ 590,000 tonnes) exceeded local demand by approximately 40-50% and all excess wheat had to be transported at high cost to the industrial areas in the north of the country (F. du Bois, personal communication). On the other hand, maize for use in animal feed rations has to be transported from inland producers to the Cape Province. This situation triggered an attempt to substitute some of the maize used in animal feeds with triticale. For this purpose triticale can be grown on some of the acreage planted to wheat, especially on marginal soils where it significantly outyields wheat.

The objectives of the breeding programme are therefore (i) to improve the yield, grain quality, grain protein content and disease resistance of spring type triticale intended for feed grain production, and (ii) to select dual purpose lines that can be used to provide grazing, hay and/or silage as well as a grain yield. (iii) Lines of which the grain products may be useful for human consumption are also being selected. To realise these objectives, crosses are made among locally adapted high yielding triticale lines that are suitable for the production of feed grain and/or forage and new accessions with good yield, disease resistance and/or quality. Some of the lines are also hybridized with new primary triticales, rye, bread wheat, durum wheat and wild wheat species with a high protein content to increase the variability within the crop.

Materials and methods

The evaluation of breeding material is done primarily on the Mariëndahl Experiment Station outside Stellenbosch, 152 metres above sea-level at 18°50'E longitude and 33°50'S latitude). The soil of the trial site is a shallow, sandy Kroonstad to Fernwood type. In addition to the routine breeding material, three identical grazing trials each consisting of ten entries (8 triticales, 1 bread- and 1 forage wheat) in a randomized block with four replications, were planted in 1993 at a sowing density of 350 kernels m^{-2}. Each plot of 10.2 m^2 consisted of six 10-metre rows spaced 17 cm apart. The entries used are listed in Table 1.

Table 1. Cultivars and breeding line used in the grazing trials.

No	Entry	No	Entry
1	"Usgen 10"	6	"Kiewiet"
2	"Usgen 14"	7	19th ITSN 70-4
3	"Usgen 18"	8	"Rex"
4	"Usgen 19"	9	"Palmiet" (bread wheat)
5	"Usgen 19-4"	10	"Alpha" (forage wheat)

The trials included an ungrazed control that was planted at the optimum planting time for triticale in the area, viz. the 11th of May, while the other two were planted on the 22nd of April. Growing conditions in the Winter Rainfall Region compel producers to plant dual purpose triticale 14-20 days earlier than those intended for seed production. The main reasons are to provide fodder during June while allowing enough time for the grazed plants to recover and to set seed before the end of the rain season. This practice necessitated planting of the control plots 2 weeks later than the grazed plots. If the ungrazed control plots are planted at the same time as the plots to be grazed once or twice, their yields tend to be low, mainly because the warmer temperatures in the early growth stages cause the plants to form fewer tillers. One of the trials was grazed once (on the 2nd and 3rd of June, 42 days after planting), while the other was grazed twice (on the 2nd and 3rd of June and on the 21st and 22nd of June, 18 days between grazings). Before each grazing (sheep), one square metre green material was cut from each plot in the twice grazed trial to determine the fresh biomass before being dried in an oven to determine the dry biomass. A top dressing of nitrogen fertilizer (30 kg N ha^{-1}) was applied after each grazing. Herbicides were sprayed but no fungicides or insecticides were used. Prior to harvest, one square metre was cut from each plot to determine the biomass yields. The plots (9.18 m^2) were harvested *in toto* with a "Wintersteiger" plot combine on the 16th of November.

In an attempt to transfer genes for high grain protein content from *Triticum dicoccoides* to triticale, a *T. dicoccoides*/ "Henoch" rye amphiploid F$_1$ was crossed with and backcrossed to the triticale cultivar USGEN 18 (with concurrent selection for high grain protein content).

Results and discussion

The routine breeding programme is relatively small and is based on a modified pedigree selection procedure. As inconsistent and inadequate funding is a major constraint of the programme, new crosses are limited to about 250 per annum and are supplemented with CIMMYT-derived segregating populations. Presently the university provides about 65% of the running costs, the Agricultural Research Council contributes about 30% and royalties account for approximately 4%. Since 1993 triticale as a crop is included on the national plant variety list and plant breeder's rights can be obtained. A primary objective is therefore to stabilize our basis of funding by promoting the production of triticale and by providing better yielding cultivars. During 1993 "Rainbow Chicken Farms (Pty) LTD", the biggest individual producer of livestock products and also the largest consumer of animal feeds in the country, announced its preparedness to buy triticale at guaranteed prices. Their main objective is to utilize triticale as a partial substitute for maize in chicken rations. For the first time since the inception of the breeding programme a stable market for triticale seed was established. With this change in the market an urgent need arose for short strawed, early maturing cultivars with high grain yields. A major concern is, however, that very low grain protein contents appears to be associated with the low input production conditions. There appears to be only very limited variation for grain protein content among the well adapted breeding lines. In 1993 protein contents were measured on 25 elite trial entries grown at 7 representative localities in the production area. Although significant genetic differences were illustrated, the average (over localities) protein contents varied only between 10.4% and 11.5% (N X 5.7; 12% mb) while the trial averages varied from 8.7% to 13.1%. The average protein contents were negatively correlated (r = -0.37) with hectolitre mass (range 66 to 75 kg/hl) and were comparable to the average ranges in protein content generally observed in wheat selections.

Two new cultivars were released during 1994. They are "Rex", a selection from CIMMYT's 21st ITYN 14 and "Kiewiet", a selection from CIMMYT's 18th

ITSN 33. Both are dual purpose cultivars which can be grazed and left for a seed harvest and both produce excellent grain yields. "Kiewiet" is highly resistant to the Russian wheat aphid which is a problem of primarily the summer rainfall regions of South Africa.

In 1993 routine evaluation of advanced breeding material for their utility as dual purpose genotypes was initiated. Towards this end the grain yields of breeding lines were compared following 0, 1 or 2 grazings during the growing season. This procedure will in future form an integral part of our routine testing programme. An analysis of variance of the yield data obtained in the 1993 grazing trials is given in Table 2. Highly significant grain yield differences occurred among the control and grazed trials as well as among entries (both within and across trials). Analysis of variance data of the wet and dry biomass yields recorded in the twice grazed trial are given in Table 3. It appears that ample genetic variation for fodder production existed among the trial entries.

Table 2. Analysis of variance for grain yield as recorded in grazed and control plots.

Source of variation	d.f.	Sum of squares	F-ratio	Sign. level
Blocks (Exps)	9	1.5633		
Exps	2	7.2396	6.94	0.0058
Entries	9	28.7524	6.13	0.0000
Entries X Exps	18	9.3828	11.89	0.0000
Entries X Blocks (Exps)	81	3.5500		

Table 3. Analysis of variance for biomass yield at the times of grazing.

Source of variation	d.f.	Sum of squares	F-ratio	Sign. level
Replications (A)	3	0.44563	0.67	0.5725
Entries (B)	9	23.16863	11.61	0.0000
Wet-dry biomass (C)	1	386.16903	1742.20	0.0000
1st & 2nd cut (D)	1	0.04323	0.20	0.6645
AB	27	16.675962	2.79	0.0001
AC	3	0.239012	0.36	0.7824
AD	3	0.652962	0.98	0.4048
BC	9	15.448776	7.74	0.0000
BD	9	3.652151	1.83	0.0728
CD	1	0.165766	0.75	0.3986
Residual	93	20.614048		

The best grain yields were obtained in the trial that was grazed once. A mean grain yield of 2.69 t ha^{-1}, compared to the 2.53 t ha^{-1} of the ungrazed trial and 2.11 t ha^{-1} of the one that was grazed twice, was obtained. In Fig. 1 the grain and biomass yields are depicted for entries and trials. Entry 7 is clearly a superior dual purpose genotype while the latest releases (entries 6 and 8) are also well suited to this application. The two wheat varieties included (entries 9 and 10) were clearly the poorest performers. However, since the trial was done on a marginal soil type, different results may be obtained on more fertile soils.

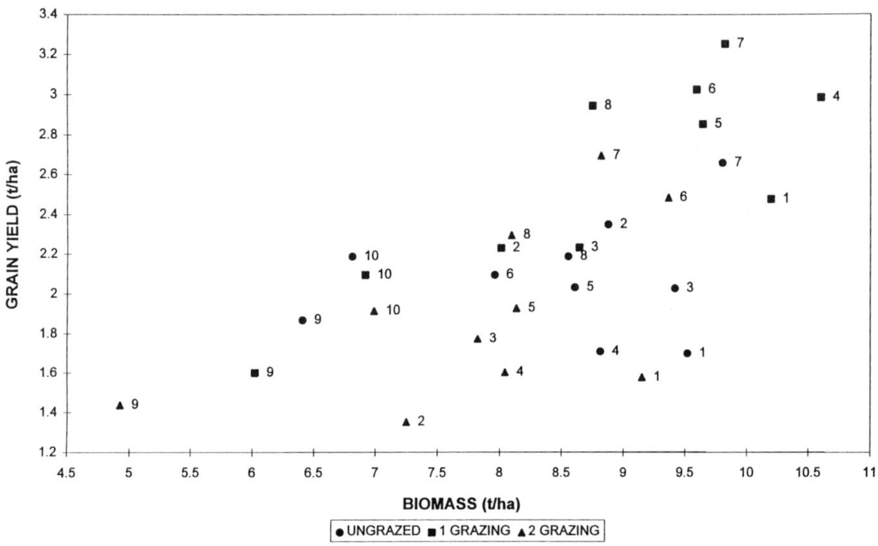

Fig. 1. Relationship between biomass and grain yield without grazing, after one grazing at 42 days and two grazings 42 and 60 days after planting. Entries 1 to 8 = triticale, entry 9 = bread wheat and entry 10 = forage wheat.

An attempt to transfer genes for higher grain protein content from *Triticum dicoccoides* to triticale has progressed to the fourth backcross. Among the segregating progeny of the advanced backcrosses, considerable variation for grain protein content and plumpness occurred, while it became apparent that the segregates with very high protein contents were invariably those plants that produced poor quality seeds. With grain protein content probably being controlled by more than one gene, each having a small effect (Levy et al., 1990), a recurrent selection procedure may be more appropriate for the further improvement of the material. Agronomically superior plants of which the grain protein contents are higher than that of the recurrent parent used during backcrossing, will therefore be selected and intercrossed randomly to create a base population for a recurrent selection programme.

Acknowledgements

The generous support of the Grain Crops Institute of the Agricultural Research Council, Rainbow Chicken Farms, Bokomo, overseas and local collaborators, participating farmers and Sensako is greatly appreciated.   We also wish to thank Dr JH Louw for useful consultations.

References

Feldman M, Avivi L, Levy AA, Saccai M, Avivi Y, & Millet E.   High protein wheat.   In: Bajaj YPS, editor. Biotechnology in Agriculture and Forestry, vol 13. Berlin: Springer-Verlag, 1990:595-614.
Marais GF.   Yield assessment of wheat breeding lines in the Winter Rainfall Region of South Africa. I. Homogeneity of the region. SAJ Plant & Soil 1985;130-34.
Scott DB.   Wheat diseases in South Africa. Dept Agric Development, South Africa, 1990: Tech Comm no 220.

REGISTRATION OF THREE NEW TRITICALE VARIETIES:- CRATO, ARRUDA AND ALTER

José Coutinho; Benvindo Maçãs; Francisco Bagulho; Conceição Gomes and João Coco
Estação Nacional de Melhoramento de Plantas, Elvas, Portugal

Abstact

Three spring triticale (triticosecale Witt.) genotypes developed by CIMMYT and selected at Estação Nacional de Melhoramento de Plantas (ENMP) were registered in Portuguese National Catalogue (CNV). Crato and Arruda in 1991 and Alter in 1993. These cultivars represent a new generation in Portugal, showing improved grain characteristics such as test weight and grain yield.
All complete triticales, they are morphologically very similar but presenting slightly differences in heading time.
Presented data was obtained from adaptation yield trials at the national network.
ENMP is responsible for maintenance of breeders seed using head-row purification and pre-basic seed production.

Introduction

Crato (a Stier line), Arruda (a Gnu line) and Alter (a Rhino line), are three complete triticale varieties included in the National Catalogue in 1991 [1] and 1993 [2]. As a consequence of a strong selection pressure, these new cultivars may be considered the fourth generation in our triticale program [3], after the implementation of the *species* into the agricultural systems of Portugal, through the varieties Armadillo, Beagle, Bacum and Arabian .
To overcome the traditional problems that the oldest varieties showed like low test weight, shriveled grain and occasional threshing difficulties (as in Beagle), new germplasm has been developed to create better gene combinations [4].

675

H. Guedes-Pinto et al. (eds.), Triticale: Today and Tomorrow, 675–678.
© 1996 *Kluwer Academic Publishers. Printed in the Netherlands.*

The breeders were successful and the new cultivars not only show better behaviour for the above handicaps, but also meet very well the specific agro-ecological conditions of Portugal, where late frosts and a lack of water after May impose a quick grain filling period.

Origin and breeding history of Crato, Arruda and Alter triticales

This raw material was introduced in Portugal in 1984/85 through the cooperative network between CIMMYT and Estação Nacional de Melhoramento de Plantas (ENMP).
In 1985 all three lines were selected in the Cereal Department (DC) at ENMP, and in 1986 they were included in a new scheme of trials. Once the behaviour of these genotypes was classified as satisfatory, they completed two more years in adaptation trials in the most important cereal areas and data was collected from 20 locations (Fig. 1 and Fig. 2).

In 1988 and 1989, they were candidates to the Portuguese National Catalogue (CNV), where after three years of trials, Crato and Arruda were released in 1991 and Alter in 1993(Table 1).

Table 1 - Results obtained in trials conducted by Centro Nacional de Protecção à Produção Agrícola (CNPPA) responsible for the   Portuguese National Calalogue (CNV).

Cultivar	Year	Nº of Trials	Yield (kg/ha)	% In Relation To C1	% In Relation TO C2	Average (kg/ha)	Average in Relation to: C1(%)	C2(%)
A R R U D A	88/89	5	4063	129.8	120.2	4214	131	119
	89/90	4	4366	132.3	117.1			
C R A T O	88/89	5	3645	116.4	107.9	4057	126	114
	89/90	4	4470	135.5	119.9			
A L T E R	89/90	4	4596	139.2	123.2	4350	112	110
	90/91	4	4180	93.3	103.9			
	91/92	4	4275	103.2	102.0			

CHECKS: - C1 - ARABIAN
C2 - BORBA

Previously, breeders seed was increased from a 1987 DC source, rogued for off-types, and delivered to the Seed Multiplication Department (SMD) at ENMP, where a scheme of head rows is maintained to provide pure seed.

Release of Crato, Arruda and Alter triticales

These triticales were proposed for release in the basis of their high grain production and yield stability (Fig.1), better test weight (Fig.2), good grain shape and with a late heading time relative to the most known triticales (more than 8 days than Bacum and Arabian and more than 2 to 4 days later than Juanilho and Borba, (Table 2)). This aspect is extremely important due to the possibility of late frosts in April, during the heading time. In the main cereal regions of Portugal, late frosts can occur (in 1994, the last one was on April 21st), and the rain stops normally after May, wich means that in the majority of the years, the grain filling period occurs under severe stress conditions [5,6].

Table 2 - Number of days between sowing date and heading time.

Varieties	86/87	88/89	89/90	90/91	Average
Arruda	132	140	136	126	133
Crato	136	140	140	126	136
Alter	130	136	136	123	132
Bacum	122	130	135	122	127
Arabian	125	130	133	122	127
Beagle	128	135	132	125	130
Juanilho	128	133	138	123	130
Borba	129	135	138	126	132

Procedure for maintaining seed stock classes

Breeders seed will be manitained by the DC. Foundation seed is produced by SMD at ENMP. Foundation, registered and certified classes of seed are recognized in the usual way for small grains. Foundation seed is produced periodically from composited head-rows purification plantings as needed.

References

1 - PPA (PO) - 27/91 - Centro Nacional de Protecção à Produção Agrícola. Divisão de Ordenamento. Lisboa, 1991.

2 - PPA (PO) - 15/93 - Centro Nacional de Protecção à Produção Agrícola. Divisão de Ordenamento. Lisboa, 1993.

3 - Bagulho, F. ; M.T. Barradas; B.Maçãs & J.Coutinho, 1994 - Some Remarks on Triticale Breeding Evolution in Portugal. a) Role played by the National Plant Breeding Station. 3rd International Triticale Symposium. 13-17 June, Lisbon - Portugal.

4 - Maçãs, B.; J.Coutinho; F.Bagulho, C.Gomes & N.Pinheiro, 1994 - Behaviour of triticale germplasm selected   at   ENMP by introgression of winter genes on spring genotypes. 3rd   International   Triticale   Symposium.   13-17   June, Lisbon. Portugal.

5 - Coutinho, J. 1991 - Estudo Preliminar de 11 Cultivares de Trigo Mole (*T. aestivum* L em Thell) representativas de várias fases do   melhoramento em Portugal. Dissertação do Curso de Mestrado em Produção Vegetal. ISA, Lisboa, 150pp.

6 - Maçãs, B., 1990 - Selecção de Cultivares de Trigo Mole (*T. aestivum* L. em Thell) para sementeiras antecipadas. Dissertação do Curso de Mestrado em Produção Vegetal. ISA, Lisboa, 109pp.

CHARACTERIZATION OF LAMB-2, A NEW TRITICALE VARIETY FOR THE HIGHLY VARIABLE MOISTURE SITUATION OF CENTRAL MEXICO

Rebeca M. Gonzalez Iñiguez and Wolfgang H. Pfeiffer
SARH-INIFAP and CIMMYT
Tte. Isidro Aleman 294
Morelia, Mich., Mexico

Abstract

The highly variable rainfall patterns in the eastern part of transversal volcanic zone in the Central Highlands of Mexico are reflected in the grain yields of small cereals grown there. Such situations demand germplasm that combines low moisture-efficiency with high moisture- and input-responsiveness to exploit the changing environmental yield potentials.

Lamb-2 was identified for release as a commercial triticale variety for drought conditions after four years of testing over a range of diverse moisture environments. It was developed from the cross Muskox/Beaguelita and, across 24 different environments, outyielded the standard triticale varieties by as much as 35%. Under monitored moisture situations, the yield of Lamb-2 responded to increased moisture. Agronomic components will be presented and discussed in context of adaptation to moisture stress.

Crop Improvement for Varying Patterns of Moisture Stress

The rainfall patterns in the eastern part of transversal volcanic zone in the Central Highlands of Mexico are highly variable. This is reflected in the grain yields of small cereals grown there. These yields can range from 1 to 6 Mgha^{-1} in different years. Such situations demand germplasm that combines low moisture-efficiency with high moisture- and input-responsiveness to exploit the changing environmental yield potentials.

Major problems in developing and identifying germplasm adapted for the eastern and central Mexican highlands are uneven and poor rainfall distribution, which result in contrasting moisture stress patterns in different years. Moisture stress can extend from pre-anthesis to post-anthesis (terminal); in some years, it can detrimentally affect yield if it occurs during the early and late growth stages, while in other years grain yields can be stable if there is stored or residual moisture. This all complicates systematic breeding. For example, pre-anthesis moisture stress affects the crop from tillering to anthesis, but not during grain-filling. Increasing rainfall and temperatures after anthesis may create a favorable environment for biotic stress [1]. Hence, biomass production will be low, plant height will be reduced, harvest index will be high, and late maturity will favor tillering

H. Guedes-Pinto et al. (eds.), Triticale: Today and Tomorrow, 679–683.

© 1996 *Kluwer Academic Publishers. Printed in the Netherlands.*

due to an expanded vegetative phase and contribute to grain yield. In contrast, moderate moisture stress around anthesis will increase in severity throughout the post-anthesis grain-filling period. High vegetative biomass production, phenotypic expression of plant height, and a low harvest index are characteristics of post-anthesis stress situations. Earliness is frequently positively associated with grain yield.

As a consequence, a testing environment in one year may have little relation to those experienced in the next [2]. A common strategy to cope with such situations is testing over a representative range of conditions and sampling spatial and temporal variation. However, testing for many crop cycles is not feasible and breeders opt to substitute temporal variation with spatial variation, assuming that testing over a wide range of test sites can ensure a parallel degree of temporal buffering capacity in germplasm [3]. Hence, data from the International Triticale Yield Trial (ITYN) were used as support criteria to identify superior germplasm for commercial release for the Central Mexican Highlands.

Performance

Lamb-2 (X65985-5M-3Y-2M-1Y-4M-1Y-0Y), developed from the cross Muskox/Beaguelita, was identified for release as a commercial triticale variety for variable drought conditions after four years of testing over a range of diverse moisture environments. The performances of Lamb-2 and check variety Eronga-83 in ITYN 18 and at different environments in the Central Mexican Highlands during 1990 to 1992 are given in Table 1.

Table 1. Perfomance of Lamb-2 and Eronga-83 in International Yield trails and different enviroments in the Central Mexican Highlands.

Data Source	Environment	Lamb-2 (1)	Lamb-2 (2)	Eronga-83 (1)	Eronga-83 (2)
18 ITYN	Across locations	5017	70	4714	69
18 ITYN	El Batán	5005	69	4199	68
18 ITYN	El Bajío (Irrigated)	11952	–	9575	–
18 ITYN	Tlaxcala	3607	70	3964	70
1990 Huamantla Yield trials	Across 4 experiments	3072	68	2901	68
1991 Huamantla Yield trials	Across 7 experiments	5612	66	5309	65
1992 Toluca yield trials	Across 4 experiments	2555	58	1906	53
1992 Huamantla Yield trials	Across 5 experiments	6476	73	6480	70

(1) Grain yield (kg/ha).
(2) Test weight (kg/hl).

Table 2. Three-year average for agronomic components for Lamb-2 and Eronga-83, measured at Cd. Obregon under full irrigation (Data source: K. Sayre, 1988, 1989, 1990 yield potential trials).

Agronomic components	Lamb-2	Eronga-83
Grain yield (kg/ha)	7,629	7,670
Biomass (kg/ha)	18,978	19,074
Straw yield (kg/ha)	12,265	12,354
Harvest index (%)	35.5	35.5
Plants/m^2	267	276
Spikes/m^2	359	307
Spikes/plant	1.4	1.2
Grains/spike	48	54
Grains/m^2	17,009	16,570
1000 grain weight (g)	39.7	41.0
Plant height (cm)	118	139
Days to heading 1) (days)	74	75
Days to maturity 1) (days)	121	121
Grain-fill duration (days)	47	45
Vegetative growth rate (kg/day/ha)	166	165
Biomass production rate (kg/day/ha)	157	158
Grain biomass production rate A 2) (kg/day/ha)	63	64
Grain biomass production rate B 3) (kg/day/ha)	163	169
Spike weight (g)	2.16	2.51
Spike growth rate (mg/spike/day)	46.3	55.3
Grain growth rate (mg/grain/day)	0.93	0.98
Hectoliter weight (kg/hl)	71.8	70.3

1) Emergence to physiological maturity.
2) Calculated from days to maturity.
3) Calculated from grainfill duration.

Data from the 22nd ITYN were used to calculate the regression of grain yield of Lamb-2 on the long-term triticale check Beagle (Figure 1), and Lamb-2 on bread wheat check Genaro 81 (Figure 2). The figures reveal environmental cropping situations where Lamb-2 outperformed the check varieties. At the line with a slope of 1.0, the grain yields of Lamb-2 and the checks are equal. All the points above this line reflect locations were Lamb-2 had higher grain yields. Data points below this line refer to higher yields of Beagle (Figure 1) or Genaro 81 (Figure 2). Both figures show that Lamb-2 had superior yields compared with the checks in high and low production environments. Across the 37 locations of ITYN 22, Lamb-2 yielded 4685 kg/ha compared with 4428 kg/ha for Eronga-83 (Pfeiffer et al. 1993).

682

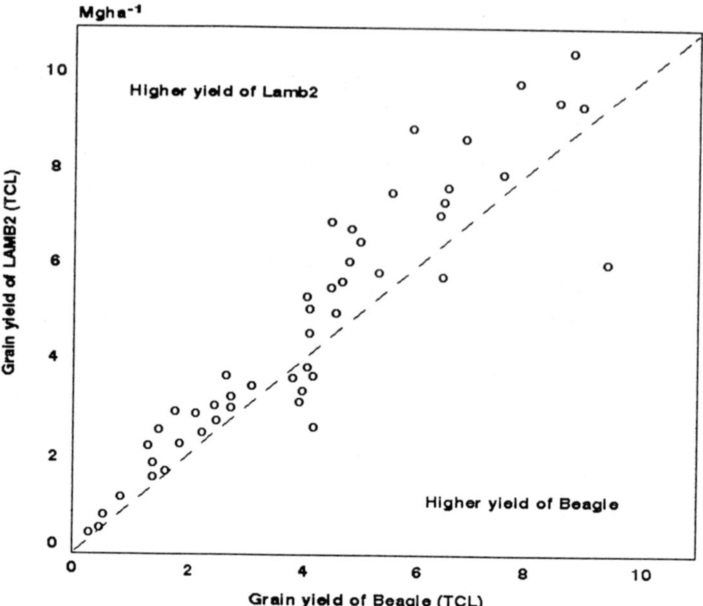

Figure 1. Yield of Lamb2 against the long term check Beagle in ITYN-22.

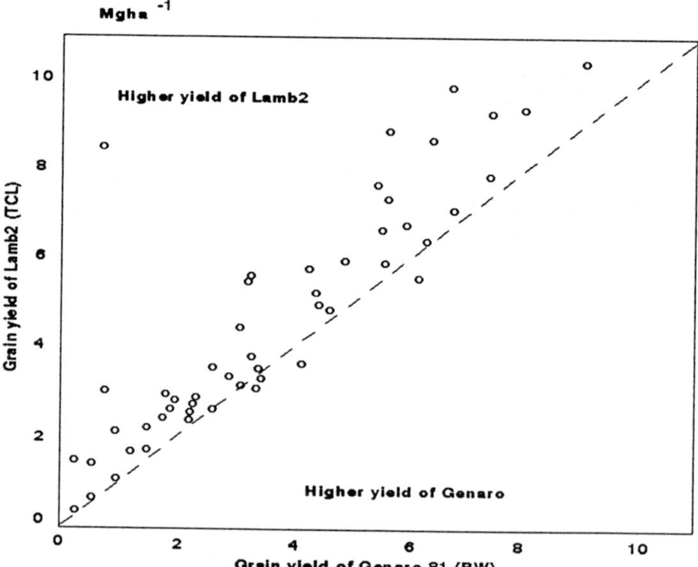

Figure 2. Yields of the triticale Lamb2 against the widely adapted bread wheat Genaro 81 from ITYN-22.

Agronomic Components

The agronomic components of Lamb-2 and Eronga-83 were measured at Ciudad Obregon under near optimal, irrigated growing conditions (Table 2). Since agronomic components were measured under no stress, differences in agronomic components between Lamb-2 and the check variety Eronga-83 have to be interpreted with caution. Compared with Eronga-83, Lamb-2 has reduced plant height, an increased number of spikes/plant with a corresponding higher number of spikes/m^2. The slightly lower number of grains/spike in Lamb-2 is overcompensated by the 16% higher number of spikes/m^2 and results in an increase in grains/m^2. This increased tillering capacity of Lamb-2, and hence better balance of the yield components, may constitute its adaptive advantage over Eronga-83 in highly variable moisture stress environments and under different types of moisture stress. Available moisture at early growth stages can be exploited, and enhanced tillering capacity will favor spikes/m^2 at maturity if crop emergence is poor. The reduced height of Lamb-2, 21 cm when compared with Eronga-83, may contribute to harvest index, particularly in terminal moisture stress situations. The adaptation of Lamb-2 to variable moisture stress situations has been confirmed by varietal releases by national programs in various countries. Lamb-2 has been released as TCL-83 in Tunisia in 1993, and has been identified as a candidate for release in Algeria and Iran.

References

1. Pfeiffer WH. Breeding for drought tolerance. In Rajaram S, Hettel GP, editors, CIMMYT Wheat Breeder's Conference, Cd. Obregon, Son., Mexico. 1989: 12-41.

2. Fox PN, Rosielle AA, Boyd WJR. The nature of genotype x environment interactions for wheat yield in Western Australia. Field Crops Res. 1985;11:387-398.

3. Pfeiffer WH, Fox PN. Adaptation of triticale. In Proceedings, 2nd International Triticale Symposium, Passo Fundo, Rio Grande do Sul, Brazil, 1-5 Oct. 1990. 1991: 54-59.

4. Pfeiffer WH, Fox PN, Corbett J, Rodriguez MT, Magaña RI. The twenty-first and twenty-second International Triticale Yield Nurseries (ITYN). El 21o y 22o Ensayos Internacionales de Rendimiento de Triticale. (ITYN). CIMMYT, Mexico, D.F., 1993.

VIII - REGIONAL TRIALS AND CROP MANAGEMENT

ADAPTATION OF TRITICALE TO SOUTHERN BRAZIL

Augusto Carlos Baier. National Research Center for Wheat
(EMBRAPA-CNPT), Passo Fundo RS, Brazil

Abstract

A yield nursery with 18 to 23 triticale genotypes and 2 wheat cultivars was
conducted annually in cooperation with other triticale research programs in
southern Brazil. Results over five years (1989-93) from 7 locations confirmed the
superior adaptation, yield potential and disease resistance of triticale.

High yields and better test weights (TW) were obtained at altitudes between 400
and 1000 m (latitude between 23° and 29° S) on acid soils with toxic aluminium.
The yield potential on acid, as well as on amended soils, is the most outstanding
feature for triticale. Low night temperatures during tillering should be considered
important for triticale to express its yield potential. At lower altitudes triticale
yielded less than wheat. TW of triticale cultivars was lower than wheat checks at
all locations.

Introduction

Small grain production in southern Brazil is limited by soil acidity and favourable
environmental conditions for diseases. Ten million ha cultivated with corn or
soybean in the summer season could be double-cropped with triticale or other
small grains in the winter to improve farmers' income and the national
nutritional status, as well as reducing soil erosion. There is an increased demand
for feed grain during the summer, prior to the corn harvest, due to the high
transportation costs associated with importing corn from the major producer
regions in Brazil and from the seaports. Increased feed prices (between 20 and
30 % prior to the new corn harvest and just after triticale harvest) are the main
factors responsible for the expansion of the area cultivated with triticale in
southern Brazil. Triticale area was estimated at 25,000 ha in 1991, 35,000 ha in
1992; 70,000 ha in 1993 and 100,000 ha in 1994.

H. Guedes-Pinto et al. (eds.), Triticale: Today and Tomorrow, 687–691.
© 1996 *Kluwer Academic Publishers. Printed in the Netherlands.*

The breeding programs conducted at the CNPT and the cooperative research organisations have been actively selecting and evaluating new triticale germplasm received from the International Maize and Wheat Improvement Center (CIMMYT). Regional evaluation has been performed by each institution or on a state-based cooperation system. Evaluation of all recommended and new triticale cultivars on a national basis has been coordinated by the CNPT.

Material and Methods

The Brazilian Triticale Yield Trial is conducted annually at 20 to 25 locations representing different edafoclimatic regions of South and Central South Brazil and includes 16 to 18 triticale cultivars and 2 or 3 wheat checks. Yield and test weight (TW) of 8 triticale and the highest yielding wheat check cultivars (Table 1) at 7 locations (Table 2) are presented and discussed. Triticale BR 1 and CEP 18-Caverá represent older "substituted" cultivars, released in 1981 and 1983, respectively. CEP 22-Botucaraí, CEP 23-Tatu, CEP 25, IAPAR 23-Arapoti, Triticale BR 4 and Embrapa 18 represent newer cultivars that contain the whole set of rye chromosomes.

Table 1. Name and pedigree of the evaluated cultivars.

Cultivar*	Pedigree
CEP 18 Caverá	TOB/8156//CC/3/Inia/4/SPY/5/M2A, (Teddy);
CEP 22 Botucarai	BGL/CIN//IRA/BGL;
CEP 23 Tatu	BGL/3/MTZTCL/Trigo//BGL/4/Nutria (Tatu)
CEP 25	B6712-171-11Y-4Y-0M-0A;
IAPAR 23 Arapoti	CIN/CIA//BGL/3/Merino (Hare)
Triticale BR 1	M2A/CML (Panda);
Triticale BR 4	BGL/CIN//MUS, B2686-0Y-6L-(25-36)FS;
Embrapa 18	Tapir/Yogui//3*MUS;
Wheat	Highest yielding wheat cultivar out of the two checks were: CEP 21 in 1989; Trigo BR 23 in 1990; Trigo BR 43 in 1991, Trigo BR 35 in 1992; and Embrapa 16 in 1993.

*Cultivars released by: CEP = Centro de Experimentação e Pesquisa da Federação das Cooperativas de Trigo do Rio Grande do Sul, RS; and BR and Embrapa = Centro Nacional de Pesquisa de Trigo, Empresa Brasileira de Pesquisa Agropecuária-EMBRAPA.

Passo Fundo, Lagoa Vermelha, Abelardo Luz, Chapecó and Campos Novos are locations where triticale showed good adaptation while São Borja, like other locations at lower altitudes, are regions where triticale traditionally did not perform well.

The experimental design used was a randomized block with 4 replications. Each plot contained 5 rows (20 cm row spacing and 40 cm between plots), 5 m long (1 m between replications). Seed rate was 400 viable seeds/m^2. Plots were fertilized with 30-90-30 kg/ha N, P, K, respectively, before sowing and 90 kg/ha of N as a

top dressing. The trials at Passo Fundo and Lagoa Vermelha were direct-drilled.
Plots were harvested at full maturity. Yield and TW are presented on an 87 % dry
matter basis and were evaluated with ANOVA and Duncan's New Multiple
Range Test.

TABLE 2. Characterization of the trial sites, institutions and years of triticale
yield trials.

State	Location	Latitude	Longitude	Altitude	Institution*
RS	Lagoa Vermelha (LV)	28°13'S	51°32'W	800m	CNPT (89-93)
	Passo Fundo (PF)	28°15'S	52°24'W	684m	CNPT (89-93)
	Cruz Alta (CA)	28°04'S	53°32'W	473m	CEP (89-93)
	São Borja (SB)	28°45'S	56°00'W	99m	IPAGRO (89-93)
SC	Chapecó (C)	26°04S	53°01'W	679m	EPAGRI (89-92)
	Campos Novos (CN)	27°24'S	51°12'W	920m	EPAGRI (89-92)
	Abelardo Luz (AL)	26°35'S	52°22'W	860m	EPAGRI (89-92)

*CNPT = Centro Nacional de Pesquisa de Trigo; CEP = Centro de
   Experimentação e Pesquisa da Federação das Cooperativas de Trigo do Rio
   Grande do Sul; IPAGRO = Instituto de Pesquisas Agronômicas da Secretaria
   da Agricultura do Rio Grande do Sul; and EPAGRI = Empresa de Pesquisa
   Agropecuária e Difusão de Tecnologia de Santa Catarina S.A.

Results and Discussion

Averaged over locations and years, the newer triticale cultivars Triticale BR 4,
Embrapa 18, CEP 22 Botucaraí, CEP 23 Tatu, CEP 25, and IAPAR 23 produced
higher yields than Triticale BR 1, CEP 18 Botucarai, and the highest yielding
wheat check of each trial. In 1993, the wheat check Embrapa 16 outyielded the
triticale cultivars at Lagoa Vermelha, Passo Fundo, Cruz Alta and São Borja.
The highest yields were obtained in Chapecó and Lagoa Vermelha and the lowest
in São Borja and in the trial on acid soil at Passo Fundo.

Average yield for the evaluated cultivars was highest in 1989 (5.3 t/ha), followed
by 1992 (4.6 t/ha), whereas in 1990 (2.8 t/ha), 1991 (2.9 t/ha) and 1993 (2.7 t/ha)
the yields were significantly lower for the average of the evaluated cultivars.

Among the triticale cultivars Triticale BR 1 had the lowest average TW while
Embrapa 18 had the highest, but it was 5 kg/hL below the average of the best-
yielding wheat check in each trial. The highest test weights were produced in
Campos Novos, Chapecó, Abelardo Luz and Lagoa Vermelha, while the lowest
occurred in Passo Fundo, Cruz Alta and São Borja. TW was highest in 1989 (75
Kg/hL), 1991 (73 Kg/hL), 1992 (73 Kg/hL) while in 1990 (66 Kg/hL) and 1993 (67
Kg/hL) it was significantly lower. High yields were associated with high TW, with
the exception of 1991 when yield was low and TW high.

Table 3. Average yield (t/ha) of triticale and the wheat check cultivars in southern Brazil (1989-1993).

Cultivar	LV	PF(N)	PF(Al)	CA	SB	C	CN	AL	Aver.
				Location*					
BR 1	3.9 a**	2.7 a	2.0 a	3.0 a	2.1 ab	3.7 a	2.9 a	2.8 a	2.8 a
BR 4	5.5 d	4.5 c	2.6 b	3.5 bc	2.0 ab	5.8 bc	4.4 c	4.9 c	4.1 d
EMB 18	5.1 cd	4.4 c	2.7 b	3.1 ab	1.9 a	5.6 bc	4.0 bc	4.2 bc	3.8 d
CEP 18	4.6 bc	3.8 bc	2.4 ab	3.4 a-c	2.4 bc	5.0 b	3.7 b	3.4 ab	3.5 c
CEP 22	5.3 d	4.5 c	2.7 b	3.3 a-c	2.0 ab	5.8 bc	4.3 c	4.8 c	4.0 d
CEP 23	5.1 d	4.4 c	2.7 b	3.5 bc	2.1 ab	6.0 bc	4.4 c	4.5 c	4.0 d
CEP 25	5.4 d	4.0 bc	2.7 b	3.4 bc	2.1 ab	6.4 c	4.4 c	4.5 c	4.1 d
IAPAR 23	5.6 d	3.7 bc	2.7 b	3.1 ab	2.1 ab	6.2 c	4.3 c	4.6 c	3.9 d
WHEAT	4.5 b	3.4 b	2.2 a	3.1 ab	2.5 c	3.8 a	3.5 b	3.3 ab	3.3 b
Average	**5.0 CD**	**3.9 BC**	**2.5 A**	**3.3 AB**	**2.1 A**	**5.4 D**	**4.0 BC**	**4.1 BC**	**3.7**

* See Table 2 for location description.

** Means followed by the same lowercase letter in columns or uppercase letter in the average line do not differ (Duncan's Multiple Range Test $p = 0.05$).

Table 4. Average TW (kg/hL) of triticale and the wheat check cultivars in southern Brazil (1989-1993).

Cultivar	LV	PF(N)	PF(Al)	CA	SB	C	CN	AL	Aver.
				Location*					
BR 1	69 a**	66 a	66 a	66 a	68 ab	70 a	71 a	71 a-c	68 a
BR 4	73 bc	70 cd	69 ab	68 a-c	69 b-d	73 ab	74 b-d	73 a-c	71 c
EMB 18	74 c	72 d	70 b	70 bc	72 e	75 b	75 cd	75 c	73 d
CEP 18	71 ab	70 bc	69 ab	67 ab	69 a-c	72 ab	73 b-d	73 bc	70 c
CEP 22	72 bc	69 bc	68 ab	67 ab	67 a	70 a	72 a-c	70 ab	69 b
CEP 23	71 ab	71 d	68 ab	67 a	68 ab	71 a	72 ab	69 a	69 c
CEP 25	73 bc	70 cd	69 b	69 a-c	70 c-e	72 a	75 d	72 a-c	71 c
IAPAR 23	73 bc	71 d	71 b	71 c	71 de	72 ab	75 d	74 bc	72 d
WHEAT	78 d	75 e	75 c	76 d	77 f	80 c	80 e	79 d	77 e
Average	**72 B**	**70 A**	**69 A**	**69 A**	**70 A**	**72 B**	**74 C**	**72 B**	**71**

* See Table 2 for location description.

** Means followed by the same lowercase letter in columns or uppercase letter in the average line do not differ (Duncan's Multiple Range Test $p = 0.05$).

Earlier triticale cultivars were susceptible to scab (*Gibberella zeae, Fusarium graminearum*). This is considered an important yield-limiting disease. To evaluate resistance, a spore suspension was inoculated into one central flower of 30 spikes of each cultivar at three different periods during the 5 years. Triticale BR 1 averaged 7.4 on a scale of 1 to 9 and was significantly more susceptible than Triticale BR 4 (6.4), CEP 25 (5.8), Embrapa 18 (6.0), CEP 22 (6.6), CEP 23 (6.2), IAPAR 23 (5.9), and CEP 18 (6.0). The newer cultivars did not have severe infection in the field or in the trials. This method, performed under field conditions, seems to be effective in evaluating and selecting for scab resistance of triticale under the climatic conditions prevailing in southern Brazil.

The severity of Helminthosporium (*Bipolaris sorokiniana*) was recorded visually in the field at three different periods during 4 years (1989-92) using a scale from 1 to 9. Analysis of variance indicated that the differences among cultivars and the interaction Year/Recording Period were highly significant. On average Triticale BR 4 (2.0), CEP 22 (2.3), CEP 23 (2.5), CEP 25 (2.9), IAPAR 23 (3.0) and Embrapa 18 (3.1) were significantly more resistant than the wheat cultivar Trigo BR 35 (4.2). CEP 18 (5.5) and Triticale BR 1 (6.4) were significantly more susceptible than Trigo BR 35.

Conclusions

At higher elevations (+400m) in southern Brazil, on acidic soil with low pH and toxic aluminium, triticale cultivars produced high yield and TW whereas, at lower altitudes triticale cultivars were out-yielded by the wheat check cultivars.

TW of all triticale cultivars was lower than wheat although some newer cultivars, such as Embrapa 18 and IAPAR 23, had plumper grains.

The resistance for scab and Helminthosporium in the newer cultivars has improved significantly.

Acknowledgments

Acknowledgments are due to Drs. Jorge Nedel, Leo Del Duca who conducted the experiments in Passo Fundo in 1993; also to Luiz Hermes Svoboda, Ari Caumo and Stanislao Dias Dávalos who conducted the experiments in Cruz Alta, São Borja and Santa Catarina, respectively; and to Antônio Zimmermann and João Américo Wordell Filho for technical assistance. The author was supported by the Conselho Nacional de Desenvolvimento Científico e Tecnológico, CNPq, Brasília.

References

1. Baier AC. Potential of triticale in southern Brasil. In: Symposium, 2, 1990. Passo Fundo. Proceedings. México: CIMMYT/EMBRAPA-CNPT/ITA, 1991, 9-13.

2. Baier AC, Nedel JL, Reis EM, Wiethölter S. Triticale: cultivo e aproveitamento. Passo Fundo: EMBRAPA-CNPT, 1994. 72p. (EMBRAPA-CNPT. Documentos, 19).

TRITICALE AT HIGH LATITUDES IN ALBERTA CANADA

Donald F. Salmon[1], Vernon S. Baron[2], J. Grant McLeod[3] and James H. Helm[1]
[1]FCDC, AARD, Lacombe, AB, Canada. [2]AAFC Res. Station, Lacombe,
AB, Canada. [3]AAFC Res. Station, Swift Current SK, Canada

Abstract

Triticale (X *Triticosecale* Wittmack L.) has the potential to become an important alternative
feed grain and fodder crop in the prairie region of western Canada. The main area for
expansion of triticale production in Alberta appears to be extended grazing (winter type) and
conserved fodder (spring type). Cooperative work between the Field Crop Development
Centre (FCDC) and Agriculture and Agri-Food Canada (AAFC) at Lacombe, Alberta has
demonstrated the value of winter triticale as a spring seeded long season grazing crop.
Similar cooperative work between the FCDC and AAFC (Lacombe and Swift Current) has
shown that the new spring triticale varieties have high whole plant dry matter production
with excellent digestibility. The mandate for the FCDC triticale program is to produce
agronomically improved spring and winter varieties which meet the needs of the Alberta
livestock industry.

Introduction

The development of a new generation of improved triticale (X *Triticosecale* Wittmack L.)
varieties with potential as a feed grain and forage crop has stimulated a renewed interest from
farmers on the Canadian prairies. In Alberta the main area for seed production of winter triticale
falls within the Parkland (100-115 frost free days) climatic zone which has good snow cover,
whereas the seed production of spring triticale is best suited to the Prairie Grassland (115-120
frost free days) climatic zone. Since the Field Crop Development Centre is located in the
Parkland, the spring triticale program stresses early maturity as a major parameter in the
development of improved cultivars. In the winter program cold tolerance and snowmould
(*Coprinus psychromorbidus sp*) resistance are the main priorities.Due to the high concentration
of beef cattle in Alberta, the high whole plant yield compared to spring barley (*Hordeum
vulgare* L.) and good feeding quality [1, 2] has resulted in forage production being the main
area of expansion for spring triticale. In a similar fashion, the primary interest in winter triticale
has been due to its value as a high quality grazing crop [3,4,5]. At the present time, evaluation
of winter and spring triticale in the breeding program is limited to testing advanced lines for
biomass production. Detailed evaluation of triticale for forage production and quality is
being carried out as a cooperative project with the forage program at Agriculture and
Agri-Food Canada (AAFC) at Lacombe and the spring triticale program at AAFC

693

H. Guedes-Pinto et al. (eds.), Triticale: Today and Tomorrow, 693–699.
© 1996 *Kluwer Academic Publishers. Printed in the Netherlands.*

Swift Current , Saskatchewan. In spring triticale the objective is to determine the quality as well as silage yield of new varieties and advanced lines compared to barley and oat. In winter triticale the objective is to identify its value compared to winter rye (*Secale cereale* L) as a frost tolerant species suitable for use in a range of cropping systems aimed at extending the grazing season beyond that of perennial forages.

Procedures

SPRING TRITICALE BREEDING

Approximately 20-50 spring crosses (conventional x reduced awn) are produced in each year with emphasis on seed development, high yield, early maturity, and sprouting tolerance. The spring wheat RL4137 has been used as a source of potential sprouting resistance and reduced awn expression. The F_1 through F_3 generation are subjected to generation advance without selection. The F_4 and subsequent generations are grown as bulks (approximately 10,000 plants per population) at 2 locations in Alberta and are subjected to modification for visual traits. The F_5 and subsequent generations are subjected to evaluation for visual and yield related parameters. Table 1 indicates the sequence of events in the breeding program.

Table 1. The sequence for the spring triticale breeding program at the Field Crop Development Centre at Lacombe Alberta.

Generation	Selection sites	Trial Type	Selection criteria
F_1	El Centro CA	Hills	Generation advance
F_2	Lacombe AB	Bulk	Maturity, reduced-awns
F_3	El Centro	Bulk	Generation advance
F_4-	2 sites AB	MB	Maturity, reduced -awn,disease,plant type.
F_5	Lacombe	Row	Uniformity, maturity,disease, plant type .
F_6-F_9	10 sites	Yield	Yield etc. plus forage evaluation, purification
F_{10}-F_{12}	10 sites	Yield	Registration trials

The advanced yield trials are conducted as a joint effort with Dr. Grant McLeod at AAFC Swift Current Saskatchewan.

WINTER TRITICALE BREEDING

Approximately 20-30 crosses are produced each year with winter hardiness as the first priority and snowmould as the second priority (Table 2). Short stature is incorporated using germplasm lines obtained from R. J. Metzger and CIMMYT. Pika is used extensively in backcrosses to incorporate winter hardiness and moderate levels of snowmould tolerance. Metzger material with Musketeer fall rye cytoplasm (*Secale cereale* L) is currently being evaluated for hardiness. Advanced lines are artificially tested for cold tolerance and snowmould after field evaluation.

Table 2. The sequence for the winter triticale breeding program at the Field Crop Development Centre at Lacombe Alberta.

Generation	Selection sites	Trial type	Selection criteria
F_1	Lacombe	Hills	Cold frame or cabinet
F_2	Lacombe	Bulk	Field selection for hardiness and SM
F_3-	Lacombe	MB	Hardiness, SM, plant type, disease, seed etc.
F_4	Lacombe	Row	Hardiness, SM, plant type, disease, seed etc.
F_5-F_6	3 sites AB	Hardiness	as in F4, plus protein
	Hermiston OR	Increase	Uniformity check and seed increase
F_7-F_9	7 sites	Yield	Yield etc. forage evaluation, purification registration.
	Hermiston OR	Increase	Uniformity check and seed increase.

In the F_7-F_9 , a site at the Crop Development Centre Saskatoon (Dr D.B. Fowler) and AAFC Swift Current Saskatchewan are used for hardiness and yield evaluation.

FORAGE

Spring triticale. Two cooperative projects are the evaluation for quality of residues of spring and winter triticale after grain harvest (straw) compared to other species and a study to determine genotype x environment effects for dough stage-forage yield and quality among current and new lines of spring types across the prairie provinces..

Winter triticale. Adaptation of winter types as spring planted pasture in monocrops and mixtures with spring cereals for subsequent use in mid and late season has been investigated for several years [3, 4, 5, 6, 7]

Discussion

SPRING TRITICALE

In 1987 the Field Crop Development Centre at Lacombe released the spring triticale cultivar Wapiti, a selection from the CIMMYT family Juanillo 90. Wapiti set a new standard for the evaluation of new varieties for seed yield and biomass production. Subsequent, cultivars that are currently in production have improved (Table 3) upon Wapiti for test weight, and yield, but Wapiti has remained popular for silage (Table 2). An increase in the number of varieties has demonstrated regional adaptation with Frank performing best in the eastern prairies and Banjo and Wapiti performing best in the central/western prairies. All of the current varieties are 10 days later in maturity than the Canadian Western Red Spring Wheat varieties and are susceptible to sprouting, thus restricting seed production to the long season dryland areas.

696

Table 3. The seed yield, biomass yield ,and maturity of current spring triticale varieties on the Canadian prairies (1987-1992).

| | Black soil zone | | | Brown soil zone | | |
Variety	Seed (%)	Biomass (%)	Maturity (days)	Seed (%)	Biomass (%)	Maturity (days)
Biggar (wheat)	100	-	104	100	-	99
Wapiti	112	108	108	115	112	101
Frank	111	109	109	106	109	102
Banjo	117	114	110	112	110	103
Cascade (oat)	-	100	-	-	100	-
Johnston (barley)	-	97	-	-	103	-

Comparison for seed yield based on Biggar wheat and Biomass yield based on Cascade oat. Insufficient data was available for AC Copia.

Until very recently, the primary source of plump seed type and earlier maturity has been the CIMMYT line Panda R 203 (ID 79Q133001). Most crosses produced from this line have been early, low tillering and low yielding. One cross with Wapiti which has been in the yield trials since 1986 , has shown some potential for yield and early maturity. One line (T124) ranges from 3-4 days earlier maturity and yields equal to or higher than the checks (Table 4).

Table 4.   Grain yield and agronomic potential of FCDC spring triticale lines in the registration trials (1992-1993).

Varieties/line	Yield (t/ha)	Mat (days)	Hgt (cm)	Ldg (0-9)	Twt (kg/ha)	Kwt (g)
Biggar (wheat)	4.85	117	83	1.9	75.3	36.3
Wapiti	5.51	124	112	2.4	66.1	44.6
Frank	5.58	124	107	2.4	68.0	39.7
Banjo	5.28	125	118	1.8	66.2	44.7
AC Copia	5.40	124	113	2.5	69.7	45.3
T124	5.71	122	111	2.1	67.3	41.5

Data derived form the cooperative registration trials coordinated by Dr. Grant McLeod AAFC Swift Current Saskatchewan.

Preliminary evaluations of the conventional and reduced-awn types for post-harvest sprouting resistance have demonstrated only small improvements in Hagberg Falling Number and dormancy. However the reduced-awn lines appear to have potential for the production of unchopped fodder. The poster " Developing Spring and Winter Triticale with Reduced-Awn Expression" discusses the potential of the reduced-awn types in the program.

WINTER TRITICALE

In 1990 Pika was released by the FCDC for production in the higher snowfall zones of Alberta. Although Pika is not as winter hardy as fall rye (Table 5), it is at least as hardy as the hardy winter wheats produced in Alberta. Pika also carries an improved tolerance to cottony snowmould. Work at the FCDC indicates that extending the hardening period for winter triticale by planting in late August improves its winter survival.

Table 5. The winter survival, seed yield, and maturity of current winter triticale varieties in western Canada.

| | Black soil zone | | | Brown soil zone | | |
| Maturity | Surv. | Yield | Maturity | Surv. | Yield | Maturity |
Variety	(%)	(%)	days)	(%)	(%)	(days)
Wintri	79	120	234	85	109	221
Pika	83	120	234	85	120	217
Musketeer (rye)	94	110	218	95	113	197
Norstar (wheat)	82	100	229	82	100	219

Yield is compared to Norstar winter wheat as 100%. Maturity is based on days after January 1.

In 1992 a series of winter triticale populations with Musketeer as the female parent were obtained from R. Metzger. The intention was to determine if these lines were hardier than conventional types. Preliminary field data from 1993 indicated that the selections range from very sensitive to hardiness levels in the Pika - Musketeer range. It was assumed that if the Cytoplasm was a major factor all lines would have been similar in hardiness since selections were from the same backcross populations. 1994 field data has shown no differential kill .

FORAGE

Spring triticale. Until recently, forage yield and general adaptation for conserved forage has been the primary concern of the triticale breeding programs. In general, spring triticale lines, typified by Wapiti, have yielded more dry matter than barley on the drier brown soils and as much or more than oats on the moister black and grey wooded soil locations.

Evidence indicates that recent selections may have undergone some indirect improvement because of associated selection criteria. The variety Carman exhibited forage quality intermediate to barley and oats at the dough stage. Data collected in 1993 (Table 6) suggests that spring lines such as Banjo, Frank, and Wapiti and unregistered lines have forage quality similar to barley. All spring cultivars have rough awns. While this trait is not necessarily detrimental to feed consumption it can restrict consumption in some feeding scenarios. The reduced - awn type should alleviate this problem.

Table 6. The yield potential and forage quality of spring triticale , barley and oat when harvested at the early dough stage.

| Species | Yield | Protein | IVDOM | NDF |
(%)	t ha^{-1}	(%)	(%)	
Triticale	16.9	8.6	67.2	44.0
Barley	12.4	10.0	68.2	46.6
Oat	16.0	11.6	61.6	48.9

IVDOM is in vitro digestible organic matter. NDF is neutral detergent fibre.

From tests conducted in all three prairie provinces, residue or straw (whole-plant minus grain) quality of spring triticale is less than oats and barley, but higher than wheat (Table 7). Straw and chaff residues are used as feedstuffs for gestating beef cows in the dry areas of the prairies.

Table 7. Quality of spring triticale, barley, oat and wheat straw.

Species	Protein	IVDOM	NDF
	----------------------------(%)----------------------------		
Triticale	4.3	45.3	71.6
Barley	4.3	47.2	73.8
Oat	5.6	49.4	71.6
Wheat	4.0	42.6	74.3

IVDOM is <u>in vitro</u> digestible organ matter. NDF is neutral detergent fibre.

Winter triticale. Winter triticale fills a role as a pasture when spring planted (Baron et al ,1992; Baron et al, 1993A; and Baron et al 1993B). Because the spring seeded winter triticale remains vegetative it will continue to regrow until limited by temperature in late October (Table 8). Yield may be increased and the grazing period initiated earlier in the season if the winter triticale is grown in a mixture with a spring cereal. Although the total number of grazing days is similar for the winter cereal and perennial forage , the grazing of the winter cereal starts later in the season and continues for an additional 4 weeks.

Table 8. Performance of a spring seeded binary combination of winter triticale and spring barley compared to perennial grass as a pasture in central Alberta.

	Triticale/barley Steers	Perennial grass Heifers
Rainfall (mm)	264	274
Initiation	July 1	May 29
Completion	Oct. 23	Sept. 9
Grazing days	115	103
Stocking rate (animals/ha)	0.38	0.38
Daily gain (kg/day)	0.94	0.74
Total gain (kg/ha)	23.6	19.3

Note: differences in gain between the crop systems are what would be expected based on the comparisons of steers and heifers.

This research has also shown that yield can be maximized if the mixture is allowed to advance until near the dough stage of the spring cereal (Baron et al 1992). While the mixture yield will be slightly less than the spring monocrop (about 10%), the reduction is more than compensated for by the regrowing winter cereal. At the same time the forage quality of the mixture is superior to a spring cereal monocrop at the dough stage due to the leafy content provided by the winter cereal. Our research has shown that the longer the winter cereal remains shaded by the spring cereal the poorer the potential for winter cereal recovery.

Systems which remove the forage initially closer to heading should provide pasture more consistently during the subsequent period. Triticale tends to survive well in the mixtures where light levels are often below the light compensation point, recovers well after the initial harvest and regrows somewhat better than rye and wheat during the late season.

Acknowledgement

The authors gratefully acknowledge the contribution of the technical staff at the Fieldcrop Development Centre (Lacombe)and Agriculture and Agri-Food Canada (Lacombe and Swift Current). The Alberta Heritage Fund is acknowledged for its contribution of funds for the work on winter cereal forage production.

Literature

ZoBell, D.R., L.A. Goonewardense, and D.F. Engstrom. 1992. Use of triticale silage in diets for growing steers. Can. J. Anim. Sci. 72:181-184

Khorasani, G.R., E.K. Okine, J.J. Kennelly, and J.H. Helm. 1993. Effects of whole crop cereal grain silage substituted for alfalfa silage on performance of lactating dairy cows. J. Dairy Sci. 76:3536-3546.

Baron, V.S., H.G. Najda, D.F. Salmon, and A.C. Dick. 1992. Post-flowering forage potential of spring and winter cereal mixtures. Can. J. Plant Sci. 72:137-145.

Baron, V.S., H.G. Najda, D.F. Salmon, and A.C. Dick. 1993A. Cropping systems for spring and winter cereals under simulated pasture: Yield and yield distribution. Can. J. Plant Sci. 73:703-712.

Baron, V.S., H.G. Najda, D.F. Salmon, J.R. Pearon, and A.C.Dick. 1993B Cropping systems for spring and winter cereals under simulated pasture: Sward structure. Can. J. Plant Sci. 73:947-959.

Baron, V.S., A.C. Dick, and E.A. de St. Remy. 1994. Response of forage yield and yield components to planting date and silage/pasture management in spring seeded winter cereal/spring oat cropping systems. Can. J. Plant Sci. 74:7-13.

Salmon, D.F., V.S. Baron, and A.C. Dick. 1993. Winter survival and yield of early-seeded winter wheat and triticale. Can. J. Plant Sci. 73:691-696.

REGIONAL TRIALS ON TRITICALE - NATIONAL LIST

Ana Paula A. Cruz de Carvalho
Centro Nacional de Protecção da Produção Agrícola. Tapada da Ajuda,
1300 Lisboa. Portugal.

Abstract

In the 1993 edition of Portuguese National List are listed 12 triticale varieties, 8 of them being of nationally bred.

The evaluation of cultivation value for the purpose of registration in the National List is based upon as experimental layout, the National Trial Network. The trial designs are randomised blocks in four replicates.

The results of the triticale National Trial Network in general, and particularly for the varieties Arabian and Borba, in what concernes grain yield (kg/ha) at 14% moisture content, hectolitre weight, weight per 1000 seeds, protein content and cycle are presented.

Introduction

An assessment of the varieties which have been submitted for registration, on the National List, since 1980 and which are currently registered encourages the use of nationally bred varieties.

The results obtained from obligatory studies forming the basis for the final analysis of registered varieties by CNPPA's (National Center for Plant Protection) Evaluation Committee are briefly presented.

An analysis of the varieties Arabian and Borba with respect to the relationship between yield and level of environmental fertility is made.

Registration of varieties on the National List

The certification of triticale seed, with few exceptions, is only allowed for varieties registered on the National List. The parties interested in certification submit their varieties for evaluation through the official authorities (CNPPA). This evaluation is made within the conditions required by Portuguese Law (Order n°. 310/91 of August 16th, Regulations n° 844/85 of November 8th and n° 43/92 of January 24th) and Community Law (Directives 70/457/CEE and 70/458/CEE).

H. Guedes-Pinto et al. (eds.), Triticale: Today and Tomorrow, 701–708.
© 1996 *Kluwer Academic Publishers. Printed in the Netherlands.*

Under article 3 of Order n° 301/91 of August 16th the evaluation of new varieties applying for registration on the National List is based upon the principle that these varieties satisfy the following conditions:

- are found to be distinct, uniform and stable;
- show an increase in the Cultivation Value and/or Use Value, compared to the check varieties known and marketed in the country.

This evaluation of new varieties includes the following tests:

- Distinctness, Uniformity and Stability (D.U.S.);
- Cultivation Value (C.V.);
- Use Value (U.V.)

For each variety, results of at least two years of trials are analysed by an Evaluation Group for Winter Cereals, which prepares proposals to be present to the CNPPA's Evaluation Committee, who, according to the Technical Regulations for Evaluation, take a final decision of rejection or addition of the variety to the National List (Fig. 1)

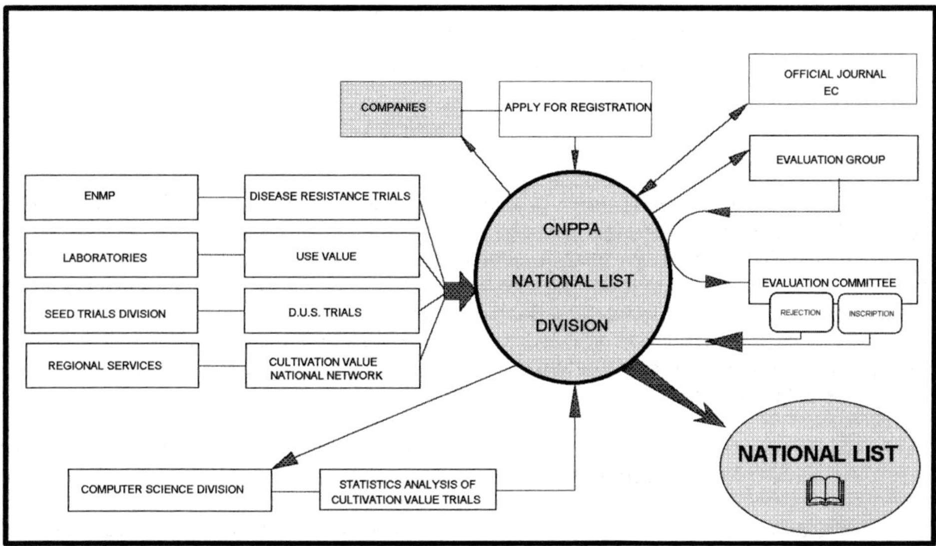

Fig.1 -Registration of triticale varieties

Trials for inscription

DISTINCTNESS, UNIFORMITY AND STABILITY

A variety shall be regarded as **distinct** if it is clearly distinguishable, in one or more important characteristic, from any other variety know in the Community. This is regardless of whether the origin of the variety is artificial or natural.

A variety shall be regarded as sufficiently **uniform** if, apart from a very few aberrations, the plants composing it are similar or genetically identical with respect to the specified characters used for identification.

A variety shall be regarded as **stable** if the variety remains true to the description of its essential characteristics after successive propagation's or multiplication's.

CULTIVATION VALUE

The evaluation of the Cultivation Value of the new varieties is based mainly on their average yield (kg/ha) at 14% moisture content, being from a minimum of eight valid trials. The Cultivation Value is expressed as a percentage of the average yield of the check varieties.

Studies of physiological and plant health characteristics are complementary, as these can affect the reliability of yield over years.

USE VALUE

The value of a variety for use relies on the analysis of samples from the valid trials and for triticale are determined as follows:

- total protein content of grain (expressed as a percentage of dry matter);
- hectolitre weight (expressed in kilograms);
- weight of 1000 seeds (expressed in grams).

National Trial Network

The experimental basis that enables the evaluation of new varieties' behaviour is the National Trial Network.

This Network which is located in places spread throughout the country, intends to represent the cropping areas of the species, in order to reflect the adaptability of new varieties not only to cultivation techniques, but also to different environmental conditions.

The National Trial Network includes two different zones - Zone 1 and Zone 2 (Fig. 2). In the first one the varieties with a vegetative cycle equal or larger than Clercal are tried, and in the second one, all the others.

The statistical design is randomised blocks with 4 replicates and the validity of trials is based in the following two parameters:

- experimental error, expressed by the variation coefficient which must be lower than 18%, and;

- "agronomic validity" - the trials which general average yield is lower than 2000 kg are submitted to a special analysis.

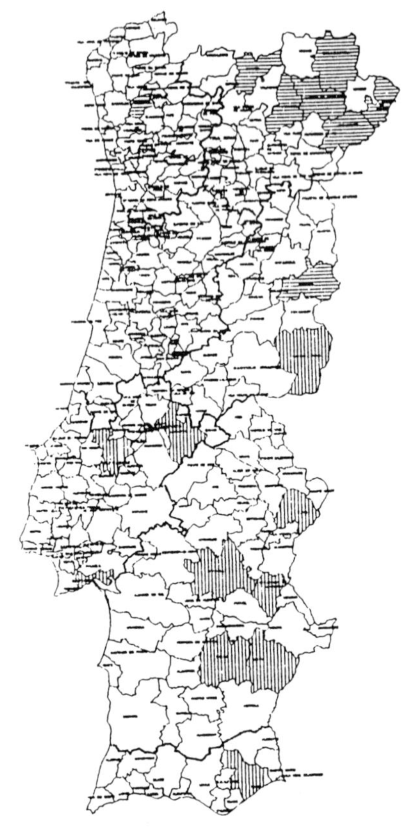

Zone 1 - trials established by the Regional Services of Agriculture of "Entre Douro e Minho", "Trás os Montes e Alto Douro" and part of "Beira Interior (Sabugal)".

Zone 2 - trials established by the Regional Services of Agriculture of "Beira Interior (Idanha-a-Nova)", "Ribatejo e Oeste", "Alentejo" and "Algarve".

Fig.2 - Triticale National Trial Network

Varieties registered in the National List

Since 1980, 26 triticale varieties have been submitted for registration. Twelve of these have been registered, eight of which were nationally bred, by the National Plant Breeding Station (ENMP). From these, 8 are nationally bred - National Plant Breeding Station (Table 1).

	Nationally bred	Foreign bred
Applications	18	8
Rejected/Discontinued	10	4
Inscriptions	8	4

Table 1 - Quantification of the registered varieties

Two of the registered varieties have long cycles, suitable only for cultivation in the centre and north of the country. The others have early to medium vegetative cycles.

The excessively early ear emergence of most triticale varieties leads farmers to sow late and this can lead to technical problems due to excessive soil moisture. Breeding varieties with later ear emergence may allow an early sowing time with minimal risk of frost damage during the flowering period. Other characteristics which breeders are attending to are increase in hectolitre weight, reduction of grain wrinkling and ease of threshing. The recent nationally bred varieties to be registered on the National List show some improvement in these characters. Table 2 shows data used as a basis for the decisions taken by the CNPPA's Evaluation Committee for the registration of the present varieties on the National List.

Varieties Arabian and Borba

These two varieties represented about 33% of the total certified seed marketed in 1992/93 in Portugal, and are now the check varieties to the Zone 2 of the National Trial Network.

The behaviour of these two varieties with respect to yield in increasing levels of environment fertility is shown (Fig. 3).

The data used to quantify the different levels of environment fertility are comprise on the general average of 36 valid trials in 7 years study of the National Trial Network.

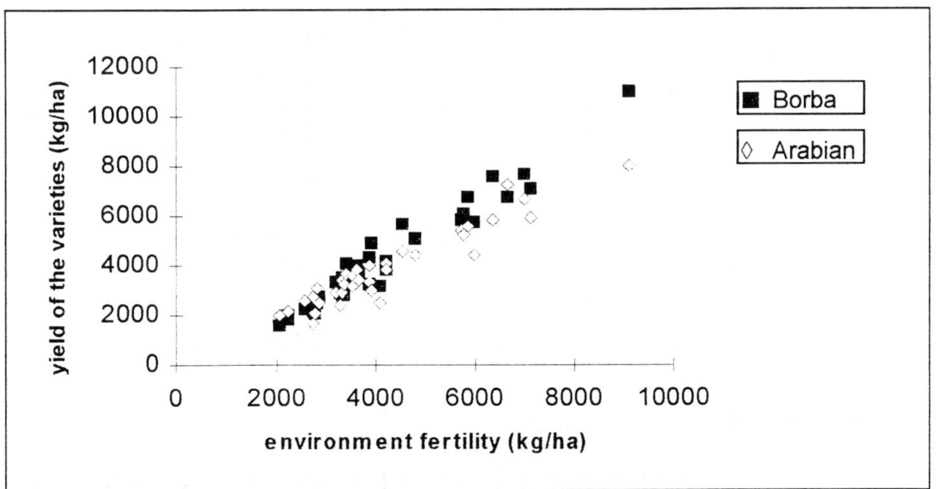

Fig. 3 -Relationship between yield of the varieties Arabian and Borba and environment fertility.

From fig. 3 is possible to verify, that the variety Arabian shows a trend to have a better yield than the variety Borba in a low fertility environment.

On the contrary, Borba achieves higher yields as the environment fertility increases, out yielding Arabian.

Varieties registered in the National List	trials	years	yield (%check variety)	hectolitre weight (kg)	1000 seeds weight (g)	protein content(%)	cycle (ear emergence)
ALTER	12	1989/90 1990/91 1991/92	109,4%	73,25 - 76,85	35,6 - 40,7	13,0 - 15,2	early-medium

check variety : (Arabian+Borba)/2
100%=3976,3 kg/ha

Varieties registered in the National List	trials	years	yield (%check variety)	hectolitre weight (kg)	1000 seeds weight (g)	protein content(%)	cycle (ear emergence)
*ARABIAN	15	1987/88 1988/89 1989/90	---	62,55 - 76,45	34,8 - 47,9	11,8 - 19,2	early
ARRUDA	9	1988/89 1989/90	130,9%	76,45 - 76,95	39,1 - 41,7	13,4 - 15,2	early

check variety : Arabian
100%=3207,0 kg/ha

Varieties registered in the National List	trials	years	yield (%check variety)	hectolitre weight (kg)	1000 seeds weight (g)	protein content(%)	cycle (ear emergence)
*BACUM	9	1984/85 1985/86	---	65,85 - 74,35	36,1 - 48,3	11,0 - 17,2	early
*BEAGLE	9	1984/85 1985/86	--	58,85 - 71,65	38,3 - 54,4	10,0 - 17,9	early
BORBA	11	1982/83 1983/84 1984/85	107,7%	71,40 - 72,90	40,9 - 51,2	10,8 - 16,6	medium

check variety : Arabian
100%=5784,8 kg/ha

* These varieties have been submetted only to DUS trials.

Table 2-Varieties registered in the National List

Variety		Years					
CLERCAL	10	1985/86 1986/87	130,5%	68,20	36,0 - 41,7	11,1	long

check variety : Centeio Regional
100%=3080,1 kg/ha

CRATO	9	1988/89 1989/90	125,1%	76,45 - 78,10	37,7 - 39,5	12,9-16,2	early

check variety : Arabian+Borba)/2
100%=3976,3 kg/ha

CUME	16	1986/87 1987/88 1988/89	106,7%	66,35 - 72,70	36,6 - 48,2	11,8 - 13,2	medium

check variety : Arabian
100%=3916,6 kg/ha

JUANILHO	15	1984/85 1985/86 1986/87	120,0%	66,90 - 72,90	46,9 - 55,7	11,8 - 13,2	early

check variety : Bacum
100%=4646,7 kg/ha

TRIMARAN	9	1990/91 1991/92	127,4%	72,25 - 73,00	32,0 - 37,8	13,3 - 14,4	long

check variety : Clercal
100%=3671,8 kg/ha

TRITANO	9	1988/89 1989/90	126,8%	71,95	42,4 - 45,7	11,7 - 14,0	early

check variety : Arabian
100%=3205,8 kg/ha

Table 2-Varieties registered in the National List

References

Carvalho P, Carneiro R. Resultados das variedades de cereais que completaram o ciclo de ensaios em 1898/90. Oeiras: CNPPA, 1991: 66-73.

CNPPA - Divisão de Ordenamento. Cereais Outono-Invernais - Regulamento Técnico de Avaliação de Variedades (VAU+DHE). Oeiras: CNPPA, 1993: 1-10: 14-18.

CNPPA - Divisão de Ordenamento. Cereais de Inverno - Resultados das variedades que completaram o ciclo de ensaios. Oeiras: CNPPA, 1993: 69-72.

DECRETO-LEI nº 301/91. «D.R.-I Série B». 187 (91-08-16).

DIRECTIVA DO CONSELHO nº 70/457/CEE. Jornal Oficial das Comunidades Europeias nº L225. (70-10-12).

Teixeira C. Cereais de Sequeiro - Resultados e decisões sobre as variedades que completaram o ciclo de ensaios. Oeiras: CNPPA, 1986: 109-111.

Teixeira C. Ensaios Experimentais - Resultados de ensaios de variedades em homologação para o Catálogo Nacional de Variedades. Lisboa: CNPPA, 1987: 7-35.

Teixeira C. Cereais Praganosos de Sequeiro - Resultados das variedades que completaram o ciclo de ensaios em 1988/89. Oeiras:CNPPA, 1990: 71-75.

QUANTITATIVE AND QUALITATIVE COMPONENTS OF ELEVEN GREEK TRITICALE CULTIVARS (X *Triticosecale* Wittmack) COMPARED TO CV. "VERGINA" (*T. aestivum* L. em. Thell.).

D. M. Gogas[1], N. Triantaphilakos[1], S.Stratilakis[1], S. Georgiadis[2], T. Adamidis[3], G. Skipitaris[4], H. Kartitsi[5], H. Tsivopulu[6] and I.Efstathiu[7]
[1]Cereal Institute - Thessaloniki, Greece; [2]A.R.S.of Xanthi-Xanthi, Greece, [3]A.R.S. of Ptolemais - Ptolemais, Greece; [4]A.R.S. of Tripolis - Tripolis, Greece; [5]A.R.S. of Serres - Serres - Greece; [6]F.C. and Pastures Institute - Larissa, Greece; [7]A.R.S. of Vardates - Vardates, Greece

Abstract

Eleven hexaploid triticale cultivars were studied for some quantitative and qualitative components and compared to the breadwheat cultivar "Vergina" which is widely cultivated in Greece.

The data from 144 experiments, carried out during the period 1976-1992 in a wide spectrum of environments, were analysed by using the method of regression and the mean square of the deviations from linear regression.

The results show that triticales exceed by 4% the yielding ability of "Vergina" and this superiority is ranged from 0% to 10%. Triticales also exhibited lower interaction with environments and years (b=1 and s^2d low) compared to "Vergina".

Seed protein percentage (Pr%) and sedimentation value (Zeleny test), the mean values of triticale varieties were not significantly different from the corresponding values of "Vergina". Two triticale varieties were found to exceed Vergina in Pr% and another one had a higher sedimentation value.

These two components varied less through years and locations in comparison to "Vergina". Finally the 1000 kernel weight mean value of triticale varieties was found to be significantly higher compared to Vergina. This component, exhibited low variation (b≈1 and s^2d low).

The results presented above and similar results published by other scientists support the belief that the time has come for triticale to be officially recognized as a commercial product and to be used in environments when the other cereals do not perform well.

Introduction

In Greece, cereals are grown on 40% of the total cultivated area and provide 18% of the gross national income because of their ability to exploit the semiarid, hilly and highland areas better than other crops. The grain quality resulting from cultivation in these areas is usually satisfactory but the yield is usually low (3000 kg/ha) in spite of the high yielding potential of the cultivars.

H. Guedes-Pinto et al. (eds.), Triticale: Today and Tomorrow, 709–718.
© 1996 *Kluwer Academic Publishers. Printed in the Netherlands.*

In these areas where the limiting factor is soil moisture, triticales exhibit better productivity (Gogas 1988, 1993) and reach an acceptable quality level. It is very interesting and useful to see the range of this superiority of triticales in comparison to other cereals and especially to bread wheat.

This work represents a first attempt to compare triticales with bread wheat through a series of experiments which were established in 15 different locations during the period 1976-1992.

Materials and Methods

Eleven hexaploid Greek triticale varieties, five of them included in the national and the EU list, were compared to the widely cultivated in Greece, Greek bread wheat cultivar "Vergina". The parameters which were used in this comparison are: yield in kg/ha, protein % of kernel dry material, sedimentation value (Zeleny test), 1000 kernel weight and their variation over years and locations.

All the comparisons were based on the data from 144 comparative yield trials, established during 1976-1992 at 15 different locations in Greece. The bread wheat cultivar "Vergina" was the long term check in all these experiments. The number of experiments used for each variety ranged from 10 to 92. Each experiment consisted of 24 new varieties randomized in a randomized complete block design, with four replications. Each plot consisted of seven rows (seven m^2), the five being threshed (5 m^2) during the harvest.

All the experimental data were analyzed by the method of regression analysis, as described by Gogas (1988). This method of analysis is a combination of the methods proposed by Finlay and Wilkinson (1963), Eberhart and Russel (1966) and Johnson *et al.* (1968). In this analysis the experimental mean was the independent and the mean of the variety was the dependent variable. The means were compared by the Z criterion and the variability by using the b value, the mean value of the characteristic and the mean square of deviations from the linear regression (s^2d).

Finally a hierarchical analysis of the data was made to find the years and locations where each variety exhibited the best performance for each of the characteristics studied.

Results and Discussion

The mean yield and the mean values of the qualitative characteristics, of the eleven triticale varieties, expressed in metric units and in % of the check used in the same series of experiments are given in Table 1. In the same table, the means for triticale as a whole, as well as the means for the check and the experiment as a whole are given. It must be emphasized that the percentage of check in the table is based on the mean of the check in the series of experiments used to calculate the means of each variety and not on the grand mean, which is the mean of all (144) experiments.

Based on this table it can be stated that:

1. Triticales as a whole yielded significantly higher than "Vergina" by 0.5%-10.02%. This confirms the estimations made by Varughese *et al.* (1986), Gogas (1988, 1993), Xinias *et al.* (1992), CIMMYT (1990, 1991, 1992), among others.

TABLE. 1. Yield and quality of eleven triticale cultivars compared to the breadwheat cultivar Vergina.

S.N.	VARIETY NAME	No. OF EXP.	YIELD MEAN (KGR/H)	YIELD % OF CHECK*	PROTEIN CONTENT MEAN (PR%)	PROTEIN CONTENT % OF CHECK*	1000 KERNEL WEIGHT MEAN (GR)	1000 KERNEL WEIGHT % OF CHECK*	ZELENY TEST MEAN (SED.VALUE)	ZELENY TEST % OF CHECK*
1	DADA	78	4285^h	102.60	13.69	101.20	37.87	106.30 –	24.67	95.90
2	NIOVI	92	4123^{gh}	100.50	14.30	105.30 –	36.35	103.90	26.20	105.30 –
3	ERATO	54	4447^f	105.08	13.00	100.00	38.44	106.50 –	21.09	99.30
4	VRONTI	30	5228^b	107.62	12.84	102.10	38.97	107.70 –	19.47	92.80
5	THISVI	30	5240^b	107.86	12.79	101.00	40.00	110.00	22.00	103.60
6	FENTRA	35	4569^d	102.86	13.41	99.50	37.89	106.30 –	23.17	106.40
7	VRITO	26	5361^a	107.82	12.76	102.10	43.60	119.90 –	16.04	79.40
8	DIONI	25	4186^g	99.79	14.11	102.00 –	36.46	102.60	24.00	89.30
9	LITO	18	4727^c	106.80	12.88	96.60	38.39	108.00 –	20.56	83.70
10	KIVELI	11	4260^{gh}	108.67	13.77	100.70	37.46	107.90	24.91	91.30
11	EKATI	10	4217^{gh}	110.02	13.63	101.00	37.60	101.30	24.90	111.70
12	TOTAL TR.	409	4513^e	104.06	13.51	101.70	38.13	106.90	23.08	98.00
13	TOTAL CHECK	409	4337^g	100.00	13.28	100.00	35.69	100.00	23.55	100
14	TOTAL EXP.	409	4443^f	102.44	13.42	101.10	35.57	105.30	22.07	93.70

* Estimated by deviding the variety mean by the check mean in the same series of experiments

– significantly higher than check

TABLE.2. The linear regression components of yield.[*]

S.N.	VARIETY NAME	N.OF EXP.	VAR. MEAN	EXP. MEAN	CHECK MEAN	b_1	b_2	b_3	r_1	r_2	r_3	a_1	a_2	a_3	$s^2 \bar{d}_1$	$s^2 \bar{d}_3$
1	DADA	78	4285	4316	4174	0.97	1.00	0.87	0.97**	0.92**	0.93**	11.64	10.73	41.87	285,262	300,666
2	NIOVI	92	4123	4135	4103	0.93	0.88	0.94	0.95**	0.92**	0.92**	26.25	49.64	23.32	243,610	338,207
3	ERATO	54	4447	4226	4232	0.95	1.05	0.71	0.96**	0.89**	0.85**	14.18	-0.66	122.00	194,203	324,618
4	VRONTI	30	5228	5116	4858	1.06	1.08	0.72	0.91**	0.86**	0.78**	-18.75	-4.03	115.81	273,195	395,021
5	THISVI	30	5240	5116	4858	1.11	1.15	0.72	0.90**	0.87**	0.78**	-41.92	-36.55	115.81	348,683	395,021
6	FENTRA	35	4569	4461	4442	1.01	0.88	0.88	0.97**	0.85**	0.88**	5.59	65.46	49.92	143,176	388,227
7	VRITO	26	5361	5230	4972	0.94	0.79	0.65	0.94**	0.70**	0.73**	47.08	142.52	158.85	135,254	399,532
8	DIONI	25	4186	4197	4195	0.92	0.77	1.01	0.97**	0.90**	0.92**	32.31	94.65	-5.79	112,564	130,288
9	LITO	18	4727	4589	4426	1.04	0.90	1.04	0.97**	0.92**	0.94**	-2.61	76.80	-34.66	198,261	361,275
10	KIVELI	11	4260	4079	3920	1.01	1.19	0.83	0.98**	0.98**	0.99**	15.24	-40.03	52.47	106,812	134,868
11	EKATI	10	4217	4052	3833	1.01	1.16	0.86	0.99**	0.98**	1.00**	12.43	-24.16	36.18	56,143	23,281
12	TOTAL	409	4513	4443	4337	0.98	0.97	0.86	0.95**	0.90**	0.90**	16.79	32.75	51.31	221,157	342,670

*: a,b,r. 1: VAR(Y)/EXP.(X)
        2: VAR(Y)/CHECK(X)
        3: CHECK(Y)/EXP.(X)

TABLE.3. The linear regression components of 1000 kernels wheight.[*]

S.N.	VARIETY NAME	N.OF EXP.	VAR. MEAN	EXP. MEAN	CHECK MEAN	b_1	b_2	b_3	r_1	r_2	r_3	a_1	a_2	a_3	$s^2\bar{d}_1$	$s^2\bar{d}_3$
1	DADA	78	37.87	37.39	35.62	0.83	0.74	0.77	0.87**	0.75**	0.80**	6.99	11.68	6.71	5.39	8.29
2	NIOVI	92	36.35	36.63	34.98	0.81	0.69	0.72	0.84**	0.69**	0.75**	6.08	12.29	8.47	7.74	11.27
3	ERATO	54	38.44	37.66	36.09	0.94	0.87	0.81	0.81**	0.79**	0.77**	3.01	7.01	5.73	10.09	9.93
4	VRONTI	30	38.97	38.97	36.20	1.01	0.46	0.74	0.80**	0.36**	0.75**	-0.25	22.26	7.36	6.03	4.33
5	THISVI	30	40.00	38.98	36.35	1.06	0.65	0.78	0.81**	0.52**	0.76**	-1.33	16.36	5.89	5.80	6.60
6	FENTRA	35	37.89	37.22	35.64	1.05	0.76	0.87	0.86**	0.70**	0.77**	-1.25	10.95	3.36	7.97	10.46
7	VRITO	26	43.60	39.08	36.35	1.08	0.84	0.70	0.71**	0.53**	0.74**	1.23	12.96	9.02	12.19	4.38
8	DIONI	25	36.46	37.14	35.54	0.83	0.72	0.76	0.82**	0.76**	0.72**	5.66	10.98	7.15	6.52	10.92
9	LITO	18	38.39	36.97	35.53	1.04	0.70	0.75	0.92**	0.68**	0.70**	0.54	13.67	7.68	4.78	15.17
10	KIVELI	11	37.46	37.63	34.73	0.83	0.71	0.76	0.91**	0.82**	0.72**	6.16	12.70	6.23	4.64	17.36
11	EKATI	10	37.60	38.67	37.10	1.18	0.70	1.52	0.61[NS]	0.96**	0.56[NS]	-7.91	11.56	-21.50	30.80	62.20
12	TOTAL	409	38.13	37.57	35.68	0.92	0.75	0.77	0.81**	0.69**	0.74**	3.45	11.46	6.74	7.91	10.30

[*]: a,b,r.   1: VAR(Y)/ EXP.(X)

2: VAR(Y)/CHECK(X)

3: CHECK(Y)/EXP.(X)

713

714

TABLE.4. The linear regression components of protein content %.[*]

S.N.	VARIETY NAME	N.OF EXP.	VAR. MEAN	EXP. MEAN	CHECK MEAN	b_1	b_2	b_3	r_1	r_2	r_3	a_1	a_2	a_3	$s^2\bar{d}_1$	$s^2\bar{d}_3$
1	DADA	78	13.68	13.62	13.52	0.96	0.82	1.03	0.96**	0.90**	0.94**	0.63	2.54	0.51	0.34	0.54
2	NIOVI	92	14.29	13.75	13.58	1.04	0.85	1.01	0.94**	0.86**	0.92**	0.05	2.74	-0.32	0.43	0.65
3	ERATO	54	13.00	13.16	13.00	0.93	0.77	1.08	0.92**	0.89**	0.92**	0.76	3.05	-1.16	0.48	0.66
4	VRONTI	30	12.84	12.80	12.58	1.03	0.92	0.97	0.94**	0.88**	0.93**	-0.27	1.30	0.13	0.39	0.41
5	THISVI	30	12.81	12.73	12.68	0.87	0.78	1.00	0.96**	0.92**	0.94**	1.71	2.93	-0.02	0.12	0.45
6	FENTHA	35	13.41	13.54	13.48	1.00	0.77	1.09	0.92**	0.86**	0.90**	-0.13	2.94	-1.23	0.51	0.82
7	VRITO	26	12.76	12.64	12.50	0.96	0.87	1.00	0.94**	0.90**	0.94**	0.68	1.95	-0.09	0.36	0.39
8	DIONI	25	14.11	14.03	13.84	0.83	0.69	1.04	0.94**	0.88**	0.93**	2.47	4.55	-0.81	0.25	0.49
9	LITO	18	12.88	13.52	13.34	1.11	1.00	0.99	0.97**	0.92**	0.94**	-2.10	-0.50	-0.01	0.42	0.63
10	KIVELI	11	13.77	13.75	13.68	1.07	0.83	1.06	0.98**	0.87**	0.92**	-1.00	2.43	-1.00	0.19	0.86
11	EKATI	10	13.63	13.66	13.49	1.02	0.86	1.04	0.98**	0.92**	0.93**	-0.34	2.04	-0.65	0.24	0.80
12	TOTAL	409	13.52	13.42	13.28	0.99	0.84	1.03	0.94**	0.88**	0.93**	0.19	2.34	-0.48	0.36	0.60

[*]: a,b,r.

1: VAR(Y)/EXP.(X)

2: VAR(Y)/CHECK(X)

3: CHECK(Y)/EXP.(X)

TABLE.5. The linear regression components of sedimentation value (Zeleny test). [*]

S.N.	VAR. NAME	N.OF EXP.	VAR. MEAN	EXP. MEAN	CHECK MEAN	b_1	b_2	b_3	r_1	r_2	r_3	a_1	a_2	a_3	$s^2\bar{d}_1$	$s^2\bar{d}_3$
1	DADA	78	24.67	22.78	25.73	1.14	0.64	1.50	0.90**	0.88**	0.87**	-1.31	8.13	-8.50	9.10	22.31
2	NIOVI	92	26.20	22.74	24.87	1.26	0.66	1.46	0.81**	0.76**	0.82**	-2.38	9.74	-8.41	22.33	28.37
3	ERATO	54	21.09	21.37	21.24	0.63	0.51	0.93	0.60**	0.83**	0.54**	7.70	10.23	1.47	19.89	38.22
4	VRONTI	30	19.47	20.57	20.97	0.93	0.68	1.15	0.93**	0.86**	0.89**	0.39	5.28	-2.63	3.43	8.60
5	THISVI	30	22.00	20.42	21.23	0.98	0.64	1.38	0.90**	0.87**	0.92**	1.91	8.49	-6.96	4.33	6.60
6	FENTRA	35	23.17	21.73	21.77	0.91	0.46	1.62	0.88**	0.77**	0.93**	3.33	13.11	-13.40	7.10	11.92
7	VRITO	26	16.04	19.58	20.19	0.68	0.41	1.37	0.88**	0.81**	0.90**	2.73	7.77	-6.64	2.92	6.92
8	DIONI	25	24.00	23.52	26.88	1.09	0.51	1.77	0.88**	0.78**	0.91**	-1.67	10.39	-14.72	10.2	19.68
9	LITO	18	20.56	22.93	24.56	1.17	0.66	1.74	0.95**	0.95**	0.98**	-6.35	4.35	-15.40	5.71	5.35
10	KIVELI	11	24.91	25.40	27.27	1.16	0.70	1.60	0.86**	0.90**	0.92**	4.64	5.91	-13.42	23.31	22.55
11	EKATI	10	24.90	22.41	22.30	1.07	0.48	0.46	0.77**	0.36^{NS}	0.45^{NS}	0.99	14.16	11.94	7.30	7.70
12	TOTAL	409	23.08	22.07	23.55	1.08	0.62	1.43	0.81**	0.80**	0.83**	-0.71	8.47	-8.09	12.36	23.13

[*]: a,b,r. 1: VAR(Y)/EXP.(X)

2: VAR(Y)/CHECK(X)

3: CHECK(Y)/EXP.(X)

TABLE 6. The best three locations where the quantitative and or the qualitative characteristics were better expressed.(Hierarchical analysis).

S.N.	VARIETY NAME	YIELD		PR %		1000 KERNEL WEIGHT		ZELENY TEST	
		VARIETY	CHECK	VARIETY	CHECK	VARIETY	CHECK	VARIETY	CHECK
1	DADA	6- 9-7	9-6-7	11- 8- 2	11-10- 8	7- 6-5	5-12- 7	11-12- 8	11-10- 8
2	NIOVI	6- 9-7	9-6-5	2- 8-15	11-12-15	6- 7-5	5-12-12	15-11- 8	11-15- 8
3	ERATO	9- 6-7	9-7-14	11- 3- 8	11-10- 8	14- 3-8	2- 9-14	8-11-10	8-10-11
4	VRONTI	9-11-6	9-7-6	10- 8-11	10-11- 8	6- 4-7	2- 1- 3	8-10- 3	8-11-10
5	THISVI	9- 6-7	9-7-6	10- 8-11	10- 8-11	6- 1-4	2- 1- 8	8-11-10	8-11-10
6	FENTRA	6- 9-5	9-6-5	2- 1- 8	1-11- 2	6- 7-3	5- 9- 6	11- 8- 2	2-11- 8
7	VRITO	9- 6-7	9-7-6	10-11- 8	10- 8- 9	5- 4-1	2- 3- 1	8-10- 3	8-10- 2
8	DIONI	9- 7-10	9-3-7	11- 8- 5	11- 8- 5	5-10-8	5- 8- 9	11- 8- 5	8-11- 3
9	LITO	6- 5-3	6-9-5	11- 2- 8	11- 8- 2	5-10-9	5- 6- 7	11- 8- 2	11- 2- 8
10	KIVELI	6- 5-9	6-5-9	11- 8- 2	11- 2- 8	5- 6-8	2- 5- 6	11- 2- 8	11-12- 8
11	EKATI	6- 5-9	6-5-9	11- 8- 3	11- 8- 3	11- 5-9	11- 5- 6	9- 6-12	8-12- 6
12	TOTAL	6- 9-7	9-6-7	11- 8- 2	11- 8- 2	6- 5-7	5- 2- 9	8-11- 3	8-11- 2

2. The differences between the variety Vrito, which is the best yielding, and Thisvi and Vronti are very small.

3. Triticale, as a whole, do not differ significantly from the check, but the varieties Niovi and Dioni exceed significantly the check, in Pr%.

4. Triticales, in general, exceed the check in 1000 kernel weight. The varieties Vrito, Thisvi, Vronti, Erato and Lito were the top varieties.

5. Triticales are not inferior in sedimentation value. Niovi is the best variety exceeding "Vergina" and most of the other triticale varieties.

From the tables 2, 3, 4 and 5, where the components of the linear regressions are shown, the following results can be extracted:

1. Triticales, as a whole, exhibit better stability in grain yield than the check (high mean yield, low b value and s^2d). Best varieties: Vrito, Vronti, Lito and Fentra.

2. The variability of 1000 kernel weight over years and locations in triticale was lower than the check. Best varieties: Vrito, Thisvi, Vronti, Erato, Fentra.

3. In the case of Pr% the b values of triticale are nearly equal to the b values of the check, but the s^2d values are lower than the s^2d values of check. Best varieties: Fentra, Niovi, Erato, Dada, Kiveli, Dioni.

4. Triticales generally show lower variability in sedimentation value and some varieties combine high mean with low s^2d and b values near to the unit. Ekati and Fentra are the best varieties for this characteristic.

From the table 6 where a part of hierarchical analysis is shown the following results can be concluded:

1. Some differentiation in the hierarchy of the best locations for yield and 1000 kernel weight appear between triticales and the check, but the best three locations are the same in the case of grain yield and only one location is the same in the case of 1000 kernel weight.

2. The best three locations are the same in the case of Pr% and nearly the same in the case of sedimentation value.

Two trends emerge from the regression and hierchical analysis. First: better stability of triticale which was also observed in the past (CIMMYT 1990, 1991, 1992, Gogas 1988, 1993). Triticales tend to show wide adaptation due to their unique genotypes.

Second: the differentiation in environmental requirements between triticale and wheat. Due to the limited space only a part of hierarchical analysis is given but a similar situation exists in the case of years.

Conclusions

The results presented above and similar results published by many other scientists support our belief that triticales have reached a high enough level of yield and quality so as to claim a considerable position in the cereal production.

The time has come for triticale to be officially recognized as a commercial product and be used in environments where the other cereals do not perform well.

References

CIMMYT. 1990. Results of 1987-88 Triticale Nurseries. Mexico, D.F.: CIMMYT.
CIMMYT. 1991. Results of 1988-89 Triticale Nurseries. Mexico, D.F.: CIMMYT.

CIMMYT. 1992. Results of 1989-90 Triticale Nurseries. Mexico, D.F.: CIMMYT.

Eberhart, S.A. and Russel, W.A. 1966. Stability parameters for comparing varieties. Crop Sc. 6: 36-40.

Finlay, K.W., Wilkinson. 1963. The analysis of adaptation in plant breeding programmes. Aust. J. Agr. Res. 14: 742-754.

Gogas, D.M. 1988. Contribution to the study of triticale stability (X. *Triticosecale* Wittmack) in Greece. Second congress on plant breeding. Thessaloniki. 1988. Proc. p.p. 25-31.

Gogas, D.M. 1993. Through years yield evolution of triticale (X. *Triticosecale* Wittmack) in Greece. 4th Congress on plant breeding. Thessaloniki 1993. Proc. p.p. 11-15.

Johnson, V.A., Shaffer S.E. and Schmidt W. 1968. Regression analysis of general adaptation in hard red winter wheat. Crop Sci. 8: 187-191.

Varughese G., Barker T. and Saari E. 1986. Triticale CIMMYT, Mexico, D.F. 32 pp. ISBN 968-6127-11-9.

Xinias, J., Gogas, D., Stratilakis, S. and Hadjilambrou, K. 1993. Yield components of six (6) new triticale (X. *Triticosecale* Wittmack) varieties. 4th Congress on plant breeding. Thessaloniki. 1993. Proc. p.p. 96-99.

TRITICALE IN MOROCCO: A PROMISING CROP

Mergoum Mohamed

Aridoculture Center, INRA, Settat, Morocco

Abstract

The severe droughts that have plagued most Moroccan regions during the 1991-92 and 1992-93 crop seasons brought focus on drought tolerance and genetic diversity in cereals. During the last decade, triticale has shown good potential and adaptation in the arid and semi-arid as well as in the sandy and high elevation areas of Morocco. This good performance of triticale as compared to the other traditional cereals is particularly, obvious in dry years. This paper illustrates the performance of triticale since 1988 as compared to the other cereals in the major growing areas of Morocco. Special attention is given to the water stressed environments. It also highlights the unique aspects of triticale and the major research accomplishments.

Introduction

Despite low and erratic rainfall, cereals have been cultivated in Morocco for centuries. While barley (<u>Hordeum</u> <u>vulgare</u> L.) is grown in about 50% of cereal growing areas, major emphasis has been placed on expanding output of bread wheat (<u>Triticum</u> <u>aestivum</u> L.) and to a lesser extent that of durum wheat (<u>T</u>. <u>turgidum</u> var. <u>durum</u>, L.). Recently, considerable effort has been made to promote triticale (<u>X</u>. <u>Triticosecale</u> Wittmack) as an alternative cereal in Morocco's dryland, sandy, and acid soils [2,3,4]. Triticale, the most recently introduced cereal crop in Morocco has good resistance or tolerance to most cereal diseases. Over the recent years, substantial progress and improvement of triticale (seed characteristics, lodging, sterility, shattering, preharvest sprouting, milling, baking and nutritional qualities) were achieved [2,3]. The objective of this paper is to illustrate the major research achievements and performance of triticale in the different cereal growing areas of Morocco.

H. Guedes-Pinto et al. (eds.), Triticale: Today and Tomorrow, 719–724.
© 1996 *Kluwer Academic Publishers. Printed in the Netherlands.*

Materials and Methods

Except for quality tests and fertilization studies, all the remaining data were collected from the breeding nurseries, particularly, advanced trials which are installed in 6 to 8 locations over most cereal growing zones of Morocco. These trials included 24 entries in a randomized block design with four replications. Each entry was planted in six rows 5 meters long plot. The results of triticale fertilization were collected from on-farm experiments conducted in two locations in the arid and semi-arid regions [1]. The layout of these experiments was similar to the previous nurseries. Quality tests of triticale flour and mixtures with wheat flour were performed according to the Chopin method (1) at the cereal technology laboratory of INRA at Rabat, Morocco.

Results and Discussions

VARIETY IMPROVEMENT

Triticale is one among several emerging technologies which are likely to have a significant and immediate contribution to cereal production in Morocco. The first triticales were introduced in the early 1960's from the International Maize and Wheat Improvement Center (CIMMYT) by the INRA breeders. Since then substantial efforts were devoted to identify advanced cultivars from CIMMYT germplasm for the different zones of Morocco. These breeding efforts resulted in the release of the first three cultivars (Beagle, Jaunillo and Drira Out Cross) in 1988. Since late 1980's, increasing hectarages have been grown to triticale. Actually, this crop is planted on more than 10000 ha. However, the released varieties have several limitations and undesirable characteristics such as seed, straw and flour qualities that need improvement (Table 1, TCL2).

Table 1. Milling and baking characteristics (Chopin) of bread wheat (BW) "Nasma", two triticales (TCL1 and TCL2) flours, and their mixtures (1988-1990).

Type of Flour and its mixture (%)	Water Content (%)	Ash (%)	Zelemy (ml)	Hagberg (s)	Water Absorption (%)
Nasma (100)	16.1	0.82	31	380	62.8
TCL1[*] (100)	15.7	0.78	21	241	51.3
TCL2 (100)	15.55	0.64	22	114	52.0
TCL1+BW (50:50)	15.95	0.79	26	317	57.0
TCL2+BW (50:50)	15.86	0.70	25	171	58.5
TCL1+BW (25:75)	15.98	0.79	30	374	60.4
TCL2+BW (25:75)	15.87	0.73	29	284	59.8

[*]TCL1 = Rhino "s"
TCL2 = Beagle (control)

In 1993, two new improved varieties were released; 'Borhan' and 'Firdaws'. These cultivars have better milling and baking qualities (Table 1, TCL1) and higher yields than the previous released ones. Today, many excellent lines that have better yield and quality are on the breeding pipelines.

QUALITY AND UTILIZATION

One of the handicap of earlier released triticales in many countries around the world is gain quality (TCL2, Table 1). The tremendous breeding efforts at CIMMYT and in many countries to overcome these obstacles has resulted in developing new lines with better quality (TCL1, Table 1). Triticale now can be used for human consumption (bread, tortia, biscuits,...etc) as well as for animals feeding (grazing, forage, grain straw,..etc). In Morocco, triticale is used for both human and animal consumption. In the case of human use, triticale is generally mixed with wheat. For animal consumption, triticale is mainly used as forage, hay, grain and straw. However, triticale straw is less acceptable compared to barley and wheat because of its stiffness.

ADAPTATION AND PERFORMANCE

Yield potential of triticale and wheat under favorable environments (irrigated and high rainfall) are similar (Table 2). Under severe drought of the arid, semi-arid and sandy soils, triticale performs well compared to the other cereals such as wheat (Table 2). During the period of 1991-1993 for instance, except triticale and to a certain extent barley, wheats grown under rainfed conditions were totally lost. Whereas, wheats did not survive severe drought and hessian fly insect (<u>Mayetiola destructor</u> say) attack, triticale lines produced on average 0.7 to 1.0 tone (t) per hectare in addition to the straw. Under similar conditions, some advanced lines in our breeding nurseries produced more than 1.4 t/ha.

Table 2. Average grain yields (t/ha) of the best 3 cultivars of bread wheat, durum wheat, barley and triticale under adverse environments from 1988 to 1992 in Morocco.

Species	Environments[*]				
	IRR.	H.R.	H.EL.	A/S.A	S. Soil
Bread wheat	59 a[**]	44 a	12 b	18 c	19 c
Durum wheat	60 a	36 c	8 c	17 c	18 c
Barley	50 b	37 c	19 a	18 b	22 b
Triticale	61 a	46 a	20 a	24 a	28 a

[*]IRR.= Irrigated; H.R.= High rainfed; H.El.= High elevation;
A/S.A.= Arid and semi-arid; and S. Soil= Sandy soil
[**]Means followed by the same letter are not significantly different at 0.05 (α) probability level.

Over many years of testing, triticale has demonstrated good resistance and tolerance to most cereal diseases and pests (Table 3). Under favorable conditions where foliar diseases such as rusts, mildew and septoria prevail, triticale is generally resistant. Similarly, in the stressed environment where Hessian Fly and root rot diseases are causing substantial yield losses, triticale is very tolerant.

Table 3. Reaction of most grown triticale, wheats and barley cultivars to most prevalent diseases and pests under Moroccan environments (1988-1992).

Disease / Pest	Species			
	Triticale	Barley	D. wheat	B. wheat
Hessian fly	R[*]	MR	S	S
Saw fly	R	MR	MR	MR
Foliar diseases	R	S	S	S
Root rot	R	-	S	MR
Viruses (BYDV)	R	S	MS	MR

[*] R= Resistant, S= sensible, MR= Medium resistant, MS= Medium susceptible.

Triticale germoplasm tested in Morocco is mainly complete hexaploid type. However, a large amount of triticale lines received from CIMMYT are substituted. Data collected from advanced lines of both types have similar performance with a slight advantage to the complete ones (Table 4). The substitute triticales are in general short and susceptible to shattering.

Table 4. Grain yield means (t/ha) of five complete and substitute triticale lines in different environments in Morocco from 1988 to 1992.

Environments	Type of Triticale	
	Complete	Substituted
Arid	25.5	27.4
Semi-arid	44.4	44.6
Sandy soil	29.9	27.8
High rainfed	51.2	48.7
Irrigated	66.7	65.9

FERTILIZATION

The most recent aspect that was considered to promote triticale was fertilization, especially, nitrogen (N) and phosphorus (P) depending on soil test levels. The on-farm field studies conducted on a shallow soil (Petrocalcic Palexeroll) deficient in both N and P and in a low rainfall zone, showed significant yield (Dry matter and grain) increases with N applications (Table 5). Similar results were obtained with P fertilization in presence of N (50 Kg/Ha of N).

Table 5. Response of triticale (means of dry matter and grain yield
(t/ha)) of cultivar "Juanillo" to separately applied Nitrogen (N)
and Phosphorus (P) from 1989 to 1991 in the dryland regions of
Morocco.

Fertilizer	Yield (t/ha)	
(kg/ha)	Dry matter	Grain
Nitrogen levels: 0	33	11
60	45	16
100	55	18
150	59	21
LSD (0.05)	0.78	0.29
Phosphorus levels: 0	43	14
(+50 Kg of N) 20	45	16
40	52	18
60	53	19
LSD (0.05)	0.51	0.18

Conclusions

Although, triticale is relatively a new cereal crop to Moroccan
agriculture, it is becoming a well established crop. Over many
years of tests, triticale has shown many advantages. Therefore, we
can conclude that:
   - Triticale has demonstrated good adaptation (yield, diseases
and insects resistance) in many parts of Morocco. Its good
performance was obvious, particularly under stressed environments
of arid and semi-arid zones; sandy soils; and high elevation areas.
   - Several improved cultivars particularly for quality are (and
will be) available for farmers to be grown for their needs.
   - Under low input conditions, triticale showed excellent
response (dry matter and grain yield) to N and P fertilizers.
   - Triticale is used for human consumption as well as for
animal feeding.
   - The future of this crop in Morocco however, will certainly
depend on many aspects related to the marketing process (prices,
utilization,...) as well as its transfer to farmers.

References

1. Mergoum M., J. Ryan, and M. Abdel Monem. 1990. Response of
      High-yielding Triticale to N and P in a Rainfed Mollisol
      and Vertisol in Morocco. p. 60. In Agronomy Abstracts.
      ASA, San Antonio, TX. USA.

2. Mergoum M. et A. Taylor, 1991. Résultats de l'amélioration
      génétique des triticales. Rapport d'Activité de
      Recherches du Programmme Aridoculture de 1990-91: 96-98.

3. Mergoum M., J. Ryan, and J. P. Shroyer. 1992a. Triticale in
   Morocco: Potential for Adoption in the Semi-Arid Cereal
   Zone. J. Nat. Resour. Life Sci. Educ. 21:137-141.

4.Mergoum M., G. A. Taylor, N. Nsarellah, and J. Ryan. 1992b.
   Triticale: An Alternative Cereal in the Drought-Prone
   Moroccan Zone. p. 69. In Agronomy Abstracts, ASA,
   Minneapolis, MN.

PERFORMANCE OF 'DRIRA OUT-CROSS 7' TRITICALE IN WEST ASIA, NORTH AFRICA AND MEDITERRANEAN EUROPE

Yau Sui K. and Miloud M. Nachit[1]
The International Center for Agricultural Research in the Dry Areas (ICARDA),
P. O. Box 5466, Aleppo, Syria
[1] CIMMTY staff based at ICARDA

Abstract

In the Mediterranean region, especially in countries of West Asia and North Africa, few studies have been conducted on the adaptation of triticale. This study aimed to find areas within the region in which triticale has the potential for commercial production. Six years' (1986/87 to 1991/92) data of the CIMMYT/ ICARDA Regional Durum Wheat Yield Trial, which included 'Drira Out-cross 7' as a triticale check, were analysed. Sites in the region were divided into six sub-regions. Entry means and standard deviations of relative yield were calculated for each sub-region. In all seasons, 'Drira Out-cross 7' was the highest yielding entry in Mediterranean Europe and in North Africa; however, its performance in West Asia was not consistent. Relative to other sub-regions, it had the poorest performance under irrigated environments in Egypt. Regression coefficients indicated that it had average response to improved conditions. However, the generally much higher standard deviations than the durum wheat entries showed that it did not have wide adaptation within sub-regions. Implications of these results on triticale breeding for the Mediterranean region were discussed.

Introduction

Triticale is a promising crop in some countries with specific, adverse growing conditions for wheat and rye. It is replacing wheat in regions with highly acid soils where aluminium and copper toxicities are common, such as Mexico, northern India and Brazil (Bushuk and Larter 1980). In Europe it is grown mainly on sandy soils previously devoted to rye.

In the Mediterranean region, triticale may be adapted to certain areas. From early work at ICARDA, several triticale cultivars (including "Drira Out-cross 7") were relesed in North Africa. However, no consistent results emerged from many attempts to compare

H. Guedes-Pinto et al. (eds.), Triticale: Today and Tomorrow, 725–729.

© 1996 *Kluwer Academic Publishers. Printed in the Netherlands.*

triticale with the dominant winter cereal crops in West Asia and North Africa (WANA), i.e., barley, durum wheat and brad wheat (Nachit 1983; Hajichristodoulou 1984; Yau 1987; Genc *et al.* 1989; Ryan *et al.* 1991). In an analysis of CIMMYT's (International Center for Maize and Wheat Improvement) 14th ITYN (International Triticale Yield Nursery), triticale entries were found to yield better than the durum wheat check in the Mediterranean region (Villareal *et al.* 1992), but differential adaptation of triticale was not investigated.

In view of the large spatial and seasonal variability in WANA, results across many environments would be needed to draw useful conclusions. The objective of this study was to compare the performance of triticale with durum wheat over years and sites to find areas within the region in which triticale has a consistent yield advantage over durum wheat.

Materials and Methods

Six seasons' (1986/87 to 1991/92) data of the CIMMTY/ICARDA Regional Durum Wheat Yield Trial for Moderate Rainfall Areas, which was assembled at and distributed to co-operators from ICARDA, Aleppo, Syria, were analysed. This trial was grown by national programmes in a randomized complete block design with three to four replicates and generally with a plot size of 4.5 m^2 (six rows 2.5 m long, spaced 30 cm apart). These were 24 entries, including five (national, long-term, improved, bread wheat and triticale) checks. The improved durum wheat check was 'Cham 1', which has wide adaptation and has been released in Algeria, Cyprus, Jordan, Saudi Arabia, Syria and Turkey. The triticale check was 'Drira Out-cross 7', which was selected from a multiple backcross programme with several plump-seeded triticale lines. It had a full set of rye cchromosomes, the type shown to be better adapted to dry areas than substituted triticale (Nachit and Tahir 1983). Durum wheat entries promoted into the trial had performed well in at least two years of yield testing in Syria. Except for the checks, entries stayed in the trial for about two to three years before being replaced by new ones.

Sites in WANA and Mediterranean Europe were divided into six sub-regions to investigate differential adaptation of triticale to specific environments. These were: West Asia 1 (Iraq, Iran, Syria and Turkey), West Asia 2 (Cyprus, Jordan and Lebanon), West Asia 3 (Saudi Arabia and the United Arab Emirates), North Africa (Algeria, Libya, Morocco and Tunisia), Nile Valley (Egypt), and Mediterranean Europe (France, Greece, Italy, Portugal and Spain). About half of the growing sites changed during the study period. Since sites within sub-regions had large differences in mean yield, relative yield (entry yield at each site was expressed as a percentage of the site mean yield) was calculated for each entry in each sub-region to give equal weight to each site (Yau and Hamblin 1994). Entries with low standard deviation of relative yields across sites are considered to be agronomically stable.

Results and Discussion

A clear pattern of differential adaptation of 'Drira Out-cross 7' was found. In all seasons, it was the highest yielding entry in Mediterranean Europe (Table 1), having more than 10% higher yield than the best durum wheat entries in four seasons. In North Africa, it also out-yielded the highest yielding durum wheat entries consistently, but the margin was not as big as in Mediterranean Europe. In contrast, it had the poorest performance in the Nile Valley, where it ranked lower than 10 and yielded below sub-region mean in three seasons.

The performance of 'Drira Out-cross 7' in West Asia was less consistent (Table 1). In sub-region 1, it yielded well in the later half of the study period but not in the first half. In sub-region 2, it ranked among the top five in grain yield for five seasons. In sub-region 3, it had higher yield than the best durum wheat entries in four of the six seasons.

The reasons why 'Drira Out-cross 7' performad well in Mediterranean Europe and North Africa but not under irrigation in Egypt are not clear. Length of the growing season, temperature, rainfall and fertility of the test sites might have played a dominant role. Although the triticale check headed earlier than the durum entries on average, its grain--filling period was longer. This character might have made it more adapted to the longer season and more fertile conditions in Mediterranean Europe. However, the same character might be disadvantageous under the heat-stress conditions in Egypt. It should be noted that the durum entries in the trial had gone through cycles of heat tolerance selection using the late-planting technique (Nachit and Ketata, 1991). In West Asia, Hajichristodoulou (1984) and Yau (1987) showed that triticale performed relatively better when rainfall was plentiful or with supplementary irrigation than in dry conditions. The large seasonal variation of climatic conditions in West Asia sub-regions 1 and 2 might have caused the inconsistent performance of 'Drira Out-cross 7' there.

The performance of 'Drira Out-cross 7' was not stable within sub-regions with few exceptions, as indicated by the generally much higher standard deviations when compared with the best durum wheat entries and the durum improved check (Table 1). This suggested that it had narrow adaptation. New triticale lines with wider adaptation would be desirable. A study is needed to determine whether this was a characteristic of 'Drira Out--cross 7' or a common character of triticale. The sub-regions might be further subdivided to find the specific countries or sites where 'Drira Out-cross 7' was well adapted. The regression coefficients of 'Drira Out-cross 7' (based on all sites in the trials) did not differ significantly from 1.0 (from 0.91 to 1.16) in all seasons, indicating that it had average response to improved conditions.

Since only one triticale was compared with many durum wheat entries, results should be interpreted with care. Besides, in trials with small plots, the taller triticale might have a competitive advantage over shorter durum wheat entries.

Table 1. Mean, rank and standard deviation of relative yield for the triticale check, the best durum wheat entry, and the durum wheat improved check for each of the six sub-regions in WANA and Mediterranean Europe.

Year and entry	W. Asia 1			W. Asia 2			W. Asia 3			N. Africa			Nile Valley			Med. Europe		
	Mean	R[#]	SD[##]	Mean	R	SD	Mean	R	SD	Mean	R	SD	Mean	R	SD	Mean	R	SD
1986/87																		
Triticale	103	9	28	107	5	17	141	1	40	117	1	36	89	23	11	113	1	18
Best durum	109	1	7	117	1	12	120	2	21	112	2	16	115	1	5	112	2	10
Durum check	104	7	14	117	1	12	97	13	19	102	10	15	105	7	9	103	9	16
Mean (kg/ha)	3870			3065			5655			3916			7208			5042		
No. of sites	10			7			2			8			5			9		
1987/88																		
Triticale	100	11	19	92	17	25	128	1	36	117	1	29	95	18	7	131	1	23
Best durum	110	1	14	126	1	3	108	3	14	109	3	13	121	1	13	119	2	20
Durum check	110	1	14	126	1	3	108	3	14	98	15	4	100	8	13	104	6	10
Mean (kg/ha)	4431			2418			5093			2981			6995			4927		
No. of sites	14			3			5			2			2			7		
1988/89																		
Triticale	96	16	24	110	4	10	102	13	21	119	1	23	100	14	18	142	1	35
Best durum	116	1	37	112	1	6	117	1	10	108	2	13	111	1	12	110	3	20
Durum check	102	9	10	112	1	6	106	8	10	108	2	13	101	9	14	108	4	20
Mean (kg/ha)	2787			2475			6069			4005			5838			4448		
No. of sites	8			7			4			9			6			6		
1989/90																		
Triticale	135	1	84	114	3	29	117	1	31	117	2	32	100	10	17	139	1	26
Best durum	124	2	20	115	2	21	117	2	9	112	3	17	117	1	12	113	3	12
Durum check	81	19	35	106	8	14	97	18	4	104	6	13	102	8	9	111	4	16
Mean (kg/ha)	2329			3327			5346			2967			5905			3301		
No. of sites	4			7			4			7			2			5		
1990/91																		
Triticale	118	1	30	107	5	34	119	2	21	117	2	24	96	15	19	121	1	18
Best durum	109	2	9	113	1	12	114	3	19	112	3	18	119	1	40	110	3	16
Durum check	104	8	10	113	1	12	105	9	23	110	4	19	108	5	20	110	3	18
Mean (kg/ha)	2378			2068			3871			4005			4539			4630		
No. of sites	8			7			3			8			6			5		
1991/92																		
Triticale	105	3	24	112	2	6	110	8	8	115	1	23	101	10	21	115	1	23
Best durum	109	1	10	113	1	22	122	1	7	114	2	23	118	1	18	110	3	9
Durum check	102	9	8	106	7	29	115	3	2	109	4	22	100	12	16	108	4	9
Mean (kg/ha)	4160			3444			5033			3560			4292			5440		
No. of sites	12			4			2			7			9			6		

[#] rank; [##] standard deviation

Acknowledgements

We thank all those co-operators who share their data with ICARDA.

References

Bushuk W, EN. Larter, 1980. Triticale: production, chemistry, and technology. In: Y. Pomeranz (Ed.), Advances in Cereal Science and Technology, pp. 115-157. Am. Ass. Cereal Chem. Inc., St. Paul.

Genc I, AC Ulger, Yagbasanlar T. Screenig for high yielding triticales to replace wheat in the Cukurova region of Turkey. Rachis, 1989; 8(1): 25.

Hajichristodoulou A. Performance of triticale in comparison with barley and wheat in a semi-arid Mediterranean region. Expt Agric 1984; 20: 41-51.

Nachit MM. The effect of clipping, during the tillering stage, on triticale. Rachis 1983; 2: 11-12.

Nachit MM, Ketata H. Selection for heat tolerance in durum wheat. In: E. Acevedo, E. Fereres, C. Gimenez & J.P. Srivastada (Ed.), Improvement and Management of Winter Cereals Under Temperature, Drought and Salinity Stresses, pp. 239-250. Inst. Nacional de Investigaciones Agrarias, Madrid 1991.

Nachit MM, Tahir M. The effect of rye chromosomes in triticale on moisture stress tolerance and its yield potential in North Africa and West Asia. In: Proc. 6th International Wheat Genectis Symposium, Kyoto, Japan 1983; pp. 925-931.

Ryan J, Abdel Monem M, Mergoum M, El Gharous M. Comparative triticale and barley responses to nitrogen at locations with varying rainfall in Morocco's dryland zone. Rachis 1991; 10(2): 3-7.

Villareal, RL, Varughese G, Adbulla OS. Advances in spring triticale breeding. Pl. Breed. Rev. 1992;8: 43-90.

Yau SK. Comparison of triticale whit barley as a dual purpose crop. Rachis 1987; 6(1): 19-20.

Yau SK, Hamblin J. Relative yield as a measure of entry performance in variable environments. 1994; Crop Sci 34 (in press).

TRITICALE CULTIVATION FOR DIVERSIFYING AND STABILISING A WHEAT DOMINATED AGRO-ECOSYSTEM

Gurnam S. Dhindsa and Jaswinder Singh
Department of Plant Breeding, Punjab Agricultural University,
Ludhiana, India

Abstract

Bread wheat, the main crop of the Indo-gangetic plains of India faces a major threat (biotic and abiotic) to stability of production on account of its monoculture. Results on performance of triticale varieties developed at Punjab Agricultural University are presented and their role in diversifying and stabilising this agro-ecosystem is discussed. Triticale performed better or at par with best wheat varieties under normal growing conditions but had a clear edge over wheat in drought stress environments. Observations on disease resistance show triticale to be highly resistant to major wheat diseases. It is proposed that triticale would have a two-fold effect in improving the stability and sustainability of the agro-ecosystem. One, the area under triticale interspersed in the wheat acreage would destroy the monoculture. Two, being itself extremely tolerant to diseases like rusts and bunts and at the same time being hardy and drought tolerant, triticale cultivation would further enhance the stability of the system.

Introduction

Bread wheat is the prime cereal crop and staple food of the Indo-gangetic plains of India. During the crop season the entire area is planted to a small number of wheat cultivars, a large proportion often being occupied by a single genotype. This type of monoculture results in an extremely fragile agro-ecosystem, which is vulnerable to both biotic and abiotic stresses. The monoculture of resistant wheat cultivars exerts selection pressure on obligate pathogens such as those causing wheat rusts. A new virulent race can multiply to epidemic proportions as a result of extensive and continuous availability of cultivar whose resistance has broken down. Moreover, natural calamities such as drought can greatly reduce yields.

H. Guedes-Pinto et al. (eds.), Triticale: Today and Tomorrow, 731–735.
© 1996 Kluwer Academic Publishers. Printed in the Netherlands.

Triticale (X *Triticosecale* Wittmack), a man-made cereal, is an inter-generic hybrid between *Triticum* and *Secale*. It has come a long way from being an experimental hybrid and is likely to assume an important role in feeding mankind.

Material and Methods

Breeding of suitable triticales for irrigated and water-stress conditions has been in progress since 1968 at Punjab Agricultural University, Ludhiana, India. After synthesis of the primary triticale lines, the procedure to improve triticale is similar to other crops, particularly wheat. The material is handled by the pedigree method of breeding after initial hybridization superior strains are evaluated in the different agroclimatic conditions. Multilocation yield trials are conducted by growing genotypes in randomised block design experiments with four replications. Observations on disease reaction, maturity and other agronomic traits are also recorded in these trials. The results, discussed below, involve the two triticale cultivars TL 419 and TL 1210 developed at Punjab Agricultural University, Ludhiana in comparison with the best wheat checks.

Results and Discussion

In developing triticale, the main objective was to arrive at a judicious combination of the two parent species in order to develop a new crop which combined high productivity with tolerance to biotic and abiotic stresses[1]. The extent to which the work at Punjab Agricultural University has been successful is discussed in view of the results of evaluation trials over the last decade.

PRODUCTIVITY

If triticale is to replace wheat in some areas it is vital that its yield potential is at least equivalent, if not better, than wheat. As a new crop, which combined two diverse genomes, triticale faced problems of meiotic instability leading to sterility and mitotic abnormalities resulting in grain shriveling [2]. Both of these problems had a direct bearing on productivity. Intensive breeding efforts have resulted in the amelioration of these problems to a considerable extent as is evident from yield data from high fertility, irrigated conditions presented in Table 1. The data from seven crop seasons and 3 locations in the Punjab State Trial show triticale cultivars to be equal or better than the best wheats.

PRODUCTIVITY UNDER LOW INPUT CONDITIONS

Any agro-climatic zone has some heterogeneity, with some areas being unfavourable

Table 1. Yield performance (Kg/ha) of triticale cultivars with best wheat check over years under normal growing conditions

	Triticale cultivars		Best wheat cultivars		Critical difference (0.05)
	TL 1210	TL 419			
1984-85	4378	3877	4019	(WL 711)	786
1985-86	4779	4129	4354	(WL 1562)	208
1986-87	2323	2536	2419	(WL 711)	509
1987-88	3774	3247	2874	(WL 711)	254
1988-89	5096	4624	4838	(PBW 154)	259
1989-90	3712	3199	3481	(PBW 154)	333
1990-91	4274	3447	3847	(HD 2329)	170
Mean	4048	3578	3960		

to crop productivity, season-to-season fluctuations, particularly in the amount of precipitation received, will also affect productivity. A crop or a cultivar which withstands these fluctuations by giving good response to poor or low input conditions can considerably improve the productivity and sustainability of an agro-ecosystem.

Table 2. Performance of triticale against wheat in rainfed conditions

Crop Season	Yield(Kg/ha)		Critical difference (0.05)
	Best triticale	Best wheat	
1980-81	3193	2580	437
1981-82	3135	2512	588
1982-83	3527	2621	680
1983-84	1759	1802	506
1984-85	1772	1748	216
1985-86	2711	2460	231
1986-87	2172	2030	304
1988-89	3411	3251	112
1989-90	1442	1171	201
1990-91	3245	2942	198
Mean	2637	2318	

Under rainfed conditions, triticale's superiority over wheat is further accentuated; the best triticale had higher yields than the best wheat cultivar in seven of the ten crop seasons (Table 2). The adaptability of triticale to diverse agro-climatic zones is evident from its higher overall means as compared to wheats when evaluated in three distinct zones for seventeen crop seasons (Table 3).

Table 3. Mean performance (Kg/ha) of best triticale and best wheat in different zones of India averaged over 17 crop season (1974-75 to 1990-91)

	North Hill Zone	North Plain Zone	Central Zone	Overall mean
Triticale	2545	4128	3872	3515
Wheat	2152	3986	3975	3371
Critical difference(0.05)	370	320	480	-

Based on: Results of All India Coordinated Wheat and Triticale Trials 1972 to 1991.

Many workers have observed that triticale has better root geometry and, therefore, greater potential to tolerate nutrient deficiencies [3]. Its better efficacy in the uptake and utilization of nutrients was also observed. Triticale is known to perform well in acid soils and is also tolerant to aluminum toxicity [4].

DISEASE RESISTANCE

The disease reaction of triticale and wheat cultivars to yellow rust, brown rust, karnal bunt, loose smut and powdery mildew, which are the diseases prevalent in this region, is given in Table 4.

Triticale emerged almost totally unscathed by these diseases whereas wheat cultivars exhibited varying levels of disease reaction, mainly with increased susceptibility. The importance of this observation cannot be over-empahsised. Sometimes, it is argued that inoculum used for screening tolerance of triticale comes mainly from the wheat crop and pathogen races possessing virulence specific to triticale are not used [3]. This, however, does not detract from the importance of triticale as a physical barrier which can prevent the spread of wheat diseases by breaking the contiguity of the crop.

NUTRITIONAL QUALITY

Acceptance of triticale as food and feed is, to a great extent, governed by its nutritional quality. Early work showed that triticale possessed high protein content [5]

Table 4. Disease reaction of triticale and wheat varieties* in 1990-91 crop season

Varieties	Yellow rust	Brown rust	Karnal bunt	Loose smut	Powdery mildew
TL 1210 (Triticale)	Free	Free	0	Free	Free
TL 419 (Triticale)	Free	Traces	0	Free	Free
PBW 222 (Bread wheat)	Moderately resistant (5S)	Moderately susceptible (20S)	(4.8% Incidence)	Susceptible	Susceptible
HD 2329 (Bread wheat)	Moderately resistant (5S)	Moderately susceptible (30S)	(44.4% incidence)	Susceptible	Susceptible

* Triticale and wheat cultivars approved by Punjab Agricultural University.

and, thus, high nutritional value. One aspect which is still far from satisfactory is the bread-making and chapati-making attributes of triticale. Presently, this is the focus of intense breeding efforts. Meantime, mixing of bread wheat and triticale flour in 50:50 proportions is recommended to avail of its better nutritional value without a substantial lowering of cooking attributes [6].

As the research for sustainable and input non-intensive alternatives in agriculture gathers momentum, triticale has much to offer and deserves to be given a serious examination for diversifying and stabilising this wheat dominated agro-ecosystem.

References

1   Gill, K.S. Perspectives of two decades of research on triticale in India. *Proceedings of 2nd International Triticale Symposium*, Oct 1-5 1990, Brazil.
2   Gupta, A.K., Priyadarshan, P.M. 1982. Triticale: Present status and future prospects. *Advances in Genetics* 21, 255-345.
3   Skovmand, B., Fox, P.N. and Villareal, R.L. 1984. Triticale in Commercial Agriculture: Progress and Promise. *Advances in Agronomy* 37, 1-42.
4   Slootmaker, L.A.Z. 1974. Tolerance to high soil acidity in wheat related species, rye and triticale. *Euphytica* 23, 505-513.
5   Gill, K.S., Sandha, G.S. and Sekhon, K.S. Breeding triticale for food and feed. *Proceedings of 4th International Congress of SABRAO,* 1981. Malaysia, 299-337.
6   Sekhon, K.S., Saxena, A.K., Randhawa, S.K. and Gill, K.S. 1980. Use of triticale for bread, cookie and chapati making. *Journal of Food Science and Technology* 17, 233.

SOME PHYSIOLOGICAL FEATURES OF TRITICALE AND THEIR CONSEQUENCES ON CROP MANAGEMENT IN FRANCE

Georges Laroche and Philippe Gate
Institut Technique des Céréales et des Fourrages
8 Avenue du Président Wilson - (Paris) France

Abstract

In France, winter triticale has confirmed its high yield potential which is at the same level as wheat in identical situations. Its resistance to eye-spot makes it a second straw cereal where it has to compete with barley. It is favoured in soils with excess of moisture (hydromorph soils, rainy climates) where it can adapt itself easily and also shows adaptation to acid soils. This cereal presents good tolerance to cold. The late maturity, which can be a handicap, is compensated for by an adaptation linked with summer conditions (high temperatures and drought).

Introduction

Triticale has shown a high yield in France and a good regularity. Whereas we thougt it was necessary to improve the knowledge of this species. The first results conduct to a large development of triticale in the country.

The potentialities of triticale

The first experiments done by ITCF and by development organizations date from the crop of 1974. A first synthesis (1) established from more than 90 trials during the period 1975-1980, essentially in breeding areas and in difficult soils, showed that tricticale and rye had a slighty better yield than wheat. But, overall, the regularity of triticale yield was far higher than that of wheat. Consequently, triticale has inherited a part of the resistance of rye.
A second more recent synthesis was done from studies carried out in the plain and in the mountain areas of the Massif Central (2) which gave the same results : very close potentials of yield between triticale (Lasko) and wheat (Arminda) with a better regularity for triticale.
A third synthesis was carried out by ITCF in Brittany and the Pays de Loire during the last five years when cultivar plots were sown on the same plots as those used for the experiments on wheat and barley. It showed that in favourable conditions (deep loams), triticale gives a high yield potential, but is a bit more irregular than wheat. Each year, a difference of at least 10 quintals per hectare is noted between barley and triticale, always in favour of triticale. But,

H. Guedes-Pinto et al. (eds.), Triticale: Today and Tomorrow, 737–741.
© 1996 *Kluwer Academic Publishers. Printed in the Netherlands.*

unlike wheat, triticale presents better adaptation in second straw (after a crop of wheat) : its yield potential is not affected, and it allows access to good levels of yield that can be compared with barley. Finally, the few measures that actually exist on the production of straw, which are very often an important criterion in breeding areas, give confirmation of a superiority in straw production relative to barley and even to the most productive varieties of wheat.

Resistance to moisture excess

Compared with wheat, triticale is more resistant to heavy rains at the end of stem elongation and during maturation. This better behaviour at the end of the cycle certainly comes from its low sensivity to diseases and from its maturity being less subject to climatic conditions.

Triticale is also less penalized by an excess of water during tillering. In fact, at the « ear at 1 cm » stage, the number of stems/m² with 3 leaves unfolded, indicates the future ear-population. We can observe that this number of stems/m² is not limited in the case of water excess in autumn or winter. On the contrary, wheats present a much higher sensivity to water excess : tillering is penalized. That can be translated into a limiting state of growth and development at the end of winter. This more regular behaviour of triticale certainly comes from rye ; some experiments carried out in conditions of contrasted hydric state (with and without excess of water) on clay soils in the North East of France, have indeed shown that rye was producing almost the same quantitiy in both cases, unlike nearly all the wheats.

Leaf appearance and tillering

As in all cereals, the leaf appearance, from the stage « 3 leaves unfolded » on the main stem, is synchronized with stem appearance. Thus, the rate of leaf appearance on the main tiller is a direct indicator of the rate of the axillary tiller appearance.

When comparing the different species of straw cereals, it is noted that on some triticales, the leaf appearance is quicker than that of wheats and almost the same as those of barley and rye. For example, an experiment carried out in the South of Parisien Bassin with early sowings (mid-October) shows that the needs of thermal time (basis 0) to form a leaf, are about 80°C for a variety of triticale such as Lasko, and about 100-110°C for a range of wheat varieties.

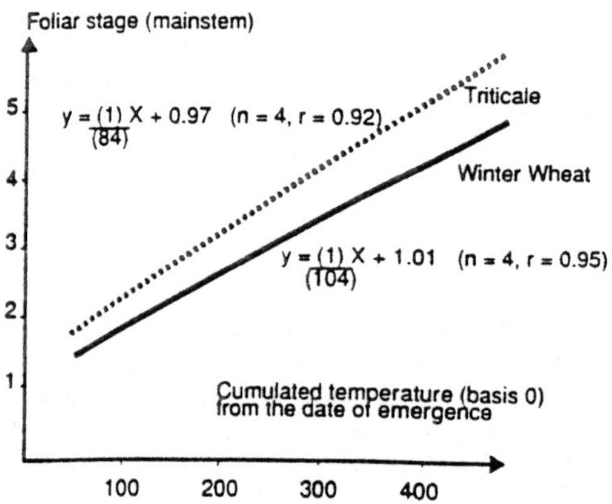

Figure 1 : Comparison of leaf rate appearance
between Winter Wheat and Triticale
(mid October sowing)

These lower needs, inherited from rye, gives it a good start to growthnd to non limiting tillering. Indeed, if we compare 2 varieties on the same earliness at stem elongation (Lasko for triticale and Arminda for bread wheat), triticale will be able to present a leaf stage more developed than wheat, with an average gain of about 2 leaves in early sowings. This quicker development enables some varieties of this species to to be more flexible with respect to late sowings and tower plast densities.

Resistance to cold

Some varieties of triticale can present a resistance to winter cold equal or superior to that of the most resistant bread wheats, thanks to its very high resistance to frost. The average of the higher levels of resistance for ryes turns around -30°C and the most resistant can fight temperatures inferior to -40°C.

Wheats, according to their varieties, resist temperatures around -20°C, -25°C and some of them up to -30°C, -32°C.

Winter barleys are the least resistant : the average level, depending on the barley variety, turns around -15°C and the most rustic cannot resist below   -20°C.

Triticales, which have various parentages, present a great genetic variability : from about -12, -15°C for the most sensitive down to -30, -35°C for the most resistant.

This criterion of resistance to winter cold can justify the development of triticales in high altitudes in contrast to barley.

Concerning the ear sensitivity to frost, resulting from the appearance of temperatures of -4°C in shelter after the « ear at 1 cm stage », we can observe, in the same way as bread wheats, some varieties are highly influenced by the precocity of the variety at the stem elongation. The earliest varieties in their growth during hard frost will show more significant damage than the later varieties.

Nevertheless, the frost conditions of 1991 revealed the existence, for varieties of the same earliness at the ear at 1 cm stage, of different levels of intrinsic tolerance. Thus, the Central and Newton varieties appeared more sensitive than varieties such as Torpedo and Magistral .

Moreover, as wheats, triticales can also be affected by low temperatures during meïosis of the pollen nother cells and probably untill fecundation. This was clearly put in evidence in high altitude zones (East of Massif Central) where variations occured in the number of grains per ear with abnormally low levels occuring in very early sowings (15-25 September).

When taking into account the present available information, it is difficult to forecast a classification of triticales in comparison to wheats for this type of risk. We can suppose that its tendency to allogamy, that is to say the crossed fecundation, allows it to compensate for fecundation instability through allogamy. On the contrary, it is thought that with the existing genetic variability, some triticales will present higher instability levels.

Resistance during maturation

The few references in France, found in dry mountain areas in the East of the Massif Central, show that triticales (Lasko and Clercal) are more resistant to thermic excess than wheats and barleys. Moreover, they also present better behaviour in conditions of water deficit. Thus, when temperature excesses remain very low but with a very high hydric deficit (-40 mm during the phase of physiological maturation), losses of weight of 1000 grains are not higher than 3 grams.

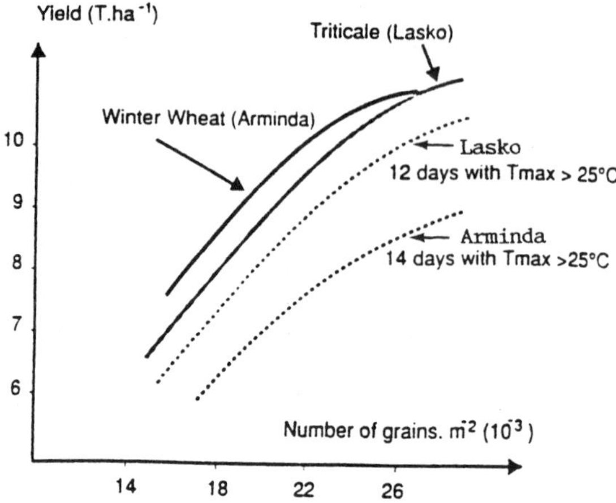

Figure 2 : Example of effect of excess
of temperature (number of days with Tmax > 25° C)
during the grain filling period

On the contrary, with similar conditions of temperatures but with a less important deficit of water (from -15 to -30 mm), losses for barley reach about 10 grams. Thus, barley avoids stress thanks to its earliness whereas triticale can tolerate it.

In the same manner, in cases of high temperature, wheats lose on average about 0,8 to 1 gram of weight per 1000 grains as soon as the day temperature is higher than 25°C (maximum temperature) during physiologic maturation. A thermic excess has fewer consequences on triticales since the middle losses stand at about 0,5 gram per day higher than this level.

These French results confirm the work of CIMMYT in a comparison of the agronomic performance of triticales with bread-wheats and durum-wheats in an international network of experiments.

The shrivelling that is sometimes noted on triticale certainly comes from rye. Indeed, in comparison to wheat, the enzymatic activities (amylolytic) last during the whole maturation of the grain.

Rythms of development (3,4)

The beginning of tillering : triticale tillering often begins sooner than that of wheat. This earliness of tillering comes from the fact that triticales have fewer needs than wheats for the growth of leaves. The beginning of the tillering stage corresponds, on average and whatever the species is, to the time when the main stem has reached 4 leaves.

Ear at 1 cm stage : the beginning of this stage has important agronomic consequences. It conditions the limits of the sowing period (3). The « ear at 1 cm » stage must be reached when the risk of an ear frost occurs in a frequency of less than 2 years out of 10.

Now the range of earliness of triticale at the ear at 1 cm stage is wide. In comparison with wheat, its ranges from a late type like Apollo to a early type like Récital.

Heading : even with the same earliness at the ear at 1 cm stage, triticale generally shows a heading stage sooner : this comes from its crossing with rye. Consequently, it presents a shorter stem elongation period than wheat.

Physiological maturity : on the same level of precocity at the heading stage, triticales reach the physiological maturity far later than wheats : this comes from its crossing with rye. Thus, compared with late wheats like Apollo, the majority of the present triticale varieties have a later physiological maturity. Wheat needs 700°C between its heading and its maturity. These needs always appear higher for triticales (about 850°C) and even far higher (about 950°C) for ryes and more reduced for barleys (about 600°C). It is to be noted that the longer these periods are, the better the behaviour of the species during maturation against hydric and thermic stress (4).

The above information enables us to present table 1 which, for a delinpated region, indicates the best periods of sowing.

TABLE 1 : BEST PERIODS OF SOWING FOR BEAUVAIS REGION

Precocity at maturity

+

Latest sowing dates		0	1	2	3	4	5	
	F			*Celtic*	*Plaisant*			
15-11	E					**Talent**	**Récital**	Slejpner = soft wheat
05-11	D			**Must**	**Thésée**	**Soissons**	**Courtot**	Lasko = triticale
05-11	C		**Rossini**	**Beauchamp**	Alamo		Torpedo	*Celtic* = barley
30-10	B		**Apollo**			Magistral		
25-10	A	**Slejpner**	**Beaver**	Lalsko	Dagro	Clercal		
	-	0	1	2	3	4	5	+ Precocity at the begining of stem elongation
		25-09	01-10	10-10	10-10	20-10	05-11	

Earliest sowing dates

References :

(1) Laroche G., 1981. Le Triticale valeur agronomique et fourragère. perspectives Agricoles n° 51, p20-29.
(2) Gate P., Demijolla P., Laroche G., Noailles D., 1990. Comportement et adaptation des céréales à paille en montagne de l'Est du Massif Central. Perspectives Agricoles, n°147, p51-64.
(3) Gate P., 1985. Une typologie des variétés pour une meilleure adaptation à la date de semis. Perspectives Agricoles n°94, p58-61.
(4) Laroche G. and al, 1994. Le Triticale, du débouché à la culture. Perspectives Agricoles, n° 188, p.I à XXXII.

SUCCESSIVE EFFECT OF HERBICIDES ON TRITICALE SEED GERMINATION AND PLANT GROWTH

Stankowski Sławomir, Maciorowski Robert
Department of Biometry, Academy of Agriculture, Szczecin, Poland

Abstract

In the laboratory experiment the successive effect of 5 herbicide variants (isoxaben, chlorsulfuron, isoproturon, chlortoluron and control) on the germination and plant growth of 4 winter triticale cultivars (Grado, Lasko, Malno and Ugo) was analysed. Germinative energy and germinating power of winter triticale seeds, obtained from plants treated with herbicide, were generaly lower, in particular for the isoproturon and chlorsulfuron variants. The reaction of triticale cultivars to herbicides was differentiated. The most sensitive were Grado and Ugo. Successive effect of herbicide treatment on the height and dry matter of seedlings was not observed.

Introduction

Results of ivestigations show different reaction of cereals on the herbicide application from stimulation to significant decrease of yield. In some papers [1, 2] is reported that very low concentration of simazine would stimulate seed germination and growth of seedlings, but in the other that is no effect of herbicide application on the seeds quality . The aim of this work is estimation of the successive effect of herbicide on seeds quality.

Material and methods

Experimental material were the grain samples taken from the field experiment conducted on heavy soil in 1993 with 5 herbicide variants (isoxaben – X-Pand, chlorsulfuron – Glean 75 DF, isoproturon – Graminon 500 FW, chlortoluron – Tolurex 80 WP and control – without herbicide) and 4 winter triticale cultivars (Grado, Lasko, Malno and Ugo). Data concerning doses and terms (Zadokscode) of herbicide application are presented in Table 1.

The successive effects of herbicides on germinative energy and germinating power and initial plant growth were determined.

743

H. Guedes-Pinto et al. (eds.), Triticale: Today and Tomorrow, 743–747.
© 1996 *Kluwer Academic Publishers. Printed in the Netherlands.*

Germinative energy and germinating power were estimated on Petri dishes in the four replicates 100 each after 4 and 7 days, respectively. Seedlings height and dry matter were determined in the germinationbeds in four replicates 25 seeds each after 12 days from the sowing (12 in Zadoks scale)

The results were worked out statistically by the analysis of variance in the completely randomized design. LSD 0.05 values were calculated by Tukey test.

Table 1. The doses and terms of herbicide application

Herbicide	Aktive substance	Dose	Term of application (Zadoks code)
X-Pand	12.5% isoksaben	1 l/ha	after sowing
Graminon 500 FW	50% isoproturon	2 l/ha	25
Glean 75 DF	75% chlorsulfuron	20 g/ha	25
Tolurex 80 WP	80% chlortoluron	2 kg/ha	25

Result and discussion

The germinative energy and germinating power of the seeds obtained from plants treated with herbicides are generaly lower than those obtained from non-treated plants (Table 2 and 3).

Table 2. The effect of herbicides and triticale cultivars on the germinative energy of seeds

Cultivar	Herbicide					Mean
	iso-xaben	isopro-turon	chlor-sulfuron	chlor-toluron	control	
Grado	89.5	81.5	86.0	86.0	92.5	87.1
Lasko	91.5	88.0	90.0	88.0	90.5	89.6
Malno	90.5	92.0	87.5	87.5	90.5	89.6
Ugo	77.5	73.0	79.0	85.5	88.0	80.6
Mean	87.3	83.6	85.6	86.8	90.4	86.7

$LSD_{0.05}$ for: cultivar – 2.05
herbicide – 2.45
interaction – 4.89

The biggest influence on the seed germination had the urea herbicide isoproturon, which decreased the germinative energy and germination power by 6.4% and 4%, respectively, as compared to the control variant. The significant herbicide x cultivar interaction indicated that differences among herbicides were not the same. Herbicides had no significant effecton germinative energy of Malno and Lasko cultivars, but reactionof Grado and Ugo was negative after isoproturon and chlorsulfuron application. The biggest decrease in germination was observed for grado, and amounted to 8%. Reaction of examined cultivars to remaining herbicides was totally different. Germination of Lasko seeds was stimulated by isoxaben, but for Malno the significant decrease was observed.

Table 3. The effect of herbicides and triticale cultivars on the germinating power of seeds

Cultivar	Herbicide					Mean
	iso-xaben	isopro-turon	chlor-sulfuron	chlor-toluron	control	
Grado	92.5	85.6	89.5	90.0	93.5	90.2
Lasko	95.0	91.0	94.0	94.0	92.5	93.3
Malno	92.5	94.0	94.0	94.5	94.5	93.9
Ugo	86.0	88.0	87.5	91.5	94.0	89.4
Mean	91.5	89.6	91.3	92.5	93.6	91.7

$LSD_{0.05}$ for: cultivar − 2.04
herbicide − 2.43
interaction − 4.87

The different reaction of triticale to chlortoluron and isoproturon − the herbicides from the same urea group − is worth mentioning.

The height and dry matter of seedlings were similar for the different herbicide variants (Table 4 and 5). The interaction among herbicides and cultivars was not found. The significant differences wer observed for the cultivars only.

Table 4. The effect of herbicides and triticale cultivars on the height of sedlings

Cultivar	Herbicide					Mean
	iso-xaben	isopro-turon	chlor-sulfuron	chlor-toluron	control	
Grado	145	145	143	152	150	147
Lasko	155	167	165	164	158	162
Malno	114	116	107	110	108	111
Ugo	121	118	109	124	118	118
Mean	134	137	131	137	133	134

$LSD_{0.05}$ for: cultivar − 6.9
herbicide − non significant
interaction − non significant

The results of prevailing investigations, connected with the influence of herbicides on seeds and young plants, are not equivocal. Lorenzoni (after Odiemah, Gaspar and Maroua [1] and Lowe and Ries [2] reported that very low concentration of simazine would stimulate seed germination and growth of seedlings. Anderson [3] and Kozanczenko [4] have not observed any significant influence of chlorsulfuron on the seed germination. Odiemah et al. [1] reported, that herbicides, applied in higher doses could increase the number of dead or abnormal seeds. The results of investigation conducted by Gournay et al. [5] and Sixto, Garcia-Baudin [6] showed that detoxication of herbicide depended on cultivar and agent.

Table 5. The effect of herbicides and triticale cultivars on the dry matter of seedlings

Cultivar	Herbicide					Mean
	iso-xaben	isopro-turon	chlor-sulfuron	chlor-toluron	control	
Grado	10.24	10.24	10.09	10.75	10.45	10.35
Lasko	8.83	9.70	9.58	9.22	9.09	9.28
Malno	7.30	7.91	7.11	7.27	6.87	7.29
Ugo	7.58	7.4	6.78	7.6	7.45	7.36
Mean	8.49	8.81	8.39	8.71	8.46	8.57

$LSD_{0.05}$ for: cultivar – 0.625
herbicide – non significant
interaction – non significant

Conclusions

1. Germinative energy and germinating power of winter triticale seeds, obtained from plants treated with herbicides, were generaly lower. The strongedt decrease in seed germination was caused by isoproturon and chlorsulfuron.
2. The reaction of triticale cultivars to herbicides was differentiated. The most susceptible were Grado and Ugo.
3. Successive effect of herbicide treatment on the height and dry matter of seedlings was not observed.

Acknowledgements

We appreciate the technical assistance of ms. Stanisława Ulasik

References

1. Odiemah M, Gaspar S, Maroua A. The influence of herbicides applications on seed quality of winter wheat. Acta Agronomica Hungarica 1988; 37:47–54.
2. Lowe LB, Ries SK. Endosperm protein of wheat seeds as a determinant of seedling growth. Plant Physiol. 1973; 48:57–60.

3. Anderson RL. Metribuzin and chlorsulfuron effect on grain of treated winter wheat (Triticum aestivum). Weed Science 1986; 34:357-436.

4. Kozaczenko H. Ocena skuteczności chlorsulfuronu (preparat Glean) w zwalczaniu chwastów oraz jego wpływu na plon jęczmienia jarego. Acta Academiae Agriculturae AC Technicae Olstenensis 1988; 45:193-201.

5. Gournay de X, Dufour L, Clair D. Essais recentsau domaine d'exposses sur les differences varietales de sensibilite aux phenilureas chez le ble tendre d'hiver. Comp. Rend. 7th Columa 1973; 2:375-382

6. Sixto H, Garcia-Baudin JM. Differentes respuestas a los herbicidas chlorotoluron e isoproturon de tres cultivars de trigo blando. Invest. Agr. Prod. Y. Prot. Veg. 1988; 3:243-252.

COMPARATIVE WEED SUPPRESSION BY TRITICALE, CEREAL RYE AND WHEAT

Deirdrie Lemerle[1] and Kath Cooper[2]

[1] NSW Agriculture, Agricultural Research Institute, Wagga Wagga, Australia

[2] University of Adelaide, Waite Campus, Glen Osmond, Australia

Abstract

Suppression of the grass weed annual ryegrass, *Lolium rigidum* by wheat, triticale and rye was compared in field trials at Wagga Wagga in 1993. Cereal rye and triticale appeared to be more competitive than wheat, with a biomass of annual ryegrass at maturity of $70g/m^2$ with triticale compared to $170g/m^2$ with wheat. Early seedling vigour, superior height and broad leaves appeared to influence the greater competitive ability of the triticale and cereal rye.

Introduction

The development of herbicide-resistant weeds and the need to reduce the dependence of farmers on herbicides, has led to the evaluation of the competitive ability of winter cereals as an additional means of weed control.

Methods

Six triticales, 200 wheats and two cereal ryes were grown in 1 x 1 m field plots at Wagga Wagga in 1993, with and without the grass weed annual ryegrass, *Lolium rigidum*. The cereals were sown at standard densities and annual ryegrass was sown immediately after the crops by broadcasting seed and raking it into the surface of the soil to achieve a density of 250 plants/m^2. Crop and ryegrass biomass were recorded at anthesis and maturity and attributes of plants which could affect weed suppression were recorded.

Results

The suppression of annual ryegrass varied with crop species and cultivar. Cereal rye and triticale were generally more competitive than the wheats. The biomass of annual ryegrass at maturity was $70g/m^2$ with triticale, compared with an average of $170g/m^2$ for the wheats. Early seedling vigour, superior height and broad leaves appeared to

H. Guedes-Pinto et al. (eds.), Triticale: Today and Tomorrow, 749–750.

© 1996 Kluwer Academic Publishers. Printed in the Netherlands.

influence the greater competitive ability of the triticale and cereal rye. Details of the suppression of annual ryegrass (i.e. dry matter at maturity) by cereal rye and triticale and wheat cv Dollarbird, and the attributes of crop growth that influence competitive ability are listed in the Table 1.

Table 1. Suppression of annual ryegrass by cultivars of triticale, cereal rye and wheat and the attributes of crop growth that influence competitive ability.

Cereal cultivar	Suppression of ryegrass ie. dry matter at maturity (g/m^2)	Cereal tillers/m^2	Cereal height (cm)	Diameter of first cereal leaf (cm)	Score of cereal habit at early jointing (1-erect, 7=floppy)	Light interception by crop at ground level at early jointing (%), ie. shading	Time of cereal head emergence (Julian days)
Tahara	65	575	120	3.83	6	90	271
Muir	70	480	130	3.65	6	92	271
Madonna	44	520	107	3.98	5	89	292
Abacus	38	760	103	3.45	4	87	278
OX83-34	37	455	125	3.89	6	79	264
OX83-50	158	360	125	3.58	6	73	278
Cereal rye Ryesun	48	825	170	3.17	6	93	264
Cereal rye 30B761	22	725	183	3.64	6	97	264
Wheat cv. Dollarbird	186	665	118	2.19	5	70	278

Conclusions

Triticale and cereal rye show great potential for weed suppression compared to wheat. Suppression of weed growth by competitive crops will decrease seed production of the weed and help reduce the cost of weed control. Competitive crops such as triticale should be grown in weedy paddocks or where resistant weeds occur.

Acknowledgments

Technical assistance of Birgitte Blater and financial support of the Grains Research and Development Corporation.

IX - TRITICALE USES

FACTORS AFFECTING TRITICALE AS A FOOD CROP

Roberto J. Peña
International Maize and Wheat Improvement Center (CIMMYT)
Apdo Postal 6-641, 06600 Mexico, D.F. Mexico

Introduction

Estimates indicate that the area dedicated to triticale production is now slightly above 2 million hectares (Pfeiffer WH, personal comm.). Triticale is used successfully in animal feeding because it is similar to, or slightly better than, other cereal grains as a source of protein and energy [1]. Although it is recognized that triticale can be used to prepare some food products [2], its utilization as a food grain is rather limited due to reasons associated with: grain compositional factors, breeding priorities, region-specific grain preferences of consumers, competitiveness with other grains, and economic, marketing, and processing aspects. Triticale could become a major crop if it were used as a human food grain in addition to an animal feed grain, particularly if it were so on a commercial scale. The objectives of this paper are to discuss, in general, grain and nongrain compositional factors associated with utilization of triticale as a food crop. There is also a brief discussion on the potential improvement of some compositional factors. More emphasis will be given to grain than to nongrain factors since there is much more documentation on the former than on the latter.

Factors Affecting Triticale as a Food Grain

As already indicated, the potential utilization of triticale as a food grain is influenced by both grain and nongrain factors. Grain factors directly affect the processing quality of the grain as well as the quality of the end-use product. On the other hand, nongrain factors affect triticale utilization by limiting the grain supply for food processing at the domestic and the commercial levels. These two groups of factors will be treated separately.

GRAIN FACTORS

The approximate chemical composition of triticale is, in general, intermediate between its two parental species, but resembling wheat more than rye, except for total sugar content, which is closer to that of rye [3-5]. Triticale, therefore, has been studied widely in relation to its milling and baking properties, the two main uses of both parental species. Additionally, triticale has been examined with respect to its potential utilization in malting

753

H. Guedes-Pinto et al. (eds.), Triticale: Today and Tomorrow, 753–762.
© 1996 *Kluwer Academic Publishers. Printed in the Netherlands.*

and brewing. Considering these as the main potential end-uses, the grain factors that more importantly influence the use of triticale for human consumption are: enzymatic activity (alpha-amylase in particular), flour milling potential, and polysaccharide (pentosans) and protein composition. Although starch plays a major role in some baking products, it is not considered a problem for the food utilization of triticale, since its properties are basically the same as those of wheat and rye.

Alpha-amylase activity (AAA). Triticale shows AAA levels generally higher than those of wheat. Triticale AAA tends to increase while in the field, due primarily to pre-harvest sprouting caused by wet conditions that prevail prior to grain harvest. High AAA is probably the most important grain-limiting factor in the utilization of triticale as a food. This is particularly true for baking because it significantly alters the functional properties of starch and, consequently, the properties of the baking system in which it is contained. In addition, the products of hydrolysis (sugars, gums) may negatively alter the quality of the end-use product. Triticale exhibits large genetic variability for AAA and pre-harvest tolerance [6,7], which has allowed breeders to select for low AAA in some breeding programs.

On the other hand, the facility of triticale to produce high AAA could be advantageous in the production af triticale malt, which can be used as additive in the food industry, or in brewing. In the latter case, triticale malt has been found acceptable in relation to amylolytic activity and wort yields. However, it is slightly high in proteolytic activity, as it results in high levels of solubilized protein, which could cause problems during fermentation and storage (protein precipitation) and in the color (dark) of the beer [8,9]. Although there is malting quality variability in triticale, breeding for this trait may be difficult because there is no methodology that allows for rapid and simultaneous screening for both protein solubilization and carbohydrate modification.

Grain milling. Due to its grain morphology (size and shape) and composition, triticale is more suitable for wheat than for rye milling. Wheat and triticale milling requires the use of both corrugated and smooth rolls to attain maximum flour extraction rates. By contrast in rye milling, smooth rolls are not used because they flake the rye middlings (due to its high pentosan content) in the reduction step, which causes a reduction in flour yield [10]. At low ash contents (<0.6%), triticale flour yields are 10 to 15% lower than those of wheat due to the occurrence of large, narrow, and shrivelled grains, which are difficult to extract flour from. Further improvement of grain plumpness should result in triticales with better milling quality.

One possible way to improve the flour yield of triticale is by milling wheat-triticale grain blends; flour yields in the range of 71 to 74.9%, (at ash contents below 0.55%) have been obtained in experimental milling (Buhler mill) from 75-25% wheat-triticale grain blends [11]. This may not be desired when the wheat supply is not limited, but may be acceptable in countries aiming to reduce wheat imports. Milling quality should not be a utilization constraint in cases where triticale is directed to the production of whole meal and high ash flour baking products.

Pentosans. Pentosans (arabinose + xylose), which are cell wall constituents, play a major role in determining the viscous properties of rye doughs, which, in turn, are very important to baking quality because they determine dough yield, stability, and volume, and partially influence bread loaf volume and crumb texture. Rye dough viscosity is influenced largely by water-soluble pentosan content. Proteins are important in rye doughs, but not to the same extent as in wheat doughs [12].

Table 1. Soluble pentosan content (%) in rye, triticale, and wheat.

Flour	Soluble Pentosans		Flour Ash (%)[b]	Viscosity of Water Extract[c]
	Grain[a]	Flour[b]		
Rye	3.89	2.4	0.97	3.15
Tcl	1.82	0.5	0.70	1.39
Wheat	2.16	0.5	0.46	1.31

a: Data from Saini and Henry [5].
b: Data from Fengler and Marquardt [13].
c: Values relative to water.

The soluble pentosan content of triticale grain and flour is similar to that of wheat and much lower than that of rye (Table 1) [5,13]. Therefore, triticale doughs in rye bread production will have inferior dough viscosity properties and baking quality as compared to those made with 100% rye flour. Triticale flour may be used, however, as a substitute of wheat flour, or of rye flour, in the production of American mixed wheat-rye bread, which uses organic acids to increase protein solubilization and, consequently, dough viscosity. In light rye bread, 60-85% of wheat flour is used in the blend [12]. Alternatively, it could be used for European rye-wheat or wheat-rye mixed breads.

Proteins. In wheat, storage proteins interact to form gluten. Gluten quantity and quality are responsible for the visco-elastic properties of the dough, which enables the production of a large variety of leavened and unleavened breads. Triticale has a storage (NaCl-insoluble) protein content considerably lower than wheat. In addition, only part of it forms gluten (Table 2); the other part corresponds, most likely, to some of the proteins inherited from rye, which do not form gluten. These differences in the amount and composition of storage proteins are the main factors responsible for the inferior bread making quality of triticale as compared to wheat; triticale breadmaking doughs have deficient visco-elasticity and poor handling properties, which yield breads with low loaf volumes and compact crumb.

Triticale shows genetic variability for gluten content and for gluten quality [14-16]. Table 3 shows the variability in SDS-sedimentation volume (a gluten quality-related parameter) and gluten content in complete and substituted triticales. Peña et al. [16] concluded that the highly significant correlation observed between SDS-sedimentation and specific loaf volume (volume/gluten protein content) (Table 4) indicates that, in addition to gluten quantity, gluten quality is a major factor influencing the bread making quality of triticale. The highest gluten contents in triticale are still 10-15% below those of wheat, a situation that is difficult to improve substantially with the present gene pool of triticale due to the presence of some rye secalins which do not form gluten. Further improvements in

Table 2. Protein solubility in 0.5M NaCl and gluten protein content of wheat, triticale, and rye flours

Flour	NaCl-sol. (%)	NaCl-insol. (%)	Gluten Prot. (GP) in Flour Prot. (%)	Difference (insol - GP) (%)
Wheat	17.7	78.2	78.5	-0.3
Tcl (S)[a]	32.4	65.6	50.5	15.1
Tcl (C)	32.5	64.2	46.4	17.8
Rye	36.7	63.0	nd	nd

a: S = substituted; C = complete. Data correspond to the mean of 8 different lines in each case.

Table 3. Gluten protein content (in flour protein) of complete and substituted hexaploid triticales.

SDS-Sedim. Group	SDS-a Sedim. (ml)	Gluten Protein in Flour Protein (%)
Complete High SDSS	13.2a 11.5-15.0	53.6a 50.0-56.5
Complete Low SDSS	6.0b 5.5-6.5	39.2b 31.6-47.4
Substituted High SDSS	12.2a 11.0-14.0	58.8a 56.6-61.1
Substituted Low SDSS	7.5b 7.0-8.5	42.3 28.2-53.6

a: Means followed by the same letter are not significantly different (a = 0.05).

Table 4. Correlation coefficients between breadmaking quality parameters in triticale.[a,b]

	SDS-Sedim. (SDSS)	Bread Loaf vol. (LV)	LV/GLU
GLU[c]	0.56**	0.19	-
SDSS	-	0.49**	-
SDSS/GLU	-	-	0.67**

a: Data from Peña et al. [16].
b: Complete (57) and substituted (38) hexaploid triticale lines were used in this study.
c: Gluten content (dry basis).

Table 5. Food uses of triticale, experiences in some major triticale producing countries (60-650 thousand hectares).

Country	Product	Proportion of Triticale Flour	Result
Australia	Breads, cookies, biscuits	100%, blend	+
Brazil	Variety breads	40-100%	+
Germany	Leavened bread	40%	+
Poland	Rye-type bread	100%	+
Russia	Rye-type bread	100%, blend	+
USA	Layer cake	50%	+

triticale gluten quantity and quality by chromosome transformations seems to be possible. A. Lukaszewski (U of C, Riverside, Cal.) has induced substitutions and translocations into some triticale genotypes, involving chromosome 1DL of bread wheat, chromosome arm carrying genes encoding for high M_r glutenins which contribute importantly to the bread making quality, and the group 1 chromosomes of triticale. Quality evaluations of some triticales carrying 1DL gene pull indicate that both gluten quality and quality can be improved (A. Lukaszewski, personal comm.). Some of these genetic stocks are being used by some breeders in an attempt to improve the gluten quality of triticale.

NONGRAIN FACTORS

In spite of its grain compositional problems, triticale has been found acceptable for human consumption mainly by partially substituting wheat and/or rye flours in baking products. Table 5 shows that diverse baking products have been successfully prepared with triticale alone, and in blends with wheat and rye flours in major triticale producing countries. The fact is, however, that triticale utilization as a food is almost nil.

In addition to grain composition factors, reasons for this are related to nongrain factors, such as breeding-related aspects, unacceptability as a food grain, lack of promotion as a food crop, and marketing and processing difficulties. Some of these factors apply globally while some others are region or population specific.

Breeding for quality. Deliberate triticale quality improvement seems to occur only in breeding programs in Poland, India, and Mexico (CIMMYT) [17-19]. Reasons for this could be: 1) no need for triticale as a food grain because the supply of traditional food grains is satisfied locally; 2) triticale quality improvement is desired, but sources for quality improvement are not available, or 3) the breeding program does not have the resources to conduct quality improvement. In the latter case, it will be very useful to form international quality nurseries that group triticales according to their potential food use, and to form an international network that can help to interchange and distribute triticale germplasm with a food orientation.

Acceptability as food grain. Despite agronomic and quality suitability, the acceptability of a new food grain for the preparation of traditional foods is not always well received. For example, Algeria has been importing large amounts of wheat for the production of bread. In attempts to reduce wheat imports, triticale was tested and found suitable in wheat-triticale blends (70-30%) for bread making, but it could not be utilized because the people seem to be unprepared to use cereals other than wheat in breadmaking [20]. Another factor that influences acceptability of triticale is the economic competitiveness of triticale with other cereals as a cash crop. For example, barley is a better cash crop than triticale in Algeria [20] and oat is a better cash crop than triticale in Michoacan, Mexico [21]. Finally, acceptability may be limited due to socio-political factors. One example is that triticale is recognized officially as a feed grain, but not suitable for food use in some parts of Europe (Krattiger A, personal comm.).

Promotion as food crop. Promotion plays a very important role. If none of the nongrain factors mentioned above were a problem for triticale utilization as a food crop then promotion would become the main limiting factor. The type of promotion required varies according to the targeted area, population, or sector. For smallholders, it should be demonstrated that triticale is a cereal that can play a role in subsistence farming, as a low--input cereal that could be used as a dual-purpose crop (food and forage) that adequately fits local crop rotations. For consumers looking for nutritious foods, it must be promoted as a good source of energy, lysine, and dietary fiber. Finally, at the end-use level, it must be demonstrated that triticale is suitable for local cereal-based foods and that it can be used for new nonconventional foods (snacks, breakfast cereal) or raw material for the food industry (starch production, malted triticale).

Marketing and food processing. A reliable grain supply is necessary for the establishment of a triticale market, but farmers often claim that they require a well established market before they set up production. Marketing also finds difficulties due to a lack of official grading factors and price for triticale. In addition, some experiences have shown that, when tested at the industrial level for milling, baking, or malting, triticale is treated equally to wheat or barley, respectively, with disappointing results. This is due to the inexperience of food processors with triticale. However, feasible processing modifications have demonstrated satisfactory results. Unwillingness of food processors to do so becomes the limiting factor in this case.

Conclusions

The utilization of triticale as a food grain is influenced by grain and nongrain factors. In spite of some important grain composition problems, triticale can be used as a food grain, mainly as a replacement for wheat and rye in proportions that will depend on the type of baking product. Further improvements, particularly on grain plumpness and gluten quantity and quality, are expected to make triticale more attractive as a food grain. In this respect, global networking among breeding programs willing to improve triticale quality could play a determining role in interchanging germplasm and in accelerating triticale quality breeding. The latter has been suggested previously (Darvey N. personal comm.), but no actions have been taken so far. This may be most likely due to economic problems faced by breeding programs worldwide. Nongrain factors, which are diverse, complex, and in many cases region- and population-specific, may be more limiting than grain factors in the utilization of triticale as a food grain.

References

1. Hill GM. Quality: Triticale in animal nutrition. In: CIMMYT. Proceedings of the Second International Triticale Symposium. Mexico, D.F.:CIMMYT, 1991: 422-27.

2. Gustafson JP, Bushuk W, Dera AN. Triticale: production and utilization. In: Lorenz KJ, Kulp K, editors. Handbook of Cereal Science and Technology. New York: Marcel Dekker Inc., 1991:373-99.

3. Heger J, Eggum BO. The nutritional values of some high-yielding cultivars of triticale. J Cereal Sci 1991;14:63-71.

4. Peña RJ, Bates LS. Grain shrivelling in secondary hexaploid triticale: I. Alpha-amylase activity and carbohydrate content of mature and developing grains. Cereal Chem 1982;59:454-58.

5. Saini HS, Henry RJ. Fractionation and evaluation of triticale pentosans: Comparison with wheat and rye. Cereal Chem 1989;66:11-14.

6. Oettler G, Mares DJ. Alpha-amylase activity and falling number in complete and substituted triticales. In: CIMMYT. Proceedings of the Second International Triticale Symposium. Mexico, D.F.:CIMMYT, 1991: 477-82.

7. Trethowan RM, Peña RJ, Pfeiffer WH. Evaluation of pre-harvest sprouting in triticale compared with wheat and rye using a line source rain gradient. Aust J Agric Res 1994;45:65-74.

8. Gupta NK, Singh T, Bains GS. Malting of triticale. Effect of variety, steeping moisture, germination and gibberellic acid. Brewers' Digest 1985; March:24-27.

9. Holmes MG. Triticale for malting and brewing. Triticale topics 1989; June:14.

10. Rozsa TA. Rye milling. In: Bushuk W, editor. Rye: Production, Chemistry, and Technology. Minnesota: American Association of Cereal Chemists Inc, 1976:111-25.

11. Peña RJ and Amaya A. Milling and breadmaking properties of wheat-triticale grain blends. J Sci Food Agric 1992;60:483-87.

12. Drew E, Seibel W. Bread-baking and other uses around the world. In: Bushuk W, editor. Rye: Production, Chemistry, and Technology. Minnesota: American Association of Cereal Chemists Inc, 1976:127-78.

13. Fengler AM and Marquardt RR. Water soluble pentosans from rye: II. Effects on rate of dyalisis and on retention of nutrients by the chick. Cereal Chem 1988;65:298-302.

14. Macri LJ, Ballance GM, Larter EN. Factors affecting the breadmaking potential of four secondary hexaploid triticales. Cereal Chem 1986;63:263-67.

15. Peña RJ, Ballance GM. Comparison of gluten quality in triticale: A fractionation-reconstitution study. Cereal Chem 1987;64:128-32.

16. Peña RJ, Pfeiffer WH, Amaya A, Zarco-Hernandez J. High molecular weight glutenin subunit composition in relation to the bread making quality of spring triticale. In: Martin EJ, Wrigley CW, editors. Proceedings of the Conference Cereals International. Victoria, Australia. Royal Australian Chemical Institute, 1991:436-40.

17. Amaya A, Peña RJ. Triticale industrial quality improvement at CIMMYT: Past, present and future. In: CIMMYT. Proceedings of the Second International Triticale Symposium. Mexico, D.F.:CIMMYT, 1991: 412-21.

18. Gill KS. Food uses of triticale in India. In: CIMMYT. Proceedings of the Second International Triticale Symposium. Mexico, D.F.:CIMMYT, 1991: 487-91.

19. Ceglinska A, Wolski T. Breeding for baking quality. In: CIMMYT. Proceedings of the Second International Triticale Symposium. Mexico, D.F.:CIMMYT, 1991: 531-35.

762

20. Benbelkacem A. Triticale research in Algeria. In: EUCARPIA-Triticale, Meeting of the Cereal Section on Triticale of EUCARPIA. Berlin. Akademie der Landwirtschaftswissenschaften der Deutschen Demokratischen Republik, 1987:451-55.

21. Carney J. Triticale production in the central Mexican highlands: smallholders' experiences and lessons for research. CIMMYT Economics Paper No. 2. Mexico, DF.: CIMMYT, 1990.

TRITICALE MALTING - AN EVALUATION OF CHARACTERISTICS AND PRODUCTION

Maria Luísa Beirão da Costa and Maria João Cabo Verde
Instituto Superior de Agronomia; Secção Autónoma de Ciência e
Tecnologia de Alimentos, Lisboa, Portugal

Abstract

Triticale cv. Manigero was malted according to a Central Composite Design with respect to time, temperature and relative humidity. Responses studied on malt were: yield (malting losses), soluble nitrogen, enzymatic activity and reducing sugars. To evaluate brewing suitability, wort characteristics (extract, viscosity, filterability, pH and colour) were determined. Triticale showed a higher enzymatic activity than barley this conducting to higher values of reducing sugars. Steeping and germination trials lead to the conclusion of possibility of malting triticale whenever conditions of germination are less intensive then those generally used for barley.

Introduction

Triticale is a challenging cereal, presenting interesting behavior from the agricultural point of view. Nevertheless as a crop it will only have a future if there are available processing methodologies that can be used.

Several processing technologies have been applied to this crop such bread baking [1,2,3] which led to poor results in the preparation of hard rolls. Also pasta products such as noodles were processed from triticale [4]. Extrusion-cooking of triticale was tried [5], resulting in acceptable products when using the whole triticale kernel tempered to 20% moisture and extruded at 350° F. Some trials have also investigated the possibility of using triticale as a malt supplement to low sugar dough [6]. The tendency of triticale grain to pregerminate suggests that it may be easily malted. Other researchers have studied the micromalting of triticale grains [7,8,9,10,11]. All of these studies showed a high α-amylase activity but also high levels of soluble nitrogen. The wort viscosity's produced from these malts showed also higher values of viscosity compared to those of barley malts, being that applications of both gibberelic acid and potassium bromate reduced malting losses and wort viscosity.

The present study was conducted to compare triticale with barley malt produced in standard conditions, aiming to a further knowledge of possibilities of malting triticale and trying to optimize malting conditions.

A response surface methodology (RSM) based on a Central Composite Experimental Design was used being the results adjusted by a second order polynomial. Responses studied on malt were yield (malting losses), soluble nitrogen, enzymatic activity and reducing sugars. To evaluate brewing ability wort characteristics like extract, viscosity, filterability, pH and colour were also determined.

763

H. Guedes-Pinto et al. (eds.), Triticale: Today and Tomorrow, 763–769.
© 1996 *Kluwer Academic Publishers. Printed in the Netherlands.*

Materials and Methods

SAMPLES DESCRIPTION AND PREPARATION

Barley cv. Elodie and triticale cv. Manigero were purchased in a commercial source. To prevent mold growth during malting process all samples were washed first in a ethanol/water mixture (70/30 V/V) containing 10g of CaCl/liter of the mixture and afterwards in a sodium azide solution (0.15g/l).

STEEPING TRIALS

In order to establish steeping curves, trials were conducted on step water at 20°C for different periods, i.e. for: 12, 24, 36 and 48 hours. The water uptake by the grain was evaluated by the increase in moisture content.

MALT PRODUCTION

The steeping phase was performed in water at 20°C/12h. Barley was malted in standard conditions, i.e. at 15°C and 20% relative humidity for 4 days.

To find the best germination conditions for triticale with respect to time, temperature and relative humidity, a central composite experimental design was established for these variables according to the matrix shown in Table 1.

Table 1. Matrix to establish central composite design for temperature, relative humidity and germination time for triticale

Factor	x1	x2	x3
Block 1	-1	-1	-1
	-1	+1	+1
	+1	-1	+1
	+1	+1	-1
	0	0	0
	0	0	0
Block 2	-1	-1	+1
	-1	+1	-1
	+1	-1	-1
	+1	+1	+1
	0	0	0
	0	0	0
Block 3	+1.68179	0	0
	-1.68179	0	0
	0	+1.68179	0
	0	-1.68179	0
	0	0	+1.68179
	0	0	-1.68179
	0	0	0
	0	0	0

The variables were tested for the following ranges:

T (x_1) - 15 - 25°C
R.H (x_2) - 20 - 30%
Time of germination (x_3) - 2 - 6 days

The points of the matrix were established by the expressions:

0-central point (middle value of the range)
+1.68179 - upper level
-1.68179 - lower level
$+1 = 0+(\Delta./2)/\sqrt{n}$
$-1 = 0-(\Delta/2)/\sqrt{n}$
Δ. - range
p = number of factors

METHODS OF ANALYSIS

Raw Materials and Malt

Samples of barley, triticale and their respective malts were tested for total nitrogen content by the kjeldhal method and moisture content by gravimetric. Starch content was evaluated in raw materials by the DMSO method [12]. Enzymatic activity was determinate by the ICC colorimetric method [13].

Germination capacity and energy of the kernels were evaluated by the method of Gupta *et al.* [8].

The malting yield is expressed in terms of malting losses in %.

Wort

The wort was tested with respect to extract yield, filterability and total reducing sugar by AOAC methods [14]. pH was determined by potentiometer. Wort apparent viscosity was evaluated using a Cannon-Fensk capillary viscometer, size 100. The kjeldhal method was used as method for soluble nitrogen in the wort.

Colour of the wort was measured in the Lovibond tintometer.

Results and Discussion

RAW MATERIALS CHARACTERIZATION

Results of analysis on the malted materials are shown Table 2. It can be seen that triticale and barley present similar germination capacity values. The nitrogen content of triticale was lower than that of barley which may be an interesting advantage for brewing. α-amylase activities were much higher in triticale then in barley. This suggests that long germination times not to be necessary for triticale.

Table 2.Analyses of the samples of unmalted raw materials

Sample	Moisture %	Total N % D.M.	Starch % D.M.	Germinat Capacity %	α-amylase activity A units*
Barley	10.5	1.89	67	94	4.8
Triticale	10	1.69	64	96	10.6

$$*A = \frac{100 \, xfx}{t2-t1} \, lg.(E_{t1}-E_{t2})$$

f = dilution factor

E_{t1} extinction value at times t1 and t2

E_{t2}

STEEPING CURVES

The results of steeping trials are shown in Figure 1. At the end of 12h triticale achieved the proper moisture content for malting, i.e. 40 %. Barley attained this value at the end of a 24h period. This can be caused by the presence of the husk which does not exist in triticale grain.

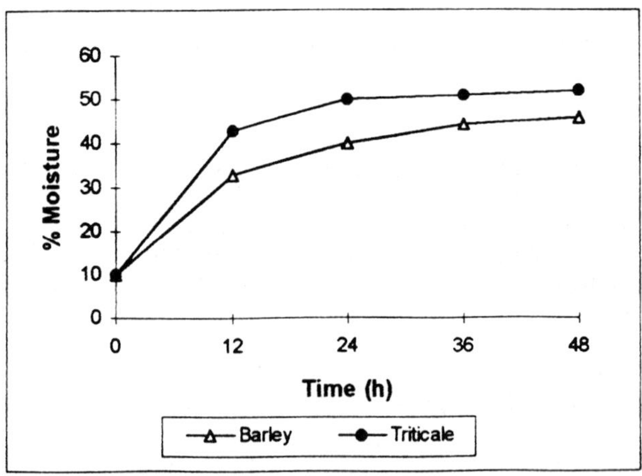

Figure 1. Steeping curves of barley and triticale at 20° C

MALT AND WORT PROPERTIES

Results for the relationship of total nitrogen to germination conditions in triticale malts, produces a response surface curve described by the following equation

$$y = 1.809 + 0.006\, x_1 + 0.021\, x_2 - 0.021\, x_3 + 0.038\, x_1^2 + 0.050 x_2^2 + 0.053\, x_3^2 - \\ - 0.035\ x_1 x_2 + 0.038\, x_1 x_3 - 0.079\, x_2 x_3$$

$$r^2 = 0.636$$

Lower levels of nitrogen content in the triticale malt were obtained for the central conditions, i.e. 20°C / 25% R.H/4 days and the total nitrogen for those is 2.1%. This is a little higher value than the one shown by barley malt (1.79%).

Nevertheless and looking also for convenient soluble nitrogen in the wort we could find a correlation that is expressed by

$$y = 1.9667 + 0.0005\, x_1 + 0.0046\, x_2 + 0.0242\, x_3 - 0.0101 x_1^2 + 0.0093\, x_2^2 \\ - 0.0154\, x_3^2 -- 0.0082\, x_1 x_2 - 0.0032\, x_1 x_3 - 0.0057\, x_2 x_3$$

$$r^2 = 0.616$$

This shows that values closer to those of barley malt (N= 1.06%) can be achieved for the lower conditions tested.

With respect to reducing sugars, malt from triticale always produced higher values than barley (9.6%). The conditions that yield lower values for triticale malt (19.5°C at 23.5% R.H.) were independent of germination time. Under these conditions the reducing sugars content was 11%. The equation correlating reducing sugars to germination condition is

$$y = 11.736 + 0.0119\, x_1 - 0.0366\, x_2 + 0.0858\, x_3 - 0.0209 x_1^2 + 0.5094 x_2^2 - \\ - 0.0566 x_3^2 - 0.0335\, x_1 x_2 + 0.0414\, x_1 x_3 - 0.0835\, x_2 x_3$$

$$r^2 = 0.435$$

Given the low r^2 this equation does not represent a statistical significant correlation; therefore it can only be used as an index of a tendency.

To obtain tritricale extract levels that are similar to those produced in barley malt, germination should be conducted at low R.H. (22%) and temperature (15,5°C) for a short period of germination (2,8 days). Extract level correlation with germination conditions is expressed by

$$y = 11.946 - 0.07\, x_1 - 0.246\, x_2 + 0.1213\, x_3 - 0.0328 x_1^2 + 0.261 x_2^2 - 0.2863 x_3^2 + \\ + 0.4981\, x_1 x_2 + 0.1398\, x_1 x_3 + 0.3793\, x_2 x_3$$

$$r^2 = 0.769$$

With respect to apparent viscosity of the wort, low temperature and R.H. conditions for a short period of germination produced best characteristics. It must be noted that triticale wort was less clear than the one produced from barley malt. The colour of triticale wort was darker than barley wort expressed in terms of yellow colour intensity as determined by Lovibond tintometer. No correlation was found between the colour and germination conditions. Nevertheless, values similar to those of barley malt wort (2.4) were obtained for the central point treatments that is 20°C/25% R.H./4 days

Malting losses on triticale were generally much higher than those on barley. Only if the grain is malted for a short time at low temperature and relative humidity it is possible to reduce those losses. Malting yields results are expressed by

$$y= 9.793 - 0.497x_1 - 0.474 x_2 + 1.265 x_3 - 0.357 x_1{}^2 - 0.393 x_2{}^2 - 0.729 x_3{}^2 + + 0.909 x_1 x_2 + +0.309 x_1 x_3 + 0.334 x_2 x_3$$

$$r^2= 0.717$$

Conclusions

From the results obtained it can be concluded that triticale can be successfully malted. Given the grain characteristics, the germination process should be shorter in time, and at lower temperatures and relative humidity than those generally used for malting barley. The best combination of conditions was T=17°C R.H.=22% during 2.8 days. Under these conditions, malting losses should be about 7%

References

1. Lorenz K. Food uses of triticale. Food Technol 1972; 26(11):66
2. Lorenz K, Welsh J, Norman K, Maja J. Comparative mixing and baking properties of wheat and triticale flour C Chem 1972;49:18
3. Ten C C, Hoover W J, Farred E P. Baking quality of triticale flours. C Chem 1973;50:16
4. Lorenz K, Dilsaver W, Lough J. Evaluation of triticale for the manufacture of noodles. J.Food Sci. 1972;37:764.
5. Lorenz K, Welsh J, Norman R, Beetner G, Frey A. Extrusion processing of triticale. J Food Sci 1974;39:572.
6. Finey K F, Shogren M D, Pomeranz Y, Bolte L C. Cereal malts in bread baking. Baker's Digest 1972;46:38.
7. Tombros S, Briggs D E. Micromalting triticale. J Inst Brew 1984;90:263-265.
8. Gupta N K, Singh T, Bains G S. Malting of triticale. Effect of varietey, steeping moisture, germination and gibberelic acid. Brewers Digest 1985;March:24-27.
9. Blanchflower A J, Briggs D E. Micromalting triticale:optimising processing conditions. J Sci Food Agri 1991;56:103-115.
10. Blanchflower A J, Briggs D E. Micromalting triticale. Comparative malting characteristics J Sci Food Agri 1991;56:117-128
11. Blanchflower A J, Briggs D E. Quality characteristics of triticale malt and wort J Sci Food Agri 1991;56:129-140.
12. Garcia W J, Wolf M J. Polarimetric determination of starch in corn with dimethyl sulfoxided as a solvent. C Chem 1972;49:298-306.

13. Perten H. A colorimetric methods for the determination of alpha-amylase activity (ICC method). C Chem 1966;43:336-341.
14. AOAC Association of Oficial Analytical Chemists. Official Methods of Analysis. 12th ed. Washington, D C, 1975.

RELATIONSHIP BETWEEN THE PENTOSANS OF TRITICALE FLOUR AND BREAD LOAF VOLUME

Małgorzata Cyran & Maria Rakowska
Department of Nutritional Evaluation of Plant Materials. Institute of Plant Breeding and Acclimatization, Radzików, 05-870 Błonie, Poland

Abstract

Physical, rheological and baking tests as well as chemical analyses, especially in respect to pentosan content were performed on six cultivars of triticale differing in baking quality and two wheat cultivars used as a standards. The pentosan content ranged from 1.90% to 2.77% of dry matter in triticale flours, whereas in wheats it was 2.36% and 1.90%, respectively for Begra wheat of high baking quality and Emika representing medium baking quality. The lowest content of pentosans in triticale cultivar Alamo was equal to that of wheat cultivar Emika. A negative relationship between flour pentosan content and bread volume, obtained by the wheat baking method was found in triticale. Mean pentosan level of triticale dry gluten was twofold higher (5.5%), than that of Begra wheat (2.7%). Falling numbers of five triticales were greater than 200, whereas in wheats the respective values exceeded 400. No significant differences were observed in falling numbers as well as in protein content among triticale samples. There was no relationship between sedimentation value and bread volume, while gluten content had an explicit impact on it. Physical and rheological tests confirmed the weaker gluten characteristics of triticale compared to wheat standards, although triticale cultivars Alamo and Vero had high baking quality in our experiment.

Introduction

The non-starch polysaccharides in cereals comprise a series of substances, which despite their relatively small contribution to the total weight of grain are of technological [1], medical [2] and nutritional importance [3]. Among these substances, especially pentosans, glycoproteins and to some extent β-glucans influence the rheological behavior of doughs and the texture of bakery products. Pentosans present in flour constitute 2-3% of dry matter in wheat [4], about 4% in triticale [5] and 5% in rye [6]. Together with glycoproteins, pentosans act as a bridge between proteins and carbohydrates in dough at the presence of oxidizing agents [7]. Beside this, pentosans were found to be a very

H. Guedes-Pinto et al. (eds.), Triticale: Today and Tomorrow, 771–777.
© 1996 *Kluwer Academic Publishers. Printed in the Netherlands.*

important regulators of water absorption and distribution in the dough, due to their unique properties such as, ability to form gels and viscous solutions as well as high water binding capacity. They neither coagulate through heat treatment, which is the case with proteins or retrograde through cooling and storage, which is the case with starch [6]. Many researchers have studied the effect of pentosans on loaf volume in wheat and rye, but opposite results have been obtained. This fact may be attributed to various methods of isolation, the purity of the pentosan preparation and the baking method used.

Therefore, the present work was undertaken to study the relationship of flour pentosan content in triticale, assayed by enzymatic digestion of starch and protein, to bread loaf volume, obtained by the wheat baking method.

Materials and Methods

Six winter triticale cultivars (Alamo, Vero, Moreno, Tewo, Ordo, Lasko) and two winter wheat cultivars (Begra and Emika) harvested in 1992 at the same location were tested. The dry matter content was determined by oven-drying at 105°C for 16h. Protein (Nx5.83) was evaluated by the Kjeldahl method (using a Kjeltec system). Minerals were assayed gravimetrically after combustion at 550°C for 6h. Soluble and insoluble dietary fiber fractions were isolated by enzymatic digestion of starch and proteins [8]. Dietary fiber fractions and dry gluten samples underwent acid hydrolysis (1N TFA at 125°C for 1h) and aldononitrile acetate derivatives of neutral sugars were obtained according to McGinnis [9] and analyzed by gas chromatography. Pentosan content was calculated from the sum of arabinose and xylose residues, expressed as polysaccharides using the factor 0.88. Milling was carried out in Brabender Quadrumat Senior laboratory mill, after conditioning of cereal grains. The mixing characteristics were analyzed in Brabender farinograph. The falling number and Zeleny sedimentation tests were also performed on whole meal. Wet gluten was separated automatically in a Glutomatic 2100. The baking test was carried out by the wheat baking method according to the following formula: flour 300g, yeast 9g, salt 3g, dough consistency 350 farinograph units. The loaves were fermented at 28°C and 75% relative humidity for 2h and then baked at 230°C for 20 min. The bread volume was measured after 24h by rapeseed displacement.

Results and Discussion

The mean grain protein content of six winter triticale cultivars was 12.0% expressed on a 14% moisture basis [Table 1] and was slightly lower, than that of Begra wheat (12.1%), but higher than the respective value for Emika wheat (11.5%). Similarly, the average protein content of triticale flour was intermediate (10.2%) between two wheat cultivars (11.6% and 9.6%). No significant differences were found in grain protein content as well as in flour protein content in triticale. Ash content ranged from 0.86% to 0.66% in triticale flours, while it was intermediate in Begra wheat (0.76%) and

Table 1. Chemical characteristic and milling flour yield of triticale cultivars compared to wheat standards

| Sample | Protein % * | | Ash % | Flour |
	Grain	Flour	Flour	yield %
Triticale				
Alamo	11.8	10.6	0.85	75.4
Vero	11.4	9.9	0.66	76.6
Moreno	11.4	10.0	0.86	73.3
Tewo	11.8	10.0	0.72	69.5
Ordo	11.8	10.0	0.76	73.9
Lasko	13.7	10.8	0.72	73.0
Wheat				
Begra	12.1	11.6	0.76	87.5
Emika	11.5	9.6	0.52	74.1

* on 14% moisture basis

lowest in Emika wheat (0.55%). Among triticale cultivars, Vero and Alamo had the best milling efficiency (76.6% and 75.4%, respectively), although Begra wheat had significantly higher milling efficiency (87.5%) and Emika wheat had lower (74.1%), than those of the best triticale cultivars.

Table 2. Falling number, sedimentation value, gluten content and gluten index of triticale cultivars compared to wheat standards.

| Sample | Falling number (s) | | Sedimentation value (ml) | Dry gluten % | Gluten index % |
	Grain	Flour	Flour		
Triticale					
Alamo	158	218	15	5.5	63.8
Vero	268	276	17	4.8	90.0
Moreno	316	341	15	4.2	71.8
Tewo	252	259	16	4.7	51.8
Ordo	84	120	15	4.1	72.6
Lasko	285	275	13	2.6	55.9
Wheat					
Begra	378	422	25	7.7	93.8
Emika	401	407	19	7.1	62.6

Table 3. Farinograph parameters of triticale cultivars compared to wheat standards

Sample	Absorption (%)	Stability (min)	Development time (min)	Resistance
Triticale				
Alamo	52.6	1.0	1.5	2.5
Vero	53.4	0.5	1.5	2.0
Moreno	55.6	1.0	1.5	2.5
Tewo	53.2	0.5	1.5	2.0
Ordo	53.2	0.5	2.0	2.5
Lasko	53.6	0.5	2.0	2.5
Wheat				
Begra	58.4	2.0	3.0	5.0
Emika	53.4	1.0	2.5	3.5

The falling number, sedimentation values, gluten content as well as gluten index are presented in Table 2. Generally, all test values obtained for triticale samples were inferior to the wheat standards. The falling numbers of four triticale cultivars exceeded 250 s, the acceptable level with respect to alpha-amylase activity [10].
No significant differences were found in sedimentation values obtained for triticale flours. The triticale cultivar Alamo had the highest dry gluten content (5.5%) but of inferior quality (gluten index 63.8%), while Vero had the lower dry gluten content (4.8%) of superior quality (gluten index 90.0%).
Physical dough tests on the farinograph confirmed the weaker characteristics of triticale dough [11], which is mostly attributable to gluten quality and quantity [Table 3].At the same consistency, the lower water absorption was found in triticale compared to wheat, whereas usually the opposite relationship is observed [12].

Flour pentosan content and baking characteristics of triticale cultivars compared to wheats are summarized in Table 4. Soluble pentosan content ranged from 0.63% to 1.04% of dry matter in triticale flours, while in wheat flours they constituted 0.75% and 0.62% for Begra and Emika, respectively. Soluble as well as total pentosan content of wheat flours was consistent with the results reported by Izydorczyk et al.[13]. The triticale flours had significantly lower total pentosan content (from 1.90% to 2.77%), than in the grain (from 5.11% to 6.38%) showing, that a considerable amount of pentosan is lost with the bran during milling. The highest bread volume (350cc) was obtained for Alamo, where the lowest pentosan content was observed, while the lowest bread volume (240cc) was obtained for Lasko and Ordo, where the highest pentosan content was found. Bread volume obtained for Begra wheat (340cc) was slightly lower, than that of Alamo and equal to that of Vero.

Table 4. Flour pentosan content (% of dry matter) and baking characteristics of triticale cultivars compared to wheat standards

Sample	Pentosans (%)			Loaf vol. (cc)	Porosity (1-8)	Crust* colour
	Insoluble	Soluble	Total			
Triticale						
Alamo	1.27	0.63	1.90	350	7	DB
Vero	1.32	0.81	2.13	340	6	DB
Moreno	1.63	0.88	2.51	320	6	B
Tewo	1.66	0.94	2.60	300	6	BB
Ordo	1.74	1.03	2.77	240		DB
Lasko	1.68	1.04	2.72	240		B
Wheat						
Begra	1.61	0.75	2.36	340	5	B
Emika	1.28	0.62	1.90	320	5	B

* B-brown; DB-dark brown; BB-bright brown

Table 5. Chemical composition of triticale dry gluten samples compared to wheat standards

Sample	Proteins (%)	Ash (%)	Pentosans (%)	Glucose Polymers (%)
Triticale				
Alamo	71.9	3.4	5.7	4.7
Vero	73.6	3.1	4.9	4.9
Moreno	69.5	3.2	5.4	3.2
Ordo	71.6	3.7	4.9	3.6
Lasko	70.5	3.3	6.4	2.3
Wheat				
Begra	79.8	3.2	2.7	4.5
Emika	72.4	2.6	3.8	5.6

The average composition of dry gluten (Table 5) was as follows: 71.4% protein, ash 3.3%, pentosans 5.5% and 3.8% glucose polymers (range 2.3%-4.9%). These glucose polymers, probably correspond to damaged starch and β-glucans.

Conclusions

The evaluation of six, winter triticale cultivars showed a negative relationship between flour pentosan content and bread volume. These results should be confirmed in a larger numbers of samples, before being used in the testing for breeding purposes. The level of pentosans found in triticale gluten samples was twice as high compared to high baking quality wheat, indicating a higher pentosan contribution to form gluten in triticale compared to wheat. Two of the triticale cultivars of distinctly lower pentosan content, showed the best baking quality characteristics.

Acknowledgments

The authors want to thank Professor T.Wolski and Ms. A.Ceglinska for their help in obtaining suitable samples of triticale and wheat and in carrying out baking quality testing in the laboratory of Laski Breeding Station.

References

1. Fincher GB and Stone BA. Cell walls and their components to cereal grain technology. In: Pomeranz Y, editor. Advances in cereal science and technology. American Association of Cereal Chemists Inc., St Paul, MN, USA,1986: 207-95.

2. Jenkins DJA and Wolever TMS. Slow release carbohydrate and treatment of diabetes. Proc Nutr Soc 1981;40:227-35.

3. Vahouny GV. Effects of dietary fiber on digestion and absorption. In: Johnson LR, editor. Physiology of the gastrointestinal tract. New York: Raven Press, 1987:1623-48.

4. Shogren MD, Hashimoto S, Pomeranz Y. Cereal pentosans: Their estimation and significance. IV. Pentosans in wheat flour varieties and fractions. Cereal Chem 1988;65:182-5.

5. Biskupski A. Jakosc konsumpcyjna ziarna. In: Tarkowski C, editor. Biologia pszenżyta. Warsaw: PWN, 1989; 352-71.

6. Meuser F and Suckow P. Non-starch polysaccharies. In: Blanshard JMV, Frazier PJ, Galliard T, editors. Chemistry and physics of baking. Spec. Publ. No 56, R Soc Chem 1986;56:42-61.

7. Shelton DR and D'Appolonia BL. Carbohydrate functionality in the baking process. Cereal Foods World 1985;30:437-42.

8. Asp NG, Johansson CG, Hallmer H, Siljestrom M. Rapid enzymatic assay of insoluble and soluble dietary fiber. J Agric Food Chem 1983;31:476-82.

9. McGinnis G. Preparation of aldononitrile acetates using N-methylimidazole as catalyst and solvent. Carbohydr Res 1982;108:284-92.

10. Amaya A and Pena RJ. Triticale indusrtial quality improvement at CIMMYT: past, present and future. Proceedings of the Second International Triticale Symposium 1990; October 1-5; Passo Fundo, Rio Grande do Sul, Brazil.

11. Schuller EH. Valuation and milling of triticale for human consumption of bread. Proceedings of the International Triticale Symposium; 1986 February 2-8; Sydney. Sydney: Australian Institute of Agricultural Science,1986.

12. Akmal Khan M and Rashid J. Nutritional and technological value of triticale. Proceedings of the International Triticale Symposium; 1986 February 2-8; Sydney. Sydney: Australian Institute of Agricultural Science, 1986.

13. Izydorczyk M, Biliaderis CG, Bushuk W. Comparison of the structure and composition of water-soluble pentosans from different wheat varieties. Cereal Chem 1991;68:139-44.

RHEOLOGICAL AND BIOCHEMICAL PROPERTIES OF MEAL FROM WINTER TRITICALE VARIETIES IN CONNECTION WITH THE RESISTANCE TO PREHARVEST GERMINATION

Stanislav Grib, M.P. Shishlov & N.P. Shishlova
Belarus Research Institute of Arable Farming and Fodders, Timiryazev,
Zhodino, Minsk Region, Belarus

Abstract

The analysis of biochemical properties (protein, starch and starch fractions, lipids, phenol combinations) and their correlations with caryopsis germination under the influence of abiotic factors (temperature, moisture, light) has shown that germination is mainly correlated with protein contents, lipid properties, and starch fractional composition. The analysis of rheograms using a permitted the identification of microviscosimeter, which was developed in BelNIIZK, the most informative characters: rheogram height - H (mm); the time the curve comes to a peak position - t (s); the area of the triangle inscribed in the rheogram - S (mm^2); and rheogram height in 60 s after coming to a peak position - H 60s (mm). The definition of amylolytic activity in 50 triticale varieties and lines using the Hagberg-Perten method as well as the viscosimeter has shown the reliability of the results obtained on the basis of these methods (r=0.917).

The evaluation of rheological and biochemical meal properties reveal correlations relations between them and resistance to preharvesting germination (PG), thus making it possible to reliably predict the reaction of a genotype to the environment.

The suggested rheological and biochemical characters are suitable for mass evaluation of resistance of genotypes to PG in the early stages of the breeding process.

Introduction

The problem of preharvesting germination in winter triticale is very acute in Belarus, due to specific weather conditions, which result in yield losses and deterioration of seed quality. The most applicable and convenient method for genotype estimation for the resistance to germination is the estimation of falling number (FN). The main limiting factor to a breeding solution for this problem is the lack of an express method and analyzers to estimate meal reological properties in F_2 and M_2 genotypes. To solve this problem, we have developed an experimental model of an electromicroviscosimeter and have obtained rheograms for different cereal varieties; the comparative analysis of rheograms has been made and correlations between FN, seed biochemical properties and the yield of winter triticale, rye and wheat varieties determined.

Materials and Methods

The seeds of the winter wheat, Berezina, the winter rye, Verasen and 50 varietal samples of winter triticale were differentiated according to yield and resistance to PG and the FN in

H. Guedes-Pinto et al. (eds.), Triticale: Today and Tomorrow, 779–784.

© 1996 *Kluwer Academic Publishers. Printed in the Netherlands.*

the meal then determined by the micromodified Hagberg-Perten method [1] as well as protein content [2]; phenol combination content [3]; lipid and starch content [4]; optical density of the iodine-starch complex at 540 and 640 nm. Meal rheological properties were estimated using an electromicroviscosimeter which makes it possible to analyze meal with minimum sample size of 750 mg. The obtained rheograms were analyzed for the following parameters: H (mm) - the height of reogram peak; t(s) - the time of coming of the curve to a peak position; S(mm^2) - the area of the triangle inscribed in the curve ascent; H 60s (mm) - reogram height in 60s after the curve comes to a peak position.

Results and Discussion

The suggested electromicroviscosimeter consists of a water bath [1] with a heat exchanger [5] and a cuvette [3] with a stirerr [4], rotated by the electrostabilized motor (Fig. 1). The boiling intensity off the water bath is regulated by a transformer. Current quantity and stirrer rotation velocity are regulated by the control unit. Changes in the strength of current by changing the stirrer rotation frequencies are controlled by the amplification unit, and are recorded by the registration unit.

The analysis of PG, rheogram parameters and biochemical characters in the samples under has revealed significant differences between these characteristics (Table 1). In triticale samples FN varied from 40 (H-1173) to 151 s (MAH 7834-27). In rye and wheat these characteristics were 172 and 159 s respectively. The height of the rheogram peak has changed from 14 (H-905-1, H-1173) to 36 mm (H-172-44, H-1178-1). In rye and wheat it was 45 and 44 mm respectively, which exceeds significantly the maximum value of this character in triticale. The time of reogram coming to a peak position varied for triticale samples from 20 (H-905-2) to 46 s (H-675). The area of the triangle varied from 75 (H-905-2) to 403 mm^2 (H-675). In rye and wheat this parameter had the maximum values of 450 mm^2 and 418 mm^2 respectively. The height of the rheogram in 60 s after coming to a peak position changed from 0 (H-905-1, H-905-2, H-1173) to 22 mm (H-1178). Figure 2 shows typical rheograms for the winter triticale samples (H-172-44, H-972-1, H-977-1), as well as rye and wheat. Slow changes in viscosity of the solution under the influence of amylases were characteristic of rye as distinct from wheat and triticale. 60 s after the rheogram came yo a peak psoition, the rheogram height in rye was 1.4 times less, in wheat 2.1 times less, and triticale 1.9, 2.8 and 3.6 times less than the peak position. This sharp decrease in viscosity of water-meal solutions in triticale under the influence of amylases is accounted for by their high activity this leading to grain germination in the spike under unfavourable conditions.The protein content in the triticale samples under study varied from 7.5 (MAH 7834-27) to 13.7% (H-157); the content of phenol combinations - from 0.179 (BAD 181) to 0.654% (BAD 192); lipids - from 1.18 (H-156) to 1.76% (L-160); starch - from 59.6 (H-158) to 70.2% (Rukh.). The analysis of biochemical properties and their correlations with caryopsis germination under the influence of abiotic factors (temperature, moisture, light) has shown that germination is mainly connected with the content of protein, lipids and starch fractional composition [5].

The correlation between FN, seed biochemical properties and yield capacity in triticale, rye and wheat was also studied. High coefficients of the paired correlations between FN and rheogram parameters are indicative of their close correlation: FN - H (r=0.877); FN - t (0.772); FN - S (0.904); FN - H 60s (0.917). There was no significant difference for the rest of pairs. For the pairs yield-protein, yield-starch and protein-starch respectively, correlation coefficients were as follows: r=-0.511; -0.013; -0.052.

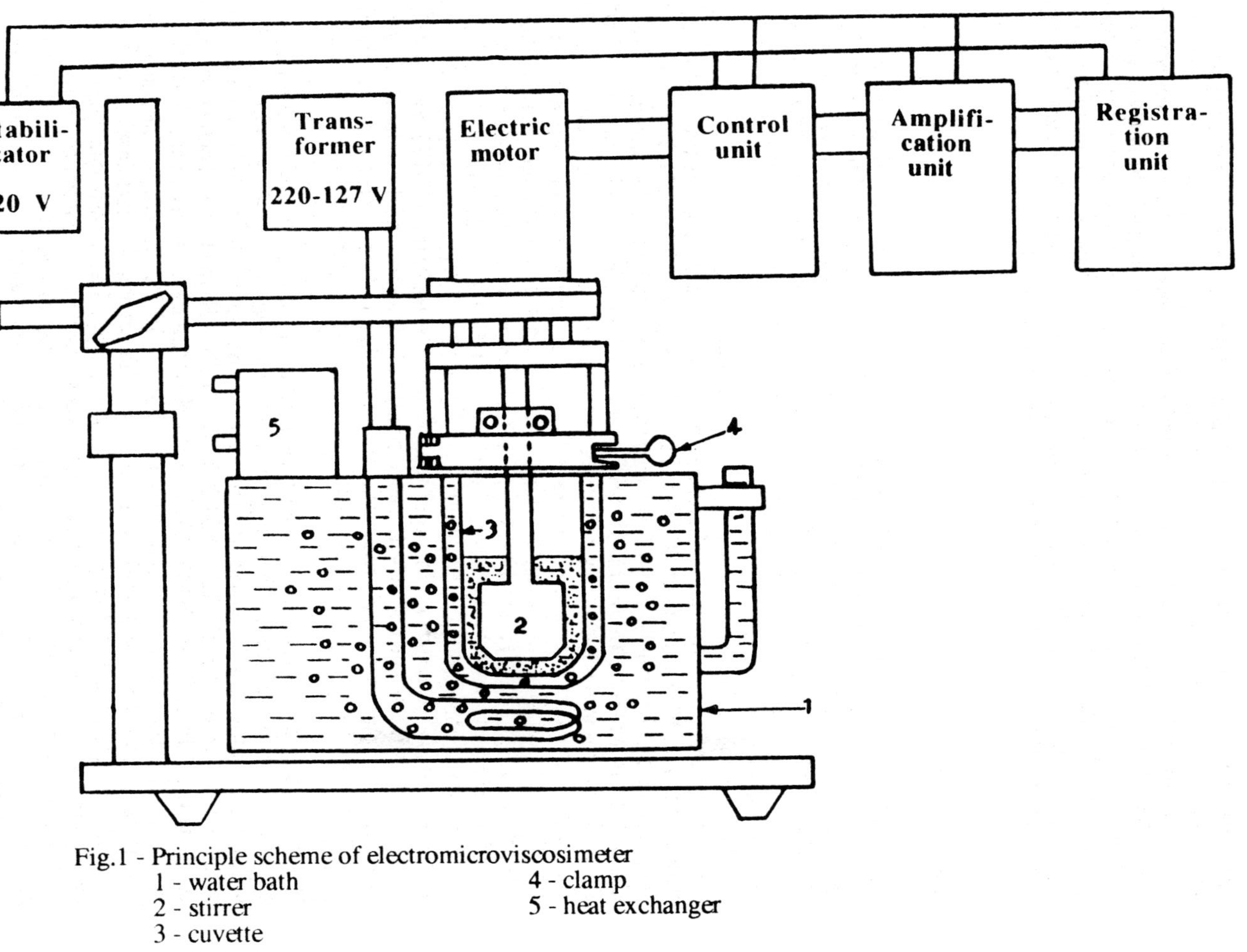

Fig.1 - Principle scheme of electromicroviscosimeter
1 - water bath
2 - stirrer
3 - cuvette
4 - clamp
5 - heat exchanger

Table 1. Rheological and Biochemical Properties of Meal from Winter Triticale, Rye Verasen and Wheat Berezina Seeds

N	Sample	Yield, t/ha	FN, s	H, mm	t, s	S, mm^2	H 60s, mm	Starch, %	Protein, %	E 540	E 640
1	Verasen	5.97	172	45	40	450	32	62.6	9.0	0.307	0.432
2	Berezina	4.58	159	44	38	418	21	67.8	12.5	0.307	0.388
3	Dar Belorussii	5.57	98	30	44	330	14	64.4	12.1	0.320	0.430
4	Rukh	3.62	72	17	26	111	7	65.4	11.9	0.310	0.410
5	H-675	6.67	129	35	46	403	19	64.8	10.9	0.346	0.486
6	H-172-44	5.69	118	36	44	396	19	66.2	11.5	0.301	0.410
7	H-172-45	3.82	100	24	36	216	13	62.4	11.9	0.343	0.450
8	H-905-1	4.51	48	14	28	98	0	64.0	12.0	0.305	0.416
9	H-905-2	7.35	41	15	20	75	0	64.8	10.7	0.328	0.446
10	H-972-1	4.70	92	25	36	225	9	69.2	10.6	0.318	0.434
11	H-972-2	4.20	92	31	36	279	15	65.0	11.1	0.293	0.389
12	H-972-3	3.72	53	20	32	160	5	65.6	11.1	0.312	0.409
13	H-977-1	6.05	73	18	32	144	5	64.0	11.0	0.308	0.435
14	H-977-2	4.96	109	22	42	231	13	64.0	11.2	0.318	0.420
15	H-977-3	5.50	85	27	38	257	11	64.2	11.0	0.325	0.455
16	H-1123	4.30	131	30	42	315	15	65.0	11.7	0.320	0.440
17	H-1161	5.44	52	19	32	152	4	64.6	11.3	0.295	0.399
18	H-1171-1	4.80	79	22	40	220	7	64.8	12.1	0.314	0.410
19	H-1171-2	6.14	126	31	44	341	15	66.6	10.4	0.317	0.441
20	H-1171-3	5.78	109	30	42	315	15	67.2	11.0	0.315	0.444
21	H-1171-4	5.57	103	24	40	240	8	64.8	12.0	0.294	0.393
22	H-1171-5	4.97	66	23	28	161	7	62.8	11.0	0.323	0.426
23	H-1178-1	5.42	125	36	44	396	22	64.6	11.1	0.336	0.457
24	H-1178-2	6.13	108	26	42	273	15	67.8	12.5	0.280	0.405
25	H-1173	5.43	40	14	22	77	0	62.0	13.2	0.286	0.396
26	H-1288	7.06	53	17	32	136	5	65.8	11.6	0.305	0.426
27	H-1290	5.99	92	25	34	213	7	68.2	9.6	0.315	0.420
28	H-1303	5.38	108	31	38	295	15	65.2	10.6	0.314	0.427
29	K-3155	5.23	50	25	34	213	9	63.4	12.4	0.320	0.427
30	MAH 7834-27	7.70	151	32	40	320	19	63.6	7.5	0.321	0.435
31	MAH 6282-3/10	6.90	69	29	32	232	10	66.8	10.0	0.315	0.456
32	MAH 10434-7/1	5.97	135	32	42	336	16	66.6	11.4	0.340	0.464
x for triticale		5.49	90	25	36	239	11	65.1	11.2	0.315	0.429

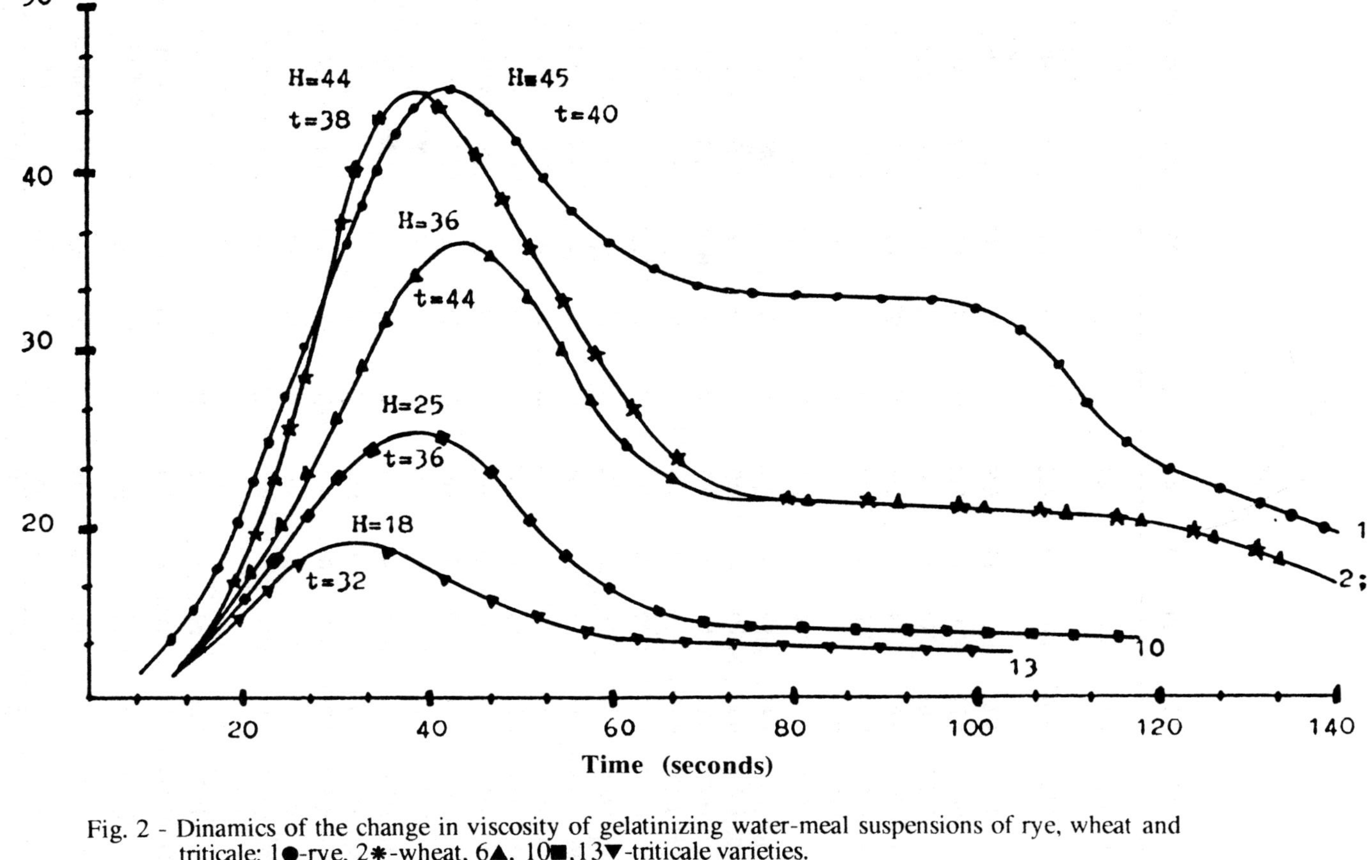

Fig. 2 - Dinamics of the change in viscosity of gelatinizing water-meal suspensions of rye, wheat and triticale: 1●-rye, 2✳-wheat, 6▲, 10■,13▼-triticale varieties.

The multiple correlation analysis between FN (dependent variable) and different variants of independent variables has shown a significant correlation for the following variants: FN - rheogram parameters, biochemical properties, yield (r=0.945); FN - rheogram parameters, biochemical properties (0.942); FN - rheogram parameters (0.926).

Conclusions

Rheological meal properties have been studied and seed biochemical composition determined in 50 varieties of winter triticale and one variety of both rye and wheat, with triticale line varying significantly in yield and resistance to preharvesting germination. Correlations between falling number, rheogram parameters and seed biochemical characters have been revealed. The most informative rheogram parameters and biochemical characters suitable for the mass evaluation of genotype resistance to PG, have been identified.

A new instrument the electromicroviscosimeter, which permits estimating of PG-resistance in small samples of F and M genotypes in the early stages of the breeding process, is recommended on the basis of its high correlation with the Hagberg-Perten method.

References

1. Berkutova, N.S. & L.G. Pogorelova, 1986. Methodical recommendations on the evaluation of cereal sprouting resistance in breeding for quality. VASKHNIL, Moscow.
2. Kizel, A.R., 1934. Practical guide to plant biochemistry. Biomedgiz, Moscow: Leningrad.
3. Harborn, J. 1968. Biochemistry of phenol compounds. Mir, Moscow.
4. Ermakov, A.I. and G.A. Lukovnikova, 1972. Methods of plant biochemical study. Kolos, Leningrad.
5. Shishlova, N.P. and S.I. Grib, 1993. Genotypical specificity of triticale seed germination with different temperature and light regimes. Plant Physiol. and Biochem., 25: 44-51.

ENVIRONMENTAL AND GENOTYPICAL INFLUENCES ON TRITICALE GRAIN QUALITY IN NORTHEAST OF PORTUGAL

José S. Oliveira[1], Benilde Mendes[1], Ana Fernando[1], Rui Fernandes[1], Henrique Guedes-Pinto[2], Olinda Pinto-Carnide[2], Valdemar Carnide[2], Bruno de Sousa[3], Fernanda Cabral[3], José Baeta[4]
[1]Grupo de Disciplinas de Ecologia da Hidrosfera, Faculdade de Ciências e Tecnologia, Universidade Nova de Lisboa, Monte de Caparica; [2]Departamento de Genética/Universidade de Trás os Montes e Alto Douro; [3]Secção Autónoma Química Agrícola/Instituto Superior Agronomia/ Universidade Técnica de Lisboa [4]Estação Agronómica Nacional/Instituto Nacional Investigação Agrária

Abstract

The effect of environmental conditions on the nutritional values of six triticale genotypes was studied by testing at five stations located in the Vila Real and Bragança districts.

The chemical composition, protein fractions, amino acid composition and the energy content of the grains were determined in order to characterize their nutritional value. The data was statistically analysed by ANOVA and for all the parameters studied the results showed highly significant differences between the five locations. No significant differences were observed between the six triticale genotypes, although, for some parameters (those related with the protein quality of the grain) interactions between genotype and environment were observed.

Introduction

Triticale is a valuable cereal crop for the North of Portugal (Guedes-Pinto *et al.*, 1982; Guedes-Pinto *et al.*, 1986); indeed, this cereal brings together some of the most interesting characteristics of wheat and rye, namely yield studied in wheat and the adaptability of rye.

Concerning the improvement of triticale, a study was developed with some triticale lines, in various regions of Trás-os-Montes (Interior/Northern Portugal) - Miranda do Douro, Montalegre, Bragança, Macedo de Cavaleiros and Vila Real. This study evaluated the effect of environmental conditions (soil, climate, water supply) on the nutritional value of triticale grain, as well as genotype effects. The influence of the interactions between genotype and environment was also studied.

785

H. Guedes-Pinto et al. (eds.), Triticale: Today and Tomorrow, 785–792.
© 1996 *Kluwer Academic Publishers. Printed in the Netherlands.*

Materials and Methods

Six hexaploid triticale cultivars or advance lines were used: EPM 293/81, EPC 17/85, EP 2/85, EP 26/85, CHD 775 and Clercal. Four randomized blocks with 10 m^2 and 400 viable seeds/m^2 were used.

2.1 - Chemical Analysis
    2.1.1 - Moisture, by drying at 103 ± 2°C, in an electric oven up to constant weight.
    2.1.2 - Ashes, by calcination at 550 ± 50°C for two hours.
    2.1.3 - Fats, by extraction using diethylether in a Soxhlet apparatus.
    2.1.4 - Total nitrogen (N), by the Kjeldahl method.
    2.1.5 - Crude protein, using the factor 6.25 to convert the nitrogen value to protein.
    2.1.6 - Cellulose, by the method of Belluci.
    2.1.7 - Calcium and magnesium, determinated by the EDTA titration method.
    2.1.8 - Phosphorus, determinated by the vanadomolybdophosphoric acid colorimetric method.
2.2 - Protein fractionation and Amino acid analysis, the extraction of protein fractions (albumins, globulins, prolamins and glutelins) and amino acid analysis of degreased milled Triticale grain were performed essentialy according to Guedes--Pinto *et al* (1990)
2.3 - Energy content, by bomb calorimetry.
2.4 - Statistical analysis, for each parameter an ANOVA - a two way analysis - was done.

Results

Tables 1-4 show some of the chemical analysis results (% moisture, calcium, magnesium and phosphorus contents are not presented because data is not significant). According to the derived results it's clearly significant that the Triticale grain obtained in Miranda do Douro (Malhadas) (all the six genotypes), show a much higher content of ash, fat and cellulose than in the other stations. However, the highest content of protein is obtained in Vila Real.

Table 1 - Ash content (% dry matter) of the Triticale grain

STATION / GENOTYPE	Malhadas	Montalegre	Bragança	Mac. Cav.	Vila Real
EP 2/85	8.56 ± 3.97	2.05 ± 0.20	2.15 ± 0.39	2.19 ± 0.08	2.82 ± 0.36
CHD 775	5.03 ± 2.51	1.64 ± 0.09	2.00 ± 0.56	1.97 ± 0.09	1.93 ± 0.29
EPC 17/85	3.59 ± 1.13	2.15 ± 0.14	2.15 ± 0.16	2.34 ± 0.12	3.28 ± 0.92
EPM 293/81	6.63 ± 4.52	1.77 ± 0.17	1.47 ± 0.18	2.40 ± 0.03	2.51 ± 0.21
EP 26/85	4.91 ± 3.74	1.92 ± 0.15	1.63 ± 0.08	2.12 ± 0.05	2.56 ± 1.16
CLERCAL	6.80 ± 4.09	2.03 ± 0.14	1.76 ± 0.07	2.14 ± 0.10	2.12 ± 0.42

Table 2 - Fat content (% dry matter) of the Triticale grain

STATION GENOTYPE	Malhadas	Montalegre	Bragança	Mac. Cav.	Vila Real
EP 2/85	4.25 ± 0.54	1.95 ± 0.13	1.48 ± 0.10	1.58 ± 0.13	2.43 ± 0.33
CHD 775	3.77 ± 1.64	3.99 ± 1.64	1.58 ± 0.12	1.78 ± 0.22	1.98 ± 0.49
EPC 17/85	3.92 ± 0.70	2.25 ± 0.28	1.84 ± 0.08	1.95 ± 0.24	2.49 ± 0.61
EPM 293/81	3.13 ± 0.84	2.04 ± 0.11	1.44 ± 0.11	1.78 ± 0.17	2.25 ± 0.61
EP 26/85	3.24 ± 0.84	2.25 ± 0.09	1.58 ± 0.20	1.62 ± 0.07	2.64 ± 1.01
CLERCAL	3.41 ± 0.67	2.09 ± 0.08	1.73 ± 0.17	1.72 ± 0.12	2.16 ± 1.02

Table 3 - Protein content (% dry matter) of the Triticale grain

STATION GENOTYPE	Malhadas	Montalegre	Bragança	Mac. Cav.	Vila Real
EP 2/85	7.48 ± 0.46	12.70 ± 1.14	8.87 ± 0.77	8.91 ± 1.05	12.20 ± 2.08
CHD 775	7.34 ± 0.54	9.41 ± 0.66	8.96 ± 1.04	10.34 ± 0.94	13.69 ± 2.04
EPC 17/85	9.51 ± 0.90	12.42 ± 0.82	12.02 ± 1.22	10.10 ± 0.76	14.87 ± 1.14
EPM 293/81	0.09 ± 0.76	10.69 ± 1.10	10.33 ± 0.65	7.79 ± 0.57	12.36 ± 2.28
EP 26/85	9.05 ± 0.67	9.75 ± 0.66	10.94 ± 1.06	9.65 ± 0.72	13.59 ± 1.97
CLERCAL	7.99 ± 0.63	11.55 ± 0.32	10.28 ± 0.37	9.40 ± 0.62	12.01 ± 1.86

Table 4 - Cellulose content (% dry matter) of the Triticale grain

STATION GENOTYPE	Malhadas	Montalegre	Bragança	Mac. Cav.	Vila Real
EP 2/85	5.88 ± 1.40	2.64 ± 0.09	2.65 ± 0.30	2.14 ± 0.14	4.70 ± 0.53
CHD 775	3.71 ± 1.61	1.91 ± 0.11	2.52 ± 0.30	2.02 ± 0.19	2.70 ± 0.12
EPC 17/85	3.68 ± 0.40	2.60 ± 0.17	2.65 ± 0.25	2.61 ± 0.33	4.23 ± 1.28
EPM 293/81	5.29 ± 2.96	2.47 ± 0.16	2.38 ± 0.27	2.24 ± 0.15	4.19 ± 0.48
EP 26/85	4.65 ± 2.77	2.38 ± 0.12	2.46 ± 0.24	2.73 ± 0.22	3.33 ± 0.67
CLERCAL	4.55 ± 1.71	2.74 ± 0.13	2.38 ± 0.25	2.47 ± 0.16	4.08 ± 0.91

The results obtained from the protein fractionation are shown in Tables 5-8. Similar to the values of protein content, Vila Real is the station that presents the highest amounts of albumin and globulin, although its prolamin content is the lowest of all. In all of the Triticale grains studied, the albumin was always found to be the highest protein fraction observed.

Table 5 - Albumin content (% dry matter) of the Triticale grain

STATION GENOTYPE	Malhadas	Montalegre	Bragança	Mac. Cav.	Vila Real
EP 2/85	3.89 ± 0.79	6.28 ± 0.60	3.14 ± 0.27	2.89 ± 0.67	5.85 ± 1.29
CHD 775	2.92 ± 0.30	3.96 ± 0.55	3.08 ± 0.53	2.93 ± 0.42	8.69 ± 2.24
EPC 17/85	4.19 ± 0.81	5.69 ± 0.95	3.71 ± 1.61	1.59 ± 0.53	6.00 ± 1.16
EPM 293/81	4.26 ± 0.98	5.06 ± 0.29	2.11 ± 0.58	3.13 ± 0.84	6.89 ± 2.49
EP 26/85	3.32 ± 0.55	4.05 ± 0.28	3.40 ± 0.76	2.24 ± 0.82	7.04 ± 1.22
CLERCAL	3.69 ± 0.28	6.33 ± 0.51	2.37 ± 1.07	2.10 ± 1.47	5.75 ± 1.91

Table 6 - Globulin content (% dry matter) of the Triticale grain

STATION GENOTYPE	Malhadas	Montalegre	Bragança	Mac. Cav.	Vila Real
EP 2/85	1.62 ± 0.21	1.89 ± 0.17	1.72 ± 0.35	1.02 ± 0.26	2.95 ± 0.94
CHD 775	1.90 ± 0.29	2.05 ± 0.22	0.84 ± 0.27	1.52 ± 0.70	2.16 ± 0.48
EPC 17/85	2.20 ± 0.25	2.70 ± 0.69	1.57 ± 0.25	1.03 ± 0.30	1.34 ± 0.26
EPM 293/81	1.70 ± 0.46	1.94 ± 0.36	1.58 ± 0.29	0.81 ± 0.18	3.04 ± 1.44
EP 26/85	1.63 ± 0.23	2.81 ± 0.21	1.28 ± 0.61	1.21 ± 0.90	2.45 ± 0.00
CLERCAL	1.70 ± 0.17	1.64 ± 0.09	1.65 ± 0.21	0.78 ± 0.44	1.88 ± 0.31

Table 7 - Prolamin content (% dry matter) of the Triticale grain

STATION GENOTYPE	Malhadas	Montalegre	Bragança	Mac. Cav.	Vila Real
EP 2/85	0.47 ± 0.06	0.42 ± 0.05	0.80 ± 0.20	0.39 ± 0.09	0.30 ± 0.14
CHD 775	1.16 ± 0.27	0.73 ± 0.15	0.42 ± 0.09	0.40 ± 0.12	0.30 ± 0.08
EPC 17/85	1.33 ± 0.39	0.68 ± 0.05	0.37 ± 0.10	0.61 ± 0.09	0.22 ± 0.03
EPM 293/81	0.73 ± 0.10	0.81 ± 0.23	0.57 ± 0.12	0.58 ± 0.12	0.22 ± 0.20
EP 26/85	0.85 ± 0.21	0.90 ± 0.20	0.79 ± 0.24	0.57 ± 0.16	0.36 ± 0.12
CLERCAL	0.69 ± 0.18	0.32 ± 0.05	0.42 ± 0.13	0.55 ± 0.12	0.39 ± 0.09

Table 8 - Glutelin content (% dry matter) of the Triticale grain

STATION GENOTYPE	Malhadas	Montalegre	Bragança	Mac. Cav.	Vila Real
EP 2/85	1.48 ± 0.15	2.32 ± 0.28	1.92 ± 0.40	1.21 ± 0.32	1.08 ± 1.19
CHD 775	1.87 ± 0.28	1.77 ± 0.23	0.30 ± 0.07	1.62 ± 0.45	2.06 ± 0.97
EPC 17/85	2.24 ± 0.62	2.43 ± 0.48	1.74 ± 0.11	2.52 ± 0.05	1.96 ± 0.28
EPM 293/81	1.78 ± 0.48	2.21 ± 0.54	0.94 ± 0.26	1.38 ± 0.37	0.97 ± 0.74
EP 26/85	2.27 ± 0.12	1.97 ± 0.33	0.78 ± 0.55	2.38 ± 0.43	2.27 ± 0.43
CLERCAL	2.19 ± 0.29	1.96 ± 0.05	0.51 ± 0.20	2.10 ± 0.44	1.75 ± 0.26

Tables 9-11 summarise the data related to the aminoacid analysis. Only the Cystine+Methionine content results were chosen as the cis+met were found to be the limiting aminoacids in all the Triticale grains studied. Vila Real and Malhadas show too low amounts of cis+met (Table 9), consequently they also show the lowests values of chemical class (Table 10) and EAA index (Table 11). Montalegre and Macedo de Cavaleiros show the highest amounts of cis+met and they also show the highest values for the chemical class and the EAA index.

Table 9 - Cis+Met content (g/16g N) of the Triticale grain

STATION GENOTYPE	Malhadas	Montalegre	Bragança	Mac. Cav.	Vila Real
EP 2/85	1.45 ± 0.48	2.62 ± 0.85	2.61 ± 0.54	2.27 ± 0.43	0.28 ± 0.07
CHD 775	1.44 ± 0.11	3.02 ± 1.47	1.65 ± 0.23	2.24 ± 0.37	0.92 ± 0.13
EPC 17/85	1.27 ± 0.12	2.26 ± 0.31	1.98 ± 0.30	2.27 ± 0.20	1.50 ± 0.23
EPM 293/81	1.21 ± 0.26	2.07 ± 0.59	1.68 ± 0.27	2.80 ± 0.54	0.90 ± 0.11
EP 26/85	1.27 ± 0.12	2.26 ± 0.31	1.98 ± 0.30	2.27 ± 0.20	1.00 ± 0.24
CLERCAL	1.25 ± 0.30	2.15 ± 0.11	1.65 ± 0.11	2.46 ± 0.27	0.80 ± 0.09

Table 10 - Chemical class of the Triticale grain

STATION GENOTYPE	Malhadas	Montalegre	Bragança	Mac. Cav.	Vila Real
EP 2/85	25 ± 8	44 ± 13	44 ± 9	39 ± 7	2 ± 0
CHD 775	24 ± 2	51 ± 25	28 ± 4	38 ± 6	8 ± 0
EPC 17/85	22 ± 2	38 ± 5	34 ± 5	38 ± 3	1 ± 0
EPM 293/81	21 ± 4	35 ± 10	28 ± 5	47 ± 9	8 ± 0
EP 26/85	22 ± 2	38 ± 5	34 ± 5	38 ± 3	1 ± 0
CLERCAL	21 ± 5	36 ± 2	42 ± 5	28 ± 2	7 ± 0

Table 11 - EAA index of the Triticale grain

STATION GENOTYPE	Malhadas	Montalegre	Bragança	Mac. Cav.	Vila Real
EP 2/85	32 ± 11	69 ± 8	85 ± 10	80 ± 12	24 ± 2
CHD 775	37 ± 2	90 ± 32	59 ± 4	66 ± 1	28 ± 3
EPC 17/85	24 ± 2	71 ± 3	67 ± 11	75 ± 5	30 ± 3
EPM 293/81	31 ± 7	70 ± 14	59 ± 5	92 ± 14	28 ± 2
EP 26/85	24 ± 2	71 ± 3	67 ± 11	75 ± 5	30 ± 2
CLERCAL	33 ± 7	67 ± 5	78 ± 9	61 ± 7	32 ± 4

To have a complete image of the results obtained in this study an ANOVA - a two way analysis - was done.

Table 12 - Results from the ANOVA - Two Way Analysis (P-value)

PARAMETER	SIGNIFICANCE		
	G	E	G x E
MOISTURE	0.3286 ns	< 0.0001 ***	0.7589 ns
CALCIUM	0.3281 ns	0.0018 **	0.2624 ns
MAGNESIUM	0.0111 *	< 0.0001 ***	0.0072 **
PHOSPHORUS	0.0316 *	< 0.0001 ***	0.0007 ***
ASHES	0.2176 ns	<0.0001 ***	0.9050 ns
CELLULOSE	0.0297 *	< 0.0001 ***	0.7533 ns
FATS	0.1386 ns	< 0.0001 ***	0.0086 **
PROTEIN	0.2706 ns	< 0.0001 ***	< 0.0001 ***
ALBUMINS	0.6002 ns	< 0.0001 ***	< 0.0001 ***
GLOBULINS	0.1323 ns	< 0.0001 ***	< 0.0001 ***
PROLAMINS	0.0002 ***	< 0.0001 ***	< 0.0001 ***
GLUTELINS	0.0079 **	< 0.0001 ***	< 0.0001 ***
AA/CHEMICAL CLASS	0.0650 *	< 0.0001 ***	0.0526 *
AA/EAA INDEX	0.0278 *	< 0.0001 ***	0.0002 ***
CIS+MET	0.1006 ns	< 0.0001 ***	0.1979 ns
ENERGY CONTENT	0.9460 ns	0.0001 ***	0.9064 ns

G - genotype; E - environment; ns - not significant; * - P < 0.05; ** - P < 0.01; *** - P < 0.001

This statistical analysis permits us to distinguish the main effects due to the genotype (**G**) and due to the environment (**E**) and to estimate the **G** x **E** interaction. Of course the results of such an analysis are only valid for the stations and genotypes assayed in one determined year. However, permits us to appraise one certain genotype in one determined region and, still, specify the regions in which the interest in the cultivation of that genotype is more evident. The results from this statistical analysis, for each studied parameter, are presented in Table 12.

Those results showed that the effect due to the environment is always greater than the effect due to the genotype. To all of the studied parameters, except to the content in calcium, highly significant differences were observed between the five stations (P<0.001).

Between the six genotypes, only the content in one protein fraction - the prolamins - has highly significant differences. The content in glutelins (another protein fraction) has, statistically, significant differences (P<0.01) but the differences on the contents in magnesium, phosphorus, cellulose, as well as the aminoacid parameters - chemical class and EAA index - are on the edge of the statistical significance (P<0.05). No differences were observed for all the other parameters.

For the **G** x **E** interactions, the parameters related with the protein quality of the grain (contents in protein, albumins, globulins, prolamins, glutelins) are the ones that showed highly significant differences. Also the differences on the content in phosphorus is highly significant. The contents in magnesium and fats showed significant differences and, once again, the differences in the pattern of aminoacid parameter - chemical class - is on the edge of the statistical significance. No differences were observed for all the other parameters.

Conclusions

Since the results obtained in this study are related to only a few number of years, they should be considered with some caution. Using the derived results from our experiments, it may be possible to achieve a number of conclusions; however, a more substancial conclusion may be obtained by a prolonged experimental period.
  - The effect due to the environment on the nutritional value of the Triticale grain, planted in several places of the Vila Real and Bragança districts, proved to be highly significant in terms of the chemical composition of the grain and in terms of the nitrogenised metabolism, namely the protein fractions and the aminoacid composition.
  - The effect due to the genotype proved to be not significant.
  - Interactions between **G** x **E** were observed mainly in the nitrogenised metabolism - in the protein content, in the protein fractions and in the aminoacid composition.
All the six genotypes showed the highest amounts of protein, albumin and globulin in Vila Real. However the best aminoacid composition was observed in Montalegre and in Macedo de Cavaleiros.

References

Guedes-Pinto, H.; Pinto-Carnide, O.; Carnide, V. - 1982 - Dois anos de ensaios de adaptação com triticales em Trás-os-Montes. Melhoramento **27**, 49-89.

Guedes-Pinto, H.; Pinto-Carnide, O.; Carnide, V.; Portela, J.; Portela, E.; Coutinho, J.; Castro, C.; Pires, A. L.; Rangel-Figueiredo, T. - 1986 - Triticale Breeding and "on farm" evaluation trials in Northeastern Portugal. Proc. International Triticale Symposium, University of Sydney, Australia 2-8 Feb. 1986: 340-344.
Guedes-Pinto, H.; Oliveira, J.F.S.; Pinto-Carnide, O.; Mendes, B.S.; Sousa, B.; Cabral, F.; Carnide, V.; Mendes, F.; Oliveira, L. - 1990 - Interacções Genótipo x Local na produção de grão. V Reunião Portuguesa sobre Triticale. ENMP, 23 of May, Elvas.
Guedes-Pinto, H.; Oliveira, J.F.S.; Pinto-Carnide, O.; Mendes, B.S.; Sousa, B.; Cabral, F.; Carnide, V.; Mendes, F.; Oliveira, L. - 1990 - Toxicidade de cobre em Triticale por aplicação foliar: I - Efeito na produção de grão. V Reunião Portuguesa sobre Triticale. ENMP, 23 of May, Elvas.
Guedes-Pinto, H.; Oliveira, J.F.S.; Pinto-Carnide, O.; Mendes, B.S.; Sousa, B.; Cabral, F.; Carnide, V.; Mendes, F.; Oliveira, L. - 1990 - Toxicidade de cobre em Triticale por aplicação foliar: II - Efeito na qualidade do grão. V Reunião Portuguesa sobre Triticale. ENMP, 23 of May, Elvas.
Oliveira, J.S. - 1977 - A qualidade dos produtos alimentares vegetais. IPVR, Vila Real, 195 pp.
Oliveira, J.S.; Mendes, B.; Mendes, F.; Oliveira, L.; Sousa, B.; Cabral, F.; Guedes-Pinto, H.; Carnide, V.; Pinto-Carnide, O. - 1990 - Toxicidade do cobre em Triticale. Efeitos sobre a qualidade das proteínas. V Congresso Nacional de Biotecnologia, 31 of October to 3 of November, Universidade do Minho, Braga.

CONCENTRATIONS OF MINERALS IN THE GRAINS OF A HIGH- AND A LOW-PROTEIN WINTER TRITICALE

Dario Fossati[1], Aldo Fossati[1] & Boy Feil[2]

[1] Station Fédérale de Recherches Agronomiques de Changins (RAC), Nyon, Switzerland
[2] Institute of Plant Sciences, ETHZ, Zürich, Switzerland

Abstract

Ten hexaploid winter triticales were grown at three sites in western Switzerland for two cropping periods. The grains were analyzed for their concentration of N. Averaged across the environments, entries 50728 and 50893, two advanced breeding lines, produced about the same grain yield but differed markedly in the concentration of grain N (Fossati, Fossati & Feil, 1993, Euphytica 71, p. 115-123). Research on wheat and other cereals has shown that concentrations of protein and minerals are often positively correlated. Therefore, it appears likely that high-protein triticales, such as entry 50728, will also exhibit elevated levels of minerals. Grain of entries 50728 and 50893, from our experimental site Changins, were used to test this hypothesis. The kernels were assayed for a number of nutritionally relevant mineral elements (P, K, Mg, Mn, Ca, Fe, Zn, and Cu) by ICP-AES. Averaged over the two cropping periods, 50728 grain was clearly higher in N, P, K, Ca, and Fe and slightly higher in Mg and Zn than those of entry 50893. There were no differences between the entries for Mn and Cu.
As expected it appear that the grain of high-protein triticales may show increased levels of many mineral elements.

Introduction

Cereal grain plays an important role in meeting the mineral requirements of the human population, either directly as food or indirectly via animal feed. The concentration of mineral elements in cereal seeds is affected by many factors including the choice of cultivar. Since triticale is a relatively new cereal and the production area is still small, very little research has been conducted on the genotypic variation in the mineral composition of triticale grain [1, 2].

H. Guedes-Pinto et al. (eds.), Triticale: Today and Tomorrow, 793–797.
© 1996 *Kluwer Academic Publishers. Printed in the Netherlands.*

There are indications that the concentrations of protein and minerals are positively interrelated, i.e. the grains of high-protein cereal cultivars are high in minerals and vice versa. This was found for wheat [3, 4] and maize [5, 6]. There are, as yet, no corresponding results for triticale.

In an earlier study, we identified two hexaploid winter triticale lines with very similar grain yields but clearly different concentrations of grain protein [7]. On the basis of observations made for cereals other than triticale, it is hypothesized that the kernels of the high-protein line are rich in minerals while those of the low-protein line have low mineral contents.

Materials and Methods

Two hexaploid winter triticale lines (entries 50728 and 50893) were compared. Details about the origin of the lines are available on request. The grains used in this study stem from our experimental site at Changins (western Switzerland) where the lines were grown for two growing periods (1989/90 and 1990/91). For detailed information on the soils, rates of fertilizer application, and conditions of growth see [7].

Seeding rate was 350 grains m^{-2}. Plots consisted of eight rows, 4.75 m long and 16 cm apart. There were five replicates. At maturity, 1 m of each of the two centre rows was harvested. The grains were dried to constant weight at 65° C for 48h. The N concentration in the ground kernels was determined by near infrared reflectometry (Technikon Infra Analyzer 400). For the determination of mineral elements, the samples were ashed at 525 °C for two hours. The residue was taken up with dilute HCl. The mineral elements were analyzed by inductively coupled plasma - atomic emission spectrometry (ICP-AES).

The remainder of the plots was harvested with a plot harvester for the assessment of grain yield.

Results and Discussion

Averaged over the two growing periods, the lines produced very similar grain yields whereas they differed markedly in the concentration of grain N (Figure 1a). In small grains, including triticale [7, 8], there is often a strong inverse relationship between grain yield and grain protein concentration, i.e. high-yielding cultivars show low concentrations of grain protein and vice versa. The data presented in Figure 1c demonstrate, however, that genotypes with similar grain yields may exhibit large differences in grain protein production.

As expected, the grains of the high-protein line displayed higher levels of minerals than did those of the low-protein line. With the exception of Mn (Figure 1i), the differences were statistically significant. The differences between the lines were relatively

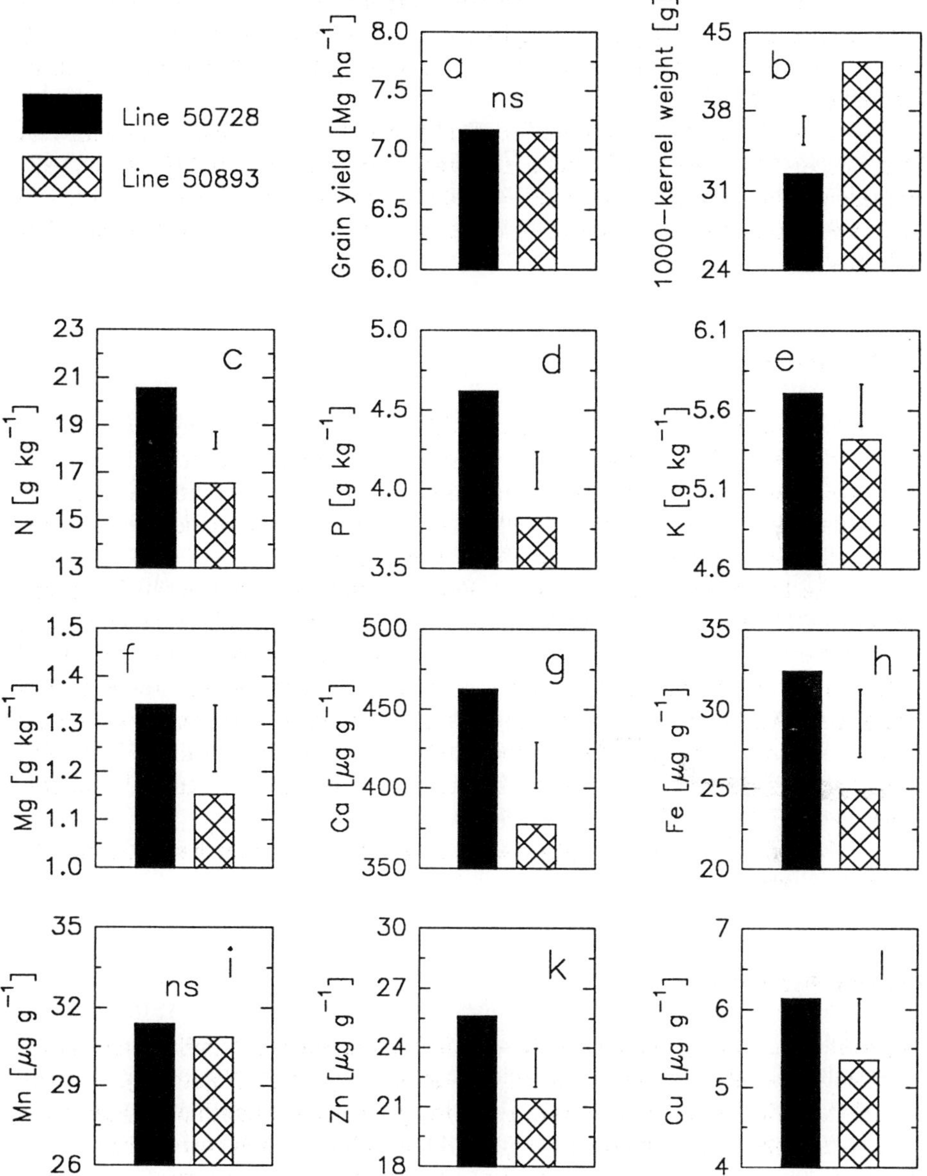

Figure 1 - Grain yield (a), 1000-kernel weight (b), and concentrations of protein (c) and various mineral elements (d - l) for two hexaploid winter triticale lines. Values are means across two crop years. Error bars indicate LSD$_{.05}$ values. ns = not significant at the P<0.05 level.

small for K, Cu and, especially, Mn (Figures 1e, 1l, and 1i, respectively). Peterson et al. [3] evaluated a collection of 27 wheat cultivars for variations in the levels of protein and 10 mineral elements. The concentrations of protein and most mineral elements in both flour and bran were found to be positively associated. In line with this, Raboy et al. [4] reported that protein and P in wheat grains are highly and positively correlated. Similar results were obtained for tropical maize [6, 9]. Long-term divergent selection for high and low levels of grain protein resulted in maize strains with high and low concentrations of P in the grains [5]. Positive relationships between the concentrations of protein and minerals have also been reported for soybean lines [10].

It appears that the physiological basis of the positive interrelations between the concentrations of protein and minerals in cereal grains is not yet understood. In ripe wheat kernels, the major proportion (50 - 80 %) of the minerals is located in the aleurone layer. At the same time, the aleurone layer makes up 87 % of total grain phytate, the major storage form of P in cereal seeds [11]. Studies of the fine structure of wheat kernels revealed that protein bodies in the aleurone layer contain P-rich inclusions, the so-called 'phytin bodies' [12]. Phytin is a carrier of cationic minerals [13]. On the basis of these observations, it is suggested that the extent of protein deposition in the aleurone layer determines the extent of phytin production which, for its part, regulates the accumulation of minerals. This may account for the observation that high-protein cultivars also show high levels of minerals.

The concentrations of minerals are relatively low in the endosperm, while they are very high in the outer portions of the grain [3, 14]. Thus, differences in whole-grain mineral concentrations among cultivars may be due, in part, to genetic variation in seed plumpness [3]. The lines tested in our experiment differed markedly in 1000-kernel weight (Figure 1b), possibly indicating that the line differences in the levels of minerals are based on differences in kernel configuration.

High levels of minerals suggest an added nutritional advantage of the high-protein triticale line 50728. However, several studies revealed that the concentrations of phytate and P are highly and positively correlated [4, 15]. Phytate is a poor source of P for simple-stomached lifestock and human beings. It is thought to act as an antinutrient, inter alias because it forms insoluble complexes with nutritionally important multivalent cations such as Ca, Fe, Cu, and Zn in the digestive tract of non-ruminants, thus making them unavailable or only partially available for absorption [16]. In evaluating the nutritional value of line 50728, these facts must be taken into account. Recent research has demonstrated, however, that the adverse effects of phytate can be alleviated by applying microbial phytase to cereal-based diets of monogastric farm animals [17].

Conclusions

Our results indicate a substantial genetic control of the mineral composition of triticale grains. As expected from observations made for cereals other than triticale, the grains of the high-protein line showed increased levels of many nutritionally relevant mineral elements.

References

1. Lorenz K, Reuter, FW Sizer C. The mineral composition of triticales and triticale milling fractions by X-ray fluorescence and atomic absorption. Cereal Chem 1974; 51: 534-542.
2. Singh B., Reddy NR. Phytic acid and mineral compositions of triticales. J. Food Sci 1977; 42: 1077-1083.
3. Peterson CJ, Johnson VA, Mattern PJ. Evaluation of variation in mineral element concentrations in wheat flour and bran of different cultivars. Cereal Chem 1983; 60: 450-455.
4. Raboy V, Noaman MH, Taylor GA, Pickett SG. Grain phytic acid and protein are highly correlated in winter wheat. Crop Sci 1991; 31: 631-635.
5. Raboy V, Below FE, Dickinson DB. Alteration of maize kernel phytic acid levels for protein and oil. J of Heredity 1989; 80: 311-315.
6. Feil B, Thiraporn R, Lafitte HR. Accumulation of nitrogen and phosphorus in the grain of tropical maize cultivars. Maydica 1993; 38: 291-300.
7. Fossati D, Fossati A, Feil B. Relationship between grain yield and grain nitrogen concentration in winter triticale. Euphytica 1993; 71: 115-123.
8. Mather DE, Poysa VW. Genetic analysis of the protein and lysine content of spring triticale. Can J Genet Cytol 1983; 25: 378-383.
9. Feil B, Thiraporn R, Geisler G, Stamp P. Genotype variation in grain nutrient concentration in tropical maize grown during a rainy and a dry season. Agronomie 1990; 10: 717-725.
10. Raboy V, Dickinson DB, Below FE. Variation in seed total phosphorus, phytic acid, zinc, calcium, magnesium, and protein among lines of Glycine max and G. soja. Crop Sci 1984; 24: 431-434.
11. O'Dell BL, de Boland AR, Koirtyohann SR. Distribution of phytate and nutritionally important elements among the morphological components of cereal grains. J Agric Food Chem 1972; 20: 718-721.
12. Morrison IN, Kuo J, O'Brien TP. Histochemistry and fine structure of developing wheat aleurone cells. Planta 1975; 123: 105-116.
13. Lott NA, Spitzer E. X-ray analysis studies of elements stored in protein body globoid crystals of Triticum grains. Plant Physiol 1980; 66: 494-499.
14. Pedersen B, Bach Knudsen KE, Eggum BO. Nutritive value of cereal products with emphasis on the effect of milling. In: Bourne GH, editor. Nutritional value of cereal products, beans and starches. Basel: Karger, 1989; 1-91.
15. Lolas GM, Palamidis N, Markakis P. The phytic acid - total phosphorus relationship in barley, oats, soybeans, and wheat. Cereal Chem 1976; 53: 867-871.
16. Maga JA. Phytate: Its chemistry, occurrence, food interactions, nutritional significance, and methods of analysis. J Agric Food Chem 1982; 30: 1-9.
17. Rimbach G, Pallauf J. Enhancement of zinc utilization from phytate-rich soy protein isolate by microbial phytase. Z. Ernährungswiss 1993; 32: 308-315.

BREADMAKING QUALITY IN TRITORDEUM: THE USE-POSSIBILITIES OF A NEW CEREAL

Juan B. Alvarez, & Luis M. Martín
Departamento de Genética, Escuela Técnica Superior de Ingenieros
Agrónomos y de Montes, Universidad de Córdoba, Córdoba, Spain

Abstract

Tritordeum (X*Tritordeum* Ascherson et Graebner) is the amphiploid derived from the cross between a South American wild barley (*Hordeum chilense* Roem. et Schulz.) and wheat. Because this amphiploid has shown agronomic characteristics of a new crop, we thought to evaluate the possible uses of this new cereal and its role in Agriculture. For this reason, in the last four years, several lines of tritordeum, along with some lines of durum and bread wheat, and of triticale, have been analyzed for breadmaking quality by the tests commonly used for evaluating quality in cereals. The results have indicated that both hexa- and octoploid tritordeum exhibit quality characteristics similar to those of bread wheat and very different to those of durum wheat. Likewise, a wide range for quality characters has been shown between the tritordeum tested lines. Although any lines analyzed has been improved for quality, the hexaploid tritordeum have exhibited baking properties slightly poorer than those of bread wheat. On this basis, we think that the role of tritordeum in the food industry could be similar to that of bread wheat, although the end-use and potential cultivated-zone are yet to be determined.

Introduction

Some part of the genetic variability lost in wheat breeding programmes can be regained by interspecific crosses and chromosomal manipulation. In general, the resulting materials exhibit a tendency to lose quality characteristics [1-2]. On this basis, the search of species which could be used for transferring some interesting characters to wheat without detrimental effects on breadmaking quality, has a great importance in any wide-crosses programme.

Recently, *Hordeum chilense* Roem. et Schulz., a South American wild barley, has been used for obtaining a new amphiploid with durum wheat (*Triticum turgidum* conv. *durum* Desf. em. M.K.), named tritordeum (X*Tritordeum* Ascherson et Graebner) [3-4]. Prior to the production of this amphiploid, Martín and Chapman [5] obtained a partially fertile octoploid form from the cross between *H. chilense* and bread wheat (*T. aestivum* L. em. Thell., cv. 'Chinese Spring'). Later, new octoploid fertile tritordeums were obtained with other bread wheat cultivars [6].

From the beginning, the hexaploid form showed promising characteristics as a new crop [7]. Later studies have confirmed these expectations [6,8]. Cubero *et al* [8] and

H. Guedes-Pinto et al. (eds.), Triticale: Today and Tomorrow, 799–805.

© 1996 *Kluwer Academic Publishers. Printed in the Netherlands.*

Martin [6] indicated that tritordeum could be used as protein source crop. Nevertheless, recent studies have showed that the protein contents between tritordeum and wheat are not significantly different, when their grain yields are similar [9].

The improvement of breadmaking quality is a very important aim in the most of cereal breeding programmes. For this reason, Alvarez *et al* [10] evaluated the breadmaking quality properties of hexaploid tritordeum. They found that the flour of this new cereal exhibited viscoelastic properties very similar to those of medium quality of bread wheat cultivated in Spain.

Because the quality properties of each parent are very different from those of the amphiploid, the search of variability for the quality characters within tritordeum has great importance in the breeding programme. So, the principal objective of the current work has been to investigate the variation of the breadmaking quality characters in tritordeum. Likewise, the end-use quality of this new cereal has been studied.

Materials and Methods

GRAIN SAMPLES

Twelve lines of primary tritordeum, six hexa- and six octoploid, were analyzed, along with their respective wheat parents. These materials were grown in 6-m^2 plots distributed in a completely randomized design with two replications.

F_3 seeds from seven crosses made from HT-31, one of the most yielding line of tritordeum developed in our Department [6], along with their parent HT-31, were analyzed.

Eight experimental lines of hexaploid tritordeum, catalogued for this work as line 1 to line 8, were analyzed, along with one bread wheat cultivar ('Yecora') used as control.

QUALITY EVALUATION

Kernel weight, test weight, and carotene content of flour were determined according to Williams *et al* [11]. Protein content was calculated from the nitrogen content by the Kjeldhal method (%N x 5.7). Dry gluten content was determined by the glutomatic system by ICC standard 137 [12]. SDS-sedimentation assays (SDSS) were performed according to Williams *et al* [11]. However, in the first trial, because of the small amount of grain, these assays were performed as in Peña *et al* [13], with a solution of 2% SDS. Because the protein content has a great influence on the SDSS volume, the quality index, obtained of the relationship between the SDSS volume and protein content, has been used for a better evaluation of protein quality [14]. Alveograph tests were performed using the Chopin method, according to the standard method ICC 121 [12]. Alveograms were evaluated in terms of overpressure (P), average length of the curve (L), and deformation energy (W) [15]. Bread loaves were calculated according to Campaña *et al* [16].

STATISTICAL ANALYSIS

Data were statistically analyzed and the least significant difference among lines was determined according to Steel and Torrie [17]. The multivariate analyses (principal component analysis -PCA- and cluster) were performed according to Kendall [18].

Table 1.- Quality characteristics of the durum and bread wheat parents and their derived tritordeums.

	Durum wheat	Hexaploid tritordeum	Bread wheat	Octoploid tritordeum
Test weight (kg/hl)	85.48 a	74.94 c	80.48 b	73.71 d
Kernel weight (g)	53.29 a	31.34 c	34.79 b	28.91 d
Carotene content (ppm)	4.59 c	9.81 a	3.87 d	7.76 b
Protein content (%)	11.19 d	14.85 a	12.09 c	13.33 b
SDSS volume (ml)	5.27 d	11.19 a	8.54 c	10.46 b
Quality index	0.47 d	0.75 b	0.71 c	0.79 a

Means followed by a same letter are not different significantly at $P > 0.5$.

Results and Discussion

In table 1 the average from the lines of hexa- and octoploid tritordeum and their wheat parents used in the first trial are shown. Data showed that both tritordeums were more similar to bread wheat than to durum wheat. As expected, the typically flat and oblong kernels of tritordeum weighed considerably less than those of bread and durum wheats. Test weight of tritordeums was also lower than those of the wheats, which is related with kernel shape. These differences were higher between the hexaploid tritordeums and their parents than between the octoploid tritordeums and them (table 1).

While Cubero *et al* [8] observed great differences between tritordeum and wheat protein contents, in this study, tritordeum showed protein content values slightly higher than wheat. However, for SDSS volume, the hexaploid tritordeums showed a values considerably higher than their durum wheat parents. These differences were less marked in the case of the octoploid tritordeums and their parents. For this index, the octoploid tritordeums exhibited the highest values, followed by the hexaploid tritordeums, bread wheats, and durum wheats. For carotene content, the hexaploid tritordeums exhibited the highest values, followed by the octoploid tritordeums and durum and bread wheats, respectively.

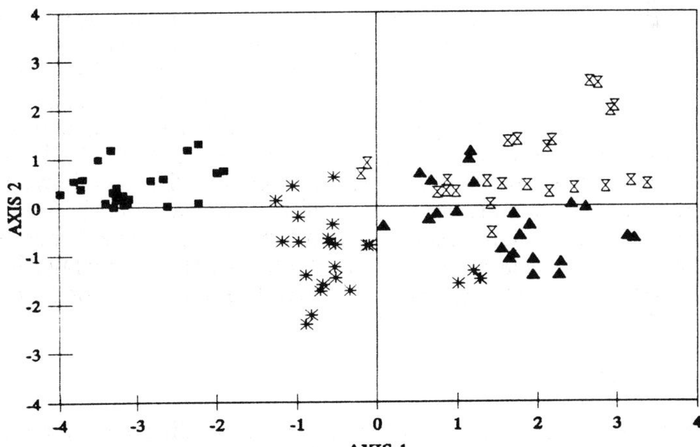

Figure 1.- Principal components analysis -PCA- (✻ Bread wheat; ■ Durum wheat; ✗ Hexaploid tritordeum; and ▲ Octoploid tritordeum).

All these parameters were evaluated in a PCA (fig. 1). There was a sharp separation of durum wheat lines from the rest. It was also observed that the material with H^{ch} genome appeared on the positive side of axis 1 and those with D genome on the negative ones of axis 2. Because the separation between the hexaploid tritordeums and their parents is higher than between the octoploid tritordeums and them, it seems to suggest that the effects of H^{ch} genome are greater in the absence of the D genome.

Table 2.- Quality characteristics of F$_3$ seeds of several crosses of hexaploid tritordeum obtained with HT-31.

	Protein content %	SDSS volume ml	Quality index	P mm	L mm	W x 10^{-4} J
HTC-174	15.70	44.00	2.80	53.10	154.90	138.90
HTC-175	15.90	58.00	3.64	50.40	154.50	143.90
HTC-227	17.10	30.00	1.75	103.50	20.00	72.70
HTC-232	18.50	38.00	2.06	46.50	145.20	99.30
HTC-233	15.80	41.00	2.59	65.40	105.80	147.40
HTC-254	18.50	41.00	2.22	66.50	128.00	131.40
HTC-262	17.40	48.00	2.74	67.50	129.00	160.50
HT-31	10.46	52.75	5.05	51.00	109.20	126.60

Similar results were obtained when several crosses of hexaploid recombinant tritordeum with one common parent (HT-31) were analyzed (table 2; fig. 2). Only the cross HTC-227 showed data different to the common pattern observed in the rest.

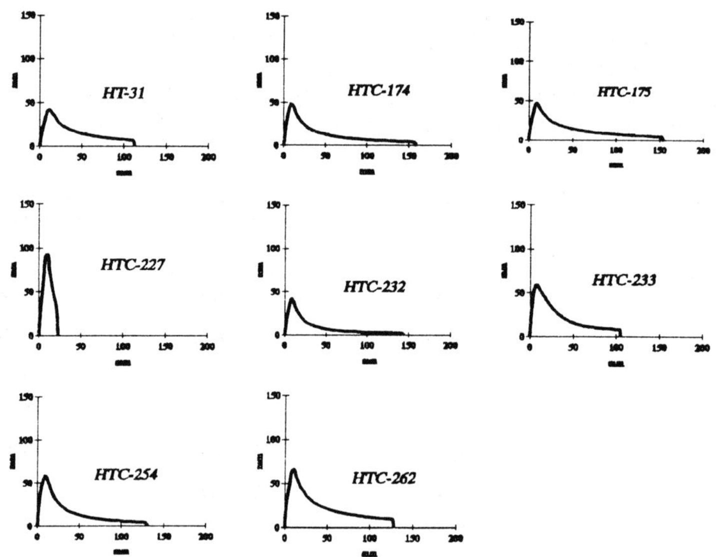

Figure 2.- Alveograms of F$_3$ seeds from several crosses of hexaploid tritordeum and their parent (HT-31).

All parameters evaluated were used in a cluster analysis to group each one of the crosses (fig. 3). This analysis emphasizes the sharp separation among the cross HTC-227 and the rest and the fact of that all crosses appeared to group by the *H. chilense* accession used in the synthesis of their another parent (H-1 for HTC-232; H-7 for HTC-233, HTC-254 and HTC-262; H-8 for HTC-175; H-11 for HTC-227; and H-16 for HTC-174). Because both H-1 and H-7 were involving in the synthesis of HT-31, this line appears between the groups synthesized from both accessions.

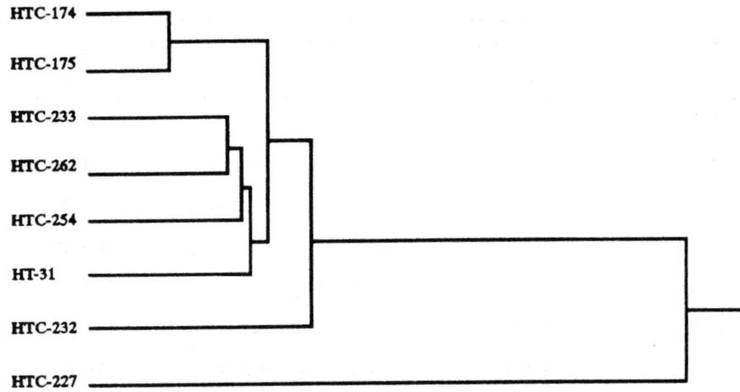

Figure 3.- Cluster analysis of several recombinant lines of hexaploid tritordeum with one common parent.

In general, tritordeum exhibits grain and flour characteristics very similar to bread wheat. Recently, the first results regarding end-use quality have been obtained. Several recombinant lines have been analyzed by baking tests. The data suggest that, in general, the tritordeums present rheological and baking characteristics were slightly poorer than those of bread wheat (table 3.). Although, no effect of the carotene content on baking quality was found, all tritordeum lines exhibited a marked yellow colour in loaf. Also, the interior texture was enough good for the most of tritordeums, with exception of the line 4 (fig. 4).

Table 3.- Baking characteristics of the lines tested.

	Weight g	Volume cm^3	Specific volume cm^3/g
Line 1	136.50	533.00	3.90
Line 2	137.50	445.00	3.24
Line 3	137.50	428.00	3.11
Line 4	132.50	474.00	3.58
Line 5	138.00	347.00	2.51
Line 6	135.00	376.00	2.79
Line 7	125.00	498.00	3.98
Line 8	135.00	426.00	3.15
cv. 'Yecora'	127.50	579.00	4.54

Conclusions

The data suggest the existence of a wide range of variation for some quality characters in the small sample of tritordeums investigated in this study. It seems the most important source of quality variability is the *H. chilense* line used in each cross. On this basis, it seems feasible to obtain a good quality tritordeum, using the broad collection of new synthesised tritordeums and well-established breeding procedures. Furthermore, an important question is that no tritordeum line used in this work, not even the line HT-31, has been deliberately selected for breadmaking quality.

We think that the role of tritordeum in the food industry could be similar to that of bread wheat, although the end-use and potential cultivated-zone of tritordeum are yet to be determined. Likewise, since *H. chilense* has shown few detrimental effect on breadmaking quality, it could be useful in transferring some interesting characters to wheat, at the same time serving to widen the genetic background of cereals.

Figure 4.- Cross-sections of loaves from several lines of hexaploid tritordeum (1-8) and one bread wheat cultivar (wheat= 'Yecora').

Acknowledgments

This work was supported by a grant No. AGR91-0844 from Comisión Interministerial de Ciencia y Tecnología (CICYT) of Spain. The first author thanks the FPI programme of the Spanish MEC for a predoctoral fellowship. We wish to thank to Prof. A. Martin and Dr. J. Ballesteros from CSIC (Spain) for the seed materials. Also, we want to thank to the ICI (Spain) for a Cooperation Project.

References

1.- Graybosch RA, Peterson CJ, Hansen LE, Mattern PJ. Relationships between protein solubility characteristics, 1BL/1RS, high molecular weight glutenin composition, and end-use quality in winter wheat germ plasm. Cereal Chem. 1990; 67: 342-9.

2.- Graybosch RA, Peterson CJ, Hansen LE, Worrall D, Shelton DR, Lukaszewski A. Comparative flour quality and protein-characteristics of 1BL/1RS and 1AL/1RS wheat-rye translocation lines. J. Cereal Sci. 1993; 17: 95-106.

3.- Martin A, Sanchez-Monge Laguna E. A hybrid between *Hordeum chilense* and *Triticum turgidum*. Cereal Res. Comm. 1980; 8: 349-53.

4.- Martin A, Sanchez-Monge Laguna E. Cytology and morphology of the amphiploid *Hordeum chilense* x *Triticum turgidum* conv. *durum*. Euphytica 1982; 31: 262-7.

5.- Martin A, Chapman V. An hybrid between *Hordeum chilense* and *Triticum aestivum*. Cereal Res. Comm. 1977; 5: 365-8.

6.- Martin A. Tritordeum: the first ten years. Rachis 1988; 7: 12-5.

7.- Martin A, Cubero JI. The use of *Hordeum chilense* in cereal breeding. Cereal Res. Comm. 1981; 9: 317-23.

8.- Cubero JI, Martin A, Millan T, Gomez-Cabrera A, de Haro A. Tritordeum: a new alloploid of potential importance as a protein source crop. Crop Sci. 1986; 26: 1186-90.

9.- Ballesteros J. Base genética del contenido proteico en tritórdeo. Ph. D. thesis, Universidad de Córdoba, Spain. 1993.

10.- Alvarez JB, Ballesteros J, Sillero JA, Martin LM. Tritordeum: a new crop of potential importance in the food industry. Hereditas 1992; 116: 193-7.

11.- Williams P, El-Haramein FJ, Nakkoul H, Rihawi S. Crop quality evaluation. Methods and guidelines. Technical Manual n° 14. 2nd rev ed. Aleppo. Syria: ICARDA, 1988.

12.- ICC. Standard Methods of the International Association of Cereal Chemists. Viena: The Association, 1986.

13.- Peña RJ, Amaya A, Rajaram S, Mujeeb-Kazi A. Variation in quality characteristics associated with some spring 1B/1R translocation wheat. J. Cereal Sci. 1990; 12: 105-12.

14.- Halverson J, Zeleny L. Criteria of wheat quality. In: Pomeranz Y, editor. Wheat: chemistry and technology, 3rd ed. St. Paul. MN: Am. Assoc. Cereal Chem., 1988: 15-45.

15.- Faridi HA, Rasper VF. The Alveograph Handbook. 2nd ed. St Paul. MN: Am. Assoc. Cereal Chem., 1987.

16.- Campaña LE, Sempé ME, Filgueira RR. Physical, chemical, and baking properties of wheat dried with microwave energy. Cereal Chem. 1993; 70: 760-2.

17.- Steel RGD, Torrie JH. Principles and procedures of statistics. A biometrical approach. New York: McGraw-Hill, Inc., 1980.

18.- Kendall MG. A course in multivariate analysis. London: Charles Griffin and Company Limited, 1972.

INFLUENCE OF FEEDING GROWING-FINISHING PIGS TRITICALE, WHEAT OR MAIZE BASED DIETS ON RESULTING CARCASS COMPOSITION, AND ON TASTE AND QUALITY CHARACTERISTICS OF PORK

Joel H. Brendemuhl, Robert O. Myer and Dwain D. Johnson
Animal Science Department, University of Florida, Gainesville, Florida, USA

Abstract

A study was conducted to determine the effect of feeding triticale (*X Triticosecale Wittmack*), soft red winter wheat (*Triticum aestivum*) or maize (*Zea mays L.*) based diets to growing-finishing pigs (25 to 100 kg) on resulting carcass composition, and on taste and quality of pork. Dietary grain source did not influence ($P > .05$) carcass lean content or quality characteristics of the longissimus muscle, such as amount of marbling, muscle texture, firmness or color. Slightly lower levels of total ($P < .05$) polyunsaturated fatty acids were noted in the backfat from pigs fed triticale or wheat based diets compared to that from pigs fed the maize based diets. Juiciness, flavor and tenderness of broiled loin chops did not differ ($P > .10$) due to dietary grain source. The feeding of triticale based diets resulted in small changes in carcass fat composition, however, there was no evidence that the taste of broiled loin chops was different from the taste of broiled loin chops from pigs fed maize based diets.

Introduction

Replacement of common feedstuffs with alternatives for pig feeding requires the assessment of not only pig performance, but also of resulting carcass characteristics and meat quality. A review by Melton[1] indicated that the feeding of diets with different feed grains to pigs affected pork flavor in some studies but not in others. Few studies have been conducted to determine whether or not maize replacement by triticale or soft red winter wheat in pig diets would influence pork flavor and quality. Therefore, this study was conducted to determine the effect of feeding triticale, soft red winter wheat or maize based diets to growing-finishing pigs (25 to 100 kg) on resulting carcass composition, and on taste and quality of pork.

Materials and Methods

Two trials were conducted each involving 18 crossbred pigs. The trials were of similar design but each utilized grain from a different crop year. Each trial involved direct comparisons of nutritionally adequate diets in which triticale, soft red winter wheat or maize was the main feedstuff. The diets within the trials were formulated to be equivalent in lysine content; soybean meal was the supplemental protein source.

H. Guedes-Pinto et al. (eds.), Triticale: Today and Tomorrow, 807–811.

© 1996 *Kluwer Academic Publishers. Printed in the Netherlands.*

Composition of the diets and a description of the grain used is given elsewhere in these proceedings[2].

The pigs were fed ad libitum the experimental diets from 25 to 100 kg average body weight per pig. After the feeding phases were terminated, all pigs were slaughtered. One pig was condemned at slaughter from the second trial for reasons unrelated to treatment. The results, therefore, are presented for 35 pigs. After chilling, each carcass was evaluated for fat firmness, backfat thickness, and longissimus muscle area, color, marbling and firmness using standard procedures[3].

Fatty acid composition was determined on subcutaneous fat (outside fat covering) obtained opposite the 8th - 10th rib section. Each subcutaneous fat sample was removed from the carcass to include all layers, and was vacuum packaged and frozen. Sample preparation and fatty acid analysis was performed as outlined previously[4]. Fatty acids were expressed as a percentage of the total sample.

A section of the loin (2nd to 10th rib) was removed from each carcass and frozen for future taste panel evaluations. For the evaluations, the loin section was thawed and divided into chops. The chops were broiled (75° C) and evaluated using a trained taste (sensory) panel following recommended procedures[5]. The taste panel evaluated the chops for flavor, juiciness, tenderness and off-flavor.

Carcass and taste panel data were determined on a per pig basis. Data were analyzed by analysis of variance procedures[6]. Hot carcass weight was used as a co--variate.

Results and Discussion

Carcass characteristics for pigs fed triticale, wheat or maize based diets are presented in Table 1. Although pigs were fed diets with grain from different crop years, no year affect was noted and thus the data for the two trials were pooled. Due to differences in slaughter weight, both within trial and between trials, hot carcass weight was used as a covariate in the statistical analysis of the data.

Table 1. Carcass characteristics of growing-finishing pigs fed diets containing triticale (TRIT), wheat or maize[a].

Item	Dietary Grain Source			
	TRIT	WHEAT	MAIZE	SE[b]
Avg backfat[ch], cm	3.1	2.9	2.9	0.1
Longissimus muscle area[ch], cm^2	36	36	36	0.2
Lean color score[dh]	2.3	2.4	2.8	0.2
Marbling score[eh]	3.4	3.9	4.0	0.4
Lean firmness score[fh]	3.5	3.1	2.8	0.2
Lean texture score[gh]	3.3	2.9	2.6	0.3

[a]Each mean is based on information from 35 animals; [b]Standard error; [c]Adjusted to a hot carcass weight of 78.9 kg; [d]Scores: 1 to 5; 2 = gray; 3 = light pink; 4 = reddish pink; [e]Scores: 1 to 10; 3 = traces; 4 = slight; 5 = small; 6 = modest; [f]Scores: 1 to 5; 2 = firm; 3 = slightly firm; 4 = slightly soft; [g]Scores: 1 to 5; 2 = fine; 3 = slightly fine; 4 = slightly coarse; [h]Means do not differ (P >.05).

Dietary treatment was found to have no effect (P >.05) on resulting average backfat thickness or longissimus muscle area of the carcasses. Dietary treatment also had no influence (P >.10) on the color, marbling, firmness or texture of the longissimus muscle.

Taste evaluations for broiled loin chops from each pig are summarized in Table 2. Loin chops were evaluated for flavor intensity, juiciness, tenderness, the presence of off-flavors and percent cooking loss. Dietary treatment was found to have no effect (P >.05) on any of the above criteria. The lack of effect of dietary treatment on palatability evaluations of the broiled loin chops agrees with that previously reported from trials with pigs that were fed triticale or wheat based diets[1]. However, Melton[1] indicates that limited research has been conducted concerning the effects of different cereal grains in pig diets on resulting pork flavor. Research with male breeder turkeys[7] has shown that cooked breast meat from turkeys fed diets containing just triticale and supplemental vitamins and minerals was more tender than cooked breast meat from turkeys fed either maize-soybean meal or triticale-soybean meal diets supplemented with vitamins and minerals. Although not significantly different (P >.10), broiled loin chops from pigs in the present study fed triticale based diets were judged slightly more tender than those from pigs fed wheat or maize.

Table 2. Taste evaluations of broiled loin chops from pigs fed diets containing triticale (TRIT), wheat or maize[a].

| | Dietary Grain Source | | | |
Item	TRIT	WHEAT	MAIZE	SE[b]
Flavor intensity, avg score[ce]	5.0	5.1	5.0	0.1
Juiciness, avg score[ce]	4.4	4.2	4.3	0.3
Tenderness, avg score[ce]	5.7	5.3	5.3	0.3
Off-flavors, avg score[de]	5.7	5.8	5.8	0.1
Cooking loss, %[e]	30	29	32	1.0

[a]Each mean is based on information from 35 animals.

[b]Standard error.

[c]Scores: 1 to 8 scale with the trait increasing with an increase in score.

[d]Scores: 1 to 6 scale with a less intense off-flavor with an increase in value.

[e]Means do not differ (P >.05).

Carcasses from pigs fed maize diets had softer fat (P <.01) than carcasses from pigs fed triticale or wheat based diets (Table 3). Fat firmness was similar (P >.10) in those carcasses from pigs fed either triticale or wheat. The softer fat observed in the maize fed pigs is the result of a higher fat (oil) content in the maize diets (3.1 %), as compared to the triticale (1.2 %) or wheat (1.3 %), and the higher degree of unsaturation in the fat (oil) of maize vs triticale or wheat. Although carcasses from the maize fed pigs were softer, it was not detrimental to carcass quality.

Fatty acid profile of the carcass fat is presented in Table 3. The fatty acid data were adjusted to an equal hot carcass weight of 78.9 kg and data from the two trials were pooled. Carcass fat from maize fed pigs contained more (P <.01) polyunsaturated, but less (P <.01) saturated fatty acids than fat from triticale or wheat fed pigs.

Table 3. Fatty acid analysis and fat firmness of carcass fat from pigs fed diets containing triticale (TRIT), wheat or maize[ab].

Item	Dietary Grain Source			
	TRIT	WHEAT	MAIZE	SE[c]
Fatty acids[d], %				
C14:0	1.5[k]	1.6[l]	1.4[j]	0.02
C16:0	32.7[n]	32.6[n]	31.4[m]	0.13
C16:1	2.0[k]	2.2[l]	1.8[j]	0.06
C18:0	13.8	14.0	13.7	0.19
C18:1	44.7[j]	45.2[k]	45.4[k]	0.22
C18:2	3.2[n]	2.6[m]	4.2[o]	0.11
Total saturates[e], %	48.3[n]	48.5[n]	46.7[m]	0.26
Total monounsaturates[f], %	47.5[j]	48.1[k]	47.9[j]	0.23
Total polyunsaturates[g], %	4.2[n]	3.4[m]	5.4[o]	0.16
Sat:Unsat ratio[h]	.94[n]	.95[n]	.88[m]	0.01
Avg fat firmness score[i]	1.8[n]	1.7[n]	2.8[m]	0.20

[a]Each mean is based on information from 35 animals.

[b]Least-square means were adjusted to an equal hot carcass weight of 78.9 kg.

[c]Standard error.

[d]Fatty acid analyses were done on backfat samples. Fatty acids are reported as a percentage of total fatty acids.

[e]Total of 12:0, 14:0, 16:0, 18:0, 20:0, 22:0, and 24:0.

[f]Total of 14:1, 16:1, 18:1, 20:1, 22:1, and 24:1.

[g]Total of 18:2, 18:3, 20:2, 20:3, and 20:4.

[h]Ratio of total saturates to total unsaturates (mono and poly).

[i]Scores: 1 to 4; 1 = firm; 2 = slightly soft; 3 = soft; 4 = very soft, oily.

[j,k,l]Means in the same row with a different superscript differ (P <.06).

[m,n,o]Means in the same row with a different superscript differ (P <.01).

Fatty acid profile of the carcass fat is presented in Table 3. The fatty acid data were adjusted to an equal hot carcass weight of 78.9 kg and data from the two trials were pooled. Carcass fat from maize fed pigs contained more (P <.01) polyunsaturated, but less (P <.01) saturated fatty acids than fat from triticale or wheat fed pigs. Pigs fed wheat diets had carcass fat that was higher (P <.06) in monounsaturated, but lower (P <.01) in polyunsaturated fatty acids than triticale or maize fed pigs. Carcass fat from triticale fed pigs was intermediate in percentage of polyunsaturated fatty acids as compared to that obtained from wheat or maize fed pigs, percentage of saturated fatty acids was similar to that from wheat fed pigs, and percentage of monounsaturated fatty acids was similar to that from maize fed pigs. The higher percentage of polyunsaturated fatty acids in carcass fat from maize fed pigs may explain the softer fat observed in those carcasses. Fatty acid profiles of the carcass fat in our study are partially in agreement with that reported by Melton[1] for triticale and maize fed pigs. In our study, carcass fat from triticale fed pigs had more (P <.01) palmitic (16:0) and palmitoleic (16:1) acids than carcass fat from maize fed pigs, but less (P <.01) linoleic (18:2) and similar (P >.10) stearic (18:0) acid percentages. Research reported by Melton[1] noted that backfat from pigs fed triticale diets had increased percentages of palmitoleic and linoleic acids, but reduced percentages of palmitic and stearic when compared with backfat from maize fed pigs. Differences in pig rearing climate, fatty acid analysis methodology and fat sampling could account for the differences observed.

Overall, feeding pigs triticale or wheat based diets as compared to maize based diets had little or no effect on carcass characteristics, carcass fat fatty acid profile or the palatability of broiled loin chops. Our research further supports that triticale is an effective replacement for maize in diets for growing-finishing pigs.

References

1. Melton SL. Effects of feeds on flavor of red meat: A review. J Anim Sci 1990;68:4421-4435.
2. Myer RO, Brendemuhl JH, Barnett RD. Synthetic amino acid supplementation of triticale and wheat based diets for growing-finishing pigs. Proceedings of the Third International Triticale Symposium; 1994 June 13-17; Lisbon.
3. National Pork Producers Council (NPPC). Procedures to evaluate market hog performance. 3rd rev. ed. Des Moines: NPPC, 1991: 16.
4. Myer RO, West RL, Gorbet DW, Knauft DA, Young CT. Performance and carcass characteristics of swine as affected by the consumption of peanuts remaining in the field after harvest. J Anim Sci 1985;61:1378-1386.
5. American Meat Science Association (AMSA). Guidelines for cooking and sensory evaluation of meat. Chicago: AMSA, 1978.
6. SAS. SAS User's Guide: Statistics. Cary: SAS Institute Inc., 1985.
7. Savage TF, Holmes ZA, Nilpour AH, Nakaue HS. Evaluation of cooked breast meat from male breeder turkeys fed diets containing varying amounts of triticale, variety flora. Poul Sci 1987;66:450-452.

SYNTHETIC AMINO ACID SUPPLEMENTATION OF TRITICALE AND WHEAT BASED DIETS FOR GROWING-FINISHING PIGS

Robert O. Myer, Joel H. Brendemuhl and Ronald D. Barnett
Animal Science and Agronomy Departments, University of Florida, Gainesville, Florida, USA

Abstract

Dietary supplementation of feed grade synthetic amino acids may be more advantageous for diets based on triticale (*X Triticosecale Wittmack*) or soft red winter wheat (*Triticum aestiuum*) than maize (*Zea mays L.*) for growing and finishing pigs (25 to 109 kg). Four trials, involving 190 pigs, were conducted to evaluate the effectiveness of supplementation of triticale and wheat based diets with synthetic lysine and threonine for growing and finishing pigs. Substantial replacement of supplemental protein (soybean meal) with synthetic lysine and threonine did not affect pig growth or carcass lean content (P>.10). In fact, complete or nearly complete replacement of soybean meal occurred for finishing pig (55 to 109 kg) diets formulated with triticale.

Introduction

Triticale (*X Triticosecale Wittmack*) and soft red winter wheat (*Triticum aestiuum*) are good alternatives to maize (*Zea mays L.*) for pig diets. Triticale, and to a lesser extent soft red wheat, contain higher levels of amino acids than maize. When these grains are utilized in pig diets, instead of maize, the amount of a supplemental protein source(s), such as soybean meal, required to meet the pig's amino acid requirements is reduced. Synthetic feed grade amino acids, such as L-lysine HCl, are becoming more available commercially and are also increasingly cost effective. Since triticale and wheat are high in amino acids, it may be possible by selective supplementation of one or two limiting amino acids to reduce the amount of or eliminate supplemental protein when these grains are used as the primary energy feedstuff in growing and finishing pig diets. The first two limiting amino acids (lysine and threonine) in triticale and wheat based swine diets are well defined[1]. Therefore, supplementation with these amino acids is relatively simple.

This study was conducted to evaluate the effectiveness of supplementation of triticale and wheat based diets with synthetic lysine and threonine for growing and finishing pigs (25 to 109 kg).

Materials and Methods

Four feeding trials involving a total of 190 pigs were conducted, two at Gainesville (trials 1 and 3) and two at Marianna (trials 2 and 4). Two different crop years of triticale (mixture of `Florida 201', `Florico', and `Sunland') and wheat (mixtures of

813

H. Guedes-Pinto et al. (eds.), Triticale: Today and Tomorrow, 813–817.
© 1996 *Kluwer Academic Publishers. Printed in the Netherlands.*

three cultivars each year) were used (year 1, trials 1 and 2; and year 2, trials 3 and 4). Representative samples of each crop were analyzed for moisture, crude protein and amino acids by conventional procedures.

Diets in which triticale or wheat served as the main feedstuff were supplemented with soybean meal to meet the requirement of the first (lysine) or third (isoleucine) limiting amino acid. Diets formulated to the third limiting amino acid were also supplemented with feed grade L-lysine HCl and L-threonine to meet the lysine and threonine requirements. Levels of other nutrients in the diets (i.e., minerals, vitamins) met or exceeded National Research Council[1] recommendations. A maize-soybean meal diet was also included as a control treatment giving a total of five dietary treatments. Composition of the experimental diets is given in Table 1.

Table 1. Composition of experimental diets, %.

Treatment	Ground grain Yr1[a]	Ground grain Yr2	Soybean meal (48%) Yr1	Soybean meal (48%) Yr2	L-Lys HCl Yr1	L-Lys HCl Yr2	L-Thr Yr1	L-Thr Yr2	Other[b] Yr1	Other[b] Yr2
				Grower diets (25 to 55 kg)						
Trit-soy	81.0	81.0	16.0	16.0	--	--	--	--	3.0	3.0
Trit-lys, thr	88.6	89.6	8.0	7.0	0.3	0.3	0.1	0.1	3.0	3.0
Wheat-soy	78.5	80.0	18.5	17.0	--	--	--	--	3.0	3.0
Wheat-lys, thr	87.6	89.6	9.0	7.0	0.4	0.4	0.1	0.1	3.0	3.0
Maize (control)	77.0	77.0	20.0	20.0	--	--	--	--	3.0	3.0
				Finisher diets (55 to 109 kg)						
Trit-soy	87.2	88.2	10.0	9.0	--	--	--	--	2.8	2.8
Trit-lys, thr	94.8	96.7	2.0	0	0.3	0.4	0.1	0.1	2.8	2.8
Wheat-soy	85.2	87.2	12.0	10.0	--	--	--	--	2.8	2.8
Wheat-lys, thr	93.7	94.8	3.0	2.0	0.4	0.3	0.1	0.1	2.8	2.8
Maize (control)	83.2	83.2	14.0	14.0	--	--	--	--	2.8	2.8

[a]Year 1 and year 2 crops, respectively, of grain used in the diets.
[b]Mineral and vitamin supplements (dicalcium phosphate, calcium carbonate, salt, pig vitamin supplement, pig trace mineral supplement, and antibiotic supplement).

Trial 1 and 3 involved 40 and 30 crossbred pigs, respectively, with an average initial weight of 25 kg. Pigs were assigned to pens of two pigs each and each of the five treatments were fed to four (trial 1) or three (trial 3) pens. Trials 2 and 4 each involved 60 crossbred pigs with an average initial weight of 28 kg. Pigs were assigned to pens of six pigs each and each treatment was assigned to two pens. The pigs were fed the experimental diets until they reached a pen average of 100 (trials 1 and 3) or 109 (trials 2 and 4) kg per pig. Feed and water were offered ad libitum. Upon termination of the feeding phase, all pigs were slaughtered. After chilling, each carcass was measured for backfat thickness and longissimus muscle area, and carcass lean content was calculated[2].

Performance (average daily gain, average daily feed intake and feed-to-gain ratio) and carcass data were determined on a per pen basis. Data were analyzed by analysis of variance procedures for a randomized complete block design. Orthogonal contrasts were done to determine significant differences due to grain type, amino acid supplementation and comparison to control.

Results and Discussion

The average nutrient composition of the triticale, soft red winter wheat and maize used in the present research is presented in Table 2. The triticale utilized was a mixture of three cultivars recently released for grain production in the southeastern United States. These triticales, like most modern triticales, are lower in crude protein and most amino acids than older cultivars of triticale. The above three cultivars, however, like other newer triticales, were plumper and of heavier test weight. The triticale averaged 35 and 2% more crude protein than the maize and soft wheat, respectively, and 52 and 15% more lysine than the maize and soft wheat, respectively. The concentrations of lysine and other nutrients of the triticale used were in general agreement with those summarized by the National Research Council[3] for the newer cultivars of triticale.

Table 2. Average nutrient composition (%) of triticale and soft red winter wheat[a].

Grain	Crude protein	Lysine	Threonine	Isoleucine	Met + Cys
Triticale[b]:					
Year 1	10.5	0.38	0.34	0.35	0.42
Year 2	11.6	0.41	0.39	0.38	0.46
Soft red wheat[b]:					
Year 1	9.4	0.29	0.31	0.32	0.42
Year 2	12.3	0.40	0.39	0.41	0.49
Maize[c]	8.2	0.26	0.30	0.26	0.36

[a]Values are expressed on an 88% dry matter basis.

[b]Average of two analyses for each year's crop.

[c]Average of four analyses.

Performance and carcass lean content data from the growing-finishing pig trials are summarized in Table 3. Data from both Gainesville trials (trials 1 and 3) were pooled since statistical analysis indicated that crop year had no effect (P>.10) on the feeding value of the triticale or wheat for the pigs. Similarly, data from the Marianna trials (trials 2 and 4) were also pooled.

For both Gainesville and Marianna trials, pigs fed the triticale or wheat based diets with reduced supplemental soybean meal and supplemented with synthetic lysine and threonine had overall average daily gains and feed-to-gain ratios that were not different (P>.10) from those obtained with pigs fed the triticale or wheat based diets with full soybean meal supplementation. These performance figures were similar to (P>.10) those of pigs fed the maize control diets. Percent carcass lean content was also not affected (P>.10) by amino acid supplementation of the diets fed.

Pigs fed the triticale based diets had average daily gains that were similar (P>.10) to that of pigs fed the wheat based diets or the maize controls in all trials; however, slightly more feed (3 to 5% more) was required per unit of weight gain in comparison to pigs fed the wheat based diets (P<.10) (Table 3). Feed-to-gain ratios of the triticale fed pigs were not different (P>.10) from that of pigs fed the maize control diets. Carcass lean content was not affected (P>.10) by dietary grain source in any of the trials.

Table 3. Performance and carcass lean content of growing-finishing pigs fed triticale or wheat based diets supplemented with synthetic amino acids[a].

| | Treatment | | | | | |
| | Triticale | | Soft red wheat | | Maize-Soy | |
Item	Soy[b]	aa's[c]	Soy	aa's	(control)	SE[d]
Trials 1 and 3 (Gainesville):						
Avg. daily gain, kg	0.83	0.82	0.79	0.83	0.83	0.03
Avg. daily feed intake, kg	2.88	2.91	2.72	2.64	2.89	0.08
Feed/unit gain[e], kg/kg	3.49	3.53	3.49	3.19	3.51	0.10
Carcass lean[f], %	50.6	49.5	51.1	51.1	48.8	0.08
Trials 2 and 4 (Marianna):						
Avg. daily gain, kg	0.94	0.93	0.94	0.94	0.94	0.02
Avg. daily feed intake, kg	3.09	3.04	2.97	3.02	3.00	0.06
Feed/unit gain[e], kg/kg	3.30	3.29	3.17	3.20	3.20	0.05
Carcass lean[g], %	44.1	44.4	43.4	43.7	44.5	0.07

[a]Seven pens per treatment with two pigs per pen; on experiment from 25 to 100 kg average body weight (Gainesville trials). Four pens per treatment with six pigs per pen; on experiment from 28 to 109 kg average body weight (Marianna trials).

[b]Soybean meal (48%) was the supplemental protein source.

[c]Synthetic lysine and threonine used to meet the requirements of the first and second limiting amino acids in the diet.

[d]Standard error; n = 7 (Gainesville trials) or 4 (Marianna trials).

[e]Triticale vs. wheat (P<.10).

[f]Adjusted to 100 kg end weight.

[g]Adjusted to 109 kg end weight.

The results indicated that a substantial amount or complete elimination of a supplemental protein source is possible when wheat or triticale is used as the grain source and supplemented with lysine and threonine in diets for growing and finishing pigs (Table 1). Previous researchers found they could eliminate supplemental protein if they added a small amount of L-lysine HCl when triticale was the grain source in diets for finishing pigs (55 to 105 kg)[4,5,6]. However, these studies utilized triticales that were higher in protein than the triticale used in the present study. The results of our present research indicated that supplementation of both lysine and threonine would be required with the new lower protein cultivars of triticale.

Recent research has shown that there may be a limit to the amount of natural protein that can be replaced with synthetic amino acids in pig diets[7,8]. There is a requirement for non-specific nitrogen that may not be met when most or all of the supplemental protein (intact protein) is replaced with synthetic amino acids. The above researchers[7,8] used lower protein cereal grains (maize and grain sorghum) and thus less intact protein was present in their diets. These diets, therefore, may have been marginally deficient in non-specific nitrogen which may explain the depressed growth of the pigs. In our study, the higher natural protein contents of triticale and soft red wheat, as compared to maize and grain sorghum, would have supplied more non-specific nitrogen which is supported by the similar pig growth across treatments.

References

1. National Research Council (NRC). Nutrient Requirements of Swine. 9th rev. ed. Washington: National Academy Press, 1988: 93.
2. National Pork Producers Council (NPPC). Procedures to evaluate market hog performance. 3rd rev ed. Des Moines: NPPC, 1991: 16.
3. National Research Council (NRC). Triticale: A promising addition to the world's cereal grains. Washington: National Academy Press, 1989: 103.
4. Hale OM, Morey DD, Myer RO. Nutritive value of Beagle 82 triticale for swine. J Anim Sci 1985;60:503-510.
5. Myer RO, Barnett RD. Triticale (`Beagle 82') as an energy and protein source in diets for starting and growing-finishing swine. Nutr Rep Internat'l 1985;31: 181-190.
6. Coffey MT, Gerrits WJ. Digestibility and feeding value of B858 triticale for swine. J Anim Sci 1988;66:2728-2735.
7. Davis DJ, Brendemuhl JH, Myer RO, Walker WR, Combs GE, Jr. Amino acid supplementation of low protein corn-soybean meal diets for 7-25 kg swine. J Anim Sci 1991;69(suppl. 1):21 (abstract).
8. Hansen JA, Knabe DA, Burgoon KG. Amino acid supplementation of low protein sorghum-soybean meal diets for 20 to 50 kilogram swine. J Anim Sci 1993;71:442--451.

NUTRITIVE VALUE AND FEED EFFICIENCY OF BROILER DIETS CONTAINING DIFFERENT LEVELS OF TRITICALE

EL-Yassin Fayez , Haj - Omar Nedal and Abboud Mousa
Dep.of Animal Prod . , Fac. of Agriculture,
University of Aleppo , Syria

Abstract

Two metabolism trials were carried out with chickens   to study the nutritive value of triticale grains . In the 1 st trial the experimental diets consisted of one of four triticale cultivars as well as maize , wheat and barley grains . In the 2nd trial iso - nitrogenous and iso - energetic diets containing 0 , 35 , 50 and 65 % triticale from either Beagle or Drira outcross 07 cultivars were introduced at the level of 100 g / bird daily .

Diets containing 0 , 35 , 50 , 65 , and 80 % triticale were compared in two feeding experiments with a diets containing maize during starting and finishing periods (1 - 48 days) or during finishing period only (22 - 48 days) .

Results from the metabolism trials on chicken showed no significant differences between triticale cultivars in digestibility , N - balance and N - intake . Nitrogen retention was significantly higher (P < 0.05) for both wheat and triticale compared with those of barley and maize , due to a higher N - intake . Adding triticale to the diet at a concentration higher than 35 % lowered the digestibility and N - retention (P < 0.05) in comparison with the control treatment . among diets containing triticale , the best N - balance and daily live weight gain were observed in group fed on a diet containing 35 % triticale grains .

A decrease in productive parameters was observed when triticale was added at a rate higher than 50 % . The best results in both experiments were obtained when triticale rate in the diet was 35 %

Introduction

Preliminary studies have shown the feasibility of growing triticale in Syria under the same condition where wheat and barley are grown (ICARDA , Ann.,Rep.,!984) . The importance of triticale as a feed crop lies in the fact that its grains can be used as replacement for maize in poultry diets . Therefore , the development of triticale growing in Syria would lead to a real saving in the consumption .

H. Guedes-Pinto et al. (eds.), Triticale: Today and Tomorrow, 819–826.

© 1996 *Kluwer Academic Publishers. Printed in the Netherlands.*

The University of Aleppo in cooperation with IDRC have carried out research project relating the development of triticale growing in Syria .

The study reported here is a part of this project . the specific objective of this study was to determine the nutritive value of the available four triticale cultivars in Syria and the possibility of including triticale grains in broiler diets.

Materials and Methods

METABOLISM TRIALS :

Two trails were carried out with thirty five 30 days old broiler chickens(Lohman) . In both trials , the birds were kept in individual cages and randomly divided into 7 groups . In the 1 st trial , each experimental group received one feed diet (ad - lib) during a 7 days preliminary and a 7 days collection period . The seven experimental diets were :

1. Triticale Beagle (Be) , 2. Triticale Drira outcross 07 (Dr) , 3. Triticale Dolphine (Do) ,

4. Triticale Juanillo (Ju) , 5. Maize , 6. Wheat , 7. Barley .

In the 2 nd trial , iso - nitrogenous , iso - energetic diets containing 0 , 35 , 50 , and 65 % triticale from either Beagle and Drira outcross 07 cultivars were used (Table 1) . The experimental diets were introduced at the level of 100 g / bird daily . Samples from diets and excreta were taken and analyzed for dry matter (DM) , organic matter (OM) , ash and nitrogen (N) .

FEEDING TRIALS :

Two feeding trials were conducted at the Moslemia Research Center to study the utilization of triticale by broilers .In the 1st trial total of 625 unsexed 1 - day old broiler chickens (Lohman) were randomly divided into 5 groups each of 125 chickens with a maximum difference in the initial mean body weight 2.1 g.

The 1 st group (control) received a diet composed of maize , soybean meal (SBM 44%cp) and Super Concentrate . Other groups received diets (Table 2) in which part of the maize and SBM were replaced with four levels of triticale grain (Mixture of 50 % Drira cv and 50 % Beagle cv) . Two feeding periods were applied , a starting (S , 21 days long) and a finishing (F . 28 days long) .

" Table 1 " Composition of the diets used in the 2 nd metabolism trial (%)

Exp. diets	Maize	SBM	Super	Triticale	
				Drira	Beagle
1. control	70	20	10	0	0
2	42	13	10	35	0
3	30	10	10	50	0
4	18	7	10	65	0
5	42	13	10	0	35
6	30	10	10	0	50
7	18	13	10	0	65

In the 2nd trial , 4 groups of 125 chicken each were fed the same control diet during the starting period . In the 2nd period (F) they received the experimental diets containing 0 , 35 , 50 , and 65 % triticale . The composition of the diets was similar to those used in the 1 st trial (Table 2) . Chickens had a 24 Hours free access to feed and water . Feed ingredients were milled but not pelted . Feed consumption and live weight were weekly recorded .

" Table 2 " Composition (% Fresh weight) ,metabolizable energy (Mcal / kg DM) and crud protein content (CP %) of experimental diets

Item		Experimental diets				
		1- control	2	3	4	5
Maize	S	67.0	36.00	22.50	9.00	0.0
	F	73.0	41.00	27.25	13.50	0.0
Triticale	S	00.0	35.00	50.00	65.00	75.0
	F	00.0	35.00	50.00	65.00	80.0
S B M	S	23.0	19.00	17.50	16.00	15.0
	F	17.0	14.00	12.75	11.50	10.0
Supper	S	10.0	10.00	10.00	10.00	10.0
	F	10.0	10.00	10.00	10.00	10.0
M E	S	3.13	3.15	3.15	3.15	3.2
	F	3.20	3.20	3.20	3.20	3.1
C P	S	21.7	21.50	21.50	21.50	20.5
	F	19.8	19.90	19.90	20.00	19.8

Data were subjected to analysis of variance and the differences between treatment means were tested for significance by Duncan s new multiple range test (Steel and Torrie ,1960)

Results and Discussion

METABOLIC TRIALS :

The chemical analysis of the grains used in the 1 st metabolic trial showed that crud protein level was 17.4 , 12.4 and 9.7 % for wheat , barley and maize, respectively , where it ranged between 10.1 - 14.1 % for triticale cultivars (Table 3) . Among triticales the higher crude protein content was for Dolphine cv . Large variation in crude protein content of triticales have also been reported by others (Zillensky and Borloug, 1971;Villegos et al ,1970; Bushuk, 1980) .
Data relating to the intake and digestibility of DM and OM of the different diets is shown in 9 (Table 4) . It is clear that the DM and OM intake was higher (P < 0.05) for the groups fed the triticale diets compared to other groups . Significant differences in this respect were also observed (P < 0.05) between the groups which received maize , wheat and barley .

The DM and   OM digestibility was significantly higher for the maize and lower for the barley diets (P < 0.05) compared to all other diets . However , no significant differences were

" Table 3 " The chemical analysis of grain used in the 1 st metabolism trial (% of DM)

Item				Triticale cultivar			
	Maize	Wheat	Barley	Be	Dr	Do	Ju
Dry matter	89.2	90.2	90.2	90.6	90.8	90.5	90.3
Organic matter	97.8	97.9	96.9	97.7	97.9	97.8	97.9
Ash	2.1	2.1	3.1	2.3	2.1	2.2	2.1
Crud protein	9.7	17.4	12.4	11.1	10.1	14.1	11.0

Each value represents the mean of four samples .

observed when the triticale cultivars were compared . Similar observation were reported by (Leterm et al ., 1990) .

" Table 4 " Dry matter and organic matter intake and digestibility of grains .

Diets	Dry matter		Organic meter	
	intake g / d .	Digestibility %	intake g / d.	Digestibility %
Maize	76.4c	84.1a	75.4c	84.9a
Wheat	79.8b	73.1c	78.2b	74.4c
Barley	71.6d	64.4d	69.3d	66.2d
Triticale cv .				
Beagle	89.6a	77.1b	87.5a	78.4b
Drira	90.2a	73.8bc	88.2a	75.3bc
Dolphine	89.7a	75.3bc	87.7a	76.3bc
Juanillo	87.4a	76.4bc	85.6a	77.7bc

a b c d : Values have different subscribes in intake or digestibility significantly differ at (P < 0.05) .

Only the triticale Beadle cv diet had a significantly (P < 0.05) higher digestibility than that of the wheat diet . The digestibility of other triticale cvs was numerically but not significantly higher than that of wheat . this is in agreement with earlier results reported by (Bragd and Sharpy., 1970) who concluded that triticale could be   replaced wheat in broiler diets without adverse effect on performance .

Data pertaining to N intake and retention   for the diets used in the 1 st metabolism trial is shown in (Table 5) . The difference in the intake of N between diets reflects the CP content of those diets . Nitrogen balance for chickens was positive on all treatments . N - retention was higher (P < 0.05) for chicken fed wheat or Dolphine cv diet than those fed other diets due to the higher intake of nitrogen . The retention of N (expressed as % of intake) was higher (P < 0.05) for chickens fed the maize diet compared to the rest of other diets . this might have been due to the higher available energy in the maize diet relative to the other diets .

Results form the 2nd metabolism trial showed no significant differences between diets containing 0 , 35 , 50 and 65 % of either triticale Beagle or Drira in relation to the intake, digestibility of DM and OM and to N - intake (Table 6) . These results are in agreement with earlier results of (Moran et al ., 1969 ; and Peterson and Aman ., 1990) who reported

" Table 5 " Nitrogen retention by chickens fed different grains in the 1 st metabolism trial .

| Item | Maize | Wheat | Barley | Triticale cultivars | | | |
				Be	Dr	Do	Ju
N - intake(g/d)	1.17 f	2.21a	1.41 e	1.67c	1.42e	2 .00b	1.50d
N - balance(g/d)	0.57de	0.90a	0.58cde	0.72bc	0.53e	0.84ab	0.60cde
N - retention (%) of intake	48.7a	40.7b	41.1b	45.9ab	37.3b	42.00ab	40.00ab

a b c d e f : Values on the same row with different subscribes are significantly different (P < 0.05) .

that the inclusion of triticale in the diet of chickens had no adverse effect on the digestibility and nutritive value of the diets .

It can be seen from (Table 6)that N balance was positive for all groups of chickens . However , the inclusion of triticale (i . e ., Beagle and Drira cvs) at a level higher than 35 % of the diet significant reduced (P < 0.05) N - retention when compared to the control diet .

When   diets containing different levels and cultivars of triticale were compared the highest N - balance was observed in group fed a diet containing 35 % of Beagle cv (P < 0.05) .

Similar observations concerning the reduction in feed efficiency were reported when triticale replaced maize and some SBM in diets fed to chickens (Myer et el., 1990) .

FEEDING TRIALS :

The diets used in both trials for all experimental groups have very similar contents of crude protein and metabolizable energy (Table 2) .

Results from the first feeding trial showed that feed consumption was influenced by the type of diet . Broiler fed the control diet consumed more than these fed diet containing triticale (Table 7) . It is obvious that the inclusion of more than 50 % triticale in the diets increased feed consumption in the 1 st feeding period   (1 - 21 days) but decreased it in the 2nd feeding period(22 - 49 days) compared to diets containing 0 or 35 % triticale .

No significant differences (Table 7) were observed in relation to the mean body live weight and growth rate of broilers when diets fed containing 0 or 35 % triticale were compared . However , the inclusion of 50 % triticale or more in the diet brought about a significant reduction (P < 0.05) in the live weight and growth rate in both feeding periods . The differences regarding the live weight decrease with age during the second feeding period and in the last week of the experiment , these differences become insignificant .

Data from the first feeding trial showed that feed conversion efficiency (kg feed/kg live weight gain) was higher (P < 0.05) when broiler received diets containing 35 % triticale compared to all other groups . However , increasing triticale level decreased the feed conversion efficiency (Table 7) .

Data relating to the feeding trial (Table 8) showed that feed consumption was higher for diets containing triticale compared to the control diets . No significant differences in the live weight and the growth rate between the experimental groups were observed when triticale was only used in the 2nd feeding period. The best feed conversion efficiency was obtained for broilers fed the control diet containing no triticale .

" Table 6 " Dry matter , Organic matter and nitrogen usage by chickens fed diets containing different levels of triticales .

| | control | Triticale cultivars | | | | | |
| | | Beagle | | | Drira | | |
Item		35	50	65	35	50	65
Dry matter :							
Intake g / d	87.10	87.40	87.70	87.20	87.60	87.30	87.00
Digestibility %	75.10	74.70	75.60	75.60	75.20	76.20	75.40
Organic matter :							
Intake g / d	82.70	82.20	83.50	82.70	82.90	82.70	82.30
Digestibility %	78.10	78.00	78.70	78.80	78.30	79.20	78.40
Nitrogen :							
Intake g /d	3.05	2.78	2.35	2.49	2.59	2.59	2.36
N - balance	1.81a	1.56b	1.40c	1.39c	1.39c	1.47c	1.27d
N - retention							
% of intake	56.3a	56.1ab	54.90b	55.80b	53.7b	56.70ab	53.80 b

a b c d : Values on the same row with different subscribes are significantly different (P < 0.05) .

" Table 7 " Growth rate , feeding conversion and mortality of broilers fed different levels of triticale .

| | Triticale level (%) | | | | |
Item	0	35	50	65	75-80
Feed consumption (g / d)					
At 1 - 21 days	55.1	49.9	53.9	58.9	57.7
At 22 - 49 days	148.9	132.3	134.2	129.4	133.8
Live weight (g)					
At 1 - 21 days	546.7a	524.9a	498.1b	458.1b	488.9b
At 22 - 49 days	2163.4	2104.6	1918.0	1906.5	1969.9
Growth rate (g / d)					
At 1 - 21 days	23.9a	22.8a	21.7b	19.7c	21.1bc
At 22 - 49 days	57.7a	56.3a	50.7b	51.5b	52.9 b
At 1 - 49 days	43.3a	42.1a	38.2b	38.0a	39.3 b
Feed conversion (kg F. / kg W.)					
At 1 - 21 days	2.30	2.16	2.48	2.98	2.73
At 22 - 49 days	2.51	2.35	2.64	2.51	2.52
At 1 - 49 days	2.47b	2.33a	2.61c	2.61c	2.58c
Mortality rate (%)					
At 1 - 21 days	4.0	4.0	3.2	0.8	0.8
At 1 - 49 days	5.0	5.8	4.1	1.7	2.5

a b c:Values on the same row with different subscribes are significantly different(P < 0.05) .

The inclusion of triticale in the diet at a level of 50 % or more reduced the mortality rate in both experiments and feeding periods compared to the diets which contained 0 or 35 % triticale (Table 7 and 8) .

" Table 8 " Growth rate , feeding conversion and mortality of broilers fed different levels of triticale during the 2nd period (22 - 49 d) .

	Triticale level (%)			
Item	0	35	50	65
Feed consumption (g / d)				
At 22 - 49 days	137.4	149.3	154.7	144.4
At 1 - 49 days	105.1	119.0	115.0	109.1
Growth rate (g / d)				
At 22 - 49 days	56.6	55.3	55.6	52.3
At 1 - 49 days	42.8	42.0	42.4	40.3
Feed conversion (kg F . / kg W .)				
At 22 - 49 days	2.42a	2.69b	2.70b	2.77b
At 1 - 49 days	2.45a	2.66b	2.71b	2.71b
Mortality rate (%)				
At 1 - 49 days	4.0	4.8	2.4	3.2

a b : Values on the same row with different subscripts are significantly different ($p < 0.05$) .

The detrimental effects as a results of increasing triticale level in the diet of chicken on body weight and /or feed efficiency were obtained by other workers. The results reported   by (Smith et al ., 1989) who fed broiler chicken from 1 to 21 days old diets in which Morrison and beagle 82 triticale cvs were substituted for 50 or 100 % of maize , indicated that feed conversion efficiency and body weight were generally reduced for chicken diets containing   triticale . Differences between cultivars were also observed . However ,(Maurice et al .,1989) concluded that triticale Florida 201 was equal to maize in promoting growth and had no detrimental effect on body weight or feed conversion . According to the same authors, the general weakness of several broiler studies giving   variable results could be attributed to the use of undenitrified triticale cvs , the lack of complete chemical analysis , and the use of different cultivars triticale grown at different locations .

On the basis of the results reported in this study , the use of triticale grain in broiler diets at a level of 35 % , while   maintaining the recommended protein - energy ratio , can be recommended .

References

Brag, D.B. and T. F. Sharpy . 1970 . Nutritive value of triticale for broiler chick diets .
    Poultry Sci . , 49 : 1022 - 1027 .
Bushuk , W. 1980. Triticale : Chemistry and Technology . Chemia i Technology Pszenrta , 24 :
    603 - 612 .
Leterme,P.,A.Thewis and F.Tahon . 1990. Nutritive value of triticale in pigs as a function of its
    chemical composition . From : Proceedings of the 2nd Inter. Triticale Symp., CIMMYT ,
    Mexico .
Maurice,D.V.,T.E. Jones,S.F.Lightsey,J.F.Rhoades and K .T.Hsu.1989.Chemical and nutritive

value of triticale (Florida 210) for broiler chickens . Appl . Agr .Res., 4 : 243 .

MoranE.T.Jr.,S.P.Lall and J .D.Summers . 1969. The feeding value of rye for the growing chick : Effect of enzyme supplements , antibiotics , autoclaving and geographical area of production . Poult . Sci . , 48 : 939 - 949 .

Myer R.O.,G.E . Combs and R . D . Barnett . 1990 . Evaluation of triticale cultivars adapted to the southeastern U . S . A . as potential feed grains for swine . Proc . of the 2nd Inter . Triticale Symp . , CIMMYT , Mexico .

Petterson,D.and P.Aman.1990 . Composition and productive value for broiler chickens of wheat , triticale and rye.Proc. of the 2nd Inter . Triticale Symp . , CIMMYT , Mexico .

Smith,R.L.,L.S.Jensen,C.S.Hoveland and W.W.Hanna .1989 .Use of pearl millet , sorghum and triticale grain in broiler diets . J . of Prod . Agric . , 2 : 78 .

Steel, R.G.D .and J.H.Torrie.1960.Principles and Procedures of Statistics .McGrow - Hill Book c : New York .

Villegas,E.,C.E .McDonald and K.A .Gilles.1970 . Variability in the lysine content of wheat, rye and triticale proteins . Cereal Chem . , 47 : 746 - 757 .

Zillinsky,F.J . and N.E.Borlaug . 1971 . Progress in developing triticale as an economic crop. CIMMYT Res . Bull . , 17 : 1 - 27 . Mexico .

EVALUATION OF TRITICALE AS A FORAGE PLANT THROUGH THE ANALYSIS OF THE CINETIC OF SOME QUALITATIVE PARAMETERS FROM STEM ELONGATION TO MATURITY.

Stefano Bocchi[1], Graziano Lazzaroni[1], Nicola Berardo[2], Tommaso Maggiore[1]
[1]Istituto di Agronomia - Università degli Studi, Milano, Italy
[2]Istituto Sperimentale per le Colture Foraggere, Lodi, Italy

Abstract

For two years a field experiment has been carried out at S.Angelo Lodigiano (Milano-Italy) with Mizar and Rigel, two common triticale varieties, grown in plots replicated four times in a randomized complete block design. Plants from each plot were sampled from stem elongation to earing every ten days, from earing to waxy maturity every three days, and from the latter to full maturity every ten days. After having measured the water content of each sample, those parameters characterizing the quality of the forage (Raw Protein, NDF, ADF, ADL , Ash) and estimating its nutritional value (MFU kg-1) were analyzed.
Due to the warmer spring in 1990 the yield was 16.3 t ha-1 at milky maturity, whereas in 1991 it was 13.2. The fiber fractions increased from stem elongation to earing and decreased afterwards. At milky-waxy maturity stage, the best one for the biomass to be insilated, the difference between the two years was not significant.

Introduction

In Italy triticale (x Triticosecale) is grown on a total area of about 6800 ha. On almost half of this surface (about 3000 ha) the crop is harvested at the milky-waxy maturity for the production of tritico-silage. Interest for this plant began in the 70's, when it was initially considered particularly suited for forage production. Preliminary studies have shown that the whole plant could be used by ruminants [1] and the grain by monogastrics [2,3,4]. The differences of production of biomass initially observed between the first varieties of triticale and the varieties of other winter cereals [5,6,7] have been gradually eliminated [8,9] thanks to new varieties.
In Italy in 1980 Francia and Martillotti [10] with trials carried out in Lazio, by comparing triticale (Mizar) with winter wheat, found higher yield of Mizar for forage cut at milky-waxy maturity and for grain and straw.

Trials carried out in the Po Valley [11] comparing triticale, barley, winter wheat, and rye for forage production in fertile soil conditions showed a higher biomass production of rye which however was characterized by a lower nutritive value than was demonstrated in barley, wheat and triticale. The same authors found a forage quality of triticale biomass that was an intermediate between rye and wheat. At harvest, carried out when the percentage of dry weight reached 35 %, the percentage of the Neutral Detergent Fiber (NDF) of the whole plant (about 35 %) was lower than for rye (37 %), but a little more

H. Guedes-Pinto et al. (eds.), Triticale: Today and Tomorrow, 827–834.
© 1996 *Kluwer Academic Publishers. Printed in the Netherlands.*

than that found in winter wheat and barley (32 %). In the stems the highest values of Acid Detergent Fiber (ADF) were found in rye (46%), these were higher than in barley (44 %), and in triticale or wheat. In the ears the highest percentages were detected in triticale and wheat (19 %) higher than those of rye (17 %) and barley (15 %). Lignin reached higher values in triticale and wheat whole plants (7.1 %) than in rye and barley (6.5 %). In more recent trials [12] where the quality of the forage obtained with different varieties of wheat, barley, rye, and triticale was compared, a high biomass production of a new variety of triticale (Rigel) was observed. Rigel forage was better than that of Mizar in terms of quality: 62.6 of NDF, 39.8 of ADF and 0.62 Milk Folder Units (MFU) of Rigel vs. 69.8, 44.7 and 0.53 respectively for Mizar.

The experiments carried out in the past showed triticale as a species suitable for environments where winter wheat and barley could have difficulties. On the contrary more recent research shows that the new triticale varieties have been strongly improved not only for biomass production, but also for the forage quality. In the fertile environments of the Po Valley the autumn-spring artificial grassland followed by corn for silage is generally composed of ryegrass (*Lolium multiflorum* L.), which can be stored in silos after harvested at earing. The main inconveniences of this type of grassland are: 1) the facility of lodging; 2) the low percentage of dry matter at the harvest stage; 3) poorer quality with the course of the development stages.
Alternative species which can be harvested at different periods must be found.
The study, the results of which are presented in this paper, took in consideration triticale by analysing the trend from stem elongation to grain maturity of the biomass accumulation and forage quality.

Materials and Methods

The trials were carried out at S.Angelo Lodigiano (Milano, Italy) during the years 1989-90 and 1990-91, on a sandy-loam, sub-acid, soil with good contents of assimilable phosphorous, low percentages of exchangeable potassium and average percentages of organic matter.

Two varieties of triticale were used: Mizar, which has been cultivated for several years, and Rigel, a more recent cv. The agrotechniques applied are shown in table 1.

Table 1. Agrotechniques of the field experiment in 1990 and 1991.

	1989-90	1990-91
Previous crop	Maize	Maize
Pre-sowing fertilization (kg ha-1 of N/P2O5/K2O)	40/120/120	40/120/120
Sowing (date)	Nov. 2	Nov. 16
Top-dress. fertilization (kg ha-1 N-NH4NO3)	26	26
(kg ha-1 N-Urea)	144	144
Weed control	Ioxynil + MCPP	

A complete randomized block design with four replications was adopted with a plot of 20 m^2, half of which was used for the sampling of plants from stem elongation to grain maturity, and half for the harvest at milky-waxy maturity stage.
In 1990 and in 1991 15 samplings and 17 samplings respectively were collected. In 1990 the sampling started on March 20th and in 1991 on April 2nd. For each sampling all the plants located on 1 linear meter completely surrounded by other plants ("competitive conditions") were harvested. The sampling was carried out every 10 days from stem elongation to heading, every 3 days from heading to milky-waxy maturity, and every 10 days until complete maturity.

After each sampling the fresh weight and the percentage of dry matter were determined. The percentage of dry matter was calculated by drying the plants in an oven at 70°C.
The dry samples pass through a mill having an opening of 0.75 mm in diameter were used to analyse Total Nitrogen, Ash, and the fiber fractions (Neutral Detergent Fiber or NDF, Acid Detergent Fiber or ADF, and Acid Detergent Lignin or ADL) according to Goering and van Soest [13].

To determine these fractions 256 samples were put in an oven at 60 °C for 24 hours and, afterward, they were analyse by spectroscopy with the NIRS system, ISI software and the PLS statistical modified model. Since the calibration patterns specifically for triticale at different development stages were not available, NIRS chose the samples to analyse following the normal procedure. A total of 120 samples was analysed (60 samples each year). Moreover, 40 samples were chemically analyzed to verify the equations of the calibration curves. The quality of the spectropical analysis was very high as shown by the correlation coefficient values related to the standard error of the cross validation equal to 0.99 , 0.93, 0.98, 0.98, 0.80 for Raw Protein, Ash, NDF, ADF, and ADL respectively.
To determine Total Nitrogen the elementary N-analyzer (modified Dumas method - Kirsten, 1983) was used. The fiber fractions were detemined, with two replications, according to Goering and van Soest [13,14]. Ash was determined by putting everything in the oven at 550 °C.

With the obtained data the nutritional value was calculated. The Milk Folder Units were calculated using the following equation:
$$(1.085 - (0.015 \; x \; \%ADF)) \, / \, 0.786$$

All the data were statistically analyzed (ANOVA) after the requested transformations for each year.

Results and discussion

For a better interpretation of the results it is appropriate to premise some information about the meteorological course. On average the temperatures recorded in autumn-winter 1989-1990 were the same as for the past 30 years. There was little rainfall in the same period. Spring growth was precocious thanks to favourable conditions, at least to April. The following year was characterized by heavy rainfall during the sowing period and low temperatures that lasted almost throughout the productive cycle so much so to retard both autumn and spring growth.

Table 2 shows the average results obtained with samplings regarding the total Dry Matter (t ha-1), percentage of Dry Matter (DM),Raw Protein (%) ,NDF (%), ADF (%), ADL (%), Ash (%), and MFU kg-^{1} of D.M. Statistical analysis has shown a general uniformity of the varieties experimented which gave different results in both years for the samplings

830

which were carried out during the milky-waxy stage of the more precocious variety (Mizar).
Table 3 shows tha analysis of variance results relative to these periods. During the harvest of Mizar, Rigel in the first year was not very different in the percentage of DM and in the production of dry matter. In 1991 there were differences only in the percentage of DM. In both years the qualitative parameters were more favorable for Rigel whose NDF, ADF and ADL are sharply inferior. These differences became even greater when we compare the two varieties at the same development stage (Table 4).

Table 2- Experimental results of the two years (1990, 1991) along the cycle.

1990 stages	Julian days	Dry Mat (%)	Dry Mat. (t ha-1)	Raw Prot. (%)	NDF (%)	ADF (%)	ADL (%)	Ash (%)	MFU N° kg-1
Stem elongation	79	14.5	1.132	32.4	43.8	19.9	3.4	9.9	1.001
	92	13.0	2.647	30.9	50.2	24.8	3.5	9.9	0.906
	101	12.5	4.335	28.6	54.1	29.4	3.5	9.9	0.819
	110	12.6	5.611	22.4	57.5	32.0	3.7	8.8	0.769
	120	14.9	8.614	16.8	59.7	36.5	4.1	7.4	0.683
earing	123	15.5	8.831	15.6	65.3	39.4	4.7	7.2	0.628
flowering	127	18.7	11.545	14.4	64.9	40.0	4.5	6.9	0.617
	130	20.7	12.699	13.2	66.8	41.2	4.8	6.5	0.594
	134	24.7	13.452	13.0	67.0	41.9	5.0	6.1	0.581
	137	25.6	14.148	12.6	67.7	42.3	5.3	6.4	0.572
milky maturity	143	27.7	15.072	10.8	67.4	42.5	6.0	6.4	0.569
milky-waxy	149	30.2	16.379	10.5	64.7	39.9	5.9	6.2	0.619
	152	32.0	16.262	10.5	62.7	37.8	5.7	5.7	0.660
physiol.	164	38.9	17.613	9.7	60.9	36.8	6.1	5.8	0.678
full maturity	171	54.1	16.140	9.4	63.9	39.3	6.7	6.1	0.630
1991									
Stem elongation	92	15.2	0.316	28.2	41.9	16.9	1.3	9.9	1.057
	102	15.3	1.031	26.4	46.8	21.6	1.5	9.6	0.968
	112	18.1	2.211	20.4	46.9	22.4	2.0	8.5	0.953
	124	15.5	3.801	17.1	57.7	31.6	3.3	8.6	0.777
	127	17.9	4.712	14.9	55.9	32.0	3.3	7.3	0.769
earing	134	16.4	5.602	13.9	62.9	36.8	4.2	7.7	0.678
flowering	137	17.1	6.167	12.3	65.9	39.6	4.6	7.3	0.625
	141	24.2	8.217	10.7	64.9	39.9	4.7	6.7	0.620
	144	24.2	8.171	10.9	67.5	41.7	5.2	6.8	0.584
	148	26.1	10.119	9.2	67.6	42.6	6.2	6.3	0.568
milky maturity	150	30.7	11.662	8.8	66.2	42.3	6.5	6.1	0.574
milky-waxy	152	32.9	13.181	8.6	63.8	40.7	6.5	6.3	0.603
	163	34.9	15.401	7.5	56.6	34.3	5.9	5.2	0.726
physiol.maturity	172	41.3	17.923	7.2	53.5	30.8	4.9	5.2	0.793
	179	54.6	16.830	7.1	51.2	29.0	3.9	5.3	0.827
	186	79.3	17.429	7.3	51.7	30.0	4.4	4.9	0.807
full maturity	198	89.7	15.268	7.9	50.5	29.6	4.3	5.0	0.815

The two varieties reached the earing stage simultaneously. When triticale forage is harvested at this stage, the percentage of DM is lower compared to ryegrass during earing (about 16 % against 20 % respectively) and has higher percentages of fiber fractions.

The results are shown in Table 2 and in Figures 1, 2 and 3 where they are expressed with third degree functions in which both the julian days and the values of the parameters examined are respectively independent and dependent variables; in the same figures the equations and the R^2 are shown.

The late spring growth reported in 1991 is highlighted in figure 1 by the arrows which indicate the main phenological stages. Except for the beginning of the earing - flowering phase, when the plants seem to temporarly stop growing, the biomass increases at an ever faster pace from stem elongation to flowering, when it reaches fluctuating values around $0.67 - 0.7$ t ha^{-1} d^{-1}.

At milky maturity the production was 16.3 and 13.2 t ha^{-1} in 1990 and 1991 respectively.

From stem elongation to earing the percentge of DM does not change much while from the earing stage to the milky - waxy maturity the increase is rapid: it starts first in 1990, but it is later in 1991, due to less favourable climatic conditions (Table 2).

At the beginning of stem elongation higher RP percentages in forage were recorded in 1990 (about 32 %) compared to what was obtained in 1991 (about 28 %); one more reason for the different rate of N uptake which reached higher values in the more temperate year.

Table 3. Dry Matter (DM - %), Dry Matter (t ha-1), Raw Protein (RP - %), NDF,ADF, ADL, Ash (%), and MFU kg-1 of Mizar and Rigel at 11° and 12° sampling, in 1990 and 1991 respectively.

cv	1990	DM (%)	DM (t ha-1)	RP (%)	NDF (%)	ADF (%)	ADL (%)	Ash (%)	MFU N°kg-1)
Mizar		28.4	15.4	10.9	72.1	46.9	6.6	6.7	0.485
Rigel		27.0	14.7	10.8	62.7	38.2	5.3	6.0	0.652
Mean		27.7	15.1	10.8	67.4	42.5	6.0	6.4	0.569
LSD		n.s.	n.s.	n.s.	3.62	2.07	0.49	0.54	0.041
	1991								
Mizar		34.2	12.4	8.7	66.3	41.6	6.7	6.3	0.586
Rigel		31.7	14.0	8.5	61.2	39.8	6.3	6.3	0.620
Mean		32.9	13.2	8.6	63.8	40.7	6.5	6.3	0.603
LSD		1.06	n.s.	n.s.	3.15	0.42	0.14	n.s.	0.029

Table 4. Dry Matter (DM - %), Dry Matter (t ha-1), Raw Protein (RP - %), NDF,ADF, ADL , Ash (%) and MFU kg-1 of Mizar and Rigel at milky-waxy maturity, in 1990 and 1991.

1990	Julian Day-1990	DM (%)	DM (t ha-1)	RP (%)	NDF (%)	ADF (%)	ADL (%)	Ash (%)	MFU N°kg-1)
Mizar	143	28.4	15.4	10.9	72.1	46.9	6.7	6.7	0.485
Rigel	149	28.6	16.6	10.5	64.1	39.7	5.8	6.5	0.622
Mean		28.5	16.0	10.7	68.1	43.3	6.3	6.6	0.554
LSD		n.s.	0.8	n.s.	4.01	3.61	0.41	n.s.	0.051
	1991								
Mizar	150	33.8	11.7	9.1	67.5	42.5	6.8	6.5	0.569
Rigel	152	31.7	14.0	8.5	61.2	39.8	6.3	6.3	0.620
Mean		32.8	12.9	8.8	64.4	41.2	6.6	6.4	0.595
LSD		2.0	1.1	0.6	3.8	3.5	0.37	n.s.	0.047

832

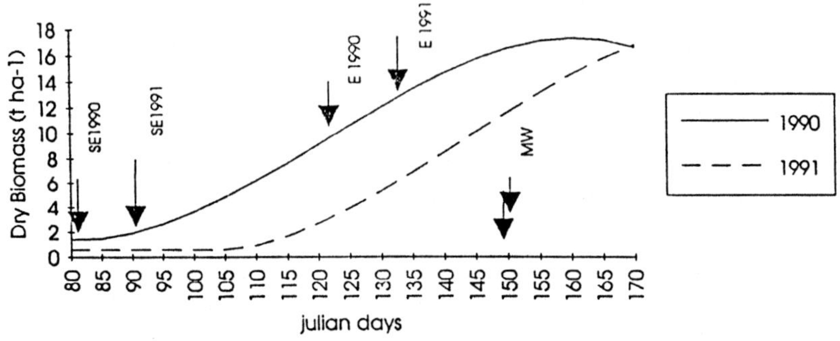

Figure 1. Total Dry Biomass produced along the cycle in 1990 and 1991.
SE = Stem Elongation; E = Earing; MW = Milky-Waxy maturity.

1990: y = 85.093548 - 2.480615 x + 0.02299988 x^2 - 0.0000634 x^3   R2 = 0.99
1991 y = 151.12495 - 3.635425 x + 0.02778667 x^2 - 0.0000634 x^3   R2 = 0.99

Figure 2. NDF percentage along the cycle in 1990 and 1991.

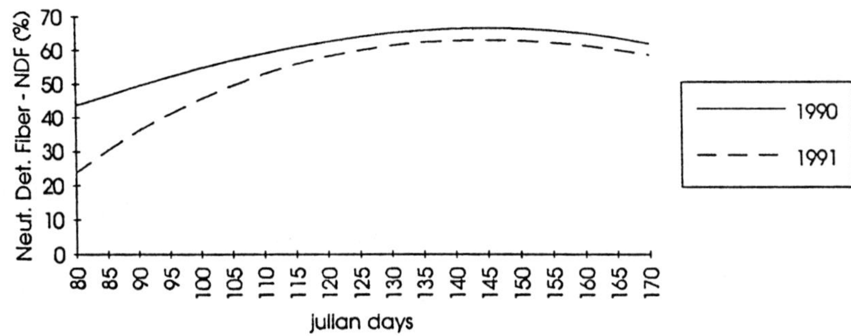

1990: y = -9.170679 + 0.56834564 x + 0.00295648 x^2 - 0.0000226 x^3   R2 = 0.92
1991: y = -187.6267 + 4.1411153 x - 0.0213039 x^2 +0.00003215 x^3   R2 = 0.78

Figure 3. ADF percentage along the cycle in 1990 and 1991.

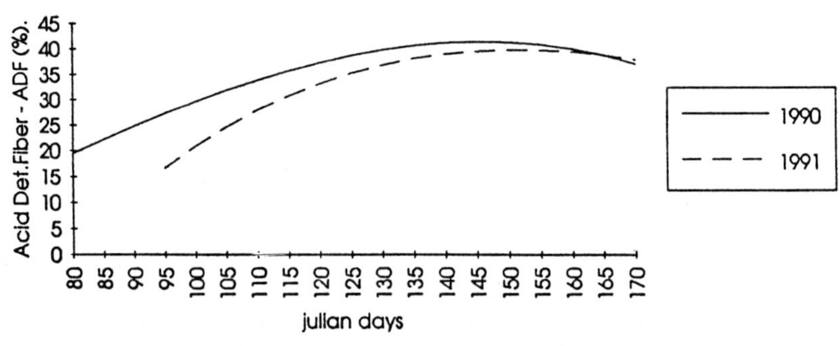

1990: y = -22.0746 + 0.31686051 x + 0.00460502 x^2 - 0.000026 x^3   R2 = 0.94
1991: y= -176.4445 + 3.3764828 x - 0.0163017 x^2 + 0.0000227 x^3   R2= 0.81

Note: for figures 2 and 3 refer to the arrows in figure 1 which indicate development stages.

In the later stages the values progressively decrease, reducing, but not eliminating the differences observed in two years: at the milky-waxy maturity the percentage of Raw Protein in the harvested forage in the two years of the study were 11 and about 9 %, respectively. In both years, NDF percentage progressively increases ranging from 43.8 and 41.9 respectively at the stem elongation of the first and the second year, up to maximum values of 67 %, generally observed at the end of the flowering-beginning of grain filling phase.Afterwards the percentage decreases to significantly lower values at the physiological maturity (3 weeks after the milky maturity in 1990, 4 weeks after in 1991).

The ADF cinetic is similar to that described for NDF: the percentages range from values less than 20 % at the beginning of stem elongation to maximum values around 43 % observed near the beginning of grain filling stage, afterwards it decreases. The percentage of ADL in 1991 follows the same cinetic. In 1990, after an initial decrease, lignin content grew during the last phases. This could be due to the metereological conditions that could have determined a low percentage of grain on the total biomass which, in turn, may have negatively affected the Ash, Protein and lignin contents (Table 2). The Milk Folder Units (MFU) kg-1 of DM from values of approximately 1 observed at the beginning of stem elongation decrease, following the ADF percentage, to reach values around 0.56 - 0.58. Triticale productivity of MFU was however high: in the most favourable year MFU can be more than 10.000 unit ha-1.

Conclusions.

A trial on biomass accumulation and the cinetic of the main qualitative parameters, from stem elongation to maturity was carried out in the Po Valley with two common varieties of triticale (Mizar and Rigel). The result was a better quality of Rigel forage at milky-waxy stage. Generally the species confirmed the high productive potentiality that, in terms of MFU ha-1, can be compared with other winter cereals. Even for triticale the highest quality of the forage is obtained by harvesting the plants between the milky-maturity and the waxy-maturity stages, that is to say with the percentage of DM ranging between 28 and 35 %. In fact, during this phase, NDF and ADF contents are regularly decreasing thanks to increasing percentages of grain in the total biomass and related increase of MFU kg-1 of DM. Late harvest is not suitable because the high percentage of DM can be problematic for forage conservation in silos. If harvested at earing, the two varieties are not diferent; they produced the same quantities which can be reached in the same environment with a ryegrass cut at earing, but with inferior quality due to the high percentages of fiber fractions.

References

1. McColay AW, Sherrod LB, Albin RC, Hansen KR. Nutritive value of triticale for ruminants. J Anim Sci 1970;32:534-539.
2. Longnecker T. Triticale comparable to sorghum for swine. Crop soils 1973;25:20-21.
3. Shimada AS, Martinez RL, Bravo FO. Studies on the nutritive value of triticale for growing swine. J Anim Sci 1972;33: 1266-1269.
4. Jessop RS, Leterme P, Wright R, Lopez J,Kolding M, Fialho E. Triticale as Forage of Grain for Animals. Proc Second Int Triticale Symp, Rio Grande do Sul, Brasil, 1-5 Oct 1990; 64-65.

5. Carnide VP, Guedes-Pinto H. Forage aptitude of primary 8xTriticale compared with rye and wheat progenitors. Proc Second Int Triticale Symp, Rio Grande do Sul, Brasil, 1-5 Oct 1990; 536-541.
6. Barnett RD, Stanley WH, Chapman WH, Stith RL. Triticale. New feed and forage crop in Florida. Sun State Agric Res 1971;Jul-Sep:12-14.
7. Mongini L, Corleto A. Possibilità produttive di alcuni cereali vernini (avena, orzo e triticale) raccolti a maturazione latteo-cerosa. Ann Fac Agr Bari 1970;XXVII:961-964.
8. Bishoi UR, Walker JC. Evaluation of existing triticale varieties for growth, yield and forage qualities. Alabama Univ Ann Res Rep 1976;6:70-88.
9. Carnide V, Ferreira A, Pires A, Guedes-Pinto H. Study of a triticale variety for forage. Genetics and breeding of Triticale, Eucarpia meeting, Clermont-Ferrand, 2-5 July 1984; 686-691.
10. Francia U, Martillotti F. Produzione, conservazione e valutazione nutrizionale di due varietà di frumento e del triticale (Mizar) a diversi stadi di maturazione. Ann Ist Sper Zoot 1980;13:153-165.
11. Gentinetta E, Giammonta M, Delogu G, Lorenzoni C, Maggiore T. Triticale ed altri cereali a semina autunnale: confronto qualitativo del prodotto della pianta intera e delle sue parti. Riv di Agron 1993;XVVII:392-396.
12. Merli D, Bocchi S, Berardo N, Maggiore T. Confronto fra cereali vernini e loglio italico per l'insilamento. L'Inf Agr 1993;33:61-68.
13. Goering HK, van Soest PJ. Forage fiber analysis. USDA-ARS Handbook 1970;379:1-20.
14. Martillotti F, Antogiovanni M, Rizzi L, Santi L, Bittante G. Metodi di analisi per la valutazione degli alimenti di impiego zootecnico. CNR-IPRA, 1987; Quaderno metodologico 8.

EFFECT OF THE SEEDING RATE ON FORAGE YIELD AND QUALITY OF TRITICALE

Valdemar P. Carnide[1], Henrique Guedes-Pinto[1], Armando Mascarenhas-Ferreira[2] and Carlos Sequeira[2].
1 - Depto. de Genética e Biotecnologia; 2 - Depto. de Nutrição Animal.
Universidade de Trás-os-Montes e Alto Douro, Vila Real, Portugal

Abstract

Studies on triticale as a cereal for grain have been carried out since 1975 in the Northeast of Portugal. Because of its high biomass production, triticale has also been studied as a forage crop since 1979. These studies were designed to investigate whether rye, the major cereal in that region which sometimes is grown as a forage crop or oats, sown in cereal-legume mixtures, could advantageously be replaced by triticale.

According to previous data, the line UTAD 36/85, from the UTAD triticale breeding programme, and the Polish cultivar Presto were selected for studying the effect of the seeding rate on forage yield and quality. The regional rye population Vila Pouca and the Portuguese oats cultivar Boa-Fé were used as testers.

Dry matter yield, forage quality (*in vitro* organic dry matter digestibility and crude protein content) and number of stems. m^{-2} were evaluated at the seeding rates of 175, 350, 525 and 700 viable seeds.m^{-2}. Over all variables, triticale line UTAD 36/85 had the highest mean of dry matter yield (8715 kg.ha^{-1}) while the regional rye population Vila Pouca had the lowest (7315 kg.ha^{-1}). At the rates of 175 and 350 viable seeds.m^{-2} the average number of stems. m^{-2} for all genotypes was similar but it increased with higher rates. The highest *in vitro* organic matter digestibility was observed on triticale cv. Presto, 66.6% at the seeding rate of 175 viable seeds.m^{-2}, and the lowest on oats cv. Boa-Fé, 52.8% at the same seeding rate. The highest crude protein content was observed for all seeding rates on the rye population while the oats cultivar showed the lowest one.

Introduction

Rye is the main cereal in Northeast Portugal and it is exploited either for grain or forage production. A triticale programme for grain production was started for this region in 1975 [1, 2]) and from 1979 onwards several studies have been carried out with triticale as a forage crop. Several triticale lines/cultivars were evaluated alone and in mixture with legumes, and as a dual purpose crop [3, 4, 5, 6].

An increase in seeding rate of triticale, wheat and rye from 50 to 100 kg.ha^{-1} produced more forage [7]. Hernando (1981) showed that in wheat, higher seeding rates corresponded to higher dry matter yields. Regarding the number of stems.m^{-2} in wheat,

H. Guedes-Pinto et al. (eds.), Triticale: Today and Tomorrow, 835–841.
© 1996 *Kluwer Academic Publishers. Printed in the Netherlands.*

it increases with the increase of seeding rate [9 10, 11]. In rye a higher seeding rate resulted in a higher dry matter production. Although the seeding rate x population interaction was not statistically significant, some populations were considerably more affected by the seeding rate than others [12]. In legume - cereal mixtures Osman *et al.* (1983) found that dry matter production was significantly improved with an increase in the seeding rate.

The aim of this study was to evaluate the effects of seeding rate and cultivar on forage yield, number of stems.m^{-2}, as well as on the nutritive value of the forage.

Materials and Methods

A three year (1990/91 to 1992/93) field experiment was conducted at Vila Real. Each year two triticales, line UTAD 36/85 from UTAD breeding programme and Polish cultivar Presto, a regional rye population, Vila Pouca, and a Portuguese oats cultivar, Boa-Fé, were seeded at four different rates - 175, 300, 525 and 700 viable seeds.m^{-2}. A complete randomized block design with four replications was used. The plots consist of 6 rows 25 cm apart in a 4 x 1.5 m plot (6 m^2). The 4 central rows were clipped manually at heading stage. Planting date was the second fortnight of October. Prior to seeding 28 kg.ha^{-1} of nitrogen, 56 kg.ha^{-1} of phosphate and potassium were incorporated into the soil. When the plants reached the tillering stage 120 kg.ha^{-1} of nitrogen were top dressed. Forage samples were collected and dried in a forced air dryer at 65°C and yields calculated on an oven-dry basis. Sub-samples of forage were ground and used for *in vitro* organic matter digestibility and crude protein content which were evaluated according to Marten and Barnes (1980) and AOAC (1975), respectively. The number of stems.m^{-2} was also recorded.

Results and Discussion

Seasonal patterns for rainfall during the main growing season (February-April) varied widely over the years (Table 1). The 1991 growing season had a limited moisture availability, specially in April once from 15th March the rainfall was only 27 mm, i. e. 66.3% of the average April rainfall of the 30 years period . In 1992 the spring was very dry and the decrease in average rainfall between February and April was 90.6%. In 1993 the situation was the reverse of 1991, i. e., February and March were dry with a 92.8% reduction in the average rainfall and April was rainy, 36.3% above the average.

Table 1 - Monthly accumulated rainfall (mm) from January to April in the period 1991-1993.

Month	1991	1992	1993	Average 1950-1980
January	88.0	3.2	46.9	163.8
February	130.6	7.2	5.6	165.5
March	161.8	11.0	14.6	133.6
April	25.9	12.0	104.8	76.9

FORAGE PRODUCTION

Forage production was significantly ($P<0.01$) influenced by all main effects (year, $F=169.9$, genotype, $F=8.8$, and seeding rate, $F=5.3$) and by year x genotype interaction ($F=8.2$). All the other interactions were not significant.

The average forage production for each genotype and seeding rate is shown in Table 2.

Table 2 - Average of forage dry matter production ($Kg.ha^{-1}$) for each genotype at four seeding rates and across three years.

GENOTYPE	SEEDING RATE (viable seeds.m^{-2})				
	175	350	525	700	Average
Triticale cv. Presto	6532	8420	8348	8218	7880
Triticale line UTAD 36/85	8220	8818	8498	9325	8715
Reg. rye pop. Vila Pouca	6922	7085	7212	8040	7315
Oats cv. Boa-Fé	8042	8400	8385	8680	8377
Average	7429	8181	8110	8566	
LSD at 5%	585				

Over all variables, triticale line UTAD 36/85 had the highest mean (8715 $Kg.ha^{-1}$) while the regional rye population Vila Pouca had the lowest (7315 $Kg.ha^{-1}$). The higher seeding rate (700 viable seeds.m^{-2}) resulted in higher forage production except for triticale Presto where the highest production was obtained at a seeding rate of 350 viable seeds.m^{-2}.

These results confirm those of Pfahler *et al.* (1982) who in rye found that a higher seeding rate results in higher productions. In triticale, wheat and rye Bishnoi (1980) observed that an increase of seeding rate produced more forage and Osman *et al.* (1983) concluded that dry matter production was significantly improved with an increase of seed rate in several forage mixtures. However in oats, Folkins and Kaufmann (1974) did not observed any significant effect on forage production.

The highest forage production was observed, for all seeding rates, on triticale line UTAD 36/85 showing at the seeding rate of 700 viable seeds.m^{-2} a production of 9325$Kg.ha^{-1}$, which was statistically different ($P<0.05$) from all the other forage productions. The triticale Presto showed the lowest production, 6532 $Kg.ha^{-1}$ at the seeding rate of 175 viable seeds.m^{-2}. This production was not statistically different ($P>0.05$) from that of the rye population Vila Pouca.

Several authors [17, 18, 19] have also found triticale productions superior to those of rye and oats. However, Carnide *et al.* (1991a,b) found that triticale cultivar Corgo had lower productions than oats and higher than rye at booting stage and similar to rye at milk dough stage.

NUMBER OF STEMS.M^{-2}

There were significant year ($F=4.6$, $P<0.05$), genotype ($F=53.1$, $P<0.001$), seeding rate ($F=27.9$, $P<0.01$), and year x genotype interaction ($F=24.1$, $P<0.01$) for the number of stems.m^{-2} while the interactions involving seeding rate were not significant for this parameter.

Regarding all variables, oats cv. Boa-Fé showed the highest number of stems.m^{-2} (724 stems.m^{-2}) while triticale EP 36/85 had the lowest (341 stems.m^{-2}) (Table 3).

Table 3 - Average number of stems.m^{-2} for each genotype at four seeding rates and across three years.

	SEEDING RATE (viable seeds.m^{-2})				
GENOTYPE	175	350	525	700	Average
Triticale cv Presto	342	453	490	605	472
Triticale line UTAD 36/85	226	298	361	478	341
Reg. rye pop. Vila Pouca	507	638	702	754	650
Oats cv. Boa-Fé	502	645	828	920	724
Average	394	363	595	689	
LSD at 5%	67				

The number of stems.m^{-2} at the seeding rates of 175 and 350 viable seeds.m^{-2} was similar but increased with higher seeding rates. These results agree with those of Folkins and Kaufmann (1974) for oats.

The highest number of stems.m^{-2} was observed at a seeding rate of 700 viable seeds.m^{-2} on the oats cv. Boa-Fé (920 stems.m^{-2}) which was statistically different (P<0.05) from the numbers observed in all the other genotypes. Triticale line UTAD 36/85 showed the lowest number of stems.m^{-2} (226 at a seeding rate of 175 viable seeds.m^{-2}), which was statistically different (P<0.05) from all other three genotypes.

The seeding rate means, averaged over years, showed that increasing the seeding rate increases the number of stems.m^{-2}. There is however, a genotype x seeding rate interaction, particularly at the lower and intermediate rates.

IN VITRO ORGANIC DRY MATTER DIGESTIBILITY

The main effects year (F=44.6, P<0.01), genotype (F=57.6, P<0.01) and seeding rate (F=3.0, P<0.05) and the interactions year x genotype (F=3.3, P<0.05), genotype x seeding rate (F=7.0, P<0.01) and year x genotype x seeding rate (F=3.8, P<0.01) were significant for *in vitro* organic dry matter digestibility.

The highest average *in vitro* organic dry matter digestibility was observed on triticale genotypes and was statistically different (P<0.05) to that observed in the other two species (rye population Vila Pouca and oats cv.Boa-Fé). However, the triticale line UTAD 36/85 at a seeding rate of 700 viable seeds.m^{-2} was not statistically different (P>0.05) from those species.

Triticale cv. Presto had the highest *in vitro* organic dry matter digestibility, 66.62% at the seeding rate of 175 viable seeds.m^{-2}, while oats cv. Boa-Fé had the lowest, 52.82%, at the same seeding rate (Table 4).

The seeding rate means, averaged over years, showed a tendency for decreasing the *in vitro* organic dry matter digestibitity with the increase of seeding rate. However, due to genotype x seeding rate interaction the behaviour of the different genotypes is variable.

Table 4 - Average of *in vitro* organic dry matter digestibility (%) for each genotype at four seeding rates and across three years.

	SEEDING RATE (viable seeds.m^{-2})				
GENOTYPE	175	350	525	700	Average
Triticale cv. Presto	66.62	64.22	62.23	59.57	63,16
Triticale line UTAD 36/85	62.95	62.98	62.63	58.95	61,88
Reg. rye pop. Vila Pouca	59.27	59.57	56.75	57.58	58,29
Oats cv.Boa-Fé	52.82	52.92	57.10	57.82	55,16
Average	60.42	59.92	59.68	58.48	
LSD at 5%	1.39				

So, in triticale Presto the increase of seeding rate decreases the *in vitro* organic dry matter digestibility while in oats cv. Boa-Fé it increases specially from the seeding rate of 350 to 525 viable seeds.m^{-2}. On population Vila Pouca it seems that seeding rate has not an important effect on this parameter.

Carnide *et al.* (1991a) reported a higher *in vitro* organic dry matter digestibility in triticale than in oats and rye. Lopez (1991) found a higher *in vitro* digestibility for triticale than for rye and barley and similar to oats. According to Jessop *et al.* (1991) triticale has interest as fodder for livestock on account of its good nutritive value.

CRUDE PROTEIN CONTENT

Crude protein content was significantly influenced by the effects year (F=43.2, P<0.05), genotype (F=100.1, P<0.05) as well as by the interaction year x genotype (F=3.5, P<0.05), while the seeding rate effect (F=1.2)and the interactions involving it (seeding rate x year, F=1.3, seeding rate x genotype, F=1.3, and seeding rate x genotype x year, F=1.0) were not significant.

The highest crude protein content was observed for all seeding rates on rye population Vila Pouca, ranging between 9.84% at the seeding rate 525 viable seeds.m^{-2} and 10.08% at 350 viable seeds.m^{-2}. Oats cv. Boa-Fé showed the lowest crude protein content ranging, between 5.04% and 5.63% at seeding rates 525 and 700 viable seeds.m^{-2}, respectively (Table 5).

Table 5 - Average crude protein content (%) for each genotype at four seeding rates and across three years.

	SEEDING RATE (viable seeds.m^{-2})				
GENOTYPE	175	350	525	700	Average
Triticale cv Presto	7.77	6.36	6.87	6.90	6.98
Triticale line UTAD 36/85	6.99	7.42	6.68	6.02	6.78
Reg. rye pop. Vila Pouca	9.97	10.08	9.84	9.85	9.94
Oats cv.Boa-Fé	5.48	5.40	5.04	5.63	5.39
Average	7.55	7.32	7.11	7.10	
LSD at 5%	0.56				

The data obtained in this study indicate that triticale line UTAD 36/85 has a great potential for forage production and can replace either the rye or the oats. Although the highest yield of dry matter in this triticale had been obtained at the seeding rate of 700 viable seed.m-2, the rate of 350 viable seed.m-2 is the one that maximizes the digestible dry matter and crude protein yields. For the regional rye population and oats cultivar the seeding rate 700 viable seed.m-2 gave the highest dry matter as well the highest number of stems.m-2.

References

1 - Guedes-Pinto H, Carnide O, Carnide V. Dois anos de ensaios de adaptação com triticales em Trás-os-Montes. Melhoramento 1982; 27: 49-89.
2 - Guedes-Pinto H., Bernard M. Étude comparative de quelques cultivars de blé, seigle et triticale dans le Nord du Portugal. I Productions de grain, de paille, de proteínes. Agronomie 1983; 3: 691-700.
3 - Carnide V, Ferreira A, Guedes-Pinto H. A comparative study of triticales lines as a forage crop. Proceedings of EUCARPIA Meeting of the Cereal Section on Triticale; 1988; 266: 591-604.
4 - Carnide V, Guedes-Pinto H, Ferreira AM, Sequeira C. Avaliação do potencial forrageiro do triticale em comparação com outros cereais. I. Corte ao emborrachamento. Pastagens e Forragens 1991a; 12: 49-59.
5 - Carnide V, Guedes-Pinto H., Sequeira C, Ferreira AM. Avaliação do potencial forrageiro do triticale em comparação com os outros cereais. II. Corte no estado de grão pastoso/pastoso duro. Pastagens e Forragens 1991b; 12: 61-74.
6 - Carnide V, Guedes-Pinto H, Ferreira AM, Sequeira C. Triticale-legume mixtures. Proceedings of the Second International Triticale Symposium; 1990 Oct 1-5; Passo Fundo, Brazil. Passo Fundo: CIMMTY 1991c; 542-545.
7 - Bishnoi UR. Effect of seeding rates and row spacing on forage and grain production of triticale, wheat, and rye. Crop Sci. 1980; 20: 107-108.
8 - Hernando, J. Comparaction de genotipos de trigo a diferentes densidades de siembra: III. Produccion y supervivencia de hijuelos. Su porte al rendimiento final. Anales del Institute Nacional de Investigaciones Agrarias 1981; 15: 73-82.
9 - Power JF, Alessi J. Tiller development on yield of standard and semidwarf spring wheat varieties as affected by nitrogen fertilizer. J. Agric. Sci. 1978; 90: 97-108.
10 - Joseph KM, Alley M, Brann D, Gravelle D. Row spacing and seeding rate effects on yield and yield components of soft red winter wheat. Agron. J. 1985; 77: 211-214.
11 - Chatha MR, Ahmad D, Gill MA. Yield of wheat cultivars as affected by different seed rates under irrigated conditions. Pakistan J. Agric. 1986; 7: 241-243.
12 - Pfahler PL, Barnett RD, Luke HH. Seasonal distribution of rye forage production in Florida as affected by population, seeding rate, location, and year. The Soil and Crop Science Society of Florida, Proceedings, 1982; 41: 61-67.
13 - Osman AE, Nersoyan N, Somarroo BH. Effects of phosphate, seed rate, seed ratio and harvesting stage on yield and quality of forage legume-cereal mixtures. Forage Res. 1983; 9(2): 127-135.
14 - Marten CB, Barnes RF. Prediction of energy digestibility of forages with in vitro rumen fermentation and fungal enzime systems. In: Pigden WC, Balch CC,

Graham M, ed. Standardization of analyticalmethodology for feeds. Ottawa, IDRC, 1980; 61-71.

15 - Association of Official Analytical Chemists (AOAC). Official methods of analysis. 12th ed., AOAC, Washington D. C., 1975.

16 - Folkins LP, Kaufmann ML. Yield and Morphological studies with oats for forage and grain productions. Can. J. Plant Sci. 1974; 54: 617-620.

17 - Lopez JR. Breeding forage and dual-purpose triticale in Bordenave, Argentina. Northern Mexico. Proceedings of the Second International Triticale Symposium; 1990 Oct 1-5; Passo Fundo, Brazil. Passo Fundo: CIMMTY 1991; 161-163.

18 - Lozano AJ. Studies on triticale forage production under semiarid conditions of Northern Mexico. Proceedings of the Second International Triticale Symposium; 1990 Oct 1-5; Passo Fundo, Brazil. Passo Fundo: CIMMTY 1991; 264-267.

19 - Wright RL, Agyare JA, Jessop RS. Selection factors for Australian grazing/dual purpose Triticales. Proceedings of the Second International Triticale Symposium; 1990 Oct 1-5; Passo Fundo, Brazil. Passo Fundo: CIMMTY 1991; 438-441.

20 - Jessop RS, Leterme P, Wright R, López J, Kolding M, Fiolho E. Triticale as forage or grain for animals. Proceedings of the Second International Triticale Symposium; 1990 Oct 1-5; Passo Fundo, Brazil. Passo Fundo: CIMMTY 1991; 64-65.

APTITUDE OF SPRING AND WINTER TRITICALES FOR DUAL-PURPOSE (FORAGE AND GRAIN) IN MEDITERRANEAN CONDITIONS

Concepción Royo, Eugenia Penella, Francesc Tribó and
José Luis Molina-Cano
Centre R + D de Universidat de Lleida, Institut de Recerca i Tecnologia
Agroalimentária, Lleida, Spain

Abstract

A nursery of thirty genotypes was grown in 1991 with the aim of testing the aptitude of triticale as a dual-purpose crop. Forage yields ranged between 2.7 and 5.2 t of DM.ha^{-1} whereas grain yields after forage removal fluctuated between 0.5 and 3.3 t.ha^{-1}.

The best ten cultivars were further tested in 1993 in a randomized complete block design experiment, with three replicates and plots of 12 m^2. In dual-purpose plots, forage was harvested at the end of tillering, after which 50 extra units of N.ha^{-1} were applied. Grain was harvested from all plots (clipped and unclipped) at maturity. Biomass samples were periodically taken to calculate growth indexes.

Winter triticale yielded 40 % more forage than spring triticale due to its higher growth. Grain yield was reduced by approximately 25 % in both winter and spring triticales by clipping. Among yield components, only thousand kernel weight was responsible for yield decrease because neither the number of spikes per m^2 nor the number of grains per spike were modified by clipping. Leaf Area Index (LAI) at anthesis and Leaf Area Duration from anthesis to maturity (LAD) were drastically diminished by cutting.

Spring triticales outyielded winter types in grain production because of their higher number of grains.spike^{-1} and kernel weight. The high tiller mortality of winter triticales led to a similar number of spikes.m^{-2} between the two groups at anthesis. Both Net Assimilation Rate between booting and anthesis (NAR) and LAD seemed to be responsible for higher grain yield production in spring triticales. The total biomass (forage + grain) produced by winter types was 6.8 t.ha^{-1} and 7.7 t.ha^{-1} for spring triticales.

Introduction

Triticale (*XTriticosecale* Wittmack) is mainly used as grain for animal feeding in the Mediterranean area. However, an increasing interest in forage and dual-purpose uses has been recently noticed. The dual-purpose practice consists in cutting (or grazing) the crop at early stages of development and then leaving the regrowth for grain production.

The potential of triticale as a dual-purpose crop has been highlighted in some recent

H. Guedes-Pinto et al. (eds.), Triticale: Today and Tomorrow, 843–849.
© 1996 *Kluwer Academic Publishers. Printed in the Netherlands.*

844

studies [1-3]. Of the two main sources of triticale cultivated today (*spring types* bred basically at CIMMYT, and *winter types* mainly developed in Eastern Europe), winter triticales are the most promising as forages [4].

Jointing or the stage when the growing points begin to elevate above ground level, is the critical stage of development for determining the time of forage removal [5]. Where the grass is being grown for grain, removal of the mainstem and primary tiller apices by early spring grazing results in reduced seed yields as the higher order tillers produce fewer and smaller seeds [6]. Forage removal usually influences grain yield depending on environmental conditions, moisture and fertility of the soil, husbandry practices, plant genotype and stage of growth at clipping [7-9]. Increases in grain yield following grazing have been associated with reduced lodging, whereas decreases in grain yield have been mainly attributed to a reduced number of spikes per m^2 at harvest [1,8,10] and to reductions in kernel weight [7]. Clipping usually modifies plant tillering, stem elongation and subsequent growth of leaves and roots [11].

The aim of this work was to compare the aptitude of winter and spring triticales for dual-purpose according to their forage and grain productions. The origin of the recorded differences were explained physiologically.

Materials and Methods

Two trials were sown at Lleida (North-East of Spain) under irrigation and high soil fertility. A nursery of 30 triticale genotypes containing winter and spring triticales was sown in 1990 in plots of 4 m^2. Forage was removed at the end of tillering, and the grain produced on the regrowth was harvested at the end of the season.

The best ten genotypes were further tested in a randomized complete block design, with plots of 12 m^2 (6 rows 20 cm apart), and three replicates. The two harvesting treatments were: (1) grain production only, and (2) cutting the crop at the end of tillering for forage and harvesting the grain produced after regrowth. Half of the plot was assigned for destructive samples of biomass. Forage and grain yields were determined by harvesting the other half (6 m^2). Both experiments were sown at the end of November at a seed density of 550 viable seeds.m^{-2}. Forage was removed at stage 31 of the Zadok's scale [12] for each genotype, leaving the apical dome intact. Fifty extra units of N.ha^{-1} were applied after cutting on dual-purpose plots. Dry matter content (DM) was calculated after drying a sample of 250 g of green forage in an oven at 70° C for 72 hours. Biomass samples were periodically taken by harvesting the plants contained in 50 cm of row length. Five randomly sampled plants per plot were selected among them in the laboratory to determine the estimates on a per plant basis. Growth indexes were calculated according to the methodology described by Ramos *et al.* [13]. Standard analyses of variance were used to analyze the data obtained. Differences between means were compared by the Duncan test.

Results

A great variability in forage and grain yield was found in the selection nursery. Forage yields ranged between 2.7 and 5.2 t of DM.ha^{-1} and grain yields fluctuated between 0.5 and 3.3 t.ha^{-1}. The best ten genotypes (five spring and five winter triticales) were further tested in 1993. Table 1 shows the forage and grain yields obtained.

Table 1. Forage and grain yields in dual-purpose plots for each genotype and mean values for both growth habits and harvesting treatments.

		Forage yield (kg.ha^{-1})	Grain yield (kg.ha^{-1})
Genotype:			
Spring:	Brumby 6	2639 b	3855 defg
	CC-7252	1889 bc	6893 ab
	CTM-19808	2766 ab	5203 cde
	SWT-1697	1424 c	5539 abc
	CTM-10816	1379 c	7022 a
Winter:	Lasko	2883 ab	4760 cdef
	LT-1	2525 b	3079 g
	MT-2	2180 bc	3607 efg
	Maliszin27/Tcle 712	3739 a	3162 fg
	CT 474.85	2894 ab	5368 bcd
Growth habit:	*Winter*	2844 a	3995 b
	Spring	2019 b	5702 a
Treatment:	*Grain*	-	6443 a
	Dual-purpose	2431	4849 b

Means sharing a common letter within a column and effect are not significantly different (P=0.05).

Table 2. Biomass components of winter and spring triticales at cuttingtime.

	Winter triticales		Spring triticales	
Crop Dry Weight (CDW) (g.m^{-2})	674.4	a	494.8	b
Weight of leaves per plant (g)	0.7	a	0.5	b
Weight of stems per plant (g)	0.7	a	0.4	b
Leaf Area Index (LAI)	7.7	a	6.0	b
Inverse of Leaf Area Ratio (1/LAR) (g.m^{-2})	89.1	a	82.7	b
Tiller number	3.6	a	1.4	b
Number of leaves in main stem	4.4	b	5.1	a
Number of leaves in tillers	10.2	a	4.2	b
Length of main stem (cm)	25.6	a	19.9	b
Length of first tiller (cm)	21.8	a	13.8	b

Differences between means followed by different letters (within a row) are significant at the 0.05 probability level.

Winter triticales outyielded spring types in forage production (approximately 140 %). The analysis of crop growth at cutting time (Table 2), revealed that spring types had slower growth rates at early stages of development. Only leaf number of the main stem was higher in spring types, as it was expected due to its superior leaf emergence rate (leaves.day^{-1}) [14]. However, the higher tillering capacity of winter triticales was responsible for the higher number of leaves per plant, Leaf Area Index (LAI), and leaf weight per plant. Winter types also had higher values for the Inverse of Leaf Area Ratio (1/LAR). These observations explain why the total above-ground dry matter yield (biomass) of winter types surpassed by 136 % the crop dry weight of spring triticales, an amount similar to the differences recorded in forage production.

Grain yield was reduced by forage removal approximately 25 %. Among yield components, thousand kernel weight (TKW) was the only once responsible for yield decrease because neither the number of spikes per m^2 nor the number of grains per spike were modified by clipping (Table 3). When grazing removes the apical dome, the tillers die but may be replaced by higher order tillers [15]. In our trial the number of tillers per plant were not significantly affected by forage removal as the terminal meristems were not removed by clipping.

Spring triticales outyielded winter types in grain production. The higher tillering capacity of winter triticales at the early stages of development did not lead to a superior number of spikes at anthesis. Most of the tillers died during stem elongation and, consequently, the number of spikes.m^{-2} did not differ between groups at anthesis. The higher grain productivity of spring types may be attributed to greater number of grains per spike and grain weight (Table 3).

Table 3 . Grain yield components for winter and spring triticales and for grain and dual-purpose treatments.

	Number of spikes.m2	Number of grains. spike^{-1}	Thousand kernel weight (g)
Growth habit:			
Winter	608 a	38,2 b	33,1 b
Spring	599 a	43,1 a	40,6 a
Treatment:			
Grain	601 a	41,4 a	38,8 a
Dual-purpose	606 a	39,8 a	35,0 b

Means sharing a common letter within a column and effect are not significantly different (P=0.05).

In growth terms forage removal reduced biomass (CDW) and LAI at anthesis, and Leaf Area Duration (LAD). The efficiency indexes (1/LAR, Net Assimilation Rate or NAR and Grain:Leaf Ratio or G) increased after cutting, as was expected.

Table 4. Growth indexes at anthesis for winter and spring triticales and both harvesting treatments.

	CDW (g.m^{-2})	LAI	1/LAR (g.m^{-2})	NAR (g.m^{-2}. week^{-2})	LAD (m^2.m^{-2} .week)	G (kg.ha^{-1}. week)
Growth habit:						
Winter	2036 a	6,3 a	340 a	25,4 b	12,1 b	446 a
Spring	2138 a	6,4 a	350 a	45,6 a	16,3 a	464 a
Treatment:						
Grain	2391 a	7,9 a	314 b	23,1 b	18,6 a	385 b
Dual-purpose	1783 b	4,8 b	377 a	48,0 a	9,7 b	525 a

Means sharing a common letter within a column and effect are not significantly different (P=0.05). See text for abbreviation meanings.

The higher capacity of spring triticales to recover after cutting may be explained in terms of NAR and LAD (Table 4).

Discussion

The wide range of forage and grain yields obtained in the 1991's nursery revealed that enough variability exists within triticale to select successfully for dual purpose. Winter triticales, that are believed to be better for forage production at advanced stages of development [4] were also better to produce forage at the end of tillering. The higher values of the traits that define growth (LAI, 1/LAR, tiller and leaf number per plant and stem length) at early stages of development in winter triticales, explain the observed differences in biomass and forage yield between the two groups of germplasm.

Grain weight was only responsible for the reductions observed in grain yield by forage removal. Yield components are sequentially defined during the plant cycle. At early jointing, when cuting was carried out, the potential number of spikes.m^{-2} and the number of spikelets.spike^{-1} (and, in consequence, the maximum number of grains.spike^{-1}) were already defined. As the apex was not removed by clipping, only the remaining yield component to be defined (grain weight) was significantly affected. The decrease in TKW may be due to the delay in heading date produced by cutting [9,7], which in Mediterranean conditions leads these plants to mature under higher temperatures, typical of late spring. Cultivar differences in response to cutting have been related to phenological development [16].

The number of spikes.plant^{-1} has been reported as the main component responsible for yield reductions in some previous studies in Mediterranean conditions [1,2]. In both works, reductions in kernel weight and specific weight were also recorded but they were too small to be statistically significant.

LAI and LAD were drastically affected by cutting. It has been already pointed out that reduced grain yield following forage harvest can be due to decreased leaf area at anthesis or leaf area duration from jointing to anthesis [2,17]. Also, reductions in dry matter yield and winter growth rates have been associated with cutting [16]. In wheat lower yields of crops defoliated early in their phenological development resulted from reduced leaf area available to support production [18].

Spring triticales yielded more than winter types after cutting. The total biomass (forage + grain) produced by winter types was of 6.8 t.ha^{-1} and for spring triticales 7.7 t.ha^{-1}. The strong tiller mortality of winter triticales led to a similar number of spikes.m^{-2} between the two groups at anthesis. The larger number of grains.spike^{-1} and kernel weight in spring types led them to surpass winter triticales in grain production.

Acknowledgments: We wish to thank the staff of the Area de Conreus Extensius of Centre UdL-IRTA for their skilled technical assistance. This work was partially supported by CICYT AGR90-0509-C02.

References

1. Royo C, Montesinos E, Molina-Cano JL, Serra J. Triticale and other small grain cereals for forage and grain in Mediterranean conditions. Grass & For Sci 1993, 48:11-17.
2. García del Moral LF. Leaf area, grain yield and yield components following forage removal in triticale. J Agron & Crop Sci 1992; 168: 100-107.
3. Scott WR, Hines SE. Effects of grazing on grain yield of winter barley and triticale: the position of the apical dome relative to the soil surface. New Zealand J Agr Res 1991; 34: 177-184.
4. National Research Council. Triticale: A promising addition to the world's cereal grains. Washington: National Academy Press, 1989.
5. Gilmore EC. Forage production of new small grains. Proceedings of the 6th pasture and forage crops short course 1971; Texas A & M Univ., College Station: 96-110.
6. Brown KR. Seed production in New Zealand grasses. 1. Effect of grazing. New Zealand J Exp Agr 1980; 8: 27-32.
7. Royo C, Insa JA, Boujenna A, Ramos JM^{a}, Montesinos E, García del Moral LF. Yield and quality of spring triticale used for forage and grain as influenced by sowing date and forage cutting stage. Field Crops Res (in press).
8. Dunphy DJ, McDaniel ME, Holt EC. Effect of forage utilization on wheat grain yield. Crop Sci 1982; 22: 106-109.
9. Poysa VW. Effect of forage harvest on grain yield and agronomic performance of winter triticale, wheat and rye. Can J Pl Sci 1985; 65: 879-888.
10. Brignall DM, Ward MR, Whittington WJ. Yield and quality of triticale cultivars at progressive stages of maturity. J Agr Sci Cambridge 1988; 111: 75-84.
11. Milthorpe FL, Davidson JL. Physiological aspects of regrowth in grasses. In: Milthorpe FL, Ivins JD, editors. The growth of cereals and grasses. London: Butterworths, 1966: 241-255.
12. Zadoks JC, Chang TT, Konzak CF. A decimal code for the growth stages of cereals. Weed Res 1974; 14: 415-421.

13. Ramos JM, García del Moral LF, Recalde L. Vegetative growth of winter barley in relation to environmental conditions and grain yield. J Agr Sci Cambridge 1985; 104: 413-419.
14. García del Moral LF, Boujenna A, Insa JA, Arbonés A, Royo C. Evaluation of a set of triticale genotypes for dual-purpose (forage and grain). Melhoramento (in press).
15. Langer RHM. How grasses grow. Studies in biology 1979; 34. London. Edward Arnold. 66 pp.
16. Andrews AC, Wright R, Simpson PG, Jessop R, Reeves S, Wheeler J. Evaluation of new cultivars of triticale as dual-purpose forage and grain crops. Aust J of Exp Agr 1991; 31: 769-775.
17. Dunphy DJ, Holt EC, McDaniel M.E. Leaf area and dry matter accumulation of wheat following forage removal. Agron J 1984; 76:871-874.
18. Winter SR, Thompson EK. Grazing duration effects on wheat growth and grain yield. Agron J 1987; 79: 110-114.

THE EFFECTS OF SOWING DATE ON THE GROWTH AND YIELD OF DUAL-PURPOSE AND GRAZING TRITICALES UNDER COOL AUSTRALIAN CONDITIONS

Joseph A. Agyare, Roy L. Wright and Robin S. Jessop
Department of Agronomy and Soil Science, University of New England,
Armidale NSW 2351, Australia

Abstract

Two field experiments examined the phasic development, tillering, dry matter production and grain yield of a number of registered and unregistered forage triticales (*Tritico secale*) compared to oats (*Avena sativa*). The experiments were conducted in Armidale (elevation 1070m) on the Northern Tablelands of New South Wales, Australia (30°S). The experiments were sown in either early or late autumn and included uncut, single and double cutting treatments to assess the forage and grain yields of each line. The oats remained vegetative longer and frequently produced more tillers than the triticales; the only exception was the triticale cv. Empat which, when sown in early autumn, produced increased tiller numbers. Dry matter production was highest with the triticales in both early and late sowings although increasing cutting frequency decreased total dry matter, particularly with the unnamed triticales. Oats showed less effect of cutting on grain recovery but the triticales in the uncut treatment always gave the highest grain yields. The results are discussed in relation to the future of grazing of dual-purpose triticales in the cooler areas of eastern Australia.

Introduction

Forage cereal crops are widely grown in the southern and eastern regions of Australia to fill winter feed gaps when pasture growth is limited by low temperatures. Oats (*Avena sativa*) has historically been the most popular cereal for this purpose but recent studies [1] have indicated that triticales (*Tritico secale*) are equivalent or superior to a range of oat varieties in forage production and grain yield with digestabilities in the range of 70-80%.

In the Northern Tablelands of New South Wales, oats intended for grazing are frequently sown from early February to early March; such oat sowings commonly suffer from leaf (*Puccinia recondita*) and stem rust (*P. graminis*) infections [2]. Newer triticales have low levels of leaf rust and have been selected for high stem rust resistance.

H. Guedes-Pinto et al. (eds.), Triticale: Today and Tomorrow, 851–857.

© 1996 *Kluwer Academic Publishers. Printed in the Netherlands.*

Variable rainfall often means that it is difficult to sow grazing cereals at the optimum time, which appears to be from late summer (February) to early autumn (late March). The objectives of this study were two-fold: firstly, to select dual-purpose triticales from triticale cultivars and newer lines for high levels of forage production and grain yield, and secondly, to compare their performance under similar conditions to widely-used oat cultivars when sown at different times in the early and late autumn.

Materials and Methods

The experiments were conducted at the Laureldale Research Station, Armidale (University of New England, latitude 30°S, longitude 150°E, altitude 1070 m, 500 km north of Sydney with a cool temperate climate). Soil type was chocolate basaltic with 70% clay content, pH 6.5 (1:5 soil:water) and a 0.19% total nitrogen content.

The experiments used a split-plot randomised complete block design with 8 main plots (6 triticales, 2 oats), 3 sub-plots (3 cutting systems) and 4 replicates. Main plot size was 6 x 1.24m and the sub-plot size was 2 x 1.2m. The 8 cultivars used were 2 released triticales (Empat, a dual-purpose longer season line and Tiga, a shorter season grazing line), 2 widely used oats (Blackbutt and Carbeen) and 4 unregistered triticales (AT40, AT41, AT43 and AT46). The three cutting systems, designed to simulate zero or repeated grazing, were uncut (dry matter and grain yield being measured at maturity), one-cut (a single cutting on 7 July, 1989 and on 17 July, 1990 followed by a dry matter and grain yield recovery harvest at maturity) and two-cut (similar to one-cut but with an additional dry matter cutting on 9 October, 1989 and on 18 September, 1990).

The experiments were sown on 3 May, 1989 (late sowing) and 22 March 1990 (early sowing); for these forage species on the Northern Tablelands. Seed rates were adjusted so that plant establishments were close to 33 plants/m of row with a row spacing of 18cm. 100kg of ammonium nitrate/ha (34% N) was applied to all plots after each cutting. In winter and early spring, three plants per treatment were randomly removed and their apical meristems viewed under the microscope to assess their stages of development, using the scale ranging from 0 (vegetative) to 12 (spikelet primordia) described by Andrews *et al.* 1991 [1]. At each cutting time, tiller numbers (i.e. all rooted shoots with two or more leaves) were counted on a 1m length row. Plants in the relevant subplot were cut to a height of 2cm above ground level; plant material was collected from the six inner rows and dried to assess dry matter production. A split plot analysis of variance was carried out for all data; in the case of tiller production where counts were involved, the analysis was carried out on square root transformed data in order to make the error variance independent of the means [3].

Results

PHASIC DEVELOPMENT

The apical development scale used in both years was the same with a score of 4 as the beginning of floral initiation (Table 1). On 10 August 1989, the two oat varieties and AT43 were the only ones which remained vegetative. By early spring of this year (September) ear formation had started in all cultivars. A similar trend occurred with the earlier sowing in 1990/91 which showed the oats to be less advanced in June and July. Although there appeared to be differences between the oats and the triticales in development patterns, the varietal means were only statistically different in 1989 (Table 1).

TILLER DEVELOPMENT

Tiller development was greatest in the oat variety Blackbutt in 1989; in the following year, with earlier sowing, Empat produced equal tiller numbers to Carbeen and these varieties gave the greatest number of tillers per plant.

Cutting frequency had no effect on the number of tillers produced per plant in 1990 whilst it had some effects in 1989 but differences were not significant. Cutting once or twice increased tillering in Blackbutt, Tiga and AT46 with little effect on the other varieties.

DRY MATTER PRODUCTION

In the first year's experiment (1989/90 with late sowing), all the triticale varieties produced significantly higher levels of dry matter than Blackbutt and Carbeen oats (Table 2). Whilst there was no significant difference between the oat varieties in dry matter production in either year, more dry matter was produced with the earlier sowing in 1990. Within the triticales, AT40, AT41 and Tiga gave the best dry matter yields with late sowing (1989) whilst Empat and AT43 produced the highest dry matters with earlier sowing in 1990.

Cutting frequency had no statistically significant effect on dry matter yields with the late sowing in 1989 (Table 2) but there was a suggestion that total dry weight production decreased with increased frequency of cutting. In 1990, this effect was more pronounced with close to 50% reductions in the unnamed AT lines. Such an effect was either much reduced or absent in the oat varieties.

GRAIN RECOVERY

Grain recovery after grazing or grain produced from the no-cut treatments was highest from the triticale lines and this was relatively little (compared to dry matter production) influenced by time of sowing (Table 3).

The oat cultivars showed less changes in grain yield following grazing compared to the triticales, particularly in 1990. With late sowing in 1989, the no-cut treatment produced the highest grain yields; the triticale lines Empat, AT41 and AT43

Table 1. Development scores for oat and triticale varieties (1989 and 1990) (See methods section for details of score system.)

Crop/cultivar	10/8/89	7/9/89	15/6/90	17/7/90
Oats				
Blackbutt	2.40	10.69	1.00	1.76
Carbeen	1.48	11.06	1.50	3.19
Triticales				
Empat	4.45	7.94	4.08	3.41
Tiga	5.73	10.81	5.83	6.83
AT40	5.05	8.63	4.50	5.59
AT41	5.45	9.31	6.10	8.00
AT43	3.90	8.19	4.50	4.76
AT46	5.28	10.75	5.58	5.25
S.E.	0.297	0.297	0.440	0.440

$P < 0.05$

Table 2. Effects of cutting frequency on dry matter production (t/ha) in 1989 (late sowing) and 1990 (early sowing).

Crop/Cultivar	1989			1990		
	Uncut	One-cut	Two-cuts	Uncut	One-cut	Two-cuts
Oats						
Blackbutt	2.96	2.45	3.33	4.65	5.09	4.29
Carbeen	2.90	2.52	2.34	5.46	4.80	3.42
Triticales						
Empat	5.66	6.03	5.10	9.80	9.32	6.82
Tiga	6.47	5.34	5.18	9.83	6.50	5.35
AT40	6.17	6.67	6.19	10.53	7.66	5.37
AT41	6.14	5.33	5.89	11.47	8.25	4.83
AT43	5.27	5.37	4.79	13.13	8.64	6.46
AT46	5.18	4.99	4.58	9.30	7.66	4.44

SE (Standard Error) of mean = 0.418   SE (Standard Error) of mean = 0.506
$P < 0.01$ $P < 0.01$

Table 3. Effects of cutting system on grain yield (t/ha) in 1989 (late sowing) and 1990 (early sowing).

Crop/cultivar	1989			1990		
	Uncut	One cut	Two cuts	Uncut	One cut	Two cuts
Oats						
Blackbutt	1.00	0.54	0.76	1.48	1.59	1.16
Carbeen	1.05	0.76	0.35	0.96	1.27	0.51
Triticales						
Empat	3.32	2.78	1.86	2.67	2.60	1.34
Tiga	1.63	1.69	1.53	1.42	1.00	0.60
AT40	2.33	2.41	1.83	1.57	1.27	0.82
AT41	3.20	2.81	2.34	2.72	1.59	0.91
AT43	3.13	2.92	1.69	3.91	2.42	1.06
AT46	1.07	1.42	0.98	0.98	0.93	0.33

SE of means = 0.205 SE of means = 0.218

$P < 0.01$ $P < 0.01$

gave greater yields than other lines. The two-cut treatment reduced grain yield compared to the single cutting in Carbeen, Empat and AT43.

With the earlier sowing in 1990, Empat, AT41 and AT43 gave the highest grain yields under the no-cut treatment. The cutting treatments again reduced grain yield in some of the triticales; this occurred with Empat (two cuts), Tiga (two cuts), AT41 (one cut) and AT43 (one cut).

Discussion

These two experiments support earlier studies [1] which indicated that the newer triticales had at least equivalent forage production to well-adapted oat varieties in the cooler temperate regions of eastern Australia. Ambient temperatures for the current study were close to those of the long-term means for 118 years for the Tablelands areas, with a wetter growing season in 1990 (20% higher than the long-term average rainfall for 110 years). This increased rainfall, combined with earlier sowing, gave rise to the greater dry matter levels in 1990 compared to the later sowing and drier conditions in 1989 close to normal rainfall.

In both years, the triticale lines (essentially spring types) appear to have only slight vernalisation requirements, especially for Tiga and Empat (1), and were more advanced in their phasic development during the winter (June - August) than the oat varieties. This rapid phasic development was generally associated with fewer tillers in both years but, with later sowing, the well-adapted line Empat was superior to the other triticales.

Whilst simulated grazing, single or double cutting appeared to have little effect on tiller production in most lines, indications were that defoliation could stimulate tillering in oats and some triticales.

The triticales produced up to double the dry matter yields of the oats in both early and late sowings and this advantage in potential grazing value was larger than those produced in earlier studies [1]. This degree of yield advantage needs to be confirmed in larger farm situations over a range of environments.

Delayed sowing with less rainfall reduced dry matter yields in all varieties, both cut and uncut. Compared to the uncut yields, the percentage reduction was greatest in the higher yielding early sown experiment in 1990 (Table 2). Blackbutt oats showed least negative effects of cutting in both years. In 1989, all species showed minimal effects of cutting on dry matter production, however in 1990, the two-cut treatment markedly reduced dry matter yields particularly in triticales which had advanced phasic development.

The ability of current and newer lines of triticale to produce larger quantities of dry matter than other winter cereal crops has also been demonstrated elsewhere [4,5]. Although the negative effects of winter grazing or cutting appear to vary between years, these could be minimised by sowing mixtures of two or more lines of different phasic development rates; early maturing spring types would provide early grazing and late maturing types with winter vegetative growth would provide late grazing [4].

The triticale varieties yielded up to three times the grain production of the oats when sown late in autumn and double the yield with an earlier sowing. There appeared to be few season x variety interactions with comparative grain yields; in both years Empat, AT41 and AT43 produced the highest grain yields from the uncut treatments.

However, there were marked differences in the varietal grain yield responses to the cutting treatments between the two experiments. As an example, the decreases in grain yield after two cuttings during growth were similar in each year for Blackbutt and Empat whilst some of the triticales showed much more severe reductions in the 1990 experiment. The lines with the largest reduction in grain yield tended to be those which were most phenotypically advanced in mid-winter (July 1990) such as AT41. Therefore, it is suggested that for optimum dry matter and grain yields in dual-purpose triticales, grazing or cutting strategy needs to be closely tied to a knowledge of their apical development patterns.

Grain yields were not high in either experiment but were similar to those in earlier studies with older varieties[1,6]. In excess of 100,000 ha of grain triticale are presently grown in Australia and newer lines are currently under assessment which are expected to be more disease-resistant and to produce high herbage and grain yields. The future of triticale as a dual-purpose crop in Australia appears promising because it has high dry matter yields, high quality herbage, the ability to produce higher grain yields on acid soils than other cereal crops and accepted markets as a stockfeed grain provided grain quality can be maintained or improved [7,8,9]

Acknowledgements

We wish to thank the Grains Research and Development Corporation (GRDC) of Australia for funding this research. We also thank Mr. J. Kruideniner and Mr. D. Wheatley for their technical assistance and Miss V. Hall for secretarial assistance.

References

1. Andrews AC, Wright R, Simpson PG, Jessop R, Reeves S, Wheeler J. Evaluation of new cultivars of triticale as dual-purpose forage and grain crops. Aust J Exp Agric 1991;31:769-775.
2. Crofts FC. Nitrogen fertilisers for balancing pasture production with animal needs. In: James BJF, editor. Intensive Utilisation of Pastures. Sydney: Angus and Robertson. 1965: 76.
3. Snedecor GW. Statistical methods. 5th ed. Iowa State University Press, Ames, Iowa, U.S.A. 1962.
4. Lozano AJ. Studies on triticale forage production under semi-arid conditions of northern Mexico. Proceedings of the Second Internation Triticale Symposium (ITS); 1990 Oct 1-5; Passo Fundo, Brazil. Passo Fundo: CIMMYT 1991.
5. Lopez JR. Breeding forage and dual-purpose triticale in Bordenave, Argentina. northern Mexico. Proceedings of the Second Internation Triticale Symposium (ITS); 1990 Oct 1-5; Passo Fundo, Brazil. Passo Fundo: CIMMYT 1991.
6. Wright RL, Jessop RS, Matheson E. Selection of dual purpose/grazing triticales suitable for cool Australian climates. Proceedings of the First International Triticale Symposium (ITS); 1986 February 2-8; Sydney, Australia; Sydney: Australian Institute of Agricultural Science Publication, 1987.
7. Hill GM. Triticale in animal nutrition. In northern Mexico. Proceedings of the Second Internation Triticale Symposium (ITS); 1990 Oct 1-5; Passo Fundo, Brazil. Passo Fundo: CIMMYT 1991.
8. Jessop RS, Leterme P, Wright R, Lopez J, Kolding M, Fialho E. Triticale as forage or grain for animals. Proceedings of the Second Internation Triticale Symposium (ITS); 1991 Oct 1-5; Passo Fundo, Brazil. Passo Fundo: CIMMYT 1990.
9. Jessop RS, Wright RL. Triticale. In: Jessop RS, Wright RL, editors. New Crops: Agronomy and Potential of Alternative Crop Species. Melbourne: Inkata Press, 1991: 175-182.

DEVELOPMENT OF TRITICALE AS A FORAGE CROP AND GRAIN FEED IN ALGERIA

Abdelkader Benbelkacem and Ali Zeghida
Experimental Station ITGC El Khroub, Constantine, Algeria

Abstract

Triticale is promoted in Algeria with the aim of reducing maize and barley imports. The crop is mainly grown for forage production (hay, green, silage and mixed with legumes), as a grain feed or for dual purpose (forage and grain).

Many new adapted spring cultivars were developed during this last decade (all cultivars are complete types). In terms of production, triticale outyielded durum wheat and barley by 11% in favourable and semi-arid environments. However, bread wheat still outyields triticale (10% increase) in the semi-humid zones.

Forage production (fresh biomass and dry matter) of triticale gave encouraging results in comparison to barley and oats, when used in mixture with vetch or peas for hay production, the amount of legume in the mixture was always higher with triticale than with oats or barley.

Triticale produces more straw, commonly fed to sheep, than other crop during dry years (frequent in Algerie) where the free market price increases drastically. Straw yield of triticale compensated largely for its low price when compared to barley straw or wheat and gave in general better economic return to farmers (1230 USD/ha for triticale vs 1120, 1080, 735 and 685 USD for bread wheat, durum wheat, barley and hay of oats respectively).

Introduction

Research efforts to improve triticale, conducted and supported predominantly by CIMMTY, have shown the great potential of this crop when introduced in the farming systems of developing countries (Carney, 1990, Skovmand *et al.*, 1984 and Varughese *et al.*, 1987). Triticale has shown a great adaptability to stressed environment including drought stress, and a good resistance to a large number of cereal diseases.

In Algeria, research on triticale was initiated in the early 1970's through the "Wheat Project" in collaboration with CIMMTY. However, this crop did not gain any interest until the late 1980's when the Ministry of Agriculture started to actively promote triticale to production through local and special projects. The area cropped to triticale has drastically increased 81500 ha in 1984 as compared to 15000 ha in 1992 mainly as forage), although, its still currently accounts for less than 0,52% of the total area grown to cereals.

859

H. Guedes-Pinto et al. (eds.), Triticale: Today and Tomorrow, 859–865.
© 1996 *Kluwer Academic Publishers. Printed in the Netherlands.*

Average productivity of triticale exceeds that of durum wheat and barley in the northern zone of the country. However, substantial yield variability is noted between the different zones (from the littoral to high plateau). On station and on farm research has consistently shown that triticale is on the average more productive and more resistant to most diseases and lodging than local and new barley or durum wheat varieties (ITGC Annual Reports 1990-1993). Results recorded in severe situations indicated that triticale performed better than barley and wheat. This is also the case this season (1993/1994), where severe early and terminal drought took place.

Triticale was originally introduced in order to reduce maize and barley imports for poultry and livestock feed. Some researchers acknowledged that declining self sufficiency levels for wheat coupled with steadily increasing food demand, triticale will ultimately play an important role in bread making (e.g. adding up to 25% of triticale to wheat flour).

DEVELOPMENT OF TRITICALE GERMPLASM FOR GRAIN AND FORAGE PRODUCTION

During the last five seasons, the development of triticale focused on three different end-uses: grain, forage and dual purpose

For grain, high yield potential, yield stability, grain quality, straw quantity and quality, drought, disease and lodging resistance are the traits for selection.

For forage, a high biomass production and forage quality are the main criteria for development.

For dual purpose, the most important characteristics are a vigourous early development, rapid regrowth capacity after grazing or clipping, a high biomass and leaf index at anthesis, a large leaf area duration from anthesis to maturity with good grain and forage quality (V. Carnide *et al.* 1985, 1990, Lopez, 1990).

The triticale materials evaluated express a wide range of genetic variation as demonstrated by different growth habits and production potential.

Thus it is possible to select new combinations with potential for different types of forage, i. e., green forage, hay making, silage or grazing in many regions of the country.

FORAGE

Triticale as a forage crop, was and is still mainly used in mixtures with legumes for hay production.

Research results have shown a clear advantage of triticale over other forage cereal types (barley, oats), in fresh biomass yield, dry matter and quality in terms of legume-triticale proportions in the final product, where the legume level, which is the main source of nitrogen, seen as the main contraint for livestock feeding, was higher than with barley or oats, without affecting the total dry matter production. Indeed, the legume (vetch and/or peas) proportion was higher in triticale (30% of the total production as compared to 15 and 18% with oats and barley respectively) (table 1).

Table 1: Legume-cereal association production (100 Kg/ha)

MIXTURES	TOT. DM	LEG. DM	GER. DM	LEG. %
VETCH-OATS	136.55	20.74	115.78	15.75
PEAS-OATS	145.05	16.11	128.87	15.82
VETCH-BARLEY	153.32	22.49	134.00	18.18
PEAS-BARLEY	143.42	22.23	121.02	17.22
VETCH-TRITICALE	136.06	47.99	88.07	32.15
PEAS-TRITICALE	130.57	38.60	91.96	29.35

When used for dual purpose: grazed/grain, triticale performs better than other cereals when it is grown for grain only, its grain yield is less affected by diseases, predators and lodging as does barley. Triticale and barley showed a better yield recovery than oats, after grazing or cutting (table 2).

Early facultative and late spring triticales used in this study gave the highest yielding cut forage/grain dual purpose. Results suggest the possibility of developping grain triticales with high potential for early forage, hay and whole crop silage.

Table 2: Dual purpose yield results (100 Kg/ha)

CROPS	GRAIN YIELD	YIELD / CUTTING
OATS	33.09	32.09 (3 Years)
BARLEY	39.45	50.35 (2 Years)
TRITICALE	48.14	50.05 (3 Years)

GRAIN

Through two decades of rechearch, major advances have been in overcoming problems of sterility, shrunken endosperm, susceptibility to lodging, to some diseases and low yields etc...

Extensive triticale materials (originating mainly from CIMMTY) possessing high yield potential, diseases resistance, good quality and plump grains have been obtained. Three cultivars are already released for seed increase (Juanillo 159, DOC 7 and Clercal), and many other newly developed lines compared favourably for various agronomic and quality characteristics with wheat are in the pipeline (LAMB 2, REH/HARE 212-11, RHINO.1.).

In general, triticale performed better than wheat under rainfed conditions in Algeria. The best triticale cultivars outyielded the best wheat cultivars in the semi-arid zone during the last two seasons (35.5 9x/ha vs 26.5 9x/ha). This superior performance is mainly due to better adaptability to this type of environment. The combined average grain yield on station and on farm trials showed a clear advantage of triticale over durum wheat, barley and bread wheat (28%, 26% and 4,5% increase respectively) (table 3).

Table 3: Highest grain yield of cereals combined over years and zones (1990/1993)- (100 Kg/ha).

	FAVORABLE ENVIRONMENT	SEMI ARID ENVIRONMENT	MEAN
DURUM WHEAT	45.72	22.00	33.86
BREAD WHEAT	56.48	26.50	41.49
TRITICALE	51.26	35.50	43.38
BARLEY	46.92	19.00	34.46
MEAN	50.09	25.75	-

However the new developed bread wheat cultivars outyielded triticale by 9%, mainly in the sub-humid environment.

Under drought conditions and regions with soil problems, complete triticale show a distinct yield superiority and appear to have advantages over wheat.

This indicates that the future of triticale as a commercial crop lies with complete type triticales in more marginal environments, while it could complement wheat production in optimal environments.

In light of these promising prospects the ministry of agriculture initiated (late 1980s) a program to promote the cultivation of triticale. In order to encourage farmers adoption, the ministry of agriculture has from the outset guaranteed a producer price that is almost the double of the barley. This makes triticale more attractive than barley for production (table 4).

Table 4: Evolution of cereal prices (Algerian Dinar/100 Kg)

Years	Durum Wheat	Bread Wheat	Barley	Triticale
1988	270	220	170	170
1989	400	300	230	130
1990	500	330	230	230
1991	540	370	230	250
1992	1025	910	470	715

Algerian Dinar (DA). 1986 1 USD - 5AD, 1992 A USD - 22 AD

Despite the availablity of local research data supporting the positive effect of triticale on poultry and ruminant productivity, the incorporation of triticale in the diet is still limited.

STRAW

Triticale straw is also widely used for on-farm animal feeding even though some farmers complained about its poor quality. It is mainly due to this quality parameter that the price of triticale straw is lower than the one of barley.

However, under rainfed conditions and in severe dry years and because of its stature the triticale straw yield is at least twice as for barley or wheat. This yield advantage (table 5) compensates for the relative lower grain price compared to wheat.

From the rapid appraisal survey done last season (1992/93) it appears that triticale fields gave a better net return to farmers (1230 US$/ha) vs 1120, 1080, 735 and 685 US$ for bread wheat, durum wheat, barley and oat hay respectively.

Table 5: Survey result

Farmers	D. Wheat		B. Wheat		Barley		Triticale		Hay
	Grain	Straw	Grain	Straw	Grain	Straw	Grain	Straw	
Souici	18	55	20	40			23	215	
Kadri	17.5	70	17	28	14	50	17	65	28.5
Bouaoun	18.2	70	21.2	65	15.7	55		90	72
Baraouia	22.3	60	25.4	58	21.6	60	13	70	65
Laaziz	16.4	58	16	49	7.3	65	28	350	100
Boucheb	16	26	22	32	12	85	15	113	40
Rahal	17	70	31	65			9	90	147
Mean	17.9	58.4	21.8	48.1	14.1	63	17.5	142	75.4
Str=Gra	58.8		5.2		19.9		20		
Yield	23.7		27		34		37.5		
Return	23750 AD		24840		16150		27000		15080

1 USD = 22 AD (Algerian Dinar)

Conclusion

One of the most critical issues in Algeria is that in the last few years food production has consistently failed to meet the domestic demand, and population is expected to increase by about 60% by the year 2000. New crops or technologies that allow the increase in food production in Algeria without substantial increase in inputs would directly adress this critical issue.

Triticale has undoubtedly a niche in the rainfed farming system of Algeria. Triticale varieties currently grown have demonstrated a resistance to major diseases and lodging. This has translated into higher and less variable yields compared to barley or wheat. In addition, triticale benefits from a relatively good guaranteed producer price. This advantage has made triticale production more attractive. Yet triticale adoption has been very slow, leading to a share (area) of less than 1% of total cereal area. One major constraint appears to be a shortage of seed.

References

1. Carney, J. Triticale production in the Central Mexican Highlands: Smallhoders' experiences and lessons for research. CIMMTY. Economics Paper nº2. Mexico, DF. 1990.

2. Skovmand, B. *et al.*, Triticale in Commercial Agriculture: Progress and Promise. Advances in Agronomy, 1984, Vol.37.

3. Varughese, G. *et al.* Triticale. CIMMTY, Mexico, DF. 1987.

4. ITGC, Annual Reports. 1990-1993.

5. Carnide, V. *et al.* Study of a triticale variety for forage. Eucarpia Meeting, Clermont Ferrand, France. 685. 1985.

6. Carnide, V. *et al.*, Triticale-legume mixtures. Proceedings of the 2nd Int. Triticale Symposium, Brazil, P 542. 1990.

7. Lopez JR. Breeding forage dual purpose triticale in Bodenave Argentina, Proceedings of the 2nd Int. Triticale Symposium, Brazil, P 161. 1990.

PROSPECTS OF TRITICALE AS A DUAL PURPOSE CROP

Jaswinder Singh, Gurnam S. Dhindsa, Govinder S. Nanda and Rohit K. Batta
Department of Plant Breeding, Punjab Agricultural University Ludhiana, India 141004

Abstract

Eight triticale genotypes were evaluated using two currently cultivated fodder oats varieties as control. The experiment comprised of 2 sets viz. grain purpose (with all the genotypes grown to maturity) and fodder plus grain purpose (all the genotypes harvested for fodder seventy days after sowing and then allowed to regenerate and set seed). The data were recorded for green fodder yield, dry matter yield, grain yield, plant height, tillers per meter, harvest index and fodder quality attributes. Distinct genetic differences were observed for most of the characters particularly number of tillers per meter in the regenerated crop after the fodder harvest. In dual purpose experiment there was a general reduction in plant height (up to 31 per cent) and a corresponding reduction in lodging tendency which is a desirable feature. The salient finding of the study was the identification of dual purpose triticale genotype like TFL-3 which gave adequate fodder yield (72% of oats) without any significant reduction in grain yield (96% of the grain purpose experiment) showing its potential as a fodder cum grain crop.

Introduction

Triticale (X *Triticosecale* Wittmack) has acquired importance as a crop owing to its tolerance to biotic and abiotic stresses, better nutritional quality and high productivity, particularly on marginal lands. In most of the countries triticale is accepted as a grain crop for feed and food and is commercially grown in Australia, Poland and Brazil etc. Many workers have investigated the potential of triticale as a forage crop. Triticale was found to have higher forage yield and protein content than wheat, rye and barley (1). The novel concept of using forage cereals first for grazing or harvesting green fodder and allowing regeneration for grain production has been in circulation for some time (2, 3, 4). The present study investigates the usefulness of triticale for this purpose.

Materials and methods

Ten triticale genotypes including 2 commercially cultivated oats checks were sown in randomized block design in 3 mt x 1.38 mt plots in November 1992 at Punjab Agricultural University, Ludhiana (India). The experiment comprised of 2 Sets. In Set 1, all the genotypes were grown to maturity and harvested for grain. In Set 2, all the genotypes were harvested for fodder seventy

H. Guedes-Pinto et al. (eds.), Triticale: Today and Tomorrow, 867–871.
© 1996 *Kluwer Academic Publishers. Printed in the Netherlands.*

days after sowing and then allowed to regenerate and set seed. Data were recorded for plant height, grain yield, harvest index, tillers/m in both the Sets and green fodder yield, dry matter yield and crude protein content in Set 2 only. The crude protein content was determined by Kjeldahl procedure. The dry matter yield/ha was extrapolated from the weight of 300 gm fresh samples dried in the oven at 70^0 C for 72 hrs. The regeneration ability of different varieties was inferred from the number of tillers per meter row length in the regenerated crop in Set 2.

Results

GRAIN YIELD

Grain yield is of prime importance both in grain purpose and dual purpose triticales. Observations on grain yield in Set 1 and Set 2 (Table 1) revealed TFL-3 as the best dual purpose triticale. In Set 2 which was grown for dual purpose of fodder and grain, no reduction in grain yield was observed in case of TFL-3. All the other genotypes recorded significant reduction in grain yield.

Table 1: Grain yield and other characteristics of triticale genotypes grown as 'grain only ' (Set 1) and dual purpose (Set 2) crop.

Genotypes		Grain yield (Kg/plot)	Tillers/m	Plant height (cms)	Harvest Index
TFL-1	Set 1	1.75	92.00	143.80	0.21
	Set 2	0.44	86.50	124.20	0.09
TFL-2	Set 1	1.90	104.00	149.80	0.22
	Set 2	0.69	76.00	120.50	0.13
TFL-3	Set 1	1.77	82.00	147.20	0.21
	Set 2	1.70	80.00	130.20	0.27
TFL-4	Set 1	1.81	105.50	142.70	0.25
	Set 2	1.30	62.00	135.30	0.24
TFL-5	Set 1	1.55	119.00	144.70	0.19
	Set 2	1.10	104.00	139.40	0.18
TFL-6	Set 1	1.32	136.50	141.90	0.16
	Set 2	0.29	77.50	104.40	0.06
TFL-7	Set 1	2.48	112.50	111.30	0.36
	Set 2	0.63	74.50	89.39	0.23
TFL-8	Set 1	1.42	106.00	124.10	0.18
	Set 2	0.29	81.00	86.30	0.06
KENT (Oats)	Set 1	1.64	78.50	153.10	0.18
	Set 2	1.25	84.00	125.90	0.19

OL-9	Set 1	1.24	73.00	162.90	0.15
(Oats)	Set 2	0.70	81.50	129.40	0.12
Mean	Set 1	1.69	105.45	142.15	0.21
	Set 2	0.84	80.70	118.50	0.16
CD (0.05)					
Individual mean		0.27	NS	13.06	0.03
Overall mean		0.08	13,54	4.13	0.01

TILLERS/M

A comparison of number of tillers/m row length in Set 1 and Set 2 in case of each genotype (Table 1) indicates their regeneration potential and suitability for the dual purpose. It was observed that this parameter did not differ significantly between the two sets for either oats or triticale genotypes individually though the count was significantly higher in Set 1 on the whole. It was noteworthy that the tiller count registered an increase after cutting for fodder in the 2 oats varieties, whereas triticale variety TFL-3 showed little change in the two Sets.

PLANT HEIGHT

Reduction in plant height particularly in tall genotypes in the regenerated crop after harvesting for fodder is a desirable attribute as the grain crop suffers greater loss on the account of lodging. Most of the genotypes, barring two (TFL-4 and TFL-5) registered significant decrease in height (Table 1). Reduction in plant height ranged from 3 to 31 per cent.

HARVEST INDEX

High harvest index in Set 2 is critical for obtaining high grain yields. The genotypes which exhibit an increase in harvest index in regenerated crop are more suitable for dual purpose. It was observed that harvest index of the oats varieties does not change over Sets whereas most of triticales show a decline (Table 1). TFL-3 was an outstanding exception as it recorded significant increase (0.21 to 0.27) in harvest index in Set 2.

FORAGE CHARACTERISTICS

Forage yield, dry matter yield and crude protein content were recorded on the fodder harvested from the genotypes in Set 2 (dual purpose). None of the triticale genotypes outstripped the oats checks for forage yield (Table 2). However, among the triticale genotypes TFL-3 fared relatively well giving up to 69% and 72% Of the forage yield of Kent and OL-9 oats varieties. In terms of dry matter yield TFL-3 was numerically superior to all other genotypes including oats check though the differences were statistically non-significant. In case of forage crude proteins the triticale genotypes as a group show superiority over oats checks.

Table 2: Forage yield and forage quality characteristics of triticale genotypes grown as dual purpose crop

Genotype	Forage yield (Kg/ha)	Dry matter yield (Kg/ha)	Crude protein (%)
TFL-1	10788	3123.91	11.7
TFL-2	11231	3260.87	9.8
TFL-3	13285	4432.37	11.0
TFL-4	9581	2306.76	10.2
TFL-5	7729	2878.38	11.0
TFL-6	11272	3784.06	11.5
TFL-7	12672	2890.46	10.7
TFL-8	10789	3097.83	10.7
KENT	18236	3366.55	10.2
OL-9	19162	3474.15	8.2
CD (0.05)	3790	NS	–

Discussion

The present study has resulted in identification of a near ideal dual purpose genotype TFL-3 The superiority of the genotype was evident from better forage characteristics, adequate regeneration ability and increased harvest index in post cut crop resulting in higher grain yield. The forage yields are obtained without a drastic sacrifice in terms of grain yield or prolonged duration. Thus overall productivity in biological as well as economic terms stands enhanced. An additional benefit of the dual purpose system is that triticale genotypes which are otherwise excellent but suffer from lodging on account of tallness can be exploited. The post cut crop showed up to 31% reduction in height as compared to ' grain only ' crop. Similar observations have been reported by other workers (3, 5).

Differential response of triticale genotypes to forage harvest before allowing them to regrow and set seed as seen in this study has also reported elsewhere (5). Thus genetic variation for this trait exists in triticale and specific breeding programs for dual purpose triticale can be taken up.

References

1. Bishnoi UR, Chitapong I, Hughes I, Nishimuta J. Quantity and quality of triticale and other small grain silage. Agronomy Journal 1978, 70 : 439-441.

2. Brown AR, Almodares A. Quantity and quality of triticale forage compared to other small grains. Agronomy Journal 1976; 68 : 264-266.

3. Poysa VW. Effect of forage harvest on grain yield and agronomic performance of winter triticale, wheat and rye. Canadian Journal of Plant Science 1985; 65 : 879-888.

4.	Brignall DM, Waid MR, Whittington WJ. Yield and quality of triticale cultivars at progressive stage of maturity. Journal of Agricultural Science, Cambridge 1988; 111 : 75-84.

5.	Royo C, Montesinos E, Molina- Cano JL, Serra J. triticale and other small grain cereals for forage and grain in mediterranean conditions. Grass and Forage Science 1993; 48 : 11-17

APICAL DEVELOPMENT IN TRITICALE GROWN FOR DUAL PURPOSE IN A MEDITERRANEAN ENVIRONMENT

A. Boujenna, J. M. Ramos, J. A. Yanez and L. F. García del Moral
Departamento Biología Vegetal, Facultad de Ciencias, Universidad de Granada, Granada, Spain

Abstract

Apical development in three triticale varieties grown for dual use (forage and grain) was studied in southern Spain during 1991. Three treatments were compared: C0, an uncut control; C30, one cutting made at Zadoks' stage 30 (pseudo stem erection); and C31, one cutting made at stage 31 (first node detectable). The cuttings significantly delayed the terminal spikelet stage, this effect being more pronounced after cutting at stage 31. The treatments, however, significantly reduced the duration from this last stage to the anthesis. The number of primordia at the terminal spikelet stage, anthesis and physiological maturity, was only significantly diminished after C31 treatment. The forage removal reduced the rate of spikelet initiation, the lowest values obtained from C31 treatment. The spikelet losses from the terminal spikelet stage to maturity were significantly higher in the treatment C30.

Introduction

Triticale develops through a series of phases, whose timing depends on genetic and environmental factors, as well as on agronomic practises. Observation of the shoot apex leads to the precise identification of its stage of development, giving thus a useful information not only for efficient cereal crop husbandry but also for assessing the effect of certain treatments on yield components and grain yield.

An adequate duration and rate of apical development is essential for cereals, because they control the final number of spikelets per ear and the number of florets per spikelet [1]. The most important phases in the apical development are the vegetative, the double ridge and the terminal spikelet stages [2]. The final number of grains per ear depends both upon the duration and rate of spikelet initiation and the proportion of spikelets that survive to set grain [3,4].

Forage removal in triticale delays development and anthesis [5], thus reducing the mean grain weight [5,6]. Spikelets per ear and floral fertility are also reduced after forage utilization, this effect being stronger in the ears of tillers [7].

In this work the development of the apical meristem in the main stem was studied in relation to the number of grains per ear in three triticale varieties grown for dual purpose in a Mediterranean environment.

873

H. Guedes-Pinto et al. (eds.), Triticale: Today and Tomorrow, 873–877.
© 1996 *Kluwer Academic Publishers. Printed in the Netherlands.*

Material and Methods

During the 1991 season in Granada (Spain), three hexaploid triticales (Tropical, Tutor and Trujillo) were sown at a rate of 300 seeds m^{-2} in six-row 10-m long plots separated by 0.5m-wide uncultivated pathways. The crops were seeded on 3 Dec. 1990 on a irrigated silty clay typic Xerofluvent soil. The experimental design was of randomized blocks with four replications. Three cutting treatments were compared: C0, an uncut control; C30, one cutting made at Zadoks' stage 30 (pseudo stem erection); and C31, one cutting made at stage 31 (first node detectable). Each cutting reduced the entire plot to a height of 4-6 cm without removing any of the growing points.

For the measurement of apical development five plants were harvested from each plot on a twice-weekly basis, and dissected in order to obtain the apex developmental stage and the total number of primordia in the main stem. The rate of spikelet initiation was obtained as the slope of the linear regression of the number of spikelet primordia againts time, for the period between the double-ridge and terminal spikelet stages. In addition, the number of spikelets per ear at complete anthesis and phisiologycal maturity were also recorded. A plot was judged to be at complete anthesis when anthers were visible at the basal spikelets in 50% of ears. A plot was assumed to be physiologically mature when 75% of the glumes of the primary ear had turned yellow in 50% of the plants. In terms of apical development, C30 treatment was applied after the double ridge stage and C31 after the lemma primordium stage.

The data were subjected to an analysis of variance, and differences between means were compared by the least significance difference test.

Results and Discussion

Apical development was divided into four phases: 1) Vegetative period, from sown to double ridge stage; 2) Spikelet initiation, from double ridge to terminal spikelet stage (when the maximum number of primordia is attained); 3) Spikelet growth, from terminal spikelet stage to anthesis; and 4) Grain growth, from anthesis to physiological maturity.

The results showed (Table 1) that cutting treatments delayed the terminal spikelet stage, this effect being more pronounced after cutting at stage 31. However, treatments significantly reduced the duration from terminal spikelet initiation to anthesis. The grain growth period was not significantly affected by cutting (Table 1). Thus, although the treatments lengthened the period from the double ridge to the terminal spikelet stage, they shortened the period from this last stage to anthesis. Therefore, the differences observed in the date of anthesis between treatments were only 5-6 days. These results are in accordance with previous findings in southern Spain [6] and can be attributed to a longer photoperiod, stronger irradiation and higher temperatures during spring in comparison with northern locations [5].

Table 1 - Duration in days of developmental stages of three genotypes of triticale as affected by two treatments of cutting.

	Vegetative period	Spikelet initiation	Spikelet growth	Grain growth
Uncut	70.8a	37.6c	48.2a	23.1a
Cut 30	70.3a	48.3b	44.2b	23.8a
Cut 31	70.7a	52.0a	39.1c	22.7a

a-b: Means followed by the same letter in each column do not differ at the P= 0.05 probability level using the LSD test.

At the double ridge stage no statistical differences were found in the number of primordia produced in the main stem between the three treatments (Table 2), because at this moment cutting had not yet been applied. After the C30 treatment, the number of primordia initiated at the lemma primodium stage (Table 2) was significantly diminished in relation to uncut plants. This effect however, was later compensated for and thus the number of primordia at the terminal spikelet stage, anthesis and physiological maturity, was only significantly diminished after C31 treatment (Table 2).

Substantial differences exist in the literature concerning the effects of clipping on primordia production in cereals. Sharrow and Motazedian [8] indicated that in winter wheat forage removal increased the number of spikelets per ear. Christiansen *et al.* [9] reported that grazing winter wheat during the autumn increased the number of grains per ear as compared with no grazing. This increase was attributed to reallocation of resources to reproductive growth instead of new tiller formation. However, García del Moral [7] reported that forage utilization in triticale did not significanlty modify the final spikelet number in the main stem ears, but drastically reduced this number in the ears of tillers. These results agree with those of Miller *et al.* [10] that forage removal decreases the number of spikelets per head in wheat and triticale.

Table 2 - Number of primordia at different stages of apical development.

	Double ridge	Lemma primordium	Terminal spikelet	Anthesis	Maturity
Uncut	13.2a	25.7a	28.1a	27.8a	24.9a
Cut 30	13.5a	22.5b	27.8a	26.8a	24.0a
Cut 31	13.1a	24.7a	25.6b	25.9b	22.5b

a-b: Means followed by the same letter in each column do not differ at the P= 0.05 probability level using the LSD test.

Forage removal reduced the rate of spikelet initiation (Table 3), the lowest value obtained from C31 treatment. However, differences between uncut and cut plants at stage 30 were not significant. This fact is probably due to an increase in the competition between the vegetative and reproductive growth for the available resources. In effect, cutting at tillering severely reduce photosyntate production when there is a high demand of assimilates for the simultaneous production of new leaf area, stem elongation and reproductive growth [7,11].

Table 3 - Production and loss of primordia in triticale as affected by two treatments of cutting.

	Primordia production rate mm d^{-1}	Losses from		
		terminal spikelet to anthesis %	anthesis to maturity %	Total %
Uncut	0.496a	3.3ab	8.4b	11.7b
Cut 30	0.478a	4.6a	14.1a	18.7a
Cut 31	0.389b	1.5b	11.5ab	13.0b

a-b: Means followed by the same letter in each column do not differ at the P= 0.05 probability level using the LSD test.

The loss of spikelets was more significant from anthesis to maturity than from the terminal spikelet stage to anthesis (Table 3), and higher in the C30 than in the C31 treatments. Losses from the terminal spikelet stage to anthesis were lower in the plants cut at C31, probably due to less production of primordia (Table 2).

Conclusions

From this study it can be concluded that forage removal lengthened the spikelet initiation period, but reduced the duration from the terminal spikelet stage to anthesis. When cutting was done at Zadoks' stage 31 the primordia initiation rate and the number of spikelets per ear were significantly reduced, but application at Zadoks' stage 30 did not modify these characteristics.

Acknowledgements

We thank Dr. J. Marinetto and M. Pelaez for technical assistance of the field trials. This work was supported by CICYT, Spain under project AGR90-0509-C02.

References

1. Whingwiri EE, Stern DR. Spikelet development and grain yield of the wheat ear in response to applied nitrogen. Aust J Agric Res 1982;31:637-47.
2. Kirby EJM, Appleyard M. Cereal plant development and its relation to crop management. In: Gallagher EJ, editor. Cereal production. London: Butterworts, 1984:161-173.
3. Kirby EJM, Appleyard M. Cereal development guide. 2nd ed. Stoneleigh: Arable Unit, National Agriculture Centre, 1986.

4. García del Moral MB, Jimenez Tejada MP, García del Moral LF, Ramos JM, Roca de Togores F, Molina-Cano JM. Apex and ear development in relation to the number of grains on the main-stem ears in spring barley (*Hordeum distichon*). J agric Sci, Camb 1991;117:39-45.

5. Royo C, Insa JA, Boujenna A, Ramos JM, Montesinos E, García del Moral LF. Yield and quality of dual purpose spring triticale as influenced by sowing date and cutting stage. Field Crop Res 1994; in press.

6. Martinez-Ochoa B, García del Moral LF, Marinetto J, Fernandez Conejo A. Producción de grano en triticale pastoreado. Agric Med 1989;119:109-18.

7. García del Moral LF. Leaf area, grain yield and yield components following forage removal in triticale. J Agron Crop Sci 1992;168:100-07.

8. Sharrow SH, Motazedian I. Spring grazing effects on components of winter wheat yield. Agron J 1987;79:502-4.

9. Christiansen S, Svejcar T, Phillips WA. Spring and fall cuttle garzing effects on components and total grain yield of winter wheat. Agron J 1989;81:145-50.

10. Miller JL, Joost RE, Harrison SA. Forage and grain yields of wheat and triticale as affected by forage management practices. Crop Sci 1993;33:1070-75.

11. Dunphy DJ, Holt E.C, McDaniel ME. Leaf area and dry matter accumulation of wheat following forage removal. Agron J 1984;76:871-74.

TRITICALE AS FORAGE IN CHINA

Y.S. Sun[1], Y. Xie[2], Z.Y. Wang[1], L. Hai[1] & X.Z. Chen[1]
[1]Institute of Crop Breeding and Cultivation, CAAS Beijing, China
[2]Beijing Municipal Burean of State farm, Beijing,China

Abstract

Triticale is a new cereal crop resulting from a cross between wheat and rye. It not only retains the high yielding performance and good quality of wheat but also combines disease resistance, tolerance and lush growth from rye. Furthermore, it desplays improved characters; such as higher protein and lysine content in the grain compared to the parent's. According to an analysis of biomass nutrient content in production and dairy cattle feeding trials, triticale has been identified as a promising silage crop compared to barley, the farmers traditional silage crop in Beijing. The relationship between biomass yield of triticale and some major agronomic measures such as sowing date, seed rate and nitrogen application was studied and defined. It indicated that sowing date is the primary factor influencing yield followed by N-application and seeding rate. A set of cultivation measures for silage triticale has been researched and released. Triticale as forage has been well accepted by farmers in China.

Introduction

Wheat and rye are from the same family, but are in different genera. Over a long evolutionary period, the wheat chromosome ploidy level developed from a diploid (BB, 2n=14) to a tetraploid (AABB, 2n=28) and then to a hexaploid (AABBDD, 2n=42). In the course of a evolution, yield and nutrition quality of wheat improved. Rye remains diploid (RR, 2n=14) with good disease resistance, stress tolerance and lush growth. Crossing rye wheat (2n=28 or 42) wheat and doubling the hybrid's chromosome number can create new, hexaploid and octoploid triticale. This crop displays a range of new characteristics due to genetic recombinations. At present research on triticale world wide is mainly concentrating on the hexaploid (AABBRR, 2n=42) and the octoploid (AABBDDRR, 2n=56) types. There are research activities on hexaploid and octoploid as well this in China. It is in this way that forage triticale was developed. The shortage of silage has been a primary constraint on the development of dairy cattle in Beijing during the late spring and early summer. A set of triticale varieties for silage which selected from crossing octoploid and hexaploid triticale. They are not only high in biomass yield but have superior

H. Guedes-Pinto et al. (eds.), Triticale: Today and Tomorrow, 879–886.
© 1996 *Kluwer Academic Publishers. Printed in the Netherlands.*

nutritional quality. We want to replace barley, traditional silage in Beijing, and bring about a great advance in agricultural and livestock production.

Material and Methods

In order to demonstrate its value as a forage crops, a series of trials comparing triticale to wheat, rye and barley were conducted at six state farms near Beijing from 1989 to 1993. In order to develop suitable component measures for triticale, a further trial was conducted between 1990-1992. In these trials the triticale cultivar H1890 was tested using a quadratic regression rotary combination design. Sowing date X1, Sowing rate X2 and Nitrogen rate X3 were chosen as decision variances, each with 5 levels (r=1.682), with the purpose of identifying the optimum component techniques with high yield, fine quality and low input. Twelve dairy cows were fed each kind of silage measuring milk yield and nutrition content over a 45 day period at Beijing.

Results and Discussion

TRITICALE CHARACTERISTIC
High Biomass, Triticale possesses a huge root system, produces lush growth and high biomass because it combines some fine characteristics from wheat and rye. Study results from multi-year experiments showed that no obvious difference (Table 1) was found for grain yield between triticale and wheat, but biomass yield of triticale was higher than that of rye and barley. As regards above ground biomass, the triticale was 30-50% higher than wheat, rye and barley. The straw to grain ratio of triticale reached 3.3, which is favourable for feed production.

Table 1. Triticale, wheat, rye and barley biomass

Crop	Plant height (cm)	Grain yield kg/mu	Above ground biomass (kg/mu)	Straw to grain ratio
Triticale H1890	145	270.5	901.7	3.3
Wheat				
Fengkang No.8	85	283.0	673.0	2.4
Rye AR132	135	210.5	661.0	3.1
Barley Cuan	110	204.2	583.0	2.8

Vigorous tillering ability and growth rate, Triticale possesses vigorous tillering ability and a high-speed growth rate. Even under low temperature conditions (3-10°C) in early spring, the crop is still able to quickly produce tillers. In normal conditions during late spring it can keep 5-6 tillers per plant on average. The number of tillers is 2-3 higher than wheat and barley. The growth rate of the crop is twice that of wheat and barley from the elongation stage to the heading stage. The hypertrophied foliage accounts for about 45% of the total above ground fresh fodder yield at heading-flowering stage. This is considerably higher than for wheat and barley (Table 2).

Table 2. Fresh fodder yield components of wheat, barley and triticale at heading stage

Crop	Plant height (cm)	Leaf (%)	Stem (%)	Head (%)
Wheat	92	30.6	55.4	14.0
Barley	105	33.6	54.2	12.2
Triticale	146	44.3	45.6	10.1

High disease resistance and adversity tolerance, Triticale is a disease resistance crop. It demonstrates good resistance to powdery mildew, bushy stunt, yellow dwarf, virus disease and leaf rust, so the crop thrives and maintains green foliage for a long time. Triticale shows high resistance to stresses, such as drought, cold and sour of basic soil. In addition, triticale can efficiently absorb water and nutrients from subsoil because it has a huge root system. In one word triticale always yields highest than wheat and barley under stress conditions.

Early maturity, Most triticale cultivars have a moderate growth period duration. They can be planted before rice, maize and sorghum. Winter cultivars are generally planted in early October and harvested in the middle of May at Beijing. There are about 140-150 days remaining in the growing season for following crops. Therefore, for following crops, the cultivars with middle and long growth duration can be chosen as components after harvesting triticale as a high yielding grain or feed. Spring triticale cultivars used for silage are planted in early March and harvested in late May or early June. In normal condition, feed triticale can win over 20-30 days for the following crops compared with wheat (harvest in middle of June). In the mean time, spring cultivars help alleviate the labor and power pressure of planting and harvesting in Autumn due to planted date regulating.

Fine Nutrition Quality,
a. Nutrition Quality of Grain, The contents of protein and lysine in triticale grain are quite high. The analysis of 508 triticale lines showed that grain protein content ranged from 10.1% to 20.2%, while the average content was 15.3%. This was about a 30% increase relative to wheat (11.5%). Lysine content of the grain was 0.51% on average, about 50%higher than for 0.33% of wheat (Table 3). The composition and content of amino acids

in the protein of triticale grain differed from other cereals. Evangelina *et al.*, from CIMMYT, found 10 kinds of essential amino acids. Except for trytophan and leucine, the contents of all essential amino acids in triticale were the highest among the tested cereals. In addition, a much great variety of amino acids and more balanced content was found in the triticale grain compared with other cereals. All facts mentioned above indicate that triticale is a fine protein feed crop.

Table 3. The contents of protein and lysine in triticale and wheat grain

Crop	Average	Protein (%) range	Number of lines	Average	Lysine(%) range	Number of lines	Test nuit
Triticale	15.30	10.10-20.22	508	0.51	0.24-0.81	307	CAAS
Wheat	12.05	9.20-13.88	28	0.33	0.20-0.48	11	CAAS
Triticale	14.8	10.00-22.50	5100	0.49	0.28-0.62	5100	CIMMYT
Wheat	12.20	6.9-22.00	12613	0.33	0.24-0.46	12613	CIMMYT

b. *Nutrition Quality of Fodder,* For silage and fresh feed crops, it is important to contain varied nutrition components in plants. In order to identify the feeding value, the nutrition components of triticale and barley were determined before silage was harvested. Table 4 shows that except for sugar, the content of all nutrition components in triticale are higher than barley. Table 5 listed the determined results for 17 kinds of essential amino acid content in triticale and barley. From the Table 5, the new feed crop triticale as compared with farmer's feed crop barley, the amino acid contents are high as well as balanced. Triticale not only supplied good balanced proteins but it also supplied plentiful energy and vitamin resources for livestock and poultry. All feeding trials conducted in 12 dairy farms of Beijing between 1988-1992 achieved good results. The results proved that triticale was fine high-yielding silage and fresh feed resource.

Table 4. Nutrition components in triticale and barley plants at flowering stage

Crop	Water %	Protein %	Fat %	Cellulose %	Sugar %	Ash %	Carroten mg/g
Triticale H1890	69.23	4.36	1.08	10.42	2.43	4.06	1.12
Barley Guan	74.22	3.44	0.89	8.26	3.01	2.43	0.78

Table 5. The contents of 17 kind indispensable amino acido in triticale and barley plants

Crop	Lys %	Asp %	Thr %	Ser %	Glu %	Gly %	Ala %	Cys %	Val %
Triticale H1890	0.34	0.64	0.33	0.28	1.00	0.35	0.53	0.12	0.42
Barley Guan	0.28	0.52	0.28	0.24	0.74	0.30	0.50	0.14	0.34

Crop	Met %	Iso %	Leu %	Tyr %	Pth %	His %	Arg %	Pro %	Total %
Triticale H1890	0.22	0.31	0.54	0.28	0.36	0.12	0.36	0.37	6.67
Barley Guan	0.18	0.24	0.46	0.24	0.31	0.10	0.31	0.34	5.52

FRESH FOODER YIELD AND COMPONENT TECHNIQUE OF TRITICALE

Highlighted Results, The Fresh fodder yield was taken for goal function and the regression model between the yield and the variances was established as following:

$$Y=4748.5-598.8x_1+141.2x_2+335.0x_3+91.0x_1x_2+10.5x_1x_3+0.5x_2x_3$$
$$-119.8x_1+37.7x_2^2+92.3x_3^2$$

F test: $F=6.5^{**}(F_{0.01}=5.00)$, $R=0.95^{**}(R_{0.01}=0.88)$

the net determination coefficients (c) showed the order of priority for the tested decision variances affecting yield is:

Sowing date (c=0.99)>N rate (0.61)>sowing rate (0.18)

In triticale management, as shown in Fig.1, sowing date and N rate have a strong effect on fresh fodder yield. Fresh fodder yield decreased as the time of sowing date was delayed. After level 0 (October 16), the yield reduction greatly increased. Fresh fodder yield increased with the N rate and sowing rate. For N rate, fodder yield increased by a big margin between level - 1.682 to level 0. Peck yield occured at level 1. However, sowing rate had small influence on fresh fooder yield.

The application of a systematic theory viewpoint comprehensively evaluted fresh fodder yield and the benefit from triticale. The optimum component techniques for triticale were identified and presented in Table 6.

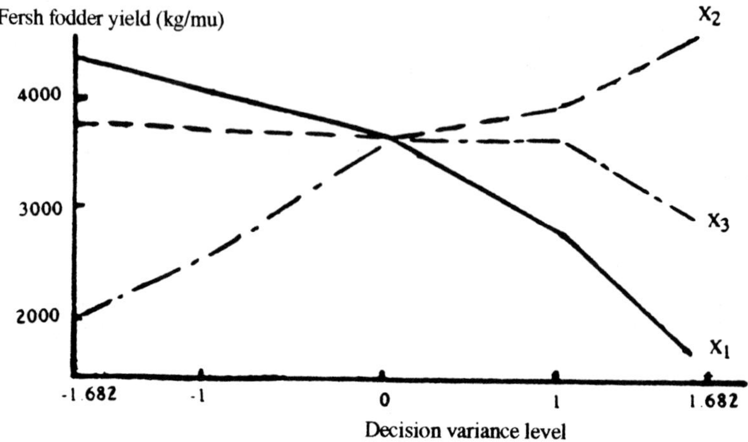

Fig. 1 - Demonstrated the relationship between the fresh fodder yield and the decision variances.

Table 6. The optimized component techniques of triticale

Yield	Optimum Combination		
level (kg/mu)	Sowing date	Sowing rate (kg/mu)	N rate (kg/mu)
4000	Oct. 1-5	12.1-14.5	6.1-8.9
3000	Oct. 11-24	13.0-14.5	3.6-7.2

APPLICATION OF TRITICALE AS FEED

Silage, Triticale was first planted at Zhong-A State Farm near Beijing in 1972 for the feed purpose. The fresh fodder yield of the crop exceeded 2300kg/mu (34.5t/ha), which was 40% over barley. Regional tests from 6 state farms in Beijing between 1989-1992 showed that the fresh fodder yield of cultivar H1890 reached 2949kg/mu(44.2/ha), which was about a 48% increase over barley (farmer's traditional feed crop). In addition, triticale was easy to harvest with machines due to its hard stalk and lodging resistance.
Normally silage triticale was harvested in mid-May. It added 20-30 day growth period for the following crop compared to wheat. Triticale-Rice cropping system tests conducted in farmer's fields between 1988-1991 showed that as a fresh fodder, the first crop of triticale

exceded 300kg/mu. The grain yield of the second crop rice reached 400-450kg/mu. Moreover, new cropping pattern needed lower labour, power and inputs, but gave higher output as than the existing cropping patterns. Experimental results verified that triticale is a better feed crop before rice.

Feeding trials indicated that quantity of milk produce by feeding triticale silage or traditional silage (barley and corn) was not different. However, the quality of milk from cows fed with triticale in better (Table 7).

Table 7. Milk quality from different source of silage

Silage	Protein (%)	Fat (%)	Lactose (%)	Water (%)
Corn	3.03	2.83	4.59	88.00
Barley	3.05	3.30	4.60	88.08
Triticale	3.14	3.38	4.67	87.03

Fresh fooder, Triticale vigorously grows from elongation stage to heading stage. Triticale fresh fodder contains varied nutrient components. Protein is about 13-19% in triticale. Due to its disease resistance, stress tolernace and lush growth, triticale has been well accepted as a silage and fresh fodder crop by farmers. It's acreage has expanded from 2000 mu in 1989 to 16000 mu in 1992 in the Beijing area.

Concentrated feed, Protein and lysine content in triticale grain is about a 10-30% and 20--30% increase, respectively, over wheat and barley. If crushed grain from triticale is substituted for barley or sorghum concentrated feeds, feed efficiency could be greatly improved. A feeding trial conducted in Heilongjiang Province in 1989 showed that the mean live weight of rabbits increased by 40% after 45 days when fed with the 45% triticale grain substituted for maize on a feed. This result was similar to those obtained in other countries, such as Ukraine, Brazil and Poland.

Hay and hay powder, The contents of protein and carotene in hay powder from triticale increased by 10-15% and 34%, respectively, compared to rye and wheat. In the Shansi province, triticale intercropped with alfalfa and other legume feeds, had a protein content of up to 18-20%. In the Southwest part of China, especially the mountain region, farmers feeding livestock with triticale straw achieved better benefits. Triticale straw has fine palatibility and helps livestock. Feeding trials conducted at Tu Mu Ji farm of Mongolia showed that growth increased by 80%, if triticale hay substituted for traditional hay. In addition, triticale hay powder can further be processed into pelletized feed and compound feed to enhance nutritions quality. In one word, triticale has become a new feed resource for livestock farming in the country although it needs testing, improvement and popularizing.

References

Bushuk, W., 1974, Proteins of Triticale and Physical Characteristics, in Triticale First Man-made Cereal, Am. Assoc. Cereal Chemi., St. Paul Minn., U.S.A., P128-136.

Charmley, E., and Greenlagh, J., 1987, Nutritive Value of Triticale for Sheep, Pigs and Poultry, Anmi. Feed Sci. Technal, 18: 19-35.

Jessop, R.S., Leterme, P., 1990, Triticale as Forage or Grain for Animal, Proc. od 2nd Int. Triticale Symp., Mexico, DF, CIMMYT. P64-65.

Müntzing, A., 1979, Triticale Results and Problems, Advances in Plant Breeding, 10: 1--103.

Wang Chongyi and Sun Yuanshu, 1986, Triticale Breeding in China, Proc. It Int. Triticale Symp. Sydney, 1986, Aust, Inst. Agric. Sci., Occasional Publication № 24. P50-59.

Sun Yuanshu and Wang Chongyi, 1990, Triticale as a New Silage for Dairy Cattle, Proc. of 2nd Int. Triticale Symp., Mexico, DF, CIMMYT. P514-515.

WINTER TRITICALE: A ROUGHAGE SOURCE FOR WINTERING RANGE COWS?

Mathias F. Kolding and Robert J. Metzger, Oregon State University and ARS-USDA, Oregon, USA

Abstract

Hay, grain yield, and nutrient production of advanced generation winter triticale harvested from a irrigated sandy soil were determined. Dry weight yield ranges and LSD (0.05) for hay was 12 to 22 Mg he (6.4) and for grain 1740 to 7030 kg he (1100). Total protein, crude fat, Nitrogen Free Extract, and Total Digestible Nutrients for hay and grain are given and discussed.

Introduction

Winter triticale (X *Triticosecale* Wittmack) for the irrigate soils in the Pacific North-West Columbia Basin, United States of America are selected for grain production. Farmers, however, allocate water to high income crops, such as potatoes and onions when anticipating water short-ages. Consequently, there may be less cereal stubble and straw fodder for ranchers to over-winter range cows. Soil conservation laws, hoewever, require growers to protect fragile sandy soils. Fall seeded forage triticale could provide winter and summer soil protection, a high quality forage prior to heading, a source of bulk hay or silage after heading, or a grain and straw source. Therefore, a study was initiated to estimate gross nutrient and dry matter production of advanced generation winter triticale hayed at the soft dough stage and harvested for grain.

Materials and Methods

This study was conducted on a Winchester sandy loam at the Oregon State University, Hermiston Agricultural Research and Extension Center, Hermiston Oregon. 18 treatments were seeded October 10, 1991 in a randomized block of 4 replications. Triticale treatments were: one private variety selected for forage, four mid-height and ten mid-tall to tall triticales, one hooded barley (*Hordeum vulgare* L.), the triticales, hooded barley mixed 50/50 with one of the triticales, and the hooded barley seeded every other row with the triticale used in the 50/50 mixture. Plots comprised four rows spaced at 30.4mm and 4.56m long. The prior crop was a weedy fallow following potatoes. 100 kg ha of nitrogen was

887

H. Guedes-Pinto et al. (eds.), Triticale: Today and Tomorrow, 887–888.
© 1996 *Kluwer Academic Publishers. Printed in the Netherlands.*

applied via the watering system just prior to stem elongation. Irrigation was done with "hand-lines" fitted with sprinkler heads. Water applications were limited to 90 percent of evapotranspiration (ET total=25mm February 1 through May 15, 1992) and stopped when plant development ranged from late boot to anthesis. Two rows were harvested for hay and two for grain. Sub.samples from each replication were mixed. A private laboratory determined moisture and nutrient values.

Results and Discussion

Hay dry matter production ranged from 12 to 22 Mg ha, [LSD (0.05) = 6.4]. Dry weight ranges and the LSD (0.0%) of protein, crude fat, Nitrogen Free Extract (NFE), and Total Digestible Nutrients (TDN) in kg ha were 830 to 1680 (460), 190 to 440 (127), 6,870 to 11,000 (3460), and 6,250 to 10,770 (3228) respectively. Pearson correlations (0.05) for plant height and total dry matter production was 0.2618, and for stem count and dry matter production was 0.1746.

Grain dry weight ranges and the LSD (0.05) of grain, protein, crude fat, NFE, and TDN in kg ha were 1740 to 7030 (1100), 240 to 760 (126), 38 to 132 (19.8), 1330 to 5869 (883), and 1550 to 5240 (830) respectively.

Advanced generation winter triticale could provide new varieties that would offer the grower a fall sown crop after onion and potato harvest. Nine of the mid-tall lines were the top nine treatments and produced over 17 Mg ha of dry matter and 8.8 Mg ha of TDN. Six of the mid-tall lines were also in the top nine for grain production. Weak correlations of height and stem count to dry matter production may point to differences in leafiness.

892

falling numbers, 589, 590, 615–618, 621, 771–774
farmers, 34
fat content, 787
fats, 790, 885
fatty acid, 808–810
feed, 42, 693
feed efficiency ratio, 632
feed grain, 669
feeding, 863
feeding values, 42, 627
fertility, 627, 628, 630, 633
field conditions, 623
FISH, 97, 155, 156, 161
flour pentosan content, 774–776
fodder, 589, 669, 693
fodder quality, 867
forages, 8, 321, 322, 391, 653, 669, 694, 697–699, 827, 833, 835–837, 840, 843, 844, 846, 847, 859–861, 869, 873, 879, 880, 887
forage crop, 867
forage production, 851, 852, 855
forage quality, 828, 860, 870
forage triticales, 851
forage yield, 870
foundation seed, 677
fresh fodder yield, 881
fresh fooder, 885
fresh fooder yield, 883
fungal diseases, 315
fungicide treatments, 658
Fusarium, 373–375, 499, 555
Fusarium spp., 504, 506, 658
Fusarium avenaceum, 505
Fusarium culmorum, 67, 505, 650
Fusarium graminearum, 505, 549, 550, 552, 553, 650, 690
Fusarium nivale, 315–317
fusarium scab, 527, 530

G × E interactions, 609, 611, 791
GA₃ insensitivity, 275
Gaeumannomyces graminis, 500, 513

galactose, 237, 238
gametoclonal variation, 384, 387
gene expression, 120, 278
gene suppression, 151, 153
gene transfer, 7
general combining ability, 598, 599
genetic control, 396, 420
genetic diversity, 243, 244, 250, 251
genetic interactions, 96
genetic markers, 149–151
genetic parameters, 603
genetic resources, 7, 243, 244, 261, 262, 266, 299
genetic resources network, 270
genetic variation, 609, 616
genome combining ability, 51
genomic DNA, 183
genomic interactions, 149, 275
genotype × N interaction, 603, 607
germination, 615
germplasm, 100, 243–247, 250, 254, 255, 258, 266, 285–287, 429, 431, 433, 571, 572, 574, 577, 579, 640, 679, 680, 860
germplasm bank, 244–246
germplasm creation and deployment, 269
Gibberella, 504
Gibberella zeae, 690
gliadins, 149, 150, 211, 413
globulin content, 788
globulins, 790
glucose, 237, 238
glucose 6-phosphate isomerase (GPI), 150
glucose polymers, 236
glutamate oxalacetate transaminase, 409
glutelins, 789–791
gluten, 755, 756, 758, 760
gluten content, 773, 774
gluten protein, 757
gluten proteins, 256
glutenins, 411, 571, 578
glycoproteins, 771

896

regenerated plantlets, 374
regenerated plants, 349
registration, 701, 702, 704, 705
relative yields, 725, 726
repetitive DNA, 155–157, 159, 161
reproductive behaviour, 292
resistance, 315, 505, 506, 508–511, 513–
    515, 527, 536, 538, 557, 655, 737,
    859, 864
resistance genes, 499
resistance to cold, 739
resistant to thermic excess, 739
restorer, 226
restorer genes, 350
restoring genotypes, 227, 230
RFLP, 7, 339, 340, 342
rheogram height, 779
rheological properties, 780
rheological tests, 771
Rhizoctonia solani, 513
Rht genes, 7, 281
Rht1, 281–283, 663
Rht1 gene, 667
Rht2, 281–283
Rht2+3, 281–283
Rht3, 281–283
Rht3 gene, 653
Rhynchosporium secalis, 67
rice, 14
root dry weights, 441
root length, 437, 440
root lesion, 557
rRNA genes, 119, 120, 123
rye, 119, 121, 124
rye cytoplasm, 77, 78
rye germplasm, 253
rye inbred, 74
rye spacer region, 123
rye-wheat comparative map, 342
ryes, 541

salt tolerance, 429, 431, 433
salt tolerant, 430
scab, 549, 553, 690

Schizaphis graminum, 67
SDS-PAGE, 383, 409, 410
Secale, 183
Secale cereale, 119, 179
Secale cereale-montanum, 179
secalin, 211
secondary triticales, 101
sedimentation values, 773, 774
seed companies, 34
seed germination, 616–618, 621, 744,
    746
seed shrivelling, 75
seeding rate, 835–840
self-fertilization, 7
selling, 35
semidwarf, 581
sensitivity to frost, 739
Septoria nodorum, 67, 315–317, 499,
    507, 650, 651, 658
Septoria nodorum resistance, 667
Septoria tritici, 67, 639, 658
Septoria tritici resistance, 253, 258
septoriosis, 535
shoot and root growth, 492
shoot apex, 873
shoot growth, 493
short straw, 663, 667
silage, 285, 861, 879, 880, 882, 884,
    885
silver nitrate impregnation, 136
silver staining, 129, 130
silver staining technique, 127
soil acidity, 445
soluble dietary fiber, 234
soluble nitrogen, 763
somaclonal variation, 8, 108, 330, 349–
    351, 353
somaclones, 508
somatic embryogenesis, 328, 329, 333
somatic embryos, 369
somatic variation, 405
Southern blot analyses, 349–351
sowing date, 851
spacer region, 122

Typhula ishikariensis, 513, 541, 543, 545

univalency, 97
use value, 702, 703
Ustilago, 508
Ustilago nuda, 67
Ustilago tritici, 67

variability, 465
vernalisation, 637
virus-like, 514
viruses, 514, 881
vitality of pollen, 195, 196, 200

water-stress, 732
waterlogged, 490
waterlogging, 489
waterlogging tolerance, 489, 490
waterlogging-sensitive, 492
weed control, 659
weed suppression, 749, 750
wheats, 14, 18, 120, 281, 283, 284, 541
wheat rye addition, 233
wheat streak mosaic virus (WSMV), 514

wheat Triticum aestivum, 119
wheat-rye hybrids, 83–85, 135
winter and facultatives genotypes, 287
winter hardiness, 663, 667
winter triticale, 535

X Triticosecale Wittmack, 119
Xanthomonas campestris, 514
xylose, 237, 238

yellow dwarf, 881
yield, 26, 63, 78, 563, 566, 567, 571, 572, 574–577, 710–712
yield capacity, 663
yield components, 607
yield per spike, 597, 598
yield potential, 18, 19, 656, 657, 687, 860, 862
yield stability, 677, 860
yield trials, 675
yielding ability, 709
yielding potential, 709

Zea mays, 379
Zeleny test, 709–711, 715